19—

D0543333

PHYSICS

Other titles in the Project

Telecommunications   John Allen
Medical Physics   Martin Hollins
Energy   Robert Hutchings and David Sang
Nuclear Physics   David Sang

Design Electronics   Bill Phillips

Biology   Martin Rowland   Geoff Hayward
Applied Genetics   Geoff Hayward
Applied Ecology   Geoff Hayward
Micro-organisms and Biotechnology   Jane Taylor

Project Director: J. J. Thompson, CBE

PHYSICS

ROBERT HUTCHINGS

MACMILLAN

First published 1990
Reprinted 1990

Published by
MACMILLAN EDUCATION LTD
Houndmills, Basingstoke, Hampshire RG21 2XS
and London
Companies and representatives
throughout the world

Typeset by Typematter Graphics, Basingstoke, Hampshire

Printed in Hong Kong

British Library Cataloguing in Publication Data
Hutchings, Robert
Physics.
1. Physics
I. Title
530
ISBN 0–333–46515–6

Contents

The Project: an introduction

The **University of Bath · Macmillan Science 16–19 Project,** grew out of a reappraisal of how far sixth form science had travelled during a period of unprecedented curriculum reform and an attempt to evaluate future development. Changes were occurring both within the constitution of 16–19 syllabuses themselves and as a result of external pressures from 16+ and below: syllabus redefinition (starting with the common cores), the introduction of AS-level and its academic recognition, the originally optimistic outcome to the Higginson enquiry; new emphasis on skills and processes, and the balance of continuous and final assessment at GCSE level.

This activity offered fertile ground for the School of Education at the University of Bath and Macmillan Education to join forces with a team of science teachers, drawn from a wide spectrum of educational experience, to create a flexible curriculum model and then develop resources to fit it. This group addressed the task of satisfying these requirements:

- the new syllabus and examination demands of A- and AS-level courses;
- the provision of materials suitable for both the core and options parts of syllabuses;
- the striking of an appropriate balance of opportunities for students to acquire knowledge and understanding, develop skills and concepts, and to appreciate the applications and implications of science;
- the encouragement of a degree of independent learning through highly interactive texts;
- the satisfaction of the needs of a wide ability range of students at this level.

Some of these objectives were easier to achieve than others. Relationships to still evolving syllabuses demand the most rigorous analysis and a sense of vision – and optimism – regarding their eventual destination. Original assumptions about AS-level, for example, as a distinct though complementary sibling to A-level, needed to be revised.

The Project, though, always regarded itself as more than a provider of materials, important as this is, and concerned itself equally with the process of provision – how material can best be written and shaped to meet the requirements of the educational market-place. This aim found expression in two principal forms: the idea of secondment at the University and the extensive trialling of early material in schools and colleges.

Most authors enjoyed a period of secondment from teaching, which not only allowed them to reflect and write more strategically (and, particularly so, in a supportive academic environment) but, equally, to engage with each other in wrestling with the issues in question.

The Project saw in the trialling a crucial test for the acceptance of its ideas and their execution. Over one hundred institutions and one thousand students participated, and responses were invited from teachers and pupils alike. The reactions generally confirmed the soundness of the model and allowed for more scrupulous textual housekeeping, as details of confusion, ambiguity or plain misunderstanding were revised and reordered.

The test of all teaching must be in the quality of the learning, and the proof of these resources will be in the understanding and ease of accessibility which they generate. The Project, ultimately, is both a collection of materials and a message of faith in the science curriculum of the future.

J.J. Thompson
January 1990

How to use this book

The aim of this book is to help and encourage you in your Advanced Level physics course. Physics is not a collection of facts and equations to be memorised. It needs to be considered as a few fundamental principles which can be applied in many different situations. At the start of the course it is important to acquire habits of study and experiment which will help your understanding of the subject. Understanding is the key; memorised facts without understanding are virtually useless. The short cut to an understanding of physics is the ability to use the principles, not the ability to put numbers into equations.

It has been assumed that in preparation for taking Advanced Level physics you have done some physics previously as part of a GCSE science course, or as a subject by itself. However, much GCSE work is covered again in this core book to ensure you have a solid grounding in the fundamentals before tackling new ideas.

Calculus has been used when necessary, but wherever an alternative explanation is easier to understand it is used instead. However, you do need to be able to find the gradient of a graph or the area beneath it and calculus is a very precise way of doing this.

The study of physics needs to be done with a notebook, pen and calculator always at hand. You will find on may occasions that a topic is introduced with a numerical problem. Where this is the case you will find more steps in the arithmetical working than usual in books at this level and there are situations where steps are regarded as simple numerical calculation. The algebra, and in some cases calculus, comes later.

In some of the simple numerical calculations you may find that units have not necessarily been given. It is the counsel of perfection to include units in every calculation and in many complex equations they can be more of a hindrance than a help. Books frequently dodge this problem by working algebraically until the last line and then doing the calculation in one step, but I hope it helps to work arithmetically, rather than algebraically, in places. You will also find that, on occasion, more significant figures have been used than are justifiable, but this has been done in order to avoid rounding up errors and, where necessary, will have been corrected in the last line of the answer.

Fewer abbreviations have been used in this book than in many at this level. This has been done to limit the number of obstacles to understanding and to try and reduce the amount of jargon. Where words are used in place of abbreviations they are written in full, 'a force of 6 newtons' for example. I am convinced that examination candidates could demonstrate their ability and knowledge better if they were prepared to express their approach to the solution of a problem in a sentence or two, rather than assuming that all the examiner is interested in is the answer. I hope you will see the benefit of being prepared to used words where they are needed rather than relying solely on symbols and abbreviations.

I wish you success and satisfaction from the hard work you will need to put into your course of study.

Learning objectives

Each chapter starts with a list of learning objectives which outlines what you should gain form the chapter. They are statements of attainment and often link closely to statements in a course syllabus. Learning objectives can help you make notes for revision, especially if used in conjunction with the summaries at the end of the chapter, as well as for checking progress.

Questions

In many Advanced Level physics textbooks the only questions are those taken from past A-level examination papers. These have their place and plenty are provided in this book at the end of the chapters for you to do at the end of a topic or for revision. You will also find in-text questions dealing with one point at a time which will help your understanding and ability to use principles as you progress through the course. You should do these questions as you read through each chapter.

Do not avoid doing questions because you are a little uncertain what to do, you will always learn something in the attempt! The following tips on how to approach questions may help. Draw a sketch diagram; list, in your own words, the facts you have been given; make use of the units of the quantities involved; calculate anything obvious even if it is not specifically asked for. Look at all your answers to see if they are reasonable; look up the correct answer in the back of the book; try a different approach.

Answers to Questions

In the study of physics everyone makes mistakes from time to time. Checking the work you do is important. You need to be alert to the possibility of going wrong and to look out for likely mistakes. Powers of ten are a common source of arithmetical nonsense, for example. If answers appear to be nonsensical than they are probably wrong, whereas if you can find two different approaches to the same problem and both give the same answer, you have strong confirmation you are correct. Checking needs to be done continuously and not just at the end of a problem. Check each line quickly for significant figures, for units and for making sense.

Answers to most questions are given in Appendix A.

Investigations

If experiments are done thoughtlessly by slavishly following instructions they simply use up time unprofitably. The investigations in this book are designed to require thought about the experimental method being used as well as to give experience of some key principles of physics.

Data analysis

These extended exercises are intended to give you some insights into the way physics is used to solve the problems of everyday life and to help you learn how to apply the principles to a problem. Hints and some answers are given in Appendix C.

Summaries

Each chapter ends with a brief summary of its content. These summaries, together with the learning objectives, should give you a clear overview of the subject, and allow you to check your own progress.

Acknowledgements

The author and publishers wish to thank the following who have kindly given permission for the use of copyright material:

The Associated Examining Board, Joint Matriculation Board, Northern Ireland Schools Examinations Council, Oxford and Cambridge Schools Examination Board, University of Cambridge Local Examinations Syndicate, University of London School Examinations Board, University of Oxford Delegacy of Local Examinations.

The author and publishers wish to acknowledge, with thanks, the following photographic sources:

Allsport *pp 45 lower, 60 lower, 73, 471;* Clive Barda *p 214;* Barnaby's Picture Library *pp 1, 26, 27, 66, 70 71 upper, 72;* Ivan Belcher *p 506 lower;* Beken of Cowes *p 153;* Bridgeman Art Library *p 140;* British Aerospace *pp 58, 88, 114, 246 right, 269;* Bristol University, Physics Department *p 566 left and right;* The British Petroleum Company plc *p 178;* British Textile Technology Group *p 526;* Cable & Wireless *p 169;* Casio Electronics Co Ltd *p 412;* The Cavendish Laboratory *pp 220, 532;* CEGB *pp 30 left, 233, 245 upper, 275, 334, 421, 445;* Computer Weekly *p 290;* Colorsport *p 45 upper;* Crown Copyright, Central Office of Information *p 543;* Electricity Council *p 510;* Ford Motor Company Ltd *pp 85, 152;* GEC *pp 245 lower, 507;* GEC Turbine Generators *p 356;* Philip Harris *p 467;* Harman UK Ltd *p 389;* IBM Corporation *p 395;* The Image Bank *pp 6, 108, 153 left, 212;* JET *p 336;* Kent Industrial Measurements Ltd *p 38 upper;* Frank Lane Picture Agency *pp 146, 122, 282, 292 lower, 385, 458;* Magnetti Marelli *p 358;* Minolta *pp 197, 201;* Moulinex Ltd *p 289 right;* NASA *pp 52, 120 upper, 127, 135;* National Physical Laboratory *p 3;* Osram GEC *p 248;* Peckham's of Stroud *pp 13, 37, 94 right and lower left, 213, 223, 343, 452;* Physics of Materials by B Cooke and D Sang *p 430 upper and lower;* The Photo Source *pp 76, 89 upper, 125, 190, 475, 521;* Polaroid *p 173 top;* Precision Metal Forming Ltd *p 30 right;* courtesy of Rolls Royce plc *pp 15, 173 centre, 506 upper;* The Royal Greenwich Observatory *pp 204 left and right;* The Royal Military School of Music *p 216;* R S Components *pp 246 left, 299, 388, 399, 405;* Salford Electrical Instruments Ltd *p 351;* Science Museum, London *pp 40, 493;* Science Photo Library *pp 38 lower, 44, 81, 94 upper left, 97, 120 lower, 154 173 lower, 192, 227, 288, 289 left 292 upper, 324, 385, 397, 426 left and right, 441, 450, 524, 531 left and right, 567 upper and lower, 574 all pictures;* Shell Photo Library *p 437;* Telefocus *pp 276, 366;* Teletron Ltd *p 539;* Tony Stone *pp 60 upper, 104;* UKAEA *pp 469, 573;* Volvo Concessionaires Ltd *pp 89, 90;* Westland Helicopters Ltd *p 57;* ZEFA Picture Library *pp 33, 35, 45 centre, 56, 71 lower, 74, 139, 160, 162, 236, 359.*

Every effort has been made to trace all the copyright holders, but if any have been inadvertently overlooked the publishers will be pleased to make the necessary arrangement at the first opportunity.

Theme 1

MEASUREMENT

Accurate measurement is central to the development of any science. The importance of measurement was apparent to ancient civilisations. Throughout history the accuracy with which measurements could be made has been improved by the use of more and more sophisticated instruments. At almost every stage, improved measuring techniques have resulted in new concepts and ideas.

Chapter 1

PHYSICAL QUANTITIES AND UNITS

> **LEARNING OBJECTIVES**
>
> At the end of this chapter you should be able to:
>
> 1. quote quantities in their correct SI units;
>
> 2. find the dimensions of any quantity.

1.1 PHYSICAL QUANTITIES

The science of physics, like all other sciences, is based upon taking measurements. All scientific theories and laws must be tested experimentally and all experiments necessitate making measurements. One of the reasons for the development of physics as a science is that the measurements which are made in physics experiments are usually more reliable than the measurements made in some other sciences. It is very difficult to make accurate measurements on, for example, aspects of human behaviour. Sociology is therefore a less precise science than physics although the scientific procedures which it uses are basically the same.

Physical quantity is a term which is used to include measurable features of many different items. The area of a playing field, the mass of a bag of sugar and the speed of an aeroplane are all physical quantities. Abstract quantities, such as love, fear and hope are not measurable in the same way. In quoting any measurement of a physical quantity two items need to be stated. The first is the **numerical value** of the quantity and the second is its **unit.** *It is important at all times to think and write the value and the unit of any quantity together.* Apart from the technical accuracy of writing in this way it will also help you to acquire a mental appreciation of the size of the quantity you are considering. For instance, if the diameter of a wire is found to be 1.46 millimetres this appears reasonable; if the diameter is found to be 1.46 metres then something is wrong. A value stated as simply 1.46 does not give this check and is meaningless.

Realising whether a particular value is possible or not is something which comes with experience, but it is only by thinking about the sizes of physical quantities that this experience is gained.

1.2 THE INTERNATIONAL SYSTEM OF UNITS (SI)

History has provided for us far more units than are necessary. If units of length only are considered, we have miles, furlongs, yards, feet, light years, inches, fathoms, nautical miles, metres, microns and Ångstrøms, to say nothing about units of length which are not now used such as chains, rods and leagues. One unit of length can, of course, be converted into another by using a conversion factor. There are a huge number of conversion factors, however, so that few people know more than a few of them. The main problem with having so many units is that familiarity with one unit under one circumstance cannot easily be transferred to a different circumstance. Measuring distance up a mountain in feet, and distance along a road in

Table 1.1 Prefixes which can be used with any SI unit

Prefix	Multiplying factor	Symbol
atto	10^{-18}	a
femto	10^{-15}	f
pico	10^{-12}	p
nano	10^{-9}	n
micro	10^{-6}	μ
milli	10^{-3}	m
centi	10^{-2}	c
deci	10^{-1}	d
deca	10	da
hecto	10^{2}	h
kilo	10^{3}	k
mega	10^{6}	M
giga	10^{9}	G
tera	10^{12}	T
peta	10^{15}	P
exa	10^{18}	E

Several of these terms are used only infrequently.

yards or miles, makes it difficult to relate one to the other when the gradient of a road is needed.

The **metric system** of units was introduced at the time of the French Revolution to rationalise the chaos of units which then existed. It has been modified over the last two hundred years and all countries now use, at least in part, the **Système International (SI)** system of units. The advantage of the SI system of units is that any quantity has only one unit in which it can be measured. In the case of length the metre is the only unit of length used, together with multiple units such as the kilometre, and submultiple units such as the millimetre. Table 1.1 gives the value of each of these multiple and submultiple units. The prefixes may be added on to any SI unit.

Be particularly careful when changing from one multiple to another to get the arithmetic the right way round. If it is not done correctly then the answer is usually nonsensical. For example, if you change 5.09 km into metres and accidentally divide by 1000 then the equation 5.09 km = 0.005 09 m is produced. A mental image of these two lengths should immediately tell you that something has gone wrong. In this particular case your equation is incorrect by a factor of a million. In your study of physics you need always to be alert to nonsensical results.

The SI system of units starts by defining seven **base units** arbitrarily. This means that there is no particular reason for the size of the unit chosen. In practice, there are historical reasons for the approximate size of the base units but, as it has become possible to measure more accurately, earlier definitions have ceased to be used. The metre, for example, was intended to be one ten-millionth of the distance from the equator to the north pole, and while this is approximately true, it is now realised that the Earth is slightly irregular in shape so that all distances from the equator to the north pole are not the same. Added to this is the fact that it is necessary to have

Fig 1.1 An atomic clock at the National Physical Laboratory, London. Although it looks nothing like a normal clock it is capable of measuring time to 1 part in 10^{13} and is used as the standard in the United Kingdom.

the standard of length within a laboratory so that comparisons can be made with it. Similarly, the second was originally based on the time the Earth takes to rotate, the day. Accurate measurements now show that the Earth wobbles appreciably and is slowing down gradually, so that all days are not of the same length.

One of the seven base units, the candela, is a unit of luminous intensity and it will not be used in this book. The other six are listed in Table 1.2, but do not make any attempt to learn the figures involved. They are needed only by standards laboratories.

Several features of Table 1.2 need to be noted. As with all measurements, it is important to distinguish between the quantity being measured and the unit in which it is being measured. Later, when definitions are being given, it will be important to define units in terms of units, and quantities in terms of quantities. Expressions such as distance per second should not be used as this is a mixture of a quantity and a unit. The quantity *distance per unit time* or the unit *metre per second* should be used. Try to acquire good habits over such matters at an early stage and you will improve your accuracy.

Table 1.2   The six SI base units which are used in this book

Quantity	Unit	Definition
mass	kilogram (kg)	The mass of the international prototype kilogram; a particular cylinder of platinum–iridium alloy kept at the International Bureau of Weights and Measures at Sèvres, near Paris.
time	second (s)	The duration of 9 192 631 770 periods of the radiation corresponding to the transition between the two hyperfine levels of the ground state of the caesium-133 atom.
length	metre (m)	The distance light travels in a vacuum in 1/299 792 458 second.
electric current	ampere (A)	The constant current which, if maintained in two straight parallel conductors of infinite length and negligible circular cross-section, when placed 1 metre apart in a vacuum would produce a force of 2×10^{-7} newton per metre of length between these conductors.
thermo-dynamic temperature	kelvin (K)	The fraction 1/273.16 of the thermodynamic temperature of the triple point of water.
amount of substance	mole (mol)	The amount of substance of a system which contains as many elementary entities as there are atoms in 0.012 kilogram of carbon 12.

It can also be seen from the table that some of the base units depend on other units. It is therefore important that units should be defined in a definite order. The kilogram is now arbitrary and does not depend on anything else. The same is true of the second, which is defined the way it is because extreme accuracy can be achieved with clocks using caesium beam tubes. Such a clock (see Fig 1.1) using the precise frequency of the oscillation within the atoms of caesium would have an uncertainty in a time measurement of less than 1 part in 10^{13} (1 second in 300 000 years).

The metre, on the other hand, is dependent on the second having been defined first. The metre was first defined in this way in 1983 when accurate measurements of the speed of light, together with the accuracy of time measurements, made previous length measurements a little uncertain. The effect of this definition is that the speed of light in a vacuum is now, by definition, 299 792 458 metres per second exactly. This also means that experiments which used to be experiments to measure the speed of light have become experiments to measure distance.

More will be written about the other three base units in later chapters but it should already be clear that the ampere cannot be defined immediately as it depends on a unit of force which we do not yet have.

1.3 DERIVED UNITS

Using the base units it is possible to construct a system of units which can be used to measure all the quantities we shall need to measure. A list of quantities and their units is given in the Appendix. Units additional to the base units are called **derived units** and there are very many of them. Some of these derived units have specific names such as newton, joule and watt; others are clearly derived units as they do not have names and just use combinations of the base units, such as metres per second and cubic metres. Many quantities can be measured in two or more equivalent units. Note that, although units such as the newton are correctly abbreviated N as in 6.7 N, the capital letter is not used when the word is written in full, so this will become 6.7 newtons. The plural form of the abbreviation, Ns, is never used because N s means newton second.

Not only is it important to know the unit which is used for each quantity but it must be realised that the sequence in which they can be defined is restricted. It is impossible to define a unit in terms of something not yet itself defined. Table 1.3 shows some of the quantities upon which physics is based. It is essential that you know how to use all of them and their units by the end of your course. They are given in a particular order so that, as each quantity is introduced, its definition depends only on quantities already defined. The quantities in bold type are the SI base units.

Table 1.3   A possible sequence for defining certain quantities and their SI units. Thorough knowledge of these 20 quantities and their units is absolutely essential by the time you finish your course

Quantity	Unit	Abbreviation	Definition (Page number)
mass	**kilogram**	**kg**	4
time	**second**	**s**	4
frequency	hertz	Hz	140
length	**metre**	**m**	4
angle	radian	rad	105
area	metre^2	m^2	–
volume	metre^3	m^3	–
density	kilogram per metre^3	kg m^{-3}	422
velocity	metre per second	m s^{-1}	36
acceleration	metre per second^2	m s^{-2}	39
momentum	kilogram metre per second	kg m s^{-1}	46
force	newton	N	51
pressure	pascal	Pa	434
energy	joule	J	83, 87
power	watt	W	86
electric current	**ampere**	**A**	4
electric charge	coulomb	C	234
potential difference	volt	V	239
resistance	ohm	Ω	241
capacity	farad	F	298

1.4 DIMENSIONS

It is sometimes convenient to express a unit in terms of base units only. Because of the multitude of names of units, and because the same unit can be expressed in different ways, a comparison of units can best be made by using this form. Similarly the quantity itself can be expressed in terms of the base quantities. These expressions are called the **dimensions** of a quantity. Use may be made of the dimensions of a quantity to check an equation for correctness or to suggest the form an equation may take. Dimensions of a quantity are placed in square brackets [ ] and it is usually sufficient to express them in terms of [M] for mass [L] for length [T] for time and [I] for electric current.

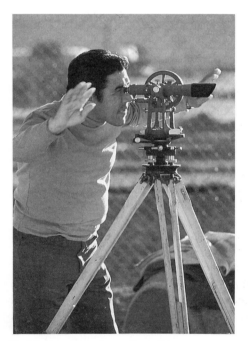

Fig 1.2 A surveyor using a calibrated telescope measures angles very accurately. The measurements can be used to calculate distances.

Example 1

Express the joule in terms of base units and, energy in its dimensions.

These two problems are very similar to one another so they are answered in parallel columns so you can see comparisons between them.

energy = force × distance

joule = newton × metre

force = mass × acceleration

newton = kilogram × metre second^{-2}

energy = mass × acceleration × distance

joule = kg m s^{-2} × m

Therefore dimensions of energy are:
$$[M \times LT^{-2} \times L]$$
$$= [ML^2T^{-2}]$$

Therefore the joule in base units is kg m^2 s^{-2}

Note that there is never any need to do both of these deductions, as conversion from the answer of one to the answer of the other is simply a matter of putting in kg as the unit of mass, m as the unit of length and s as the unit of time. Do be careful to distinguish between m, metre, the unit of length, and [M] the dimension of mass. This is one good reason for ensuring that you always use the square brackets [ ] around any dimension.

QUESTIONS

1.1 Find the dimensions of the following quantities:
frequency, force, pressure, work, power, potential difference, resistance, capacitance.

1.2 Express the following units in terms of their base units:
hertz, newton, pascal, joule, watt, volt, ohm, farad.

Example 2

Bernouilli's equation relates the pressure difference p between two points along a pipe with their height difference y, the density ρ of the fluid flowing along the pipe, its velocity v and the acceleration of free fall g. Show that each term in the equation has the same dimensions and find the dimensions of the constant k.

$$p + \rho g y + \tfrac{1}{2}\rho v^2 = k$$

PHYSICAL QUANTITIES AND UNITS

Quantity	Defining equation	Dimensions
pressure	$\dfrac{\text{force}}{\text{area}} = \dfrac{\text{mass} \times \text{acceleration}}{\text{area}}$	$\dfrac{[\text{M L T}^{-2}]}{[\text{L}^2]} = [\text{M L}^{-1}\text{T}^{-2}]$
density	$\dfrac{\text{mass}}{\text{volume}}$	$\dfrac{[\text{M}]}{[\text{L}^3]}$
acceleration	$\dfrac{\text{change in velocity}}{\text{time}}$	$\dfrac{[\text{L T}^{-1}]}{[\text{T}]} = [\text{L T}^{-2}]$
velocity	$\dfrac{\text{distance}}{\text{time}}$	$\dfrac{[\text{L}]}{[\text{T}]}$

Dimensions of

1st term p	2nd term $\rho g y$	3rd term $\frac{1}{2}\rho v^2$
$[\text{M L}^{-1}\text{T}^{-2}]$	$[\text{M L}^{-3}] \times [\text{L T}^{-2}] \times [\text{L}]$	$[\text{M L}^{-3}] \times [\text{L T}^{-1}]^2$
$[\text{M L}^{-1}\text{T}^{-2}]$	$[\text{M L}^{-1}\text{T}^{-2}]$	$[\text{M L}^{-1}\text{T}^{-2}]$

The dimensions of each term are the same and the equation is therefore dimensionally correct. The constant k must have the same dimensions as well. Constants do often have dimensions and therefore units; they can also be dimensionless as the $\frac{1}{2}$ was in this example, so the dimensions of $\frac{1}{2}\rho v^2$ are the same as the dimensions of ρv^2. Note that it is not possible to add quantities which are not dimensionally the same. You simply cannot add kilograms to seconds; there is no meaning to any answer you might get. This is equally true for all units, however they may seem to be nearly equal to one another. It is impossible to add them unless they are exactly equal. The analysis on the equation only shows that it is a possible equation; it does not and cannot check whether or not the equation is correct. There is no way in which dimensional analysis can find dimensionless constants.

QUESTIONS

1.3 When a body moves through a fluid there is a resisting force acting on the body. In the case of a sphere of radius r moving with constant velocity v through a liquid of density ρ, the force F is given by the equation

$$F = k\rho r^2 v^2$$

Show that k is a dimensionless constant.

1.4 Which of the following equations are dimensionally correct? If an equation is dimensionally incorrect it cannot be the correct equation. This provides a good way of checking the accuracy of equations, but remember it cannot take into account any dimensionless constants.

(a) $T = \dfrac{1}{2\pi}\sqrt{\dfrac{l}{g}}$ where T is a time, l a length and g an acceleration

(b) $F = mv^2 r$ where F is a force, m a mass, v a velocity and r a distance

(c) $E = mv^2$ where E is an energy, m a mass and v a velocity

(d) $I = \dfrac{ML}{12} + \dfrac{ML^2}{4}$ where I is a moment of inertia, L a length and M a mass

(e) $c = \sqrt{\dfrac{p}{d}}$ where c is a velocity, p a pressure and d a density

1.5 Show that the unit of momentum given in Table 1.3 as kg m s^{-1} is the same unit as the N s.

1.6 Express the unit of force and the unit of charge in terms of the SI base units. Using the equation

$$F = \frac{q_1 q_2}{4\pi \, \varepsilon_0 \, r^2}$$

where F is the force acting between two charges q_1 and q_2 when placed a distance r apart, express the unit of ε_0 in terms of the base units.

1.7 If velocity v, energy E, and force F, had been chosen as the fundamental physical quantities, what would be the dimensions of mass, length and time?

SUMMARY

- All physical quantities should be quoted with their numerical value and their unit.
- The SI system of units will be used exclusively throughout the course.
- The six base units used are kilogram, second, metre, ampere, kelvin and mole.
- 20 key quantities and their units are given in Table 1.3 in a correct sequence.

EXAMINATION QUESTIONS

1.8 What are the base SI units of mass, length and time?
Name the SI units of (i) energy, (ii) frequency.
The Stefan constant σ is measured in $W\,m^{-2}\,K^{-4}$.

Theoretically $\sigma = \dfrac{a\,k^4}{h^3\,c^2}$

where a is a dimensionless constant; k, the Boltzmann constant, has units $J\,K^{-1}$;
c, the speed of light, has units $m\,s^{-1}$.
What are (i) the units, (ii) the dimensions, of the constant h?

(UCLES 1983)

1.9 **(a)** Write a word equation to define electrical potential difference.
Hence or otherwise derive an expression for the volt in terms of base units.
(b) Explain what is meant by homogeneity in a physical equation.
Show that the equation below is homogeneous:

$$P = V^2/R$$

where P is the power dissipated in a circuit element of resistance R and V is the potential difference across the circuit element. **(9)**

(ULSEB 1988)

1.10 **(a)** Define *power*.
(b) Why is it that, in the SI system of units, power **cannot** be defined using the equation

power = potential difference $\times$ current? **(2)**

(UCLES 1988)

1.11 **(a)** Explain briefly what is meant by the *dimensions* of a physical quantity. Why is knowledge of the dimensions important? **(2)**
(b) State the dimensions of *energy*, and of the *moment of a force about a point*, and comment briefly on them. **(4)**
(c) The fundamental dimensions commonly used to analyse electrical quantities are mass (M), length (L), time (T) and current (I). Derive the dimensions of (i) *potential difference* and (ii) *inductance* in this system. **(3)**

(O & C 1983)

1.12 The unit $J\,s^{-1}$ can be used as the unit of power instead of the watt. Give a unit for each of the following quantities using appropriate combinations of metre (m), second (s), ampere (A), joule (J), and volt (V) only.
Quantity: energy; pressure; electrical charge; electrical resistance. **(4)**

(UCLES 1988)

1.13 **(a)** How do you check a formula for dimensional consistency? Why does this method of checking not give a definite confirmation that an equation is correct?
(b) A unit for μ_0 the permeability of a vacuum, is $kg\,m\,s^{-2}\,A^{-2}$. Use this unit, and a unit for ε_0 (Question 1.6) to decide which one of the following relations between $\varepsilon_0\,\mu_0$ and c, the speed of light in a vacuum, is dimensionally consistent:

$$\varepsilon_0\,\mu_0 = c^2;\ \varepsilon_0\,\mu_0 = c;\ \varepsilon_0\,\mu_0 = c^{-1};\ \varepsilon_0\,\mu_0 = c^{-2}.$$

(UCLES 1982)

1.14 **(a)** State the SI units of mass, length, time and current. **(2)**
Explain how the method of dimensions may be used to check physical formulae. What are the limitations of such dimensional analysis? **(3)**
(b) The speed v of ocean waves, g the acceleration of free fall and λ the wavelength, are related to one another by the equation $v = k\sqrt{(g\lambda)}$ where k is a dimensionless constant. Show that this equation is dimensionally consistent.
An ocean wave has a speed of $18.0\ m\,s^{-1}$ and a wavelength of 200 m. Calculate
(i) the wave frequency, (ii) the value of k. **(5)**
(Take the value of g, the acceleration of free fall, as $10\ m\,s^{-2}$.)
(c) Another ocean wave has a speed of $21.6\ m\,s^{-1}$. Calculate
(i) its wavelength, (ii) its wave frequency. **(4)**
(d) Deduce the dimensions of the following quantities in terms of mass, length, time and current:
(i) resistance
(ii) magnetic flux density
(iii) capacitance. **(6)**

(UCLES 1986)

1.15   The drag coefficient C_D of a car moving with a speed v through air of density ρ is given by

$$C_D = \frac{F}{\frac{1}{2}\rho v^2 A}$$

where F is the drag force exerted on the car and A is the maximum cross-sectional area of the car perpendicular to the direction of travel. Show that C_D is dimensionless (that is, it does not have a unit).   (3)

(UCLES 1987)

1.16   The dimensions of torque are the same as those of energy. Explain why it would nevertheless be inappropriate to measure torque in joules. State an appropriate unit.   (2)

(UCLES 1985)

PHYSICAL QUANTITIES AND UNITS

Chapter 2

MEASUREMENT TECHNIQUES

LEARNING OBJECTIVES

At the end of this chapter you should be able to:

1. find systematic and random uncertainties;

2. calculate the uncertainty in an experiment;

3. plan an experiment to give reliable readings;

4. use graphs accurately.

2.1 MEASUREMENT UNCERTAINTY

Whenever a measurement of a physical quantity is made, some measuring instrument has to be used to make that measurement. The instrument may be as ordinary as a ruler or as sophisticated as a modern mass spectrometer for chemical analysis. In using the instrument the experimenter has to make use of his own skill to obtain as accurate a reading as possible. Built into the instrument however is a limit of accuracy within which the experimenter is working. The result of this is that the readings which the experimenter takes have a degree of **uncertainty**. Physical quantities cannot be measured exactly with any instrument. An accurate clock might measure a time interval to a millionth of a second but even this is not quite exact. If a person is using a ruler, the reading will probably be taken to the nearest millimetre. The reading might then be stated as 208 ± 1 mm. This implies that the person taking the reading thinks that the best value is 208 mm but that the value will not fall outside the range from 207 mm to 209 mm. The ±1 mm is called the uncertainty of the reading. Sometimes people refer to this as the **error** in the reading but the word error seems to imply that a mistake has been made. This is not the case. Any reading *always* has an uncertainty. Many modern instruments hide this fact by giving digital displays of their readings.

A digital ammeter stating that the current is 568 mA seems to be stating that the current is exactly 568 mA. Correctly, the current in this case cannot be stated with any greater accuracy than to the nearest milliamp. It may be the case however that the uncertainty is greater than 1 mA. The meter itself may be fluctuating around the 568 mA figure; the meter may have been badly calibrated in the first place; the meter may have been misused; the current itself may be changing. All of these factors may make it more suitable to quote this reading as, say, (568 ± 20) mA. This raises another point about readings. If the uncertainty in this reading is ±20 mA, the figure 8 in the 568 mA is so uncertain that it is not worth while recording it. The reading then becomes (570 ± 20) mA.

This way of writing the value of the current introduces doubt about the number of significant figures being used. It is preferable to state the current in amperes as

$$(0.57 \pm 0.02) \text{ A}$$

or in standard form as

$$(5.7 \pm 0.2) \times 10^2 \text{ mA}$$

Either of these statements makes it quite clear that the uncertainty is in the final figure of the reading, the 7 in this example. Since the readings are themselves uncertain, the size of the uncertainty is even more difficult to be sure about. There is therefore almost never any need to give uncertainties to more than one significant figure.

The size of an uncertainty needs to be considered together with the size of the quantity being measured. The example given above of the ruler measuring a distance of 208 mm with an uncertainty of 1 mm may be an acceptable uncertainty for a particular experiment. However, an uncertainty of 1 millimetre will not be acceptable if the actual distance you are trying to measure is itself only 2 mm. In other words:

(208 ± 1) mm is a fairly accurate measurement
(2 ± 1) mm is highly inaccurate

In order to compare uncertainties, use is made of **absolute, fractional** and **percentage** uncertainties. For the reading (208 ± 1) mm:

1 mm is the absolute uncertainty
1/208 is the fractional uncertainty (= 0.0048)
0.48% is the percentage uncertainty

Note that the absolute uncertainty has units, mm in this case, but that the fractional and percentage uncertainties are ratios and therefore are dimensionless. As we usually require uncertainty to only one significant figure, the two values given above would be used as 0.005 and 0.5%.

QUESTIONS

2.1 State both the fractional and the percentage uncertainties if:
   (a) a distance of 7.84 m is measured to the nearest centimetre;
   (b) a time of 10.03 seconds is measured to the nearest 0.02 s;
   (c) a mass of 6000 kg is measured to the nearest 5 kg.

2.2 What percentage error is introduced by taking:
   (a) π to be 22/7;
   (b) π^2 to be 10;
   (c) 1 year to be $\pi \times 10^7$ s;
   (d) g to be 10 m s^{-2} when its value is 9.807 m s^{-2};
   (e) the speed of light to be 3×10^8 m s^{-1}, when its value is 2.998×10^8 m s^{-1}.

Note that these are errors rather than uncertainties because there is no need to make these approximations. It can, nevertheless, be useful to use approximations such as these when checking calculations, or when only approximate answers are required.

2.2 SYSTEMATIC AND RANDOM UNCERTAINTIES

Uncertainty can be of two types: **systematic** or **random**. A systematic uncertainty will result in all the readings taken being faulty in one direction. Using a stopclock which is running fast will result in all time readings being too big; using an ammeter with a zero reading of − 0.2 A will result in all readings taken being 0.2 A too small. Calibration errors result in systematic uncertainty and experimenters are somewhat too fond

Fig 2.1 (a) This student is introducing a large systematic error into his experiment because of parallax. In (b) he has the ruler closer to the mercury level he is reading and also has his eye on the same level. A mirror placed behind the mercury ensures greater accuracy if the image of the mercury level is aligned with the level itself.

of assuming that any systematic uncertainty must be due to the apparatus they are using. This is not necessarily so. Systematic uncertainty can be introduced into an experiment by poor experimental technique. The experimenter shown in Fig 2.1(a) will be reading all his length readings incorrectly because his eye is not on the same level as the mark on the ruler. He is introducing a parallax error which will affect all of his readings in the same way.

Systematic uncertainty is difficult to estimate and eliminate, but there are some simple procedures which can reveal it. Two ammeters placed in series must have the same current flowing through them. If their readings are not the same then there must be a systematic error in one of them. Use a third meter if you are not sure which one is in error. Two thermometers must read the same value in the same environment. Reducing systematic error can be done by using good experimental technique and by varying the instrumentation being used. Systematic uncertainty can never be eliminated by taking repeated readings with the same built-in faults.

Random uncertainties, as their name implies, result in a scatter of readings about a mean value. They have an equal chance of being positive or negative. Random uncertainty results from the inability of the observer to repeat his actions precisely. If the period of oscillation of a pendulum is being measured, an experimenter might be timing 50 swings. There are several things which cannot be reproduced exactly each time:

- the reaction time on the stopclock might vary a little;
- the start of the first swing and the end of the fiftieth swing may not be noted exactly;
- the same starting amplitude may not be used each time so that although the period of oscillation is nearly independent of the amplitude there is a small influence of amplitude on period;
- the hand of the stopclock may not be viewed from quite the same angle, so introducing random parallax uncertainty.

There are standard statistical methods for handling random uncertainties. These can give standard deviations for a series of readings, but when the number of readings is not large it is useful to have a method to obtain the approximate value of the uncertainty without doing a formal statistical analysis. Work through the following example to see one method of doing this.

Example 1
From the set of data given in Table 2.1, obtain values for d, the diameter of a wire, and T, the period of swing of a pendulum.

Table 2.1 Readings of the diameter of a wire and of 50T

	Diameter of wire /mm	Time for 50 swings of pendulum /s
	0.83	104.3
	0.83	104.5
	0.85	104.5
	0.83	104.5
	0.85	104.2
	0.86	104.7
	0.85	104.4
Average	0.843	104.44

First obtain the differences between the average value and the individual values. Ignore positive and negative signs. This gives the results of Table 2.2 which is normally the only table drawn up.

Table 2.2   Table 2.1 with the differences added

Diameter of wire /mm	Difference /mm	Time for 50 swings of pendulum /s	Difference /s
0.83	0.013	104.3	0.14
0.83	0.013	104.5	0.06
0.85	0.007	104.5	0.06
0.83	0.013	104.5	0.06
0.85	0.007	104.2	0.24
0.86	0.017	104.7	0.26
0.85	0.007	104.4	0.04
Average 0.843	**0.011**	**104.44**	**0.12**

The average difference is a measure of the uncertainty of the reading, so giving these values:

$$\text{diameter} = (0.84 \pm 0.01) \text{ mm} \qquad 50T = (104.44 \pm 0.12) \text{ s}$$
$$T = (2.089 \pm 0.002) \text{ s}$$

Note that in each case one significant figure has been dropped when quoting the final values. The uncertainty in the diameter measurement is much larger than the uncertainty in the period. This becomes apparent if percentage, rather than absolute, uncertainties are given.

Fractional uncertainty in diameter reading $= 0.01/0.84 = 0.012$
Percentage uncertainty in diameter reading $= 1.2$
Percentage uncertainty in $50T$ $= 0.1$
Percentage uncertainty in T $= 0.1$

The percentage uncertainty is not affected by dividing by 50. This is why a large number of swings need to be used in such a timing experiment. The absolute uncertainty in measuring the time for 50 swings is not much different from the absolute uncertainty for measuring the time for one swing. This gives only a small percentage uncertainty for $50T$ and corresponding accuracy for T.

QUESTION

2.3 The following readings of time interval were taken for a capacitor to discharge to 30% of its fully charged value:

67.3 s, 71.6 s, 68.4 s, 68.9 s, 63.6 s, 68.0 s, 70.2 s,
69.5 s, 70.6 s, 69.4 s, 69.0 s, 70.8 s

What is the time interval for this discharge and what is its fractional uncertainty?

2.3 ACCURACY AND PRECISION

These two words are often taken to mean the same thing. It is possible however to have readings taken with great precision which are not accurate. This will happen if there is a systematic error. Similarly it is possible to have readings which are accurate but not very precise. This will occur if there is only a small random error in the readings taken.

A precise reading will be taken to a large number of significant figures, but do be careful to use instruments of appropriate precision. It would be quite unsuitable to measure out food for a recipe using a balance capable of measuring to a milligram. It would take hours to get 100.000 grams of

Fig 2.2 Great experimental accuracy can be achieved with some modern equipment. Here a Rolls Royce engineer is using an electron microscope to make measurements on a specimen piece of material.

butter into a mixing bowl – and then some of it would stick on the wooden spoon. Similar absurdities can happen in a physics laboratory, resulting in students spending a long time taking unsuitable readings. While the aim in doing any experiment is to be as accurate as possible, there is no point in taking some readings to a very much higher precision than other readings in the same experiment. Ideally for each of the readings taken, the fractional uncertainty of each reading should be the same. In practice, things do not work out ideally but the fractional uncertainty for each reading should be known, at least approximately. Consider the following data which can be used with the equation stated to find the Young modulus E of a steel wire of length l and diameter d:

		Percentage uncertainty
Length of wire (l)	3.025 ± 0.005 m	0.17
Diameter of wire (d)	0.84 ± 0.01 mm	1.2
Mass supported by wire (M)	5.000 ± 0.002 kg	0.04
Extension caused (e)	1.27 ± 0.02 mm	1.6
Acceleration of free fall (g)	9.81 ± 0.01 m s^{-2}	0.10

$$E = \frac{4\,Mgl}{\pi e d^2}$$

It is clear from the percentage uncertainty column that two of these values are less certain than the others. But the overall uncertainty is also affected by d being squared. d is effectively used twice and so the error in d^2 is twice that in d. If any improvement is to be made to the above experiment therefore, the first reading which needs to be improved is that of d. A more accurate micrometer could be used, for example. After this, steps would need to be taken to improve the value of e. The experimental details of this experiment are given in Chapter 23. The value used for constants such as π is taken to as many significant figures as are needed to prevent the introduction of error as a result of using that constant. This is also true for physical constants such as g, the acceleration of free fall. Approximating g to have the value 10 m s^{-2} automatically introduces a 2% error into all your results. This may be acceptable in an experiment in which the overall uncertainty is 20% but it is not acceptable if, as in this case, an uncertainty of about 4% is achievable. If greater accuracy than this is needed then a book of constants should be used to enable you to determine the value of g for your locality.

It is necessary with readings such as those for the measurement of E, given above, to find the resulting uncertainty in E as a consequence of the uncertainty in each of the separate readings. When an expression involves only the multiplication or division of quantities, all that is needed to find the overall uncertainty is to add the percentage uncertainties of all the individual terms. In this case, therefore, the total uncertainty is

$$0.17 + 2(1.2) + 0.04 + 1.6 + 0.10 = 4.31\%$$

As stated above, this value is only an estimate and so only one significant figure is used and the value will be given as 4%. The equation gives the value of E as

$$\frac{4 \times 5.000 \times 9.81 \times 3.025}{3.1416 \times 0.00127 \times (0.00084)^2} = 210\ 819\ 704\ 710$$

This answer, taken straight from a calculator, must not be left as it is, because there are far too many significant figures, no units are given and there is no statement of uncertainty. An uncertainty of 4% of this is 8.4×10^9 giving

$$E = 210\ 800\ 000\ 000 \pm 8\ 400\ 000\ 000$$

This is still much too clumsy, so standard form needs to be used and the value of E is given as

$$E = (2.11 \pm 0.08) \times 10^{11}\ \text{Pa}$$

the pascal (Pa) being the SI unit in which the Young modulus is measured.

There is always an element of personal judgement in such results. It can easily be argued that since there are always likely to be more, rather than fewer, uncertainties to be taken into account, this result should be given as

$$E = (2.1 \pm 0.1\ ) \times 10^{11}\ \text{Pa}.$$

In making this calculation there are several points at which more significant figures are being used than is necessary. π need not have been substituted as 3.1416 for instance. However, with calculators available for easy calculation there is no harm done in the middle of a calculation if too many figures are used. Too few figures, however, must be avoided. At the conclusion of the calculation the corrected number of significant figures must be given. If you are in any doubt about the number of figures to use, then you will not usually be very far out if you use 3 significant figures.

In the above explanation all the terms in the expression were multiplied or divided. When addition or subtraction occurs in an expression then the absolute uncertainty must be found for that term. The rules for uncertainties are therefore:

addition and subtraction	ADD absolute uncertainties
multiplication and division	ADD percentage uncertainties
powers	MULTIPLY the percentage uncertainty by the power

Example 2

Find v and the uncertainty in v if

$$v = \frac{(a - b)\, d^2}{q\, \sqrt{T}}$$

and $a = (1.83 \pm 0.01)$ m, $b = (1.65 \pm 0.01)$ m, $d = (0.001\,06 \pm 0.000\,03)$ m $q = (4.28 \pm 0.05)$ s and $T = (3.7 \pm 0.1) \times 10^3$ s.

First calculate the percentage uncertainties in each of the 4 terms:

$(a - b)$	$= (0.18 \pm 0.02)$m	11%
d	$= (0.001\,06 \pm 0.000\,03)$ m	3%
q	$= (4.28 \pm 0.05)$ s	1.2%
T	$= (3.7 \pm 0.1) \times 10^3$ s	3%

The uncertainty in $(a - b)$ is now very large, although the readings themselves have been taken carefully. This is always the effect when subtracting two nearly equal numbers. The uncertainty in d^2 will be twice the uncertainty in d; the uncertainty in $\sqrt{T}$ will be half the uncertainty in T because a square root is a power of $\frac{1}{2}$. This gives:

uncertainty in $v = 11 + 2(3) + 1.2 + \frac{1}{2}\,(3) = 19.7\% \approx 20\%$

This gives $v = (7.8 \pm 1.6) \times 10^{-11}\,\text{m}^3\,\text{s}^{-1}$, a rather uncertain result which would be better expressed as:

$$v = (8 \pm 2) \times 10^{-11}\,\text{m}^3\,\text{s}^{-1}$$

There are some circumstances where the uncertainty in the final value is best found by working the problem through twice, once with the readings as taken and once with the limiting values which will give the maximum result. Equations containing trigonometrical ratios, or exponentials, or equations in which some of the terms appear both on the top and the bottom of the expression, such as

$$f = \frac{uv}{u + v}$$

are best dealt with this way.

QUESTIONS

2.4 Two lengths are recorded as (1.873 ± 0.005) mm and (1.580 ± 0.005) mm. What is the maximum possible value of the sum of the two lengths and what is the maximum possible value of the difference between the two lengths? What is the fractional uncertainty in both the sum of the two lengths and the difference between the two lengths?

2.5 In a simple pendulum experiment to determine g the equation used is

$$T = 2\pi \sqrt{\frac{l}{g}}$$

where T, the period is found to be (2.16 ± 0.01) s when the length l of the pendulum is (1.150 ± 0.005) m. Find the value of g and its uncertainty.

2.6 The volume of liquid V flowing through a pipe of radius r in time t is given by

$$\frac{V}{t} = \frac{\pi r^4 (p_1 - p_2)}{8 \eta L}$$

where p_1 and p_2 are the pressures at each end of the pipe, L is its length and η (eta) is a physical constant called the viscosity of the liquid. Use the following readings to determine η together with its uncertainty

$$
\begin{aligned}
r &= (0.43 \pm 0.01) \text{ mm} \\
p_1 &= (1.150 \pm 0.005) \times 10^5 \text{ Pa} \\
p_2 &= (1.000 \pm 0.005) \times 10^5 \text{ Pa} \\
L &= (5.5 \pm 0.1) \text{ cm} \\
V &= (10.0 \pm 0.1) \text{ cm}^3 \\
t &= (4.0 \pm 0.1) \text{ s}
\end{aligned}
$$

(These figures are based on typical figures for the flow of water through a hypodermic needle. The effect of the r^4 term means that needle size is much more important than thumb force in controlling the rate of flow of liquid through the needle.)

2.4 EXPERIMENT PLANNING

Careful design of an experiment can also play a large part in increasing the reliability of results. Many physical experiments are of the cause and effect type. They are designed to answer questions such as: 'What will happen to the length of a wire if the tension in the wire is increased?' Here the tension is the cause and the extension is the effect of that cause. It is of the utmost importance in such experiments to ensure that all other factors are kept constant. In this case the temperature of the wire must remain constant throughout the experiment. What other factors might change during the course of the experiment? How would you keep them constant?

The design of individual experiments varies from one topic to another and many experimental details are given in different places throughout this book, but there are some techniques which are common to many experiments and it is sensible to be aware of them and to work to establish good experimental practice. Some of these, together with some common faults, are listed here.

1. Support apparatus firmly. It is impossible to obtain reliable readings of force, for instance, if the piece of apparatus is wobbling. A firm support will also make it more difficult for apparatus to be damaged by falling over.

2. Make certain that the measuring instruments that you are using are suitable for the readings you are trying to make. An ammeter reading up to 30 mA will not be suitable to measure a current of 1 mA. Similarly a metre rule might be suitable for measuring the length of a wire but unsuitable for measuring its extension, when a vernier scale will probably be needed.

3. Repeat readings where possible. Random uncertainty can be reduced by taking many readings using the same apparatus, and systematic uncertainty can be reduced by using different apparatus and procedures. It is often possible to increase and then to decrease a quantity, taking readings in both directions. A particular example of this would be in examining how the resistance of a resistor varies with temperature. The temperature can first be raised, while values of resistance are measured and then lowered. If the two sets of readings are markedly different then perhaps the resistor has been damaged by the high temperature. If they are very similar to one another then an

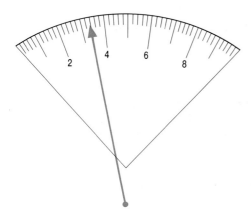

Fig 2.3 A common mistake with scales leads to this scale reading being given as 2.7. A quick check that it must be between 3 and 4 enables the correct value of 3.4 to be obtained.

Table 2.3   A table should have plenty of space for extra columns and extra lines

M /kg	x /m	y /m	$y - x$ /m	
0.100				
0.200				
0.300				
0.400				
0.500				
0.600				
0.700				

average can be taken. It is likely that the temperature of the resistor on the way up is recorded too low, while on the way down it may well have been recorded too high. Why?

Avoid parallax error when reading from a scale. See Fig 2.1. A mirror placed behind a scale can be used to ensure that the person taking the reading is directly in front of it. When viewed from the correct position the person will see an image of his eye directly behind the scale, and if there is a pointer over the scale he will not be able to see the image of the pointer.

5. Be careful to read scales correctly. Make certain you know what the smallest division on the scale is and just check that the reading you have taken is sensible. 2.7 is *not* sensible for the scale reading in Fig 2.3.

6. Be honest in taking readings. Do not decide in advance what you are going to find for a particular reading. By all means check that the reading is approximately what you expect. That is a good way of avoiding mistakes, but if you think that there is an established pattern in the figures you are recording and then slavishly follow it, it is likely that you will have to make only small early corrections to the actual figures but that the size of these corrections will be of an ever increasing size. You will also miss any real change that does take place in the pattern. There are examples in scientific history where small deviations from expected results have been absolutely crucial in establishing new theories. Einstein's theory of relativity is one such example.

7. Check for mistakes. Do not go blindly on when it should be clear that there is something drastically wrong. Try to find the cause of the mistake. If the mistake occurs in only a few, out of many, readings, then eliminate those readings from your calculations.

8. When performing counting experiments of any kind, make certain that you start counting from zero and not from one. If you are counting dots on a ticker tape from 1 to 10, then although you have counted up to 10 you will only have included 9 time intervals.

9. When measuring length, as, for instance, in measuring the difference in height between the two levels of liquid surfaces in a manometer, measure both levels from the same fixed horizontal surface. See Fig 2.4. This gives much greater accuracy than trying to hold a ruler in mid air and the zero alongside one of the liquid levels. Remember also that the difference in height between the levels of the liquid surfaces in a manometer gives only a difference in pressure. It does not give the total pressure in a system. Whether an absolute value of a quantity is required or the difference between two values of the quantity is often of crucial importance.

10. Take readings promptly. You will be more likely to keep constant those features which you want to keep constant if you do not delay taking readings unnecessarily. A battery will not have as much time to go flat; room temperature will not have as much time to alter; a resistor will not have as much time to heat up, etc. There can be some contradictory demands here, so each experiment must be taken on its merits and you will have to use some personal judgement. Experiments involving heating are often the exception to this advice as it is often necessary to wait until the whole system under test has acquired the same temperature.

11. Record readings actually taken on a prepared table. Do not do any arithmetic before recording these values. Do not use odd scraps of paper. Leave plenty of room for extra columns you may need and allow plenty of lines for more readings. Your table should look like Table 2.3 as you commence taking readings.

12. Use a graphical presentation of your results where possible, as explained in the next section.

2.5 GRAPHICAL REPRESENTATION

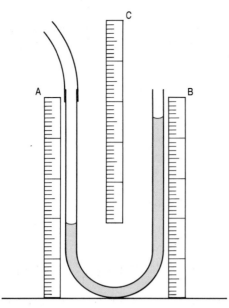

Fig 2.4 If the height difference between the liquid levels is required take two readings from A and B and substract. A single reading from C will have a large uncertainty.

A graph can be used to show very clearly the dependence of one quantity upon another. A single line graph can only be used to relate two quantities to one another, so again everything else must remain constant. Normally the cause is plotted on the *x*-axis. This is called the independent variable. The effect of that cause is plotted on the *y*-axis. Because the effect is dependent upon the cause, this is called the dependent variable. If time is one of the variables it is usually plotted along the *x*-axis.

If one quantity varies with another in a complex way then the graph will show this complexity. Often, however, a graph is used to establish a pattern in the variation; that is, to establish a law which can be used to predict what will happen on another occasion. In order to do this it is usually necessary to plot a straight line graph. The reason for this is that one curve can look very like another even though the algebraic equation that they are obeying may be different. A parabola can appear like a circle if only a portion of both are drawn.

The generalised equation of a straight line graph is

$$y = mx + c$$

where *m* is the gradient of the graph and *c* is the intercept on the *y*-axis.

If the graph from a series of readings is a straight line then it is possible to find the gradient and the intercept directly. Consider the graph drawn in Fig 2.5 which has been plotted to show distance against time. The gradient of this graph will give the speed and the intercept the distance from a fixed point at the start of the measurements. This gradient is given by:

$$\frac{67.5 - 27.0}{20 - 0} = 2.03 \text{ m s}^{-1}$$

The intercept on this graph is 27 m. This cannot be found to any greater certainty than to the nearest 0.5 m.

Note that this way of calculating the gradient makes it quite clear which points have been used to find the gradient and that the values on the *y*-axis have been estimated to a tenth of a small square. The gradient is quoted to 3 significant figures. A common mistake when calculating gradients is to quote inappropriate numbers of significant figures. Often this is because the calculation is done with too small a figure on the bottom of the division. It is not good practice to hunt for apparently convenient points on the graph. In this case it would not be correct to use the values of *d* when *t* is 12 and 7 to get a gradient of $(50 - 40)/5 = 2$. This has ignored the actual position of the line, and the accuracy of such a method is low.

If the relationship between two quantities being plotted is known, but is not a linear relationship, then it is still possible to plot a straight line graph by carefully choosing what to plot. Consider the gas law equation $pV = nRT$. If an experiment is being done to measure how the volume *V* of a fixed mass of gas varies with the pressure *p* at a constant temperature, then a straight line graph can be obtained by plotting $1/V$ against *p*. This graph will have the gradient $1/nRT$ and will pass through the origin if the gas is obeying this gas law.

Consider as an example an experiment on an amplifier connected to a series of different loudspeakers. The output resistance of the amplifier is *r* and the resistance of a loudspeaker is *R*. The power output *P* is measured for each different loudspeaker for a standard setting on the amplifier volume control. A straight line graph can be used to find the e.m.f. *E* of the amplifier's output and its output resistance, since the relationship between these terms is known:

$$P = \frac{E^2 R}{(R + r)^2}$$

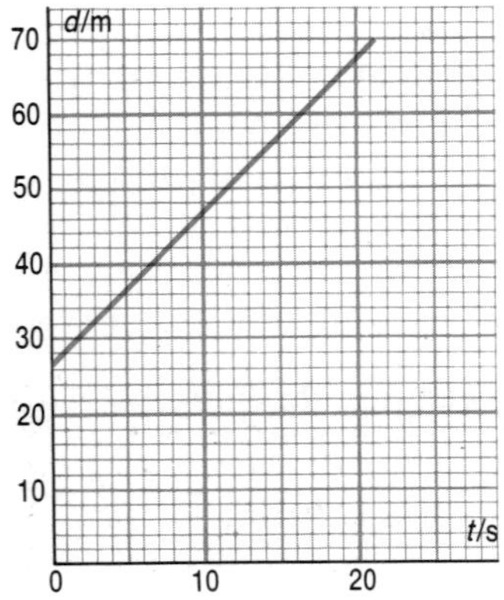

Fig 2.5

The equation has first to be re-written in the form $y = mx + c$. Note that the first term contains only a single variable, the second term a constant m and a variable, while the third term contains only variables. Here we can have any combination of R and P in the first term, anything in the second term and only a combination of r and E in the third term. It is necessary to try various possibilities until one is found which satisfies these requirements, e.g.

$$P(R+r)^2 = E^2 R$$

$$(R+r)^2 = \frac{E^2 R}{P}$$

$$(R+r) = E\sqrt{\frac{R}{P}}$$

$$R = E\sqrt{\frac{R}{P}} - r$$

If therefore R is plotted against $\sqrt{R/P}$, the gradient of the graph will be E and the intercept will be $-r$.

The experimental readings will therefore look something like those in Table 2.4.

Table 2.4 How the power out of an amplifier varies with the resistance of the loudspeaker to which it is connected.

R /Ω	V /V	I /A	P /W	R/P /Ω W^{-1}	$\sqrt{R/P}$ /$\Omega^{1/2}$ W$^{-1/2}$
2.0	2.90	1.40	4.06	0.493	0.702
3.0	3.90	1.27	4.95	0.606	0.778
4.0	4.79	1.19	5.70	0.702	0.838
8.2	7.35	0.90	6.62	1.24	1.113
12.1	8.80	0.73	6.42	1.88	1.37
16.4	9.83	0.60	5.90	2.78	1.67
30.5	11.8	0.40	4.72	6.46	2.54
51	12.8	0.25	3.20	15.93	3.99

The values of resistance have been measured using an ohmmeter. Readings of current I, and potential difference V, for the loudspeaker have also been taken. The remaining columns have been calculated from these readings. The plotted graph is shown in Fig 2.6.

The gradient of this graph is found to be 15.2 and the intercept is –9.0. This gives

$$E = 15.2 \text{ V and } r = 9.0 \ \Omega$$

Note that it is only numerical values which are plotted on graph paper. That is why the axes are labelled the way they are. R/Ω means the resistance in ohms divided by ohms, to give a figure with no units to be plotted. This graph has shown the relationship to be valid. The data could have been analysed in a different way. A graph plotted of power against resistance could have been used to find the maximum power which could be transferred from the amplifier to the loudspeaker. Inspection of the data shows this to be between 6 and 7 W.

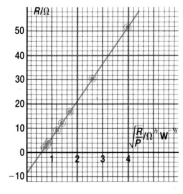

Fig 2.6

One other feature of the advantages of using graphical analysis is that, as here, several of the readings taken produce points which are rather too close together. Additional work could be done using other loudspeakers to increase the number of points in the higher regions of the resistance. Another point that becomes clear is that the ammeter being used is rather too insensitive when the resistance is large. It would be better to use a meter reading to 0.1 mA for the later readings.

Often the equation relating the two quantities being plotted is not known. In this case a log graph can be plotted. Suppose you are trying to find a relationship between the diameter of a piece of fuse wire and the current needed to cause it to melt. First of all you must have the fuse wire under controlled conditions, probably in a fuse holder. Then you will be able to measure the fusing current by increasing the current slowly until the fuse cuts off the current, and record the maximum ammeter reading. A table can be drawn up of current I against diameter d.

Assume that I is proportional to d to a power n, i.e.

$$I = kd^n \text{ where } k \text{ is a constant}$$

Taking logs of this equation gives

$$\log I = \log k + n \log d$$

The equation is now of the form required to plot a straight line graph. Log I can be plotted on the y-axis against log d on the x-axis. n will be the slope of the graph and the intercept will be log k.

Table 2.5 gives a series of values obtained for this experiment.

Table 2.5 The data for a logarithmic graph.

d/mm	I/A	$\log(d/\text{mm})$	$\log(I/\text{A})$
0.4	5.5	−0.398	0.740
0.5	8.4	−0.301	0.924
0.6	10.3	−0.222	1.01
0.7	13.5	−0.155	1.13
0.8	16.2	−0.097	1.21
0.9	21.0	−0.046	1.32

When these points are plotted, the graph is as shown in Fig 2.7. Note here how the position of the points does not lie very well along the line. This shows that the readings themselves were not very reliable. Whenever this happens, the graph plotted must be the best straight line through the points. If the points on a graph lie on a curve then the line drawn should be a smooth line rather than a zig-zag line joining the points. The spread of points around the line indicates the random nature of the readings themselves.

For this graph the gradient is 1.6, showing that

$$I = d^{1.6}$$

An estimate of the uncertainty can be obtained from any graph by considering the limits between which the line may possibly have been drawn. In this case the uncertainty was estimated to be 1.6 ± 0.1. This is a difficult experiment in which to get reliable results and the high uncertainty reflects this.

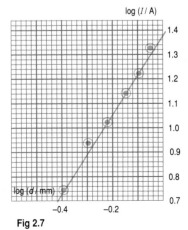

Fig 2.7

MEASUREMENT TECHNIQUES

Table 2.6

t/s	$v/\text{m s}^{-1}$
0	3.7
0.5	5.5
1.0	7.5
1.5	9.1
2.0	11.1
2.5	13.0
3.0	14.8
3.5	16.8

Table 2.7

f/Hz	Z/Ω
100	931
200	1020
300	1150
400	1310
500	1500
600	1690
700	1900

Table 2.8

Separation d/m	Force of repulsion F/N
0.050	2.03
0.055	1.53
0.060	1.17
0.065	0.92
0.070	0.74
0.075	0.60
0.080	0.49

2.7 What would you plot on the x-axis and the y-axis if you wanted to obtain straight line graphs from the following equations, if values are known for the variables and the graph is to be used to find the constants.

(a) $\dfrac{1}{u} + \dfrac{1}{v} = \dfrac{1}{f}$ f is constant

(b) $f = \dfrac{mu}{m+1}$ f is constant

(c) $I = \tfrac{1}{2} M (R_1^2 + R_2^2)$ R_1 and M are constants

(d) $v = \omega \sqrt{A^2 - x^2}$ A and ω are constants

2.8 The readings in Table 2.6 were obtained from an experiment to determine the acceleration of a vehicle. The velocity v of the vehicle at times t were recorded. Plot a graph and use it to find:
(a) the acceleration;
(b) the velocity when $t = 0.8$ s;
(c) the velocity when $t = 4.0$ s;

Why is interpolation, that is, using the graph to find an intermediate value as in (b), more reliable than extrapolation, that is, using the graph to find a reading beyond the end of the graph as in (c)?

2.9 The impedance Z of an electrical circuit is given by

$$Z = \sqrt{R^2 + 4\pi^2 f^2 L^2}$$

where R and L are constants and the frequency f is variable. Use the data of Table 2.7 to plot a suitable graph which will enable you to find values of R and L.

2.10 The force of repulsion F between two magnets placed in line with one another, N pole to N pole, varies with the separation d of their centres in the way shown in Table 2.8.
Find the law relating the repulsive force to the distance, assuming it is of the form $F = kd^n$.

SUMMARY

- For a reading $a \pm \delta a$, δa is the absolute uncertainty, $\delta a / a$ is the fractional uncertainty and $100 \delta a / a$ is the percentage uncertainty.
- When quantities are added or subtracted, the uncertainty in their total is the sum of the absolute uncertainty in each.
- When quantities are multiplied or divided, the fractional uncertainty in their total is the sum of the fractional uncertainties in each.
- If $b \pm \delta b$ and $c \pm \delta c$ are two values to be combined then

     if $y = b - c$ $y \pm \delta y = (b - c) + (\delta b + \delta c)$

           $x = b \times c$ $\dfrac{\delta x}{x} = \dfrac{\delta b}{b} + \dfrac{\delta c}{c}$

           $z = b^n$ $\dfrac{\delta z}{z} = n \dfrac{\delta b}{b}$

EXAMINATION QUESTIONS

2.11 The reading of a speedometer fitted to the front wheel of a bicycle is directly proportional to the angular velocity of the wheel. A certain speedometer is correctly calibrated for use with a wheel of diameter 66 cm but, by mistake, is fitted to a 60 cm wheel. Explain whether the indicated linear speed would be greater or less than the actual speed and find the percentage error in the reading.

(UCLES 1984)

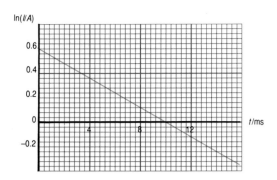

2.12 In performing an experiment to determine the Young modulus of steel, a student recorded the following values:

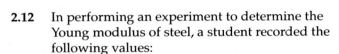

> length of wire (3.255 ± 0.005) m,
> diameter of wire (0.63 ± 0.02) mm,
> force applied to wire (26.5 ± 0.1) N,
> extension of wire (1.40 ± 0.05) mm.

(i) Calculate the value which these readings give for the Young modulus and calculate the uncertainty in the value. (8)

(ii) Discuss ways in which the experiment could be improved so as to obtain a more reliable value for the Young modulus. (3)

(UCLES 1988)

2.13 (a) Distinguish between a *systematic error* and a *random error* in the measurement of a physical quantity. Give one example of each type of error.

(b) For each of your examples describe one way in which the uncertainty of the measurement could be reduced. (8)

2.14 A quantity y varies with time according to the equation

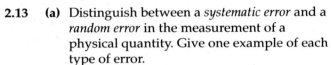

$$y = Be^{-kT} \quad \text{(i)}$$

where k and B are constants.

(a) In order to test whether the quantities y and t in a particular situation are related as in equation (i), it is usual to plot a graph of ln y against t. Why is this?

(b) One situation where this equation applies is when the current decays in a d.c. circuit with a large inductance after the power has been switched off. The line graph, represents the decay of current with time in such a circuit.

Obtain a value for k from the graph.
Sketch a graph showing how the current I varies with time t.
What does the constant B of equation (i) represent in this case? Mark its significance on the sketch graph you have just drawn. The length of time it takes for the current to decay to $1/e$ or 0.37 of its original value is called the time constant of the circuit. In what way does increasing the value of k in equation (i) change the time constant? (12)

(ULSEB 1988)

Chapter 3

SCALARS AND VECTORS

> **LEARNING OBJECTIVES**
>
> At the end of this chapter you should be able to:
>
> 1. distinguish between a scalar and a vector;
>
> 2. add, subtract and multiply vectors;
>
> 3. put a vector into its components.

3.1 SCALARS AND VECTORS

When dealing with physical quantities in Chapter 1, it was stated that all such quantities have a numerical size and a unit. Some physical quantities also have direction. Any physical quantity which has direction is called a **vector**. A physical quantity which does not have direction is called a **scalar**. To specify a scalar, only its value and its unit need be given. To specify a vector the three terms value, unit and direction must all be given. The numerical size, the value, of a vector is often called its **magnitude**.

The way in which the direction of a vector is given varies from one situation to another, but should at all times be stated explicitly even when it seems obvious. It might be given as in the direction of motion, or as upwards or downwards, or as due north, or as north 50° east. It should not be given as vertical, or horizontal, or along the axis, or parallel to the surface, because in each of these cases there is more than one possibility; 'vertically' may be vertically upwards or vertically downwards. The important point is that the direction must be clear. Often the simplest way of clarifying the direction of a vector is to draw a diagram. Scale diagrams are often useful to show vectors, in which case the length of the line drawn represents the magnitude of the vector, and the direction of the line shows the direction of the vector. Fig 3.1 shows a typical vector scale drawing and also indicates that vectors will be printed in bold italics.

Table 3.1 gives a list of many of the physical quantities you will be using and divides them up into scalars and vectors. There are some fine distinctions which need to be made in deciding what words to use and hence into which category a quantity is placed. If you are in doubt about any of the terms, look up the definition of that term by using the index.

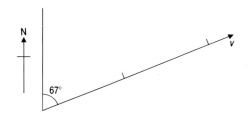

Fig 3.1 A vector showing the velocity **v** of 21.8 m s⁻¹ in a direction of N 67° E. Each unit of length represents 10 m s⁻¹.

3.2 VECTOR ADDITION

The addition of scalars is no problem; the two quantities, which must have the same unit, are added together by the normal rules of addition. A mass of 6.4 kg added to a mass of 8.3 kg gives a total mass of 14.7 kg.

Adding two vectors together is a different matter because of their directions. If they happen to be in the same direction then direct addition is possible, but usually they are not in the same direction and it is possible to have a vector of magnitude 5.3, when added to a vector of magnitude 7.0, giving a total vector of anything between 1.7 and 12.3.

Table 3.1 Scalar and vector quantities

Scalar	Vector
length	displacement
speed	velocity
time	acceleration
volume	force
mass	weight
energy	momentum
frequency	torque
density	moment
pressure	electric current
power	magnetic flux density
charge	electric field
capacity	
potential difference	
temperature	

In order to add vectors, use is made of the **parallelogram law**. This is illustrated in Fig 3.2, where vector A is being added to vector B to give vector R. It is frequently convenient to draw only half of the parallelogram, in which case the second vector is added on to the first nose-to-tail, Fig 3.3. It can be seen from Fig 3.4 that $A + B$ produces the same result as $B + A$.

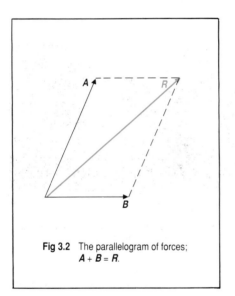

Fig 3.2 The parallelogram of forces;
$A + B = R$.

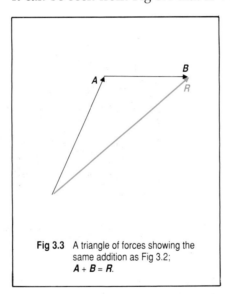

Fig 3.3 A triangle of forces showing the same addition as Fig 3.2;
$A + B = R$.

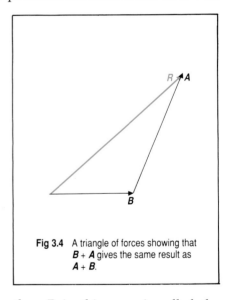

Fig 3.4 A triangle of forces showing that $B + A$ gives the same result as $A + B$.

Fig 3.5 In a high wind a person leans into the wind when walking. Gravity still acts vertically downwards but the force the ground exerts on the person has components forwards and upwards. The forward component can be equal and opposite to the force of the wind.

The result of adding two vectors together, R in this case, is called the **resultant** vector. Since so many physical quantities are vectors it is essential that you can either find a resultant by scale drawing or by calculation. Calculation of a resultant usually requires the use of the cosine equation for a triangle. This is

$$a^2 = b^2 + c^2 - 2bc \cos A$$

where a, b and c are the lengths of the three sides of the triangle and A is the angle between the sides of length b and c .

This process of addition can be extended to finding the sum of as many vectors as are wanted. Fig 3.6(a) shows the various forces acting on a tractor when it is being transported on a lorry. The forces act at many different points on the tractor, but nevertheless the resultant force on it can be found by adding all the forces together as shown in Fig 3.6(b) where the resultant R is given by

$$A + B + C + D + E + F = R$$

SCALARS AND VECTORS

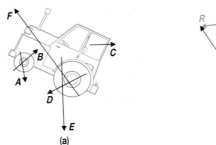
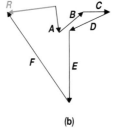

(a)   (b)

Fig 3.6 The sum of many vectors can be obtained by drawing the vectors to scale in this nose to tail way. All the forces on the tractor give a resultant force **R**.

The same resultant is obtained whatever the order in which the forces are added. This procedure can be extended into three dimensions if required, but the geometry of the diagram can then be quite difficult to work with.

If the result of adding three vectors together is zero then a **triangle of vectors** is obtained. If more than three vectors are added together and give a total which is zero, then the figure obtained is called a **polygon of vectors**.

Fig 3.7 There will be a large number of forces acting on each of these tractors but they will be fixed to the trailer and the resultant force on a tractor must give it the same acceleration as the lorry.

3.3 VECTOR SUBTRACTION

The process of vector subtraction is, in principle, similar to vector addition. The process does however produce some surprising results. It is worthwhile first considering ordinary scalar subtraction.

Consider a simple scalar problem.

Example 1
A ball at one moment has a kinetic energy of 11 J. Later its kinetic energy has increased to 18 J, find the increase in its kinetic energy.

The problem is almost too trivial; the answer is

$$18\,J - 11\,J = 7\,J$$

Note that the problem could have been done by finding what needed to be added to the first value to obtain the final value, i.e.

$$11\,J + 7\,J = 18\,J$$

This illustrates the way which is used to subtract vectors. Again there are two possible ways of looking at the problem.

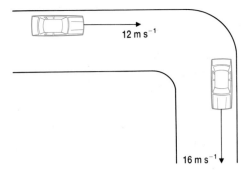

Fig 3.8 A car changing its speed by 4 m s⁻¹ as it goes round a corner changes its velocity by a very different amount.

Example 2

A car goes round a right-angled corner so that its velocity changes from 12 m s⁻¹ due east to 16 m s⁻¹ due south, see Fig 3.8. What is its change in velocity?

The first point to make is that here we are *not* dealing with scalar quantities and that therefore the answer is *not* 4 m s⁻¹. Just subtracting 12 m s⁻¹ from 16 m s⁻¹ finds the change in the speed of the car but the change in its velocity is affected very much by the change in direction. To find the change in velocity we can use one of two methods as illustrated by Figs 3.9 and 3.10. Both methods give the same answer. In the first method the vector (*–start velocity*) is added to the vector (*final velocity*). In the second method, often the easier method in practice, use is made of the fact that the change in velocity is the vector which has to be added to the start velocity to get the final velocity.

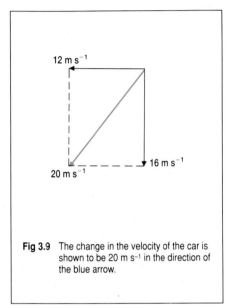

Fig 3.9 The change in the velocity of the car is shown to be 20 m s⁻¹ in the direction of the blue arrow.

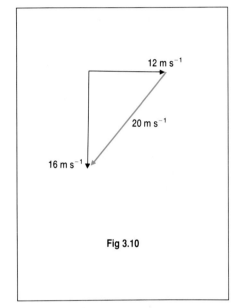

Fig 3.10

In Fig 3.9 vector –12 m s⁻¹ has been added to vector +16 m s⁻¹ to give a value of the change in the velocity of 20 m s⁻¹ in a direction of S 36.9° W.

In Fig 3.10 the required change of velocity is the vector which needs to be added on to the start velocity to obtain the final velocity.

This can be expressed in equation form in the following two ways

First method:- $(B - A) = B + (-A)$

Second method:- $A + (B - A) = B$

QUESTIONS

3.1 Show that for any two vectors A and B, the vector $(B - A)$ is equal to $-(A - B)$.

3.2 Find the change in velocity of a yacht if it changes its velocity from 5 m s⁻¹ due N to 3 m s⁻¹ due W.

3.3 Find the change in velocity of a tennis ball if it approaches a tennis racquet at 30 m s⁻¹ and leaves the racquet in the opposite direction at 40 m s⁻¹.

3.4 An aircraft travels through the air with a velocity of 200 m s⁻¹ due east. It is found to be travelling over the ground with a velocity of 220 m s⁻¹ in a direction 20° S of E. The difference between these two velocities is due to the wind velocity. What is the wind velocity?

SCALARS AND VECTORS

3.5 The Moon is travelling around the Earth in a circle at a constant speed of 1000 m s^{-1}.
  (a) What angle does it turn through in its orbit in 1 second if it takes 28 days for 1 complete revolution?
  (b) What change in velocity takes place in 1 second?
  (c) In what direction does this change in velocity occur?

The solution to question 3.5 is referred to in the chapter on motion in a circle, Section 8.2. Similar problems to these can be constructed in terms of any vector and not just velocity.

3.4 COMPONENTS OF VECTORS

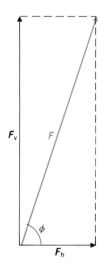

Fig 3.11 Force **F** is shown resolved into two components: F_h is the horizontal component and F_v is the vertical component.

In some circumstances it is useful to split a vector up into two parts. These parts are called the **components** of the vector. Very often the components are taken horizontally and vertically, in which case the two parts are called the horizontal component of the vector and the vertical component of the vector respectively. (If working in three dimensions, three components are often used.) Fig 3.11 shows how in two dimensions vector **F** is split into its two components F_h and F_v. The two components, taken together, have exactly the same effect as the single vector **F**.

The magnitudes of the two components are given by

$$F_h = F \cos \phi$$
$$F_v = F \sin \phi$$

In these two statements bold italic type is not used as the magnitude only is being given for the vectors. There is a real danger when using diagrams such as this that too many vectors will be used. The components are used *or* the vector itself. All three must not be used together.

It is sometimes convenient, especially in a complex problem, to resolve all the forces. In Fig 3.12 the four vectors, **A**, **B**, **C** and **D** have been resolved into their vertical and horizontal components. The sum of all four vectors is therefore equivalent to the single vertical vector

$$A_v + B_v - C_v - D_v$$

together with the single horizontal vector

$$A_h + B_h + C_h - D_h$$

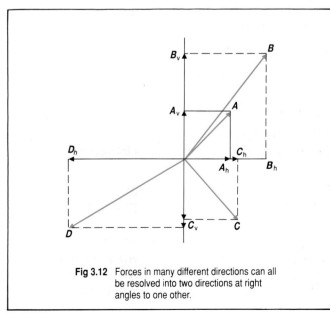

Fig 3.12 Forces in many different directions can all be resolved into two directions at right angles to one other.

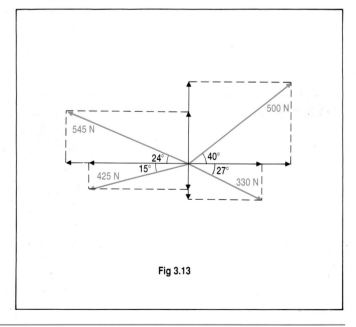

Fig 3.13

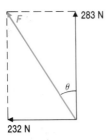

Fig 3.14

Example 3
Find the resultant of the four forces shown in Fig 3.13.

In the horizontal direction to the left the total force is given by:

$$545 \cos 24° + 425 \cos 15° - 500 \cos 40° - 330 \cos 27°$$
$$= \quad 498 \quad + \quad 411 \quad - \quad 383 \quad - \quad 294 \quad = 232 \text{ N}$$

In the vertically upwards direction the total force is given by:

$$545 \sin 24° + 500 \sin 40° - 425 \sin 15° - 330 \sin 27°$$
$$= \quad 222 \quad + \quad 321 \quad - \quad 110 \quad - \quad 150 \quad = 283 \text{ N}$$

Using the Pythagoras theorem enables the resultant force F to be found, Fig 3.14.

$$232^2 + 283^2 = F^2$$
$$F = 366 \text{ N}$$

The direction of F is given by

$$\tan \theta = 232/283$$

The resultant force F therefore has a magnitude of 366 N inclined in a direction 39° to the left of vertically upwards.

DATA ANALYSIS

Forces in frameworks
Figs 3.15 and 3.16 are photographs of two very common types of structure. In civil engineering terms they are called pin-jointed structures. The design of such structures must depend on the use to which the structure is to be put and the loading forces which are applied to it. The strength, and therefore the size, of each of the struts in such a structure need to be

Fig 3.15 The construction of a steel pylon can be seen to be a series of triangles.

Fig 3.16 A roof under construction shows the extensive steel framework consisting of a series of girders built from triangular frames.

calculated so that the most economical design can be constructed. This exercise deals with a similar, but simplified, structure of a roof which is constructed out of triangles with sides of length 3 m, 4 m and 5 m. Fig 3.17 shows *all* the forces acting on each of the joints in the framework of the roof. The forces in the struts are either forces of compression or forces of extension and they always act in line with the strut. Other external forces are applied to the structure as shown. At each joint the total force has to be

SCALARS AND VECTORS

zero if the structure is to be in equilibrium. (Further details concerning equilibrium are given in Chapter 6. This exercise is basically concerned with the addition of force vectors.)

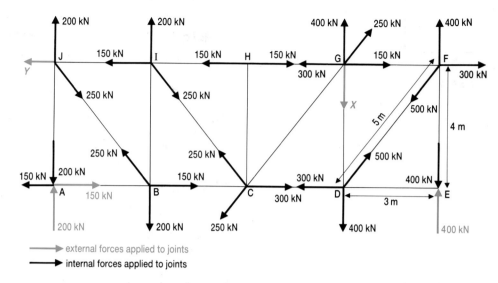

Fig 3.17 The forces acting on the joints in a steel roof structure can be calculated by working from joint to joint. The resultant (total) force on each joint must be zero.

Using the information given on the diagram, answer the following questions.

(a) Explain how the diagram shows that joints E and H are in equilibrium.
(b) Explain similarly why joint A is in equilibrium.
(c) Draw triangle of force diagrams to show that the total force acting on each of joints B, C, D, F and I is zero.
(d) Draw the polygon of forces diagram for joint A.
(e) Draw the triangle of forces for joint J. Use the fact that the total of all the forces at the joint must be zero and hence find the value of Y.
(f) Draw the polygon of forces for joint G and use it to find X.
(g) Which of the struts are in tension and which are in compression?
(h) Which 4 struts are, for this loading, redundant? That is, which struts are neither in compression nor in tension?

This exercise could have been reversed. You could have been given all the external loadings, drawn the force diagrams one by one around the structure and hence found all the internal forces applied to the joints. This reverse process is carried out by a structural engineer (or a computer) in the design process to determine the particular steel struts to use.

3.5 MULTIPLYING VECTORS

There is much more to multiplying vectors than is being stated here. Further detail can be found in books on vector analysis.

To multiply a vector by a scalar is straightforward. If vector F is multiplied by scalar t then the product is a vector of magnitude Ft in the same direction as F.

Example 4

Multiply a force of 6.0 N acting in a direction due north by a time of 15 s.

You will see in Chapter 7 that this alters the momentum of a body by an amount of 6.0 N × 15 s = 90 N s in a direction due north.

Multiplying a vector by a vector can give either a scalar product or a vector product. The following examples give one of each.

Example 5

Multiply a force of 650 N in a direction due east by a velocity of 30 m s⁻¹ in the same direction.

You will see, also in Chapter 7, that this calculation is needed to find the power output of the machine producing the force. Power is a scalar quantity and the power output has no direction; in this case it has a value of 19 500 W.

If there had been an angle θ between the two vectors of force F and velocity v then the resulting power would have been $Fv \cos \theta$.

Example 6

A current of 3.2 A flows vertically upwards through a wire of length 0.045 metres. The wire is at a angle of 30° to a magnetic field of flux density 0.089 T. Find the force on the wire.

The equation relating these quantities is $F = BIl \sin \theta$ (Chapter 17), where F is the force, B is the magnetic field, I is the current, θ is the angle between B and I, and l the length. Clearly two vectors have to be multiplied together and the result in this case is

$$
\begin{aligned}
F &= 0.089 \times 3.2 \times 0.045 \times \sin 30° \\
&= 0.0064 \text{ N}
\end{aligned}
$$

This force acts in a direction at right angles to the plane containing B and I. The rule for multiplying two vectors A and B together to get a vector product is that

the magnitude of the total vector is $AB \sin \theta$
the direction is at right angles to the plane containing A and B

where θ is the angle between the vectors.

QUESTIONS

3.6 How much work is done when a force of 3.0 kN moves a distance of 0.36 m
(a) in the direction of the force,
(b) in a direction at right angles to the direction of the force,
(c) in a direction at 40° to the direction of the force?

3.7 Using the equation given in Example 6, find the force which a magnetic flux density of 0.085 T exerts on a wire of length 0.063 m when it is carrying a current of 4.3 A if
(a) the wire is at right angles to the field,
(b) the wire is in the same direction as the field,
(c) the wire is at an angle of 25° to the field.

SCALARS AND VECTORS

- Scalars have magnitude only. They can be added, subtracted, multiplied and divided by normal arithmetic.
- Vectors have direction as well as magnitude. They must be added or subtracted by the parallelogram rule (see Fig 3.18). They can be multiplied in which case they can give

  a scalar as in   **force** × **displacement** = work
  or a vector as in **magnetic flux density** × **current** × length = **force**

- Vectors can be resolved into two vectors at right angles (see Fig 3.19).

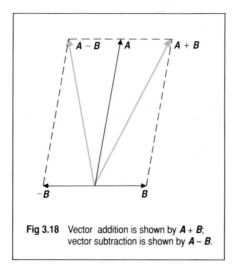

Fig 3.18 Vector addition is shown by **A** + **B**; vector subtraction is shown by **A** − **B**.

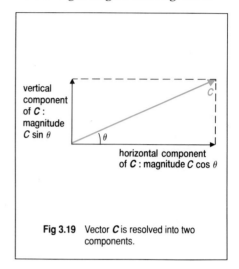

Fig 3.19 Vector **C** is resolved into two components.

Fig 3.20 All the forces acting on a yacht determine its angle to the vertical.

EXAMINATION QUESTIONS

3.8 Explain what is meant by (i) a scalar, (ii) a vector.
For each of the following state whether it is a scalar or a vector:
force, length, mass, momentum, kinetic energy, electric charge, amplitude, displacement, half life, moment of a force.
(UCLES 1985)

3.9 Two forces A and B act on an object. A is a force of 8.5 N acting vertically down. The resultant force on the object is a horizontal force of 6.3 N. Find B.

3.10 A particle moves in a circle of radius r with uniform angular speed. At one instant the particle is at A, and its instantaneous velocity is v in the direction AA'; at a time of δt later the particle has moved to B, a distance $r\,\delta\theta$ along the arc (where $\delta\theta$ is a small angle), where its velocity is v in the direction BB'.
  (a) Draw a vector diagram to show δv, the change of velocity that occurs during this time.
  (b) Show that the magnitude of δv is approximately equal to $v^2\delta t/r$.
  (c) Hence find the magnitude and direction of the acceleration of the particle.
(UCLES 1985)

3.11 A child pushes a toy box 4.0 m along the floor by means of a force of 6.0 N directed downwards at an angle of 37° to the horizontal. How much work does the child do?

Theme 2

NEWTONIAN MECHANICS

Sir Isaac Newton published his classic work Principia Mathematica just over 300 years ago in 1687. It has had a profound effect on the development of science since then and the principles put forward in it are still used a great deal today.

Chapter 4

MOTION

LEARNING OBJECTIVES

At the end of this chapter you should be able to:

1. relate distance travelled, velocity, time and acceleration for any uniformly accelerated motion;

2. use distance–time and velocity–time graphs;

3. calculate the momentum of an object.

4.1 SPEED AND VELOCITY

The speed of a car is often referred to as being, say, 120 kilometres per hour. This does not mean that the car must travel 120 kilometres or that it is on a journey that takes one hour. The expression implies a rate of movement of, in this case, 33.3 metres per second, or 33.3 millimetres per millisecond or even 0.0333 millimetres per microsecond. All of these figures, when put into SI units, will be 33.3 m s^{-1}. This illustrates the point that the speed of an object can be found at a particular moment, and this is called the **instantaneous speed** of the object. Instantaneous speed can be defined by the equation:

 instantaneous speed = $\dfrac{\text{small distance travelled by the object}}{\text{small time taken}}$

when the small time taken gets closer and closer to zero.

Those of you who are familiar with calculus will see that this is similar to one of the basic calculus ideas, differentiation, in which the gradient of a graph is found, see Fig 4.1. Indeed the instantaneous speed of an object can be defined as the gradient of the distance–time graph for the object.

The **average speed** of an object is defined by the equation

 average speed = $\dfrac{\text{total distance travelled}}{\text{total time taken}}$

If an object travels at a constant speed for a whole journey, then the instantaneous speed will be constant and equal to the average speed. In practice such journeys are unusual so it is much more reliable to write carefully which speed is being referred to. For example, 'the speed at the start is . . .' or 'the speed after 20 seconds is . . .' The statement 'speed is . . .' should be avoided; it is much too vague and leads to mistakes.

In many cases not only is the speed of an object known but also the direction in which it is travelling. This therefore implies that the quantity being dealt with is a vector. To make it quite clear when a vector is being used, a different term is applied. **Velocity** is always a vector and the **instantaneous velocity** of an object is defined by the equation:

 instantaneous velocity = $\dfrac{\text{small distance travelled in a stated direction}}{\text{small time taken}}$

as the small time approaches zero. This is a rather clumsy definition and so if **displacement** is the word used to mean the distance moved in a stated

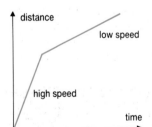

Fig 4.1 The gradient of a distance–time graph gives the speed.

direction, then we can define instantaneous velocity by the rather briefer statement:

instantaneous velocity is the rate of change of displacement

As with speed we can also use the term **average velocity**. This is defined by the equation:

$$\text{average velocity} = \frac{\text{total distance travelled in a stated direction}}{\text{total time taken}}$$

$$= \frac{\text{displacement}}{\text{total time}}$$

$$= \frac{\Delta x}{\Delta t}$$

The symbol Δ, the Greek letter capital delta, is read as 'the change in'. So average velocity can be read as 'delta x divided by delta t' or as 'the change in x divided by the change in t'. The symbol δ, the small Greek letter delta, is used to mean 'the small change in'. Instantaneous velocity can therefore be written

$$v = \frac{\delta x}{\delta t} \quad \text{as } \delta t \text{ approaches zero}$$

Using calculus terminology this is written

$$v = \frac{dx}{dt}$$

Measurement of velocity

Measurement of speed and/or velocity is often carried out by measuring a distance and dividing by the corresponding time. Standard methods for doing this in the laboratory may only involve the use of a metre rule and a stopwatch. Another method involves the use of a ticker-timer which makes a mark on a piece of paper moving at the required speed every 1/50 of a second. Electronic methods for measuring the time interval can also be used. These involve the object whose speed is required, interrupting a beam of light to a photocell. The time that the photocell is not working is the time it takes for the object to pass through the beam, see Fig 4.2. Multiflash photographs can be used, provided there is a scale for measuring the distance, see Fig 4.6.

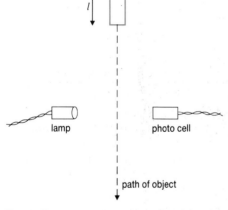

Fig 4.2 The time taken for an object of length *l* to pass between the lamp and the photocell can be used to calculate the speed of the object.

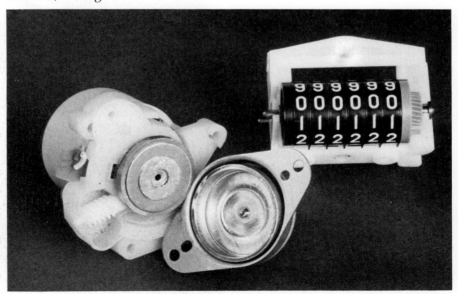

Fig 4.3 A car speedometer. The rotation of the aluminium disc is directly proportional to the car's speed.

Fig 4.4 An electromagnetic flow meter used for measuring the speed of an electrically conducting liquid. There is nothing inside the meter to impede the liquid's flow.

A car's speedometer uses a different method altogether. Here an aluminium disc is rotated at a rate proportional to the speed of the car, just under a magnet. The magnet can rotate but is restrained by a spring. The angle through which the magnet turns is proportional to the speed of rotation of the disc, see Fig 4.3.

The speed of a liquid in a pipe can be determined in several different ways. A small propeller fitted inside the pipe will rotate at a speed proportional to the speed of the liquid. The propeller will however cause resistance to the flow of the liquid, and modern methods, with liquids which conduct electricity, rely on creating an electrical potential difference across the fluid when it flows through a magnetic field. One of these instruments is illustrated in Fig 4.4 and is described in Section 19.3. If the liquid does not conduct electricity then two ultrasonic beams can be passed diagonally across the pipe. The difference in the time taken for each beam to cross the pipe with the flow or against the flow is a direct indication of liquid velocity. The instrument can be calibrated to give an output reading in m s^{-1} or, if needed, in litres per second.

Example 1

Find the average speed and the maximum speed shown on the ticker tape in Fig 4.5. The dots are printed every 1/50 s and the tape is marked in centimetres.

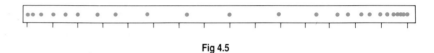

Fig 4.5

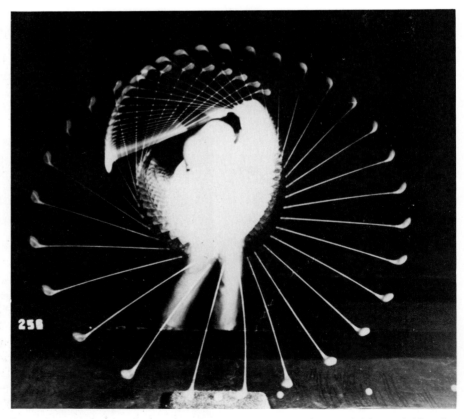

Fig 4.6 By making measurements on this photograph you are asked in question 4.1 to find the speed of the golf ball and the golf club.

MOTION

The average speed is total distance/total time.

$$
\begin{aligned}
\text{total distance} \quad &= 15.0 \text{ cm} \\
\text{total time} \quad &= (23/50) \text{ s} \quad = 0.46 \text{ s} \\
\text{average speed} \quad &= 15.0/0.46 \quad = 32.6 \text{ cm s}^{-1} \quad = 0.326 \text{ m s}^{-1}
\end{aligned}
$$

The maximum speed can be obtained less accurately; the maximum distance between any two dots is 1.9 cm. This gives the maximum speed as $1.9/0.02 = 95$ cm s^{-1}. The maximum speed is an instantaneous value of the speed. Here we are calculating for a time interval of 0.02 s. In order to work with any smaller time interval it would be necessary to plot a graph of distance against time and find its maximum gradient.

QUESTION	
	4.1 Find the speed of **(a)** the golf ball, **(b)** the golf club in Fig 4.6. The scale is 1 to 25 and the multi-flash photographs were taken at the rate of 120 per second.

INVESTIGATION

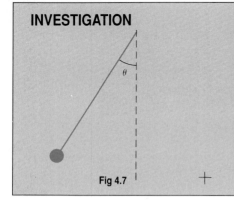

Fig 4.7

Find how the maximum speed of a pendulum bob varies with the angle θ through which it is swinging, see Fig 4.7.

1. Set up a pendulum bob so that the angle through which it swings can be measured.

2. Arrange a photocell so that the shortest time taken for the pendulum bob to interrupt a beam of light can be measured on a timer. Find the maximum speed of the bob.

3. Plot a graph of maximum speed v_{max} against angle of swing θ.

4. Plot a graph of kinetic energy against θ for a single swing.

4.2 ACCELERATION

An object whose velocity is changing is said to have an acceleration. **Acceleration** is defined as the rate of change of velocity, and since velocity is a vector then acceleration must also be a vector. This can be written using corresponding equations to those given above for velocity.

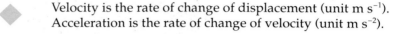

$$
\text{average acceleration} = \frac{\Delta v}{\Delta t} \text{ or}
$$

$$
\text{instantaneous acceleration} = \frac{\delta v}{\delta t} \text{ as } \delta t \text{ approaches zero} = \frac{dv}{dt}
$$

The number of equations here can confuse, but in practice both velocity and acceleration can be taken to mean instantaneous velocity and instantaneous acceleration unless it is stated clearly that this is not so. This means that the two important definitions are similar:

Velocity is the rate of change of displacement (unit m s^{-1}).
Acceleration is the rate of change of velocity (unit m s^{-2}).

Determination of acceleration practically is usually done by the same methods as were given above for velocity. In this case however it is always necessary to find two velocities, one at the beginning and one at the end of a time interval Δt. Acceleration can be measured directly using instruments called accelerometers. These depend for their function on the fact that a force is needed to cause an acceleration. There is a magnificent example of an accelerometer in the National Railway Museum in York, see Fig 4.8. The carriage was called a dynamometer and was used in Victorian times to find the power output of an engine, so that its efficiency could be found if the rate of burning of coal was known. A huge spring is used to connect an engine to the dynamometer carriage. If the spring is extended more than the distance needed to overcome friction then the engine is accelerating. The spring has to be calibrated so that the extension produced, which is plotted directly on a chart recorder, can give a graph of acceleration against time.

Fig 4.8 The large flat horizontal area in the photograph is a huge spring which was used to connect a railway engine to its carriages so that the pull of the train could be measured.

4.3 GRAPHICAL REPRESENTATION OF MOTION

Many graphs can be drawn to show how a body moves. They are particularly useful for indicating motion, and in very many cases time will be on the x-axis of the graph. This is one way of overcoming the difficulty of presenting movement on a static piece of paper. The same problem is encountered with showing moving waves on paper, and again graphs are resorted to. Whenever a graph is drawn or used, it is essential to be certain what is being plotted as the same information plotted on two different graphs can appear to be very different. Consider the movement of a tube, or underground train in between two stations. The motion of the train consists of an acceleration, then a period of constant velocity followed by slowing down to a stop. The slowing down is sometimes called a **deceleration**, which is a negative acceleration. If a velocity–time graph is plotted for the movement it appears to be very different from the corresponding distance–time graph, although they are giving exactly the same information. Figs 4.9 and 4.10 show two such graphs, plotted for idealised circumstances. It has also been assumed that the track is straight so that we are correctly dealing with velocity and not speed.

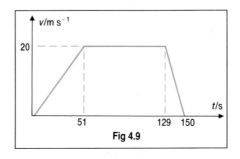

Fig 4.9

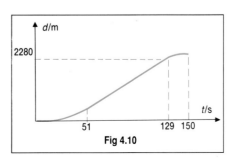

Fig 4.10

4.2 Draw a third graph showing the same information as is contained in the above two graphs but plotting acceleration against time.

4.3 What acceleration is shown by the graph in Fig 4.11?

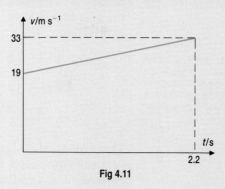

Fig 4.11

With many ways of presenting the same information, the question arises as to how to obtain information from any of the graphs. Two features of graphical work should be clear already from the basic definitions of velocity and acceleration. Because velocity is the rate of change of displacement, velocity is the gradient of a displacement–time graph. Similarly, because acceleration is the rate of change of velocity, acceleration is the gradient of a velocity–time graph. Displacement can also be obtained from a velocity–time graph.

Consider the graph shown in Fig 4.9. When the tube train is travelling at a constant velocity of 20 m s^{-1} it is travelling for 129 – 51 = 78 seconds. During this time therefore it has a displacement of 20 m s^{-1} × 78 s = 1560 m. This value, 1560, is also the area of the rectangle beneath the flat portion of the graph. This can be extended to any shape of graph. The area beneath any velocity–time graph always represents the displacement.

4.4 UNIFORMLY ACCELERATED MOTION

In many practical applications the acceleration of an object is constant. That is its velocity is increasing at a steady rate. This motion is called **uniformly accelerated motion** and is shown in Fig 4.12 for an object moving in a straight line. Here an object at time $t = 0$ has a velocity u. After moving for time t its velocity has increased to v. Assume during this time its acceleration is a and its displacement is s.

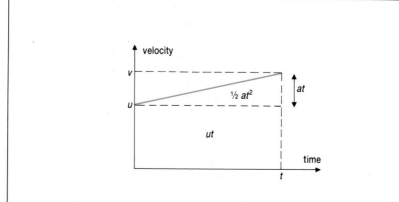

Fig 4.12 Graph showing uniform acceleration from velocity u to velocity v in time t. The area beneath the graph is the distance. The gradient of the graph is the acceleration.

Since the acceleration is constant its value is given by

$$a = \frac{v - u}{t} \tag{1}$$

Therefore the increase in velocity is at as marked. The area of the bottom rectangle is shown as ut and the area of the triangle is $\frac{1}{2}$ of $at \times t$. This gives the area beneath the graph and hence the displacement s in three different ways:

$$s = ut + \tfrac{1}{2}at^2 \tag{2}$$

$$s = vt - \tfrac{1}{2}at^2 \tag{3}$$

$$s = \frac{(u + v)t}{2} \tag{4}$$

This last equation is the average velocity × the time.

If t is eliminated from any two of the above equations the relationship between u, v, a and s can be found as

$$v^2 = u^2 + 2as \tag{5}$$

These equations of motion apply *only* to a system undergoing a constant acceleration. Mistakes are frequently made by applying them to situations where this is not the case. Of the five quantities u, v, a, s and t, one is omitted from each of the equations.

Example 2

Using the details from the tube train example above, show that the total displacement is 2280 m.

First section, acceleration $\quad s_1 = \dfrac{(u + v)t}{2} = \dfrac{20 \text{ m s}^{-1} \times 51 \text{ s}}{2} = 510 \text{ m}$

Second section, constant
velocity $\quad\quad\quad\quad\quad\quad s_2 = 20 \text{ m s}^{-1} \times 78 \text{ s} \quad\quad\quad = 1560 \text{ m}$

Third section, deceleration $\quad s_3 = \dfrac{(u + v)t}{2} = \dfrac{20 \text{ m s}^{-1} \times 21 \text{ s}}{2} = 210 \text{ m}$

Total displacement $\quad\quad\quad\quad = 510\text{m} + 1560\text{m} + 210\text{m}$
$\quad\quad\quad\quad\quad\quad\quad\quad\quad\quad = 2280 \text{ m}$

QUESTIONS

4.4 Find the distance travelled by a train in the first minute of its journey if it has a constant acceleration of 0.32 m s^{-2}.

4.5 A ball approaches a tennis racquet with a velocity of 21 m s^{-1}. The tennis racquet gives it an average acceleration of 3500 m s^{-2} for 0.020 s in the opposite direction to its initial velocity. What is the velocity of the ball after leaving the racquet and how far does the ball travel while undergoing this acceleration?

4.6 A speeding motorist is travelling at a constant speed of 40 m s^{-1} on a motorway. He passes a stationary police car which immediately accelerates at a rate of 2.5 m s^{-2}, to a constant speed of 50 m s^{-1}. Draw a sketch graph showing the variation of speed against time for each car and find
(a) the time it takes for the police car to catch the motorist;
(b) the distance each will have travelled by that time.

DATA ANALYSIS

Accelerometer graphs

Copy two sketches similar to Fig 4.13 on to a piece of A4 paper and add axes for velocity against time and distance against time as shown in outline in Fig 4.14. Use the same time scale for each of the graphs. The acceleration graph given is taken from an accelerometer on a train starting from rest. The actual acceleration fluctuates a great deal because of vibration and irregularities of traction, friction and track. To overcome the difficulty of a

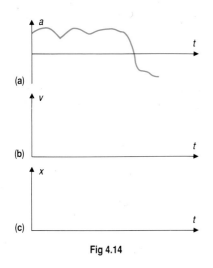

(a)

(b)

(c)

Fig 4.14

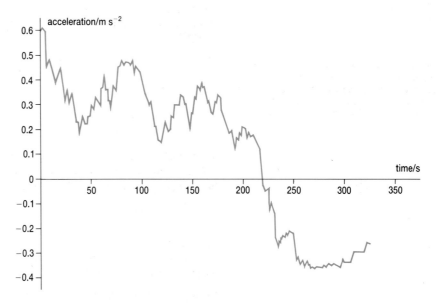

Fig 4.13 Graph of acceleration plotted against time for the 300 seconds of a train journey taken from a dynamometer like the one shown in Fig 4.8.

rapidly changing acceleration, draw a smooth graph, superimposed on the given acceleration graph, and use the smoothed out graph for taking further readings.

Using approximate numerical values:
(a) Plot the velocity–time graph from the acceleration–time graph.
(b) Find the final speed of the train.
(c) Find the maximum speed of the train.
(d) Using the velocity–time graph plot the corresponding distance–time graph.
(e) What is the total distance travelled?

4.5 FALLING BODIES

An object is said to be in **free fall** if the only force acting on it is the gravitational pull of the Earth. It then has an acceleration downwards relative to the surface of the Earth of approximately 9.81 m s^{-2}. This value varies from place to place on the Earth. It decreases as the height above sea level increases, and it also decreases as you get closer to the equator. Uneven distribution of matter within the Earth also affects the value. As changes caused by these factors are small, the downward acceleration caused by free fall is often taken to be a constant and given the symbol g. True free fall will only occur in a vacuum, where air resistance will be zero.

Consider a ball being hit so that it starts with an upward velocity of 26.1 m s^{-1}. If wind resistance is neglected the ball is in free fall from the moment it ceases to be in contact with the bat. Its acceleration will have a value of 9.8 m s^{-2} (to 2 significant figures) in a downward direction. This will have the effect of reducing its upward velocity by 9.8 m s^{-1} during each second of its movement. Values of velocity and time are therefore

time/s	0	1	2	3	4	5
velocity/m s^{-1}	26.1	16.3	6.5	–3.3	–13.1	–22.9

A velocity–time graph will therefore be a straight line going from positive to negative values. There is no discontinuity when the ball reaches the top of its flight and stops momentarily after 2.66 seconds. The acceleration throughout is equal to 9.8 m s^{-2}, even when the ball is stationary. The velocity–time graph will look like Fig 4.15.

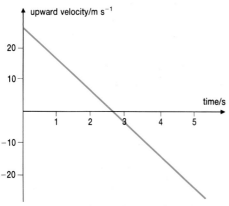

Fig 4.15 An object in free fall has its upward velocity reduced by a fixed amount each second. The acceleration is –9.8 m s^{-2} so the gradient of the graph is –9.8.

The distance the ball will travel upwards is given by the area of the triangle,

$$\tfrac{1}{2} \times 2.66 \times 26.1 = 34.7 \text{ m}.$$

The ball will fall back to the level it started from when the negative area beneath the graph equals this positive area. This will be after a further 2.66 seconds, making the total time 5.32 seconds.

If during the time the ball is in mid-air the ball also has a horizontal velocity then its vertical velocity is totally independent of its horizontal velocity. In the absence of air resistance its horizontal velocity will be constant. In the photograph, Fig 4.16, the tennis ball, before being hit by the

Fig 4.16 Before the tennis ball is hit by the racquet it can be seen to have a path which is parabolic. After being hit the path is still a parabola, but the much larger forward velocity makes this less apparent.

racquet, is not only accelerating under gravity but also has a small forward velocity. This independence of vertical and horizontal motion is important in determining the path of any projectile.

Example 3

Assuming that the maximum speed a person can run is 11.0 m s^{-1}, and that for maximum distance a long jumper should take off at an angle of 45°, find the maximum theoretical value of the world long jump record in terms of g.

Assuming that the long-jumper can change his velocity to an angle of 45° without changing his speed, a vector diagram of his take off is given by Fig 4.17.

For vertical motion, velocity upwards has to change from 11 sin 45° m s^{-1} to −11 sin 45° m s^{-1}. Therefore $\Delta v = 2 \times 11 \sin 45° = 15.6$ m s^{-1}.

Time jumper is in the air is therefore 15.6/g
Horizontal distance travelled in this time = (15.6/g) × 11 cos 45°
= 121/g

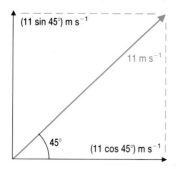

Fig 4.17 The horizontal and vertical components of the velocity of a long-jumper on take off at 45°.

(11 sin 45°) m s^{-1}

11 m s^{-1}

45°

(11 cos 45°) m s^{-1}

MOTION

Fig 4.18 Bob Beamon's world record jump in the Olympic Games in Mexico in 1968. Note the height he achieves in trying to reach the theoretical optimum take off angle of 45°.

In Helsinki, Finland where g is 9.827 m s^{-2} this is 12.31 metres.

At sea level in Mexico, where g is 9.794 m s^{-2} this is 12.35 metres.

In Mexico City, at an altitude of 2000 m, where $g = 9.787$ m s^{-2}, the distance becomes 12.36 metres.

The other, and larger, effect of jumping at high altitude is that the wind resistance is lower in the less dense air.

In fact the world long jump record was set in Mexico City by Bob Beamon in the 1968 Olympic Games. It is a distance of only (!) 9.76 metres, so athletes have some way to go before they reach the theoretical maximum. In practice it would not be possible to take off at an angle of 45° without some springboard assistance. Gymnasts are able to travel considerably further than long jumpers. This is not only because they have something to vault over but also because they do use springboards. Even the thickness of the rubber in the soles of the shoes of a long jump athlete makes a considerable difference and causes difficulties for the authorities laying down the rules of the competition.

QUESTIONS

Fig 4.19 For heights up to 10 m air resistance makes very little difference to the movement of the diver.

Fig 4.20 Taken at the World Championship Ski Jumping in 1989.

4.7 How long does a high diver take to reach the water from a 10.0 m diving stage? With what speed will she reach the water?

4.8 A football is kicked at a speed of 20 m s^{-1} at a launch angle of 45°. Find its range assuming negligible air resistance.

4.9 A stunt rider leaves a ramp at an angle of 30° and lands at the same height after travelling a horizontal distance of 36 m. What was his take off speed?

4.10 A cricket ball is thrown with a velocity of 30 m s^{-1} at an angle of 42° to the horizontal. Find
   (a) the time it takes to reach its maximum height;
   (b) the total time taken to return to its original level;
   (c) the maximum height reached;
   (d) the horizontal distance travelled.

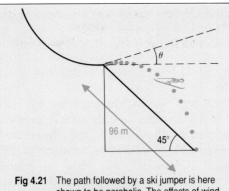

Fig 4.21 The path followed by a ski jumper is here shown to be parabolic. The effects of wind resistance are being ignored.

4.11 A ski jumper lands 96 m from his take off point. See Figs 4.20 and 4.21. The slope is at an angle of 45° and the jumper is in the air for 4.3 s. Making the rather sweeping assumption that air resistance is negligible find

(a) the horizontal distance travelled;

(b) the horizontal component of the velocity at take off;

(c) the vertical distance from take off to landing;

(d) use $s = ut + \frac{1}{2}at^2$ to find u the vertical component of the take off velocity; (Be careful to use signs so that you are consistent with directions. s is down, u is up and a is down.)

(e) the angle of take off, θ;

(f) the speed of take off.

4.6 MOMENTUM

The **momentum** of a body is defined as the product of its mass and its velocity.

Momentum = mass × velocity

It is a vector quantity, being the product of a scalar (mass) and a vector (velocity). The direction of the momentum is always the same as the direction of the velocity. The unit of momentum does not have a special name. The obvious derived unit to use is kg m s^{-1}, namely the unit of mass multiplied by the unit of velocity. It can be shown that this unit is exactly the same as the N s, and this is often used. The importance of the momentum of a body will be discussed in Chapters 5 and 6. At this stage it is useful to get some idea of the momentum of different objects.

Example 4

Find the momentum of the Earth in its orbit around the Sun. The mass of the Earth is 6.0×10^{24} kg and the radius of its orbit is 1.5×10^{11} m.

The direction of the momentum will be in the direction of the Earth's velocity; this is, at a tangent to its orbit.

The Earth takes a year to rotate round the Sun so the velocity of the Earth in this direction is given by

$$\frac{\text{circumference of orbit}}{\text{period}} = \frac{2\pi \times 1.5 \times 10^{11} \text{ m}}{365 \times 24 \times 60 \times 60 \text{ s}}$$

$$= 3.0 \times 10^4 \text{ m s}^{-1}$$

$$(= 30 \text{ km s}^{-1})$$

The momentum of the Earth is therefore

$$= 6.0 \times 10^{24} \times 3.0 \times 10^4 \text{ m s}^{-1}$$

$$= 1.8 \times 10^{29} \text{ N s}$$

QUESTIONS

4.12 Find the momentum of a car of mass 1200 kg travelling with a velocity of 40 m s^{-1} due east.

4.13 Find the momentum of an ion of mass 7.3×10^{-26} kg, travelling with a velocity of 4.2×10^5 m s^{-1} in a direction parallel to a magnetic field.

4.14 By how much does the momentum of the Earth change over a period of (a) 6 months, (b) a year?

4.15 An animal of mass 50 kg changes its velocity from 3.0 m s^{-1} due north to 15 m s^{-1} due west in order to escape from a predator. What is the change in the momentum of the animal? (The answer is *not* 600 N s.)

SUMMARY

- Velocity is the rate of change of displacement (unit m s^{-1}). It is a vector. The magnitude of the velocity is the speed. Speed is a scalar.
- Acceleration is the rate of change of velocity (unit m s^{-2}). It is a vector.

For constant velocity $\quad v = x / t$

For constant acceleration

$$v = u + at$$
$$s = ut + \tfrac{1}{2} at^2$$
$$s = vt - \tfrac{1}{2} at^2$$
$$s = \frac{(u + v)\, t}{2}$$
$$v^2 = u^2 + 2\, as$$

- Momentum is the product of the mass and the velocity of a body. It is a vector.

EXAMINATION QUESTIONS

4.16 A motorist travelling at 13 m s^{-1} approaches traffic lights, which turn red when he is 25 m away from the stop line. His reaction time (i.e. the interval between seeing the red light and applying the brakes) is 0.70 s and the condition of the road and his tyres is such that the car cannot slow down at a rate of more than 4.5 m s^{-2}. If he brakes fully, how far from the stop line will he stop, and on which side of it?
(UCLES 1982)

4.17 A tennis ball is dropped from the hand, falls to the ground and bounces back at half the speed with which it hit the ground. Draw a velocity time graph of its motion. Mark the point on the graph which corresponds to the ball hitting the ground.
Indicate how, from the graph, **(a)** the distance the ball falls, and **(b)** the distance the ball rises, can be found. (5)
(ULSEB 1984)

4.18 Water flows at a constant rate of 8.0×10^{-6} m^3 s^{-1} through a horizontally mounted tube of internal radius 1.0 mm from which it emerges.

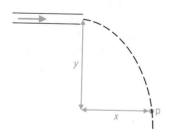

(a) Write down the horizontal and vertical components of the acceleration of the water after leaving the tube. (Neglect air resistance.) (2)
(b) Calculate the speed of water emerging from the tube. (2)
(c) A point P on the water stream is a distance y below the tube and a horizontal distance x from the tip of it. Show that $y = kx^2$. Calculate the value of k.
(O & C 1988)

4.19 The diagram shows speed time graphs for three spherical objects made of the same material but with differing radii, a_1, a_2, a_3, falling through a column of liquid.

(a) Explain why the speeds tend to a constant value and why these values differ.

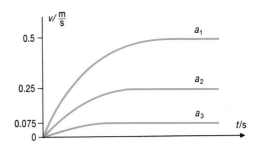

(b) If the constant speed of a falling sphere is proportional to its radius squared, find a_2, given that $a_1 = 2.0$ mm. (6)
(ULSEB 1982)

4.20 A tennis ball is served 2.5 m above the ground at an angle of 5° above the horizontal direction with an initial speed of 30 m s^{-1}.
(a) When will it hit the ground? **(b)** How far will it travel? (Neglect air resistance.)

4.21 **A** and **B** are two vectors acting at right angles. Draw labelled diagrams to show the vector addition ($A + B$) and the vector subtraction ($A - B$). State the magnitudes of the resultants and their directions with respect to vector **A**.
A beam of electrons travelling at a uniform speed of 1.50×10^7 m s^{-1} in vacuum enters the gap between two plane, parallel deflection plates along a line PQ, midway between the plates. The plates are 40 mm long. The beam emerges from the plates with a velocity of 1.60×10^7 m s^{-1} at an angle of 20.4° to the original direction.
(a) Draw, to scale, a labelled vector diagram to show the change of velocity Δv of an electron during its path between the plates. Use your diagram to find the magnitude and direction of Δv.
(b) How long does the electron take to travel between the plates? Hence find its acceleration during the deflection.
(UCLES 1983)

4.22 In the absence of air resistance with what angle to a horizontal ground must an object be thrown from ground level so that it travels a distance before landing equal to twice its maximum height?

4.23 An ice hockey puck has a mass of 0.40 kg and at one instant has a velocity of 30 m s^{-1}. After a time interval of 0.0030 s its velocity is 45 m s^{-1} in the opposite direction. Find the change in the momentum of the puck and its average acceleration during the time interval.

Chapter 5

NEWTON'S LAWS OF MOTION

> **LEARNING OBJECTIVES**
>
> At the end of this chapter you should be able to:
>
> 1. quote and use Newton's three laws of motion;
>
> 2. draw and use free body force diagrams;
>
> 3. distinguish between mass and weight.

5.1 INTRODUCTION

Newton's laws form a principal part of the foundation stones of classical physics. An understanding of their meaning is vitally important for all students of physics and you are well advised to spend time giving considerable thought to this chapter. Gaining the necessary understanding will come only with effort and with the application of the laws to practical examples.

Newton published his laws in 1687 and they have transformed science and engineering in the 300 years since then. Einstein's theory of relativity has modified the laws so that they are now regarded as an approximation to the theory of relativity, but that approximation is so precise that in practice it is only necessary to consider differences from Newton's laws when speeds approaching the speed of light are being used.

To simplify things at the start, assume that any object being considered
- has a constant mass
- is not rotating
- is in surroundings which are not accelerating.

5.2 NEWTON'S FIRST LAW

In formal language this law, translated from Newton's original Latin, states:

 Every object continues in its state of rest or uniform motion in a straight line unless it is compelled to change that state by external forces acting upon it.

Another way of putting this would be to say that any object has a constant velocity unless it is acted upon by a resultant external force. This effectively defines force. A force is necessary to accelerate an object: or, a force is that influence which, acting alone, can cause an object to be accelerated.

There are practical situations where it is easy to apply Newton's first law but there are other occasions where the law seems to contradict common sense and these often cause difficulty.

Consider the horizontal forces of **drive** and **drag** acting on a car travelling at a constant velocity of 30 metres per second (about 70 mph), see Fig 5.1. The total force on the car really is zero. That is, the force which drives the car forward exactly equals the drag force caused by friction and wind resistance. If the two forces are not equal in size and opposite in

drive   drag

Fig 5.1 A car will have a constant velocity if the drive force on the car is equal and opposite to the total drag force on the car. (Drag to include road drag and wind resistance.)

Fig 5.2 A car will slow down if the drag force is larger than the drive force.

direction, say the drive force is larger than the drag force, then the car will accelerate. If drag is greater than drive then the car will decelerate, see Fig 5.2. This is made use of in braking when the drag force is increased considerably. It is only when they are equal in size that the car has a constant velocity.

The same analysis can be carried out for a person travelling inside a vehicle. Consider a woman travelling at a constant velocity of 300 metres per second in an aeroplane. The seat on which she is sitting supports her with a force equal and opposite to her weight. She needs **nothing** to move her forward. There is **no** forward force: just two forces balancing to give **zero** total force and therefore no acceleration. She did need a forward force on her when accelerating on the runway. That was when she was conscious of the seat pushing her, but in mid-flight it is just like sitting on a chair in the garden (apart from an uncomfortable and unnecessary vibration from the engines shaking the aeroplane). That is why, if a stewardess pours her a drink, there is no need to hold the glass some way in front of the bottle to allow for the movement of the aeroplane. It is also one reason why drinks are not served during take-off!

Fig 5.3 shows a drink being poured into the glass from the bottle held by the stewardess. Why is the liquid already in the glass not horizontal? In which direction is the aeroplane accelerating?

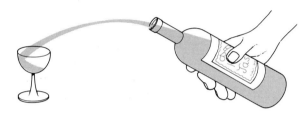

Fig 5.3 Simple actions become difficult to perform when taking place in an accelerating system.

Newton realised that there is no difference at all between the forces acting on an object travelling at a constant velocity and those on a stationary object. This was developed by Einstein as one principle of his theory of relativity. He stated that the laws of physics take exactly the same form for all systems which have a constant velocity relative to one another.

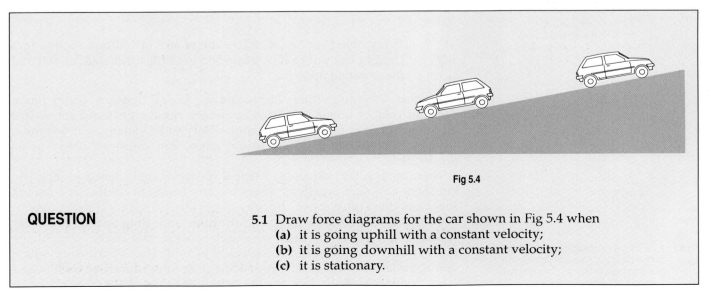

Fig 5.4

QUESTION

5.1 Draw force diagrams for the car shown in Fig 5.4 when
(a) it is going uphill with a constant velocity;
(b) it is going downhill with a constant velocity;
(c) it is stationary.

NEWTON'S LAWS OF MOTION

Fig 5.5 shows the application of Newton's first law to some practical situations.

5.3 NEWTON'S SECOND LAW

The first law defined force. If a fixed force is applied to two different masses, one large and one small, they will not have the same acceleration. The second law deals with the size of the effect that a force has on an object.

The second law in formal language states:

 The rate of change of momentum of a body is proportional to the total force acting on it and occurs in the direction of the force.

Consider an object of constant mass m being pushed by a constant force F so that its velocity increases from u to v in time t. Newton's second law gives

$$\frac{\text{change in momentum}}{\text{time}} \propto \text{Force}$$

$$\frac{mv - mu}{t} \propto F$$

$$\frac{m(v - u)}{t} \propto F$$

$$m \times a \propto F$$

where a is the acceleration.

As an equation, this becomes $F = k \times m \times a$ where k is a constant which has no unit but whose size depends on the unit to be chosen for force. This needs now to be done.

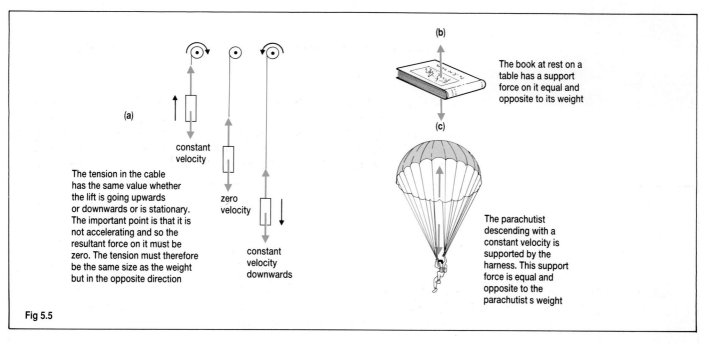

(a)

The tension in the cable has the same value whether the lift is going upwards or downwards or is stationary. The important point is that it is not accelerating and so the resultant force on it must be zero. The tension must therefore be the same size as the weight but in the opposite direction

constant velocity

zero velocity

constant velocity downwards

(b) The book at rest on a table has a support force on it equal and opposite to its weight

(c) The parachutist descending with a constant velocity is supported by the harness. This support force is equal and opposite to the parachutist s weight

Fig 5.5

One **newton** is the force which causes a mass of one kilogram to have an acceleration of 1 metre second^{-2}.

Substitution into the above equation gives 1 newton = k × 1 kg × 1 m s^{-2} and so k must be 1 for this system of units. An equation which must be thoroughly learnt therefore is

FORCE in newtons	=	MASS in kilograms	×	ACCELERATION in metre second^{-2}

As a direct result of the way the newton has been defined, k has been made equal to 1. It makes use of the equation straightforward but it must be remembered that the simplified equation is valid only if the correct units are used. It is a frequent cause of mistakes to substitute values with incorrect units into the above equation.

The equation must be used with care in several respects:

- The units used must always be newton, kilogram and metre second^{-2}.
- The force must always be the total force acting on the object. If there are several separate forces acting on an object then they must be added together to find the total force. The total force is the vector sum of all the forces acting on an object and is often referred to as the **resultant** force.
- Both the total force and the acceleration are vectors and these are **always and without exception** in the same direction. If you have the solution to a problem in which the force and the acceleration are not in the same direction then something has gone wrong!
- It is important to realise that acceleration is caused by force. Accelerations do not cause forces.

Fig 5.6 An astronaut undergoing increasingly high acceleration. All parts of the astronaut's face must have large forces on them to cause their acceleration. Note: It is not the acceleration which causes these forces, the forces cause the acceleration.

The equation is a mathematical summary of both of Newton's first and second laws of motion for a body of constant mass. Besides giving the acceleration of an object for a known force, it also shows that if the force is zero then the acceleration will also be zero.

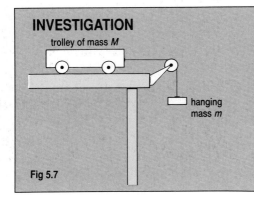

INVESTIGATION

Fig 5.7

The arrangement shown in Fig 5.7 is often suggested as an experiment to verify Newton's second law. The acceleration is usually found by using a ticker timer and ticker tape attached to the trolley.

1. Set up the apparatus so that the acceleration a of the trolley may be found.

2. Vary the mass m of the hanging mass so that a graph may be plotted of a against m. In doing this m should vary from being much less than M to a value greater than M.

3. Explain why your graph is not a straight line.

NEWTON'S LAWS OF MOTION

5.4 NEWTON'S THIRD LAW

This law is a measure of the genius of Newton. Since the days of Newton no one has ever found any situation in which the law does not apply. Much of Einstein's theory of relativity is based upon it and also the law of conservation of momentum which can be proved from it. In recent years it has become usual to write the law in a different form from that which Newton himself used, because in its original form it was often misunderstood. Formally it is now written as

 If body A exerts a force on body B, then body B exerts an opposite force of the same size on body A.

The two forces must also be of the same type. That is, if one is an electrical force then the other must be electrical too. Note that the two forces are acting on two different bodies: here, one of them acts on A and the other on B. This law applies in **all** circumstances.

The two bodies may be enormous stars or minute parts of atoms: one may be huge and the other may be tiny. The bodies may be stationary or moving or accelerating. In particular, it is important to realise that it is not necessary to have equilibrium in order for the law to be true. Newton's third law is **always** true; not just when things are nicely balanced. If a dart is stuck in a dartboard then

the force the dart exerts on the dartboard is equal and opposite to the force the dartboard exerts on the dart

If the dart is just arriving at the dartboard and is decelerating then at each moment during the deceleration it is still true that

the force the dart exerts on the dartboard is equal and opposite to the force the dartboard exerts on the dart

Statements like this can be made whenever forces exist. Newton realised that force is a mutual influence between two objects and that it is impossible to have a single force. Forces always exist in pairs. In order to become used to the ideas associated with this law it is recommended that clear force diagrams are always made and that the forces are clearly labelled to show what object is exerting the force. Care needs to be taken to draw on any diagram the correct force out of the pair of forces being considered. If the wrong one is used then it will be pointing in exactly the wrong direction. For instance, in the case of the dart hitting the dartboard, the force the dart exerts on the dartboard is a forward force; this force would only be shown on a diagram showing the forces on the dartboard. The force the dartboard exerts on the dart is a backward force which causes the deceleration of the dart; this force would only be shown on a force diagram for the dart.

In Fig 5.8 one pair of forces is shown. The gravitational force the Moon exerts on the Earth is shown on the force diagram of the Earth and the equal and opposite force which the Earth exerts on the Moon is shown on the force diagram of the Moon.

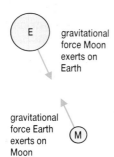

Fig 5.8 A pair of gravitational forces which, according to Newton's third law of motion, must always be equal and opposite.

gravitational force Moon exerts on Earth

gravitational force Earth exerts on Moon

5.5 TYPES OF FORCE

Some words used in physics have a meaning which is more closely defined than the word's meaning in everyday use. There is a multitude of words in the English language which represent force. Some examples of these words are: push, pull, hit, tension, knock, shove, effort, load, strength, power, vigour. In a science it is essential to be careful in the use of words so that when a word is used its meaning is clear. For instance, in the above list, some of the terms are scientifically inaccurate. Power means work done per unit time; it does not mean force. Other words in the list are simply descriptions of particular situations where forces occur; tension, effort and load come into this category.

When dealing with types of force however we find, surprisingly, that outside the nucleus of atoms there are only two possible types of force; these are:

- electromagnetic force
- gravitational force

Electromagnetic forces exist between moving or stationary charges. Since all atoms have charged particles within them, it is electromagnetic forces which bind atoms together in solids and liquids. On some occasions the electrical nature of a force is important, sometimes the magnetic nature is important. In these cases there is not usually any problem in pin-pointing where the force exists. In the vast majority of mechanics problems however, it is the electromagnetic forces between atoms which are of prime importance. Whenever the atoms of one object are close to the atoms of another object, there will be a contact force between them. *All* forces of contact are electromagnetic forces. In the list given above push, pull, hit, shove, knock, effort and load are all examples of contact forces. Tension is also a contact force, but is used in a rather special way involving internal electromagnetic forces between atoms in a string as well as the contact force between the string and the object to which it is attached.

Gravitational forces exist between any two masses and can usually be neglected unless one of those masses is very large. The gravitational force which a car exerts on a caravan is negligible: the electromagnetic force of contact which the car exerts on the caravan is the force which pulls the caravan along. In practice the only gravitational force which usually concerns us is the gravitational attraction of the Earth.

5.6 WEIGHT

One of the results of the work of Newton was the introduction of the concept of gravity. The application of Newton's third law to gravitation is dealt with in more detail later, but the link between gravity and weight needs to be considered here. Gravity is a very mysterious force. How one mass exerts a force on another when there is no contact between them and nothing in the space between them is the mystery. Because gravity is such a familiar force, its strangeness is often overlooked. What can be stated here by the application of the third law is that, as always, forces exist in pairs so that:

the gravitational force the Earth exerts on you is equal and opposite to the gravitational force you exert on the Earth

In other words because both you and the Earth have mass, you and the Earth are pulled together by gravitational attraction.

Weight is the word which is often used for the gravitational force which the Earth exerts on an object. Further consideration will be given to the term in Section 9.5.

The relationship between the weight of an object and its mass becomes much clearer if it is realised that weight is always a force and will therefore always be measured in newtons. Mass is always measured in kilograms.

Consider a ball of mass m falling freely towards the Earth, being pulled down by its weight w but with no other force acting on it (Fig 5.9). Using the second law in this situation gives

force	=	mass	×	acceleration
in newtons		in kg		in m s^{-2}
w	=	m	×	g

Fig 5.9 A ball in free fall with only its weight acting on it.

where g is the acceleration of free fall of the ball. (Note: g is *not* gravity.)

NEWTON'S LAWS OF MOTION

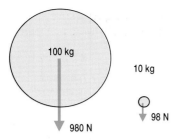

Fig 5.10 A large force of gravity on a object of large mass gives rise to the same acceleration as the small force of gravity on the object of small mass.

If *g* is measured experimentally its value is found to be approximately 9.8 m s⁻². The above equation can therefore also be written

$$\frac{w}{m} = 9.8 \text{ newtons per kilogram}$$

Fig 5.10 shows two masses in free fall; each has 9.8 N of gravitational force acting on it per kilogram of mass. The large mass has a large force acting on it and this gives it the same acceleration as the smaller force acting on the smaller mass.

Example 1

The gravitational force acting on a golf ball on the Earth is 0.430 N. The same golf ball if on the Moon would have a gravitational force on it of only 0.0725 N. Find the mass of the golf ball and the acceleration of free fall on the Moon.

Consider the ball falling freely on the Earth:

force in newtons	=	mass in kilograms	×	acceleration in m s⁻²
0.430	=	*m*	×	9.8
m	=	0.430/9.8 = 0.0439 kg		

When the ball is on the moon its mass is still 0.0439 kg so now:

force in newtons	=	mass in kilograms	×	acceleration in m s⁻²
0.0725		0.439	×	*a*

Acceleration of free fall on the moon = 0.0725/0.0430 = 1.65 m s⁻².

5.7 FRICTION

Consider a stone resting on a road. If the contact between the road and the stone is examined closely it can be seen that the two rough surfaces make close contact only at relatively few places. Where contact is made, the road will exert a force on the stone as shown in Fig 5.11(a).

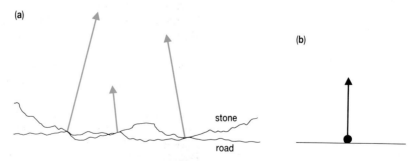

(a)

(b)

stone

road

forces road exerts on stone: stone stationary

Fig 5.11 **(a)** A stationary stone on a flat road has various forces exerted on it by the road at points of contact **(b)** The resultant of these forces is vertically upwards.

The sum of all these forces is shown using a different scale in Fig 5.11(b) and this single force is the contact force which the road exerts on the stone.

If however the stone happens to be sliding across the road to the left say, Fig 5.11 changes to Fig 5.12. Now there are rather more forces in the direction opposite to the direction of travel than in the forward direction – Fig 5.12(a). This results in the contact force which the road exerts on the stone being tilted as shown in Fig 5.12(b). The tilted force can be considered as being the vector sum of a horizontal and a vertical component.

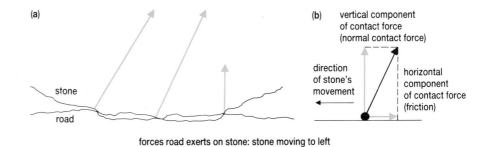

(a)

stone

road

(b) vertical component
of contact force
(normal contact force)

direction
of stone's
movement

horizontal
component
of contact force
(friction)

forces road exerts on stone: stone moving to left

Fig 5.12 **(a)** When the stone moves to the left the forces the road exerts on it are likely to be both upwards and to the right. **(b)** The resultant of all these forces is tilted to the right. It has components vertically upwards and horizontally to the right.

The horizontal component of this force is called the **frictional force**, and the vertical component is called **normal contact force**. For two surfaces where friction can be neglected the only force will be the normal contact force.

Friction itself is usually a component of a contact force. The distinguishing feature of friction is that it is in the opposite direction to the direction of motion.

Friction is often regarded as a nuisance but it is often crucially necessary. There are times when friction needs to be minimised but without friction everything would collapse – literally. If two surfaces could exert forces on each other only at right angles to the surface, i.e. if there was no friction, then there would be no nails, no nuts and bolts, no glue, no screws, no sewing, no knitting, no fabrics, no buildings, no cars, no people. The list is virtually endless. Friction is an extremely useful force. Chaos usually follows even the reduction in friction which takes place on roads during a snowfall.

(NOTE: The normal contact force is sometimes called the normal reaction but this term is being avoided here because in the old form of Newton's third law the word reaction was also used. Sometimes these two terms were the same thing and sometimes they were not. Much confusion was caused.)

In solving problems concerned with Newton's laws the following routine is strongly recommended.

1. State clearly which object is being considered.
2. Draw a free body sketch of that object only.
3. Mark on the sketch the weight of the object.
4. Mark on the sketch all the points where the object touches anything else and draw in the forces at these points. Label all forces clearly.
5. Decide which direction to call positive for the total force and acceleration.
6. Apply Newton's second law equation.

If this routine is followed you will find that it can be used to solve all problems. You will not need to use a different approach when more complicated situations arise and you will be far less likely to make unnecessary mistakes. In use the routine is direct, reliable and quick. Supposed short cuts in solving these problems lead to many mistakes being made.

The essential need is for complete free body force diagrams but not for complicated pictures.

Fig 5.13 Less friction than usual causes many problems!

Example 2

A digger is exerting an upward force of 10 000 N on a load of mass 800 kg in its bucket. Find the acceleration of the load.

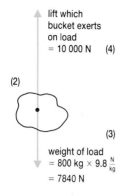

lift which
bucket exerts
on load
= 10 000 N   (4)

(2)

(3)

weight of load
= 800 kg × 9.8 $\frac{N}{kg}$
= 7840 N

Following the numbered sequence suggested in the margin notes:
(1) Consider the forces acting on the load.
(5) Consider the upward direction as positive.
(6) Force = mass × acceleration

Force	=	mass	×	acceleration
N		kg		m s^{-2}
10 000 N – 7840 N	=	800 kg	×	a
2160 N	=	800 kg	×	a

$$a = \frac{2160 \text{ N}}{800 \text{ kg}}$$

$$= 2.7 \text{ m s}^{-2}$$

This problem is rather trivial, but it is amazing how often the working is done in such a way as to give, for this question, an answer of 12.5 m s^{-2}. What method is used to obtain this answer? Why is the method and answer incorrect?

QUESTIONS

Fig 5.14 A helicopter has to be able to control its vertical height and its horizontal velocity and acceleration. It does this by altering its attitude. If it is stopping its nose is lifted up so that the rotor forces air downwards and forwards. The force this air exerts on the rotor is therefore upwards and backwards.

5.2 A helicopter of mass 5000 kg is rising with an acceleration of 2.4 m s^{-2}. What thrust must be exerted on it by its rotor?

5.3 An aeroplane of total mass 30 000 kg needs to be able to take off within a distance of 1200 m on a runway which is 2200 m long. The extra distance is necessary for safety reasons. If the take off and landing speeds of the plane are both 60 m s^{-1} find
(a) The constant acceleration needed to reach take off speed in 1200 m.
(b) The resultant force on the aeroplane during this acceleration.
(c) The thrust provided by the engines if the drag due to ground and wind resistance is taken to have a constant value of 10 000 N.
(d) On landing the mass of the aeroplane has decreased to 25 000 kg. (Why?) What minimum force is necessary to stop it on the runway?

5.4 A lift has a mass of 1800 kg and the rope supporting it exerts an upward force on it of 15 000 N. What is the acceleration of the lift? If this lift is travelling **upwards** at a speed of 3.8 m s^{-1} how long will it take to stop?

(**Note:** When the acceleration in this problem is found to be downwards there is a tendency to assume that the lift is travelling downwards. Although this might be the case it is also possible, as here, for the lift to be decelerating while moving upwards. The total force on the lift and the acceleration of the lift **must** be in the same direction but the acceleration is **not** necessarily in the direction of the velocity.)

Example 3

A man of mass 100 kg now gets into the lift in Question 5.4. If the rope now exerts a force upwards on the list of 20 000 N, find
(a) the force the lift exerts on the man;
(b) the acceleration of lift and man.

This is a two part problem: **two** force diagrams are therefore needed. Consider the forces acting on firstly the man and secondly the lift.

There is a good example of the use of Newton's third law here. F is the force which the lift exerts on the man. It is equal and opposite to the force which the man exerts on the lift and is therefore also labelled F.

Consider the upward direction as positive.

For the man
force on = mass × acceleration
man   of man   of man

$(F - w)$ = m × a
$F - 980$ N = 100 kg × a

For the lift
force on = mass × acceleration
lift   of lift   of lift

$(T - W - F)$ = M × a
$20\,000$ N $- 17\,640$ N $- F = 1800$ kg × a

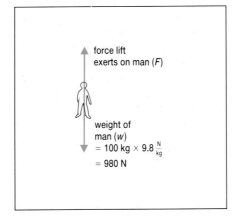

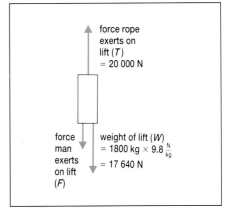

This gives two equations with two unknowns and they can be solved by adding them to give:

$$F - 980 + 20\,000 - 17\,640 - F = 100\,a + 1800\,a$$
$$1380 = 1900\,a$$
$$a = \frac{1380}{1900} = 0.73 \text{ m s}^{-2}$$

Substituting this back gives $F - 980 = 100 \times 0.73$
$$F = 73 + 980$$
$$= 1050 \text{ N (to 3 sig. figs)}$$

So: **(a)** the force the lift exerts on the man is 1050 N and **(b)** the acceleration of the lift and the man, which must be the same since they always have the same velocity, is 0.73 m s^{-2}.

The upward force the lift exerts on the man is greater than his weight, so he accelerates upwards. It is this larger than usual force acting upwards which is the force which we find strange when a lift starts off. Note that the weight of the man is still 980 N, although if he were to be in the unlikely position of standing on some scales while in the lift, the scales would read 1053 N.

DATA ANALYSIS

Fig 5.15 The European Airbus A320

Passenger loading of a European Airbus

Use the data below concerning a European Airbus (which is shown in Fig 5.15) to answer the questions which follow.

Mass of aeroplane, including crew and all equipment	42 000 kg
Capacity of fuel tanks	18 000 kg
Maximum number of passengers	150
Average mass of a passenger and baggage	100 kg
Average use of fuel	5.0 kg km^{-1}
Safety reserve of fuel at end of journey	3000 kg
Take off speed	75 m s^{-1}
Length of runway used	1500 m

(a) What is the safe range of the aeroplane?
(b) How much further could it travel if, at the end of its scheduled flight, the airport it intended to land at was closed by poor weather conditions?
(c) What is the maximum total mass of the aeroplane, passengers and fuel at the start of a flight?
(d) When the plane is taking off, what is its acceleration while on the runway? (assumed constant)
(e) What force is necessary to cause this acceleration?
(f) Runways are always longer than is needed for take off. What braking force would be needed if, just before leaving the ground, the pilot realised that something had gone wrong and that he needed to stop in a further 1200 m?

   Sometimes aeroplanes find it necessary to take off from shorter than normal runways. They then have to reduce the number of passengers they carry or the amount of fuel carried. (They are not allowed to reduce their safety margin of fuel.)
(g) What total mass can be accelerated to take off speed in a distance of only 1200 m? What is the disadvantage to such a take off?
(h) Complete Table 5.1 to show what range the aeroplane has when carrying different numbers of passengers.

Table 5.1   Data for Airbus with different operating conditions

	1500 m take off	1200 m take off	1200 m take off	1200 m take off	1400 m take off
number of passengers	150	130	110	90	
mass of aeroplane	42 000 kg	42 000 kg	42 000 kg	42 000 kg	42 000 kg
mass of passengers	15 000 kg	13 000 kg	11 000 kg	9000 kg	
total mass of fuel	18 000 kg				18 000 kg
total mass	75 000 kg				
mass of fuel in reserve	3000 kg	3000 kg	3000 kg	3000 kg	3000 kg
usable mass of fuel	15 000 kg				15 000 kg
range					

(i) In the final column find the number of passengers which may be carried for the full range of the aeroplane if the length of the runway available is 1400 m.
(j) What other physical factors will affect the range of the aeroplane and what would you do as an airline operator to make operational efficiency as high as possible, without reducing safety standards?

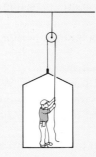

Fig 5.18 A builder's cradle which enables a workman to move up and down the face of building. Various safety devices are omitted from the diagram.

Fig 5.16 The forces on each part of the train must enable each part to have the same linear acceleration.

5.5 A local train consists of two tractor units of mass 10 000 kg (these are coaches with engines) and an unpowered coach, of mass 8000 kg. The horizontal forces acting on the three coaches are shown in Fig 5.17 and consist of:

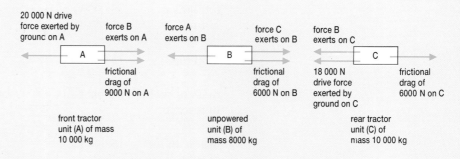

Fig 5.17 The forces acting on the three coaches of a train.

(a) drive forces of 20 000 N on A and 18 000 N on C;

(b) drag forces due to air resistance and friction of 9000 N on A and of 6000 N on both B and C;

(c) forces which act by contact between adjacent coaches.

Find the acceleration of the train and the magnitude and direction of the force which A exerts on B and the force which C exerts on B.

Fig 5.19 Note how the right knee of this high jumper is bent as he is just about to take off. He exerts a downward force on the ground much larger than his own weight so the ground exerts an upward force on him greater than his own weight.

5.6 A model helicopter, of mass 5.0 kg, rises with constant acceleration from rest to a height of 60 m in 10 s. Find the thrust which is exerted by the rotor blades during the ascent.

5.7 A man in the builder's cradle (Fig 5.18) is lifting himself upwards (there are safety devices not shown to prevent him falling). He has a mass of 80 kg and the cradle has a mass of 30 kg. He is pulling on the rope with a force of 600 N.
Draw free body diagrams of
(a) the man
(b) the cradle.
You can assume that the tension in the rope is the same at both ends of the rope, so that it exerts both a force of 600 N upwards on the man and a force of 600 N upwards on the cradle.
Find the acceleration of the man and the force which the man exerts on the floor of the cradle.

5.8 The take off acceleration of a high jumper (Fig. 5.19) is 17.0 m s^{-2} upwards. His mass is 67 kg.
(a) What force is the ground exerting on the man?
(b) What is the ratio of this force to his weight?
(c) Explain how it is possible for this force to be larger than the high jumper's weight.

SUMMARY

Newton's laws
- Every object continues in its state of rest or uniform motion in a straight line unless it is compelled to change that state by external forces acting on it.
- The rate of change of momentum of a body is proportional to the total force acting on it.
- If body A exerts a force on body B, then body B exerts an opposite force of the same size on body A.

Force
- A force is necessary to accelerate an object. The direction of the acceleration is always in the direction of the force. Force is measured in newtons. 1 newton is that force which, acting alone, will give a mass of 1 kilogram an acceleration of 1 m s^{-2}.

Force in newtons	=	Mass in kilograms	×	Acceleration in metre second^{-2}

- Forces outside the nucleus are either gravitational or electromagnetic. All contact forces are electromagnetic forces. The component of a contact force which is tangential to the surface is called the frictional force: the component of the force which is normal to the surface is called the normal contact force.

- Near the surface of the Earth the gravitational pull on a mass is approximately 9.8 newtons per kilogram. The gravitational pull on an object is called the object's weight.

EXAMINATION QUESTIONS

5.9 State *Newton's second law of motion* and describe how you would verify it experimentally. A rubber ball of mass 0.12 kg is thrown vertically upwards at 12 m s⁻¹. The diagram represents the velocity–time graph of the motion of the ball from its starting point until its return to the same point.
Estimate the accelerations at X, Y and Z.

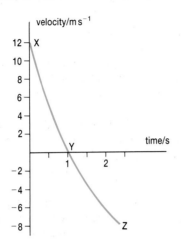

Calculate the change in momentum of the ball and the average force on the ball
(a) between X and Y, **(b)** between Y and Z.
Explain why your calculated values of accelerations and forces change during the ball's motion.

(UCLES 1984)

5.10 A barge of mass 10 000 kg, initially at rest, is pulled along a canal by two horizontal cables inclined at 30° to the bank as shown below. The tension in each cable is 2000 N. Calculate

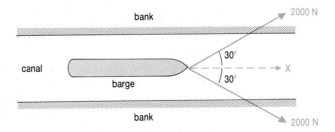

(a) the initial force on the barge in the direction X,
(b) the initial acceleration of the barge in this direction.
Explain why the acceleration of the barge is unlikely to remain constant, and sketch a likely speed–time graph for the motion of the barge. **(9)**

(AEB 1987)

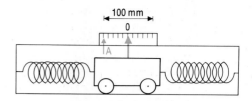

5.11 The sketch shows a mass–spring system which is to be attached to a vehicle such as a car in order to measure the acceleration. A trolley of mass 2.0 kg is free to move from left to right but constrained by two similar springs. When the mass is in equilibrium, each spring is extended by 50 mm, and the force required to displace the mass either to right or the left is shown below.

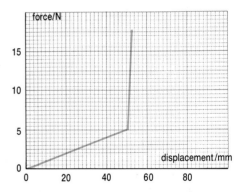

(a) Explain why the mass is displaced when the vehicle accelerates. **(3)**
(b) Use the graph to determine the force acting on the mass when the pointer is at A. What is the magnitude and direction of the acceleration of the vehicle when the pointer is in this position? **(5)**
(c) Even when the vehicle is accelerating uniformly, the pointer will not come to rest for some time.
   (i) Explain why this is so, and sketch and label a graph which shows how the displacement of the pointer is likely to vary with time from the instant the acceleration commences.
   (ii) State two factors which determine the time for the pointer to come to rest, and suggest a method which could be used to bring the pointer quickly to rest without changing the reading obtained. **(7)**

(d) (i) Make a rough numerical estimate of the average acceleration of a car as it accelerates from rest, showing how the estimate is made and giving the value of any quantity you use in your calculation. (60 mile $h^{-1} \approx 25$ m s^{-1}.) Explain whether the mass–spring system shown is suitable for measuring an acceleration of this magnitude.

(ii) Explain two ways of changing the range of the instrument. What effects would the changes you propose have on the response of the instrument to rapid changes in acceleration? (7)

(e) Suggest an explanation of the change in slope of the graph which takes place when the displacement is 50 mm. (3)

(AEB 1987)

5.12 A large cardboard box of mass 0.75 kg is pushed across a horizontal floor by a force of 4.5 N. The motion of the box is opposed by
(i) a frictional force of 1.5 N between the box and the floor, and
(ii) an air resistance force kv^2, where $k = 6.0 \times 10^{-2}$ kg m^{-1} and v is the speed of the box in m s^{-1}.

Sketch a diagram showing the directions of the forces which act on the moving box.
Calculate maximum values for **(a)** the acceleration of the box, **(b)** its speed. (6)

(ULSEB 1984)

5.13 What is meant by the statement that velocity is a *vector* quantity?
State how a velocity–time graph may be used to find
(i) the acceleration of a body;
(ii) the distance the body has travelled in a given time.

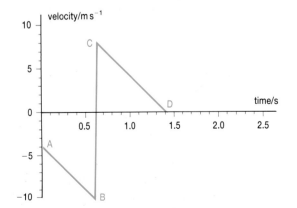

The figure is a velocity-time graph for the first 1.4 s of the motion of a ball, of mass 0.080 kg, which is thrown vertically downwards at 4.0 m s^{-1}. It hits the ground and rebounds directly upwards, taking a negligible time to bounce.

(a) How far does the ball travel before first hitting the ground?

(b) Calculate the change in velocity as the ball rebounds.

(c) How high does the ball rise after rebounding?

(d) Between C and D what is
(i) the acceleration of the ball,
(ii) the change in momentum of the ball,
(iii) the force on the ball?

(e) At what time does the ball next reach the ground?

(f) What will be the upward velocity of the ball on its second rebound if at each instant of bouncing the ratio downward velocity/upward velocity has a constant value?

(g) Sketch the extension of the graph line shown to illustrate the continued motion of the ball up to a time of 2.5 s.

(UCLES 1987)

5.14 (a) State Newton's laws of motion. Explain how the *newton* is defined from these laws. (5)

(b) A rocket is propelled by the emission of hot gases. It may be stated that both the rocket and the hot gases each gain kinetic energy and momentum during the firing of the rocket. Discuss the significance of this statement in relation to the laws of conservation of energy and momentum, explaining the essential difference between these two quantities. (5)

(c) A bird of mass 0.5 kg hovers by beating its wings of effective area 0.3 m^2.
(i) What is the upward force of the air on the bird?
(ii) What is the downward force of the bird on the air as it beats its wings?
(iii) Estimate the velocity imparted to the air, which has a density of 1.3 kg m^{-3}, by the beating of the wings.
Which of Newton's laws is applied in each of (i), (ii) and (iii)? (8)

(ULSEB 1983)

5.15 A car is travelling at speed v along a straight, level road towards a gate which may close automatically at any time. A visual warning is given a short period T before the gate closes. The car's maximum deceleration under braking is a. Show that there is a maximum safe speed given by

$$v_{max} = 2Ta$$

above which the car may be able neither to stop nor to pass (without accelerating) before the gate closes.
Calculate a suitable value for T if $a = 3.0$ m s^{-2} and speeds up to 30 m s^{-1} are to be safe.

(UCLES 1985)

5.16 A transport plane is to take off from a level landing field with two gliders in tow, one behind the other. Each glider has a mass of 1200 kg, and the friction force or drag on each may be assumed to be constant and equal to 2000 N. The tension in the towrope between the transport plane and the first glider is not to exceed 10 000 N.
   (a) If a velocity of 40 m s^{-1} is required for take off, what minimum length of runway is needed?
   (b) What is the tension in the towrope between the two gliders while they are accelerating for the takeoff?

5.17 State Newton's laws of motion.
   When a body moves through a fluid, a retarding force due to turbulence may be experienced. In the case of a sphere of radius r moving with speed v through a stationary fluid of density ρ which is at rest, this force is given by $F = k\rho r^2 v^2$ where k is a constant.
   (a) Show that the constant k is dimensionless.
   (b) By relating the retarding force to the transfer of momentum between the sphere and the fluid, explain why F is proportional to $\rho r^2 v^2$.
   When spherical raindrops fall through still air, all but the smallest experience a retarding force given by the equation above. It is found that drops of a given radius approach the ground with an approximately constant speed, which is independent of the height of the cloud in which they were formed. Explain this observation by reference to Newton's laws. Find an expression for this terminal speed v_t in terms of the constant k, the radius r of the drop, its density r_w, the density r_A of the air and the acceleration of free fall g. (Neglect the buoyancy of the air.)
   The terminal speed of a raindrop of radius 1 mm is approximately 7 m s^{-1}. In freak storms, hailstones with radii as large as 20 mm may fall. Estimate the speed with which such stones strike the ground.
   (Take the density of water as 1×10^3 kg m^{-3} and the density of ice as 9×10^2 kg m^{-3}.)

(UCLES 1986)

5.18 The figure shows the important parts of a rolling mill which is used for reducing the thickness of slabs of hot steel. The slab is compressed as it moves through the space between the two rollers which rotate as indicated.

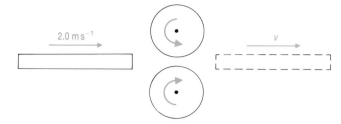

A hot slab of metal 20 mm thick, 0.50 m wide and 6.0 m long enters the mill at a speed of 2.0 m s^{-1}, and passes through in a time of 3.0 s. In the rolling process the width of the slab is not changed but its thickness is reduced to 16 mm.
   (a) Calculate the speed at which the slab leaves the mill.
   (b) Calculate the momentum of the slab as it leaves the mill, given that the density of steel is 8.0×10^3 kg m^{-3}.
   (c) Deduce the average force acting horizontally on the rollers.

(UCLES specimen)

5.19 **(a)** Define acceleration. (2)
   Use your definition to explain why it is that the velocity of a body may be in a different direction from its acceleration. (2)
   (b) Discuss whether the resultant force on a body may or may not be in the same direction as its acceleration. (2)
   (c) The moving head of an electronic printer has a mass of 0.20 kg and moves along the line of print in a jerky motion. After printing each character, the head accelerates sideways under a force of 10 N and then immediately decelerates to rest under a force of 30 N ready to print the next character. The characters are 2.5 mm apart.
   (i) Sketch a graph to show how the velocity of the head of the printer varies with time. (2)
   (ii) Calculate the time taken for movement between characters. (7)
   (iii) An additional time of 8.0 ms is required for the printing of each character. Find the maximum number of characters that can be printed each second. (2)
   (iv) What physical problems might require the printer to run at a slower speed than the value you have calculated? What changes could be made to increase the printing speed? (5)

(UCLES 1988)

Chapter 6

EQUILIBRIUM

> **LEARNING OBJECTIVES**
>
> At the end of this chapter you should be able to:
>
> 1. calculate the turning effect of forces;
>
> 2. establish whether or not a rigid body is in equilibrium;
>
> 3. understand the term centre of gravity.

6.1 EQUILIBRIUM OF A POINT OBJECT

In discussing Newton's laws in Chapter 5 a simplification was made by assuming that the object under consideration was not rotating. Another way of making that same assumption is to assume that every part of the object concerned has the same velocity. In practical terms this is not too difficult to imagine, and usually does not involve very much error. A child travelling on a scooter at 4 m s^{-1} can be visualised easily, although if greater accuracy is required it might be necessary to take into account the fact that parts of the child and wheels are moving with different velocities.

If we wish to make the simplification that all of an object is travelling with the same velocity, one way of doing it is to assume that the object is a **point object**. A point object is idealised. It has mass but no size. Point objects are not necessarily small. In gravitational problems the Earth can be considered as a point object. If no information is given about the size of an object then you have to assume that it is being regarded as a point object.

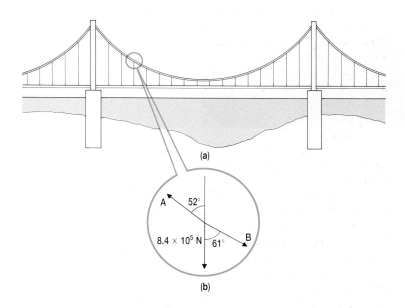

Fig 6.1 Each point of support on the main cables of a suspension bridge must be in equilibrium.

A point object is said to be in **equilibrium** if the resultant force acting on it is zero. If there is zero resultant force acting on it then it cannot have any acceleration. It may however be moving with a constant velocity. The condition for zero resultant force on an object is that the vector sum of all the forces acting on it must be zero.

Example 1

A vertical suspension bridge cable supports a load of 8.4×10^5 N. The main cable, to which it is attached, is pulled so that the higher part is at an angle of 52° to the vertical and the lower part is at 61° to the vertical. Find the tensions *A* and *B* in the cable, see Fig 6.1(a). Fig 6.1(b) shows an enlarged view of the point at which the vertical cable is attached to the main cable.

The point of suspension X must be in equilibrium so the resultant force acting on it must be zero. The sum of the three forces shown in the vector diagram, Fig 6.2 must be zero. A closed triangle will therefore be the result.

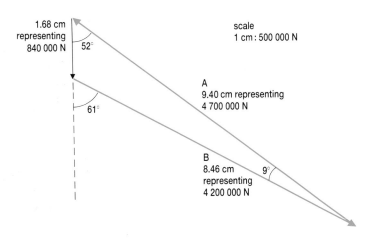

Fig 6.2   The sum of the three forces acting on each point of support must be zero. Since the angles can be measured and the load supported is known, the triangle enables the tensions **A** and **B**, to be found.

Finding the values of *A* and *B* can be done either by a scale drawing of the triangle of forces or by calculation. In this case the calculation requires the use of the sine rule, (see Appendix B).

The sine rule gives

$$\frac{8.4 \times 10^5 \text{ N}}{\sin 9°} = \frac{A}{\sin 119°} = \frac{B}{\sin 52°}$$

$$A = \frac{\sin 119° \times 8.4 \times 10^5 \text{ N}}{\sin 9°} = 4.7 \times 10^6 \text{ N}$$

$$B = \frac{\sin 52° \times 8.4 \times 10^5 \text{ N}}{\sin 9°} = 4.2 \times 10^6 \text{ N}$$

Why are these values likely to have a high uncertainty?

Fig 6.3   A bridge must be designed so that either it is in equilibrium or it will return to an equilibrium position quickly after being distorted in some way either by a load or by adverse weather conditions.

6.2 TURNING EFFECTS

At the beginning of the chapter it was stated that a point object is in equilibrium if the resultant force acting upon it is zero. If the object is not a point object then further consideration is necessary because such an object may rotate.

Consider the forces acting on the water wheel shown in Fig 6.4. The wheel makes contact with two objects: the water and the axle about which it rotates. Each of them exerts a force on the wheel and in addition there is the gravitational force which the Earth exerts on the wheel, its weight.

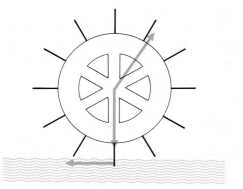

Fig 6.4 The forces acting on a water wheel give it no linear acceleration but they do cause rotation.

Since there is no linear acceleration of the wheel, the resultant force acting on it must be zero. A vector diagram, Fig 6.5(a), can be used to find the force which the axle exerts on the wheel. This force is at an angle to the

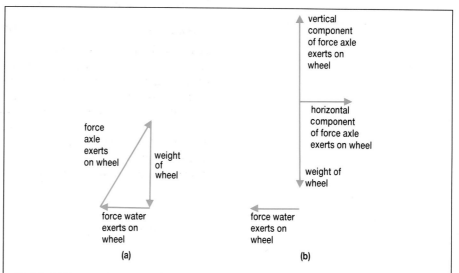

Fig 6.5 **(a)** The sum of the forces acting on the water wheel is zero so a closed triangle of forces is obtained. **(b)** Here the diagram is re-drawn with the force that the axle exerts on the wheel resolved into vertical and horizontal components.

vertical and it can be resolved into a horizontal and a vertical component as shown in Fig 6.5(b). The weight of the wheel is equal and opposite to the vertical component of the force that the axle exerts on the wheel, and the force that the water exerts on the wheel is equal and opposite to the horizontal component that the axle exerts on the wheel.

Despite there being zero force on the wheel, the wheel rotates. This is because of the turning effect of the water. The water exerts a horizontal force on the wheel at a distance from the axle of the wheel. The turning effect of a force is called its moment. The size of the moment which a force exerts depends not only on the size of the force but also on the distance of the line of action of the force from the axis of rotation. **Moment** is defined by the equation:

Moment = force × perpendicular distance between axis of rotation and line of action of force.

The unit for moment is newton metre (N m) and by convention this is not written as joule (J). Moment is a vector whose direction is, again by convention, taken along the axis in the direction of a right-hand screw.

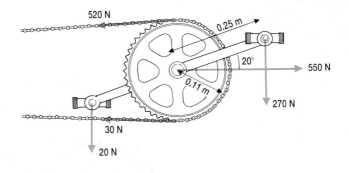

Fig 6.6 The forces acting on the pedal wheel of a bicycle.

Example 2

Find the total moment of all the forces about the axis of a chain wheel and pedals of a bicycle, when the forces exerted on it are as shown in Fig 6.6. The chain wheel has a radius of 0.11 m and the forces exerted on the pedals are at a distance of 0.25 m from the axis. The line of the pedals is 20° from the horizontal.

Note the directions of these forces. They are forces on the pedal wheel and the chain is pulling the pedal wheel to the left in the diagram. The axle prevents this wheel moving backwards relative to the bicycle.

As the pedal arm is at an angle to the horizontal, the perpendicular distance in metres of the applied force from the axis is 0.25 cos 20°.

Moment exerted by upper foot	=	270 N × 0.25 cos 20° m	= 63.4 N m
Moment exerted by lower foot	=	−(20 N × 0.25 cos 20° m)	= − 4.7 N m
Moment exerted by upper part of chain	=	−(520 N × 0.11 m)	= − 57.2 N m
Moment exerted by lower part of chain	=	30 N × 0.11 m	= 3.3 N m
Moment exerted by axle	=	550 N × 0	= 0

Total moment = (63.4 − 4.7 − 57.2 + 3.3) N m = 4.8 N m clockwise

In the absence of any other forces acting on the chain wheel and pedals, this total moment will cause the wheel to increase its rate of rotation. Note that attention has to be paid to the direction of the moment. Clockwise moments are regarded as positive and anticlockwise moments as negative. The largest force considered exerts no moment because it is exerted by the axle and therefore the perpendicular distance from the axle is zero.

 If a system is in equilibrium then the sum of the clockwise moments equals the sum of the anticlockwise moments. This is known as the *principle of moments.*

Forces are frequently applied to an object to turn it without there being any resultant force on the object. This is done by applying two equal and opposite forces to the object, often at equal distances from the axis of rotation. A coil in an electric motor has driving forces acting on it as shown in Fig 6.7. The resultant force acting on the coil is zero as there is an upward force of 200 N to cancel out the downward force of 200 N. The coil will therefore have zero linear acceleration. However, because the two forces are not in line with one another there is a moment acting on the coil and it rotates. A pair of equal forces such as this is called a **couple**. The moment provided by a couple is equal to the product of one of the forces and the perpendicular distance between them. The moment of a couple is called the **torque**. The terms torque and moment of a force are often regarded as having the same meaning but a moment can be supplied by a single force and may result in there being a non-zero resultant force on the object. The term torque always implies that only the turning effect is being considered.

The distinction between the terms is illustrated by Fig 6.8. The moment of the force in Fig 6.8(a) about O is

$$30 \text{ N} \times 0.40 \text{ m} = 12 \text{ N m}$$

This single force would cause a linear acceleration of the rod as well as a turning effect.

In Fig 6.8(b) the equal forces acting in opposite directions give a resultant force of zero so there is no linear acceleration. They are called a **couple**. The torque of the couple is a turning effect only and has a value of

$$(30 \text{ N} \times 0.20 \text{ m}) + (30 \text{ N} \times 0.20 \text{ m})$$
$$= 12 \text{ N m as before}$$

In some problems where a torque is applied to an object, the forces causing that torque are difficult to measure. A good example of such a

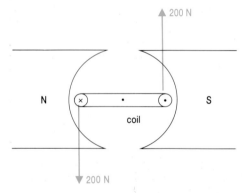

Fig 6.7 A couple acting on the coil of an electric motor.

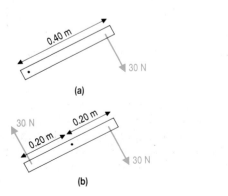

Fig 6.8 **(a)** A moment of 30 N × 0.40 m = 12 N m.
(b) A couple of 2 × 30 N × 0.20 m = 12 N m.
A couple has a resultant linear force of zero.

situation is when a screwdriver is being held in the hand to turn a screw. The hand then exerts many thousands of tiny frictional forces on the screwdriver handle, wherever contact is made between the handle and the hand, but does not exert any resultant force to accelerate the screwdriver in a straight line. The sum of the moments provided by all of these forces is often referred to as, simply, the **torque** on the screw.

QUESTIONS

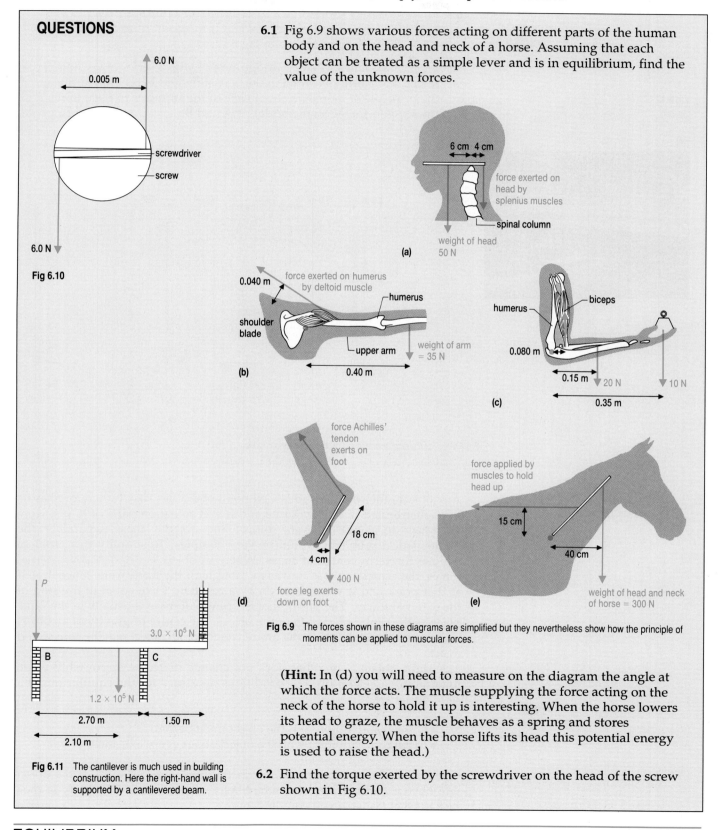

Fig 6.10

6.1 Fig 6.9 shows various forces acting on different parts of the human body and on the head and neck of a horse. Assuming that each object can be treated as a simple lever and is in equilibrium, find the value of the unknown forces.

Fig 6.9 The forces shown in these diagrams are simplified but they nevertheless show how the principle of moments can be applied to muscular forces.

(**Hint:** In (d) you will need to measure on the diagram the angle at which the force acts. The muscle supplying the force acting on the neck of the horse to hold it up is interesting. When the horse lowers its head to graze, the muscle behaves as a spring and stores potential energy. When the horse lifts its head this potential energy is used to raise the head.)

6.2 Find the torque exerted by the screwdriver on the head of the screw shown in Fig 6.10.

Fig 6.11 The cantilever is much used in building construction. Here the right-hand wall is supported by a cantilevered beam.

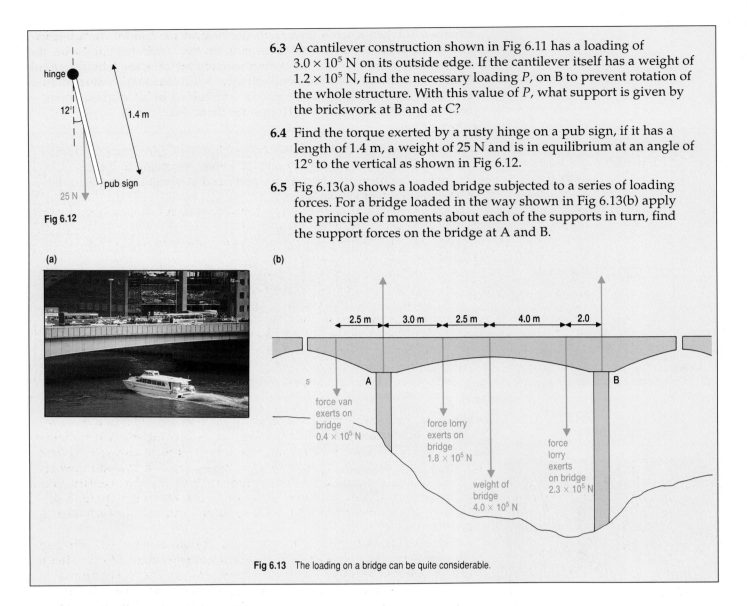

6.3 A cantilever construction shown in Fig 6.11 has a loading of 3.0×10^5 N on its outside edge. If the cantilever itself has a weight of 1.2×10^5 N, find the necessary loading P, on B to prevent rotation of the whole structure. With this value of P, what support is given by the brickwork at B and at C?

6.4 Find the torque exerted by a rusty hinge on a pub sign, if it has a length of 1.4 m, a weight of 25 N and is in equilibrium at an angle of 12° to the vertical as shown in Fig 6.12.

6.5 Fig 6.13(a) shows a loaded bridge subjected to a series of loading forces. For a bridge loaded in the way shown in Fig 6.13(b) apply the principle of moments about each of the supports in turn, find the support forces on the bridge at A and B.

Fig 6.12

(a)

(b)

hinge

12°

1.4 m

pub sign

25 N

2.5 m 3.0 m 2.5 m 4.0 m 2.0

S   A   B

force van exerts on bridge 0.4×10^5 N

force lorry exerts on bridge 1.8×10^5 N

weight of bridge 4.0×10^5 N

force lorry exerts on bridge 2.3×10^5 N

Fig 6.13 The loading on a bridge can be quite considerable.

6.3 EQUILIBRIUM OF A RIGID OBJECT

Any object other than our idealised point object has size. It therefore has its mass distributed over its volume and forces acting on parts of it. If it were necessary to consider every bit of the object's mass separately the mathematical problems would be very complex. To simplify this problem we use the term **centre of mass**. When an object is acted on by external forces, the centre of mass moves as though all the mass were concentrated at that point and it were acted on by a resultant force equal to the sum of the external forces. The position of the centre of mass is usually in the same place as the centre of gravity of the object. The **centre of gravity** of an object is the point through which the entire weight of an object may be considered to act.

A **rigid object** is one that does not change its shape appreciably when undergoing acceleration. The conditions necessary for the equilibrium of a rigid object are:

- the resultant force on the object must be zero
- the resultant torque on the object about every axis must be zero.

The first of these conditions means that the centre of mass of the object will have a constant velocity, which may often be zero. The body is then said to be in translational equilibrium.

The second condition is necessary for the object not to be changing its rate of rotation. An object rotating at a constant rate is said to be in rotational equilibrium and zero torque is required for this to be the case. Note that rotational equilibrium is, strictly, not a true equilibrium as the individual particles within the object do have acceleration. Complete equilibrium exists when there is zero resultant force on an object and it does not rotate.

INVESTIGATION

Without recourse to major surgery, find the mass of one of your own hands. You can of course extend this investigation to finding the mass of an arm or even a leg if you so wish.

Also find the position of your own centre of gravity when you are in a normal standing or lying position.

Use the results of the two investigations to answer the following questions.

- How is it possible for an athlete to jump clear over the bar in a high jump competition and yet for the centre of gravity of the athlete to go under the bar?
- An artificial leg is lighter than a real leg. How would this need to be taken into account by a physiotherapist in planning the rehabilitation of someone who has had to have a leg amputated?

6.4 CENTRE OF GRAVITY AND STABILITY

Fig 6.14 Earthquake damage in Mexico City in 1985.

When a civil engineer designs a structure, such as a bridge, or a building, he or she must make certain that the structure is in equilibrium at all times. It is clear that the structure must not accelerate so it must always have zero resultant force and zero resultant torque acting on it. It is also necessary that the structure should be in equilibrium at all stages of its construction. This frequently causes structural engineers problems as the strength of a building when completed is usually greater than its strength during construction.

Making certain that the structure is in equilibrium is, however, only part of the problem. Any structure is subjected to variable external forces after it has been constructed. These forces can be very large. The most dramatic of these forces are earthquake forces. During a severe earthquake, forces can be sufficiently large to cause enormous damage. In Fig 6.14 the photograph shows a building which has been seriously damaged by an earthquake. In areas of the world where earthquakes are common, modern buildings are designed to withstand earthquakes up to a certain magnitude and to move in one piece thereafter. This reduces the amount of injury caused by falling floors or masonry. Why is it not possible to build a totally earthquake-proof structure?

Fig 6.15 Snow loading on houses can be very large at times throughout the life of a house.

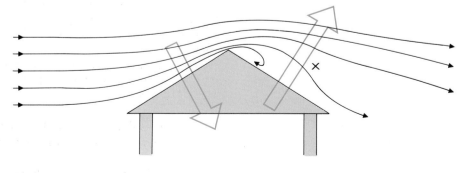

Fig 6.16 Wind blowing across a roof increases the loading on the side of the roof facing the wind and reduces the loading on the side away from the wind.

EQUILIBRIUM

More usual variable forces acting on a building are those due to wind and weight of snow. Both of these forces can be surprisingly large at some time or other during a building's lifetime. It may be that a house of floor area 50 m² has to be able to support a depth of snow of 0.60 m, see Fig 6.15. For snow of density 100 kg m⁻³ this is a total mass to be supported of 3000 kg. Wind forces can be as large as this and can also act upwards! A stream of air flowing across a roof causes a reduction in the pressure in region X, Fig 6.16. The normal pressure underneath the roof causes the roof to be lifted and in a strong gale, roofs of houses can be lifted off completely, Fig 6.17. Wind forces can also result in oscillations being set up in a structure. The size of these oscillations can increase until damage is caused (see Section 10.5).

Fig 6.17 Wind damage can be extensive. Here the roof of an appartment in Texas has been blown off by hurricane Gilbert.

A structure therefore must not only be in equilibrium but must be able to return to its original position of equilibrium if some outside cause moves it. **Stability** is the term used to describe this.

◆   A body is said to be in stable equilibrium if, when it is displaced a small amount, it will return to its original position.

The degree of stability of an object can be found by measuring the angle through which it must be rotated before its centre of gravity lies outside its

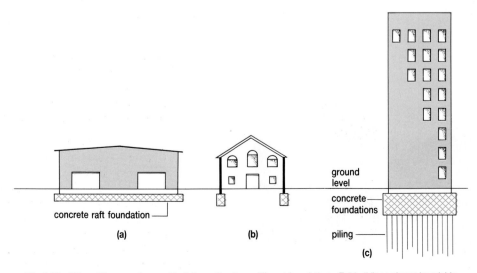

Fig 6.18 Different types and ages of buildings often have different foundations. Tall buildings almost invariably have piles pushed far down into the ground until they reach rock. Houses are usually built on strip foundations whereas factories and schools are increasingly being built on concrete raft foundations.

EQUILIBRIUM

Fig 6.19 Its wide wheelbase and low centre of gravity make a racing car very stable.

base and it topples over. This will depend on its shape and how it is supported. Fig 6.18 shows three buildings in equilibrium: (a) is the most stable and (c) is the least stable.

Figs 6.19-21 show some more examples of stability and its relation to the position of the centre of gravity. The wide base of the racing car means that it has to be tilted to a much larger angle than a saloon car before it will topple over.

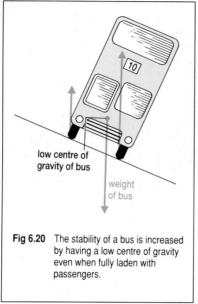

Fig 6.20 The stability of a bus is increased by having a low centre of gravity even when fully laden with passengers.

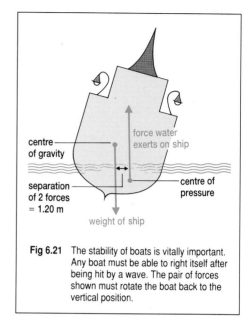

Fig 6.21 The stability of boats is vitally important. Any boat must be able to right itself after being hit by a wave. The pair of forces shown must rotate the boat back to the vertical position.

The low centre of gravity of the bus means that it will not topple until at a much greater angle than would be the case for a uniform rectangular block of the same dimensions, Fig 6.20.

When a ship keels over it is vital that the couple acting on it restores it to its original, upright position. The stability of ships is a subject on which a great deal of thought has to be given. In all cases the shape of the keel of the ship is constructed so that the force which the water exerts on the ship, which can be considered to act through a single point called the centre of pressure, lies outside of the weight of the ship, acting through the centre of gravity, (Fig 6.21).

QUESTIONS

6.6 Find the torque acting on the ship in Fig 6.21. The weight of the ship is 5.0×10^7 N.

6.7 Sketch a graph to show how the height of the centre of gravity of the bus above the ground varies with the angle of the bus to the vertical. Show both positive and negative values of the angle, and your graph should show clearly how the normal position of the bus is stable and has the least potential energy.

6.8 A dam is constructed out of a rectangular block of concrete whose dimensions are shown on Fig 6.22. What depth of water can be held back by the dam before it topples about point A? The density of concrete is 2200 kg m^{-3} and the density of water is 1000 kg m^{-3}. Start this question by drawing a force diagram for a one metre length of the dam when it is on the point of toppling over. The average pressure of the water may be considered to act at half the depth of water. (This assumption is not strictly accurate but it over-estimates the moment the water exerts on the dam. If you can use integration you should be able to find the correct moment.)

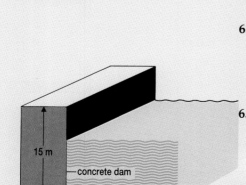

Fig 6.22

Fig 6.23 An earth dam in Holland. The sea on the left is higher than that in the lake on the right.

Dams using this principle are very common. They are often made of earth instead of concrete and so must be made with sloping rather than vertical sides. The dykes in The Netherlands are earth dams, Fig 6.23. They have to be made so that some impermeable material is incorporated within them so that they do not allow the passage of water through them or under them. If you have answered question 6.8 correctly, you will find that the depth of water which can be retained is about 18 m. Does this answer make sense?

DATA ANALYSIS

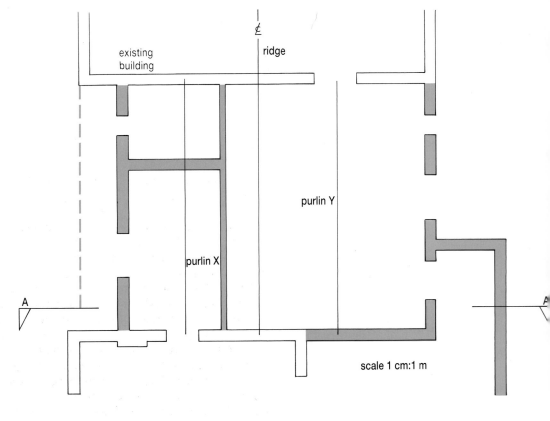

Fig 6.24 Building plans of a village hall extension. Scale 1cm: 1m

Building design

The following exercise uses data from the planning application made by a small village who had recently purchased an old school building and wished to extend the building to make it suitable for use as a village hall.

Fig 6.24 is a plan of some of the existing building together with the proposed extension shown in blue. The exercise shows how a structural engineer calculates the size required for certain key supporting beams, called purlins. ℄ is the symbol used for centre line and is the line of the highest point of the roof, the ridge.

Fig 6.25 is a section through the building; A and A′ on this diagram correspond to A and A′ on Fig 6.24.

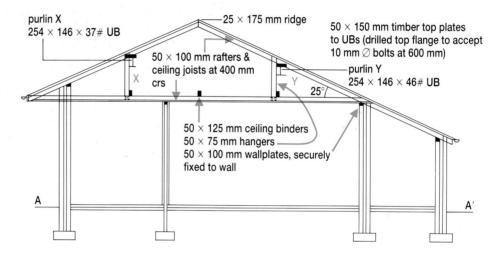

SECTION AA′

Scale 1 cm:1 m

Fig 6.25

Look at the details given on Fig 6.25 together with the following explanation of terms.

254 × 146 × 37# UB is an I-shaped steel beam (Universal beam) of depth 254 mm, width 146 mm and using a thickness of steel which gives it a mass of 37 kg m^{-1}. This is used as purlin X. Purlin Y is the other universal beam and is 254 mm deep, 146 mm wide and is made of thicker steel so it has a mass of 46 kg m^{-1}.

The rafters and the ceiling joists are shown on the section and are constructed from 50 mm × 100 mm softwood and placed so that their centres are 400 mm apart (400 mm crs). 50 mm × 150 mm wood is fixed to the top flange of the universal beam to accept bolts of 10 mm diameter every 600 mm along the length of the timber. Binders stop the ceiling joists from warping and hangers stop them sagging. The whole roof is supported by the purlins and by wall plates fixed to the walls.

The structural engineer makes the following loading calculations by using his experience and data supplied by manufacturers and planning authorities.

Fig 6.26 Buildings must be stable during construction as well as when finished.

Loading (per square metre)

Pitched roof :	slates	21.0 kg m^{-2}
	felt and slates support	7.5 kg m^{-2}
	rafters	5.5 kg m^{-2}
		34.0 kg m^{-2}
	equivalent to	330 N m^{-2}

Loading due to weight of roof $= \dfrac{330 \text{ N m}^{-2}}{\cos 25°} = 370 \text{ N m}^{-2}$

Imposed load (snow) $= 720 \text{ N m}^{-2}$

Wind load: This is worked out to an empirical formula which involves the height of the roof above the ground, the pitch (angle) of the roof, the altitude of the building and the region of the country in which the building is to be built. When the formula is applied in this case it gives the wind load as 1250 N m^{-2}.

Loading due to weight of ceiling	joists	5.5 kg m^{-2}
	fibreglass	2.0 kg m^{-2}
	plasterboard	16.0 kg m^{-2}
		23.5 kg m^{-2}
Plus loading due to storage in loft		76.5 kg m^{-2}
		100.0 kg m^{-2}
	equivalent to	980 N m^{-2}

Summary of loading

Total loading	Roof	370 N m^{-2}
	Imposed load	720 N m^{-2}
	Wind load	1250 N m^{-2}
		2340 N m^{-2}
	+ Ceiling	980 N m^{-2}

Now the structural engineer has the data to work out the loading on each purlin.

Purlin X

Loading due to roof = 2340 N m^{-2} × 2.5m $= 5850 \text{ N m}^{-1}$
Loading due to ceiling = 980 N m^{-2} × 1.3 m $= 1270 \text{ N m}^{-1}$

How much a purlin distorts as a result of being uniformly loaded depends on a term called the bending moment M. This is given by

$$M = \frac{Wl^2}{8}$$

where W is the load per metre of length and l is the length of the purlin.
This gives

$$M = \frac{7120 \text{ N m}^{-1} \times (7.6\text{m})^2}{8} = 51\,400 \text{ N m}$$

From this value another series of calculations enables the size of the beam needed to be found.

Answer the following questions which will test if you have understood the principles of the calculations and if you understand the physics of moments and equilibrium.

(a) Why is 34.0 kg m^{-2} equivalent to 330 N m^{-2}?

(b) Why is the loading due to the weight of the roof found by dividing 330 N m^{-2} by cos 25°?

(c) What depth of snow would give this imposed load? (Snow density is about 0.1 × density of water.)

(d) What does 'empirical' mean? Why is wind loading different for different houses?

(e) In calculating the loading on purlin X, why is the length of 2.5 m used?

(f) Why is a length of 1.3 m used to find the loading due to the ceiling?

(g) Why does the ceiling loading of 1270 N m^{-1} have to be added to the roof loading of 5850 N m^{-1} to obtain the total loading on the purlin of 7120 N m^{-1}?

(h) Find M, the bending moment for purlin Y. Is your result compatible with the size of the purlin actually used?

SUMMARY

- The moment of a force is a turning effect and is the product of the force and the perpendicular distance between the axis of rotation and the line of action of the force.
- A couple is a pair of forces not in line with one another. They are equal in magnitude and opposite in direction. The moment of a couple is called a torque, and has the value of the product of one of the forces and the perpendicular distance between the forces. A torque produces a turning effect but no linear acceleration.
- A point object is in equilibrium if the vector sum of all the forces on it is zero.
- An extended object is in equilibrium if the sum of the forces and the sum of the torques on it are both zero.
- The centre of gravity of an object is the point through which the entire weight of a body may be considered to act.

EXAMINATION QUESTIONS

6.9 The repair manual for a car states that wheel nuts must be tightened by a torque of 60 N m. If a spanner of length 20 cm is used on such a wheel nut, what perpendicular force must be exerted to achieve the correct torque?

6.10 AB represents the raised bonnet of a car, freely hinged at A. The bonnet is of mass 12 kg and its weight acts through the point G.

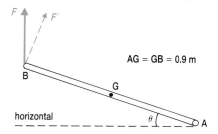

(a) Calculate the value of the vertical force F that will just maintain the position of the bonnet.

(b) Instead of applying a vertical force to the bonnet, a force F' is applied perpendicular to the surface of the bonnet.
  (i) Calculate F' for $\theta = 24°$.
  (ii) Hence explain why it is easier to raise the bonnet by applying a force in the direction of F' than by applying a force in the direction of F. (8)

(ULSEB 1988)

6.11 The drawing shows part of the force diagram for the spine of a person bending over, with the back horizontal. The spine is considered as a rod pivoted at its base. The various muscles of the back are equivalent to a single muscle producing a force T as shown. W is the force that the upper part of the body exerts on the spine.

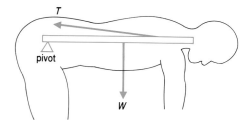

(a) Explain why, for equilibrium, the value of T is large (typically several times W).

(b) For equilibrium, a force P at the pivot is necessary. Draw a triangle of forces to show the equilibrium of the spine under the action of forces T, W and P. Comment on the size of P relative to W.

(UCLES 1988)

6.12 (a) Coplanar forces F_1, F_2, F_3 act on a point mass A, as shown in Fig (a). State, or show on a labelled diagram, the condition for the mass to be in equilibrium.

(b) Coplanar forces F_4, F_5, F_6, F_7, which do not all pass through the same point, act on a body B, as shown in Fig (b). State, or show on labelled diagrams the conditions for the body to be in equilibrium. (3)

(UCLES 1987)

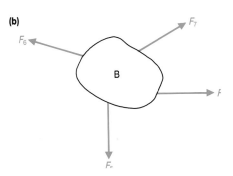

6.13 (a) (i) Explain what is meant by a *couple*. (3)
  (ii) Show that the torque (moment) of a couple is independent of the points of application of the forces constituting the couple if the magnitude of the forces and their separation remains constant. (4)

(b) An 850 kg car travelling at 20 m s^{-1} on a straight level road is brought to a standstill in 60 m by uniform braking. Calculate:
  (i) the mean retarding force acting on the car; (4)
  (ii) the energy dissipated in the brakes. (3)

(c) The brake system of each wheel of the car is shown in the diagram. When the brake pedal is depressed, hydraulic fluid causes the two brake linings to rub against the brake drum. Calculate, using the data given in the figure and the result of (b) (i):
  (i) the angular velocity ω of each wheel before the brakes are applied; (3)

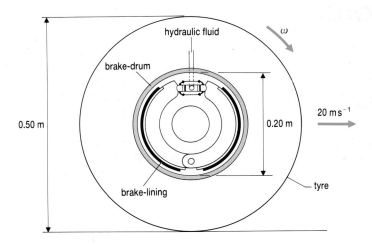

(ii) the mean tangential force exerted by each brake lining on the inner surface of a brake drum during braking. (Assume that during braking the braking force is applied equally on all four wheels.)   (6)

Chapter 7

MOMENTUM, WORK, ENERGY AND POWER

7.1 CONSERVATION OF MOMENTUM

Momentum

Momentum was defined (Section 4.6) as the product of mass, m and velocity, v. It is a vector quantity in the same direction as the velocity. In equation form this is therefore $p = m \times v$ where p is the symbol used for momentum.

Momentum was used by Newton in his second law which states that the rate of change of the momentum of a body is proportional to the total force acting on it (Section 5.3). Using SI units for the quantities we therefore get for a constant force F the equation

$$\frac{mv - mu}{t} = F$$

or $\qquad \Delta p = F \times t$

where Δp is the change in the momentum. This equation shows that the product of the constant force acting on an object and the time for which it acts gives the change in the momentum of the object. It also indicates that a unit which can be used for momentum is the newton second, N s.

Example 1

A ship of mass 6 000 000 kg has a constant force applied to it of 200 000 N. How long will it take to reach a speed of 5 m s⁻¹ if it is starting from rest?

Since the product force × time gives the change in the momentum, we need to know the final momentum p of the ship.

$$p = m \times v = 6\,000\,000 \text{ kg} \times 5 \text{ m s}^{-1} = 30\,000\,000 \text{ N s}$$
$$= F \times t = 200\,000 \text{ N} \times t$$

The time taken is therefore 150 seconds.

Questions such as this are rather artificial, as there is little chance of keeping such a force constant. The resistance to movement provided by the water will vary during the acceleration. The question does however give a guide to the time taken, and indicate the minimum time taken if the force applied by the propeller to the ship is 200 000 N. Drag through the water will, in practice, cause the time to be longer. Another point to note with this type of question is that it is not necessary to find the acceleration. As a general rule, if a question is a force–time question then immediately consider the momentum of the object.

Impulse

In many practical applications of momentum the force applied will not be constant. Consider a ball, of mass 0.20 kg, travelling with velocity 28 m s^{-1} directly towards a wall. It hits the wall and bounces off in the opposite direction with a velocity of 20 m s^{-1}.

The change in the ball's momentum is

final momentum away from wall – initial momentum away from wall
= (0.20 kg × 20 m s^{-1}) – (0.20 kg × –28 m s^{-1}) = 9.6 N s

and therefore the wall has exerted a force on the ball in such a way as to provide the same effect as a constant force of 9.6 N would do if applied for 1 second. This could however have been achieved by a constant force of 96 N for 0.1 s, or a force of 48 N for 0.2 s, or any similar combination as shown in Fig 7.1. In graphical terms the area beneath the force–time graph has to be 9.6 and the shape of the graph, in practice, is much more likely to be as shown in Fig 7.2.

This area is called the impulse. **Impulse** is the area beneath a force–time graph and its unit is the newton second. The impulse applied to an object will be a measure of the change in the momentum of the object. For a constant force it is the product force × time. For a varying force it is the average force × time. Using calculus notation it is $\int F dt$, so we get the equation:

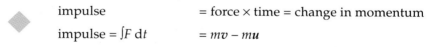

impulse = force × time = change in momentum
impulse = $\int F\, dt$ = $mv - mu$

where v is the final velocity and u the starting velocity.

Often the word impulse is used when the force is applied for a brief time, but there is no reason why it cannot be applied when longer times are involved.

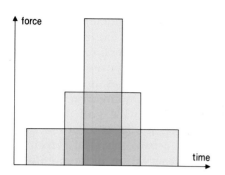

force

time

Fig 7.1 Each of these force–time graphs will cause the same change in the momentum of the object on which the force acts. The area beneath each graph is the same.

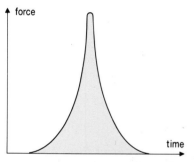

force

time

Fig 7.2 In practice when one object collides with another the force–time graph is more likely to be of this shape.

QUESTIONS

Fig 7.3 The collision between a golf club and a golf ball gives the ball a high acceleration for a short time. This is possible because of the large force exerted by the club. The force is sufficient to squash the ball appreciably.

7.1 What driving force is necessary if a car of mass 1200 kg is to reach a speed of 30 m s^{-1} in 15 s?

7.2 What braking force is necessary if the car in question 7.1 is to stop in 5 s?

7.3 A ball, initially at rest, is struck with an average force of 2600 N for 1.2×10^{-3} s. What is the final velocity of the ball if its mass is 0.047 kg? What impulse is given to the ball?

7.4 A pilot ejected from an aeroplane experiences an acceleration of 12 g for 0.25 s. If the mass of the pilot is 70 kg, find the average total force acting on him and the change in his velocity.

7.5 What impulse is given to an α particle of mass 6.7×10^{-27} kg if it is ejected from a stationary nucleus at a speed of 3.2×10^{6} m s^{-1}? What average force is needed if it is ejected in approximately 10^{-8} s?

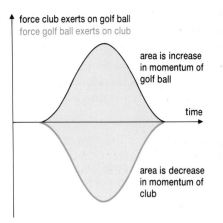

force club exerts on golf ball
force golf ball exerts on club

area is increase in momentum of golf ball

time

area is decrease in momentum of club

Fig 7.4 Force–time graphs showing how if one object gains momentum another object loses an equal amount of momentum.

Conservation of momentum

An object can only undergo a change in its momentum if a force is exerted on it. Using Newton's third law however, we know that if an object exerts a force on something then there must be an equal and opposite force on the object itself. This will change its own momentum. An example should help to make this clear. The head of a golf club strikes a golf ball horizontally. A graph showing the way in which the force which the club exerts on the golf ball varies with time may be as shown in Fig 7.4.

The impulse given to the golf ball will increase the momentum of the golf ball and this will be the area beneath the graph. During the brief moment that this change in momentum is taking place the force which the club exerts on the golf ball is, at all instants, equal and opposite to the force which the golf ball exerts on the club. The club therefore has its momentum changed by exactly the same amount but in the opposite direction. In other words, the club transfers some of its momentum to the golf ball. Any momentum it loses, the golf ball gains. The total momentum of the two objects remains constant. This is an example of a basic physical principle called the conservation of momentum. The **conservation of momentum** states:

 the total momentum of a system remains constant provided no external forces act on the system

The conservation of momentum is a deduction from Newton's laws.

The example with the golf ball and club can be extended to cover any objects. If anything loses momentum then something else gains it. The total momentum of the universe remains constant. Under all circumstances no one has ever found any way of destroying momentum. Einstein's theory of relativity uses this fundamental law. It applies to sub-atomic particles and to vast galaxies.

Example 2

A lorry of mass 10 000 kg collides with the back of a car of mass 1500 kg, see Fig 7.5. Immediately before the collision the lorry was travelling at 30 m s^{-1}, and the car at 12 m s^{-1}. Find the speed of the vehicles immediately after the collision if they remain jammed together.

All momenta are considered in the direction of travel.

Momentum of car before
collision   = 1500 kg × 12 m s^{-1} = 18 000 N s

Momentum of lorry before
collision   = 10 000 kg × 30 m s^{-1} = 300 000 N s

Therefore total momentum before collision

= total momentum after collision   = 318 000 N s

Therefore   (10 000 kg + 1500 kg) × v = 318 000 N s

v = 318 000 / 11 500

v = 27.7 m s^{-1}

= 28 m s^{-1} to 2 sig. figs

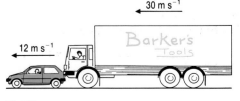

30 m s^{-1}

12 m s^{-1}

Barker's Tools

Fig 7.5

MOMENTUM, WORK, ENERGY AND POWER

7.6 A 2000 kg van and a 1500 kg car both travelling at 40 m s^{-1} in opposite directions collide head on and lock together. What are their speed and direction immediately after the collision?

7.7 A man of mass 60 kg, standing on a friction-free surface, throws a ball of mass 1.5 kg with a horizontal velocity of 25 m s^{-1}. What will be the recoil speed of the man?

7.8 An aeroplane of total mass 50 000 kg is travelling at a speed of 200 m s^{-1}. If a passenger of mass 100 kg then walks towards the front of the aeroplane at a speed of 2 m s^{-1}, what change in the speed of the aeroplane does this cause?

7.2 WORK

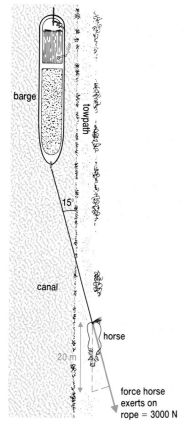

barge

towpath

15°

canal

horse

20 m

force horse
exerts on
rope = 3000 N

Fig 7.6 The work done by the horse is *not* 3000 N × 20 m because the 20 metre movement is not in the direction of the force. The work done is 3000 N × 20 cos 15° m = 5800 J.

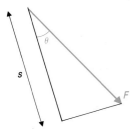

θ

s

F

Fig 7.7 The work done by the force is *Fs* cos θ.

Work is a word with a large number of meanings. It can be a noun or a verb. There is often a problem when a common English word is used as part of a strict scientific vocabulary. The difficulty arises because some of the common meanings conflict with the scientific meaning. Work in the scientific sense is precisely defined. The definition of **work** is given by the equation:

work = force × distance moved in the direction of the force

Distance moved in a stated direction is called the displacement, so the equation can be written:

work = force × the component of the displacement in the direction of the force.

Work is a scalar quantity, although obtained by multiplying two vectors. The unit of work is 1 newton × 1 metre. This unit is named the joule (J). 1 **joule** is the work done when a force of 1 newton moves its point of application through a distance of 1 metre.

Example 3
Find the work done by a crane when it exerts a force of 3000 N on a load and lifts it 20 m.

Work done = force × distance moved in the direction of the force.
Work done by crane = upward force crane exerts on load × distance load moved up.
$$= 3000 \text{ N} \times 20 \text{ m} = 60\ 000 \text{ J}$$

This is an extremely straightforward problem. Note how a difference occurs in the next example.

Example 4
Find the work done by the horse in Fig 7.6 if it moves a distance of 20 m and the tension in the rope is 3000 N.

Here the force is the same as in Example 3, but although the distance is still 20 m, it is not now in the direction of the force.

The work done by the horse = 3000 × 20 cos 15° = 58 000 J.
The distance the horse moves in the direction of the force is 20 cos 15°.

Example 4 shows that when the force F, is acting at an angle θ to the direction of the displacement s, (see Fig 7.7), the work done W is given by the equation:

$$W = Fs \cos \theta$$

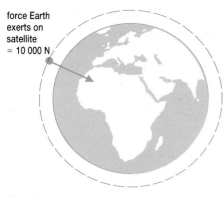

force Earth exerts on satellite = 10 000 N

Fig 7.8

It should also be noted that work done can be negative. In the example given above of the horse doing work by pulling on the rope, the wording could be arranged differently to obtain a negative answer. For instance, it would be correct to state that 'the work which the rope does on the horse is –58 000 J'. This is because the force the rope exerts on the horse is in nearly the opposite direction to the displacement. Making certain that the correct sign is given to work done needs care. It is important to state clearly what is doing work on what, and not merely to say 'work done ='.

Example 5

Find the work done by the Earth on a satellite going round it in a circular orbit, Fig 7.8.

Work done on satellite	=	force Earth exerts on satellite	×	Distance satellite moves in direction of force
	=	10 000 N	×	0
	=	0		

No work is done because there is no movement in the direction of the force.

QUESTIONS

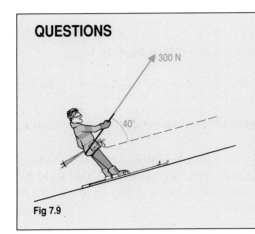

Fig 7.9

7.9 Find the work done by gravity when a diver of mass 60 kg falls through a height of 11 m.

7.10 How much work is done by the winding engine in raising 1000 kg (1 tonne) of coal out of a coal mine which is 300 m deep? Why does the winding engine not have to do any work to raise the lift itself?

7.11 A skier is dragged up a slope by a ski tow which exerts a force on the skier of 300 N at an angle of 40° to the ground, Fig 7.9. What work is done by the tow when the skier moves 200 m?

7.12 What work is done on an electron in a cathode ray tube if the force acting on it is 3.2×10^{-14} N and the force causes the electron to move 5 cm in the direction of the force?

Work done by a varying force

In all of the previous examples and questions the forces being considered have had a constant value. In many practical situations the force being exerted on an object will vary. Work in the scientific sense is done so often in everyday life that normally it is not thought about, let alone calculated. All of the following actions involve doing work:

> Eating
> Breathing
> Walking
> Shutting a door
> Lifting a baby out of a pram
> Turning the page of a book
> Kicking a football
> Stretching a spring

The individual forces involved in these actions are many and varied but all of them involve muscle action in some way. As an example of a varying force consider a muscle in your arm. The maximum tension in a muscle depends on its cross-sectional area and also on its length. The maximum tension is usually about 30 newtons per square centimetre and occurs when the muscle's length is only slightly changed from its resting length. A graph showing the way in which the force varies with the distance the muscle contracts is shown in Fig 7.10.

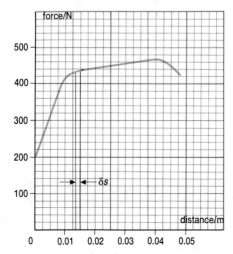

Fig 7.10 Graph showing how the force exerted by a muscle varies with the distance the muscle contracts.

MOMENTUM, WORK, ENERGY AND POWER

To calculate the work done by this muscle in contracting by 0.040 m, it is necessary to find the sum of the force × the distance moved for all the small distances δs. The problem is therefore one of finding the area beneath the graph. Using calculus notation the work done is $\int F ds$. The area beneath the graph in this example, found by counting squares, represents 20.8 J. An average force of 416 N moving the same distance of 0.050 m would give the same work done. Average forces are often quoted when the actual force varies considerably, in order to simplify problems.

The work done to stretch a spring

The extension caused to a spring in use is proportional to the force causing the extension. The force per unit extension is called the **spring constant**, k. A graph showing force plotted against extension is shown in Fig. 7.11.

The shaded area gives the work done W in stretching the spring. The work done is given by

$$W = \tfrac{1}{2} Fe$$

Since $k = F/e$, the work done can also be written as $\tfrac{1}{2} ke^2$.

The work done to compress a gas

In engines gases expand and contract at different parts of the cycle of movement. When a gas expands it does work on its surroundings; when it is compressed work is done on it by its surroundings. Inside an engine it will normally be a piston which is used to compress a gas. The piston is part of the surroundings of the gas. As a gas is compressed its pressure normally rises, and so an increasing force has to be used by the piston to reduce the volume of the gas further.

Fig 7.12 shows a piston, of area of cross-section A, compressing a gas contained in a cylinder. Consider the piston move a very small distance δx at a constant speed.

If the pressure of the gas is p then

the force exerted by the piston on the gas = pA
the work done by the piston on the gas = $pA\,\delta x$

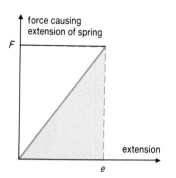

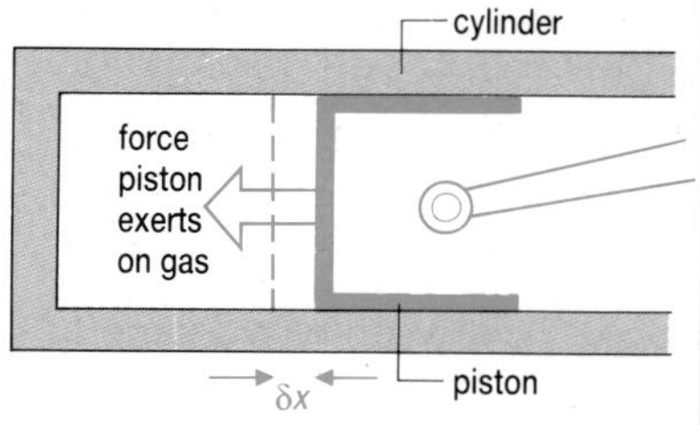

Fig 7.11 Graph of the force against extension for a spring not stretched beyond its elastic limit.

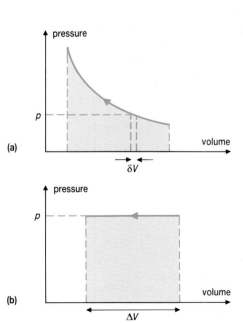

(a)

(b)

Fig 7.14

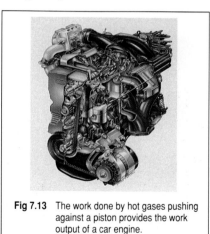

Fig 7.12 The gas in the cylinder is being compressed by the piston moving to the left.

Fig 7.13 The work done by hot gases pushing against a piston provides the work output of a car engine.

$A\,\delta x$ however is the small change in volume δV of the gas in the cylinder, so the small amount of work done by the piston on the gas δW is given by

$$\delta W = p\delta V$$

If this is put in graphical terms, the total work done is the area beneath the graph of p plotted against V (Fig 7.14(a)). In calculus notation this is

$$W = \int p\,\mathrm{d}V$$

It is possible to compress a gas at constant pressure. In this case the area required for the work done is rectangular and is the pressure × the change in the volume, Fig 7.14(b).

QUESTIONS

7.13 Estimate how much work is done by the muscle referred to in Fig 7.10, if it contracts by a distance of only 0.02 m?

7.14 How much work is done to stretch a spring, of force constant 190 N m^{-1}, a distance of 1.7 cm?

7.15 The suspension springs on a car require a force of 100 N to depress them 1.0 cm. How much work needs to be done to compress them by 5.0 cm?

7.16 How much work is done by a gas if it expands from a volume of 1.0×10^{-3} m^3 to 3.7×10^{-4} m^3 at a constant pressure of 8.2×10^5 Pa?

7.17 A gas expands in such a way that pV^2 remains constant. If it starts with a volume of 6.0×10^{-3} m^3 and pressure of 3.0×10^4 Pa and ends with a volume of 13.0×10^{-4} m^3, find the work done by the gas. Either plot a graph or integrate.

7.3 POWER

 Power is defined as the rate at which work is done

Average power is therefore given by the equation:

$$\text{average power} = \frac{\text{work done}}{\text{time interval taken}}$$

The unit of power is the joule per second. This is given the name **watt**. Note that although watts are used frequently in electrical work, the unit itself is not defined from electrical terms. Any power is measured in watts, and the definition of the watt as the joule per second is a purely mechanical unit.

Example 6
Estimate the power of a suitable motor for lifting yachts of up to 2000 kg out of the water.

In answering questions which require estimates it is often necessary to make some assumptions. In the answer these assumptions must be clearly stated. Answers should generally be given to only one significant figure.

Assume that the distance the yacht has to be lifted is 5 metres.

Assume that the time allowed should not be greater than 5 minutes.

The largest force which the hoist will need to be able to exert will occur when the yacht is out of the water. This will be the weight of the yacht, a force of 2000 kg × 9.8 N.

Total work done = Force × distance = 2000 kg × 9.8 N kg^{-1} × 5 m = 98 000 J
Time taken = 300 s

$$\text{power required} = \frac{\text{work done}}{\text{time taken}} = \frac{98\,000 \text{ J}}{300 \text{ s}} = 330 \text{ W}$$

The motor will have to be rated higher than this because it will not be 100% efficient. A suitable motor would be one rated at 500 W.

MOMENTUM, WORK, ENERGY AND POWER

7.18 A car of mass 1200 kg has an output power of 58 kW when travelling at a speed of 30 m s^{-1} along a flat road. What power output is required if it is to travel at the same speed up a hill with a gradient of 10%? (This is a gradient of 1 in 10. The angle of the slope is tan^{-1} 0.10 = 5.7°)

7.19 A motor used to pull on a cable is drawing in the cable at a speed of 4.8 m s^{-1}. If the force being exerted on the cable is 8500 N, what is the power output of the motor? What is the general equation linking power P, force F and velocity v?

7.4 ENERGY

When work is done on an object, different things may happen to it. Work cannot be done without movement, so the object *must* move. The movement may result in there being an acceleration, or a deceleration, or no change in velocity. It may result in the object rising, or in it having its shape altered. It may result in the object's temperature rising or a chemical reaction or even the production of an electrical potential difference across it. In stretching a spring, work is done on the spring. The spring can then itself do work when it returns to its original shape. In other words, when stretched it stores the ability to do work.

 The stored ability to do work is called **energy**.

Energy is measured using the same unit as work, namely the joule, and can be calculated using the same equation as that used to define work, namely: force × distance moved in the direction of the force. Work is done when energy is transferred; the two terms are closely connected. The energy stored in a compressed spring may be used to do work at any time. The electrical energy stored in a capacitor, an electrical component designed to store electrical energy, may be released in a fraction of a microsecond. The chemical energy in a lump of coal may have been stored for many millions of years before it is released.

There is much political discussion nowadays about how quickly we are, or should be, using energy. Much of the energy which is being used at a rapid rate in the world now is not renewable. One of the reasons why we are so profligate in our use of energy is that, at present, it is *very* cheap. 1 joule of electrical energy is supplied to you at home for about 0.000 002p, i.e. you can buy 500 000 J for only 1p. 500 000 J is enough energy to lift you 1000 metres, and that will put you on the top of Snowdon, all for 1p! It is not really surprising therefore, at this cost, that we are willing to waste a lot of electrical energy. There will be much greater economic incentive to find forms of renewable energy when the cost of energy is higher. In discussions about energy use it is important to use quantitative information to get some idea of the problems involved. Table 7.1 gives some possible energy values.

7.5 KINETIC ENERGY

Kinetic energy is referred to in Table 7.1. This term, which you are probably familiar with, is always associated with movement. Energy is defined in Section 7.4, as the stored ability to do work. The **kinetic energy** of an object is its stored ability to do work as a result of its motion. It can have **translational kinetic energy** as a result of its linear movement and it may also have **rotational kinetic energy** as a result of rotation.

Fig 7.15 Concorde just after take-off. Maximum power is required on the runway to accelerate the aeroplane when it has maximum mass, and is sustained to enable the aeroplane to gain potential energy and further kinetic energy.

Table 7.1 Energy values of certain systems

Description	Order of magnitude of energy stored/J
The Big Bang at the formation of the universe	$\sim 10^{70}$
Rotational energy of the galaxy	$\sim 10^{50}$
Energy released by the Sun up to the present	10^{45}
Binding energy of the Earth – Sun system	10^{33}
Kinetic energy of the Moon as it orbits the Earth	10^{28}
Energy radiated by the Sun in one second	10^{27}
Energy received at the Earth from the Sun in a year	10^{25}
Annual wind energy	10^{22}
Annual energy use by people	10^{21}
Annual energy dissipation by tides	10^{20}
Energy released by the largest nuclear bomb ever made	10^{18}
Annual electrical energy supplied by a large power station	10^{16}
Energy released in burning a tonne of coal	10^{11}
Kinetic energy of Concorde at cruising speed	10^{10}
Output energy of a car in using 1 litre of petrol	10^{7}
Daily intake of food energy for an adult	10^{7}
Energy supplied by a loaf of bread	10^{6}
Maximum kinetic energy of a top 100 m athlete	10^{4}
Maximum kinetic energy of a tennis ball in play	10^{2}
2 beats of a human heart	1
Kinetic energy of a fly flying	10^{-3}
Electrical energy stored in a 0.02 µF capacitor at 10 V	10^{-6}
Energy released by a radioactive atom decaying	10^{-13}
Kinetic energy of a single electron passing through 1 V	10^{-19}
Kinetic energy of vibration of a single atom	10^{-21}

Example 7

Find the kinetic energy of a train, of mass 300 000 kg, when travelling at 50 m s^{-1}.

The kinetic energy must be found from the amount of work the train can do against a force which stops it. We will assume, at first, that the force has a constant value of 120 000 N. With this force acting on the train in the opposite direction to its velocity the acceleration of the train is given by:

$$\text{acceleration} = \frac{\text{force}}{\text{mass}} = \frac{-120\,000\ \text{N}}{300\,000\ \text{kg}}$$

$$= -0.40\ \text{m s}^{-2}$$

Using the equation $v^2 = u^2 + 2as$ (Section 4.4) we get:

$$0 = 50^2 + (2 \times -0.40 \times s) = 2500 - 0.80\,s$$

$$s = \frac{2500}{0.80} = 3125\ \text{m}$$

where s is the distance travelled while stopping.

The work done by the train against the stopping force is

$$\text{force} \times \text{distance} = 120\,000\ \text{N} \times 3125\ \text{m} = 375\,000\,000\ \text{J}$$

This work is done by using its kinetic energy. The kinetic energy of the train at the start, that is its stored ability to do work as a result of its motion, is 375 000 000 J.

MOMENTUM, WORK, ENERGY AND POWER

7.20 If the stopping force in the above example had been 60 000 N instead of 120 000 N, find:
   (a) the distance the train takes to stop;
   (b) the work done by the train against the stopping force in coming to rest;
   (c) the kinetic energy of the train.

If you have followed through Question 7.20 correctly then you will have found that with this different force to slow down the train, the work that it is capable of doing as a result of its motion is still 375 000 000 J. If the problem is repeated with any size of force, or a force of varying size, the work that can be done by the object is always 375 MJ. That is, the kinetic energy of the train is not dependent on the way its stored ability to do work is used, but only on the mass and speed of the object. This is important in the design of cars or other transport systems to reduce the injury caused in accidents.

Fig 7.16 The work done by muscles as a sprinter starts a is race partly changed into the kinetic energy of the sprinter.

Accident analysis

Consider the kinetic energy of a person travelling in a car on a motorway to be 36 000 J. Note that this is about four times greater than the maximum possible kinetic energy of a sprinter. Table 7.2 gives values of the constant force necessary to stop the person in different distances.

Table 7.2   Variation of force with stopping distance

Force/N	Stopping distance/m
10	3600
100	360
1000	36
10 000	3.6
100 000	0.36

In all cases the work done by the person against the stopping force is 36 000 J. In the first line of Table 7.2, the example is given of a very small force being exerted over a very long distance. The person pushes forwards against the seat and the floor of the car with a total force of 10 N. Using Newton's third law, the seat and floor exert a backwards force of 10 N on the person. In the second line, the force of 100 N applied would correspond to a normal braking force. The third line would correspond to a fairly dramatic emergency stop, while stopping in 3.6 m, the fourth line, would be a serious accident. A person experiencing a force of 10 000 N applied would probably survive it if the force were applied by a seat belt, but not if applied to the head by a steering wheel or windscreen. A force of 100 000 N would stop the person in a distance of 36 cm and would kill anybody. If therefore the distance for stopping can be increased then the force necessary to stop is reduced. Most manufacturers of cars now build into their cars collapsible sections which crumple and hence increase the distance a person can travel if involved in an accident, see Figs 7.17 and 7.18. The forces involved in a collision are far from being constant. Fig 7.19 shows how in a collision in which the average stopping force is 6000 N, the maximum stopping force is over 15 000 N. It is the maximum stopping force and how that force is applied which determine how serious an accident is, as it is this force which determines whether bones are broken or not.

Fig 7.17 The box-like structure of the passenger compartment is made stronger than the engine compartment. In an accident this gives the passengers extra distance in which they can stop so reducing the size of the force on them.

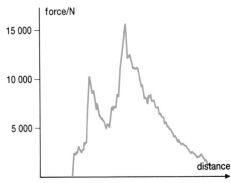

Fig 7.18 The effect of the structure described in Fig. 7.17 is seen here. The passenger compartment is undamaged even after considerable damage to the engine compartment.

Fig 7.19 Graph to show how the force exerted on a person, when an accident occurs, can have a low average value but still cause injury as a result of there being large forces applied at certain places.

Equation for kinetic energy (E_k)

So far in this section, problems have been worked out from basic principles. It is often convenient to be able to find the kinetic energy of a body from its mass m, and velocity v. If a constant force F is used to slow it down from an initial velocity v to a stop, then its acceleration a is given by

$$a = -\frac{F}{m}$$

Using $\qquad$ (final velocity)^2 = (initial velocity)^2 + $2as$

therefore gives $\quad 0 = v^2 - 2\,\dfrac{F}{m}\,s$

or $\qquad\qquad v^2 = \dfrac{2\,Fs}{m}$

where s is the distance taken to stop.

That is, $Fs = \frac{1}{2}mv^2$

Since Fs is the work done by the object against the stopping force as a result of its motion, it follows that its kinetic energy is given by:

 $\qquad$ Kinetic energy of a body, $E_k = \frac{1}{2}mv^2$

Using calculus it is possible to show that this equation is valid whether or not the force is constant. In this case the work done is the change in the kinetic energy, so

$$Fs = \tfrac{1}{2}mv^2 - \tfrac{1}{2}mu^2$$

QUESTIONS

7.21 A lorry of mass 10 000 kg has a kinetic energy of 1.13×10^6 J. What is the speed of the lorry and what speed must a car of mass 1200 kg have in order for it to have the same kinetic energy as the lorry?

7.22 During take off an aeroplane's speed changes from zero to 40 m s^{-1} while travelling 1000 m along a runway. If its mass is 20 000 kg find the driving force which needs to be applied to it.

7.23 Fishing line is usually sold by stating the breaking force of the line. What breaking force line is required if a 2.0 kg fish swimming at 1.8 m s^{-1} is to be stopped in 0.40 m?

Potential Energy

Potential energy (E_p) is defined as the stored ability of a body to do work as a result of its position or shape. When work is done on an object and the object accelerates, then the object stores an increasing amount of kinetic energy. An object does not necessarily accelerate however when work is done on it. When a lift, of total weight 8000 N, is being pulled up at a steady velocity of 2 m s^{-1} there is no increase in the kinetic energy of the lift while the constant velocity is maintained. Work is being done on the lift and the lift in its raised position can do work on something else when it moves down, so the lift gains potential energy. The potential energy of the lift at a height of 12 m above the ground is the work the lift can do when it moves down. In this case the work done is 8000 N × 12 m = 96 000 J; so the potential energy of the lift 12 m above the ground is 96 kJ. It is important when considering potential energy to state the zero reference point of the energy. In this case, the 96 kJ is the increase in the potential energy of the lift as it goes upwards 12 m from the ground. We are in effect taking the potential energy on the ground to be zero. This is an arbitrary decision however, and has the effect that if in this case the lift could descend into a basement then its potential energy there would have a negative value.

In general, if a body of mass m is near the Earth where the acceleration of free fall is g then the weight of the body is mg. If the body is at a height h above the reference zero of potential energy then the work which the body can do as a result of its position, its potential energy, is given by:

$$\text{potential energy, } E_p = \text{force} \times \text{distance} = mgh$$

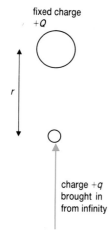

Fig 7.20 Work has to be done to bring a charge +q from a large distance to a point near a fixed charge +Q. The work done per unit charge is called the potential of the point.

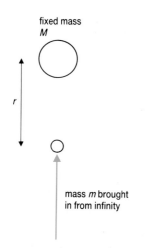

Fig 7.21 The principle illustrated in Fig 7.20 can be applied to gravitational potential but the gravitational potential is negative.

Other forms of potential energy

There are other situations in which a body can have work done on it but not increase its velocity. If a spring is extended at a constant velocity by applying a suitably increasing force to it, then work is being done on the spring and the spring can release this energy at a later date, when it contracts. The energy stored by the spring, if it obeys Hooke's law is given by:

$$E_p = \tfrac{1}{2} k e^2$$

If an electrical charge +q is brought from an infinite distance to a distance r from a fixed charge +Q, then because the two charges repel one another work has to be done, see Fig 7.20. The electrical potential energy is a form of potential energy because the system stores the ability to do work as a result of the position of the charges. See Section 15.2.

If a mass m is brought from an infinite distance to a distance r from a fixed mass M, then again there will be a change in the potential energy of the system. This will not be quite the same as there is attraction between the two masses due to gravitation, see Fig 7.21. The potential energy stored in this system is therefore negative. It is called the gravitational potential energy. See Section 9.3.

Chemical energy is a form of potential energy. When coal burns, carbon atoms combine with oxygen atoms to form carbon dioxide molecules. This reorganisation of the atoms reduces the stored electrical potential energy and energy is emitted mostly in the form of infra-red radiation.

7.24 What gain in potential energy takes place for a high jumper if she has a mass of 48 kg and raises her centre of gravity by 1.28 m? What must be the minimum speed of her centre of gravity at take off?

7.25 A cricket ball is thrown upwards with a kinetic energy of 47 J. If its mass is 0.156 kg, what maximum height can it reach?

7.26 In a pinball machine a force of 24 N is used to compress a spring a distance of 0.080 m. Find the potential energy stored in the spring and hence find the maximum speed which it can produce in a ball of mass 0.15 kg.

7.27 What energy is possibly available from a tidal barrage scheme in one cycle of ebb and flow if the area of water behind the barrage is 30 km², the tidal depth is 10 m and the density of sea water 1030 kg m⁻³? If a cycle of ebb and flow takes 12 hours 40 minutes, what average power does this represent? What practical problems are bound to occur if maximum energy output per cycle is to be achieved?

Potential

Potential is a term applied to a place or position, rather than to a particular object. The **potential of a point** is defined as the potential energy per unit mass of an object at the point. A potential diagram is like an energy contour map. In fact a geographical contour map can be turned into a gravitational potential energy diagram very simply. This is done in Fig 7.22 where the individual contour lines are renumbered to give the potential with reference to a zero of potential at mean sea level. An object of mass 12 kg at a point 70 m above sea level has a potential energy of $12 \times 9.8 \times 70 = 8232$ J (8200 J to 2 significant figures). Unit mass at this point would have $9.8 \times 70 = 686$ J of gravitational potential energy. The potential of the 70 m contour line is therefore 686 J kg⁻¹. The potential near the Earth's surface is gh, if the zero of potential is taken at the Earth's surface.

7.7 CONSERVATION OF ENERGY

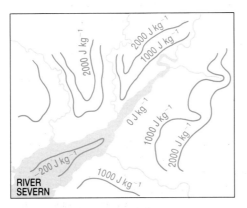

Fig 7.22 A potential map. Such a map is effectively a map showing contour lines. The potential is determined by the height above the reference zero.

During the nineteenth century Joule, among others, carried out a series of extremely careful experiments to measure many of the effects of doing work on a system. He took about forty years altogether and during that time he increased the accuracy of his results as well as increasing the variety of experiments performed. Many of his experiments necessitated measuring small temperature rises and he designed his own thermometers to measure rises in temperature to a hundredth of a degree. The type of experiment he did was to exert a known force for a known distance on a paddle wheel used to stir water. He measured the rise in the temperature of the water. He knew the work done to stir the water and he measured the heat produced in the water. He found that the same amount of work always produced the same heating effect no matter how the work on the water was done.

In Section 7.5 it was calculated how much work a train could do against the force which is being used to stop it. When this work is done the effect is the heating of the train's braking system and the rails. The kinetic energy of the train is being reduced to zero while it stops. Joule's experiments lead us to accept that all of this loss of kinetic energy by the train is transferred into increased kinetic energy of molecules in the train and its surroundings. That is, the loss of kinetic energy of the train exactly equals the gain in thermal energy of other objects. The total energy thus remains constant. This is an example of the principle of conservation of energy, one of the most important principles in physics. The principle of **conservation of energy** states that:

 Energy may be transformed from one form to another but it cannot be created or destroyed, i.e. the total energy of an isolated system is constant.

In this statement the system referred to is said to be isolated. This means that if an energy calculation is being carried out then *all* the energy within the system must be taken into account. No energy must be allowed to escape. Another statement of the principle is 'there is no change in the total energy of the universe'. The principle is dealt with in greater detail in Chapter 25.

7.8 COLLISIONS

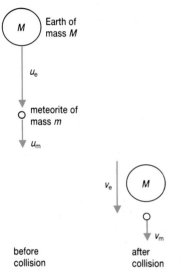

Fig 7.23 The collision of the Earth with a meteorite.

A collision, when spoken about in everyday speech, usually implies a car hitting something unintentionally. This is a very restricted meaning of the word and needs to be expanded here. On a microscopic scale molecules are colliding with each other very frequently. On a cosmic scale exploding stars can be considered as a collision. In between these two extremes there are many occasions when one object exerts a force on another object and affects the kinetic energy of both.

When a collision occurs, the principles of conservation of momentum and conservation of energy can be applied. Consider the collision shown in Fig 7.23 where the Earth, of mass M, travelling with velocity u_e in the direction shown collides with a small meteorite, of mass m travelling with velocity u_m in the same direction. u_e needs to be larger than u_m for the collision to occur at all. Assume that after the collision the Earth moves with velocity v_e and the meteorite moves with velocity v_m, both still in the same direction.

This problem has been simplified by having the objects always moving in the same direction, because when the principle of conservation of momentum is applied, it must be remembered that momentum is a vector.

The two principles give the following equations for the collision:

1. Total momentum before collision = total momentum after collision
   $$Mu_e + mu_m = Mv_e + mv_m$$

2. Total energy before collision = total energy after collision
   Kinetic energy (+ other energies) = Kinetic energy (+ other before collision energies) after collision

   or $\frac{1}{2}Mu_e^2 + \frac{1}{2}mu_m^2 = \frac{1}{2}Mv_e^2 + \frac{1}{2}mv_m^2 + \Delta E$

   where ΔE is the change in the forms of energy other than kinetic energy.

It should be clear from this that whereas it is straightforward to apply the principle of conservation of momentum to a collision, provided that you remember to consider the directions of the momenta, it is less easy to apply the principle of conservation of energy. This is because the law of conservation of energy does not apply just to the kinetic energy of the objects but to all forms of energy. In this example kinetic energy must be lost. If you see a shooting star you are seeing a collision between the Earth's atmosphere and a meteorite. You see the light emitted from the hot meteorite. Electromagnetic radiation energy is being produced by the meteorite and so some of the original kinetic energy is being used to produce this radiation.

Fig 7.24 The camera taking this photograph was pointed at the sky at night and the shutter opened. As the Earth rotated each star is seen as a short line on the photograph. Whilst the shutter was open a meteor has collided with the Earth's atmosphere. This provides the long track. Note the different colours of stars as a result of differing surface temperatures.

Elastic collisions

Some collisions take place in which the total kinetic energy of the objects being considered remain constant. These collisions are called **elastic collisions**. To be doubly sure they are sometimes called perfectly elastic collisions, or totally elastic collisions. The fact that all the kinetic energy remains as kinetic energy implies that no other energy is produced in the collision. They must be perfectly silent collisions, for example, as no sound could be produced. Elastic collisions are a theoretical perfection, although on an atomic or nuclear scale the approximation of a collision to an elastic collision is a very good one. In the laboratory a nearly perfect elastic collision may be obtained by attaching magnets to the riders on a linear air track, so that as the riders approach one another the two magnets repel one another, see Fig 7.25. An entire collision can then take place between the

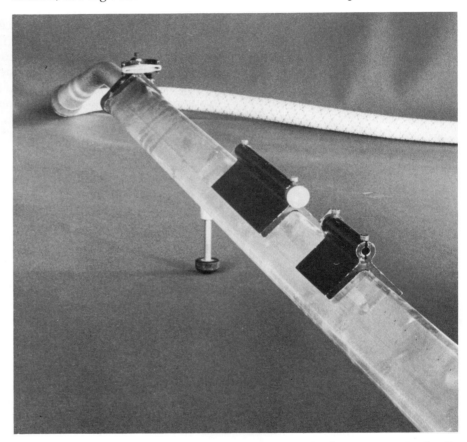

Fig 7.25 The trolleys on a linear air track glide with very little friction. By placing disc magnets on each trolley with like poles facing almost perfect collisions can be achieved. The trolleys bounce, totally silently, on the magnetic fields, without any direct contact.

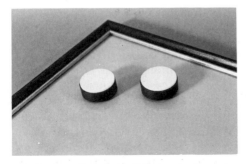

Fig 7.26 In two dimensions collisions can be observed by sliding magetic pucks towards one another across a friction free table.

two riders as each bounces on the magnetic field of the other and do not make physical contact at all. Fig 7.26 shows an elastic collision taking place in two dimensions. In this case a magnetic puck slides across a friction-free table and collides with the magnetic field of another puck.

Example 8

In a nuclear reactor, high-speed neutrons are slowed down by letting them hit the carbon atoms of the moderator. The moderator is so called because it moderates the speed of the neutrons. In atomic mass units the mass of a neutron is 1 and the mass of a carbon atom is 12. What fraction of its kinetic energy does it retain if a neutron collides elastically and head on with a stationary carbon atom? How many such collisions does the carbon atom need to undergo to reduce its energy to a millionth of its initial value?

MOMENTUM, WORK, ENERGY AND POWER

before
collision u $U = 0$

after collision

Fig 7.27

This expression shows that in the particular case of head on elastic collision, the velocity with which the bodies approach one another equals the velocity of their separation. It is worthwhile remembering that this applies to an elastic collision. Bouncing a ball on the ground illustrates this. If the ball approaches the ground at a particular speed and then it separates from the ground at the same speed, there is a perfect collision. The ground itself barely moves!

Let m be the mass of the neutron and M be the mass of the carbon atom. Using the velocity symbols given on Fig 7.27 the conservation principles give:

1. Total momentum before collision = total momentum after collision
   $(m \times u) + 0$ $= (m \times v) + (M \times V)$
2. Total k.e. before collision   = total k.e. after collision
   $\frac{1}{2} mu^2 + 0$ $= \frac{1}{2} mv^2 + \frac{1}{2} MV^2$

Solving these equations to eliminate V, a term we do not know and are not asked to find gives some tricky working. The equations are deceptively similar. First cancel the $\frac{1}{2}$ on both sides of the k.e. equation and then collect all the m terms on the left-hand side of the equations and all the M terms on the right-hand side:

$$m (u^2 - v^2) = MV^2$$

$$m (u - v) = MV$$

Divide one equation by the other and both m and M cancel to give:

$$\frac{(u^2 - v^2)}{(u - v)} = V$$

Since $(u^2 - v^2) = (u - v)(u + v)$

$$\frac{(u - v)(u + v)}{(u - v)} = V$$

$$u + v = V$$

$$u = V - v$$

Now eliminate V from the equations:

$$m (u - v) = MV \text{ and } u = V - v$$

$$m (u - v) = M (u + v) \text{ and reorganising to find } v$$

$$mu - Mu = mv + Mv$$

$$(m - M) u = (m + M) v$$

$$v = \left(\frac{m - M}{m + M} \right) u$$

So the kinetic energy of the neutron after the collision is

$$\frac{1}{2} m v^2 = \frac{1}{2} m \left(\frac{m - M}{m + M} \right)^2 u^2$$

Any kinetic energy lost by the neutron is gained by the carbon atom in this elastic collision. Substituting in the numerical values of mass gives:

$$\frac{\text{kinetic energy of neutron after collision}}{\text{kinetic energy of neutron before collision}} = \frac{\frac{1}{2} mv^2}{\frac{1}{2} mu^2} = \frac{m \left(\frac{m - M}{m + M} \right)^2 u^2}{mu^2}$$

$$= \left(\frac{m - M}{m + M} \right)^2 = \left(\frac{1 - 12}{1 + 12} \right)^2 = \left(\frac{-11}{+13} \right)^2 = \frac{121}{169} = 0.716$$

After two head on collisions the kinetic energy remaining will be $(0.716)^2 \ldots$ after n collisions the kinetic energy remaining will be $(0.716)^n$

This must be less than 10^{-6}. One way of finding n is just multiplying 0.716 over and over again on your calculator until the value is small enough. You will have to do 42 calculations but, though tedious, it is easy enough with a calculator with a built-in constant multiplier. A better way is to use logarithms. If you take logs of both sides you get:

$$(0.716)^n = 10^{-6} \text{ for the kinetic energy to fall to a millionth}$$
$$n \lg 0.716 = -6$$
$$n \times -0.145 = -6$$

$$n = \frac{6}{0.145} = 41.4$$

and 42 such collisions will be necessary to get the kinetic energy of the neutron down to less than a millionth of its original value. When this is done in practice, the neutrons do not always hit the carbon atoms head on, so many more than 42 collisions will be needed. Much of the output energy of a nuclear reactor is obtained by this mechanism. In time, the temperature of the carbon atoms would rise as a result of the collisions with the neutrons. This is prevented in a Magnox reactor by blowing a stream of carbon dioxide through the moderator. The hot carbon dioxide supplies the thermal energy output of the reactor.

QUESTION

7.28 A neutron of mass 1.67×10^{-27} kg, moving with a velocity of 2.0×10^4 m s^{-1}, makes a head on collision with a boron nucleus of mass 17.0×10^{-27} kg, originally at rest. Find the speed of the boron nucleus if the collision is perfectly elastic.

Inelastic collisions

If a collision takes place which is not elastic then it is called an inelastic collision. An **inelastic collision** is one in which kinetic energy is changed into other forms of energy. The example given above of the Earth colliding with a meteorite is an inelastic collision. So is a ball bouncing on the ground, because in reality balls do not bounce perfectly. Note that in this type of collision, although there is a net loss of kinetic energy there is never any loss of momentum. The law of conservation of momentum is therefore used to solve problems of inelastic collision. The fraction of kinetic energy lost in an inelastic collision may vary considerably. In some nearly elastic collisions only a small fraction of the total kinetic energy is lost, whereas if a brick falls on a muddy building site all its kinetic energy may be lost. Often, after a collision, the two colliding objects remain stuck together. These are sometimes called completely inelastic collisions. This does not imply that they have lost all their kinetic energy but that they both must have the same velocity after the collision.

Example 9

In a motorway accident a car of mass 1200 kg travelling at 40 m s^{-1}, runs into the back of an unloaded lorry of mass 3000 kg travelling at 25 m s^{-1}. How much kinetic energy does the car lose in the crash?

Momentum before collision = Momentum after collision
$$(1200 \times 40) + (3000 \times 25) = (1200 + 3000) \times v$$

where v is the common speed of car and lorry after the crash.

$$48\,000 + 75\,000 = 4200v$$
$$v = \frac{123\,000}{4200}$$
$$= 29.3 \text{ m s}^{-1}$$

Kinetic energy of car before crash $= \frac{1}{2} \times 1200 \text{ kg} \times (40 \text{ m s}^{-1})^2 = 960\,000$ J
Kinetic energy of car after crash $= \frac{1}{2} \times 1200 \text{ kg} \times (29.3 \text{ m s}^{-1})^2 = 520\,000$ J
Therefore the car loses 440 000 J in the crash.

The remaining 520 000 J of kinetic energy of the car will be lost immediately after the crash as the two vehicles skid to a stop. If the lorry had been more heavily loaded, the car would have lost more of its kinetic energy, up to a maximum of 585 000 J. Why cannot the car lose more than 585 000 J in the collision itself, whatever the mass of the lorry? It is interesting to note that if the driver of the car is 1/20 of the mass of the car, he has 1/20, or 22 000 J of the kinetic energy to lose. If the distance he moves during the collision is 2 m then the force the seat-belt exerts on him will have an average value of 11 000 N. As this is about 20 times his weight, this force will certainly cause injury – but he will probably survive!

QUESTIONS

7.29 A car travelling at a velocity v is struck from behind by a car of the same mass travelling with velocity $2v$. If the cars lock together what will be their common velocity?

7.30 Two blocks of mass 0.25 kg and 0.20 kg are moving towards one another on a linear air track with velocities of 0.60 m s^{-1} and 0.40 m s^{-1} respectively. The blocks collide and stick together. Find **(a)** their final velocity and **(b)** the loss of kinetic energy in the

Fig 7.28 This is a false colour photograph of a classic physics experiment. Alpha particles (green) stream upwards through a cloud chamber. One of them (shown yellow) scatters a nucleus (red) in the nitrogen gas. The alpha particle rebounds from the much heavier nitrogen nucleus.

Collisions in two dimensions

The principles applied in the earlier collision questions in this chapter were all in one-dimensional problems. When objects move on a plane then, although the mathematics gets more involved, the same principles can still be applied. Note has to be taken that momentum is a vector and therefore the conservation of momentum usually has to be applied in two directions at right angles. If the collision is elastic then there is conservation of kinetic energy as well. Since kinetic energy is a scalar, arithmetical addition of kinetic energies is possible. The following example illustrates the procedure to be applied. Since many different facts may be supplied for these problems, it is necessary to examine carefully what is given and what is to be found. A diagram stating the given masses, speeds and directions is indispensable.

Example 10

A photograph is taken of a forked track in a cloud chamber and the angles are measured to give Fig 7.29.

It is known that an α particle travelling at 3.7×10^6 m s^{-1} coming from the left-hand side is causing the fork by hitting a hydrogen atom at A. The hydrogen atom is the higher of the two right-hand arms of the fork and the recoil α particle is the lower. Find the speeds of the hydrogen atom and the α particle after the collision and find whether or not the collision is an elastic collision.

The masses of the hydrogen atom and the α particle are 1 and 4 respectively in atomic mass units. These units and the theory of the cloud chamber are dealt with in Chapter 28. A conversion factor to change all the masses into kilograms can be included, but since it will appear in every term of all the equations used it can immediately be cancelled out.

Apply the law of conservation of momentum in the initial direction of travel of the α particle.

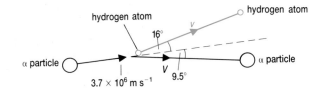

Fig 7.29 A diagram showing an α particle colliding with a hydrogen atom.

Momentum to right before collision   = momentum to right after collision:
$(4 \times 3.7 \times 10^6 \text{ m s}^{-1})$ $= (1 \times v \cos 16°)$ $+ (4 \times V \cos 9.5°)$

Momentum upwards before collision = momentum upwards after collision:

$$0 = (1 \times v \sin 16°) \quad - (4 \times V \sin 9.5°)$$
$$14.8 \times 10^6 = 0.961 \, v \qquad\qquad + 3.95 \, V$$
$$0 = 0.276 \, v \qquad\qquad - 0.660 \, V$$

This last equation gives

$$V = \frac{0.276}{0.660} \, v \qquad = 0.418 \, v$$

Substituting back into the first equation gives
$$14.8 \times 10^6 = 0.961 \, v + 3.95 \times 0.418 \, v = 2.61 \, v$$

$$v = \frac{14.8 \times 10^6}{2.61} \qquad = 5.67 \times 10^6 \text{ m s}^{-1}$$

and $V = 0.418 \, v \qquad = 2.37 \times 10^6 \text{ m s}^{-1}$

The kinetic energy before the collision, in arbitrary units because the mass is not in kilograms, is

$$\tfrac{1}{2} \times 4 \times (3.7 \times 10^6)^2 = 2.74 \times 10^{13} \text{ arbitrary units}$$

In the same units the kinetic energy of the recoil α particle is

$$\tfrac{1}{2} \times 4 \times (2.37 \times 10^6)^2 = 1.12 \times 10^{13}$$

and of the hydrogen atom is

$$\tfrac{1}{2} \times 1 \times (5.67 \times 10^6)^2 = 1.6 \times 10^{13}$$

The total kinetic energy after the collision is the same, to the degree of uncertainty we can expect, and so there has been an elastic collision.

QUESTION

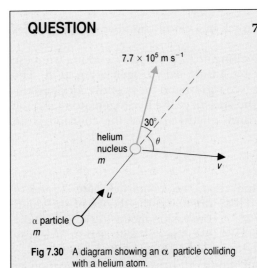

Fig 7.30 A diagram showing an α particle colliding with a helium atom.

7.31 An α particle, travelling with speed u, collides with a stationary helium nucleus and the nucleus moves with velocity $7.5 \times 10^5 \text{ m s}^{-1}$ at an angle of 30° to the original direction (Fig 7.30). Write down the equations for conservation of momentum in the original direction and at right angles to this. Write down the equation for conservation of kinetic energy assuming a perfectly elastic collision. You can use this equation to find θ. Solve the equations to find u, and v, the speed of the α particle after the collision.

Explosions

An explosion can be considered in the same way as a collision but with an increase in the kinetic energy of the system. When a rifle fires a bullet there is a recoil in the rifle. The law of conservation of momentum can be applied to this situation. The momentum of the gun backwards equals the momentum of the bullet forwards. This is a good example of Newton's third law in action. The rifle exerts a force on the bullet and the bullet exerts an equal and opposite force backwards on the rifle. Since these two forces are always equal and opposite and must be exerted for the same length of time, it follows that the momenta of the two parts are also equal and opposite. The same argument cannot however be applied to the energies of the rifle and the bullet. In this case, although the forces are equal and opposite the distance the bullet travels while the force is applied is much greater than the distance the rifle travels, and so the kinetic energy of the bullet is much greater than the kinetic energy of the rifle. They have the same momenta, but not the same kinetic energy. It is a pity it works out this way! There would be a good incentive not to use any guns if both the energies and the momenta were the same because then the person firing the gun would get just as much kinetic energy as his target. The following questions illustrate this on both a large scale and on an atomic scale.

QUESTIONS

7. 32 A cannon of mass 3000 kg fires a cannonball of mass 50 kg with a horizontal velocity of 85 m s⁻¹. Find the recoil velocity of the cannon and the ratio

$$\frac{\text{energy of cannonball}}{\text{energy of cannon}}$$

7.33 An α particle of mass 4 u is emitted from a nucleus of mass 226 u. If energy E is released as kinetic energy, find the kinetic energy of each particle.

INVESTIGATION

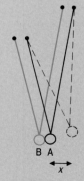

Fig 7.31 Pendulum bob A falls to collide with pendulum bob B.

In this investigation you are asked to find how the velocity of an object A changes when it makes an elastic collision with another object B and how the change in velocity depends on the mass of B.

If a linear air track is available then it should be used with A of fixed mass but with B having its mass changed by the addition of Plasticene. The sliders on the track should have magnets fitted with opposing poles so that they repel one another. The velocity of A when it approaches B should always be the same, and the timing device for the track should be used to find the velocity of A after the collision.

A graph of velocity of A after the collision should be plotted against the mass of B.

If a linear air track is not available then the experiment can be done by using a pair of pendulums as shown in Fig 7.31. It is then more difficult to obtain an elastic collision but steel ball bearings can be used for A and B, and the velocity can be found from the fact that for small swings the distance x is proportional to the velocity.

Energy and momentum of balls

Table 7.3 Data showing typical ball speeds and striker speeds for different sports

Ball	Ball mass /kg	Ball velocity / m s⁻¹ before	Ball velocity / m s⁻¹ after	Striker velocity /m s⁻¹ before	Striker velocity /m s⁻¹ after	Impact time /ms
cricket ball (hit from rest)	0.16	0	39	31	27	1.4
football (free kick)	0.42	0	28	18	12	8.0
golf ball (drive)	0.046	0	69	45	32	1.3
hand ball (serve)	0.061	0	23	19	14	1.4
squash ball (serve)	0.032	0	49	44	34	3.0
tennis ball (serve)	0.058	0	51	38	33	4.0

Using data from Table 7.3 answer the following:
(a) Find the kinetic energies of each ball just after it is struck.
(b) Find the momentum of each ball just after it is struck.
(c) Find the loss of momentum of each striker as it strikes the ball.
(d) Find the impulse which each striker gives to each ball.
(e) Find the mass of (i) the cricket bat, (ii) the golf driver and (iii) the tennis racquet.
(f) Find the loss in kinetic energy of the golf driver as it hits the golf ball.
(g) Why are the masses found in question 5 not the true masses?
(h) Which of the collisions is most nearly elastic?
(i) What average force is exerted by each striker while it is in contact with the ball?

SUMMARY

- Work is the product of a force and the distance moved in the direction of the force.
  Momentum is the product of a body's mass and velocity.
  Energy is the stored ability to do work.
  Power is the rate of doing work.

	Shown on a graph by	Can cause a change in the	Unit
Impulse	Area under force-time graph	momentum $mv - mu$	N s
Work	Area under force-distance graph	kinetic energy $\frac{1}{2}mv^2 - \frac{1}{2}mu^2$	J

- In all collisions momentum is conserved.
- In elastic collisions kinetic energy is conserved; in inelastic collisions energy is not conserved.
- The potential energy of a body of mass m raised a height h near the Earth's surface is mgh.

EXAMINATION QUESTIONS

7.34 State Newton's three laws of motion. (3)
Explain, in terms of each of these laws, how a helicopter is able to hover in flight. (4)
Most helicopters have a second (tail) rotor. Why is this? (2)
Assume that the helicopter's main rotor blades give a vertical velocity v to a cylinder of air of cross-sectional area equal to that swept out by the blades. Show that the power required by the helicopter to hover is $\pi l^2 \rho v^3$ where ρ is the density of the air and l is the length of the rotor blades. (5)
Check that the power formula is correct dimensionally. (3)
By what factor must the power increase for the helicopter to hold up a load equal to its own weight? (3)   (O & C 1985)

7.35 **(a)** Explain how a space vehicle can change its speed or its direction of travel by ejecting high speed gas particles.
(b) An advertisement hoarding has an area of 10 m², and wind blows directly at it with a velocity of 5 m s⁻¹. Given that the density of air is 1.3 kg m⁻³, estimate the force experienced by the hoarding stating any assumption you make. (8)
(AEB 1986)

7.36 How would you attempt an experimental verification of the principle of conservation of momentum? (6)
A body A of mass m moving with velocity u makes a *perfectly elastic* head-on collision with an identical body B which is initially at rest. Describe in words the motion of the bodies after the collision. (2)
The elastic collision mentioned above is one in which the bodies become temporarily compressed and remain in contact for a short time. On the same axes of velocity against time, sketch labelled graphs of the velocity of A and the velocity of B. The time axis should extend from a time before the bodies come into contact to a time after they separate: mark on this axis the time t_c at which they first touch, the time t_0 at which they suffer maximum compression, and the time t_s at which they separate. (3)
Explain why both bodies have the same velocity at t_0 (the time of maximum compression). What is this velocity? Hence find, in terms of m and u, the total kinetic energy of the bodies at t_0 and again at a time after they have completely separated. Account for the difference between these energies. (6)
(UCLES 1987)

7.37 Define *linear momentum* and state the conditions under which it is conserved. Outline an experimental method of investigating this conservation law. Discuss briefly **two** other examples of conservation laws in physics. (8)
By considering a head-on collision between two isolated bodies, show how the conservation of linear momentum is a direct consequence of Newton's laws of motion. (3)
The diagram shows a container of mass 45 kg floating in deep space. An astronaut, looking into it, observes an object of mass 15 kg, floating inside the container, explode into two fragments A and B of masses 5.0 kg and 10 kg respectively. These move apart in the directions shown in the figure. Initially, the astronaut, container and object have no relative motion. The impulse from the explosion on each fragment is 10 kg m s⁻¹.

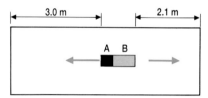

The fragments adhere to the walls of the container on impact. Describe the motion of the fragments and container. Draw graphs of position and velocity of the container, relative to the astronaut, against time for the first 5 seconds after the explosion. (9)
(O & C 1984)

7.38 **(a)** State *Newton's first law of motion* and show it leads to the concept of force. (2)
(b) Newton's second law states that 'the rate of change of momentum of a body is proportional to the resultant force acting on it'.
Show how this law, together with a suitable definition of the unit of force, leads to the relationship *force = mass × acceleration* for a body of constant mass. (3)
(c) Together with these two laws, Newton's third law can be used to derive the principle of conservation of momentum. State the third law and show this derivation. (3)
(d) A ship of mass 12 000 000 kg is moving backwards with a velocity of 0.50 m s⁻¹ towards a dockside. In order to stop the ship, the engines are ordered full ahead.

(i) Calculate the initial kinetic energy of the ship. (2)

(ii) Assuming that viscous forces are negligible, calculate the value of the constant force which must be exerted on the ship if it is to stop in a distance of 15 m. (3)

(iii) How long will it take the ship to stop under these conditions? Explain qualitatively how the result of this calculation would be affected by viscous forces. (5)

(iv) Explain how the law of conservation of momentum applies to this example. (4)

(UCLES 1988)

7.39 A simple pendulum of length l is pulled to one side through a small angle θ and then released, as illustrated below. When the string is vertical the metal bob of mass m strikes a smooth vertical metal surface S and rebounds at such a speed that one-half of the kinetic energy before impact is lost.

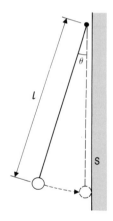

(a) Derive expressions in terms of m, l, θ and g (as appropriate) for:
   (i) the time t that elapses between release and the instant when the metal bob strikes the surface S; (2)
   (ii) the bob's kinetic energy immediately before striking S; (3)
   (iii) the magnitude of the bob's linear momentum immediately after striking S. (3)

(b) Describe briefly what will be heard by an observer, if the pendulum rebounds repeatedly along its original line of approach. (2)

(OXFORD 1989)

7.40 A projectile leaves the ground at an angle of 60° to the horizontal. Its initial kinetic energy is E. Neglecting air resistance, find in terms of E, its kinetic energy at the highest point of the motion. (3)

(UCLES 1986)

7.41 (a) Newton's third law of motion can be expressed in the form 'to every action there is an equal and opposite reaction'. By means of suitable diagrams identify clearly any action and reaction pairs of forces in the following situations and in each case describe the physical nature of each pair.
   (i) An object that is falling vertically towards the surface of the Moon.
   (ii) An object resting on the ground.
   (iii) An alpha particle at the instant it rebounds off a gold nucleus in an alpha particle scattering experiment. (8)

(b) A sub-atomic particle of mass xm and speed v collides elastically with a nitrogen nucleus of mass $14m$. The paths taken by the sub-atomic particle and the nitrogen nucleus are shown below. Assume that the nitrogen nucleus was at rest before the collision.

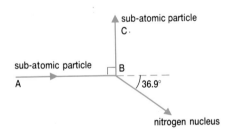

   (i) Write down equations applying the conservation of momentum principle (1) in the direction AB, (2) in the direction BC.
   (ii) Write down an equation linking the energy of the two particles before the collision with their energy after the collision.
   (iii) Deduce the value of x.
   (iv) What do you think the particle is? (10)

(ULSEB 1986)

7.42 The Grand Coulee Dam is 1270 m long and 170 m high. The electrical power output from generators at its base is approximately 2000 MW. What rate of water flow is necessary to produce this amount of power?

MOMENTUM, WORK, ENERGY AND POWER

7.43 The engine of a motor boat delivers 30 kW to the propeller while the boat is moving at 10 m s^{-1}. What would be the tension in the tow rope if the boat were being towed at the same speed?

7.44 **(a)** 'With few exceptions, mankind derives all its energy ultimately from the Sun'.
   (i) Discuss briefly three methods by which energy is derived from the Sun, explaining the role of the Sun in each process. Your choices should be as diverse as possible.
   Classify each energy source you mention as renewable or non-renewable.
   (ii) State two energy resources which are independent of solar energy.
   (iii) Indicate the approximate percentage of Britain's energy need which is supplied by each resource to which you have referred in parts (i) and (ii). (10)

(b) The solar constant at the top of the Earth's atmosphere is 1.37 kW m^{-2}. Explain the meaning of this statement.
   Discuss qualitatively the feasibility of siting a 600 MW solar power station on land, at latitude 51°N, in close proximity to an area of high population density.
   (The maximum conversion efficiency from solar energy to electrical energy is about 10 per cent.) (8)

(ULSEB 1988)

Chapter 8

CIRCULAR MOTION

LEARNING OBJECTIVES

At the end of this chapter you should be able to:

1. work with angles in radians;

2. define angular velocity and period;

3. use the term centripetal acceleration correctly;

4. see how Newton's laws apply equally well to circular motion as to linear motion.

8.1 ANGULAR VELOCITY

If an object is spinning around then it is not possible to state a unique velocity with which it is travelling, as different parts of the body are travelling with different velocities. To overcome this problem the term angular velocity is used. Angular velocity is determined by the rate of rotation of the body and could be measured in revolutions per second or

Fig 8.1 Circular motion of a cyclonic storm.

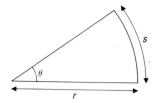

Fig 8.2 Definition of an angle.

revolutions per minute. Formally the **angular velocity** of a body is defined as the rate of change of its angular displacement. The conditions stated in Section 4.1 for linear velocity also apply to angular velocity, but with angular displacement replacing linear displacement. The angular velocity of a rigid rotating body is the same for all points of the body.

Measuring an angle

At first sight the measurement of an angle appears to be straightforward. A protractor marked in degrees is used and placed on the angle to be measured. However using a calibrated instrument does no more to define an angle than does using a watch define time.

The way an angle is defined is shown in Fig 8.2. The angle θ is *defined* by the equation

$$\theta = \frac{\text{arc length } (s)}{\text{radius of arc } (r)}$$

where s is the curved distance along the arc of radius r. This equation is probably more frequently used in the form

$$s = r\theta$$

Since θ is a ratio of two lengths it will not have dimensions. To indicate that the angle has been measured in this way the unit angle is given the name radian. One **radian** is the angle when the arc length is the same as the radius of the arc. A protractor measuring in radians looks like a protractor measuring in degrees but with a different scale (Fig 8.3). Note that since

$$\pi = \frac{\text{circumference of a circle}}{\text{diameter of circle}}$$

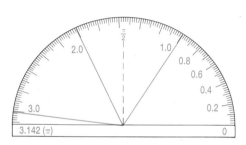

Fig 8.3 A protractor calibrated in radians.

it follows that the circumference of a circle = $2\pi r$, where r is the radius of the circle. This gives the angle in radians, for one revolution of $2\pi r/r = 2\pi$ rad.

Table 8.1 gives conversion factors between the various units of angular measure. You should take particular care always to measure angles in radians in this chapter. The advantage of using angular measure in radians is that it avoids the need to keep using conversion factors as the following example shows.

Table 8.1

Revolutions	Right angles	Radians	Degrees
1	4	2π	360
$\frac{3}{4}$	3	$\frac{3\pi}{2}$	270
$\frac{1}{2}$	2	π	180
$\frac{1}{4}$	1	$\frac{\pi}{2}$	90
$\frac{1}{6}$	$\frac{2}{3}$	$\frac{\pi}{3}$	60
$\frac{1}{2\pi}$	$\frac{2}{\pi}$	1	$\frac{180}{\pi}$
$\frac{1}{8}$	$\frac{1}{2}$	$\frac{\pi}{4}$	45
$\frac{1}{12}$	$\frac{1}{3}$	$\frac{\pi}{6}$	30
$\frac{1}{360}$	$\frac{1}{90}$	$\frac{\pi}{180}$	1

Example 1

If a needle of a record player rests on a groove of radius 12 cm, what is the length of groove which has moved past the needle when the record has turned through an angle of 0.45 radians?

$$\text{Since } 0.45 \text{ rad} = \frac{\text{arc length}}{\text{radius of arc}} = \frac{\text{arc length}}{12 \text{ cm}}$$

$$\text{arc length} = 0.45 \text{ rad} \times 12 \text{ cm} = 5.4 \text{ cm}$$

If you are not already familiar with measuring angles in radians, it is recommended that to gain familiarity with the unit you answer all the parts of Question 8.1 and that you work out Table 8.1 for yourself, with this book shut, and then check your values against those in the table!

8.1 Change the following angles into radians. Do not use a calculator. You can leave π in any of the answers.

(a) 2 revolutions

(b) 135°

(c) 5 right angles

(d) 50 revolutions

(e) 10°

(f) the angle the minute hand of a watch rotates through in a day.

Fig 8.4

Angular velocity

Angular velocity was defined formally at the beginning of the chapter as the rate of change of angular displacement. It is usually given the symbol ω, which is the small letter omega from the Greek alphabet (Ω is the capital omega). Its dimensions are time^{-1}, since the radian has no dimensions. The unit of angular velocity is normally quoted as radian per second.

 The angular velocity is the angle turned through per unit time.

The rate of change of angular displacement can be found for a small angle $\delta\theta$ turned through in a small time δt. See Fig 8.4.

The angular velocity ω is given by

$$\omega = \frac{\text{small angle turned through}}{\text{small time taken}} = \frac{\delta\theta}{\delta t}$$

and since $\delta\theta = \dfrac{\text{arc length}}{\text{radius}} = \dfrac{\delta s}{r}$

$$\omega = \frac{\delta s}{\delta t \times r}$$

But $\dfrac{\delta s}{\delta t}$ is the linear velocity v and so by substitution:

$$\omega = \frac{v}{r} \text{ or } v = r\omega$$

Example 2

A train is travelling on a track, which is part of a circle of radius 600 m, at a constant speed of 50 m s^{-1}. What is its angular velocity?

The angle the train turns through in unit time is the angular velocity.

$$\text{angular velocity } \omega = \frac{50 \text{ m s}^{-1}}{600 \text{ m}} = 0.083 \text{ rad s}^{-1}$$

Note that although the unit, m, cancels out in the division, the unit of angle, rad, is still included in the answer to make it clear that it is an angle turned through in a second and that the angle itself is measured in radians.

Example 3

A washing machine spins its tub at a rate of 1200 revolutions per minute (rpm). If the diameter of the tub is 35 cm find

(a) the angular velocity of the tub;

(b) the linear speed of the rim of the tub.

(a) $1200 \text{ rpm} = \dfrac{1200}{60}$ revolutions per second $= \dfrac{1200}{60} \times 2\pi \text{ rad s}^{-1}$

Angular velocity $= 40\,\pi$ rad s^{-1} = 126 rad s^{-1}

(b) Linear speed $v = r\omega = \dfrac{0.35}{2}$ m $\times$ 126 rad s^{-1} = 22.0 m s^{-1}

Note that the unit, rad, is not written in the answer, as a linear speed is required there.

Period

The **period** T of circular motion is the time taken for one revolution

Period is related to the angular velocity by the equation

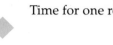

$$\text{Time for one revolution } T = \frac{\text{Angle turned through in 1 revolution}}{\text{Angular velocity}}$$

$$T = \frac{2\pi}{\omega}$$

(Compare this equation with time = distance/velocity. Here 2π is the angular distance and ω is the angular velocity.)

QUESTIONS

8.2 A clock has a second hand which is 5.0 cm long, a minute hand which is 5.0 cm long and an hour hand which is 4.0 cm long. Find the angular velocity of each of the hands and the linear speed of the tip of each hand. Assume that there is no jerkiness in their movement.

8.3 What is
  (a) the angular velocity of the Earth in its orbit around the Sun?
  (b) the angular velocity of the Earth about its axis?
  (c) the angular velocity of the Moon in its orbit around the Earth?
  (d) the angular velocity of the Moon about its axis?

8.2 CENTIPETAL ACCELERATION

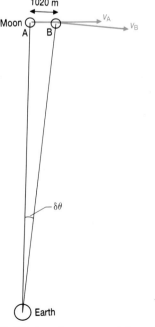

Fig 8.5 A greatly exaggerated diagram showing the angle the Moon moves through in its orbit in one second. ($\delta\theta$ should be much smaller.)

If you look at Question 8.3 you should notice that there are two types of rotation involved. The Earth in its orbit around the Sun is an object travelling in a nearly circular path. This is referred to as circular motion and the assumption is usually made that the size of the object travelling in the circle is negligible compared with the radius of the orbit. The other type of rotation is illustrated by the Earth rotating about its own axis. This is referred to as rotation. Rotation is rather more difficult to deal with than circular motion because different parts of the object have different speeds. Detailed analysis of rotation is not dealt with in this book.

When an object moves along a circular path at a constant speed its velocity is always changing because the direction in which the object travels is changing. It was indicated in Chapter 3 that some surprising results occur when changes in a vector quantity are calculated. It is now necessary to return to some of these results to find the acceleration of a body in circular motion. Consider this question, done as an example:

Example 4

Find the acceleration of the Moon. It travels at a constant speed of 1020 m s^{-1} (to 3 significant figures) and takes 27.3 days for a complete revolution of the Earth. The question is a rewording of Question 3.5, which you may have already done.

First find the radius of the orbit r
circumference of orbit = speed $\times$ time $\quad$ = 1020 m s^{-1} $\times$ 27.3 $\times$ 24 $\times$ 60 $\times$ 60 s
$\qquad\qquad\qquad\qquad\qquad\qquad\qquad$ = 2.41 $\times$ 10^9 m
radius of orbit = circumference/2π $\quad$ = 2.41 $\times$ 10^9 m/2π
$\qquad\qquad\qquad\qquad\qquad\qquad\qquad$ = 3.83 $\times$ 10^8 m

The angle through which the moon moves in one second ($\delta\theta$) is given in radians as $1020/r = 2.66 \times 10^{-6}$ rad.

Fig 8.6 The Moon as seen from a satellite in orbit about the Earth. The Moon travels in a nearly circular orbit around the Earth. Its period of rotation is 27.3 days.

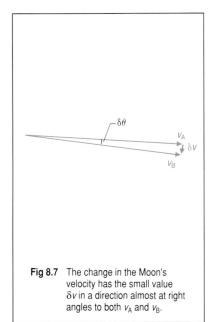

Fig 8.7 The change in the Moon's velocity has the small value δv in a direction almost at right angles to both v_A and v_B.

A vector diagram can now be drawn to find the change in the velocity of the Moon in one second, Fig 8.7. The change in velocity (δv) is the velocity which has to be added to the velocity at the start to obtain the velocity after one second. It has a small value and even in this extended triangle the angle $\delta\theta$ has had to be exaggerated. The direction of the change in velocity is towards the centre of the circle, at right angles to the individual velocities. Using the vector diagram gives

$$\delta v = v\delta\theta = 1020 \times 2.66 \times 10^{-6} = 2.72 \times 10^{-3} \text{ m s}^{-1}$$

This has made the very good assumption that the vector diagram is a sector of a circle and that therefore the angle at its centre $\delta\theta$ is by definition of an angle in radians, $\delta v/v$.

Since the change in velocity of the Moon in one second is 2.72×10^{-3} m s^{-1}, in a direction towards the centre of the orbit, then this must be its acceleration, as acceleration is defined as the change in velocity per unit time, i.e.

acceleration of the Moon = 2.72×10^{-3} m s^{-2} towards the Earth

This example shows how the acceleration of an object may be found in a particular case when an object is travelling in a circle at a constant speed. The acceleration's direction is always changing as the object moves, but is always directed towards the centre of the path in which it travels and always has the same magnitude. The acceleration towards the centre for an object travelling in a circle is called the **centripetal acceleration**. Note three important points concerning centripetal acceleration:

1   the rate at which the object travels in its circular path is constant;

2   the direction of the acceleration at any moment is in a straight line;

3   the value of the acceleration can be found directly from the definition of acceleration.

There is nothing basically different about a centripetal acceleration from any other acceleration. We can apply Newton's laws to a centripetal acceleration in exactly the same way as we can to other accelerations.

CIRCULAR MOTION

8.4 Assuming that the Earth goes round the Sun in a circle of radius 1.50×10^{11} m, find the centripetal acceleration of the Earth. You should know how long it takes for the Earth to go once round the Sun!

Example 4 works from first principles to find the acceleration of the Moon in its orbit around the Earth. To establish a general formula which can be used to calculate centripetal accelerations we need to do the same problem algebraically. The following terms will be used in the proof:

v the constant speed of the body travelling in the circular path
v_A the velocity of the body at A (v is its magnitude)
v_B the velocity of the body at B (v is its magnitude too)
r the radius of the circular path
δt the small time it takes to get from A to B
$\delta\theta$ the small angle turned through in this time
δs the small distance travelled during this time
δv the small change in velocity during this time.

Fig 8.8 shows the path under consideration and Fig 8.9 is the vector diagram to enable the change in velocity to be found. It is crucial to see that the angle between v_A and v_B is the same in both diagrams.

$$\text{In Fig 8.8} \quad \delta\theta = \frac{\delta s}{r}$$

$$\text{In Fig 8.9} \quad \delta\theta = \frac{\delta v}{v}$$

$$\text{Eliminating } \delta\theta \text{ gives} \quad \frac{\delta s}{r} = \frac{\delta v}{v}$$

But $\delta s = v\delta t$ since v is the speed travelled for a time δt, so

$$\frac{v\delta t}{r} = \frac{\delta v}{v}$$

and therefore $$\frac{\delta v}{\delta t} = \frac{v^2}{r}$$

Since $\dfrac{\delta v}{\delta t}$ is the acceleration, we get

$$\text{centripetal acceleration} = \frac{v^2}{r} \text{ in the direction of } \delta v$$

which is towards the centre of the circular path.
Since also $v = r\omega$ (see Section 8.1) we can write

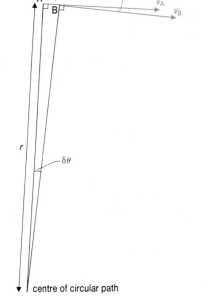

δs
A
B
$\delta\theta$
v_A
v_B
r
$\delta\theta$
$\delta\theta$
centre of circular path

Fig 8.8

$\delta\theta$
v_A
v_B
δv

Fig 8.9 The change in the velocity is in a direction towards the centre of the orbit.

Centripetal acceleration $= \dfrac{v^2}{r} = r\omega^2 = v\omega$ towards the centre of the circular path.

8.5 The maximum speed of the blades on rotary lawn mowers is restricted to avoid danger from flying stones. If the rate of rotation of the blade in one particular model is 3500 revolutions per minute and the blade has a radius of 0.23 m, find

(a) the angular velocity of the blade;

(b) the linear velocity of the tip of the blade;

(c) the centripetal acceleration of the tip of the blade.

8.6 An electric motor has a rotor which rotates with an angular velocity of 50 revolutions per second. If the rotor has a diameter of 12.0 cm, find the centripetal acceleration of the rim of the rotor. How does the centripetal acceleration of the rotor vary through the body of the rotor from its centre to its rim?

8.7 A centrifuge is required to give an acceleration of 1000g to a particle at a distance of 8.5 cm from the axis of rotation. Find the necessary angular velocity of the centrifuge.

8.3 EXAMPLES OF CIRCULAR MOTION

Having found the acceleration of an object travelling along a circular path at a constant speed, we are in a position to consider the application of Newton's laws. This needs to be done in exactly the same way as explained in Chapter 5 where the following guidelines were written:

In solving problems concerned with Newton's laws the following routine is strongly recommended.

(1) State clearly which object is being considered.

(2) Draw a free body sketch of that object only.

(3) Mark on the sketch the weight of the object.

(4) Mark on the sketch all the points where the object touches anything else and draw in the forces at these points. Label all forces clearly.

(5) Decide which direction to call positive for the total force and acceleration.

(6) Apply Newton's second law equation.

If this routine is followed you will find that it can be used to solve all problems. You will not need to use a different approach when more complicated situations arise and you will be far less likely to make unnecessary mistakes. In use the routine is direct, reliable and quick. Supposed short cuts in solving these problems lead to many mistakes being made.

The essential need is for complete free body force diagrams and not for complicated pictures.

Exactly this procedure needs to be followed in the case of circular motion. Circular motion really is no different from any other type of motion. In *all* motion the resultant force on the object controls its acceleration and the resultant acceleration is always in the same direction as the resultant force.

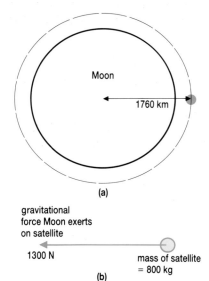

(a)

gravitational
force Moon exerts
on satellite

1300 N   mass of satellite
= 800 kg

(b)

Fig 8.10 The force on a satellite in orbit around the Moon.

Example 5

A satellite of mass 800 kg is orbiting the Moon in a circular path of radius 1760 km Fig 8.10(a). The weight of the satellite is 1300 N. Find the speed of the satellite and its period of rotation.

The force diagram is very simple in this example, Fig 8.10(b). The satellite is touching nothing, so the only force exerted on it is its weight, that is the gravitational attraction which the Moon exerts on it.

$$
\begin{array}{ccccc}
\text{FORCE} & = & \text{MASS} & \times & \text{ACCELERATION} \\
\text{in newtons} & & \text{in kilograms} & & \text{in metre second}^{-2}
\end{array}
$$

$$1300 \text{ N} = 800 \text{ kg} \times \frac{v^2}{r}$$

$$1300 \text{ N} = 800 \text{ kg} \times \frac{v^2}{1.76 \times 10^6 \text{ m}}$$

$$v^2 = \frac{1.76 \times 10^6 \text{ m} \times 1300 \text{ N}}{800 \text{ kg}} = 2.86 \times 10^6 \text{ m}^2 \text{ s}^{-2}$$

$$v = 1690 \text{ m s}^{-1}$$

$$\text{Hence } \omega = \frac{v}{r} = \frac{1690 \text{ m s}^{-1}}{1.76 \times 10^6 \text{ m}} = 9.61 \times 10^{-4} \text{ rad s}^{-1}$$

$$T = \frac{2\pi}{\omega} = \frac{2\pi \text{ rad}}{9.61 \times 10^{-4} \text{ rad s}^{-1}} = 6540 \text{ s}$$

The satellite actually takes longer to travel around the Moon than it would to travel around the Earth, despite the Earth being larger. This is because the Earth would exert a larger force on it and so there would be a much larger acceleration, with a consequent much larger speed.

QUESTION

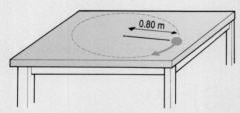

Fig 8.11   A ball of mass 0.300 kg, attached to a string, rotating on a friction free surface. The ball has a constant speed of 3.45 m s⁻¹.

8.8 Fig 8.11 shows a mass of 0.300 kg rotating in a circular path of radius 0.80 m on a friction-free table. It is attached by a string to a peg at the centre of the circle. Draw a free body force diagram for the mass and find the force which the string exerts on the mass when the mass is moving at a constant speed of 3.45 m s⁻¹.

(Your free body force diagram should have three forces shown on it. The mass is touching two objects, the table and the string, and each will exert a force on it. There is the weight of the mass as well. In this case, because the table is friction free, the weight and the force the table exerts on the mass are equal and opposite and so will cancel out. This leaves the force which the string exerts on the mass equal to the resultant force on the mass.)

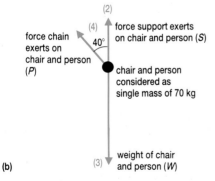

(a)

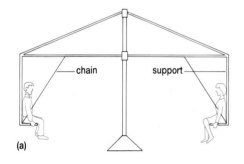

(b)

Fig 8.12   (a) A roundabout at a fair and (b) the free body force diagram for the chair and a person on such a roundabout.

Example 6

Fig 8.12(a) shows a roundabout in which a person sitting in a chair rotates in a circle of radius 2.3 m once every 9.0 s. The support for the chair (Fig 8.12(b)) is kept vertical by a chain attached to the chair at an angle of 40° to the vertical. The total mass of the chair and person is 70 kg. Find the tension in the chain and the support.

Follow the routine suggested for the solution of these problems:

(1) The chair and the person are considered together as a single object.
(2), (3) and (4)
(5) In the horizontal direction towards the centre
(6)
$$
\begin{array}{ccccc}
\text{FORCE} & = & \text{MASS} & \times & \text{ACCELERATION} \\
\text{in newtons} & & \text{in kilograms} & & \text{in metre second}^{-2}
\end{array}
$$

$$P \cos 50° = 70 \text{ kg} \times r\omega^2$$

$$= 70 \times 2.3 \text{ m} \times \left(\frac{2\pi}{T}\right)^2$$

$$= \frac{70 \times 2.3 \times 4\pi^2}{(9.0)^2}$$

$$= 78.5 \text{ N}$$

$$P = \frac{78.5 \text{ N}}{\cos 50°}$$

$$P = 122 \text{ N}$$

Fig 8.13 Some people enjoy being flung upside down at the top of a vertical circle.

Now repeat (5) and (6) but in the vertical direction

FORCE in newtons	=	MASS in kilograms	×	ACCELERATION in metre second^{-2}
$(P \cos 40° + S - W)$	=	70 kg	×	0

The weight of an object of mass 70 kg is 70 kg × 9.8 N kg^{-1} = 686 N.

$$122 \cos 40° + S - 686 = 0$$
$$S = 686 - 122 \cos 40° = 593 \text{ N}$$

The data for this problem were given to 2 significant figures so the working has been done to 3 significant figures to avoid introducing arithmetical error. The final answer should be given to two significant figures. So

tension in chain = 120 N; tension in support = 590 N

Example 7
Fig 8.13 is a photograph of an amusement park ride in which a person of mass 63 kg rotates in a vertical circle of radius 6.6 m. The time taken for one revolution is 3.2 s. Find the force which is exerted by the structure on the rider when the rider is

(a) at the bottom of the circle,
(b) at the top of the circle,
(c) half way up

The weight of the rider = 63 × 9.8 = 617 N.

The acceleration of the rider is at all times towards the centre and has a value given by:

$$\text{acceleration} = r\omega^2 = 6.6 \times \left(\frac{2\pi}{3.2}\right)^2 = 25.4 \text{ m s}^{-2}$$

Fig 8.14 shows three free body force diagrams for the person when in the three different positions under consideration.

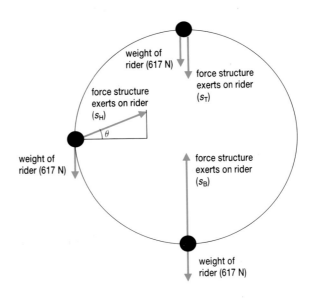

Fig 8.14 Three free body force diagrams for a person sitting in the amusement park ride shown in Fig 8.13.

CIRCULAR MOTION

(a) At the bottom the force the structure exerts on the rider is upwards and is larger than the weight of the rider. This gives a resultant upward force to cause the upward acceleration towards the centre of the circle

$$\text{FORCE} \quad = \quad \text{MASS} \quad \times \quad \text{ACCELERATION}$$
$$\text{in newtons} \qquad \text{in kilograms} \qquad \text{in metre second}^{-2}$$

$$(S_B - 617 \text{ N}) = \qquad 63 \text{ kg} \qquad \times \qquad 25.4 \text{ m s}^{-2}$$

$$S_B \qquad = (63 \times 25.4) + 617 \quad = \qquad 2220 \text{ N}$$

(b) At the top the force the structure exerts on the rider might be upwards, if there is a strap underneath him holding him up, or it might be downwards, if it is exerted on his back. If the wrong direction is chosen for the diagram then the result will come out as a negative value. It helps however to think out the direction carefully. In this case he will be pushing outwards on the structure, but that is a force on the structure **not** a force on the rider. The rider here must have the structure exerting a downward force on him because the gravitational force by itself is not large enough to give him a downward acceleration of 25.4 m s^{-2}.

$$\text{FORCE} \quad = \quad \text{MASS} \quad \times \quad \text{ACCELERATION}$$
$$\text{in newtons} \qquad \text{in kilograms} \qquad \text{in metre second}^{-2}$$

$$(S_T + 617 \text{ N}) = \qquad 63 \text{ kg} \qquad \times \qquad 25.4 \text{ m s}^{-2}$$

$$S_T \qquad = 63 \times 25.4 - 617 \qquad = \qquad 986 \text{ N}$$

(c) At the side the problem is more difficult. The sequence used to draw the free body force diagram is as follows:

1. The acceleration is known to be towards the centre and will still have the same value of 25.4 m s^{-2}.
2. The resultant force must therefore be towards the centre.
3. The weight does not change; it is always 617 N and acts downwards.
4. The force which the structure exerts on the rider when half way up, S_H, has been drawn so that when it is added vectorially to the weight, the resultant of the two forces acts towards the centre.

Now apply Newton's laws in the vertical and the horizontal directions

$$\text{FORCE} \quad = \quad \text{MASS} \quad \times \quad \text{ACCELERATION}$$
$$\text{in newtons} \qquad \text{in kilograms} \qquad \text{in metre second}^{-2}$$

Vertical:

$$(S_H \sin \theta - W) = 63 \text{ kg} \times 0$$
$$\text{so} \quad S_H \sin \theta \qquad = W \qquad = 617 \text{ N}$$

Horizontal:

$$S_H \cos \theta = 63 \text{ kg} \times 25.4 \text{ m s}^{-1} = 1600 \text{ N}$$

$$\text{dividing gives} \quad \frac{S_H \sin \theta}{S_H \cos \theta} \quad = \quad \frac{617 \text{ N}}{1600 \text{ N}} \quad = 0.386 = \tan \theta$$

$$\theta \qquad = 21°$$

$$S_H = \frac{617 \text{ N}}{\sin 21°} \qquad = 1720 \text{ N}$$

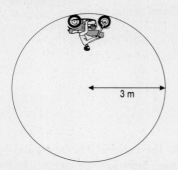

Fig 8.15 A motor cyclist inside a spherical cage and riding upside down.

8.9 Work through the roundabout problem, Example 6, again, but with a time of rotation of 6.0 s. What would happen to the force which the support exerts on the chair and the person if the time for one rotation were only 3.0 s?

8.10 A motor cycle stunt rider travels around the inside of a sphere of radius 3.0 m, see Fig 8.15. What is the minimum speed he must have to be able to remain in contact with the sphere when he is upside down?

8.11 In an amusement park a passenger of mass 60 kg travels upside down in a carriage at the top of a circle of radius 6.0 m at a speed of 12.3 m s^{-1}, Fig 8.16. Draw a force diagram showing the forces acting on the passenger. Find the magnitude and direction of the force which the carriage exerts on the passenger.

Fig 8.16 The seat belts are not essential in rides such as this one at Alton Towers. The downward acceleration is considerably larger than the acceleration of free fall so that the seat of the cab has to push the rider downwards when at the top of the loop.

8.12 A pilot of mass 75 kg who has been diving vertically downwards with a velocity of 180 m s^{-1} pulls out of his dive by changing his course to a circular path of radius 1200 m. If he maintains a constant speed what will be his maximum acceleration? What is the maximum force which his seat exerts on him.

8.13 A cyclist of mass 35 kg travelling at a constant speed of 11 m s^{-1} rounds a corner of radius 17.0 m. Draw a force diagram showing the forces acting on the cyclist. At what angle to the vertical should the cyclist be?

8.14 A coin will rest on a long playing record rotating at 45 rpm provided that it is not more than 10 cm from the centre of the record. How far away from the centre may it be placed if it is to remain on the record when rotated at 33.3 rpm?

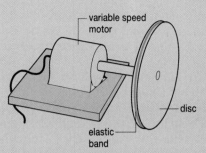

Fig 8.17 Possible experimental arrangement for investigation on tension in a rotating elastic band.

If an elastic band is placed on a rotating disc the tension in it will provide sufficient force for it to rotate with the disc only up to a critical angular velocity. At angular velocities above this critical angular velocity the elastic band will not stay on the disc. You are asked to show that the critical angular velocity, ω_c varies with r, the radius of the disc according to the relationship

$$\omega_c{}^2 = A - \frac{B}{r}$$

Fig 8.18 illustrates a possible arrangement for the experiment. Varying sizes of Meccano wheels could be used for the disc, and some trial and error is necessary to obtain an elastic band of suitable length for the wheels in use and the range of angular speeds of the motor. The rate of rotation of the motor must be able to be varied and measured. Measurement is most conveniently done using a calibrated stroboscopic lamp but it can also be done using a rev counter.

Use the straight line graph you have plotted to find A and B.

It can be shown that

$$A = 4\pi^2 k / M$$

where k is the spring constant of the elastic band (force per unit extension for a single strand of elastic), and M is the mass of the elastic band. Measure M and use your value of A to calculate k. Why is this method for finding k likely to be less accurate than finding k by a direct method?

The Behaviour of aircraft tyres when rotating at high angular speeds

When a tyre is rotated at high speed its diameter increases. (If the tyre has been retreaded there is a danger that the retread can separate from the tyre body in the same way that the elastic band did in the investigation. Evidence for this is often to be seen as strips of rubber littering the hard shoulder of motorways.) The problem is particularly important on aircraft as the undercarriage needs to be made as small and light as possible and yet to be safe and strong. This usually results in several wheels being used

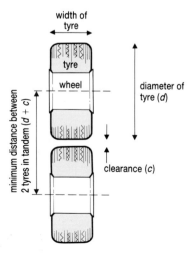

Fig 8.18 The undercarriage of a modern jet. Note the arrangement of the wheels and the hydraulic jacks for raising and lowering the undercarriage.

Fig 8.19 Arrangement of tyres in tandem on the under-carriage of an aeroplane. When rotating at high speed the tyres stretch. Clearance between wheels must be sufficient so that they do not hit one another.

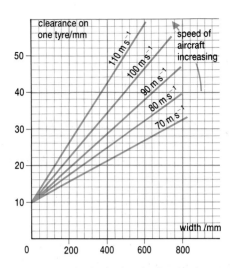

Fig 8.20 Graphs showing how the diameter of a tyre varies with the rate of rotation.

on each undercarriage and the spacing of the wheels needs to be as close as practicable. In particular, when wheels are in tandem, as shown in Fig 8.19, any increase in the diameter of a wheel could make them rub against one another.

British Standard 2M45 Part 1 gives a graph showing how much clearance must be allowed for at different tyre widths and speeds. It is reproduced in Fig 8.20, and shows that a clearance of at least 10 mm must always be used but that the wider the tyre or the faster the take off speed the more clearance must be given.

Use data from the graph to answer the following questions. You might find it easier to imagine that the tyre is being tested on a stationary test rig so that the only factors you need to consider are the ones involved with circular motion. In fact the horizontal velocity of the aircraft along the runway does not make any difference to the force necessary for circular motion, although it does alter the loading on the wheel.

(a) Consider first a tyre which has a diameter of 0.90 metre and a width of 400 mm. What will be its angular velocity at each of the speeds shown?
(b) What will be the centripetal acceleration of the tread of the tyre at each of these speeds?
(c) What clearances are required for this tyre at each of these speeds?
(d) What force is necessary to hold each kilogram of tread on the tyre when the tyre is travelling at these speeds?
(e) If a particular aircraft has a clearance between two wheels in tandem of 60 mm, plot a graph to show how the maximum take off speed possible varies with the width of tyre used.
(f) List some advantages and disadvantages of using a tyre of greater width.
(g) For a given take off speed, how does the centripetal acceleration of the tyre tread vary with the diameter of the tyre?
(h) The graph has been used to provide data for a tyre of diameter 0.90 m. It can be used for tyres of any diameter. Suggest a reason why the graph need not be different for different diameter tyres?
(i) The very large forces which are necessary to hold the tyre in one piece are supplied by steel wires embedded in the rubber of the tyre. Sketch how this wire could be arranged within the tyre and show the direction of the forces acting on the wire.

SUMMARY

- Angle θ in radians is defined by the equation $\theta = \dfrac{\text{arc length } s}{\text{radius of arc } r}$.

- Angular velocity ω is the number of radians turned through per unit time.

- The time for one revolution T, the period, is therefore given by

$$T = \frac{2\pi}{\omega}$$

  $\theta \times r = s$ by definition of θ
  therefore $r\omega = v$ where v is speed.

- A body travelling at a constant speed v in a circular path has an acceleration towards the centre of its path. This acceleration is called a centripetal acceleration and its value is given by

$$\text{centripetal acceleration} = \frac{v^2}{r} = r\omega^2 = v\omega$$

- The total force on such a body must, as always, act in the direction of the acceleration and the magnitude of the total force can be found by using $F = m \times a$.

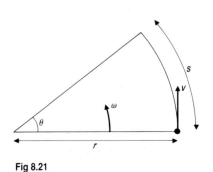

Fig 8.21

CIRCULAR MOTION

EXAMINATION QUESTIONS

8.15 The diagram shows a section of curtain track in a vertical plane. The curved section, CDE, forms a circular arc of radius 0.75 m and the point D is 0.25 m higher than B. A ball bearing of mass 0.060 kg is released from A, which is 0.50 m higher than B. Assume that rotational and frictional effects can be ignored and that the ball bearing remains in contact with the track throughout the motion.

    (a) Calculate the speed of the ball bearing (i) at B, (ii) at D. (4)

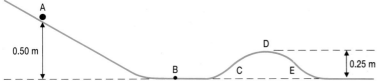

    (b) Draw a diagram showing the forces acting on the ball bearing when it is at D and calculate the force the track exerts on the ball bearing at this point. (4)

                                 (JMB 1986)

8.16 Show that a particle travelling with a constant speed v in a circular path of radius r has an acceleration v^2/r directed towards the centre of the circle. (5)

    (a) An electron of mass m and charge $-e$ has a speed v. It travels in a uniform magnetic flux density B applied in a direction perpendicular to its motion. Show that its path is a circle and find an expression for the frequency of the circular motion. Show that this frequency is independent of the radius of the circle. (6)

    (b) The vehicles of a fairground ride are supported by light cables with their upper ends at a radius R from the axis of rotation. The centre of mass of the vehicle is at a distance l from the upper point of support. Explain why, when the ride is rotating with angular velocity ω, the cables are inclined at an angle θ to the vertical, as shown in the diagram. (4)

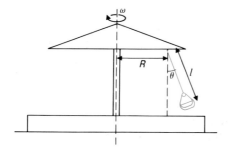

Show also that

$$\omega^2 = \frac{g \tan \theta}{R + l \sin \theta} \qquad (2)$$

Calculate the rate of rotation, in rev s^{-1}, for which $\theta = 15°$, given that $R = 3.0$ m and $l = 2.5$ m. (3)

8.17 **(a)** Explain why a particle moving with a constant speed along a circular path has a radial acceleration.
The value of such an acceleration is given by the expression v^2/r, where v is the speed and r is the radius of the path. Show that this expression is dimensionally correct. (6)

    (b) Explain, with the aid of clear diagrams, the following.

      (i)   A mass attached to a string rotating at constant speed in a horizontal circle will fly off at a tangent if the string breaks.

      (ii) A cosmonaut in a satellite which is in a free circular orbit around the Earth experiences the sensation of weightlessness even though he is influenced by the gravitational field of the Earth. (7)

    (c) A pilot 'banks' the wings of his aircraft so as to travel at a speed of 360 km h^{-1} in a horizontal circular path of radius 5.0 km. At what angle should he bank his aircraft in order to do this? (5)

                              (ULSEB 1982)

8.18 A bobsleigh rises up the side of an ice track when it follows a horizontal circular path at speed. Diagram (a) shows a cross-section of the ice track and (b) is a free body diagram showing the forces which act on the bobsleigh.

(a)
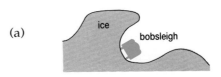

(b)
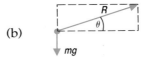

    (i)   Explain why the kinetic energy of the bobsleigh is conserved but not its linear momentum.

(ii) What effect does the horizontal component of the push, R, of the ice track on the bobsleigh have on the motion of the bobsleigh?

(iii) Calculate the value of $\tan \theta$ for a speed of 25 m s^{-1} if the radius of the circular path the bobsleigh follows is 20 m.   (8)

(ULSEB 1987)

8.19 Explain the circumstances under which a particle moving with a uniform speed can experience an acceleration.   (3)

An aircraft is travelling at a constant speed of 180 m s^{-1} in a horizontal circle of radius 20 km. A plumbline, attached to the roof of the cabin, settles at an angle ϕ to the vertical while the aircraft is turning.

(i) Find the centripetal acceleration of the aircraft.

(ii) Name the forces which act on the bob of the plumbline, and draw a labelled diagram to show the directions of these forces, and of their resultant. (Indicate the direction of the centre of the circle on your sketch.)

(iii) Find the angle ϕ.

(iv) Show by means of a simple sketch of the cross-section of the aircraft and its cabin how the plumbline is orientated with respect to the aircraft.   (9)

(UCLES 1985)

8.20 (a) The *gravitational potential energy* of a body of mass m at a height h above the surface of the Earth can be expressed as mgh. Give the *unit* and the *dimensions* of this quantity.   (2)

(b) Describe an experiment to investigate how the radial force acting on a body moving at a constant speed in a circle depends on the radius of the circle and the mass and speed of the body.   (6)

(c) The bob of a conical pendulum has a mass of 50.0 g and is attached to a fixed support at O by an inextensible string OA of length 1.00 m. The bob moves with constant angular speed in a horizontal circle with centre C. The angle maintained by the string to the vertical OC is 30°, as shown below.

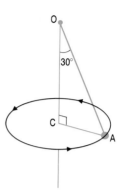

Calculate

(i) the speed of the bob,

(ii) the tension in the string, and

(iii) the linear momentum of the bob at one instant.   (7)

(d) At some point in the motion of the conical pendulum the bob splits into two parts. One part remains attached to the string. The act of splitting involves a negligible force. Discuss qualitatively the motion of

(i) the part of the bob still attached to the string, and

(ii) the detached part, up to the moment before it hits the floor of the laboratory.   (5)

CIRCULAR MOTION

Chapter 9

GRAVITATION

LEARNING OBJECTIVES

At the end of this chapter you should be able to:

1. quote Newton's law of gravitation;

2. understand the term gravitational field;

3. understand the term gravitational potential and be able to relate it to gravitational field;

4. apply Newton's law of gravitation and principles of circular motion to satellites and other bodies;

5. calculate an escape speed;

6. understand why the measured value of the acceleration due to gravity varies from place to place on the Earth's surface.

As stated in Section 5.6, gravitational attraction is a strange and remarkable force about which little is known. It is very difficult to do experiments on gravitational attraction because the force is so small between masses in the laboratory. The one case where the gravitational force is large is when it is due to the pull of the Earth, but for any particular laboratory it is then not possible to change the pull significantly on a given mass. We have electrical switches that enable us to switch on and off electrical fields so we can easily see their effect. There is no such thing as a gravity switch – which is a pity! The mystery of gravity deepens when you ask yourself how can the Earth pull on the Moon even though the Moon is 400 000 km away from the Earth and there is a vacuum between them?

9.1 NEWTON'S LAW OF GRAVITATION

In Section 8.2 the acceleration of the Moon was calculated to be 2.60×10^{-3} m s^{-2} in a direction towards the Earth. This calculation was first done by Newton, using different units. He compared the acceleration of the Moon with the acceleration of objects falling on the Earth – traditionally with the acceleration of an apple falling from a tree in his garden. In modern units these were the figures he might have obtained:

	Distance from centre of Earth	Acceleration
Moon	3.84×10^8 m	2.72×10^{-3} m s^{-2}
Apple	6.39×10^6 m	9.81 m s^{-2}

At first sight there does not seem to be anything special about the numbers, but looking for patterns and connections between figures was, and still is, one of the basic aims of scientists. Use a calculator to find the connection between these numbers *before reading on*.

Fig 9.1 Apollo 8 mission to the Moon, Dec 21 1968. A thrust of 3.3×10^7 N sends the vehicle carrying Borman, Lovall, and Anders to orbit the moon. They were the first astronauts to leave the influence of the Earth and enter the gravitational pull of the Moon.

In order to find connections such as this, the ratio of the two values should be found:

$$\frac{\text{distance of Moon from the centre of the Earth}}{\text{distance of apple from the centre of the Earth}} = 60.1$$

$$\frac{\text{acceleration of Moon}}{\text{acceleration of apple}} = 2.77 \times 10^{-4} = \frac{1}{3610} = \frac{1}{60.1^2}$$

To put this another way, the Moon is 60.1 times further away from the centre of the Earth but its acceleration is only 1/3610 of the acceleration of the apple. Now note that $60.1^2 = 3610$ and you have the basis of Newton's inverse square **law of gravitation** which states that:

> Every particle of matter in the universe attracts every other particle with a gravitational force that is directly proportional to the product of the masses of the particles and inversely proportional to the square of the distance between them.

This gives $\quad F \propto \dfrac{m_1 \, m_2}{r^2}$

By putting in a constant of proportionality G called the **gravitational constant** we get the equation form of Newton's law of gravitation:

$$F = \frac{G \, m_1 \, m_2}{r^2}$$

where F is the gravitational force, and has the same magnitude on each particle, m_1 and m_2 are the masses of the two particles and r is the distance between them. The value of G can be found experimentally to be 6.67×10^{-11} N m^2 kg^{-2}.

Fig 9.2 Optical photographs of the Orion Nebula, a bright cloud of gas and dust where stars are in the process of being born.

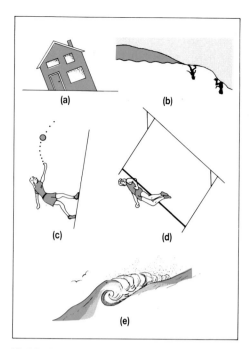

Fig 9.3 These sketches show normal situations but they are drawn without the normal convention that vertically downwards is the way objects fall. Which direction is down for each situation?

The insight provided by Newton's law of gravitation is not so much that he found a mathematical connection between the acceleration of the apple and the acceleration of the Moon, but that he realised that the force controlling the movement of the apple is of the same nature as the force controlling the movement of the Moon: that gravity is a truly universal force. Gravity exists because the bodies have mass. Gravitational pull is so familiar as weight in everyday life that its peculiarities are simply accepted as being normal. Fig 9.3 shows activities taking place at different places on the Earth, as seen from space. They look odd because the normal convention for vertical has not been followed. Work out for each one the direction of the pull of gravity and suggest where each may be taking place.

That the force of gravity is usually a very tiny force can be readily shown. If one magnet moves past another then there is an appreciable attraction or repulsion between them. If you walk past a house you feel no attraction or repulsion at all, yet there *is* an attraction; always an attraction, never a repulsion. The reason that you do not notice the attraction is that the force is so very small. For a house of mass 200 000 kg and a person of mass 60 kg, the force of attraction if the person is 10 metres from the house is only 0.000 008 N. This is a force which it is quite impossible to notice, but was the basis of the first attempt to measure G, in 1740. If a plumb bob is suspended near one side of a mountain then it is pulled very slightly out of vertical by the gravitational force the mountain exerts on it, as illustrated in a very exaggerated way in Fig 9.4.

A large gravitational force will only exist if at least one of the masses is large. In practice this usually means the Earth. The Earth, on the other hand, can hardly be considered to be a particle. Newton was able to show that the gravitational force exerted on or by any uniform sphere is the same as if the entire mass of the sphere were concentrated at its centre. The force of attraction between two masses which are not spheres can be difficult to calculate. You may assume, in gravitational problems, that any masses may be considered as point masses unless you are specifically told not to.

QUESTIONS

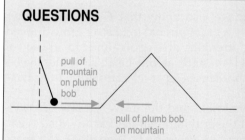

Fig 9.4 An exaggerated sketch of the way a mountain can pull a plumb bob out of the vertical.

9.1 Find the force, to two significant figures, which

(a) A molecule of mass 3.0×10^{-25} kg exerts on another molecule of the same mass when at a distance apart of 2.4×10^{-9} m.

(b) The Earth, of mass 6.0×10^{24} kg, exerts on a mass of 1 kg at its surface. The Earth has a radius of 6.4×10^{6} m.

(c) The Sun, of mass 2.0×10^{30} kg, exerts on the Earth. The Earth–Sun distance is 1.5×10^{11} m.

9.2 The mountain referred to above and shown in Fig 9.4 may be considered as a point mass of approximately 4×10^{12} kg at a distance of 3000 m from a plumb bob of mass 2 kg. Work to only 1 significant figure.

(a) Find the force the mountain exerts on the plumb bob.

(b) The Earth has a mass of 6×10^{24} kg and a radius of 6×10^{6} m find the force the Earth exerts on the plumb bob.

(c) Draw a vector diagram showing these forces on the plumb bob, and hence find the angle of deflection θ.

(d) If the angle of deflection is measured experimentally then this question, worked through from the end back to the beginning, can be used to obtain the value of the mass of the Earth even though the value of G is not known. Write down the algebraic expression to find the mass of the Earth.

GRAVITATION

9.2 GRAVITATIONAL FIELD

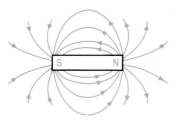

Fig 9.5 A typical drawing of a magnetic field.

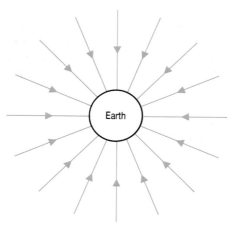

Fig 9.6 The shape of the gravitational field surrounding the Earth.

When the magnetic field surrounding a magnet is drawn, the concentration of field lines gives an indication of the strength of the field. The direction of the field lines shows the direction in which the North Pole of a compass would point at places around the magnet, Fig 9.5. The same type of diagram can be drawn to show the gravitational field around any body. Fig 9.6 is a drawing of the gravitational field surrounding the Earth. It shows spherical symmetry because the Earth is very nearly a uniform sphere. Both of these field diagrams are really three-dimensional diagrams. The magnetic field diagram shows repulsion as well as attraction, but the gravitational diagram only shows attraction because gravitation is always an attractive force. The diagram of the Earth's gravitational field clearly shows that the gravitational pull on an object is always directly towards the Earth, and that it gets stronger the closer the body is to the Earth's surface. Fields are therefore used to show the strength of a force and the direction in which it acts. This needs to be expressed more formally.

The **gravitational field strength** at a point is defined as the gravitational force per unit mass acting at that point

Gravitational field strength is a vector having direction as well as magnitude, and so addition of gravitational fields must be done by vector addition.

The symbol used for gravitational field strength is g so the equation defining g is

$$g = \frac{\text{gravitational force}}{\text{mass}} = \frac{F}{m}$$

g has units of N kg^{-1}.

Since from Newton's second law we also know that $F = ma$, it follows that an acceleration g is produced by a gravitational field strength g. Indeed the units of gravitational field strength are equivalent to those of acceleration. The gravitational field strength at the Earth's surface has a numerical value of 9.8 N kg^{-1} and this causes an acceleration due to gravity of 9.8 m s^{-2} (see Sections 5.6 and 9.5).

From the definition of gravitational field and the universal law of gravitation we can work out the value of the gravitational field at any distance r from the centre of the Earth (or from any other body if needed). If a mass m is a distance r from the centre of the Earth, which has a mass M (Fig 9.7) then

$$\text{Force of attraction} = \frac{GmM}{r^2}$$

$$\text{Gravitational field strength} = \frac{\text{force of attraction}}{\text{mass}} = \frac{GmM}{r^2 m} = \frac{GM}{r^2}$$

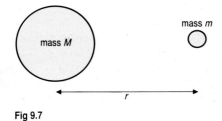

mass m

mass M

r

Fig 9.7

Example 1

Assuming the Earth to be a sphere, find its mean density from the following data:

Gravitational field strength of the Earth, $g = 9.83$ N kg^{-1}

Radius of Earth, $r = 6.37 \times 10^6$ m

$G = 6.67 \times 10^{-11}$ N m^2 kg^{-2}.

GRAVITATION

$$g = \text{force of attraction per unit mass} = \frac{GM}{r^2}$$

$$M = \frac{gr^2}{G} = \frac{9.83 \times (6.37 \times 10^6)^2}{6.67 \times 10^{-11}} = 5.98 \times 10^{24} \text{ kg}$$

Volume of Earth $= \frac{4}{3} \pi r^3 \qquad = \frac{4}{3} \pi (6.37 \times 10^6 \text{ m})^3$
$$= 1.083 \times 10^{21} \text{ m}^3$$

Mean density $= \dfrac{5.98 \times 10^{24} \text{ kg}}{1.083 \times 10^{21} \text{ m}^3}$

$$= 5520 \text{ kg m}^{-3}$$

Variation of g with distance from the centre of the Earth

From the surface of the Earth upwards the value of g decreases following an inverse square law. Within the Earth the value of g follows a complex variation because the density of the Earth is not constant. The Earth has a core which is much denser than its crust. From measurements taken of the speeds with which shock waves from earthquakes travel, it is possible to calculate the values of g at different distances from the centre of the Earth. These values are used to obtain the graph in Fig 9.8.

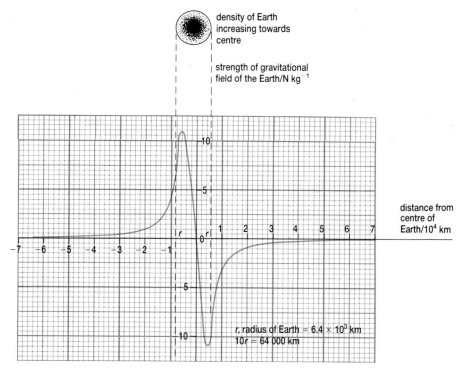

Fig 9.8 A graph showing how the Earth's gravitational field changes with distance from the centre of the Earth.

The reason for drawing this graph in this way is to illustrate two points. The first is to show that the gravitational field strength at the centre of the Earth is zero. This is because any matter at the centre would be pulled equally in all directions, giving a resultant force on it of zero. The second reason is that the graph shows that the direction of the field is in opposite directions on opposite sides of the Earth. On the diagram the field on the right-hand side acts towards the left, and so is shown as having a negative value; the field on the left-hand side acts towards the right and is shown as positive.

Gravitational field pattern of Earth/Moon system

As stated earlier, the total gravitational field strength at a point is the vector sum of the individual fields caused by different bodies. Finding the field for even two bodies, such as the Earth and the Moon, can be a lengthy process, but the pattern produced by this process is much as would be expected. Fig 9.9 is drawn to a scale of $1 : 7.5 \times 10^9$ and the grey lines show the gravitational field pattern of both the Earth and the Moon. The diagram shows how the Earth's field dominates this region of space, with the Moon's field having an influence only comparatively close to itself. The field of the Sun has not been included.

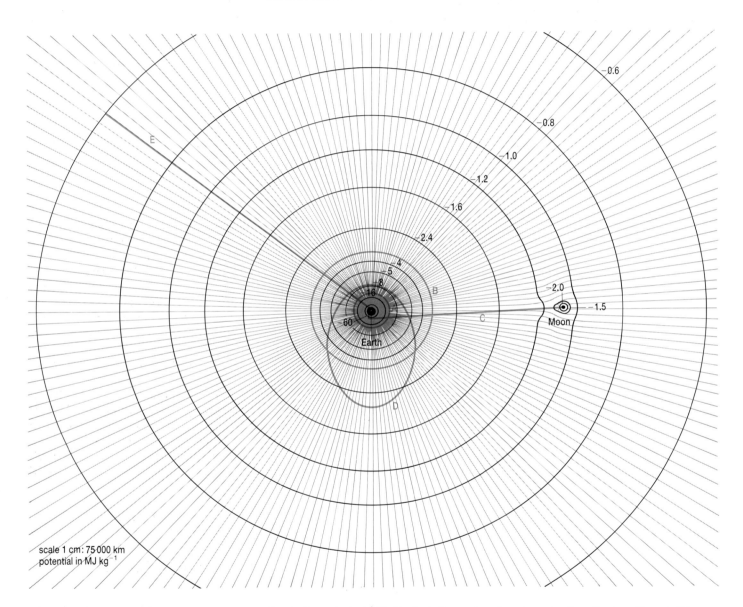

scale 1 cm : 75 000 km
potential in MJ kg^{-1}

Fig 9.9 The gravitational field of the Earth–Moon system, shown in grey together with values of the potential at all points within the system, shown in black. Some satellite paths are shown in blue.

QUESTION	9.3 The mass of the Earth is 6.0×10^{24} kg and the mass of the Moon is 7.4×10^{22} kg. Find the position of the gravitational neutral point, X, between the Earth and the Moon if the distance between their centres is 3.8×10^8 m. Show that it divides the distance between the Earth and the Moon roughly in the ratio 9 : 1.

GRAVITATION

Because the gravitational field strength is a vector quantity it is difficult to make calculations of speed and energy of objects moving in the field. Whenever a field theory is used it is therefore useful to define a scalar property of the field at each point, and this property is called the potential of the field. In this case we define the **gravitational potential** of a point in the field as the work done on unit mass in moving it to the point from a point remote from all other masses. This is often stated in more abbreviated form as: the work done to move unit mass from infinity to the point. The definition itself implies that the zero of potential energy is taken to be when the mass is at infinity. Later, in Section 15.3, the electrical potential of a point will be defined in a similar way as the work done in moving unit charge from infinity to the point. Gravitational potential, like gravitational field strength, is a property of a point in the field and does not depend on the mass at the point.

The symbol used for gravitational potential is V, and its unit is J kg^{-1}.

Fig 9.10 A NASA photograph of a satellite over the Moon which also shows the Earth, 400 000 km away.

Example 2

Using the graph in Fig 9.8 find the work done on unit mass to move it from A, a distance of $10r$ from the Earth, to B, on the Earth's surface.

The mass would fall from A to B without any work being done on it by force other than gravity and would in the process gain kinetic energy. In order for it not to gain kinetic energy, work would have to be done *by* the mass on an external force. In other words the work done *on* the mass is negative. Work is always the force × the distance moved in the direction of the force. Here the force per unit mass is plotted against distance and so the area beneath the graph gives the work done per unit mass, and it will have a negative value. The area can be found from the graph by counting squares. Each small square is 2000 km × 0.5 N kg^{-1} = 1000 kJ kg^{-1}. The area is approximately 60 small squares and the value obtained is approximately $-60\,000$ kJ kg^{-1}.

The gravitational potential difference between A and B is therefore $-60\,000$ kJ kg^{-1}.

Note that if A had been much further away from the Earth than $10r$, it would not have made much difference to the potential difference. This is because the gravitational force is so small at large distances from the Earth that very little work needs to be done in order to move the mass. We need to be able to find the potential at a point without having to plot a graph and count squares. This can be done satisfactorily only by the use of calculus.

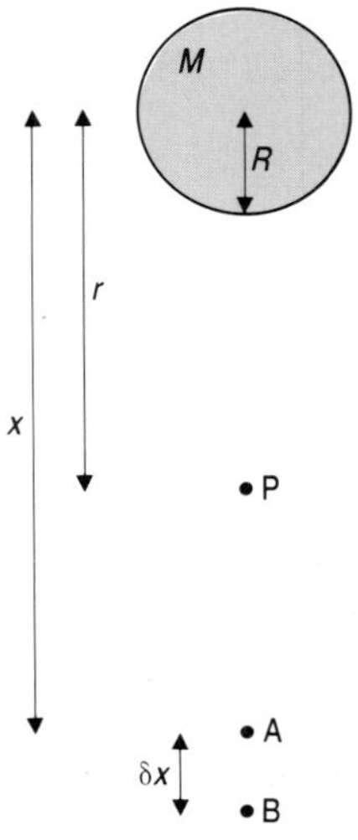

Calculation of the gravitational potential due to a spherical mass

Fig 9.11 shows a spherical mass of radius R and mass M. If unit mass is to move at constant speed from A to B, a small distance dx, then it must have a force applied to it equal and opposite to the gravitational attraction.

Force applied to unit mass = GM/x^2 to the right

Work done in moving unit mass from A to B = $\dfrac{GM}{x^2} \times \delta x$

Fig 9.11

Total work done in moving unit mass from P to infinity

$$= \int_r^\infty \frac{GM}{x^2} \, \mathrm{d}x = \left[-\frac{GM}{x} \right]_r^\infty$$

$$= \left[-\frac{GM}{\infty} \right] - \left[-\frac{GM}{r} \right]$$

$$= \frac{GM}{r}$$

Potential at infinity is therefore GM/r higher than at P. Since the definition of gravitational potential implies that the potential at infinity is zero, the gravitational potential at P, a point near a spherical mass, is given by

$$V = -\frac{GM}{r}$$

The potential at the surface of the spherical mass is therefore $-\frac{GM}{R}$.

QUESTIONS

9.4 Find the gravitational potential difference between a point on the Earth's surface and a point 1 metre above the Earth's surface. Do this calculation by the two apparently different ways stated:

(1) by use of the basic definition of potential

(2) by use of the formula just obtained, the data given in Example 1 and a calculator with enough significant figures displayed (or some algebra).

You should get the same answer by both methods of course – always a good check on a solution.

9.5 Find the gravitational potential at the Earth's surface. At what distance from the centre of the Earth does the gravitational potential fall to one half of this value? What is the maximum potential which an object can gain as a result of moving a large distance from the Earth?

9.6 A piece of rock far out in interplanetary space is at rest relative to the Sun. Under the influence of the Sun's gravitational pull it begins to fall towards the Sun along a straight radial line. Assuming that the Sun has a definite radius of 7.0×10^8 m, with what speed will the rock hit the Sun? The mass of the Sun is 2.0×10^{30} kg.

The relationship between potential and field

The answer to Question 9.4 is 9.83 J kg^{-1}. The reason this number keeps occurring is because it gives the gravitational force acting on a mass of one kilogram at the surface of the Earth. In Question 9.4 you were asked to find the difference between the potential which a 1 kg mass has on the Earth's surface and that at a height of 1 m. To find this you can simply take the gravitational force acting on the 1 kg mass, 9.83 N, and multiply it by the distance moved, 1 m. The work done is 9.83 joule on the 1 kg mass.

Put in symbols this gives the change in gravitational potential ΔV as the gravitational field strength g multiplied by the distance moved $-\Delta x$

i.e $\Delta V = -g \, \Delta x$

The – sign arises because the gravitational field is in the opposite direction to increasing potential. If the equation is written in the form

$$g = - \frac{\Delta V}{\Delta x}$$

we see that the gravitational field strength is numerically equal to the gravitational potential gradient. This is equally true for non-uniform fields.

◆ The gravitational field is minus the potential gradient: $g = - \dfrac{dV}{dx}$

9.4 SATELLITES

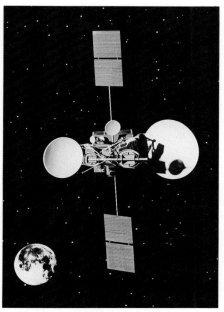

Fig 9.12 A geostationary satellite orbits the Earth once per day over the equator so it appears to be stationary. Used now for international communications and first suggested by Arthur C. Clarke.

Any satellite is kept in its orbit by the gravitational attraction of the body about which it is rotating. Many man-made satellites are in circular orbits around the Earth and once in orbit they do not need rocket motors to keep them in orbit. Provided they are sufficiently far away from the Earth's surface to have very little air resistance, they can remain rotating around the Earth for many years. Some of them are expected to remain in their orbit for millions of years. Besides man-made satellites there are also many naturally occurring satellites. All of the planets are satellites of the Sun. The Moon is a satellite of the Earth. Satellites may have circular or elliptical orbits. Circular orbits are easier to handle mathematically and there are many satellites which are in orbits which are approximately circular. As far as the Sun's planets are concerned, only Pluto has a markedly elliptical orbit. The Earth's orbit has a maximum radius of 1.52×10^{11} m and a minimum radius of 1.47×10^{11} m.

In order to be in a circular orbit, a satellite has to have a particular velocity for the distance it is from the Earth. Some satellites are placed in orbits directly above the equator and travel in an easterly direction so that they rotate about the Earth once every day. They are called geostationary satellites because to an observer on the ground, also rotating with the Earth once per day in the same direction, they seem to be stationary. The following example shows how to find the height at which a geostationary satellite must be placed.

Example 3

A communications satellite is to be placed in a circular geostationary orbit. Find its speed and height above the Earth's surface.

All such problems depend for their solution on using Newton's laws of motion.

The satellite is acted on by a single force, the gravitational attraction of the Earth. Using the Universal law of gravitation we get

$$\text{gravitational force towards the Earth} = \frac{GmM}{r^2}$$

where m is the mass of the satellite, M is the mass of the Earth and r is the distance of the satellite from the centre of the Earth. See Fig 9.13.

The centripetal acceleration of the satellite is also towards the Earth and has the value $r\omega^2$, where ω is the angular velocity of the satellite.

Apply Newton's second law:

$$\text{force} = \text{mass} \times \text{acceleration}$$

$$\frac{GmM}{r^2} = m \times r\omega^2$$

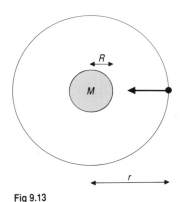

Fig 9.13

Since $\omega = 2\pi$ radians in one day:

$$\omega = \frac{2\pi \text{ rad}}{60 \times 60 \times 24 \text{ s}} = 7.27 \times 10^{-5} \text{ rad s}^{-1}$$

$$\frac{GmM}{r^2} = m \times r \times (7.27 \times 10^{-5})^2$$

$$r^3 = \frac{GM}{(7.27 \times 10^{-5})^2}$$

$$= \frac{6.67 \times 10^{-11} \times 5.98 \times 10^{24}}{(7.27 \times 10^{-5})^2}$$

$$= 7.55 \times 10^{22} \text{ m}^3$$

$$r = 4.23 \times 10^7 \text{ m}$$

Since R, the radius of the Earth, is 6.4×10^7 m, the height of the satellite above the surface = $(4.23 - 0.64) \times 10^7 = 3.6 \times 10^7$ m, or 36 000 km to 2 significant figures.

The speed of the satellite is therefore $r\omega =$

$$(4.23 \times 10^7 \text{ m}) \times (7.27 \times 10^{-5} \text{ rad s}^{-1}) = 3100 \text{ m s}^{-1}$$

The above example includes the equation showing how Newton's second law is applied to a satellite in a circular orbit, namely

$$\frac{GmM}{r^2} = m \times r\omega^2$$

To cancel out the mass of the satellite from this equation means that we are assuming that mass, treated from the point of view of an object's reluctance to accelerate, its inertia, and mass treated from the point of view of gravitational attraction are the same thing. The fact that results obtained from such equations agree in practice to a very high order of accuracy confirms the belief that inertial mass and gravitational mass are identical.

Kepler's Law for the period of rotation of a satellite

The equation

$$\frac{GmM}{r^2} = m \times r\omega^2$$

can be used in very many ways depending on the data given and required. In terms of T, the period of rotation, this gives

$$\frac{GmM}{r^2} = mr \times \left(\frac{2\pi}{T} \right)^2$$

since $\omega = \dfrac{2\pi}{T}$

Rearranging: $GMT^2 = 4\pi^2 r^3$

or $T^2 = \dfrac{4\pi^2}{GM} r^3$

or $T^2 \propto r^3$

Historically this relationship was discovered for the planets before the time of Newton. A Danish astronomer, Tycho Brahe, spent virtually every fine night for twenty years plotting the position of the planets using an astrolabe, an accurate three-dimensional protractor. Johannes Kepler analysed the results of Brahe and showed that

(period of planet)$^2 \propto$ (mean radius of planet's orbit)^3

Kepler died in 1630 twelve years before Newton was born. Newton's use of his laws of gravitation and motion to deduce Kepler's law of planetary motion gave strong experimental confirmation of the laws.

QUESTIONS

9.7 The Earth is at a distance of 1.50×10^{11} m from the Sun. Find the distance from the Sun of Venus, Mars and Neptune given that

1 year on Venus = 0.615 Earth years
1 year on Mars = 1.88 Earth years
1 year on Neptune = 165 Earth years

9.8 In 1978 a moon of Pluto was discovered. The radius of its orbit was estimated as 5×10^7 m and it rotated around Pluto in 4.8 days. Use this information and the value of G to find the mass of Pluto.

Energy of a satellite in orbit

Potential energy

A satellite of mass m in a circular orbit of radius r about a large mass M has a potential energy, E_P, given by

$$E_p = \frac{-GmM}{r}$$

as shown in Section 9.3. Fig 9.14 is a scale drawing for a satellite orbiting the Earth at a height of 400 km. At that height, although the satellite is effectively outside the Earth's atmosphere, the value of g is still

$$\frac{9.8 \text{ N kg}^{-1} \times (6.4 \times 10^6 \text{ m})^2}{(6.8 \times 10^6 \text{ m})^2} = 8.7 \text{ N kg}^{-1}$$

This has been found by direct use of the inverse square law of gravitation.

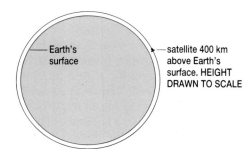

Earth's surface — satellite 400 km above Earth's surface. HEIGHT DRAWN TO SCALE

Fig 9.14 This diagram is drawn to scale. The satellite is in a virtual vacuum at its height of 400 km yet the gravitational pull of the Earth is still large.

Kinetic energy

Using force = mass × acceleration for the satellite gives

$$\frac{GMm}{r^2} = m \times \frac{v^2}{r}$$

so
$$mv^2 = \frac{GMm}{r}$$

and the kinetic energy $E_k = \frac{1}{2}mv^2 = \frac{1}{2}\frac{GMm}{r}$

This shows that the numerical value of the kinetic energy E_k when in a circular orbit is half the value of the potential energy. They have opposite signs however.

Total energy

The total energy $E = E_k + E_p$

$$= \frac{GMm}{2r} + \left(\frac{-GmM}{r}\right) = -\frac{GMm}{2r}$$

The total energy therefore has the same numerical value as the kinetic energy but the opposite sign. This means that if, as a result of atmospheric friction, the satellite loses energy, it will move to a position nearer to the Earth, r is therefore smaller, and the kinetic energy increases. It is similar to a ball rolling down a hill. The ball loses potential energy, some of this potential energy is changed into kinetic energy, but not all of it. As a result of friction there is a small increase in the temperature of the ball; its internal energy increases. In the case of the satellite half of its loss of potential energy is used against friction, and the other half is used to increase the kinetic energy of the satellite.

The velocity of a satellite in a circular orbit can be given in terms of the value of the gravitational field strength g, it is in. We then get

$$mg = \frac{GMm}{r^2} = m\,\frac{v^2}{r}$$

where v is the velocity of the satellite. Therefore

$$v^2 = gr$$
$$v = \sqrt{gr}$$

and for this case
$$v = \sqrt{(8.7\ \text{N kg}^{-1}) \times (6.8 \times 10^6\ \text{m})}$$
$$= 7700\ \text{m s}^{-1}$$

A satellite orbiting near to the Earth would have a velocity of 7900 m s^{-1} and would take 85 minutes to orbit the Earth. It is impossible to have a satellite going around the Earth faster than this unless it can keep its rocket motors going all the time to increase the force on it *towards* the Earth, see Fig 9.15. In practice this is impossible as there would be too great a need for fuel. With the existing pull of gravity it is never likely to be possible to reach Australia from the United Kingdom in less than 40 minutes!

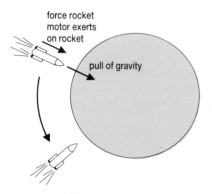

Fig 9.15 This shows the only way a rocket could go round the Earth in a shorter time than 85 min. It would have to keep its rocket motor on in a direction to exert a force on the rocket towards the centre of the Earth.

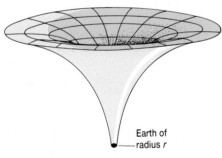

Fig 9.16 A sketch of the gravitational potential well of the earth.

Speed of escape

This term is often called the **escape velocity** but since only the magnitude of the velocity is needed and not its direction, it is more accurately called the **escape speed**. The speed of escape of a satellite is not what it appears to be. That is, it is not the speed of a satellite when it escapes from the Earth. Indeed, if it were possible to keep a rocket motor going for a long time, it would be possible to escape from the Earth at any speed. The escape speed of the Earth is 11.2 km s^{-1}. The time taken for astronauts going to the Moon, a distance of 380 000 km, was about 3 days, or 260 000 seconds. Their average speed was therefore only 1.5 km s^{-1}, well below the escape speed, and yet they escaped!

The term escape speed is in many ways unfortunate. To begin with, it is never possible to escape from the pull of the Earth's gravitational field. If you go far enough the pull of the Earth's gravity will certainly be very small but it is not zero. The term can best be considered in conjunction with the term **gravitational well**. This is illustrated in Fig 9.16. where the gravitational potential is plotted on a three-dimensional sketch and shows how the potential falls as the Earth is approached.

The potential goes on rising for ever as the distance from the Earth increases, but the slope gets less and less. The depth of the gravitational potential well is GM/r. Now consider a piece of matter, stationary relative to the Earth and a long way from it. It is pulled very gently by the Earth and accelerates towards the Earth a little. As it approaches the Earth it gets faster and faster and will eventually collide with the Earth. Its speed on collision *is* the escape speed, because if it bounced perfectly back off the Earth it would retrace its path and just stop when it was a great distance from the Earth. If only the journey outward from the Earth is considered it is possible to think of the escape speed as the speed needed at the Earth, if something is to be thrown off the Earth. The speed of an object when it is thrown upwards does, of course, fall as it rises. A rocket at the Earth, moving upwards at 11.2 km s^{-1}, will gradually slow down as it rises and will just have enough kinetic energy to reach the top of the gravitational well. A similar rocket travelling at only 10 km s^{-1} does not have enough kinetic energy to reach the top of the gravitational well, and so will eventually fall backwards. One with an upward speed of 12 km s^{-1} has more than enough kinetic energy and so will have some kinetic energy remaining when it has reached the top of the gravitational well.

The escape speed can be obtained by realising that the kinetic energy the mass must have at the Earth's surface needs to be equal to the gain in the potential energy in escaping from the gravitational well. Working with unit mass we get:

kinetic energy per unit mass	$=$	gain in potential in leaving gravitational well
$\frac{1}{2} \times 1\,\text{kg} \times v^2$	$=$	$\dfrac{GM}{r} \times 1\,\text{kg}$
v^2	$=$	$\dfrac{2GM}{r}$

and since

$\dfrac{GMm}{r^2}$	$=$	mg
$\dfrac{GM}{r}$	$=$	rg
v^2	$=$	$2rg$
v	$=$	$\sqrt{2rg}$

The escape speed v, is $\sqrt{2rg}$ and is therefore $\sqrt{2}$ times the speed needed for a circular orbit. The kinetic energy in orbit is therefore half the kinetic energy needed to escape.

QUESTIONS

9.9 What are the potential and kinetic energies of a satellite of mass 600 kg when the satellite is in a circular orbit around the Earth of radius 8.00×10^6 m?

9.10 What is the work done on a satellite of mass 600 kg in order to place it in a circular orbit around the Earth of radius 8.00×10^6 m?

9.11 As a result of air resistance a satellite of mass 105 kg loses energy slowly over a period of time and falls from an orbit of radius 7.5×10^6 m to one of radius 7.0×10^6 m. Find the changes in the potential and kinetic energies of the satellite during this time and hence find the total loss of energy. What are the initial and final values of the speed of the satellite?

9.5 WEIGHT

Weight has been referred to on many occasions so far in this book, particularly in Section 5.6. Unfortunately weight is a word which is often used to mean different things, and some of these uses are extremely misleading if not actually incorrect. Added to the problem of knowing exactly what is meant by the term weight is the problem of knowing the meaning of such terms as weightless, weightlessness, apparent weight and effective weight. Because of the confusion it is worthwhile re-stating some terms about which there should *not* be any confusion.

Mass

A measure of the inertia of a body, that is, its reluctance to undergo linear acceleration. It is measured in kilograms. If a force is applied to a body of unknown mass m, and causes it to have an acceleration a, and the same force when applied to the standard kilogram gives it an acceleration a_s, then

$$\frac{m}{1\,\text{kg}} = \frac{a_s}{a}$$

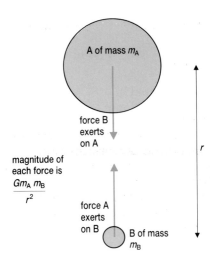

A of mass m_A

force B exerts on A

magnitude of each force is $\dfrac{Gm_A\,m_B}{r^2}$

force A exerts on B

B of mass m_B

r

Fig 9.17

For example, if m is given an acceleration three times the acceleration of the standard kilogram by the same force, then its mass is one-third of a kilogram.

The mass of an object does not vary from place to place. It is just as difficult to stop a moving hammer on the Moon as it is on the Earth, so it will be as effective in hammering a nail into a piece of wood on the Moon as it is on the Earth. Mass is a scalar quantity.

Gravitational attraction

This is a force and so will be measured in newtons. It is a force of mutual attraction between two masses, equal and opposite forces acting on each of them as shown in Fig 9.17. The magnitude of the force is given by

$$F = \frac{Gm_A m_B}{r^2}$$

Gravitational attraction is a vector.

Because there is no confusion about these terms it is preferable to use them whenever possible. Mass should always be the term used for a quantity measured in kilograms despite the fact that in everyday speech a bag of sugar is said to have a weight of one kilogram. Similarly when there is a force on an object due to the gravitational attraction of the Earth then it is more precise to mark it 'gravitational attraction of the Earth'. This has the advantage that it is easy to realise that the gravitational attraction of the Earth on an object has a fixed value near a point on the Earth's surface. The gravitational attraction does not depend on how the object is moving or accelerating.

Weight

 The weight of an object is the gravitational force exerted on the object.

Normally the only appreciable gravitational force on an object is that due to the Earth, but there may be circumstances in which other bodies are important. If someone is on the Moon then the gravitational force which the Moon exerts on him is the major contribution to his weight. The terms apparent weight and effective weight will not be used.

The gravitational pull the Earth exerts on a mass of 1 kg can be found using Newton's Universal law of gravitation:

$$F = \frac{GmM}{r^2} = \frac{6.67 \times 10^{-11} \times 1 \times 5.98 \times 10^{24}}{(6.37 \times 10^6)^2}$$

$$= 9.83 \text{ N}$$

This gives the gravitational field of the Earth, used earlier, as 9.83 N kg^{-1}.

The value varies a small amount from place to place on the Earth's surface as the Earth is a slightly flattened sphere and the distribution of matter within the Earth is not uniform. The northern hemisphere, for instance, has far more land area than the southern hemisphere. Accurate measurements of the gravitational pull are used to identify different densities of materials beneath the Earth's surface. These measurements are used in prospecting for oil, coal and other minerals.

The acceleration due to gravity (g)

If the force of gravitational attraction is the only force on a kilogram mass then the mass is said to be in free fall. This would be true if it were falling in a vacuum. It would then have an acceleration g towards the centre of the Earth given by

$$
\begin{array}{lcl}
\text{force} & = & \text{mass} \times \text{acceleration} \\
9.83\,\text{N} & = & 1\,\text{kg} \times g \\
g & = & 9.83\,\text{m s}^{-2}
\end{array}
$$

This value for the acceleration due to gravity is the acceleration of the mass towards the centre of the Earth. It is therefore not exactly the same as the acceleration of the mass towards the surface of the Earth. As the surface is going round in a circle once per day, the surface of the Earth has an acceleration at all points except at the North and South Poles. Vector diagrams must be used to relate the various accelerations to one another, Fig 9.18. It can be seen from these diagrams that if g is measured in the laboratory a value other than 9.83 m s^{-2} should be expected since the

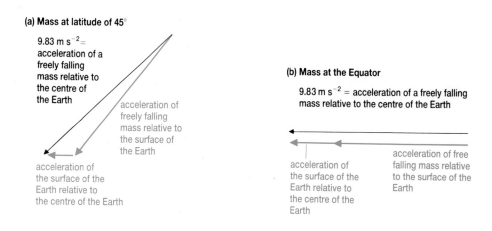

(a) Mass at latitude of 45°

9.83 m s^{-2} = acceleration of a freely falling mass relative to the centre of the Earth

acceleration of freely falling mass relative to the surface of the Earth

acceleration of the surface of the Earth relative to the centre of the Earth

(b) Mass at the Equator

9.83 m s^{-2} = acceleration of a freely falling mass relative to the centre of the Earth

acceleration of the surface of the Earth relative to the centre of the Earth

acceleration of free falling mass relative to the surface of the Earth

Fig 9.18 Vector diagrams showing that the acceleration of free fall towards the surface of the Earth is not the same as the acceleration of free fall towards the centre of the Earth.

measurement will be made relative to the surface of the Earth. Accurate values of g are now found by measurements taken directly on an object falling freely in a vacuum. They are needed because all force calibrations rely on using accurate balances and the relation between mass and weight depends on knowing an accurate value for g.

Example 4

(a) Find the measured acceleration of free fall at the equator, given that the gravitational field at the equator is 9.830 N kg^{-1}.

(b) Find the weight of a 100 kg mass at the equator and find also the reading when it is placed on a sensitive spring balance.
(Radius of the Earth = 6.37×10^6 m.)

(a) The acceleration of a point on the surface of the Earth at the equator

$$
\begin{aligned}
&= r\omega^2 \\
&= (6.37 \times 10^6\,\text{m}) \times \left(\frac{2\pi}{24 \times 60 \times 60}\ \text{rad s}^{-1} \right)^2 \\
&= 0.034\ \text{m s}^{-2}
\end{aligned}
$$

since the Earth rotates 2π radians in one day.

Using the vector diagram, Fig 9.18(b), we have

measured acceleration of mass relative to surface of Earth	+	acceleration of surface of Earth relative to centre of Earth	=	acceleration of mass relative to centre of Earth

measured acceleration of free fall	+	0.034	=	9.830

measured acceleration of free fall = $9.830 - 0.034 = 9.796$ m s^{-2}

(b) Weight of 100 kg mass at the equator = $100 \text{ kg} \times 9.830 \dfrac{\text{N}}{\text{kg}} = 983.0$ N

A force diagram for the mass when suspended by the spring balance is given in Fig 9.19. Applying Newton's law to the mass, which itself has an acceleration in rotating once per day gives

force in newtons	=	mass in kilograms	×	acceleration in m s^{-2}
$(983 - B)$	=	100	×	0.034
B	=	$983.0 - 3.4$	=	979.6 N

The reading on the balance is therefore 979.6 N, as it supplies this force on the mass. It can be seen from this example therefore that the reading given by the spring balance is not quite equal to the weight of the mass. The difference between the two is small however and can, in most cases, be neglected.

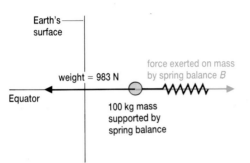

Earth's surface

weight = 983 N

force exerted on mass by spring balance B

Equator

100 kg mass supported by spring balance

Fig 9.19 A mass of 100 kg supported on a spring balance at the equator.

Weightlessness

This term should be more readily understood after working through Example 4. In that example it was shown that the reading on a spring balance holding a mass is not quite the same as the weight of the mass. The difference between the two forces causes the centripetal acceleration of the mass. There will always be a difference between the reading on the spring balance and the weight of the object if the spring balance is itself accelerating. In the unlikely event of someone standing on bathroom scales while going down in a lift, they would notice that the reading on the scales would be low while accelerating downwards, normal when travelling downwards with constant velocity, and high when stopping. The difference between the weight and the reading on the balance is greater when the acceleration is greater. The reading would become zero if the rope broke!

For a satellite in orbit around the Earth the acceleration of the satellite is equal to the acceleration of free fall. An object placed on a spring balance therefore falls freely with the spring balance and the spring balance exerts zero force on it. It is the fact that a spring balance reads zero that gives rise to the idea of weightlessness. In fact, as was shown above, there is normally a difference between the balance reading and the weight; in a satellite it is much more dramatic.

9.12 The asteroid Ceres has a radius of 550 km and a mass of 7.0×10^{20} kg. What would be the Cerian weight of a person with a mass of 90 kg and what would be the acceleration due to gravity on the surface of Ceres?

9.13 The mass of Jupiter is 1.90×10^{27} kg and it has a radius of 7.14×10^{7} m. Assuming that the planet is spherical, find the gravitational field at the surface of Jupiter and hence the weight of a 15 kg mass on its surface.

Jupiter rotates on its axis with a period of 9 h 50 min (35 400 s).

(a) What is the centripetal acceleration of an object at the equator?

(b) What force is necessary to cause this acceleration for a 15 kg mass?

(c) Draw a free body diagram for the 15 kg mass when it is (i) at the pole, (ii) at the equator.

(d) Find the support force which the planet exerts on the mass at both the pole and the equator.

Fig 9.20 A camera pointing at the pole star clearly shows the rotation of the earth.

Satellite motion

Fig 9.9 needs to be used with this exercise.

The grey lines on the figure show the gravitational field near the Earth due to the Earth and the Moon. (The gravitational field due to the Sun in this region is not negligible but its value does not change appreciably over the distance covered by the figure and so it has not been considered.)

The black lines show the gravitational potential at all points. These can best be thought of as contour lines.

The blue lines on the figure show the paths of various satellites. You are asked to ignore the movement of the Moon when answering the questions. It clearly does move and this adds an additional complication to any of these problems in practice.

Answer the following questions assuming the mass of any satellite remains constant:

(a) A is rotating at a constant speed in a circle of radius 30 000 km and B is similarly rotating in a circle of radius 120 000 km; find the ratio of the period of B to the period of A.

(b) Find the kinetic energy per unit mass of A and hence its speed.

(c) Find the gain in potential of C as it goes to the Moon.

(d) C has a mass of 3000 kg. How much potential energy does it gain in reaching the Moon?

(e) If C started out with kinetic energy of 2.1×10^{11} J, how much kinetic energy has it left when it reaches the Moon? What is its speed at the Moon?

(f) D has a mass of 80 kg. Find the change in the potential energy of D between its point of closest approach to the Earth D_1 and its point of furthest distance from the Earth D_2.

(g) D is travelling with a speed of 7000 m s^{-1} when at point D_1. What is its speed at point D_2?

(h) E is off on a journey to Mars. It starts with a speed of 13 km s^{-1}. Plot graphs on the same horizontal axis to show how
   (i)   its potential energy
   (ii)   its kinetic energy
   (iii) its velocity
change over the first half a million kilometres.

(i) Sketch the shape of the field and the lines of equi-potential on a larger scale for the area near the Moon.

SUMMARY

- Newton's law of universal gravitation states that the gravitational force F, between two point masses m_1 and m_2 separated by a distance r is given by

$$F = \frac{Gm_1m_2}{r^2}$$

where G, the gravitation constant, can be found experimentally to have the value 6.67×10^{-11} N m^2 kg^{-2}.

- The gravitational field strength at a point is the force acting on unit mass at that point. It is a vector.

- The gravitational field strength at the surface of the Earth g is 9.83 N kg^{-1}.

- Using the law of universal gravitation gives $g = GM/r^2$.

- Gravitational potential at a point in a gravitational field is the work done in moving unit mass from infinity to the point. It is a scalar and its value is given by

$$V = -\frac{GM}{r}$$

- The gravitational field is equal to minus the potential gradient. The minus sign indicates that the potential falls when moving in the direction of the field.

- The escape speed for any satellite is given by

$$v = \sqrt{\frac{2GM}{r}} = \sqrt{2gr}$$

where M is the mass of the planet, r its radius and g the gravitational field at the surface of the planet.

- The weight of a body on the Earth is the gravitational force exerted on it by the Earth. The weight of an object varies with its distance from the Earth but it is not altered by any acceleration the body may have.

EXAMINATION QUESTIONS

9.14 **(a)** A body of mass *m* travels at constant speed in a circular path of radius *r*. It takes time *T* to complete one revolution.
  (i) Write down expressions in term of *m*, *r* and *T* for the speed, the acceleration, the angular velocity, the kinetic energy and the momentum of the body. (6)
  (ii) Which of these quantities change during a revolution and which remain constant? (3)
  (iii) On a sketch show the directions of the acceleration and the momentum at a particular instant. (2)
  (b) What is the acceleration of the Moon? The Moon may be considered to travel about the Earth in a circular orbit of radius 3.82×10^8 m and period 2.36×10^6 s. Why does the Moon not fall and hit the Earth?
  By considering the acceleration of free fall at the Earth's surface, show how the magnitude of the Moon's acceleration is consistent with Newton's inverse square law of gravitation. (Radius of the Earth = 6.36×10^6 m.)

  (UCLES 1988)

9.15 The gravitational field strength on an astronaut travelling in a space vehicle in low Earth orbit is only slightly less than if he were standing on Earth.
  (a) Explain why the force is only slightly less.
  (b) Explain why, when travelling in the space vehicle, the astronaut appears to be 'weightless'. (4)

  (ULSEB 1988)

9.16 What do you understand by the term *gravitational field*; define *gravitational field strength*.
  Show that the radius *R* of a satellite's circular orbit about a planet of mass *M* is related to its period as follows:

  $$R^3 = \frac{GM}{4\pi^2}\ T^2$$

  where *G* is the universal gravitational constant. (8)
  The diagram shows two graphs of R^3 against T^2; one is for the moons of Jupiter and the other is for the moons of Saturn. *R* is the mean distance of a moon from a planet's centre and *T* is its period. The orbits are assumed to be circular.

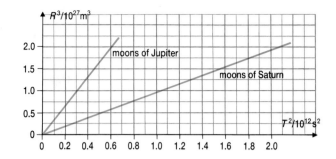

  The mass of Jupiter is 1.90×10^{27} kg.
  (i) Why are the lines straight?
  (ii) Find a value for the mass of Saturn.
  (iii) Find a value for the universal gravitational constant, *G*. (8)

  (ULSEB 1987)

9.17 A mass of 2 kg is at a point P, a height 3 m above the surface of the Earth. Taking the gravitational potential at the surface of the Earth to be zero, state
  (a) the gravitational field strength at P,
  (b) the gravitational potential at P,
  (c) the gravitational force acting on the mass,
  (d) the gravitational potential energy of the mass. (6)

  (UCLES 1988)

9.18 **(a)** Define *gravitational potential* at a point.
  (b) As a spacecraft falls towards the Earth, it loses gravitational potential energy. What becomes of the lost potential energy
    (i) when the spacecraft is falling freely towards the Earth well away from the Earth's atmosphere,
    (ii) when the spacecraft is falling through the Earth's atmosphere at a constant speed?
  (c) (i) Calculate the gravitational potential difference between a point on the Earth's surface and a point 1600 km above the Earth's surface.
    (ii) Calculate the minimum energy required to project a spacecraft of mass 2.0×10^6 kg from the surface of the Earth so that it escapes completely from the influence of the Earth's gravitational field.
    (Radius of Earth = 6400 km; universal gravitational constant = 6.7×10^{-11} N m² kg⁻²; mass of Earth = 6.0×10^{24} kg)

  (AEB 1987)

9.19 Find the speed of a satellite which orbits the Moon near the Moon's surface. What is the kinetic energy per unit mass of the satellite? (5) Mass of Moon = 7.4×10^{22} kg; radius of Moon = 1.74×10^6 m.

(UCLES 1988)

9.20 (a) Explain what is meant by *gravitational field strength*. In what unit is it measured? Starting with Newton's law of gravitation, derive an expression for g, the acceleration of free fall on the surface of the Earth, stating clearly the meaning of each symbol used. (Assume that the Earth may be considered as a point mass located at its centre.) (5)

(b) g may be found by measuring the acceleration of a freely falling body. Outline how you would measure g in this way, indicating the measurements needed and how you would calculate a value for g from them. (6)

(c) At one point on the line between the Earth and the Moon, the gravitational field caused by the two bodies is zero. Briefly explain why this is so.
If this point is 4.0×10^4 km from the Moon, calculate the ratio of the mass of the Moon to the mass of the Earth. (Distance from Earth to Moon = 4.0×10^5 km.) (7)

(ULSEB 1983)

9.21 In order to send a space vehicle to the Moon, it is usual first to place the vehicle in a circular 'parking' orbit near to the Earth and then to increase its speed in an appropriate direction so that it escapes from the influence of the Earth and becomes a satellite of the Moon.

(a) Determine the speed v of the vehicle when it is in parking orbit close to the surface of the Earth.
($g = 10$ m s^{-2}; radius of orbit = 6.4×10^6 m) (2)

(b) Describe how, if at all, the rotation of the Earth affects the energy required to place the vehicle in the parking orbit. (3)

(c) In the parking orbit, friction with the outer layers of the Earth's atmosphere causes a gradual reduction in total energy, but it is observed that the vehicle actually moves faster. Explain this paradox. (4)

(UCLES 1986)

9.22 A space station is in a circular orbit around the Earth. An object is released at rest inside the station. To an observer inside the station it appears to be weightless. Explain the meaning of the word *weightless*. Describe the motion of the object and the space station as viewed by an observer on Earth. (4)
Show how it is possible to determine the mass of the Earth from measurements of the gravitational constant G, the period T and the radius r of the space station's orbit around the Earth. Suggest a direct method of measuring the height of the space station above the Earth's surface. (6)
Assuming the Earth to be a perfect sphere, explain why an object would weigh less at sea level at the Equator than at either Pole. Calculate the surface speed of rotation that the Earth would require for an object at the Equator to be weightless. The radius of the Earth is 6.4×10^6 m. (6)
Consider a planet which has the same average density as the Earth but twice the radius. In terms of the same quantities on the Earth, find

(a) the value of the gravitational field strength at its surface, and

(b) the speed of a satellite just skimming the surface. (4)

(O & C 1987)

9.23 There is no agreement in textbooks on the definition of *weight*. Three definitions are commonly found, namely

(a) weight is the force of gravity on an object,

(b) weight is the mass of the object multiplied by the value of the acceleration of free fall with respect to the surface of the Earth,

(c) weight is the force a body exerts on its support when it is at rest on the Earth's surface.

Explain the significance of the differences between these definitions and explain how they result in three different values for the weight of a 1 kg mass at the Equator. (8)
Calculate the acceleration with respect to the centre of the Earth of an object on the Equator and stationary with respect to the surface of the Earth. Discuss how the acceleration of free fall with respect to the Earth's surface varies with latitude. You may consider the Earth to be a uniform sphere of radius 6.38×10^6 m. (6)
There is probably a planet similar to the Earth somewhere in the Universe whose day is only three hours long. Describe some of the physical differences which might arise as a result of this much increased angular velocity. (6)

(UCLES 1985)

Theme 3

OSCILLATIONS

Oscillations are of considerably more importance than is generally recognised. The oscillation of a pendulum of a clock is apparent; the oscillation of atoms within a solid is hidden. The oscillation associated with wave motion can be appreciated when a boat in a harbour gently rises and falls as the waves travel past it, but the oscillation associated with light waves cannot be perceived directly by our senses. There is an association with oscillation for all waves whether they are sound waves or electromagnetic waves such as light or radio waves. Maxwell's theory of electromagnetic radiation in 1864 was one of the outstanding achievements of the Nineteenth Century. It led immediately to the prediction of radio waves, and these were first produced by Hertz in 1887. The use of oscillations and waves in communication was not new. All communication by sight and by hearing makes use of wave oscillation. Radio waves greatly extended the range and speed of communication.

Chapter 10

FREE AND FORCED OSCILLATION

LEARNING OBJECTIVES

At the end of this chapter you should be able to:

1. define and use the terms frequency, period, displacement and amplitude;

2. describe the conditions necessary for simple harmonic motion and calculate its period;

3. calculate the energy of a body in simple harmonic motion;

4. distinguish between free, damped and forced oscillations;

5. describe examples of resonance.

10.1 PERIODIC MOTION

Fig 10.1 A clock from 1810 clearly showing the controlling pendulum.

In many practical examples of motion the moving object starts at one place and finishes somewhere else. Many such examples were given in Chapters 4 and 5. In this chapter the type of motion which will be considered is that in which there is no overall displacement of the object and the moving object finishes up where it started. One example of this type of motion would be the movement of the string of a guitar. The string is plucked, it moves backwards and forwards for a time, and then the motion dies out leaving it where it was at the beginning. Of course the guitar may have moved bodily during this time, the guitarist might be on a moving train, but here that will not be our concern. We need to examine the motion of the vibrating string. **Oscillation** is the word used to indicate that this type of movement is taking place. The motion is said to be **periodic motion**. There are many types of periodic motion. Some of them are easy to analyse while others can be very complex. Fig 10.2 shows certain familiar situations and since on the printed page it is impossible to show movement, graphs have been drawn to show how the displacement of the object might vary with time. The displacement can be a linear displacement or an angular displacement. For all of the examples given there is a clearly defined time interval which is the length of time for one complete oscillation. This time interval T is called the **period** of the oscillation. Note that one period is a complete cycle back to the same starting point. For example a clock pendulum which takes 1 second to swing from one side to the other has a period of 2 seconds since it will take another second to swing back to its starting point. Confusingly, such a pendulum is sometimes called a second pendulum.

 The **frequency** f of any periodic motion is the number of oscillations per unit time.

The time for one cycle is therefore $1/f$.

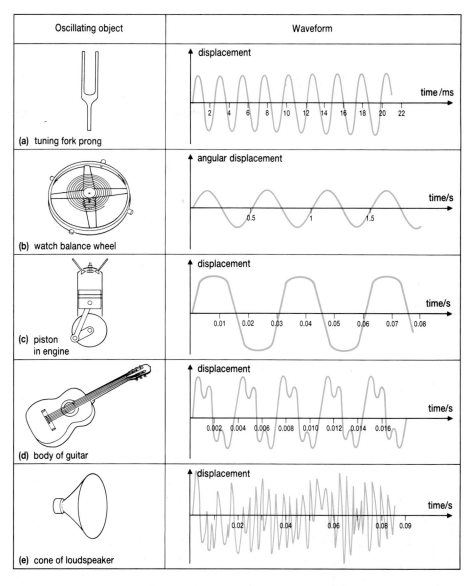

Oscillating object	Waveform

(a) tuning fork prong

(b) watch balance wheel

(c) piston in engine

(d) body of guitar

(e) cone of loudspeaker

Fig 10.2 Common examples of oscillatory movement.

This time is the period T of the oscillation so we have in general

$$T = \frac{1}{f}$$

◆ Period $= \dfrac{1}{\text{frequency}}$

Using SI units the frequency is measured in hertz (Hz). 1 Hz is 1 cycle per second.

The term **displacement** has already been used. It is the distance the object has moved from its rest position in a stated direction. It is a vector. **Amplitude** is the magnitude of the maximum value of the displacement. It is a scalar. Be careful not to introduce an incorrect factor of 2 when dealing with amplitude. For example if a guitar string moves a maximum of 4 mm from its rest position in one direction and then 4 mm from its rest position in the other direction, it will move 8 mm between stops. Its amplitude is nevertheless 4 mm and not 8 mm. These terms are illustrated in Fig 10.3 for a pure oscillation. This oscillation is one which has the same shape as a sine

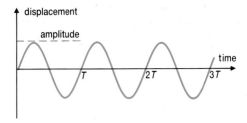

Fig 10.3   A pure sine wave.

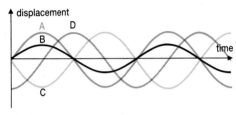

Fig 10.4

wave and is called a **sinusoidal oscillation**. The vibration of the prongs of a tuning fork has this shape, when taken over a short time interval. If a longer time interval is taken then it is seen that the amplitude of the tuning fork gradually decreases.

A problem arises in obtaining an equation giving the displacement as a function of time. The equation must have a sine function in it to give the sinusoidal shape but it is only possible to take the sine of an angle. The problem is how can an angle be associated with a linear to and fro motion? The problem can be overcome by realising that a sine wave repeats itself every 2π radians. We need therefore to associate one cycle of an oscillation with an angle of 2π radians. This also gives rise to a way of relating the pattern of movement of two separate oscillations. If the two oscillations are in step with one another they are said to be in **phase** with one another. Oscillations are said to be in antiphase if they are always moving in opposite directions, for instance if one moves up when the other moves down. This is a phase difference of half a cycle so the oscillations are said to be π radians out of phase with one another. Fig 10.4 shows an oscillation A, which is in phase with oscillation B, in antiphase to oscillation C, and out of phase with oscillation D by a quarter of a cycle, $\pi/2$ rad.

 The equation required for a sinusoidal oscillation will be of the form
$$x = x_0 \sin \omega t$$

where x is the displacement, x_0 is the amplitude, t is the time and ω is a constant. ω is chosen so that when t equals the period T of the oscillation:

$$\omega t = \omega T = 2\pi$$

i.e. $\omega = \dfrac{2\pi}{T} = 2\pi f$

and hence $x = x_0 \sin 2\pi f t$

Angular frequency is a name sometimes given to ω, but in less formal discussion concerning a pure sinusoidal motion it is not given a name at all and is just referred to as 'omega'. The term $2\pi f t$ has no dimensions but it has a unit, the radian, as does an angle and so it is now possible to take the sine of this expression and to plot the required graph. If you are not familiar with this particular piece of mathematics you really have no alternative than to work at the understanding of the measurement of angles in radians and their sines and cosines. The problem will not go away if you ignore it! It will recur in dealing with waves and also with alternating currents in electricity. It will help if you look at Example 1 to see how a sine wave graph can be plotted and then do the following questions. There are some more similar questions on the same topic in Question 19.1.

Example 1
Plot a graph showing a sinusoidal oscillation with an amplitude of 0.0006 m and a frequency of 2500 Hz. This could be the oscillation pattern of a loudspeaker coil when used to produce a high pitched note. What is the value of ω for this oscillation?

$$\omega = 2\pi f = 2\pi \times 2500 = 15\ 700\ \text{s}^{-1}$$

The graph to be plotted is $x = x_0 \sin 2\pi f t$ where $x_0 = 0.0006$ m and $f = 2500$ Hz. The numerical value of x is given by $x = 0.0006 \sin(2\pi \times 2500t)$ and care needs to be taken when plotting graphs with such awkward numbers.

Tabulated values are shown in Table 10.1. The first four lines of values are put in because this is how graphs are often started. Values for t of 1, 2, 3, 4 etc. are useless. They simply produce 0 (or error messages) on a calculator. This is because these times are long enough for thousands of cycles of the oscillation to have taken place and all the readings which have

FREE AND FORCED OSCILLATION

been calculated are after a whole number of oscillations. The later lines are the sensible ones where short time intervals of under one period are used. It is convenient to start with the time for one cycle and then to work with convenient fractions of a cycle.

Table 10.1

t	$2\pi \times 2000t$	$\sin(2\pi \times 2000t)$	x
0	0	0	0
1	$5\,000\pi$	0	0
2	$10\,000\pi$	0	0
3	$15\,000\pi$	0	0
0.000 4	2π	0	0
0.000 1	$\pi/2$	1	0.0006
0.000 05	$\pi/4$	0.707	0.000424
0.000 15	$3\pi/4$	0.707	0.000424
0.000 2	π	0	0
0.000 25	$5\pi/4$	−0.707	−0.000424
0.000 3	$3\pi/2$	−1	−0.0006
0.000 35	$7\pi/4$	−0.707	−0.000424

Once one cycle of the oscillation has been drawn it may be repeated as often as is required, see Fig 10.5.

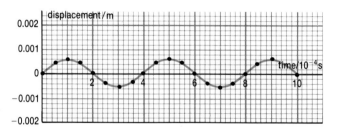

Fig 10.5

QUESTIONS	**10.1** Plot graphs to show the following sinusoidal oscillations

10.1 Plot graphs to show the following sinusoidal oscillations
(a) $x = x_0 \sin 2\pi ft$, where $x_0 = 3.0$ m and $f = 2.0$ Hz
(b) $x = x_0 \sin 2\pi ft$, where $x_0 = 0.2$ m and $f = 50$ Hz
(c) $x = x_0 \sin 2\pi ft$, where $x_0 = 0.000\,04$ m and $f = 5000$ Hz
(d) $x = x_0 \cos 2\pi ft$, where $x_0 = 0.2$ m and $f = 50$ Hz

10.2 Draw four more varied waveforms, similar to those drawn in Fig 10.1, each showing a possible pattern of oscillation.

10.2 SIMPLE HARMONIC MOTION

Fig 10.6 A mass oscillating on a long spring undergoes nearly simple harmonic motion.

A person standing on a bench with a mass attached to the end of a long spring can make the mass move with a sinusoidal oscillation, Fig 10.6. If the amplitude of the oscillation is 1.20 metres and the period of the oscillation is 3 seconds, then the frequency of the oscillation is $\frac{1}{3}$ Hz. The equation giving the value of the displacement of the mass at any time t is

$$x = x_0 \sin 2\pi ft = 1.20 \sin\left(\frac{2\pi t}{3}\right)$$

This graph is plotted in Fig 10.7(a). Since this is a displacement–time graph, the gradient of the graph at any time gives the velocity of the mass at that time. The tangent to the graph is plotted when $t = 6$ s and the gradient is found to be $1.3/0.5 = 2.6$ m s^{-1}. This is the maximum value of the velocity as the graph is at its steepest gradient where it crosses the time axis. Where the graph is at its maximum or minimum the gradient is zero and so the velocity is zero. By making a few more calculations, and by realising that the velocity is frequently negative, it is possible to plot another graph, this time one of velocity against time. This is done in Fig 10.7(b). This process can be repeated. The gradient of the velocity–time graph is the acceleration and this can also be plotted against time. This is

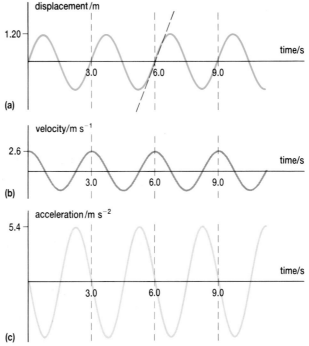

Fig 10.7 These three graphs show how the displacement, the velocity and the acceleration vary with time for the mass on the spring shown in Fig 10.6.

found to have a steepest gradient of 5.4 and hence the acceleration of the mass has a maximum value of 5.4 m s^{-2}. The acceleration–time graph is plotted in Fig 10.7(c). A fact crucial to pure sinusoidal oscillation can be seen from a comparison of these graphs. The acceleration–time graph is the same shape as the displacement–time graph but is of opposite sign. Put mathematically this can be written:

$$\text{acceleration} \propto -\text{displacement}$$
$$a \propto -x$$
$$a = -\omega^2 x$$

where ω^2 is a constant. The constant is written in this way because any number squared has a positive value and it is essential that the acceleration and the displacement themselves have opposite signs. (As you may imagine, the ω factor is the one which was introduced earlier.)

Motion of this pattern is called simple harmonic motion, or SHM for short. **Simple harmonic motion** is defined as motion taking place in which the acceleration of an object is proportional to the displacement of the object from a fixed point and in the opposite direction to the displacement.

With simple harmonic motion defined in this way the equation

$$a = \frac{d^2x}{dt^2} = -\omega^2 x$$

is the starting point for a study of SHM. This equation is called a differential equation because differentiating one factor in the equation x twice over gives another term in the equation a. A rigorous solution to this equation will not be given here but a check will be made on one of the solutions to the basic equation.

FREE AND FORCED OSCILLATION

Assume that the value of x given by

$$x = x_0 \sin \omega t$$

is a solution to the equation, then

$$v = \frac{dx}{dt} = x_0 \, \omega \cos \omega t$$

and

$$a = \frac{d^2x}{dt^2} = -x_0 \, \omega^2 \sin \omega t$$

but since

$$x = x_0 \sin \omega t$$

it follows that

$$a = -\omega^2 x$$

showing that our solution is valid. There are other solutions to this equation. If your calculus is good enough you will be able to show that

$$x = x_0 \cos \omega t$$
$$x = x_0 \cos (\omega t + \theta)$$

and

$$x = x_0 \cos \omega t + x_1 \sin \omega t$$

are also solutions to the basic equation

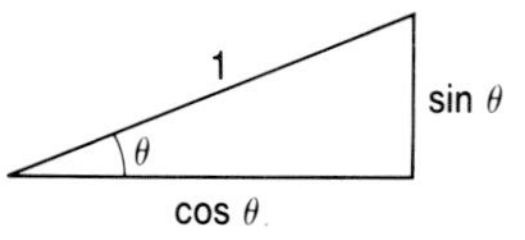

So far the displacement and the velocity of the body undergoing simple harmonic motion have been found in terms of the time. Sometimes the velocity is required in terms of the displacement. This can be done by the use of the Pythagoras equation. If a right-angled triangle with a hypotenuse of 1 is drawn, Fig 10.8, then the opposite side is $\sin \theta$ and the adjacent side is $\cos \theta$. Pythagoras therefore gives, for any angle:

$$\sin^2 \theta + \cos^2 \theta = 1$$

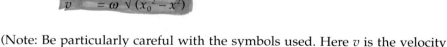

Fig 10.8

So for the angle ωt

$$\sin^2 \omega t + \cos^2 \omega t = 1$$

and

$$\left(\frac{x}{x_0}\right)^2 + \left(\frac{v}{x_0 \, \omega}\right)^2 = 1 = \frac{x^2 \, \omega^2 + v^2}{x_0^2 \, \omega^2}$$

so

$$x_0^2 \omega^2 = x^2 \, \omega^2 + v^2$$
$$v^2 = \omega^2 \, (x_0^2 - x^2)$$
$$v = \omega \sqrt{(x_0^2 - x^2)}$$

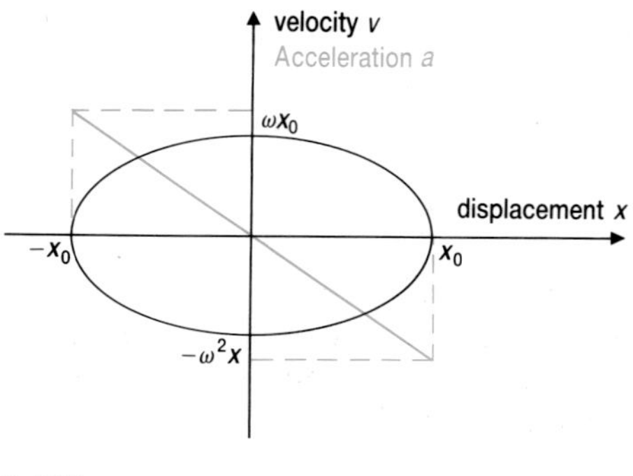

(Note: Be particularly careful with the symbols used. Here v is the velocity and this can be written $\dot{x}$; a is the acceleration and this can be written $\ddot{x}$. In some books a is used for the amplitude and the solution to the basic equation becomes $x = a \sin \omega t$.)

A graph of velocity against displacement is plotted in Fig 10.9. Note that the displacement has a maximum value of x_0 and that for any value of the displacement the velocity can be either positive or negative. On the same horizontal axis is also plotted a graph of acceleration against displacement showing the basic fact about SHM that the acceleration is proportional to the displacement but in the opposite direction. A table of the equations it is useful to be able to quote is given in the summary of this chapter.

Fig 10.9

Fig 10.10 The oscillation of the sea causing tides is something which controls the lives of sailors. In Pembrokeshire, where this photograph was taken, the tide can be 10 m from high to low water – an amplitude of 5 m.

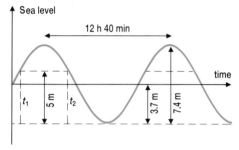

Fig 10.11 The graph shows how the sea level varies with time at Immingham. The zero of the sea level axis is taken at the mid-tide level.

Example 2

At Liverpool the tide is nearly sinusoidal and has a period of 12 hours 40 minutes. On a day when the height of high tide above low tide is 7.4 m find:

(a) The length of time within a cycle during which the water is more than 5 metres above low tide. A ship's captain always knows the depth of water his ship requires and he needs to be able to do this type of calculation from tide charts.

(b) The rate at which the tide is rising or falling when the water is 5 metres above low tide.

Fig 10.11 is a graph of sea level plotted against time for the tide. Sea level is taken from the mean sea level, so low tide will be a sea level of −3.7 m and high tide will be a sea level of +3.7 m. The period of the cycle is 760 min, or 45 600 s. Since

$$T = \frac{2\pi}{\omega}$$

$$\omega = \frac{2\pi}{45\ 600\ \text{s}}$$

This gives the equation of the displacement as

$$x = x_0 \sin \omega t$$

where
$$\omega = \frac{2\pi}{45\ 600\ \text{s}} = 0.000\ 137\ 8\ \text{s}^{-1}$$

and
$$x_0 = 3.7\ \text{m}$$

The two times t_1 and t_2 are times when the tide is at a height of 1.3 m above mean sea level, so

$$1.3\ \text{m} = 3.7\ \text{m} \sin \omega t_1$$

$$\omega t_1 = \sin^{-1}\left(\frac{1.3}{3.7}\right) = 0.359\ \text{rad}$$

$$t_1 = \frac{0.359}{0.000\ 137\ 8} = 2610\ \text{s}$$

Using the symmetrical nature of the graph gives t_2 as 2610 s before the time of half a period:

$$t_2 = 22\ 800 - 2610$$
$$= 20\ 200\ \text{s}$$

The time the sea level is more than 5 m above low tide

$$= t_2 - t_1$$
$$= 17\ 600\ \text{s}$$
$$= 4\ \text{h}\ 53\ \text{min}$$

The rate of rise of the tide when $t = t_1$ is the velocity of the oscillation and is given by

$$v = \omega \sqrt{(x_0^2 - x^2)}$$
$$\omega = 0.000\ 137\ 8 \qquad x_0 = 3.7\ \text{m} \quad \text{and}\ x = 1.3\ \text{m}$$
$$v = 0.000\ 137\ 8 \sqrt{3.7^2 - 1.3^2}$$
$$= 0.000\ 137\ 8 \times \sqrt{12.0}$$
$$= 0.000\ 48\ \text{m s}^{-1}$$
$$= 0.029\ \text{m min}^{-1} = 29\ \text{mm min}^{-1}$$

FREE AND FORCED OSCILLATION

10.3 Find the maximum values of the velocity and the acceleration of the system undergoing sinusoidal motion as given in Example 1. These values seem large but they are typical values for the behaviour of a loudspeaker. The very rapid oscillations need large forces to be applied to the moving coil within the loudspeaker.

10.4 A neap tide at Immingham, on the east coast, is one of the most accurately sinusoidal tides in the United Kingdom. On a day when low water occurred at midnight, the height of the sea above the low water mark at subsequent times had the values given in Table 10.2.

Table 10.2 Height of sea above low water at different times of day at Immingham.

Time of day /hours	Height of sea above low water level/m
00.00	0.00
01.00	0.32
02.00	0.96
03.00	1.76
04.00	2.50
05.00	3.01
06.00	3.20
07.00	2.98
08.00	2.46
09.00	1.79
10.00	0.93
11.00	0.32
12.00	0.00

Plot a graph to show the sinusoidal variation of water depth with time and answer the following questions about the oscillation.
(a) For how long is the water more than 2.0 m above low water level?
(b) For what fraction of the total time is the water lower than 1.0 m above low water level?
(c) What is the rate at which the sea level is rising at 04.00 hours?
(d) A rule of thumb used by navigators is called the rule of twelfths. It says that in the first hour after low water the sea level rises by one-twelfth of the tide height, in the second hour it rises by two-twelfths, in the third hour by three-twelfths, in the fourth hour by three-twelfths, in the fifth hour by two-twelfths, and in the last hour before high tide by one-twelfth, i.e. 1 2 3, 3 2 1. What error in tide rise is made at each time by using this rule at Immingham?
(e) What is the amplitude of the tidal wave?
(f) What is the period of the wave?
(g) What mathematical equation gives the height of water above the low water level, assuming that the rise and fall is sinusoidal?

In Example 2 and Questions 10.3 and 10.4 it was assumed or stated that the oscillation under consideration was simple harmonic motion. Once this is known it is not difficult to find the value of ω from the period or frequency and hence find anything required. In the case of many oscillations however it needs to be proved that simple harmonic motion is taking place and to find ω from the proof. In the following example this is done for a mass oscillating vertically on a spring.

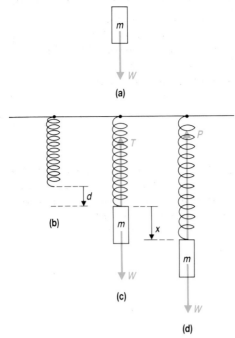

(a)

(b)

(c)

(d)

Fig 10.12

Example 3

A mass m is oscillating vertically on a spring whose spring constant is k. (The spring constant is the force required for unit extension, $F = kx$, and using SI units will have the unit N m^{-1}.) Show that the mass undergoes simple harmonic motion and find the period of the oscillation. The mass of the spring may be neglected in comparison with the mass m.

Fig 10.12 shows several different situations relevant to this problem. (a) shows the mass in free fall with only its weight acting on it. Under these circumstances

$$\text{force} = \text{mass} \times \text{acceleration}$$
$$W = mg$$

(b) shows the unstretched spring
(c) shows the mass in equilibrium when the spring is stretched a distance d.
   Under these circumstances

$$T = kd \text{ and } T = W$$
so $$kd = mg$$

(d) shows the mass in its displaced position so that the total spring extension is $d+x$ and the mass is not in equilibrium:

$$P = k(d + x)$$

It is necessary in this type of problem to be careful to indicate which direction is being considered as positive. Here consider downwards as being positive since x is measured downwards and apply Newton's second law.

Resultant force downwards	=	mass × acceleration downwards
$W - P$	=	$m \times a$
$mg - k(d + x)$	=	ma
$mg - kd - kx$	=	ma

but since

$$kd = mg$$
$$mg - mg - kx = ma$$
$$a = -\frac{kx}{m}$$

This shows that the acceleration is proportional to the displacement and in the opposite direction to the displacement. The motion is therefore simple harmonic motion and

$$\omega^2 = k/m$$

$$\omega = \sqrt{\frac{k}{m}}$$

This means that $$T = \frac{2\pi}{\omega} = \frac{2\pi}{\sqrt{\dfrac{k}{m}}}$$

$$T = 2\pi \sqrt{\frac{m}{k}}$$

It also means that an equation for displacement is

$$x = x_0 \sin \left(\sqrt{\frac{k}{m}} \, t \right)$$

and for velocity is

$$v = x_0 \sqrt{\frac{k}{m}} \cos \left(\sqrt{\frac{k}{m}} \right) t$$

FREE AND FORCED OSCILLATION

10.5 A light spring of unstretched length 0.060 m is extended to a length of 0.075 m when a mass of 200 g is placed on it. It is now given a small displacement so that the mass starts to oscillate. Find the period of the oscillation. Which of the three pieces of information given is not needed? Explain in physical terms why this piece of information is not required.

The simple pendulum

The word simple here is used to distinguish a pendulum, in which a point mass oscillates through a small angle on the end of a piece of string, from a compound pendulum in which mass is distributed at different distances from the axis of rotation. Fig 10.13 shows the difference and you should

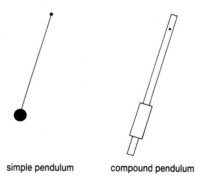

simple pendulum   compound pendulum

Fig 10.13 A simple pendulum is one in which all the mass is at a point some distance from the point of support. In a compound pendulum the mass is distributed at different distances from the point of support.

appreciate that the pendulum in a grandfather clock is, in practice, a compound pendulum. Fig 10.14 shows the forces acting on the pendulum bob of a simple pendulum when the string is at a small angle θ to the vertical. One of these forces is the weight of the bob and the other is the tension in the string. Because they are not equal and opposite to one another there will be an acceleration in the direction at right angles to the string. The diagram shows mg split into two components, one of them at right angles to the string and the other in line with it. Newton's second law can be applied to the motion at right angles to the string, taking care to assign positive signs to directions to the right since x is measured from the centre to the right.

$$\text{force to right} = \text{mass} \times \text{acceleration to right}$$
$$-mg \sin \theta = ma$$

For small angles $\sin \theta \approx \theta = \dfrac{x}{l}$

$$-mg \, \frac{x}{l} = ma$$

$$a = -\frac{g}{l} x$$

Since the acceleration is proportional to the displacement and in the opposite direction the motion is simple harmonic motion and

$$\omega^2 = \frac{g}{l}$$

$$T = \frac{2\pi}{\omega} = 2\pi \sqrt{\frac{l}{g}}$$

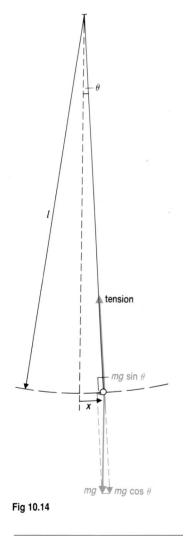

Fig 10.14

FREE AND FORCED OSCILLATION

The equation for the period of a simple pendulum is an approximate one because of the approximation of $\sin \theta$ to θ in radians. This is a good approximation only for small angles. For $\theta = 0.1$ rad ($5.73°$), $\sin \theta = 0.0998$. In other words the error in making the approximation is only 0.2%.

QUESTION

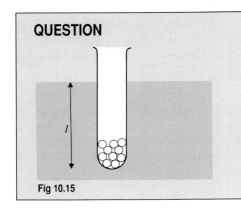

Fig 10.15

10.6 A loaded test tube floats in a liquid with a length l submerged, Fig 10.15. By calculating the resultant force on the test tube when it is displaced downwards by a small distance x, show that the motion is simple harmonic motion and show that the period of the oscillation T is given by

$$T = 2\pi \sqrt{\frac{l}{g}}$$

10.3 ENERGY IN SIMPLE HARMONIC MOTION

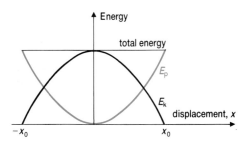

Fig 10.16 Energy alternates between kinetic energy and potential energy as any object oscillates.

In Section 10.2 the velocity of a mass undergoing simple harmonic motion was found in terms of its displacement. The relevant equation was

$$v = \omega \sqrt{x_0^2 - x^2}$$

This enables the kinetic energy E_k of the oscillating mass to be found:

$$E_k = \tfrac{1}{2} mv^2 = \tfrac{1}{2} m\omega^2 (x_0^2 - x^2)$$

The maximum kinetic energy occurs when $x = 0$, as it is at the centre of the oscillation that the velocity is greatest.

So the maximum kinetic energy $= \tfrac{1}{2} m\omega^2 x_0^2$

If the potential energy is taken to be zero at the centre of the oscillation then the maximum kinetic energy must also be the total energy of the oscillation. If no energy is lost during an oscillation then the total energy must remain constant and any loss of kinetic energy must be converted into potential energy.

This is shown graphically in Fig 10.16. The value of the potential energy at any displacement is given by

$$E_p = \text{total energy} - E_k$$
$$= \tfrac{1}{2} m\omega^2 x_0^2 - \tfrac{1}{2} m\omega^2 (x_0^2 - x^2)$$
$$= \tfrac{1}{2} m\omega^2 x_0^2 - \tfrac{1}{2} m\omega^2 x_0^2 + \tfrac{1}{2} m\omega^2 x^2$$
$$E_p = \tfrac{1}{2} m\omega^2 x^2$$

DATA ANALYSIS

Energy in SHM

When a mass oscillates on the end of a spring it interchanges its kinetic and potential energies as explained in Section 10.3. The potential energy however is of two forms: as it rises and falls it gains and loses gravitational potential energy; as the spring stretches and contracts it gains and loses potential energy too. This spring potential energy is greatest when the mass has least gravitational potential energy. It is quite difficult to show all of these different energies in graphical form, largely because the zero of potential energy is arbitrary. Zero potential energy is usually taken at the centre of the oscillation but of course the spring is stretched at this point and so does store some spring potential energy. This implies that the gravitational potential energy at the centre of the oscillation must be negative. In this exercise you are guided to use some data concerning a spring in order to plot graphs to show the interplay between the various

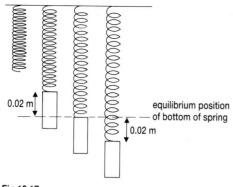

Fig 10.17

types of energy. Assume that there is no overall loss of energy to friction. The oscillation is shown in Fig 10.17.

oscillating mass (m)	3.0 kg
spring constant (k)	1200 N m^{-1}
amplitude of oscillation (x_0)	0.020 m

By answering the questions in order and completing the following partial table you should be able to sketch the final graphs. It will help to realise that the graphs in Fig 10.16 apply in this case for total energy, kinetic energy and total potential energy.

	Top	Centre	Bottom
Displacement/m	0.02	0	−0.02
Kinetic energy/J	1	2	3
Spring potential energy/J	10	11	12
Gravitational potential energy/J	14	13	15
Total potential energy/J	4	5	6
Total energy/J	7	8	9

(a) Find the extension which would be produced in the spring by placing the mass on the spring but not letting it oscillate. This gives the centre of the oscillation.
(b) Use the equation worked out in Example 3 to find ω.
(c) Find the maximum velocity.
(d) Find the maximum kinetic energy.
(e) Fill in numerical values for 1 to 9 in the table.
(f) Find the potential energy stored in the spring in each of the three positions. See Section 7.2 if this causes problems.
(g) Fill in numerical values for 10 to 12 in the table.
(h) Using 5 and 11 find 13.
(i) What increase in gravitational potential energy takes place between the centre and the top?
(j) Fill in numerical values for 14 and 15 in the table.
(k) Sketch, on the same displacement axis, graphs showing each of the five energies used.

10.4 DAMPED OSCILLATION

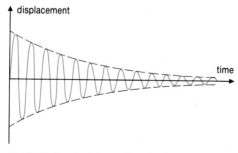

Fig 10.18 If a simple harmonic motion is subjected to frictional forces, the amplitude of a freely oscillating object gradually decreases. This is called damped simple harmonic motion.

So far in this chapter friction has been largely ignored. The oscillations which have been referred to have been assumed to be of a fixed amplitude. In practice this is not usually the case unless energy is being continuously supplied to the oscillating system. The balance wheel of a watch will cease to oscillate unless some energy can be supplied to it from the watchspring. The effect of frictional forces is to reduce the total mechanical energy of the oscillating system. This is a common everyday experience. A note played on a piano is loud immediately after it has been played and then gradually fades away; a church bell, similarly, is loud as the clapper strikes it and it too fades away. Life would be very noisy if this were not the case! A knock against a table causes the table to reverberate; this reverberation also dies away, often after many hundreds of vibrations.

A damped oscillation is shown in Fig 10.18. An oscillating system in which friction has an effect is said to be a **damped** system. The damping may be light damping, in which the amplitude is reduced in a large number of oscillations, or heavy damping, in which oscillation may not

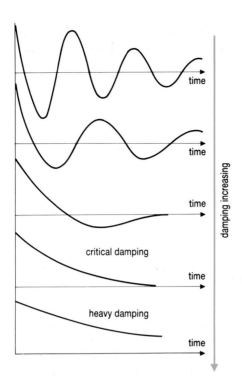

Fig 10.19 If the damping is increased, oscillation does not occur at all.

occur at all. Not only does friction have an effect on the amplitude but it also has the effect of reducing the frequency slightly.

Fig 10.19 shows the way in which the degree of damping affects the oscillation. The condition for what is known as **critical damping** is that for more than critical damping the system ceases to oscillate. It returns to, but does not cross, the mean position. The degree of damping of a mechanical system is important. Too little damping results in a large number of oscillations: too much damping leads to there being too long a time when the system cannot respond to further disturbances. This is illustrated well

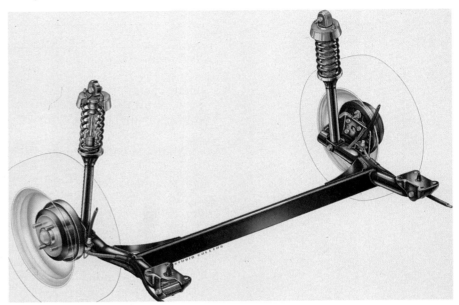

Fig 10.20 The rear suspension of a car. Damped oscillation is provided by the springs and shock absorbers.

by the trouble which car manufacturers take with the suspension of cars. The suspension is the link between the wheels and axles of a car and the body and the passengers, and consists of a spring which is damped by a shock absorber. If a car's shock absorbers are badly worn then the car becomes too bouncy and this is uncomfortable.

A good suspension is one in which the damping is slightly under critical damping as this results in a comfortable ride and also leaves the car ready to respond to further bumps in the road quickly. Fig 10.21 shows that by the time the car has reached A the shock absorbing system is ready for

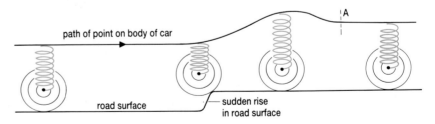

Fig 10.21 The springing of a car's suspension is damped so that when a car goes over a bump the passengers in the car quickly and smoothly regain equilibrium.

another bump. A very heavily damped shock absorbing system would still have a compressed spring by the time A is reached and so would not be able to respond to further bumps. So long as there are bumps on a road then these must have an effect on a passenger in a car. The shock-absorbing system can only reduce the forces applied. It cannot eliminate them because, clearly, in Fig 10.21, the passenger must rise eventually by the height of the bump.

FREE AND FORCED OSCILLATION

The suspension system of the human body is illustrated in Fig 10.22. The skier's body moves over the bumpy snow smoothly while his or her thighs

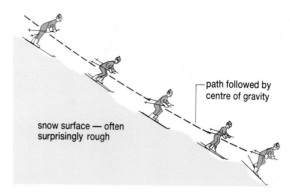

path followed by centre of gravity

snow surface — often surprisingly rough

Fig 10.22 The shock absorbing system of the human body is visible here. As the bumpy snow moves the skis up and down the knees flex and straighten so that the upper body travels smoothly.

Fig 10.23 A skier on bumpy snow uses his knees as shock absorbers.

and calves act like a damped spring. This is why skiing instructors are reputedly always saying 'Bend zee knees'. A new development in suspension is that racing cars and hydrofoils can now be fitted with active suspensions. These involve computer operated hydraulic systems in which bumps or waves are sensed, and the suspension system is adjusted accordingly. This would, if applied to the system shown in Fig 10.21, lift the spring as it reached the bump in the road. Active suspensions reduce the amount of passenger movement dramatically. A hydrofoil (Fig 10.24) is lifted clear of waves. Recent developments use computer control of the angle of the waves through the water to achieve remarkably stable movement of the craft.

Fig 10.24 A hydrofoil can be controlled so that little oscillation takes place. The angle of the foils can be computer controlled to damp out unwanted oscillations.

10.5 FORCED OSCILLATION AND RESONANCE

In many mechanical systems, vibration is a nuisance and potentially dangerous. Vibration can arise when an object is repeatedly displaced from its equilibrium position by some external force. If the repeated application of the force happens to be at the natural frequency of the object then **resonance** is said to occur and the vibration can build up to dangerous levels. The Tacoma Narrows bridge disaster of 1939 was caused by the bridge being too slender for the wind conditions in the valley. A more tragic accident took place in Angers, France in 1850 when soldiers marching over a bridge caused sufficient vibration to break the bridge and over 200 of them were killed. A more mundane example of resonance is the way in which the bodywork of a bus can vibrate at a particular engine speed.

Resonance can occur when an object with a natural frequency f_0 has a force acting upon it with a frequency f. (f is called the driver frequency). The amplitude of the oscillation of the object depends on how much it is damped and on the values of f and f_0, as is shown in Fig 10.25.

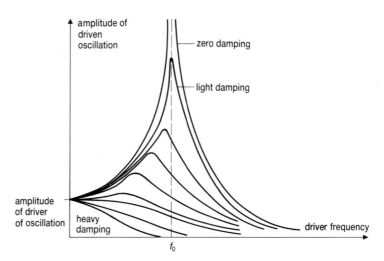

Fig 10.25   Resonance and the effect of damping on resonance.

Resonance is not only a phenomenon for mechanical oscillations. It can equally well occur in electrical circuits. These will be dealt with in more detail in Chapter 20 but at this stage the following examples are worth quoting. The electrons in a radio receiving aerial are forced to vibrate by the radio wave passing the aerial. If the aerial is the correct length for the particular frequency being used then the amplitude of the oscillation is larger and a larger signal is passed by the aerial to the receiver, where the circuitry again uses resonance to isolate and amplify the required frequency. A microwave oven also makes use of resonance. The microwaves are produced with a frequency which is the same as a natural frequency of vibration of water molecules. When some food, which always contains some water molecules, is placed in the oven the water molecules resonate, absorbing energy from the microwaves and consequently heating up. The microwaves do not heat up the containers in which the food is placed because plastic and glass do not contain any water molecules. Manufacturers of microwave ovens have to be very careful to ensure that nobody can be irradiated with microwaves as water molecules in a person's body would also resonate. They make it impossible to open the door of an oven without automatically switching off the microwaves and also they ensure that there is a conducting cage around any microwave region once the door is shut. This absorbs microwaves.

Increasing use is being made of a phenomenon called magnetic resonance. Here strong, varying radio frequency electromagnetic fields are used to cause the nuclei of atoms to oscillate. In any given molecule there will be many resonant frequencies, and whenever resonance appears energy is absorbed. The pattern of energy absorption can be used to detect the presence of particular molecules within any specimen and biochemists are using the technique to study complex molecules and the part they play in biological processes. Magnetic resonance is also being used instead of X-rays as an imaging system. Fig 10.26 is a computer enhanced photograph of

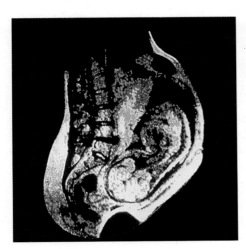

Fig 10.26   False colour magnetic resonance image showing a lateral view of a woman's abdomen during the eighth month of pregnancy. The foetus is in the normal position with head downwards.

FREE AND FORCED OSCILLATION

a foetus taken using magnetic resonance. One major advantage of magnetic resonance used in this way is that no ionising radiation is involved.

INVESTIGATION

When clothes are being dried in a spin drier or a washing machine there is often a problem of excessive vibration if the clothes are all on one side of the rotating drum. In this investigation you are asked to find how the degree of vibration depends on the rate of rotation of the drum. It is unsuitable in a school laboratory to do this with an actual washing machine working on the mains but a simple alternative with a small electric motor can be used. The motor needs to be fixed through a rubber mounting to a rigid support and to have an eccentric mass mounted on its axle as shown in Fig 10.27. It is convenient if the end casing of the motor has a definite mark or some writing on it so that

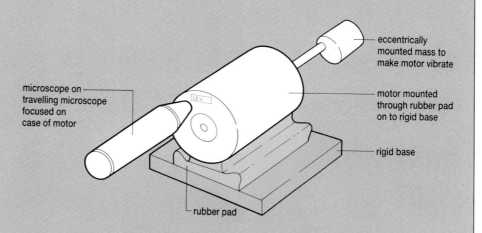

eccentrically mounted mass to make motor vibrate

microscope on travelling microscope focused on case of motor

motor mounted through rubber pad on to rigid base

rigid base

rubber pad

Fig 10.27

when it is viewed through a travelling microscope the writing seems to be blurred with the vibration which occurs. The extent of the blurring can be measured quite easily with a travelling microscope provided that the motor is well illuminated. It should be measured over a range of speeds of the motor. It is worthwhile carrying out a preliminary experiment to observe the vibration of the motor and to try to use a range of speeds within which there is a speed at which resonance occurs. If a stroboscopic lamp is used to find the speed of rotation of the motor, a graph of amplitude of vibration against rate of rotation can be plotted. If time permits, a further investigation could be made into how the position and mass of the eccentric mass affect the amplitude of the vibration.

SUMMARY

- The time for one complete oscillation is called the period T.
- The number of oscillations per unit time is called the frequency f. $f = 1/T$.
- The angular frequency ω is given by $\omega = 2\pi/T = 2\pi f$.
- The maximum value of the displacement is called the amplitude.
- A body moves with simple harmonic motion if its acceleration is always directed towards a point and has a magnitude proportional to the displacement from that point.

- For a body moving with simple harmonic motion and with the displacement given by

$$x = x_0 \sin \omega t$$

the velocity and the acceleration are given:

	in terms of time	in terms of displacement
velocity	$v = x_0 \omega \cos \omega t$	$v = \omega \sqrt{(x_0^2 - x^2)}$
acceleration	$a = - x_0 \omega^2 \sin \omega t$	$a = - \omega^2 x$

- The kinetic energy of a body undergoing SHM is $\frac{1}{2} m\omega^2 (x_0^2 - x^2)$
- The potential energy of a body undergoing SHM is $\frac{1}{2} m\omega^2 x^2$
- A system with a natural frequency f_0, if forced to oscillate at frequency f, will oscillate with a large amplitude when $f = f_0$. This effect is known as resonance.
- An oscillating system is said to be damped if the amplitude of the oscillation decreases with time.

EXAMINATION QUESTIONS

10.7 A mass on the end of a light helical spring is given a vertical displacement of 3.0 cm from its rest position and then released. If the subsequent motion is simple harmonic with a period of 2.0 s, through what distance will the bob move in **(a)** the first 1.0 s, **(b)** the first 0.75 s? (6)

(UCLES 1988)

10.8 **(a)** Define *simple harmonic motion*, and explain the meaning of the terms *amplitude, period* and *frequency*. (4)

(b) Diagram (a) shows a vertical spiral spring. Its lower end is attached to a fixed horizontal surface, and the upper end carries a platform. The force required to produce unit compression in the spring is k.

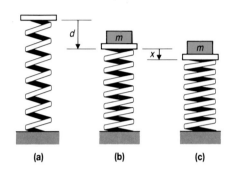

(a)   (b)   (c)

(i) Diagram (b) shows the equilibrium position when a body of mass m has been placed on the platform. Draw a labelled diagram showing the forces acting on the mass and write an equation which relates m and d. (3)

(ii) The mass is then pushed down through a short distance and released, and vertical oscillations take place. Diagram (c) shows the mass at a distance x below its equilibrium position. Deduce an equation for the acceleration of the mass at this position, and hence show that the motion of the mass is simple harmonic. Give an expression for the period of the oscillations. You may assume that the mass of the spring and the platform is negligible. (5)

(iii) Using the same time axis, sketch graphs showing how the displacement of the mass and its velocity change with time from the moment of release. (3)

(iv) If $m = 6.0$ kg and $k = 1000$ N m^{-1}, calculate the maximum amplitude of the oscillation if the mass is just to remain in contact with the platform at all times. Explain carefully how you obtain your answer. (2)

(v) Determine the energy of oscillation in the situation of (iv) and sketch a graph of the variation of kinetic energy with time for one complete cycle. (4)
(Assume that the acceleration of free fall, $g = 10$ m s^{-2}.)

(AEB 1986)

10.9 **(a)** Show that the time period, T, of small amplitude oscillations of a simple pendulum of length, l, is given by

$$T = 2\pi \sqrt{\frac{l}{g}}$$

where g is the acceleration of free fall. (4)

(b) (i) Describe how you would set up a simple pendulum of length 1.00 m in a laboratory and determine accurately the time period of small amplitude oscillations.

(ii) Discuss whether or not it would be possible to detect the effect on the time period of this pendulum if its length were increased by 0.02 m. (8)

(c) A simple pendulum of length 1.00 m has a bob of mass 0.20 kg. It is set into motion by a single sharp sideways blow to the bob which causes the pendulum to oscillate with an angular amplitude of 5.0°. Calculate

(i) the maximum vertical displacement of the bob from its rest position,

(ii) the energy acquired by the bob from the initial blow,

(iii) the maximum speed of the bob.
Neglect the effect of damping. (6)

(d) Explain briefly the mechanism by which damping would cause the amplitude of the oscillations to decrease. (2)

(JMB 1986)

10.10 A horizontal plate is vibrating vertically with simple harmonic motion at a frequency of 20 Hz. What is the maximum amplitude of vibration so that fine sand on the plate always remains in contact with it? (3)

(UCLES 1986)

10.11 State the relationship between the force on a body and the distance of the body from a fixed position when the body is executing simple harmonic motion about that position. (2)

Show that a body of mass m suspended by a light elastic string for which the ratio of tension to extension is λ will execute simple harmonic motion when given a small vertical displacement from its equilibrium position. Find the period of the motion for the case $m = 0.1$ kg and $\lambda = 20$ N m^{-1}.   (7)

A second 0.1 kg mass is attached to the first by a light inextensible wire and hangs below it. The system is allowed to come to rest, and at a time $t = 0$ the wire is cut. Calculate the position, velocity and acceleration of the first 0.1 kg mass at time $t = 1.05$ s, assuming no resistance to motion.   (6)

Give expressions for the kinetic and potential energy of the system at time t. Show that the total energy is independent of time. Outline qualitatively what would happen to the total energy of such a system set oscillating in the laboratory.   (5)

(O & C 1983)

10.12   The motion of a piston in a certain car engine is approximately simple harmonic with amplitude 40.0 mm. The frequency of oscillation is 120 Hz. Find (a) the maximum acceleration, (b) the maximum speed, of the piston.   (4)

(UCLES 1983)

10.13   (a)   Explain what is meant by simple harmonic motion (s.h.m.).   (2)
(b)   A particle executes s.h.m. of amplitude a and period T.
   (i)   Explain the meaning of the terms *amplitude* and *period*.
   (ii)   Write down expressions in terms of a and T (as appropriate) for the speed and the acceleration of the particle when at a displacement x from the midpoint of its oscillations.   (4)
   Hence derive an expression involving a and T for the total energy associated with the simple harmonic motion of a particle of mass m.
(c)   A loudspeaker cone, sounding a pure note of frequency 2.5 kHz, executes s.h.m. of amplitude 2.0 mm. Calculate:
   (i)   the maximum speed of the cone;   (3)
   (ii)   the maximum acceleration of the cone;   (3)
   (iii)   the mean power required to maintain the cone's oscillations in the presence of a mean damping force of 0.30 N.   (6)
(d)   Describe an experiment (using normal laboratory equipment) to confirm that the cone's oscillations are simple harmonic.   (5)

(OXFORD 1987)

10.14   A sealed metal tube of uniform cross-sectional area A is loaded with lead shot to a total mass M of shot plus tube. The tube floats upright in liquid of density ρ. Show that small amplitude vertical oscillations of the tube are simple harmonic and find an expression for their period.   (3)
For a tube of length 300 mm, diameter 50 mm and total mass 0.50 kg, floating in water, calculate
(a)   the length of tube above the surface of the water when floating in equilibrium, and
(b)   the period of small amplitude vertical oscillations.
   Neglect effects due to viscosity, surface tension and the motion of the fluid.   (4)
(c)   Discuss qualitatively the effect on the oscillations of replacing water by
   (i)   a less dense liquid such as ethanol,
   (ii)   a much more viscous liquid such as motor oil.   (4)
The oscillations of the tube are maintained by coupling it to a simple harmonic vibrator which can be driven at all frequencies in the range 0.1 Hz to 10 Hz. Sketch graphs of how you would expect the amplitude of the oscillations of the tube, in water and in motor oil, to vary with variations in the frequency of the vibrator. (Density of water = 1000 kg m^{-3})   (4)

(O & C 1984)

10.15   The suspension of a car may be considered as a spring under compression combined with a shock absorber which damps the vertical oscillations of the car.
(i)   Draw a sketch graph to illustrate how the vertical height of the car above the road will vary with time after the car has just passed over a bump when the shock absorber is faulty so that it provides less damping than normal.
(ii)   Draw the corresponding sketch graph assuming that the shock absorber is operating normally.

(UCLES 1989)

10.16   A long pendulum performs small oscillations and the position of the bob at 0.10 s intervals is shown below. The diagram shows exactly one half cycle of its motion.

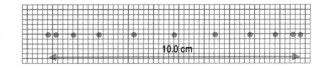

(a) (i) On graph paper plot the displacement of the bob from its equilibrium position against time.
(ii) Explain how you would show that the variation is sinusoidal.
(b) Give values for the amplitude and the period of the oscillation.
(c) Use your answers to (b) to calculate the maximum speed of the bob. (8)

(AEB 1986)

10.17 (a) Explain the meaning of (i) *free oscillation*, (ii) *forced oscillation*.
Give one example of each.
(b) The diagram shows a simple pendulum which oscillates with simple harmonic motion between the points A and D. The bob travels through a distance of 20.0 cm in moving from A to D, and the time for one complete oscillation is 2.0 s.

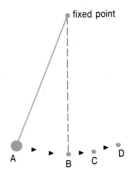

fixed point

A   B C   D

(i) What is the acceleration of the bob at B, the midpoint of the oscillation?
(ii) Calculate the acceleration of the bob at A.
(iii) The bob is released from rest at the point A. How long after its release will it reach the point C midway between B and D?

10.18 A rubber ball, initially at rest, is dropped from a height of 2.5 m onto a horizontal floor. It rebounds several times to a height which you may assume is unchanged after each bounce.
(a) Calculate how long the ball takes to reach the floor for the first time. (3)
(b) Sketch graphs showing how the following properties of the ball vary with time for the first three bounces: (i) height above the floor, (ii) velocity, (iii) acceleration. Where possible indicate numerical values on the axes. (6)
(c) Discuss the relationship between kinetic energy, potential energy and total energy of the ball during the motion. Include a

consideration of the short time during which the ball is in contact with the floor. (4)
(d) State three ways in which the motion, although periodic, differs from simple harmonic motion. (3)
A second ball, less elastic than the first, is now used. Each time the ball hits the floor it rebounds with 0.95 times the speed with which it hits the floor. Starting from the same initial conditions as the first ball, find how many bounces the ball makes before its rebound height is less than 1.5 m. (4)

(O & C 1985)

10.19 Describe one mechanical and one electrical experiment you could perform to show damped oscillations. (10)
Illustrate the difference between free oscillations and damped oscillations by drawing sketch graphs of **(a)** displacement against time, **(b)** amplitude against number of oscillations. (5)

(UCLES 1984)

10.20 Sketch a set of graphs on the same axes to show how the amplitude of forced oscillations of a resonant system varies with the driving frequency for
(i) very light damping, (ii) moderate damping, (iii) heavy damping. (3)

(UCLES 1985)

10.21 (a) What is meant by *resonance* in vibrating systems? (2)
(b) Describe a laboratory demonstration of resonance. State the apparatus used, the procedure followed, and describe the results obtained. (6)

Chapter 11

WAVE MOTION

LEARNING OBJECTIVES

At the end of this chapter you should be able to:

1. define and use the terms frequency, period, wavelength, displacement and amplitude when applied to progressive waves;

2. relate the frequency and the wavelength to the velocity of the wave;

3. distinguish between transverse and longitudinal waves;

4. describe reflection and refraction as wave phenomena;

5. use the cathode ray oscilloscope for quantitative measurements;

6. list characteristics of the various parts of the electromagnetic spectrum.

11.1 PROGRESSIVE WAVES

Wave phenomena are familiar to all readers if only in the form of water waves. However, even if only water waves are considered it is apparent that there is a great deal of variety of wave pattern. The ripples which raindrops make in a puddle are very different in speed, shape and size from a tidal wave or the waves breaking on a beach, but there are some similarities between these examples which stem from their wave nature. It is the similarities which concern us in this chapter. What similarities do all waves show? What phenomena other than water waves show wave properties?

Fig 11.1 Sea waves are rather unusual waves. Most waves do not have the possibility of becoming unstable as this wave has.

All waves involve a disturbance from an equilibrium position and the disturbance travels from one region of space to another. A wave is said to be propagated because the pattern of the disturbance occurring at one place at one time is repeated at a later time at a different place. The complex movement of the electrons in the wire of a radio transmission aerial sets up electric and magnetic fields in the space surrounding the aerial. A fraction of a second later these changing fields cause the electrons in a receiving aerial to copy the pattern of movement of the electrons in the transmitting aerial. The link between the two aerials is an electromagnetic wave, the radio wave, which has been propagated from the transmitter to the receiver. Microwaves, infra-red, light, ultra-violet, X-rays and γ rays are other forms of electromagnetic waves.

Waves are said to be **mechanical waves** if the wave has a material substance, called the **medium**, through which to travel. Examples of mechanical waves include water waves and also such waves as sound waves, shock waves as in earthquakes and waves in strings, springs and rods.

In many examples of wave travel, for instance when a ripple is made on the surface of a pond, energy from the source is being dispersed to the space surrounding the source. A wave which distributes energy in this way is said to be a **progressive** wave. As the wave is transmitted this energy is spread over a larger and larger area and so the energy associated with unit area drops. The following example illustrates the way in which the power per unit area may be calculated; it also shows how sensitive the eye is.

Example 1

A 100 W light bulb is 10% efficient. (That is, 90% of its output is invisible infra-red radiation and only 10% is visible light.) A person can see the light with the naked eye from a distance of 20 km on a dark night. If the area of the pupil of the person's eye is $\frac{1}{2}$ cm^2, find the power of the light which the eye of the person is receiving.

$$\text{Power of light produced} = 10\% \times 100 \text{ W} = 10 \text{ W}$$

This power is assumed to be spreading out uniformly in all directions, so by the time it has reached a distance of 20 km from the bulb the power is spread over the surface area of a sphere of radius 20 km. This is shown, in an exaggerated way, in Fig 11.2.

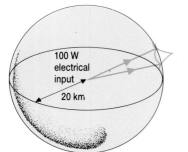

$$\text{Surface area of sphere} = 4\pi r^2 = 4\pi \times (20\,000 \text{ m})^2 = 5.0 \times 10^9 \text{ m}^2$$
$$= 5.0 \times 10^{13} \text{ cm}^2$$

$$\text{Power entering the eye} = \frac{1}{2} \times \frac{10 \text{ W}}{5.0 \times 10^{13}}$$
$$= 1 \times 10^{-13} \text{ W}$$

As indicated above, waves vary enormously in their complexity. The purest form of wave is one in which the oscillation of the source is a simple harmonic motion. Some of the terms introduced for simple harmonic motion can therefore also be applied to wave motion. These are

period *(T)*	the time for one complete oscillation
frequency *(f)*	the number of oscillations per unit time
displacement *(y)*	the vector distance a particle is moved from its rest position
amplitude *(a)*	the maximum displacement
angular frequency *(ω)*	2π multiplied by the frequency

Another connection between wave motion and simple harmonic motion concerns energy. The **intensity** of a wave depends on the energy passing through unit area in unit time. For waves, as with simple harmonic motion,

Fig 11.2 The power from a light bulb spreads out over a sphere.

Fig 11.3 Ripples from raindrops behave independently of one another.

the energy associated with the oscillation is proportional to its amplitude squared (see Section 10.3). Put in equation form becomes

◆   intensity = k (amplitude)^2 where k is a constant

This readily becomes apparent at sea. A wave one metre high can be produced by a light wind, and life on board ship is not too uncomfortable. A wave five metres high has 25 times the energy associated with it and will only be produced by a gale. It makes life on board very uncomfortable.

Wave diagrams

There are some insuperable problems with drawing diagrams of waves. A wave is essentially a moving pattern and a piece of paper can only have fixed images on it. What is written and drawn to represent waves is therefore only a second best. Also a wave is usually three-dimensional, so representing it on a two-dimensional page must mean that the diagram is incomplete. Wave diagrams have to be carefully drawn, labelled and interpreted if they are to be useful. They also have to rely on the reader being able to visualise the pattern of movement from a variety of different diagrams.

Fig 11.3 is a photograph of a puddle with raindrops falling into it. The ripples from the raindrops spread out in circles on the two-dimensional surface of the water. The ripple of each raindrop moves completely independently of all the other raindrops so the pattern is one of overlapping circles. A diagram showing this looks like Fig 11.4 and is rather confusing. Concentrating on one ripple, Fig 11.5, simplifies the

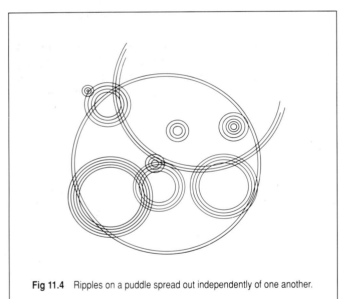

Fig 11.4   Ripples on a puddle spread out independently of one another.

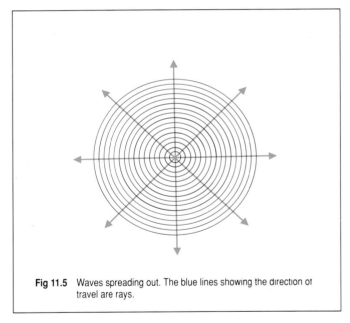

Fig 11.5   Waves spreading out. The blue lines showing the direction of travel are rays.

diagram and enables the addition of circles and arrows to be made to indicate that as time proceeds the circles get larger. A line showing the position of the crest of a wave is called a **wavefront**. A wave moves in a direction at right angles to the wavefront and lines drawn to show the movement are called rays. This diagram therefore is showing the wavefronts at different times, and rays starting from the source of the wave disturbance and spreading outwards. It is nevertheless difficult to show that the larger the circle the smaller is the amplitude of the wave. This is done in Fig 11.6 where a series of graphs are drawn of the displacement and the wave is shown in cross-section. Here it should be clear that the disturbance is of only short duration and that as the wave spreads out from the centre its amplitude is reduced. In order to show specific features of a

WAVE MOTION

wave it is often necessary to draw more than one diagram or to draw composite diagrams like Fig 11.6 in which several graphs are super-imposed.

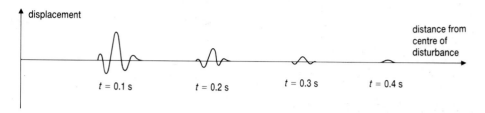

Fig 11.6   A composite diagram showing how the amplitude of a wave pulse reduces as it gets further from the source.

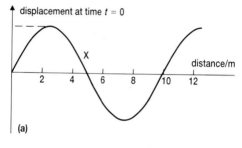

(a)

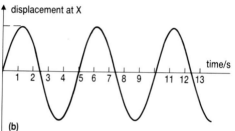

(b)

Fig 11.7   **(a)** A displacement–distance graph and **(b)** A displacement–time graph for the same wave.

A wave is a disturbance which changes both in space and in time. A graph representing a pure, continuous wave can therefore show either how the displacement changes with distance, or how the displacement varies with time. Figs 11.7(a) and 11.7(b) are graphs for the **same** wave. Fig 11.7(a) is a graph of displacement against distance and shows that one complete wave cycle occupies a distance of 10 metres. The graph shows the wave at a particular moment in time, a graphical photograph of the wave. Fig 11.7(b) shows how the displacement varies with time at a particular place. At the start there is zero displacement at point X. As time elapses the displacement at X increases to a maximum as the crest of the wave passes, falls to zero, continues to a minimum as the trough of the wave goes past, and then returns to zero to complete one cycle after 5 seconds. The period of this wave is 5 seconds, its frequency is 0.2 Hz and its angular frequency ω is $2\pi/5$ rad s^{-1} = 1.26 rad s^{-1}. Fig 11.8 is a composite graph showing this information in a series of superimposed graphs. Graphs such as this can be useful but are quite difficult to draw accurately. When sketching graphs of

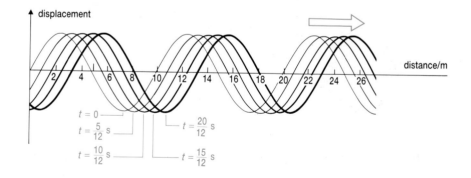

Fig 11.8   A composite graph showing movement of a wave.

wave motion it is usually better to sketch them fairly quickly and to turn the page round so that you are not drawing them from underneath the wave but from the side. Try to get your hand holding the pen oscillating smoothly from side to side before you actually bring pen to paper and then, as you start drawing the wave, move your whole arm downwards across the paper while still maintaining the oscillation of your hand. It takes a bit of practice! The graphs in this book are drawn using a stencil so you cannot and will not be expected to draw them as well as this. Teachers and examiners are aware of the problem. They only expect the particular features of the wave to be drawn which you wish to highlight.

Wave speed

Fig 11.7(a) shows that a complete cycle of the wave occupies a distance of 10 m. This distance is called the wavelength and for a pure wave is the distance between adjacent crests. In general, **wavelength** λ is defined as the smallest distance between two points which are in phase with one another. It can be seen from Fig 11.7(b) that it will take 5 seconds for a complete cycle and so the speed is 2.0 m s^{-1}. The speed of a wave c is the speed with which the crests of the wave move. It is also the speed with which energy is transferred. It is not the speed with which the particles within the wave move. In general from the basic definition of speed as

$$\text{speed} = \frac{\text{distance}}{\text{time}}$$

for one cycle the distance travelled is the wavelength and the time is the period so

$$\text{speed} = \frac{\text{wavelength}}{\text{period}}$$

and since $\text{period} = \dfrac{1}{\text{frequency}}$

$$\text{speed} = \text{frequency} \times \text{wavelength}$$

$$c = f\lambda$$

Example 2

A radar wave is an electromagnetic wave whose speed is 3.00×10^8 m s^{-1}. If its frequency is 1.64 GHz find its wavelength and the value of ω for the wave.

$$\text{Since} \quad c = f\lambda$$
$$3.0 \times 10^8 \text{ m s}^{-1} = 1.64 \times 10^9 \text{ Hz} \times \lambda$$
$$\lambda = \frac{3.0 \times 10^8 \text{ m s}^{-1}}{1.64 \times 10^9 \text{ Hz}} = 0.183 \text{ m}$$
$$\omega = 2\pi f = 2\pi \times 1.64 \times 10^9 \text{ Hz}$$
$$= 1.03 \times 10^{10} \text{ rad s}^{-1}$$

Wave equations

The equation relating displacement for a pure wave to the distance along the wave is of a sinusoidal pattern. The equation is similar to those dealt with in Chapter 10 for simple harmonic motion. In this case the graph is one of displacement against time and the angle whose sine is found is 2π multiplied by the fraction of wavelength x/λ. This gives the equation

$$y = y_0 \sin 2\pi \, \frac{x}{\lambda}$$

of which a graph is drawn in Fig 11.9(a).

If the displacement at a particular point is required as a function of time then the equation is the same as the equation for simple harmonic motion, namely

$$y = y_0 \sin 2\pi \, \frac{t}{T}$$

This graph is drawn in Fig 11.9(b).

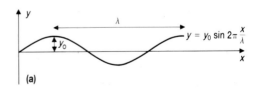

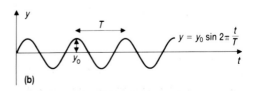

Fig 11.9 (a) Graph of the equation $y = a \sin (2\pi x / \lambda)$.
(b) Graph of the equation $y = a \sin (2\pi t / T)$

Mathematically the variable y depends on two other variables x and t. You will probably not be familiar with handling three variables and certainly it is not possible on two-dimensional paper to draw graphs showing three variables, but you may be able to see that the combined equation

$$y = y_0 \sin 2\pi \left(\frac{x}{\lambda} - \frac{t}{T} \right)$$

is the equation of a graph of y against x and the graph is different at different times. This is, in fact, the equation of a wave moving to the right.

If the substitutions

$$T = \frac{1}{f} \quad \text{and} \quad \lambda = \frac{c}{f}$$

are made then this equation can be written in many alternative forms, e.g.:

$$y = y_0 \sin 2\pi \left(\frac{x}{\lambda} - ft \right)$$

$$y = y_0 \sin 2\pi f \left(\frac{x}{c} - t \right)$$

$$y = y_0 \sin \frac{2\pi f}{c} (x - ct)$$

There is little point in memorising such equations because there are so many of them and usually one does not know in advance which form is likely to be required. It may be useful to remember one of them; it may well be necessary to switch from one form to another by means of direct substitution.

QUESTIONS

11.1 What is the range of frequencies of visible light? At one end of the spectrum the wavelength of red light is 700 nm and at the other end of the spectrum the wavelength of violet light is 400 nm. Light travels with a speed of 3.0×10^8 m s^{-1} in a vacuum.

11.2 Draw a diagram to illustrate two waves which are **(a)** $\pi/4$ radians out of phase and **(b)** $7\pi/4$ radians out of phase. Is there any need to distinguish between the two diagrams?

11.3 What is the mathematical equation of a wave with an amplitude of 5 mm, a wavelength of 600 mm and a speed of 340 m s^{-1}?

11.4 At a distance of 20 m from a point source of waves the amplitude of a wave is 1.6 mm and it has an intensity of 4.4×10^{-3} W m^{-2}. Find the intensity and amplitude of this wave at distances of 40 m and 100 m from the source.

11.5 Draw a diagram, similar to Fig 11.4, to show the wave pattern produced when the source of waves is itself moving. Use the following data.

wave speed	= 40 cm s^{-1}
frequency	= 50 Hz
speed of source of waves	= 25 cm s^{-1}

Find the apparent wavelength and frequency of the waves in front of and behind the moving source. How does the diagram change if the speed of the source of the waves is changed to
(i)   40 cm s^{-1}
(ii) 60 cm s^{-1}?

(a)

(b)

Fig 11.10 **(a)** A transverse wave set up on a slinky spring.
(b) A longitudinal wave set up on a slinky spring. Instead of having the normal wave shape, as shown in the transverse wave photograph, the longitudinal wave has a series of compressions and rarefactions.

The direction in which the displacement takes place within a wave motion affects the properties of the wave. In one type of wave, called a **transverse wave**, the particle movement is at right angles to the direction of propagation of the wave. In the other type of wave, the particle movement is in the same direction as the direction of propagation. Both of these wave types can be illustrated using a slinky, a long flexible steel coil which in use rests on a smooth table. Wave energy can be transmitted by a slinky and, for illustration, each of the turns of the coil can represent a particle of the medium through which a wave is travelling.

The transverse wave is illustrated by the spring in Fig 11.10(a). At the left-hand side a turn of the coil is being oscillated sideways, perpendicular to the direction of propagation. Since the turns are connected to one another the second coil copies the pattern of movement of the first, the third copies the second and so on down the slinky. Each time the movement pattern is repeated however, there is a small time delay so the second coil is slightly out of phase with the first. There is a slight phase lag in the vibration of the second coil with respect to the first. It is this delay which gives rise to the wave propagation and it is a characteristic of all progressive waves that adjacent particles have the same pattern of movement after a small delay.

If however the first coil is pushed and pulled in a direction parallel to the length of the slinky, then a series of compressions and rarefactions is produced in the coil. A compression is a region where the turns of the coil are closer together than average, and a rarefaction is where they are further away than average. Again, as a result of each coil repeating the pattern of movement of the one next to it with a small time delay, it is found that both the compressions and the rarefactions propagate themselves forward in the direction of wave travel as is shown in Fig 11.10(b). This type of wave is called a **longitudinal** wave. Fig 11.11 can help in the understanding of

Fig 11.11 Diagram to show longitudinal wave motion. A slot needs to be made out of a piece of A4 paper and the book moved down underneath the slot.

longitudinal wave motion. A slot needs to be prepared out of a piece of paper according to the instructions given on Fig 11.12. (The slot will be needed again to view a different type of wave when you are reading Chapter 13.) Place the slot on top of the book so that a small portion of the wave pattern at the bottom of the page shows through the slot. Then slide the book towards you, keeping the sheet of paper still. The movement of the pattern shows the propagation of the wave to the right. If you concentrate your attention on a place where the lines are close together, a

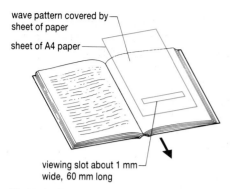

wave pattern covered by sheet of paper

sheet of A4 paper

viewing slot about 1 mm wide, 60 mm long

Fig 11.12 Diagram showing how to use the longitudinal wave pattern.

compression, it is easy to see the compression moving to the right, followed by a rarefaction, followed by a compression . . . If however you concentrate on just one line you will see that it is simply oscillating sideways. Each line represents one of the individual particles which make up the medium through which the wave travels. You can adjust the frequency and speed of this longitudinal progressive wave by pulling the book towards you at different speeds.

All electromagnetic waves are transverse waves. The displacement in the case of electromagnetic waves is a variation in the electric and magnetic fields, and more will be written about these waves in Section 11.7. Other transverse waves are water waves and waves on strings. Sound waves, on the other hand, are longitudinal waves. There are several situations in which both types of wave can exist together, but they do not usually travel at the same speed. The shock wave from an explosion or from an earthquake consists of both longitudinal and transverse waves. Transverse shock waves (shear waves) reduce in amplitude quickly when they travel through a liquid. They are said to have rapid **attenuation**. Use is made of this as a way of prospecting for oil. If the shock wave from an explosion is found to contain much stronger longitudinal than transverse waves, then there is good likelihood that the wave has travelled through a liquid. Test borings can then be made and an oil company hopes it does not discover water! There is data analysis exercise related to this after Section 11.6.

The speed with which different waves travel can be determined theoretically and it is usually found that there is a factor which has the effect of increasing the speed of the wave. This factor is associated with the strength of the elastic coupling between particles in the medium through which the wave travels. There is also a factor which reduces the speed; it is associated with the inertia of the moving particles. For instance, if the speed of a transverse wave along a string is considered, increasing the tension in the string increases the strength of the coupling between the particles and hence also increases the speed of the wave. Increasing the mass per unit length of the string has the effect of increasing the inertia of the oscillating parts of the wave and hence of reducing the wave speed. Question 11.6 tabulates the equations for the speed of several different types of waves for reference and asks you to check that they are dimensionally correct.

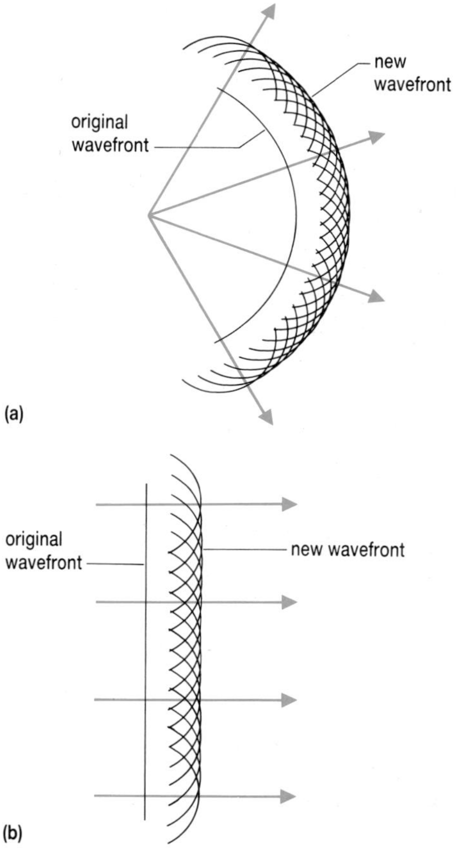

original wavefront

new wavefront

(a)

original wavefront

new wavefront

(b)

Fig 11.13 **(a)** The formation of a new circular wave front on a circular wave.
(b) The formation of a new wave front on a plane wave.

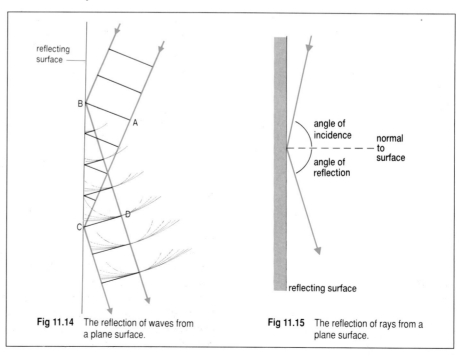

reflecting surface

B

A

C

D

angle of incidence

normal to surface

angle of reflection

reflecting surface

Fig 11.14 The reflection of waves from a plane surface.

Fig 11.15 The reflection of rays from a plane surface.

11.6 Use the data given in Table 11.1 to show that the right-hand sides of all the wave equations have the dimensions of length per unit time. (If you have not studied electrical quantities you will not be able to check the last of these equations yet.)

Table 11.1   Equations giving the speed of different types of wave

Type of wave	Equation for speed of wave
Transverse wave on a string	$c = \sqrt{\dfrac{T}{m}}$
Longitudinal wave on a spring	$c = \sqrt{\dfrac{kl}{m}}$
Longitudinal wave in a rod	$c = \sqrt{\dfrac{E}{\rho}}$
Longitudinal sound wave in a gas	$c = \sqrt{\dfrac{\gamma p}{\rho}}$
Transverse ripples on a pond	$c = \sqrt{\dfrac{2\pi\sigma}{\rho\lambda}}$
Transverse electromagnetic wave	$c = \sqrt{\dfrac{1}{\varepsilon_0 \mu_0}}$

Symbols used		
	c –	wave speed
	T –	tension
	m –	mass per unit length
	k –	force per unit extension
	E –	Young modulus (see section 23.1)
	l –	length of spring
	ρ –	density
	p –	pressure
	γ –	dimensionless constant
	σ –	surface tension (unit J m^{-2})
	λ –	wavelength
	ε_0 –	permittivity of free space (see section 16.1)
	μ_0 –	permeability of free space (see section 18.7)

11.3   REFLECTION

Fig 11.5 shows the way in which waves spread out from a source with circular wavefronts and rays perpendicular to the wavefronts. The way in which successive wavefronts are built up from preceding ones is shown in both parts of Fig 11.13 in which each point on an existing wavefront is considered as a source of a new disturbance. The sum of all these disturbances becomes the new wavefront. Fig 11.13(a) shows this for a circular wavefront in which the new wavefront is also a circle but of larger radius. If the procedure is continued, the radius of curvature becomes greater and greater so that at a large distance from the source the wavefronts become virtually straight and the rays become parallel lines, Fig 11.13(b). This method of constructing the new position of a travelling wave is making use of a technique called Huygen's construction and it can be used in situations which are more involved than those with circular wavefronts. Fig 11.14 shows a parallel wavefront approaching a plane surface and being reflected from it. After reflection, the intensity of the wavefront is appreciable where it has been drawn with thick lines but can be shown mathematically to be near to zero elsewhere. The construction shows that triangles ABC and DCB are exactly the same size and therefore the angle of incidence is equal to the angle of reflection. This is shown for rays in Fig 11.15 and illustrates the law of reflection that the angle of incidence is equal to the angle of reflection when a wave is reflected, see Section 12.2. This fact is made use of in aerial dishes. Fig 11.16 is a photograph of one such dish and Fig 11.17 shows how a parallel beam of radio waves from a satellite positioned above the equator can be reflected to a single spot by the parabolic dish. The aerial is placed at this spot to receive

Fig 11.16 A dish aerial. The parabolic shape of the aerial reflects radio waves to a sharp focus.

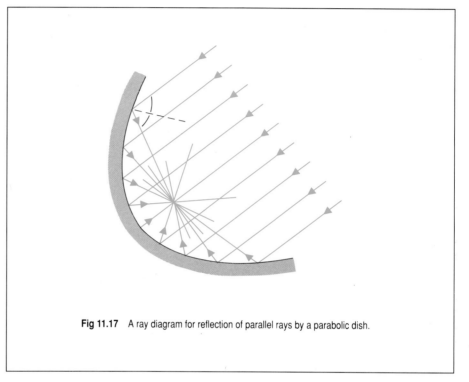

Fig 11.17 A ray diagram for reflection of parallel rays by a parabolic dish.

the largest intensity of radiation. At all points on the dish the angle of incidence is equal to the angle of refraction. For domestic use these dishes can be 1.20 m in diameter and the wavelength of the radio waves is typically of the order of half a metre.

QUESTIONS	
	11.7 On a diagram copied from Fig 11.17 show typical positions of the wavefronts before and after reflection.
	11.8 A *radar kilometre* is a unit of time. It is the time taken for a radar pulse to travel a distance of 1 kilometre to an object and to be reflected back. How long is a radar kilometre?

11.4 REFRACTION

The same procedure can be performed if, instead of being reflected a wave enters a region in which it has a slower speed. This might be when light passes from air into glass or where a water wave travels into shallower water. Assume that the change in speed takes place suddenly at a boundary. Fig 11.18 shows this and illustrates the well known fact that light is refracted as it enters a dense material. The wavefronts are shown progressing towards the boundary and continue with a smaller wavelength after refraction. Here not all of the wavefronts are shown so that the construction of one of them may be drawn.

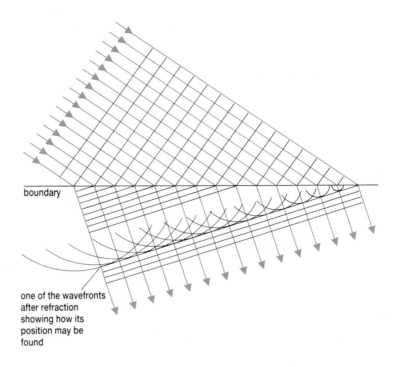

Fig 11.18 The refraction of waves.

Using a simplified diagram, Fig 11.19, shows that in triangles ADC and ABC the angles at B and D are right angles because rays are always at right

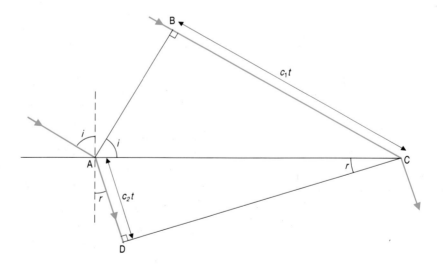

angles to wavefronts. If the wave takes time t to travel from B to C then the other end of the wave also takes time t to travel from A to D. This gives

$$BC = c_1t \text{ and } AD = c_2t$$

where c_1 is the speed of the wave in material 1 and c_2 is the speed of the wave in material 2.

The ratio $\dfrac{c_1}{c_2}$ is constant, so for a ray travelling from one material into another the ratio of

◆ $\dfrac{\sin i}{\sin r}$ is a constant

For light this is known as Snell's law of refraction. If light travels from a vacuum (material 1) into a transparent substance (material 2) then

$$\frac{\sin i}{\sin r} = \frac{c}{c_2}$$

where c is the speed of light in a vacuum.

◆ The ratio $\dfrac{\text{speed of light in a vacuum}}{\text{speed of light in a transparent material}}$ is known as

the refractive index n of the transparent material.

QUESTION	11.9 What is the refractive index from material 1 to material 2 for the refraction shown in Fig 11.18? If the waves had been travelling in the opposite direction the pattern would have been exactly the same. It is a basic property of waves that they are always reversible. What therefore would be the refractive index from material 2 to material 1 for the refraction shown in Fig 11.18? More detail about refraction, in terms of rays, is given in Section 12.3.

11.5 POLARISATION

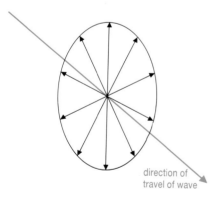

Fig 11.20 In an unpolarised transverse wave oscillations may take place in any direction at right angles to the direction in which the wave travels.

As a result of the transverse nature of the vibration, transverse waves have an additional property which is not possessed by longitudinal waves. The movement of particles in transverse mechanical waves is at right angles to the direction of propagation of the wave. In three dimensions this still leaves many possibilities for the direction of particle movement as Fig 11.20 shows. All of the double-headed arrows in this figure are drawn at right angles to the direction of propagation and show possible directions for particle movement. The direction of particle movement for a string, for example, does not necessarily have to be up and down, it can equally well be from side to side or slanting. Frequently oscillation takes place in a transverse wave in many different directions and the wave is then said to be **unpolarised**. If the oscillation does take place in only one direction then the wave is said to be **polarised** in that direction.

Whether or not a transverse wave is polarised depends to a large extent on how it is produced. Electromagnetic waves are transverse waves and can therefore be polarised. Indeed it is the fact that they can be polarised which enables us to be confident that they are transverse and not longitudinal waves. If a radio wave is transmitted from a vertical aerial then the wave will have vertical polarisation and any receiving aerial must also be positioned vertically. Use is made of the ability to polarise radio waves for television transmission. The main transmitters in the United Kingdom send out signals which are horizontally polarised, but because of the contours of the land there are some areas which obtain less than the minimum field strength which is defined for a television service. To improve reception in these areas there are about 1000 low power transmitters each covering a limited area. Most of these low power transmitters are vertically polarised so that there is no danger of the signal which they transmit causing interference with the signals from the main transmitters.

The same procedure is used for radio transmission, though many areas of the country are now served by mixed polarisation transmitters. The following investigation may not work as intended in your area. It will nevertheless let you find out if you are in an area of mixed polarisation and can also show how the strength of the received signal depends on the direction of the aerial.

INVESTIGATION

Use a portable radio to plot a graph which shows how the intensity of a received radio signal varies with the angle between the plane of polarisation of the radio wave and the direction of the aerial.

The investigation is easier if the portable radio has an FM band and a movable aerial. If the radio is rotated about a vertical axis then the loudness of the radio changes not because of polarisation but because of the direction of the transmitter from the radio. If the aerial is rotated about a horizontal axis however, the change in loudness is due to the polarisation of the wave.

Set up the radio so that maximum signal is being received and arrange so that you can measure the angle of rotation of the aerial as you rotate the aerial about a horizontal axis. Some detector must be used to measure the volume of noise which the radio is producing. This could be a decibel meter or a microphone connected to an oscilloscope. Relating the reading from the oscilloscope or decibel meter to intensity can be difficult and therefore you should be careful how the vertical axis of your final graph is labelled.

A similar investigation may be carried out with microwaves which, from a laboratory transmitter, are also polarised.

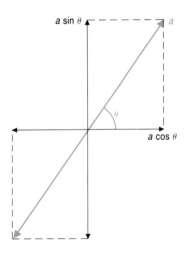

Fig 11.21 A wave of amplitude a can be resolved into two components at right angles to one another. The amplitude of the horizontal component is $a \cos \theta$ and the amplitude of the vertical component is $a \sin \theta$.

The above investigation shows that if a receiving aerial is not quite in line with the transmitting aerial, the intensity of the received signal does not suddenly drop to zero. Another illustration of the same thing is that you do not need to reposition the television aerial on the roof of your house if it is twisted a little in a gale. Nor does it become impossible to watch television when gusty wind conditions are causing the aerial to wobble. A wave oscillation can be resolved into two components at right angles to one another as is shown in Fig 11.21. Here a wave of amplitude a is oscillating so that its plane of polarisation is at an angle θ to the horizontal. This wave can be considered as two separate waves; one of amplitude $a \cos \theta$ which is horizontally polarised and one of amplitude $a \sin \theta$ which is vertically polarised.

Light from a hot object, such as the Sun or a lamp filament, is unpolarised because it is emitted totally at random from the atoms of the hot object. If unpolarised light is passed through a perfect vertically polarising filter then the half of the light which is vertically polarised light will be allowed through and the other half, which is horizontally polarised, will be absorbed. Many crystals have the effect of polarising light and, in *polaroid*, crystals of quinine iodosulphate are used in which long molecules are aligned so that light polarised in one direction only can pass through. A sheet of *polaroid* can be used to detect the presence of polarised light, in which case it is called an analyser. There are other natural means of producing polarised light. Light from blue sky is partly polarised and so is light reflected from a shiny surface.

This is why *polaroid* is used in sunglasses. It reduces glare, (see Fig 11.22) because the light reflected from a shiny horizontal surface, say the sea, is partially horizontally polarised. If the sunglasses are vertically polarised then most of the reflected light is absorbed by the sunglasses. Anglers find these glasses useful since it enables them to see into the water once the

Fig 11.22 The glare visible on one of these photographs has been eliminated by placing a *Polaroid* filter on the camera before taking the second photograph.

Fig 11.24 **(a)** Polarised light was used to take this photograph. The stress pattern within each turbine blade is clearly visible.
(b) The same technique using crossed polaroids is used here to photograph a plastic model of a crane hook. Note how there is no strain in the hook to the right of the load.

strong reflected light is reduced in intensity. The reduction in intensity of the reflected glare is shown in Fig 11.23. It happens that if the reflected ray

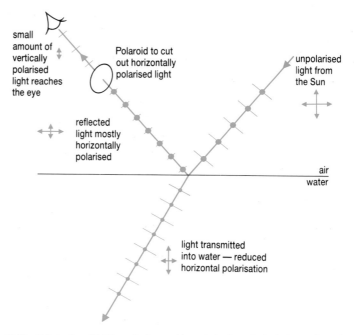

Fig 11.23 Polarisation of light by reflection enables polarised spectacles to be used to reduce glare.

and the refracted ray are at right angles to one another then the reflected ray is completely horizontally polarised. The condition for this to occur is for the angle of incidence θ_b called the Brewster angle, to be given by

$$n = \tan \theta_b,$$

where n is the refractive index of the material. See Question 12.5.

Use is often made in applications of polarisation of a property called optical activity. When polarised light is passed through some materials the plane of polarisation is rotated. The plastic of which transparent rulers are made has this property, so does Sellotape and sugar solution. Fig 11.24 shows photographs taken using this effect. The photographs are taken using an arrangement of crossed polaroids shown in Fig 11.25. If the object is not present then no light is able to pass through the analyser because it is

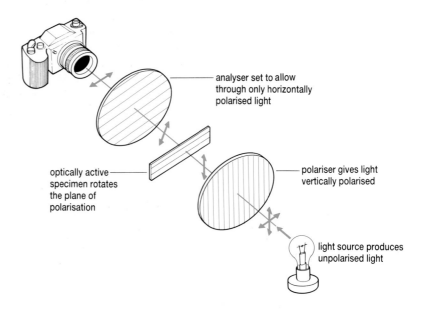

Fig 11.25 The arrangement needed to take the photographs shown in Fig 11.24.

optically active specimen rotates the plane of polarisation

analyser set to allow through only horizontally polarised light

polariser gives light vertically polarised

light source produces unpolarised light

set to transmit horizontally polarised light and the polariser is only letting through vertically polarised light. With the object in place between the crossed polaroids, any optical activity of the object rotates the plane of polarisation so that the analyser is able to pass some light. The colours in the photographs arise because different wavelengths of light are rotated by different amounts and also because different parts of the object under test are under different strains and this alters the amount of optical activity.

How much polarisation a material causes can be affected by the presence of an electric field. This is becoming an increasingly used effect. Liquid crystal displays (LCDs) are usually polarised. Rotating a pair of polarised sunglasses over the display on a calculator will show the effect. The figures will appear and disappear twice per complete revolution of the sunglasses. The effect electric field has on the polarising ability of a material is known as the Kerr effect and it can be used to interrupt a beam of light very rapidly. If a high frequency potential difference is applied across a cell of liquid placed between crossed polaroids, then light will not pass through the cell when the potential difference is zero but will when the field is set up. It is possible to obtain pulses of light with a duration as little as a picosecond (1 ps = 10^{-12} s) using this technique.

11.6 EXPERIMENTAL METHODS

Some general principles concerning measurement of wave phenomena should be appreciated. For instance, if the speed c of a wave motion is required, one way of proceeding is to find both the wavelength λ and the frequency f of the wave motion in separate experiments and then use the equation

$$c = f\lambda$$

If, on the other hand, a direct method of distance travelled divided by time taken is to be used, then it is essential in planning the experiment to be aware of the problems which are likely to be encountered. In a thunderstorm it is apparent that sound waves travel more slowly than light waves: the flash of lightning may be seen several seconds before the sound of the thunder is heard. This may be used to find the speed of sound, but only if it is assumed that light travels instantaneously.

The speed of light

For many years it was assumed that light travelled instantaneously simply because no one could find any way in which a time delay could be observed. However, as early as 1676 Roemer, by observing the moons of Jupiter, was able to estimate the speed of light with quite remarkable accuracy. His biggest problem was in knowing the radius of the Earth's orbit. In modern units the problem is illustrated by Table 11.2 which shows how either very large distances or very small times must be used. By using astronomical distances Roemer was able to give its first measured value and modern direct methods can be performed in school laboratories by methods which effectively measure fractions of microseconds.

Table 11.2 Numerical values indicating the problem involved in measuring the speed of light.

Distance measured/m	Time elapsed/s	
300 000 000 000	1000	Roemer (1676)
60 000 000	0.2	Satellite (modern)
300 000	0.001	Radar (modern)
30 000	0.000 1	Fizeau (1849)
30 000	0.000 1	Michelson (1878)
30	0.000 0001	Laboratory rotating mirror (modern)

However the speed of sound or of light is measured, it must be interrupted so that some marker is present in the succession of waves. For the measurement of the speed of sound using thunder and lightning, the marker is the thunderclap. For an echo method also, it will be the sudden bang which can be heard directly and again after reflection from a wall. For Roemer's method to find the speed of light, it was the disappearance of a moon of Jupiter as it passed behind the planet, and for Fizeau a rotating wheel chopped a light beam up into equal light and dark parts.

The speed of light has now, arbitrarily, been defined as 299 792 458 m s^{-1}. This now means that all experiments which previously had been thought of as experiments to measure the speed of light are now regarded as experiments to measure distance! Nowadays this can be done very accurately and the present definition of the metre is the distance light travels in a time of

$$\frac{1}{299\ 792\ 458} \text{ s}$$

The oscilloscope

The structure and principles of the cathode ray oscilloscope will be dealt with in detail in Sections 28.4 and 28.5, but some aspects of its use are essential when dealing with sound waves. The first thing to realise about an oscilloscope is that it is always measuring an electrical input. If therefore it is to be used to measure some feature of a sound wave, then it must be used in conjunction with a transducer of some sort. A **transducer** is a device which can convert one form of energy into another form, and here all that is necessary as a transducer is a microphone which can convert a sound wave into an electrical signal of the same pattern. The next important point is that while the pattern on the screen is often stationary and horizontal distances can be measured on the screen, these distances do not correspond with any distances on a wave being measured but are times. What might appear as a wavelength on the screen is a period of the wave. The trace is a graph of displacement against time. Finally, remember that the displacement in a longitudinal wave is in the direction of propagation of the wave, so that although the wave may look on the screen as if it were a transverse wave, it may in reality be a longitudinal one.

Frequency and period

Frequency and period are readily measured with a calibrated oscilloscope but the accuracy is not usually very high. The following example shows how this may be done.

Example 3

The stationary pattern on the screen of an oscilloscope which is graduated in centimetre squares appears as shown in Fig 11.26. The calibration for the time base, which is applied to the X-plates, is also shown. Find the frequency and period of the oscillation.

The calibration of the time base gives the information that a time of 1.0 milliseconds is taken for the trace to move 1 centimetre sideways across the screen.

The pattern shows that 5 complete cycles occupy 8.2 centimetres. 5 cycles take 7.8 milliseconds so

$$T = \frac{8.2 \text{ ms}}{5} = \frac{0.0082 \text{ s}}{5}$$

$$\text{so } f = \frac{1}{T} = \frac{5}{0.0082 \text{ s}} = 610 \text{ Hz}$$

The speed of sound

The speed of sound can conveniently be determined using an oscilloscope as a clock to measure small time intervals. The experiment usefully illustrates several features of an oscilloscope but is not particularly accurate because the accuracy of the calibration of the time base on an oscilloscope is limited by the width of the screen and on whether the speed of the spot across the screen is constant.

An oscilloscope with a sweep output needs to be used. This output is a pulse given out at the start of each sweep of the spot across the screen, and leads connect this output to an amplifier and loudspeaker as shown in Fig 11.27. The sound output of the loudspeaker can be a series of low frequency

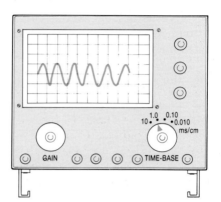

Fig 11.26 The trace on the screen of a calibrated oscilloscope enables the frequency of the oscillation to be found.

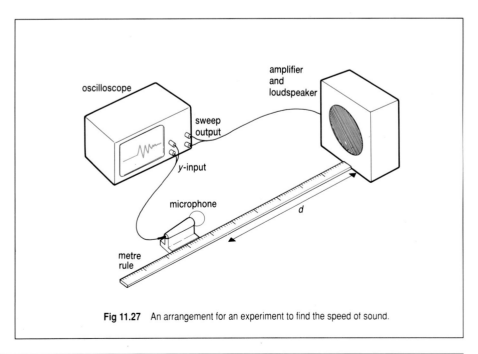

Fig 11.27 An arrangement for an experiment to find the speed of sound.

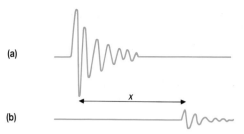

(a)

(b)

Fig 11.28 The screen patterns obtained in the speed of sound experiment.

clicks or, as the frequency is raised become more of a whistle. Since the distance the sound is to travel is something less than a metre, the time which needs to be measured is of the order of a few milliseconds. The spot on the oscilloscope screen needs to take a few milliseconds to traverse the screen. If the time base speed is set to something between 1 and 0.1 ms cm^{-1} then minor adjustments can be made as required. The sound output of the loudspeaker is picked up by a microphone, a distance d_0 from the loudspeaker, connected to the Y-input of the oscilloscope and the Y-sensitivity is adjusted to obtain a trace on the screen of suitable height. This should give a trace similar to Fig 11.28(a).

The microphone is then moved different distances d away from the loudspeaker. As this is done the pattern of the trace, Fig 11.28(b), decreases in height because the microphone receives less volume of sound from the loudspeaker. The pattern also starts at a later time on the horizontal trace because in the extra time that the sound takes to reach the microphone the spot has travelled further horizontally. A graph can be plotted of x, the extra distance across the screen, against d and the gradient $\Delta x/\Delta d$ can be used to deduce the speed of sound. To do this it must be appreciated that the time it takes for the sound to travel any extra distance to the microphone is measured by using the calibration of the time base velocity of the spot as it moves a distance x.

Frequency

Frequency can be determined accurately using an altogether different technique. If the time base is switched off and the external X-input is used, then a 50.0 Hz input can be applied from a low voltage a.c. supply. The mains supply is accurately controlled at 50.0 Hz so that electric clocks maintain their time accurately. (In times of heavy demand there can be a small fall in the frequency during the day but this has to be made up for by running slightly fast throughout the night to get clocks correct by the following morning.) If now a frequency of 250 Hz, for example, is placed across the Y-plates, a pattern similar to Figs 11.29(a) or 11.29(b) is seen. The likelihood is that the pattern will not be perfectly stable but will be drifting from one of these patterns to the other. Figures like this are called Lissajou's figures and a stable pattern is achieved whenever there is a simple, whole number relationship between the two frequencies being used. In this case the ratio is 5 to 1. Patterns like those shown in Figs 11.29(c) and (d) can be

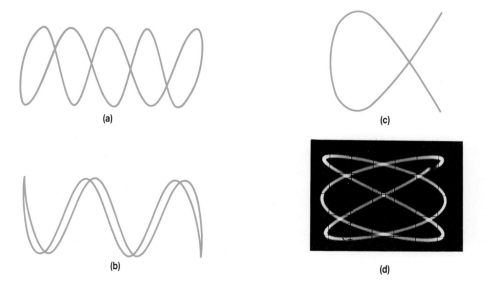

(a)

(c)

(b)

(d)

Fig 11.29 Lissajou's figures. Both (a) and (b) are for a frequency ratio $f_Y / f_X = 5/1$ (c) is for a frequency ratio $f_Y / f_X = 3/2$. (d) is a photograph taken from the screen of an oscilloscope for a frequency ratio of 5/3.

used when the frequency relationship is more awkward. The advantage of this method for comparing frequencies is that much greater accuracy can be achieved. If the patterns are stable then the frequency relationship is exact. If there is a drift from one pattern back to the same pattern after a measured time interval, then that time is the time taken for one of the sources to gain or lose one cycle on the other. This can be used as a way of measuring frequency to accuracies as high as ±0.01 Hz.

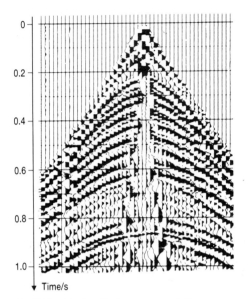

Time/s

Fig 11.30 Seismic reflection survey result. Each vertical trace represents the output from a single seismometer.

Seismic surveying

When the structure near the surface of the Earth is surveyed in prospecting for oil or minerals, one frequently used method is that of seismic reflection surveying. The process can be very complex because the strata in the Earth's crust are by no means regular and also the quantity of data which is usually received is very large. The graphical results of one particular seismic survey are shown in Fig 11.30. Each of the vertical traces shows output of a seismometer and it has been found easier to analyse the data produced if pulses are shaded in on one side.

Fig 11.31 Seismic shot firing in central Abu Dhabi.

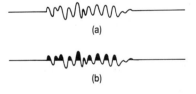

(a)

(b)

Fig 11.32 (a) The output from a single seismometer with a pulse of waves going past it. (b) The way the seismometer trace is printed.

An output from a seismometer, shown in Fig 11.32(a), is printed as shown in Fig 11.32(b). It is interesting to note that if you look at Fig 11.30 you will see that one of the seismometers failed to work.

Some of the principles behind the practice of seismic reflection surveying are explained and used in this exercise. The data have, however, been simplified.

In a place where there is a horizontal change in rock type at a certain depth, an explosion is set off at source S, in Fig 11.33. Eight detectors D_1 D_2 D_3 D_4 D_5 D_6 D_7 and D_8 are arranged as shown and these detect vibrations from the explosion a short time after the explosion. The traces received from the eight detectors are printed alongside one another in Fig 11.34 and you can see how the arrow shape of the actual seismic survey is developing even with only 8 detectors.

WAVE MOTION

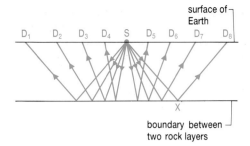

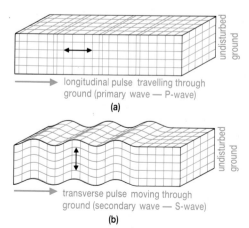

Fig 11.33   The arrangement of seismometers near S, the source of an explosion.

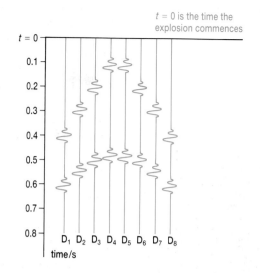

Fig 11.34   The output from eight seismometers printed alongside one other.

Fig 11.35   (a) Longitudinal waves (P-waves) travelling through rock.
(b) Transverse waves (S-waves) travelling through rock.

The rock through which the waves are travelling is known to have a density of 2700 kg m^{-3} and in rock of this density the speed of P-waves is 3.1 km s^{-1}. P-waves are longitudinal waves as shown in Fig 11.35(a) and are the waves responsible for the pulses shown in Fig 11.34. S-waves are transverse waves, shown in Fig 11.35(b), and always arrive after the P-waves.

Answer the following questions taking data from the diagrams where necessary.

(a) What other simple route can P-waves take to get from S to a detector apart from by the route shown in Fig 11.33? Identify each pulse on Fig 11.34 with the route by which the P-wave arrives at the detector.

(b) What are the distances SD_8 and SXD_8? Use these distances to find the depth of the boundary between the two rock layers.

(c) Check your answer by repeating this with a different detector. (Why is it not sensible to use D_1?)

(d) The speed of a P-wave, v_p is given by

$$v_p = \sqrt{\frac{A}{\rho}}$$

where A is a constant and ρ is the density of the rock. Find the value and the unit of A in this case.

(e) How will the traces change if, separately, the following additional factors have to be taken into account?
  (i)   The S-waves, travelling at 1.8 km s^{-1} are added.
  (ii) An extra layer of rock halfway down causes partial reflection.
  (iii) Double reflections are possible.
  (iv) The rock boundary is not horizontal.
  (v)   Some refraction takes place at an intermediate layer as a result of density changes.

When in practice all of these factors can occur together you perhaps can appreciate the problems of interpreting a seismic chart.

(f) How can you see from Fig 11.34 that the sensitivity of all the seismometers is not the same? In what respect is the amplitude shown for each pulse received drawn incorrectly?

WAVE MOTION

(g) When the wave reaches the boundary layer, some of the P-wave is transmitted downwards, with a speed of 5.7 km s^{-1}. Show that this is not possible at angles of incidence on the boundary greater than 33°. Draw a diagram to show another possible path for a P wave from S to D_8. (This question involves idea of critical angle which is dealt with in Section 12.3.)

(h) List other factors, besides those mentioned in question 5, which will affect the traces.

11.7 THE ELECTROMAGNETIC SPECTRUM

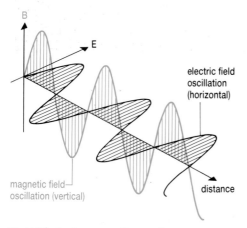

Fig 11.36 An electromagnetic wave shown as interlocking electric field and magnetic field oscillations at right angles to one another.

The electromagnetic spectrum covers a vast range of different waves, all of which travel with the same speed in a vacuum and which have the characteristics of transverse electric and magnetic waves. The electric and magnetic waves are both essential for the transmission of the energy; they interlock with one another and the electric wave is always at right angles to the magnetic wave as shown in Fig 11.36. This figure is a rather artificial diagram so it may help to visualise an electromagnetic wave if Fig 11.37 is examined. This figure shows how an electromagnetic wave originating from point P appears in cross-section. The electric field is shown in the

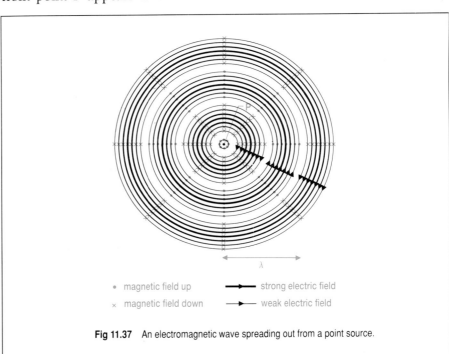

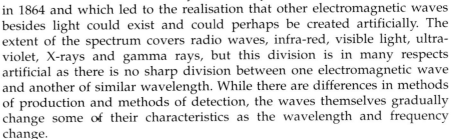

Fig 11.37 An electromagnetic wave spreading out from a point source.

plane of the paper and the magnetic field is always at right angles to the paper. The oscillation of the magnetic field creates the electric field and vice versa so that they sustain one another.

It was the mathematical analysis of this system which Maxwell achieved in 1864 and which led to the realisation that other electromagnetic waves besides light could exist and could perhaps be created artificially. The extent of the spectrum covers radio waves, infra-red, visible light, ultra-violet, X-rays and gamma rays, but this division is in many respects artificial as there is no sharp division between one electromagnetic wave and another of similar wavelength. While there are differences in methods of production and methods of detection, the waves themselves gradually change some of their characteristics as the wavelength and frequency change.

This can be illustrated by considering just that small part of the total spectrum which we call light. Light does have a precise meaning. It is that part of the electromagnetic spectrum to which the retina of the human eye is sensitive, although even here there is a gradual start and finish since the sensitivity of the human eye varies with wavelength. Around about 400 nm wavelength, deep violet light is seen and gradually the colour changes until at a wavelength of beyond about 700 nm a deep red colour fades away as the sensitivity of the eye falls. It is clear from Fig 11.38 that the

Fig 11.38 The visible spectrum.

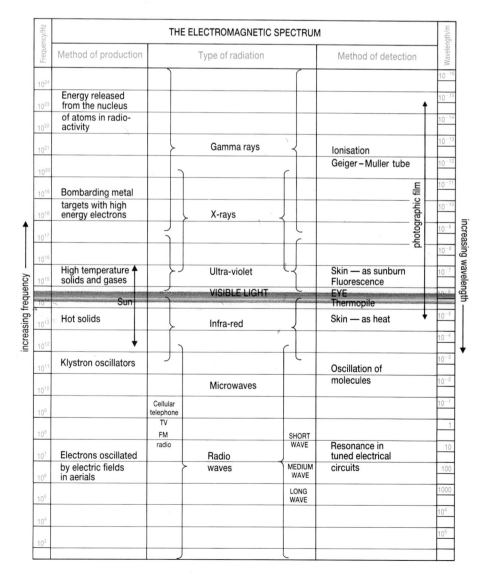

Fig 11.39 The electromagnetic spectrum.

visible spectrum contains the well known colours, but it is misleading to count the number of colours in the visible spectrum. Since one merges gradually into another there must be an infinite number of colours in the visible spectrum. The same is true of the entire electromagnetic spectrum: the labels attached to any section must be used with caution. Fig 11.39 shows the complete spectrum and the overlapping between the different regions results from this problem. Note that for all the waves the frequency multiplied by the wavelength has the constant value of 3.00×10^8 m s^{-1} and that the scales used are logarithmic. It is impossible to cover such a vast range of different values without using a logarithmic scale.

A small section of the whole spectrum is given in Table 11.3 just to illustrate how subdivision is made by the broadcasting authorities in allocation of different frequencies. It is possible to have different transmitters broadcasting on the same frequencies if they are far enough apart from one another or if they have different polarisations. In the UK the main television transmitters are horizontally polarised while small local transmitters, set up for regions where the main signals are weak, are vertically polarised. This ensures that there is no interference between them. The table covers Band IV UHF (Ultra High Frequency) and shows that each transmitter has available a range of frequencies for both the vision and the sound signals. Conventionally the sound signal is transmitted at a frequency of 6 MHz above the frequency used for the vision signal, and the width of each channel is 8 MHz. In the table the main transmitters, Divis with an output power of 500 kW and Brougher Mountain with an output power of 100 kW, each transmit four programmes and are horizontally polarised. The small stations, Glenelly Valley which has an output power of only 12.5 W and Killowen with an output power of 150 W, are vertically polarised. These stations also transmit four programmes and can use the same frequencies as the main transmitters.

Table 11.3   UHF Band IV frequencies and some of the Northern Ireland transmitters using these frequencies. (H) indicates horizontal polarisation and (V) vertical polarisation.

UHF Band IV.

| Channel | Frequencies in MHz | | Transmitter and Programme |
	Vision	Sound	
21	471.25	477.25	Divis, C4 (H), Killowen C4 (V)
22	479.25	485.25	Brougher Mountain, BBC1 (H)
23	487.25	493.25	Glenelly Valley, ITV (V)
24	495.25	501.25	Divis, ITV (H): Killowen ITV (V)
25	503.25	509.25	Brougher Mountain, ITV (H)
26	511.25	517.25	Glenelly Valley, BBC2 (V)
27	519.25	525.25	Divis, BBC2 (H): Killowen, BBC2 (V)
28	527.25	533.25	Brougher Mountain, BBC2 (H)
29	535.25	541.25	Glenelly Valley, C4 (V)
30	543.25	549.25	
31	551.25	557.25	Divis, BBC1 (H), Killowen BBC1(V)
32	559.25	565.25	Brougher Mountain, C4 (H)
33	567.25	573.25	Glenelly Valley, BBC1 (V)
34	575.25	581.25	

- Waves may be classified as mechanical or electromagnetic. Any progressive wave transmits energy from one place to another by a cyclical movement of the medium in which the wave travels but without there being any resultant transmission of the medium itself. In the case of electromagnetic waves no medium is required, but the cyclical movement then takes place in electric and magnetic fields.
- For longitudinal waves the particles move in the direction of energy transmission, whereas for transverse waves the particles travel at right angles to the direction of energy transmission.
- Only transverse waves can be polarised.
- The amplitude of a wave is the maximum movement of a particle from its rest position. The intensity of a wave is proportional to the amplitude squared.
- Waves which are out of step with one another are said to be out of phase. One cycle is 2π radians so two waves in antiphase are π radians out of phase.
- Wavelength for a progressive wave is the shortest distance between two particles which have no phase difference between them.
- Frequency is the number of oscillations per unit time.
- Speed = frequency × wavelength.
- The wave equation for displacement as a function of distance at one moment of time is

$$y = a \sin 2\pi \; \frac{x}{\lambda}$$

- The wave equation for displacement as a function of time at one place is

$$y = a \sin 2\pi \; \frac{t}{T}$$

- Reflection does not alter the speed of a wave and the angle of incidence equals the angle of reflection
- Refraction is caused when wave speed changes. At a boundary between two regions

$$\frac{\sin i}{\sin r} = \frac{\text{speed of wave before refraction}}{\text{speed of wave after refraction}}$$

for light

$$\frac{\sin i}{\sin r} = \frac{\text{speed of light in a vacuum}}{\text{speed of wave in a transparent material}} = \text{refractive index}$$

EXAMINATION QUESTIONS

11.10 (a) With the help of diagrams, explain the difference between a *transverse wave* and a *longitudinal wave*. (4)

(b) In a laboratory demonstration two different sources of radiation of wavelength 3 cm are used, one producing microwaves (short electromagnetic waves), the other producing high frequency sound waves. Suitable receivers are provided for both radiations.
Describe experimental tests that would confirm that:
(i) the electromagnetic radiation consists of transverse waves; (5)
(ii) the wavelength of one of the radiations is 3 cm. (5)

(c) The diagram illustrates part of an oil exploration test in which a beam of longitudinal(compressional) waves, generated at the Earth's surface, is directed through various strata in the Earth's crust. The frequency of the waves is 75 Hz. Using the information provided in the figure, calculate:

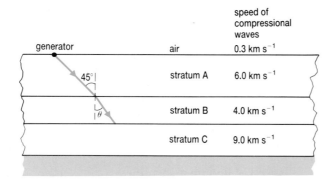

	speed of compressional waves
air	0.3 km s⁻¹
stratum A	6.0 km s⁻¹
stratum B	4.0 km s⁻¹
stratum C	9.0 km s⁻¹

(i) the wavelength of the waves in stratum A; (3)
(ii) the angle θ to the vertical at which the waves enter stratum B. (3)
Explain, with the aid of a suitable calculation, why the waves do not enter stratum C. (4)

(d) (i) Reproduce the diagram to show the subsequent paths of the waves within strata A and B, taking account of both refraction and reflection. (4)
(ii) Discuss whether the waves will eventually emerge from stratum A into the atmosphere. (2)

(OXFORD 1986)

11.11 Calculate the critical angle for sound waves travelling from air into water, across a plane boundary. The speed of sound in air is 340 m s⁻¹ and the speed in water is 1500 m s⁻¹.
Someone standing on the side of a swimming pool shouts to a friend swimming under water. Give one reason why the swimmer is unlikely to hear the shout. (5)

(ULSEB 1986)

11.12 (a) The graph shows how particle displacement varies with distance for a sound wave at one instant in time. From the graph, determine

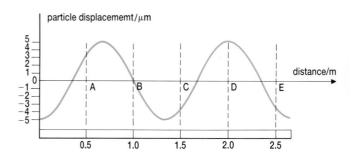

(i) the wavelength of the wave
(ii) the amplitude of the wave
(iii) the phase difference between the vibration of the particle at A and that of the particle at B.

(b) Explain what is meant by *compression* and *rarefaction* when applied to a sound wave travelling in air.

(c) At this instant a compression occurs at B.
(i) Use the graph to explain how such a compression arises.
(ii) Copy the graph and indicate on your copy the positions of another compression and a rarefaction. (10)

11.13 The diagram shows a loudspeaker, L, which emits a continuous sound of frequency 400 Hz, and a line to represent the positive x direction. The graph shows the displacements of the air masses along x from their undisturbed positions at one instant.

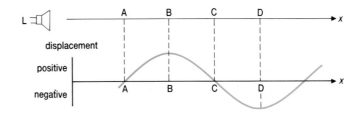

(a) At which of the four points, A, B, C, D is the instantaneous *pressure* at its peak value? Explain your answer.

(b) Calculate the time interval required for the sound to travel the distance AD. (6)

(ULSEB 1986)

11.14 What is meant by the *polarisation* of a wave? Explain whether or not (a) sound waves, and (b) electromagnetic waves can be polarised. (4)

(ULSEB 1981)

11.15 When two polarisers P and Q are placed so that their polarising directions are parallel, the amplitude of the emergent beam is *A* (see Fig).

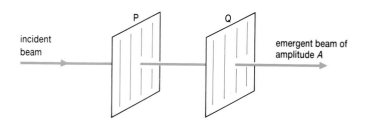

Through what angle must Q be rotated so that the amplitude of the emergent beam is reduced to *A*/2?

What will be the corresponding fractional reduction in the intensity of the emergent beam? (5)

(UCLES 1988)

11.16 A beam of plane-polarised microwaves is incident upon an aerial which is initially positioned to give maximum response. In a storm, the aerial is rotated about the direction of the incident waves until it makes an angle of 30° to the plane of polarisation, as shown below.

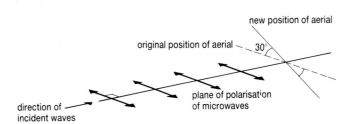

Calculate the percentage reduction in the amplitude of the signal now received from the aerial. (3)

(UCLES 1986)

11.17 An oscilloscope is used to measure the time it takes to send a pulse of charge along a 200 m length of coaxial cable and back again. The diagram shows the appearance of the oscilloscope screen, A indicating the original pulse and B the same pulse after reflection.

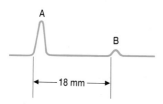

If the time base speed is set at 10 mm μs⁻¹, calculate the speed of the pulse along the cable. (3)

(ULSEB 1982)

11.18 (a) Describe with the aid of suitable diagrams how sound waves are propagated in air. Describe how you would measure the speed of sound in free air.

(b) A small loudspeaker emits sound energy uniformly into a hemispherical region in front of itself. If the total power of the sound emitted is 80 mW, what is the sound intensity, (energy per unit area) at a distance of 3 m in front of the loudspeaker? What would be the distance from the loudspeaker at which the sound intensity was half of this value? You may assume that the loudspeaker behaves as a point source of sound energy. (17)

(ULSEB 1988)

11.19 (a) Place the visible, infra-red, ultra-violet and X-ray regions of the electromagnetic spectrum in order of increasing frequency. (4)

(b) Explain two situations in which an accurate knowledge of the speed of electromagnetic waves in a vacuum is useful. (4)

(c) A small source of parallel light is situated 80 cm from the axis A of a rotating mirror R (see diagram). Light is reflected from mirror R on to a small fixed mirror M situated 1.4 × 10³ m from A. When light is reflected back to R, the mirror has rotated slightly so that the beam is finally located at a point L, alongside S, where the distance SL is 3.22 cm.

(i) Calculate the angle θ between the light beams SA and AL.

(ii) Calculate the angle of rotation of the mirror R which gives rise to θ. (4)

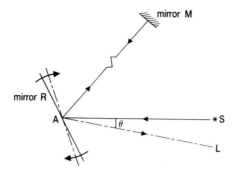

(iii) Hence, determine a value for the speed of light if the mirror is rotating at 340 revolutions per second.　(4)

(iv) The distances AM and AS can be measured to ± 10 m and ± 0.1 cm respectively and the speed of rotation to ± 2 rev s^{-1}. What is the maximum possible uncertainty in SL such that the speed of light may be quoted with an uncertainty of no more than 3%? Comment on your answer with reference to its experimental feasibility.　(6)

Chapter 12

OPTICS

LEARNING OBJECTIVES

At the end of this chapter you should be able to:

1. quote and use the laws of reflection and refraction;

2. describe the conditions under which total internal reflection occurs;

3. define and find experimentally the focal length and power of a lens;

4. use ray diagrams and the lens equation to find the position of images;

5. calculate magnification and angular magnification;

6. describe the eye, the camera and the telescope as optical instruments;

7. explain long and short sight and their correction.

12.1 INTRODUCTION

In Chapter 11 the propagation of energy in the form of waves was considered. If waves are created at a point source then they spread out from the source in concentric circles as shown in Fig 12.1. In three dimensions the circles become spheres, and when using waves of small wavelength such as light there is very little deviation from the wavefronts being perfect spheres. This implies that the directions in which the waves travel, as shown by the blue lines on Fig 12.1, are always in straight lines. For light it is more usual to draw lines representing the direction in which the light is travelling rather than the wavefronts. These lines are called light rays and in a uniform transparent material light rays travel in straight lines.

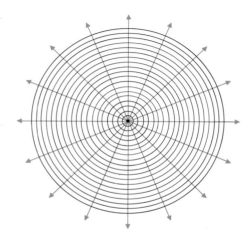

Fig 12.1 Waves spreading out from a point source. The blue lines show the direction of travel and for light are called light rays.

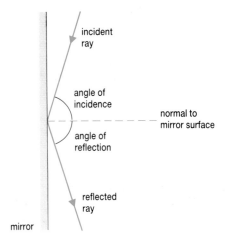

Fig 12.2 **(a)** A diagram drawn to scale showing light rays from a blue dot entering the pupil of an eye. Note how parallel the rays seem at the eye.
(b) The same diagram but with the width of the eye exaggerated.

A diagram showing the passage of light rays is called a ray diagram. Ray diagrams often have to be exaggerated to make it clear what is taking place. Consider the most straightforward ray diagram: the one for an unaided eye viewing a blue dot a distance of 15 cm from the eye. If drawn full scale, the diagram will appear like Fig 12.2(a). A more usual, easier to draw, diagram of the same situation is shown in Fig 12.2(b). The width of the cone of rays entering the eye is increased for clarity but the first diagram is the accurate one and indicates two points well. First, the amount of energy actually entering the eye is only a minute fraction of the energy emitted by the blue dot, and that itself is minute in the first place. The eye can detect light when the power entering the eye is as little as 10^{-13} W.

The second point illustrated by Fig 12.2(a) is that the light entering the eye is nearly parallel; it diverges very slightly. Objects placed at any distance from the eye between infinity and about 10 cm can be seen in clear focus by a young person with normal vision. This means that the eye is capable of receiving and focusing light only if it is somewhere between being parallel and diverging by the small amount shown here. The unaided eye cannot focus light which converges on to it, nor can it focus light which diverges more rapidly than shown.

Not only does the eye adjust to these slight differences of divergence but the brain receives a signal which it can interpret as an estimated distance of the object from the eye.

QUESTION

12.1 What is the angle of divergence in Fig 12.2? The diameter of the pupil is 3 mm and the distance of the object is 100 mm. (This question can be done very easily if you use angular measure in radians.)

12.2 REFLECTION

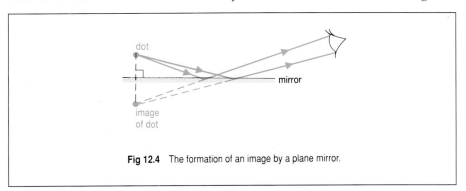

When a light wave travelling in one material strikes a different material, it can either be absorbed or transmitted or reflected. If the surface interface is smooth, a reflection called a specular reflection occurs in which the angle of incidence is equal to the angle of reflection as shown in Section 11.3 and also in Fig 12.3. A reflection such as this can give the eye a false impression of where an object is.

This is the principle of image formation and is shown in Fig 12.4. Light from the blue dot object can spread out from the object and enter the eye directly, but with the mirror present each ray obeys the law of reflection and the angle between the two rays shown remains the same after reflection as it was before. As far as the eye is concerned this reflected light is

Fig 12.3 Reflection of a light ray by a mirror

Fig 12.4 The formation of an image by a plane mirror.

diverging and will be focused in the same way as the direct light was focused. The brain will interpret the received light as having come from a place on the far side of the mirror as far away from the mirror as the object was in front of it. Two dots can then be seen and the second one is said to be an image of the first. There is no light at this image and so it is called a virtual image. A screen placed at the position of a virtual image will show nothing.

QUESTION

12.2 Draw diagrams to explain each of the following:
  (a) writing seen in a mirror is back to front;
  (b) parallel mirrors can produce an infinite number of images;
  (c) two mirrors at right angles can produce three images and the middle one of the three is not back to front;
  (d) two mirrors at right angles will reflect light back parallel to its original direction, (in three dimensions this becomes a commonly used reflector for the back of a bicycle or for a radar reflector carried on the mast of many boats);
  (e) if you remain stationary and a mirror is moved towards you at a speed of 30 m s^{-1} then your image moves at a speed of 60 m s^{-1}. (This principle is used in a radar speed detector. The mirror is the car and the radar source remains stationary. The image of the radar source is moving at a speed twice that of the car towards the detector.)

12.3 REFRACTION

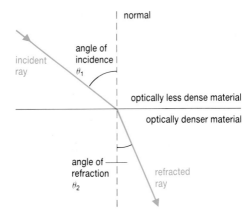

Fig 12.5 Refraction of a light ray at the surface between two transparent materials.

Light can also be transmitted at a boundary between two transparent materials and a change of direction often results. This is called refraction and is shown in Fig 12.5. It was shown in Section 11.4 that

$$\frac{\text{sine of angle of incidence}}{\text{sine of angle of refraction}} = \frac{\text{speed of light in less dense material}}{\text{speed of light in more dense material}}$$

or $\dfrac{\sin \theta_1}{\sin \theta_2} = \dfrac{c_1}{c_2}$

where c_1 and c_2 are the speeds of light in the less optically dense and more optically dense materials respectively. This can be written

$$\frac{\sin \theta_1}{\sin \theta_2} = \frac{c_1}{c_2} = {}_1n_2$$

where ${}_1n_2$ is the refractive index for light travelling from material 1 into material 2

If the less optically dense material is a vacuum this becomes

$$\frac{\sin \theta_1}{\sin \theta_2} = \frac{c}{c_2} = n$$

where n is called the (absolute) refractive index and c is the speed of light in a vacuum. For light this is known as **Snell's law of refraction**. Using the principle of the reversibility of waves which was mentioned in Question 11.9 we can also write

$$\frac{\sin \theta_2}{\sin \theta_1} = \frac{c_2}{c_1} = {}_2n_1 = \frac{1}{{}_1n_2}$$

Optical density has no connection with mass/volume. The optical density of a transparent material is related only to the speed with which light travels through it. The more optically dense the material the slower light travels through it.

Table 12.1 gives refractive indices for different transparent materials, and since the refractive index varies with the colour of light being used they are all quoted for yellow sodium light. It can be seen from the table that very little refraction takes place when light travels from a vacuum into air. It consequently makes little difference if light enters some glass from a vacuum or from air.

Fig 12.6 White light entering the prism is either reflected or refracted. The spectrum appears because blue light is slowed down more in the glass than red light. Blue light is therefore refracted more.

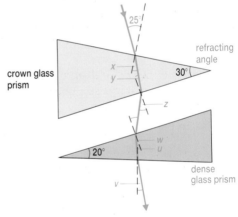

Fig 12.7 Refraction of a light ray by two prisms.

Table 12.1 Refractive indices of a selection of different materials for yellow sodium light

Glasses (typical values)	
crown	1.52
light flint	1.58
dense flint	1.66
extra dense flint	1.80
Other solids	
diamond (C)	2.417
ice (H_2O)	1.309
quartz (SiO_2)	1.544
rock salt (NaCl)	1.544
rutile (TiO_2)	2.62
Liquids	
water	1.33
ethanol	1.36
turpentine	1.47
Gas	
air	1.0003

Example 1

A ray of sodium light is shown in Fig 12.7 passing through two prisms, one made of crown glass and having a refracting angle of 30° and the other made of dense flint glass, having a refracting angle of 20°. If the angle of incidence on the crown glass prism is 25°, find the path of light through the prisms and the angle of deviation produced by them.

The principle behind this example is used in all good quality lenses. By making a lens with one component made out of dense glass and one component made out of crown glass it is possible to produce a lens which does not disperse (separate) white light into its colours. In practice many more than two components are used in camera lenses where it is particularly important that the focal length of the lens is the same for all colours.

Questions such as this are simple in principle but often somewhat tedious in practice. They depend on applying Snell's law over and over again and on using the geometry of the figure.

At the first surface the angle of incidence is 25° and the angle of refraction is x:

$$\frac{\sin \theta_1}{\sin \theta_2} = \frac{\sin 25°}{\sin x} = 1.52$$

$$\sin x = \frac{\sin 25°}{1.52} = 0.2780$$

$$x = 16.14°$$

In the triangle at the top of the prism the angle sum of a triangle gives

$$(90° - x) + 30° + (90° - y) = 180°$$

$$30° = x + y$$

giving $$y = 13.86°$$

The light now emerges from the crown glass prism so

$$\frac{\sin y}{\sin z} = \frac{1}{1.52}$$

$$\sin z = 1.52 \sin y = 1.52 \sin 13.86° = 0.3461$$

$$z = 21.35°$$

The angle of incidence on the dense glass prism must also be 21.35° since the faces are parallel. The procedure is exactly the same for the dense glass prism:

$$\frac{\sin 21.35°}{\sin w} = 1.66$$

$$\sin w = \frac{\sin 21.35°}{1.66} = 0.2193$$

$$w = 12.67°$$

which gives u

$$u = 20° - 12.67° = 7.33°$$

$$\frac{\sin u}{\sin v} = \frac{1}{1.66}$$

$$\sin v = 1.66 \sin u$$
$$= 1.66 \sin 7.33° = 0.2118$$

$$v = 12.22°$$

The light emerges at an angle of 12.2° to the normal of the last face as shown. Its deviation is found by realising that if the prisms had not deviated the light at all, the emergent ray would have been at an angle of 25° − 30° + 20° = 15° to the final face. As it is, the angle is 12.2° so the deviation is 2.8°.

Note how, with small angles, it is essential to work with 4 significant figures to keep 3 significant figures valid. At the last stage this becomes an answer to only 2 significant figures because of the subtraction.

QUESTIONS

12.3 Find the deviation produced by a crown glass prism of refracting angle 60° on a beam of sodium light which strikes it at an angle of incidence of 35°.

12.4 Find the maximum refracting angle which a crown glass prism may have if sodium light is to pass through it with two refractions and no reflection. What deviation will be caused?

12.5 A beam of light strikes a water surface and some of it is reflected by the surface and some is refracted. Find the angle of incidence of the beam of light if it is found that the refracted and the reflected rays are at right angles to one another.

Total internal reflection

If a wave is leaving a medium in which it is travelling slowly then Snell's law can be applied only within a certain range of angles, as was hinted at in Question 12.4. Figure 12.8 shows a beam of light leaving a glass block and being refracted away from the normal as expected. The relationship between the angle of incidence i and the angle of refraction r is given by the equation

$$\frac{\sin i}{\sin r} = \frac{1}{n}$$

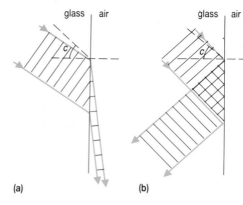

glass | air   glass | air

(a)   (b)

Fig 12.8 **(a)** Refraction of a wave emerging from a transparent material. The angle of incidence is less than the critical angle.
(b) Total internal reflection for a similar situation but when the angle of incidence is just greater than the critical angle.

From this equation it can be seen that if i is too large then $n \sin i$ will be greater than 1 and no value of r can be found, since it is not possible to have the sine of any angle greater than 1. The mathematics of this situation suggest therefore that refraction is not possible if i is greater than the value which makes $n \sin i = 1$. This value of i is called the **critical angle C** and we get

$$n \sin C = 1$$

$$\text{or} \quad n = \frac{1}{\sin C}$$

Not only does the mathematics of this situation make refraction impossible but so does Huygens's construction. In Fig 12.8(a) a wave is shown approaching the boundary at an angle of incidence which is 5° less than the critical angle. The waves produced after leaving the boundary generate a new wavefront as expected. In Fig 12.8(b) the angle of incidence is increased to 5° above the critical angle and now no new wavefront is set up. More detailed analysis of this situation shows that no wave energy can escape through the boundary but that all the energy is reflected. This is called **total internal reflection**. Because this reflection is total reflection with no energy lost at the reflecting surface, it is possible to send a light wave along a glass fibre and for the light to be reflected millions of times without any loss of brightness. This is done in optical fibre systems where the glass used is extremely pure to avoid absorption of the energy between reflections. Whereas normal glass, like sea-water, becomes difficult to see through if it is much more than a few metres thick, the glass used for optical fibre manufacture will allow light to pass through several kilometres without fading so much that it cannot be boosted by amplifiers. In fact, if sea-water were as pure as the glass used in optical fibres it would be possible to see down to the bottom of all the oceans of the world.

Fig 12.9 is a cross-section of an optical fibre of the type used to transmit messages. Because of the high frequency of light it is possible to send vast amounts of digital information down a single optical fibre. A fibre of diameter only 0.6 mm can have enough digital information sent along it to transmit many thousands of telephone calls simultaneously. The diameter of optical fibres is being reduced as the technology for making them

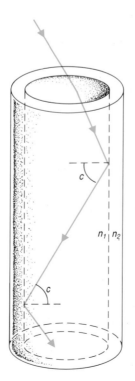

Fig 12.9 Total internal reflection enables light to be reflected many thousands of times in an optical fibre.

Fig 12.10 Optical fibres trap light inside them. The light is reflected along the fibre and can only escape at the end.

improves. Optical fibres of only a few micrometres (1 μm = 10^{-6} m = 1 micron) are now being used and they have a refractive index which varies between the centre and the circumference in a specially controlled way, see Fig 12.10.

Example 2
An optical fibre has a core made of a glass with a refractive index of 1.472 and a cladding made of glass of refractive index 1.455. What is the critical angle for the boundary between the two glasses?

The refractive index from the denser glass to the less dense glass is given by

$$n = \frac{1.472}{1.455} = 1.0117$$

since
$$n = \frac{1}{\sin C}$$

$$\sin C = \frac{1}{n} = \frac{1}{1.0117}$$

$$\sin C = 0.988\,45$$
$$C = 81.28°$$

QUESTIONS

12.6 The critical angle for light going from a certain glass to air is 41.3°. What is the refractive index of the glass?

12.7 Light is travelling along an optical fibre of refractive index 1.52. What must be the refractive index of the cladding if the critical angle is to be 85°?

12.8 The slowest speed with which light can travel along an optical fibre is when it zig-zags along the fibre making an angle of incidence equal to the critical angle whenever it is reflected. See Fig 12.9. The fastest speed is when it travels along the axis of the fibre. If the critical angle for one such fibre is 87.0° and the refractive index of the core is 1.520, find how long the fibre can be if the difference in the time taken by the two routes is not to be greater than 1 ns (1 ns = 10^{-9} s)

12.4 LENSES

Snell's law of refraction applies equally well at curved surfaces as at plane surfaces. Two curved surfaces close to one another form a lens. At one of the surfaces light travels from air into glass and at the other surface the light travels out of the glass into the air. The effect on a parallel beam of light is shown in Fig 12.11(a) for a biconcave lens and in Fig 12.11(b) for a biconvex lens.

The analysis of the precise direction the rays of light have on emerging from the lens is difficult, and manufacturers take a great deal of care in designing and making lenses for cameras. Their problems stem from the fact that light is not always parallel when it enters the lens, it does not always come from straight in front of the lens, it travels through different thicknesses of glass and it may be of any colour.

The detailed analysis necessary to design a camera lens is out of place here, but by making an assumption the main features of lens behaviour can be understood. The assumption which can be made is that the lens has spherical surfaces close to one another. Such lenses are called thin spherical lenses.

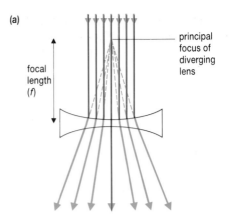

(a)

focal length (f)

principal focus of diverging lens

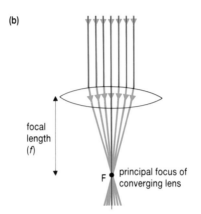

(b)

focal length (f)

F principal focus of converging lens

Fig 12.11 **(a)** The effect on a parallel beam of light of a diverging lens.
(b) The effect of a converging lens on the same beam.

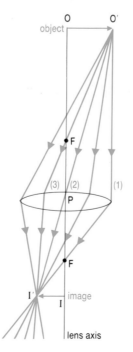

Fig 12.12 Image formation by a converging lens.

If light from a single distant point enters a thin lens then after refraction all the light will pass through a single point if it is converging, or will seem to have come from a single point if it is diverging.

It has been assumed, in drawing Figs 12.11(a) and 12.11(b) that the lenses are thin lenses. In these diagrams each point from which the light diverges or to which it converges is called the **principal focus**. The distance of the principal focus from the centre of the lens is called the **focal length** of the lens (f). The power of a lens is defined by the equation

$$\text{power of a lens} = \frac{1}{f}$$

The unit used for power is the **dioptre**. A lens has a power of 1 dioptre if its focal length is 1 m. A 2 dioptre lens has a focal length of 0.5 m.

The ability of a thin lens to collect light from a point and to refract it to another point is the essential requirement for image formation and is shown in Fig 12.12. Here blue light starts from the tip of the object. The light entering the lens is not parallel and it is not arriving from directly in front of the lens, yet the lens has brought all the light to a point. The concentration of blue light at this point will result in a blue point on a screen or on a piece of film. Every point on the object results in a corresponding point on the screen and so an image is built up. This type of image is called a **real** image. It should be noticed that the image is not at the principal focus. An image is only formed at a distance equal to the focal length from the lens when the object is at an infinite distance from the lens. Fig 12.12 also illustrates another feature of the behaviour of a thin lens, namely that light passing through the centre of the lens is not deviated because the faces of the lens are parallel at the centre.

The three rays labelled in this diagram are useful if the position of an image is to be found using ray diagrams. They are:
1. A ray parallel to the axis of the lens is refracted so that it passes through the principal focus.
2. A ray through the centre of the lens is undeviated.
3. A ray first passing through the principal focus on the near side of the lens becomes parallel to the lens axis after refraction.

The linear magnification produced by a lens is given by

$$\text{linear magnification} = \frac{\text{height of image}}{\text{height of object}}$$

This is also equal to the ratio $\dfrac{\text{image distance}}{\text{object distance}}$

because the undeviated ray (2) makes two similar triangles O'OP and I'IP, so

$$\text{linear magnification} = \frac{\text{height of image}}{\text{height of object}} = \frac{\text{image distance}}{\text{object distance}}$$

Example 3
Find by drawing the focal length of an overhead projector lens if it is able to produce an image 1.20 m high from an object 20 cm high when the lens to screen distance is 1.80 m.

The first assumption to make is that the lens can be considered to be a thin lens. Next we assume that the image to lens distance is a straight line, although usually with an overhead projector there is a mirror between them which changes the direction of the light by a right angle. The sequence required for this problem is:

OPTICS

(1) Choose a suitable scale. The horizontal and vertical scales can be different if need be. Here a scale of 1 cm representing 20 cm is used.

(2) Draw in the lens axis and mark the position of the lens and the screen (Fig 12.13).

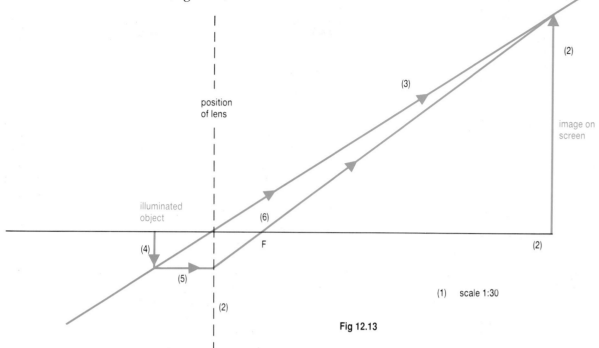

Fig 12.13

(3) Draw the undeviated ray through the centre of the lens. This is often the crucial step. The tip of the object, the centre of the lens and the tip of the image must always lie on a straight line.

(4) Find the position of the tip of the object – knowing that it must be 1 cm high on the scale drawing.

(5) Draw in a ray from the tip of the object which starts off parallel to the axis and then passes through the principal focus on its way to the image on the screen.

(6) Measure off the focal length – using the scale. Here this distance is 1.3 cm representing a focal length of 26 cm.

The accuracy of ray diagrams such as this is limited, particularly when object and image distances are very different from one another.

To calculate the solution to the problem in Example 3 it is necessary to use the equation

$$\frac{1}{u} + \frac{1}{v} = \frac{1}{f}$$

where u is the object distance, v is the image distance and f is the focal length of the lens.

In Example 3 we know that the magnification m is given by

$$m = \frac{120 \text{ cm}}{20 \text{ cm}} = \frac{v}{u} = \frac{180 \text{ cm}}{u}$$

so

$$u = 30 \text{ cm}$$

This gives

$$\frac{1}{30} + \frac{1}{180} = \frac{1}{f}$$

$$\frac{6+1}{180} = \frac{1}{f} = \frac{7}{180}$$

$$f = \frac{180}{7} = 25.7 \text{ cm}$$

Practical problems in which real lenses, which may be very thick at their centre, are treated as if they are thin lenses, give surprisingly accurate results.

12.9 A camera has a focal length of 50 mm. It is used to take a photograph of an object 80 mm high placed 450 mm from the lens. Find
(a) the power of the lens;
(b) the distance the film needs to be placed from the lens;
(c) the height of the image;
(d) the linear magnification.

12.10 An enlarger lens has a focal length of 100 mm and is used to obtain a picture 200 mm long from a negative 35 mm long. How far must the negative be placed from the lens?

Example 4
This example shows what happens if an object is placed closer to the lens than the principal focus.

An object 4 cm high is placed 6 cm from a converging lens of focal length 8 cm. Find the position and magnification of the image in Fig 12.14.

(1) Scale of 1 : 2 chosen.
(2) Axis, object and position of lens and principal focus drawn in. (There is a principal focus on both sides of the lens.)
(3) The ray passing through the centre of the lens is undeviated.
(4) The ray parallel to the axis as it approaches the lens passes through the principal focus. This ray will never cross the other ray drawn in (4).

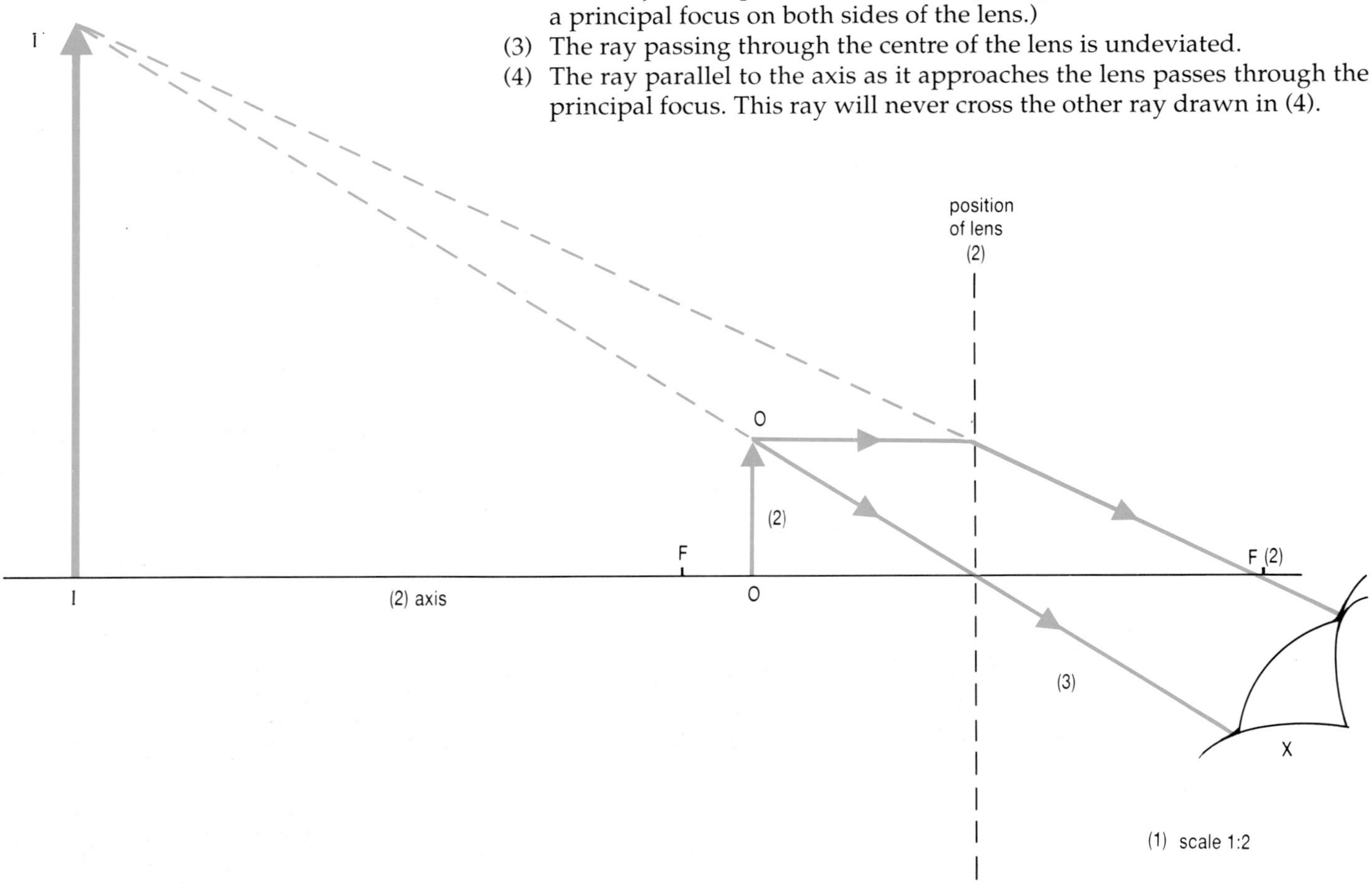

Fig 2.14

No real image can be formed by this lens arrangement but if an eye is at X, as shown, the eye receives diverging light and it will be able to focus the light. The eye/brain system will deduce that the light appears to have come from I'. A virtual image of O' is formed at I' and a virtual image II' is formed of OO'. This is the optical arrangement for a magnifying glass.

The diagram shows the image to be a distance of 24.6 cm from the lens and to have a height of 16.4 cm. The magnification is therefore

$$\frac{16.4 \text{ cm}}{4 \text{ cm}} = 4.1$$

If the same problem is done by calculation we get

$$\frac{1}{u} + \frac{1}{v} = \frac{1}{f}$$

$$\frac{1}{6} + \frac{1}{v} = \frac{1}{8}$$

$$\frac{1}{v} = \frac{1}{8} - \frac{1}{6} = \frac{3-4}{24} = -\frac{1}{24}$$

$$v = -24 \text{ cm}$$

The negative sign is always necessary for the term $1/V$ when a virtual image is formed. A negative sign is also needed on the term $1/f$ if the lens is a diverging lens. This is called the 'real is positive' sign convention.

$$\text{The magnification} = \frac{v}{u} = \frac{-24}{6} = -4.0$$

The discrepancy between the two methods is due to minor imperfections which are unavoidable in drawing the diagram.

QUESTIONS

12.11 What happens to the image in Example 4 (Fig 12.14), if the object is moved back until it is 8 cm from the lens?

12.12 Find, by drawing, the position and magnification of the image formed by a diverging lens of focal length 10 cm when an object 3 cm high is placed 12 cm from the lens. Check your drawing by calculation.

INVESTIGATION

Fig 12.15 The same view taken on cameras using lenses of different focal lengths. The numbers quoted are the focal lengths in millimetres.

The aim of this investigation is to prepare a graph which can be supplied to a customer who wishes to purchase a telephoto lens for his camera. Fig 12.15 is a series of photographs taken on a 35 mm camera with lenses of different focal lengths.

20 mm

28 mm

50 mm

85 mm

135 mm

210 mm

300 mm

600 mm

OPTICS

First find a thin lens which has any one of the focal lengths shown here. This can be done by holding the lens so that the image of a distant object is formed on the screen. The image distance is then the focal length of the lens. The magnification now has to be found for your thin lens arrangement. It will be necessary to know a dimension of your distant object.

Find also the angle of view knowing that the width of the image on the film is 36 mm. The part of the ray diagram you need to find the angle of view is shown in Fig 12.16.

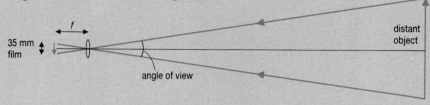

Fig 12.16

The magnification produced by the camera of the same focal length used to take the photographs will not be the same because the object distance will be different.

In the photograph supplied for the 50 mm focal length lens the magnification of the distant mountain is 0.000 044. Use this information to find the magnification in every other case. Prepare a table using these headings.

Focal length of lens/mm	Angle of view/degree	Magnification

Why is the magnification of the trees not the same as the magnification of the mountain? Why is angle of view a more useful quantity than magnification in this case?

Plot a graph of angle of view against focal length.

12.5 THE EYE

Fig 12.17 is a plan view of a person's right eye with rays showing the optical situation when a person with normal sight is viewing a distant object. The function of the various components is as follows.

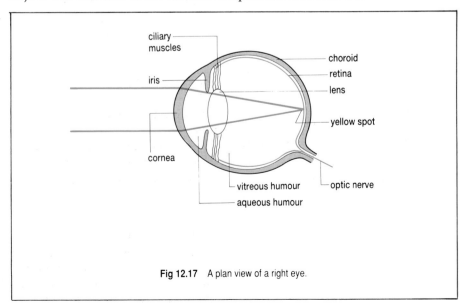

Fig 12.17 A plan view of a right eye.

The choroid is the outer casing of the eye. It is the white of the eye and is approximately spherical with a diameter of about 2.5 cm. At the front of the eye there is a bulge where the choroid becomes transparent. This region is called the cornea and is the area through which light passes into the fluids of the eye and the lens. The area of the hole through which light is allowed in is controlled by the iris. The hole itself appears black and is the pupil; the iris is the coloured part of the eye. The light-sensitive region of the eye is the retina and it is black to prevent internal reflections. The centre of the retina is its most sensitive region and this is called the yellow spot. The part of the retina where the optic nerve leaves the eye is not sensitive to light and is called the blind spot. One of the most remarkable features of the eye is the way the retina itself can adjust its sensitivity to light. The range of sensitivity of the eye is of the order of 10^9. (Full sunlight to full moonlight is an intensity ratio of about 10^6.) The ciliary muscles control the focusing of the eye and there are other muscles, attached to the choroid but not shown, which enable the whole eye to be moved.

The basic optics of the eye are very similar to that of a single converging lens producing a small real image as in a camera. There are some important differences however. Whereas in a camera light passes through the lens and emerges into the air, before hitting the film, light entering the eye through the cornea passes through the aqueous humour, the lens and finally the vitreous humour to fall on the retina. This series of transparent materials all play their part in focusing the light on the retina and the initial refraction at the cornea is particularly important.

A simple demonstration of its importance is given by opening your eyes when swimming under water. The presence of the water in contact with the cornea reduces the refraction at this surface to near zero and results in you being unable to focus on any object. Using goggles underwater enables you to see clearly because air is again present against the cornea surface.

The operation of the lens in the eye is different from that of the lens in a camera. The eye lens modifies the degree of convergence of the light passing through it by altering its shape rather than its position. When an object close to the eye is viewed, the ciliary muscles which surround the eye lens, contract. This squeezes the lens into a thicker shape so causing the lens to have greater power. If a distant object is viewed, the muscles relax and the lens has less power. This is shown in Fig 12.18 in an exaggerated diagram. Sensing the degree of muscular effort necessary is another way of estimating the distance of an object.

The degree of flexibility of the lens decreases with age and gradually the closest point which a person can see clearly in focus (the near point) recedes. A person whose eye lens becomes opaque is said to have a cataract. An operation can be done to remove a person's eye lens. This allows light to reach the retina so vision is restored but the eye cannot then adjust for different distances of objects viewed. A person who has had such an operation therefore needs several pairs of spectacles for different object distances.

Another interesting point about the image on the retina is that, as in a camera, it is upside down. The brain interprets 'top' as being anything which causes an image on the bottom of the retina and vice versa.

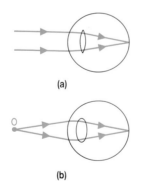

(a)

(b)

Fig 12.18 When a person looks at a distant object the eye lens is thin as shown in (a). The same person can focus on a close object by making the eye lens thicker as in (b).

Short sight

A person whose range of distinct vision is restricted to seeing only close objects clearly is said to be short sighted. Such a person has a longer eyeball than is necessary and requires a diverging lens to correct the defect. The defect and its correction are shown in Fig 12.19.

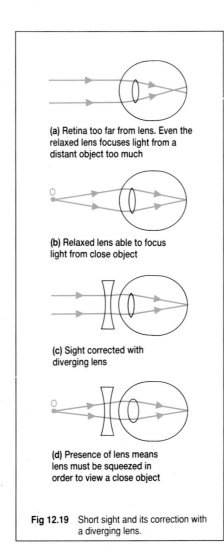

(a) Retina too far from lens. Even the relaxed lens focuses light from a distant object too much

(b) Relaxed lens able to focus light from close object

(c) Sight corrected with diverging lens

(d) Presence of lens means lens must be squeezed in order to view a close object

Fig 12.19   Short sight and its correction with a diverging lens.

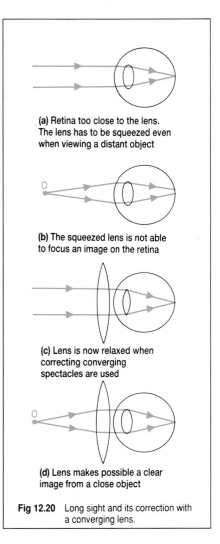

(a) Retina too close to the lens. The lens has to be squeezed even when viewing a distant object

(b) The squeezed lens is not able to focus an image on the retina

(c) Lens is now relaxed when correcting converging spectacles are used

(d) Lens makes possible a clear image from a close object

Fig 12.20   Long sight and its correction with a converging lens.

Long sight

This is the opposite defect to short sight. The eyeball is shorter than is required so light from an object close to the person cannot be focused and even for viewing distant objects it is necessary to squeeze the eye lens. Additional focusing is provided by spectacles which must have converging lenses. The defect and its correction are shown in Fig 12.20.

12.6 OPTICAL
        INSTRUMENTS

The projector, the enlarger and the camera

These have been referred to already. In essence they are single converging lens devices which produce real images on a screen from an illuminated object. In the case of a projector and an enlarger, the image is usually larger than the object, whereas in a camera a smaller image is produced. Fig 12.21 is a diagram of a camera with a ray diagram superimposed on top of it for the situation when the camera is being used to take a team photograph. The shutter is not shown on the diagram. It can be situated either just behind the iris controlling the aperture or, if it is a focal plane shutter, just in front of the film. The focal plane shutter is always used on cameras with interchangeable lenses so that when lenses are changed the shutter is unaffected.

The great variety of commercial instruments cannot be described here but the basic optics is the same for them all. The differences, of which there are many, arise from different requirements of users, different qualities and different prices.

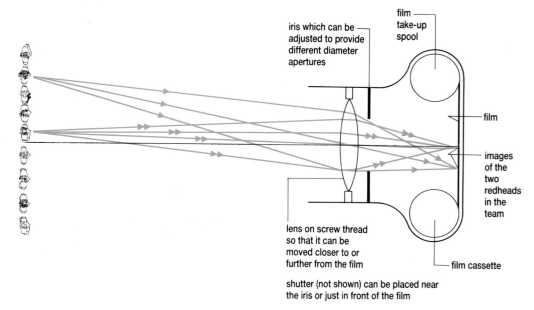

Fig 12.21 A camera in use to take a team photograph.

Labels in figure:
- iris which can be adjusted to provide different diameter apertures
- film take-up spool
- film
- images of the two redheads in the team
- lens on screw thread so that it can be moved closer to or further from the film
- film cassette
- shutter (not shown) can be placed near the iris or just in front of the film

Fig 12.22 An exploded diagram of a Minolta 7000 camera. The construction of the compound lens from many separate lenses is clearly visible, as is the optical arrangement of mirrors and prisms for the viewfinder.

The following exercise should give you some insight into the operation of a camera and enable you to use the theory of the thin lens.

DATA ANALYSIS

The camera

This information is supplied with a single lens reflex camera.

Shutter speed range	1/1000 second – 2 seconds
Focal length of lens	50 mm
Maximum usable diameter of lens (maximum aperture)	25 mm = focal length/2
The aperture of a camera is often written	$f/2$
Minimum aperture	$f/22$
Elements in the lens (number of separate pieces of glass in its construction)	6
Angle of view	47°
Accompanying graph	Fig 12.23

OPTICS

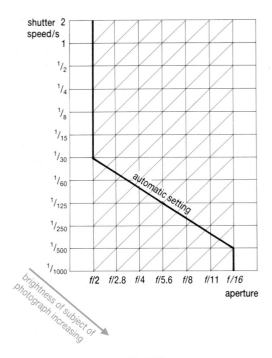

Fig 12.23

Answer the following questions:
(a) The lens moves in and out to focus on objects at distances between 0.45 m and infinity. How far must the lens be able to move?
(b) The scale for focusing around the screw thread of the lens is shown in Fig 12.24. Redraw the scale and put the intermediate distances in their correct places. You can assume that the distance on the scale is proportional to the distance the lens moves towards the film.

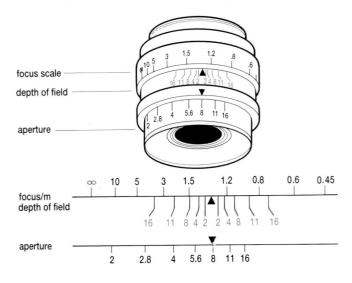

Fig 12.24

(c) The illumination of the film depends directly on the area of the hole through which the light passes. The diameter of the hole is called the aperture and it is usually measured as a fraction of the focal length. Show that $f/5.6$ gives twice the illumination of $f/8$. What would the value 5.6 be, if measured to, say, 4 significant figures?

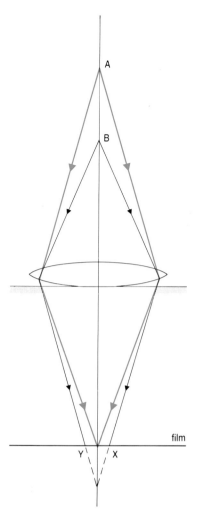

Fig 12.25 The image of A is sharply in focus. B is not in focus, but if the distance XY is small enough B may appear to be in focus.

(d) Show that each of the blue lines on the graph corresponds to the same exposure. That is, a photograph taken of the same subject will cause the same total amount of light energy to fall on the film.

(e) A photograph is taken using the automatic exposure setting of 1/125 s at $f/5.6$. The photographer then takes another photograph with the illumination received by the exposure meter thirty two times brighter. What exposure and aperture will the automatic setting then use?

(f) Fig 12.25 shows the camera receiving light from two objects at different distances. Object A is in focus on the film but object B is not. This is a normal situation for taking a photograph since not all of the object being photographed can be at the same distance from the camera. In practice, therefore, much of any photograph is slightly out of focus. If the distance XY on Fig 12.25 is less than 0.1 mm then object B also appears to be in focus. Explain, using a similar diagram to Fig 12.25, why it is that a photograph taken with a smaller aperture will have a greater range of objects apparently in focus. The distance within which objects seem to be in focus is called the **depth of field**.

(g) Using a diagram of the lens to image region and calculation, find the depth of field when the camera is focused on an object at 0.8 m and is using an aperture of $f/11$.

(h) Explain how to use the depth of field scale shown in Fig 12.24.

(i) Why do you think that some of the information found in this question by using the thin lens formula is not entirely reliable?

The telescope

The basic refracting telescope is a two lens instrument. The first lens, the objective, produces a small real image of a distant object. The second lens, the eyepiece, is used as a magnifying glass to view this first image.

The magnification produced by a telescope is not of particular importance. As with a magnifying glass the actual magnification can vary considerably without the user being very much aware of the fact that it is varying. Apart from the quality of the lenses in a telescope, which can be of overriding importance, two other factors are crucial. One of these is the size of the objective lens. An objective with a diameter of 12 cm has an area which is 9 times greater than an objective of 4 cm diameter. This will make the image 9 times brighter so much more detail can be seen, provided the lenses are of comparable quality. The second factor is the **angular magnification**.

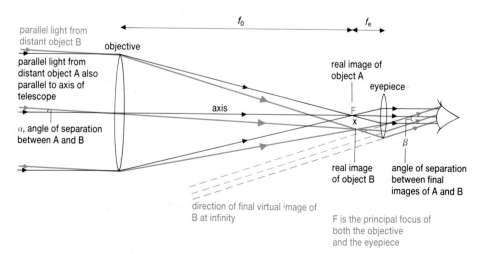

Fig 12.26 Ray diagram of a refracting astronomical telescope.

This is the magnifying quantity which really matters and is defined by

angular magnification =

$$\frac{\text{angle subtended by the image}}{\text{angle subtended by the object when the instrument is not used}}$$

Fig 12.26 is a ray diagram for a telescope. It shows parallel light (in black), from a distant object A, arriving at the objective lens parallel to the axis of the telescope. The light is brought to a focus at the principal focus and then spreads out until it reaches the eyepiece. The eyepiece is normally placed its focal length away from the principal focus of the objective, so parallel light emerges from the eyepiece and the distance between the lenses is the sum of their focal lengths. These rays show clearly how the telescope concentrates light into an observer's eye.

At the same time other parallel light, from a different distant object B, enters the telescope at a small angle α to the axis. If these two objects are viewed without the telescope then they appear to have an angle of separation of α. It may be two stars separated by an angle α. This is called the angle subtended by the two stars at the observer. The light from both A

(a)

(b)

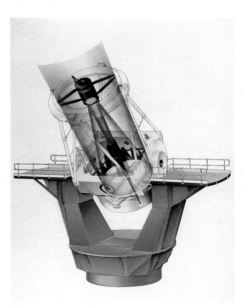

Fig 12.27   A photograph (a) and an artist's impression (b) of the William Herschel Telescope at the Royal Greenwich Observatory in the Canaries. Light is reflected from the main mirror, whose diameter is 4.2 m, to a secondary mirror and back down through a hole in the main one to the focus.

OPTICS

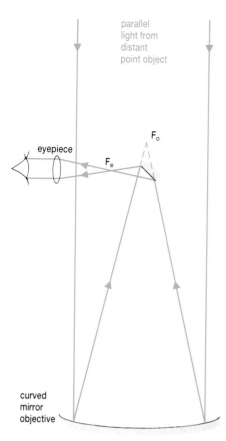

parallel
light from
distant
point object

eyepiece

F_e

F_o

curved
mirror
objective

Fig 12.28 Ray diagram of a different arrangement sometimes used in a reflecting astronomical telescope.

and B is parallel but light from A is not parallel to light from B. On entering the telescope the light from B is brought to a focus at X in the focal plane of the objective lens. When it passes on to the eyepiece it too becomes parallel before entering the eye. If it is now at an angle β to the axis the angular magnification of the telescope is given by

$$\text{angular magnification} = \frac{\beta}{\alpha}$$

$$\text{For small angles } \tan \beta = \frac{FX}{f_e} \approx \beta$$

$$\text{and} \qquad \tan \alpha = \frac{FX}{f_e} \approx \alpha$$

$$\text{angular magnification} = \frac{\beta}{\alpha} = \frac{\dfrac{FX}{f_e}}{\dfrac{FX}{f_o}} = \frac{f_o}{f_e}$$

Therefore if the eyepiece has a focal length of 2 cm and the objective a focal length of 220 cm, the angular magnification will be

$$\frac{220 \text{ cm}}{2 \text{ cm}} = 110$$

If such a telescope were to be used to view the Moon, which subtends an angle of approximately $\frac{1}{2}°$ at an observer on the Earth, the Moon would appear to be 55° wide. It would also appear to be upside down.

By using prisms it is possible to reduce the length of a telescope and to obtain an image the right way up. This is done in binoculars and the values 8 × 40 on a typical pair of binoculars indicate that the angular magnification is 8× and the diameter of the objective lens, which controls the light-gathering power of the instrument, is 40 mm.

Large telescopes are all reflecting telescopes. They have objective curved mirrors rather than objective lenses and because of this they do not suffer from chromatic aberration. All light is reflected by a mirror in the same way whereas different colours of light are refracted different amounts. An optical diagram for a reflector is similar to that for a refractor but with the diagram folded back on itself, Fig 12.28.

QUESTIONS

12.13 What power spectacles are required for a person who is short sighted and who can comfortably view objects between 80 cm and 15 cm? What distances will he be able to view comfortably when wearing the spectacles?

12.14 A telescope with an objective lens of diameter 30 cm and a focal length 2.00 m can be used with eyepieces of focal length 4.0 cm, 2.0 cm or 1.0 cm. What angular magnification can be achieved with each eyepiece and how much brighter will the image of a star appear when using the telescope rather than with the naked eye? What advantages might there be in using the longer focal length eyepieces? If such a telescope were to be used to take a photograph of a distant object, where would the photographic film be placed?

12.15 Explain why **(a)** telescopes commonly have objective lenses with as large a diameter as possible, **(b)** telescopes are sited on the top of mountains and **(c)** instruments for making astronomical observations in the ultra violet and X-ray regions of the electromagnetic spectrum are usually placed in satellites.

SUMMARY

- For reflection the angle of incidence is equal to the angle of reflection.
- For refraction

$$\frac{\sin \theta_1}{\sin \theta_2} = {_1}n_2 = \frac{\text{speed of light in material 1}}{\text{speed of light in material 2}}$$

- Total internal reflection occurs for light travelling towards a less dense medium if the angle of incidence is greater than the critical angle (C).

$$n = \frac{1}{\sin C}$$

- The principal focus of a lens is the point to which parallel light, parallel to the axis, is refracted to, or from, after passing through the lens.

$$\text{lens power} = \frac{1}{f}$$

$$\frac{1}{u} + \frac{1}{v} = \frac{1}{f}$$

$$\text{magnification} = \frac{\text{height of image}}{\text{height of object}} = \frac{v}{u}$$

v is negative for virtual images.
f is negative for diverging lenses.

- A long sighted person has an eyeball which is too short and requires a converging lens for correction
- A short sighted person has an eyeball which is too long and requires a diverging lens for correction.
- Angular magnification =

$$\frac{\text{angle subtended by an image}}{\text{angle subtended by the object if the instrument is not used}}$$

- Angular magnification of telescope when the final image is at infinity

$$= \frac{f_{\text{objective}}}{f_{\text{eyepiece}}}$$

EXAMINATION QUESTIONS

12.16 **(a)** Explain briefly Huygens' method for constructing wavefronts.
A parallel beam of light is projected on to the surface of a plane mirror at an angle of incidence of about 70°. Draw a diagram showing clearly how Huygens' method can be used to determine the direction of the reflected beam.   (6)

(b) What is meant by *critical angle*? Under what conditions will a wave be totally reflected on meeting a boundary between two media, both of which will allow the passage of the wave?   (4)
A beam of light travelling through a transparent medium A is incident on a plane interface into air at an angle of 20°. If the speed of light in the medium is 60% of that in air, calculate the angle of refraction in air.
When the beam is shone through another transparent medium B and the incident angle is again 20°, it is found that the beam is just totally reflected at a plane interface with air. Calculate the speed of light in B as a percentage of the speed of light, c, in air.   (8)

(ULSEB 1983)

12.17 A narrow parallel beam of white light strikes the midpoint of the side of a 60° prism at an angle of incidence greater than 30°. Draw a diagram showing what happens to the light as it passes into the prism and out of the far side. Explain, with references to your diagram, the meaning of the terms *refraction*, *deviation* and *dispersion*.   (7)

(ULSEB 1988)

12.18 A right-angled isosceles prism is to be used to deviate a beam of light through 90° as shown in the diagram below. What is the minimum value of the refractive index of the material of the

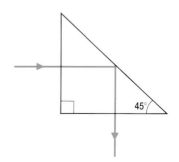

prism so that this may be achieved by total internal reflection?   (4)

(UCLES 1988)

12.19 Define the terms (i) refractive index, (ii) critical angle.   (2)
Deduce the relationship between them.   (2)
Describe briefly a method of determining the refractive index of glass.   (5)
A plane wavefront is incident at an acute angle on a plane interface between two transparent media. Show how the resulting change of direction can be explained on the basis of the wave theory of light.   (3)
A parallel beam of light travelling through water is incident on its horizontal surface with the air. The speed of light in water is 2.25×10^8 m s^{-1}. Calculate the minimum angle the beam can make with the horizontal if the light is to escape into the air, where the speed of light is 3.00×10^8 m s^{-1}.   (3)

(UCLES 1987)

12.20 **(a)** A thin converging lens L of focal length f forms, on a screen S, a sharp image I of a small object O.

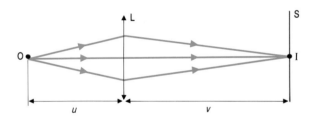

(i) State the relationship between the focal length f, the object distance u and the image distance v.   (1)

(ii) If the distance from the object O to the screen S is fixed at 1.00 m and the focal length f of the lens is 160 mm, it is possible to adjust the lens so as to obtain *two* positions at which an image I is focused on the screen. Find the two values of u for which a sharp image of O is formed on the screen.   (4)

(iii) If the object O is a small circle of radius 4.0 mm, what is the radius of the image for the two values of u in (ii)?   (3)

(b) **(i)** The normal eye can form a sharp image

(b) (i) The normal eye can form a sharp image on the retina of objects at distances in the range from infinity to 250 mm (the nearest distance of distinct vision) from the eye. How is this achieved in the normal eye? (2)

(ii) A short-sighted person cannot see clearly objects beyond a point called the far point. Draw a diagram to illustrate this defect. (4)

(iii) The far point of a short sighted person is 2.0 m from the eye. What power of spectacle lens is required to correct the defect so that very distant objects can be seen clearly? (2)

(NISEC 1989)

12.21 A thin converging lens of focal length 50 mm is to be used as a magnifying glass with the observer's eye close to the lens. If the observer can see images clearly anywhere between 250 mm from the eye and infinity, determine
(i) the range of possible object distances,
(ii) the corresponding range of magnifying powers. (8)

(JMB 1986)

12.22 A lamp and a screen are 80 cm apart and a converging lens placed midway between them produces a focused image on the screen.
A thin diverging lens is placed 10 cm from the lamp, between the lamp and the converging lens. When the lamp is moved back so that it is 30 cm from the diverging lens, the focused image reappears on the screen. What is the focal length of the diverging lens? (5)

(ULSEB 1982)

12.23 The diagram shows the paths of two rays of light from the tip of an object B through the objective, O, and the eye lens, E, of a compound microscope. The final image is at the near point of an observer's eye when the eye is close to E. F_O and F_O' are the principal foci of O and F_E is one of the principal foci of E. The diagram is not drawn to scale.

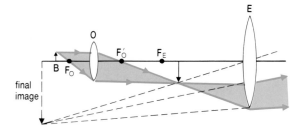

(a) Explain why (i) the object is placed just to the left of F_O, (ii) the eyepiece is adjusted so that the intermediate image is to the right of F_E. (4)

(b) In this arrangement, the focal lengths of O and E are 10 mm and 60 mm respectively. If B is 12 mm from O and the final image is 300 mm from E, calculate the distance apart of O and E. (4)

(JMB 1987)

12.24 A convex lens is used to cast an image of the full moon on to a screen.

(a) Assuming the moon to be in the centre of the field of view of the lens, draw a ray diagram showing clearly how the image is produced. Explain your method of construction. (6)

(b) The moon is 3.5×10^3 km in diameter and is 3.8×10^5 km away from the Earth.
(i) Calculate the angle in radian measure that it subtends at the eye of an observer on Earth.
(ii) If the lens has a focal length of 0.30 m, what is the diameter of the image of the moon produced on the screen?
(iii) If the observer views the image on the screen from a distance of 0.25 m, what angle will it subtend at his eye?
(iv) Calculate the angular magnification that has been achieved by using this lens.

(c) Explain carefully the effect on the angular magnification of using a lens of much longer focal length. (9)
A much greater angular magnification can be achieved by using two converging lenses, one as an objective and the other as an eyepiece. If a converging lens of focal length 0.050 m were placed 0.050 m beyond the screen (now removed) and the observer viewed the moon through both the eyepiece and the objective lenses, what would be the new angular magnification?
Apart from greater magnification, name one other difference between the image produced in this case and when only one lens was used. (3)

(ULSEB 1984)

12.25 (a) (i) Explain with the aid of a ray diagram what is meant by the principal focus and focal length of a converging (convex) lens.
(ii) Draw a ray diagram showing how a converging lens can be used to produce an upright and magnified image of a real object. (6)

(b) Explain the meaning of the term angular magnification as applied to an astronomical telescope. (2)

(c) The diagram, which is not drawn to scale, shows an optical system consisting of two thin converging lenses arranged coaxially. Lens A has a focal length of 40 mm and lens B a focal length of 375 mm. An object O of height 5.0 mm is placed 50 mm from A, and a real image I_1 is formed. This image then acts as an object for B_1 and a virtual image I_2 is formed 250 mm from B as shown.

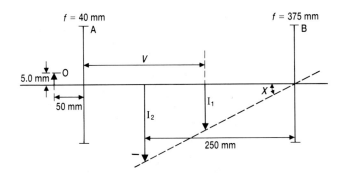

(i) Determine the position of image I_1 by calculating the distance V.

(ii) Calculate the distance between I_1 and the lens B, and deduce the angle X subtended by the final image at B.

(iii) Calculate the angle Y which would be subtended by the object placed 250 mm from an unaided eye. Determine the ratio of X to Y.

(iv) Calculate the linear magnification produced by the lens system.

(v) How would you arrange lenses A and B to form a simple astronomical telescope in normal adjustment? Calculate the angular magnification of the telescope. (13)

(AEB 1986)

12.26 (a) (i) An object is photographed several times with the same camera under conditions of constant illumination. If the f-number of the lens aperture is reduced for consecutive photographs explain why the exposure time must also be correspondingly reduced in order to allow the same amount of light to fall on the film each time.

(ii) If, in (i), the exposure time is 1/60 s at $f/11$, estimate what exposure time is required at $f/2.8$.

(b) Explain what is meant, in photography, by (i) *depth of field*, (ii) *depth of focus*. By considering point objects on the axis of a converging lens, explain with the aid of two separate ray diagrams how each of these arises.

With the aid of another diagram and an axial poit object as above, show qualitatively how **either** the depth of field **or** the depth of focus changes as the lens aperture is changed.

(JMB 1986)

Chapter 13

WAVE INTERACTION

13.1 BEATS

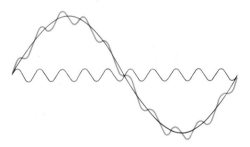

Fig 13.1 Two waves, one of frequency ten times the other, are added together to give a total displacement shown by the blue wave.

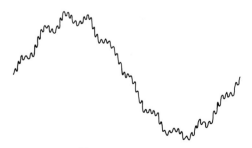

Fig 13.2 Here a third wave of higher frequency has been added to the two shown in Fig 13.1. One important property of waves is that they can be added to one another, with all sorts of surprising effects.

When waves from two sources arrive at the same point, the effect at that point is in general different from the effect which either of the waves would have made on its own. In this chapter we shall consider some situations where the resultant effect has important or interesting consequences. In order to do this the first thing which is needed is some way of establishing how waves can be added to one another. The photograph, Fig 11.3, shows one fact about waves which we make use of all the time without realising it – namely that two waves can travel through one another without affecting each other at all. The ripples in the photograph are concentric circles. There might be some unusual disturbance where two ripples meet but passing through another ripple does not alter the shape and height of a ripple. Sound waves criss cross a room when a party is taking place in it – but it is still possible to hear a record and to know that the sound from it is the same as it would have been if all the other sound in the room were not present.

Light waves also can travel through one another. The complexity of the light wave pattern within any space can be huge. Light waves of different frequencies, different polarisations and travelling in different directions in three dimensions may be travelling through the space, but each one does not get distorted by the presence of all the other light waves. Because of this we are able to make sense of light waves when they enter our eyes.

A simple demonstration with an oscilloscope shows another feature of wave behaviour. If a microphone is connected to the Y-input of the oscilloscope, and the time base and Y-sensitivity are suitably adjusted, then the pattern of any sound wave vibration can be seen. If now two sounds are made simultaneously, the pattern which appears on the screen is simply the sum of the two sounds separately.

This can be seen clearly if pure tones are used. Fig 13.1 shows two separate wave patterns, one of a frequency ten times the other and with smaller amplitude. It also shows the wave pattern obtained when both of the waves are present. This can be extended. Fig 13.2 shows a wave in which three frequencies are present.

This placing of waves one on top of another is known as **superposition**. The principle of superposition of waves is important both theoretically and

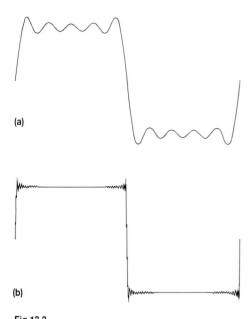

(a)

(b)

Fig 13.3

practically. One practical application is in the use of optical fibres where thousands of messages, in the form of light waves, may be passed down a single fibre simultaneously. At the other end of the fibre all of the messages can be unscrambled from the light waves.

Another interesting and important effect appears if waves of a series of increasing frequencies are added together. Fig 13.3(a) shows the result of adding together waves of frequency f, $3f$, $5f$, $7f$ and $9f$ and of amplitude 1, $\frac{1}{3}$, $\frac{1}{5}$, $\frac{1}{7}$ and $\frac{1}{9}$ respectively. If this is continued to $99f$ then quite an accurate square wave is produced as shown in Fig 13.3(b). The high frequencies are always important in the transmission of square waves. It is the high frequencies which are responsible for the sharpness of the leading and trailing edges of the square wave.

If two sound waves of similar amplitude but slightly different frequencies are listened to, a rhythmic pulsing of the intensity of the waves is heard. This is known as beating and is a direct result of superposition of waves. Fig 13.4 (a) and (b) show the variation of pressure with time for two waves of the same amplitude. At first sight the two graphs appear to be identical, but if you look carefully you will see that at X the waves are in

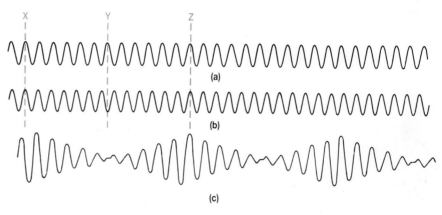

Fig 13.4 Two waves of nearly equal frequency added to one another give a beating effect of increased and decreased amplitude.

phase with one another but that by Y they are out of phase. At Z they are back in phase. Between X and Z wave (a) has had 12 oscillations whereas wave (b) has had 13 oscillations. The sum of these two waves is shown in Fig 13.4(c). Where (a) and (b) are in phase with one another, the resultant amplitude of (c) is large. Where (a) and (b) are in antiphase, that is exactly out of phase, the resultant amplitude is zero.

If f_1 is the frequency of the first wave, f_2 is the (higher) frequency of the second wave and t is the time between beats then

the number of oscillations of the first wave is $f_1 t$ and
the number of oscillations of the second wave is $f_2 t$.

But the second wave oscillates once more than the first wave so

$$f_2 t - f_1 t = 1$$
or $$(f_2 - f_1)t = 1$$

so $$(f_2 - f_1) = \frac{1}{t}$$

But t = period of beats and $t = \dfrac{1}{f}$ where f is the frequency of the beats, giving

$$f_2 - f_1 = f$$

where f is the frequency of the beats.

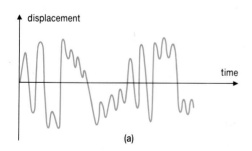

displacement

time

(a)

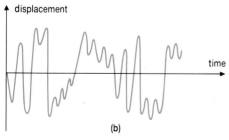

displacement

time

(b)

Fig 13.5 Sound and antisound can be added together to give zero amplitude.

WAVE INTERACTION

The beat frequency is the difference between the frequencies of the waves causing the beating.

Beats are used to detect small differences in frequency between two waves. They are also used for tuning musical instruments. On the higher frequencies a piano has two or three strings for each note. These strings must be accurately in tune with one another or a honky-tonk sound is produced. If the note is played, a person with a trained ear can listen for any beating which takes place between the sound from two strings and then adjust the tension in the strings to eliminate beating.

Another use of superposition of waves is increasingly being used to reduce noise pollution. It involves making antisound. In a noisy environment a microphone picks up the sound and an amplifier is used which reverses the sound, as shown in Fig 13.5. If the antisound is played through headphones it can cancel out the original sound in the same way that the two beating waves cancel out when they are out of phase.

13.2 STATIONARY WAVES

The term stationary waves sounds contradictory. Stationary in this context does not mean that nothing is moving but simply that the positions of the crests and troughs of the wave are not moving. An alternative name for stationary waves is **standing waves**. When a string on a violin vibrates, parts of the string are moving to and fro but some parts are fixed. A finger on the left hand of the violinist is placed on a string to stop it moving at one point and the bridge over which the string passes also keeps a point of the string fixed, Fig 13.6. In the form of its simplest vibration the string appears as shown in Fig 13.7 and the maximum amplitude of the wave is in

Fig 13.6 The frequency of vibration of the strings of the 'cello' or a violin is controlled by the length of string used, the tension in the string and its mass per unit length.

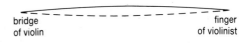

bridge of violin   finger of violinist

Fig 13.7 The movement of a string on a violin.

Fig 13.8 The formation of a stationary wave. The smaller waves have the same amplitude and are travelling in opposite directions. The large wave is the stationary wave and results from adding the other two waves together.

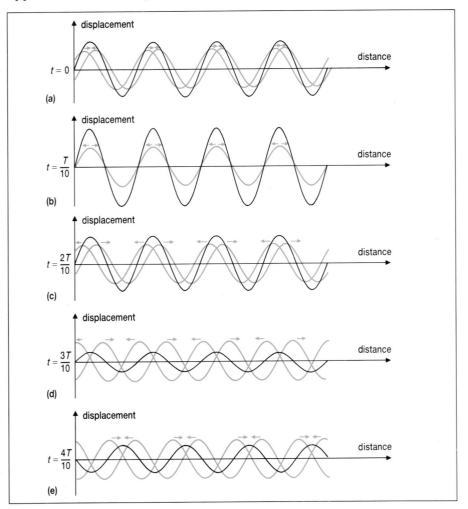

WAVE INTERACTION

the centre of the string. This is one example of a stationary wave. A characteristic of a stationary wave is that there are some parts of the wave where the amplitude is always zero. These points are called **nodes**. Halfway between nodes are points where the amplitude is a maximum. These points are called **antinodes**. Between any two nodes all the points on the wave are in phase with one another.

Stationary waves are set up as a result of the superposition of two waves of the same amplitude and frequency travelling at the same speed in opposite directions. Before considering how this comes about in practical situations, it is worthwhile examining Fig 13.8 which shows the formation of a standing wave. The blue waves and the grey waves are travelling in opposite directions, and the bold black wave is the stationary wave which results from adding the other two waves together. Note that the distance between two nodes is a half a wavelength and not a full wavelength. If the patterns of the stationary wave are superimposed on one another then the sequence of movement is as shown in Fig 13.9 and in the photograph, Fig 13.10. Transverse stationary waves are easy to see when they are set up on a string.

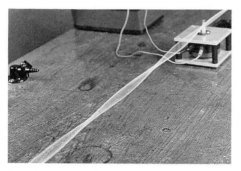

Fig 13.10 By causing this string to oscillate rapidly at a particular frequency a stationary wave has been set up. The nodes and antinodes can be seen clearly.

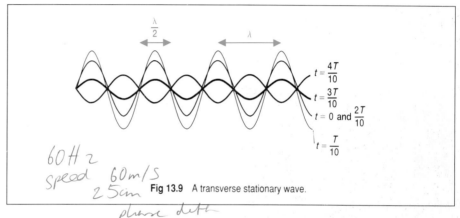

$t = \dfrac{4T}{10}$

$t = \dfrac{3T}{10}$

$t = 0$ and $\dfrac{2T}{10}$

$t = \dfrac{T}{10}$

60 Hz
speed 60 m/s
25 cm
phase diff

Fig 13.9 A transverse stationary wave.

The following investigation will enable you to see how the frequency of the vibration depends critically on the tension in the string and its length.

INVESTIGATION

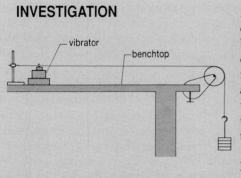

Fig 13.11 Experimental arrangement to investigate stationary waves on a string.

Attach a piece of light string or twine to a vibrator and pass the other end of the string over a pulley (which must have little friction at its bearings) to a tensioning weight as shown in Fig 13.11. Some experimentation must be carried out to obtain a suitable weight but the mass used as a tensioning weight will normally be around 100 g. The aim of the investigation is to find the series of different frequencies which will give 1, 2, 3, 4... etc. antinodes. These frequencies can be used to find the speed of the transverse waves along the string. The following table headings should be used.

Number of antinodes	Distance between nodes	Wavelength (λ)	Frequency (f)	Speed
1				
2				
3				

The investigation can be done in a more dramatic way if a stroboscopic flashing unit is used to find the frequency. It will also enable you to see more clearly exactly how the string is moving when it has a stationary wave vibration.

The second part of the investigation is to find out how the frequency of vibration is determined by the tension in the string. In this case the number of nodes and antinodes must be kept constant and the frequency adjusted for each different tension. A series of values of the frequency f and the corresponding tension T is obtained, and the problem is to find what relationship there is between f and T. This can be done using a log graph in the following way:

Assume that the relationship is

$$f = k\,T^n$$

where k is a constant and n is an unknown power. Taking logs of both sides of the equation gives

$$\lg f = \lg k + n \lg T$$

If this is compared with the equation of a straight line graph

$$y = c + mx$$

it can be seen that if $\lg f$ is plotted against $\lg T$ then a straight line graph will result in which the intercept on the y-axis is $\lg k$ and the slope of the graph is n, the power to which T has to be raised.

Fig 13.12 With the large number of strings on a harp the total force between the top and bottom is considerable.

Stationary waves are set up in all stringed instruments and these instruments can be tuned by adjusting either the length or the tension of the string. Instruments with a large number of strings, such as a piano or a harp, have such a large total force between the top and the bottom of the instrument that they need to have a very strong construction, Fig 13.12. The pitch of the note produced is also affected by the mass per unit length of the string. In practice the oscillation of a string is not as simple as has been indicated here. A complex series of oscillations is possible within any string simultaneously and this gives instruments their characteristic sound. The complexity depends on such factors as the tension and mass per unit length of the string, but it also depends on the skill of the player, the shape and quality of the instrument and the way in which the string is set in motion. A plucked string sounds different from a bowed string. This is partly because the vibrations start more suddenly and die away more quickly when a string is plucked than when it is bowed. Bowing a string is encouraging it to vibrate and it takes longer (more milliseconds) to set the vibration going.

The range of frequencies present in any complex vibration are called **harmonics**. The lowest frequency harmonic is called the **fundamental** and the others are called **overtones**. It can be seen from Fig 13.13(a) how the different harmonics can be set up in a string of fixed length, tension and mass per unit length. The amplitude of the harmonics usually decreases as the frequency rises. All these vibrations taking place in a string cause the instrument structure to vibrate and emit sound waves. The sound waves are as complex as the string vibrations, and the sum of all the harmonic oscillations gives rise to a sound wave like that shown in Fig 13.13(b).

The construction of musical instruments is as much an art as a science, but it is an art based on scientific principles. A Stradivarius violin has such a high value not only because it is antique but because the skill in shaping its sound box cannot quite be reproduced today. The balance of tone from all of the harmonics present in the oscillating string is dependent on the sound box. The shape of a wave from an instrument is called its **quality** and it depends on many factors. If electronic music is to be made artificially it is necessary to use a harmonic synthesiser. This is a series of oscillators which can be adjusted in amplitude and then added together to get any

WAVE INTERACTION

wave shape required. The exercise on the synthesiser enables you to see how differently shaped waves can be obtained by adding pure waves together, and also to see how a wave can be analysed into its harmonics.

Up to this point stationary waves have been considered only for transverse waves, but longitudinal stationary waves are also possible. The particles at a node in a longitudinal stationary wave do not move: the particles at an antinode have maximum amplitude. As with all longitudinal waves the direction of the oscillation is parallel to the direction in which the wave is moving. If a closed tube is to have the right conditions within it for a stationary wave to be set up, it must have two sound waves of equal frequency travelling in opposite directions. This can be achieved by reflecting the sound wave from the end of the pipe. In musical terms an organ pipe with a closed end is said to be a stopped pipe.

Fig 13.14 can be used to see how the particles of air move in an organ pipe when it is sounding the third harmonic. The diagram is best looked at

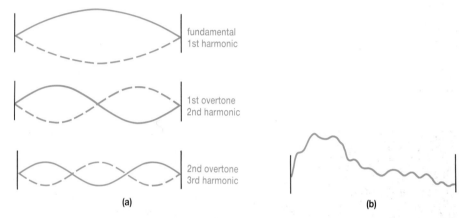

Fig 13.13 (a) The harmonic oscillations of a string on a harp.
(b) The resulting oscillation of a sound wave from a harp string.

through a narrow slit which is moved slowly down the page. The slit needs to be about 5 cm long and preferably not more than 1 mm wide. Fig 13.15 is a summary of Fig 13.14. It shows the extent of the displacement of air particles which takes place.

Fig 13.14 This diagram can be used to show a longitudinal stationary wave. To see the effect move a piece of stiff card with a slot of no more than 1 mm cut in it down over the diagram.

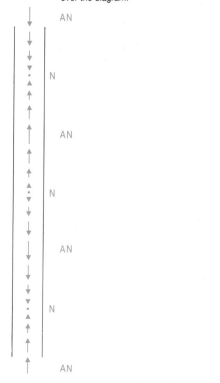

Fig 13.15 Particle movement within a pipe in which a stationary longitudinal wave has been set up.

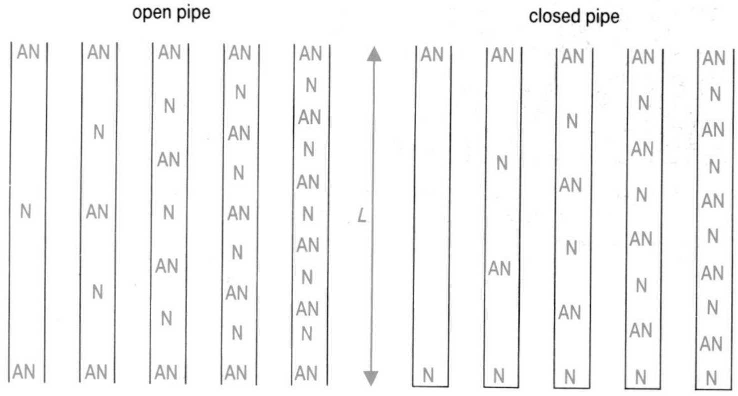

	open pipe					closed pipe				
frequency	f_0	$2f_0$	$3f_0$	$4f_0$	$5f_0$	F_0	$3F_0$	$5F_0$	$7F_0$	$9F_0$
wavelength	$2L$	$\dfrac{2L}{2}$	$\dfrac{2L}{3}$	$\dfrac{2L}{4}$	$\dfrac{2L}{5}$	$4L$	$\dfrac{4L}{3}$	$\dfrac{4L}{5}$	$\dfrac{4L}{7}$	$\dfrac{4L}{9}$
harmonic	fundamental	2nd	3rd	4th	5th	fundamental	3rd	5th	7th	9th
possible frequency/Hz	128	256	384	512	640	64	192	320	448	576
			(a)					(b)		

Fig 13.16 Harmonic frequencies in (a) open and (b) closed pipes.

A surprising fact is that a sound wave can also be reflected from the open end of a pipe so that almost any pipe can have a stationary wave set up within it. Whereas the closed end of a pipe is always a node, the open end of a pipe is always an antinode. Instruments such as the flute and the recorder are open pipes within which a stationary wave may be set up with an antinode at either end. The pitch of the note produced can be varied either by blowing in different ways to obtain different harmonics or by varying the length of the pipe being used by covering or uncovering holes placed at intervals along the length of the pipe. Fig 13.16(a) shows how different frequencies can be produced from an open pipe. It can be seen that the length of the pipe, L, is equal to half the wavelength of the fundamental (λ_0). Other possible wavelengths are $\frac{1}{2}\lambda_0$, $\frac{1}{3}\lambda_0$, $\frac{1}{4}\lambda_0$, $\frac{1}{5}\lambda_0$ . . . etc. Part of the skill of a woodwind or brass musician is being able to control which of the possible modes of oscillation actually dominates. The diagram has been constructed for an open pipe of length approximately 1.3 m. This could be for a ceremonial trumpet similar to the one shown in the photograph in Fig 13.17. Orchestral trumpets look shorter because they are coiled. They can also produce more notes than a ceremonial trumpet because the length used can be altered.

Fig 13.17   These ceremonial trumpeters play instruments of a fixed length. They can therefore only play certain tunes with a very restricted number of notes. Each note corresponds to a different harmonic in the available length of tube in the trumpet.

Fig 13.16(b) shows the corresponding frequencies for a stopped pipe. The lowest frequency is lower than for an open pipe but the even harmonics are not present in a stopped pipe.

QUESTIONS

13.1 The frequency f generated by a vibrating spring is given by the equation

$$f = \frac{1}{2l}\sqrt{\frac{T}{m}}$$

where T is the tension in the string, l is the length of the string and m its mass per unit length. On a particular piano two adjacent low notes use the same tension in wire of mass per unit length 0.030 kg m^{-1}. One of the wires is 0.87 m long and it vibrates at a frequency of 55 Hz. What is the tension in the string? The other string is slightly longer at 0.91 m. What is the frequency of its vibration and what is the beat frequency produced when both of the notes are played together?

WAVE INTERACTION

13.2 Instruments such as the clarinet use air columns which have one end open and the other end, by the reed, effectively closed. What is the effective length of a clarinet that has a fundamental frequency of 147 Hz?

13.3 What are the possible frequencies of a note if it beats at a rate of 4 Hz when heard together with a tuning fork of frequency 440 Hz? How could you tell which was the correct one?

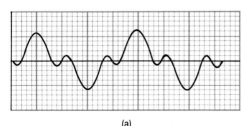

(a)

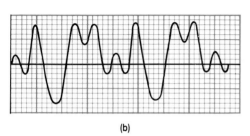

(b)

Fig 13.18

Synthesisers

An electronic synthesiser can be a very complex electronic machine but the principle it uses can be demonstrated quite simply. By combining together waves of different frequencies and amplitudes it is possible to form any regularly repeating pattern. Some of these patterns are shown in Figs 13.1 to 13.3. In a synthesiser, many amplifiers are used; the outputs from them are combined in mixer circuits to give a signal which can be used as the input to a power amplifier which drives the loudspeaker.

When plotting waves in answering the following questions, do not be over-careful with the graphs. They are easier to draw if the amplitude is not too large, and you never need to draw more than one cycle of the lowest frequency because the pattern will repeat itself. In many cases you will only need half a cycle as the other half of the cycle will be a mirror image of the first half. What you do need to be more careful with is the adding of the waves. The intention is to add or subtract displacements together for each value of θ. Look for places where one of the waves is zero; the sum will at that point be equal to the other wave.

(1) Sketch 3 graphs superimposed on top of one another of
   (a) $y_1 = \sin \theta$,   (b) $y_2 = \frac{1}{3} \sin 3\theta$ and   (c) $y_3 = \frac{1}{2} \cos 2\theta$
(2) Add graphs (a) and (c) together.
(3) Subtract graph (b) from graph (a).
(4) Add graph (c) to your answer to (3).

These graphs show some of the basic shapes which can be achieved. They also show how different frequencies can be observed in the resulting wave pattern. Without drawing any more waves, what happens to the pattern in (2) if $y_3 = \frac{1}{4} \cos 2\theta$ and to the pattern in (3) if $y_2 = \sin 3\theta$?

Now look at the wave patterns shown in Fig 13.18. One of these patterns shows the variation in pressure when a recorder is being played and the other when a violin is being played. If you wanted to synthesise the sound for each of these instruments, which frequency or frequencies would you expect to have to use a large amplitude for? Give your answer in terms of the fundamental frequency.

In order to make synthesised sound acceptable to listen to, many more frequencies need to be used than have been suggested here. Besides adding to the frequencies it is also necessary to control other features of the note being produced. List some of these additional features.

Many synthesisers have a whole range of special sound effects which they can produce. List some of these effects and outline how you would need to modify the basic signal to produce the desired effect.

Interference in sound

If two loudspeakers are arranged so that they are about half a metre apart and are connected to the same signal generator producing an output frequency of around 1 kHz, then at different positions different intensities of sound are heard. At its simplest the waves spread out in a circular pattern from each loudspeaker. Fig 13.19 shows the pattern at one instant

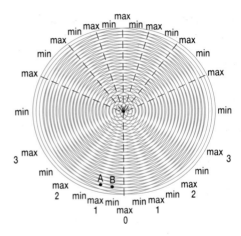

Fig 13.19 Two wave sources close to one another give this pattern of constructive and destructive interference.

but it must be appreciated that the pattern is continually being generated at the sources and spreading outwards and dissipating as it gets further from the sources. It can be seen from this figure that there are places, e.g. at A, where a compression (a crest) caused by one loudspeaker meets a compression from the other loudspeaker. If the principle of superposition is applied at these places then the amplitude of the resultant wave will be double that due to one wave by itself, and as these waves move past the hearer the intensity will be large. There are other places, e.g. at B, where a compression from one loudspeaker meets a rarefaction (a trough) from the other. At these places application of the principle of superposition will give zero resultant amplitude, and when the hearer is at these places no sound will be heard. If this is done in a laboratory then reflections of sound from the walls disturb the pattern. Having two ears also means that if one ear is at a place where the intensity should be zero the other ear is probably not, so that in practice merely a reduction in sound intensity is heard rather than no sound at all. If you move around the room the change from low intensity to high intensity is quite marked. This effect is called **interference** and the pattern is called an interference pattern. The term **path difference** is often used when dealing with problems on interference and is the difference in the distance one wave has to travel from its source to the observer compared with the other. The path difference can be measured in metres but it is often more convenient to measure it in wavelengths. Its use can be illustrated using Fig 13.19. From one source to A is 23 wavelengths; from the other source to A is 24 wavelengths, so the path difference is 1 wavelength. From one source to B is 23 wavelengths; from the other source to B is $23\frac{1}{2}$ wavelengths, and so the path difference is $\frac{1}{2}$ wavelength. We are not usually interested in the actual path length. The interesting features arise because of the path differences. In this example, for instance, a maximum intensity of sound will be heard whenever the path difference is a whole number of wavelengths. Cancellation of one wave by the other will occur whenever the path difference is an odd number of half wavelengths, e.g. $\frac{1}{2}\lambda$, $1\frac{1}{2}\lambda$. This is shown on the diagram as a series of maximum and minimum amplitudes. The numbers indicate the number of wavelengths of path difference.

The interference pattern changes as the distance separating the sources is varied and also as the frequency generated by the source is changed.

QUESTION

13. 4 Draw a series of circles similar to Fig 13.19 but with the sources 6 wavelengths apart. Use your diagram to find how the distance between positions of maximum intensity is affected by
(a) increasing the distance of the observer from the loudspeakers;
(b) increasing the distance between the loudspeakers;
(c) increasing the wavelength of the sound waves (decreasing the frequency produced by the signal generator).

Interference in light

If a similar demonstration to the one for sound is set up using lamps instead of loudspeakers, there is no possibility whatsoever of seeing any interference pattern. There are several reasons for this. The sound waves from the loudspeaker had a fixed wavelength; light waves from a domestic lamp have many different wavelengths. The wavelength of a sound wave may be around 40 cm; the wavelength of light is of the order of a million times smaller.

If monochromatic light is used instead of white light then only one wavelength is present and if the two sources are placed close to one another so that the interference pattern spreads out then it should be possible, one would think, to observe an interference pattern which would be visible under a microscope. Early experiments to detect interference in light were carried out in this way in order to establish whether light was a wave motion or not, but because no interference was seen it was assumed, by Newton among others, that light was not a wave motion. It was in 1800 that Thomas Young carried out an experiment which did show interference in light. The difference between his experiment and the others which preceded it was that he used light from a single source and split it into two. This proves critical because two separate sources of light are said to lack coherence.

A polarised, monochromatic source of light has a fixed frequency and the oscillations take place in one plane, but the light still has a complicated nature. Although the average brightness of the lamp is constant, within short time-scales the amplitude varies because the pulses of light, which are generated as the electrons within atoms lose energy, do not occur in any set pattern. This results in light waves from any source having minor imperfections in them as shown in Fig 13.20. The chance of two waves from different sources having these imperfections all occurring at the same

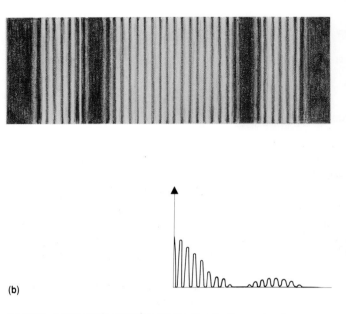

Fig 13.20 Even a monochromatic light wave has frequent interruptions to its smooth oscillation. These interruptions make interference between two different light sources impossible in practice.

instant is virtually zero. The frequency is very high, around 10^{15} Hz, so even if an imperfection occurs only once in 10^8 waves there will still be 10^7 imperfections per second. If, on the other hand, light from a single source is split into two, the two parts will have all of their imperfections occurring simultaneously. The light in these two parts is said to be **coherent** light. In light which is coherent there is a constant phase relationship between the two parts.

Fig 13.21(a) shows how Young's experiment can be repeated in a modern laboratory. It consists of a monochromatic light source such as a sodium lamp or a laser with a narrow slit placed in front of it. The light from the

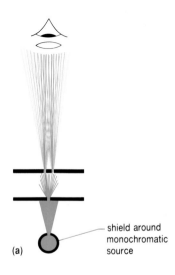

shield around monochromatic source

(a)

(b)

Fig 13.21 (a) The optical arrangement for the Young's slits experiment.
(b) The interference pattern formed by the double slit arrangement.

narrow slit spreads out and illuminates a double slit held accurately parallel to the single slit some distance in front of the single slit. If the light emerging from the double slit is viewed through a travelling eyepiece, an image similar to the one shown in Fig 13.21(b) can be seen. Suitable dimensions for the experiment are:

width of slits	0.2 mm
separation of double slits s	0.5 mm
distance of lamp to single slit	10 cm
distance of single slit to double slit	20 cm
distance of double slit to eyepiece D	30 cm

If you do this rather difficult experiment you are advised to start with the overall distance from lamp to eyepiece rather shorter than is suggested here, find the interference pattern, which will be small and bright, and slowly rotate the double slit for maximum clarity. This will occur when the slits are all parallel to one another. Then you can increase all the distances to increase the width of the fringes, that is, the separation of the light and dark regions of the interference pattern. As the fringe width increases the whole pattern gets dimmer and more blurred so it is necessary to work in a dark room and to shield the lamp to cut out scattered light. Because the fringes are blurred, the fringe separation should be calculated by measuring the distance across as many individual fringes as possible and then dividing by the number of fringes.

This experiment can be used to find the wavelength of the mono-chromatic light. To see how this can be done consider Fig 13.22(a) which shows the path which light takes from each of the double slits to the first bright fringe away from the centre of the fringe pattern. A problem arises with drawing this diagram in that the horizontal distance from the slits to the eyepiece is many centimetres whereas the slit width and the fringe width are both about a millimetre. The diagram as drawn therefore exaggerates the fringe and slit widths considerably. This means that the triangle APB is really even more elongated than is shown. Its two long sides are very nearly parallel to one another and this enables approximations to be made which are very good approximations.

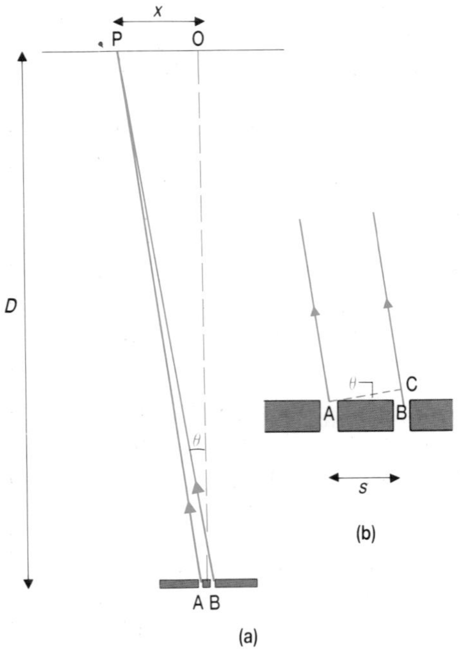

For the first bright line at P the distance BP must be one wavelength longer than the distance AP. The extra distance is BC and is shown on a larger scale in figure 13.22(b).

Because the slit width is so small compared with D, APO can be regarded as a right-angled triangle. This makes the two triangles ABC and APO to be nearly the same shape so it follows that

$$\frac{AB}{BC} = \frac{AO}{OP}$$

giving $\quad \dfrac{s}{\lambda} = \dfrac{D}{x} \quad$ giving

◆ $\quad \lambda = \dfrac{sx}{D}$

A fringe width of about a millimetre would be expected for $D = 1$ metre.

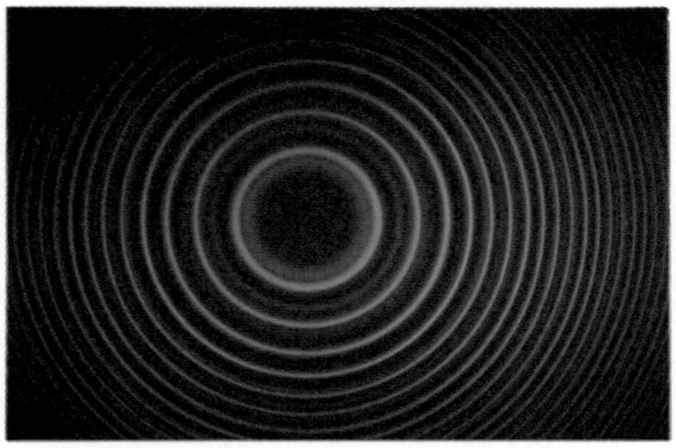

Fig 13.22

Fig 13.23 Interference is shown using 3 wavelengths (2 blue and 1 green) of the mercury spectrum. The sharp rings are formed by interference caused by multiple reflections between two accurately parallel mirrors, in this case 3.12 mm apart.

Example 1

Two loudspeakers on a radio are a distance of 40 cm apart. What frequencies will possibly be inaudible to a person whose ears are 14 cm apart and who is placed directly in front of the radio at a distance of 150 cm? See Fig 13.24. The speed of sound is 334 m s^{-1}.

If AE_2 is half a wavelength further than AE_1, then because the diagram is symmetrical BE_1 is also half a wavelength longer than BE_2. This means that

WAVE INTERACTION

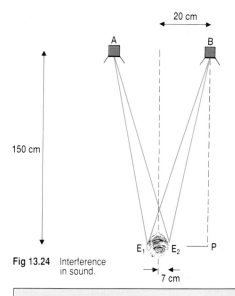

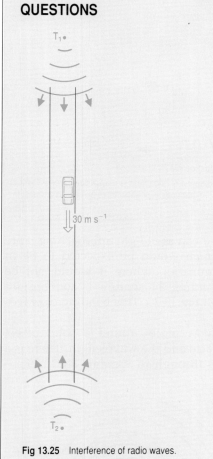

there is a possibility that the sound waves arriving at each ear are in antiphase and will destructively interfere. Using the Pythagoras theorem gives

$$BP^2 + E_1P^2 = E_1B^2$$
$$150^2 + (20 + 7)^2 = E_1B^2 = 22\,500 + 729 = 23\,229$$
$$E_1B = \sqrt{23\,229} = 152.4 \text{ cm}$$
$$\text{and} \quad BP^2 + E_2P^2 = E_2B^2$$
$$150^2 + (20 - 7)^2 = E_2B^2 = 22\,500 + 169 = 22\,669$$
$$E_2B = \sqrt{22\,669} = 150.6 \text{ cm}$$
$$\text{since} \quad BE_1 = AE_2$$
$$AE_2 - BE_2 = 152.4 \text{ cm} - 150.6 \text{ cm} = 1.8 \text{ cm}$$

If 1.8 cm is half a wavelength, destructive interference occurs.

This gives $\qquad \lambda = 3.6 \text{ cm} = 0.036 \text{ m}$

so inaudible frequency $= \dfrac{c}{\lambda} = \dfrac{334 \text{ m s}^{-1}}{0.036} = 9300 \text{ Hz}$

Fig 13.24 Interference in sound.

QUESTIONS

13.5 If laser light of wavelength 632.8 nm is shone through a pair of narrow slits a distance of 0.420 mm apart, what will be the separation of the spots on a screen placed at right angles to the initial direction of the light at a distance of 3.00 m from the slits?

13.6 Two radio beacons are located on an east – west line 6.0 km apart. They are emitting radio waves of frequency 1.00×10^5 Hz in phase with one another. The navigator of a ship finds that he is receiving signals from the two beacons which are in anti-phase to one another (that is out of phase by π radians). Draw a diagram to show several possible positions of the ship relative to the beacons. The speed of radio waves is the same as the speed of light, 3.00×10^8 m s^{-1}. How can other beacons be used to establish the exact position of the ship?

13.7 Fig 13.25 shows a car being driven at a speed of 30 m s^{-1} along a straight road between two radio transmitters. The transmitters are sending out the same programme, using a frequency of 1.50 MHz. The radio is heard to fade and strengthen regularly. What is the period of this regular fading?

This problem can be considered either as an interference problem or as a stationary wave problem. You should be able to see from a diagram why these two approaches are equivalent.

Fig 13.25 Interference of radio waves.

13.4 DIFFRACTION

If any wave meets an obstruction then the effect on the wave depends on the relative sizes of the obstruction and the wavelength. An echo can be caused when a sound wave meets a large object. Bats use reflected ultrasonic waves to detect objects around them and because the wavelength of these ultrasonic waves is small the quality of the reflected signal is good enough for a bat to be able to detect a moth in flight. The same is true of light waves being reflected from the pages of this book; because the wavelength of the light is very small compared with the size of the print,

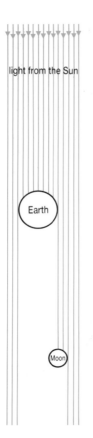

light from the Sun

small details can be seen sharply in focus. The formation of shadows takes place because light travels in straight lines, and when a large obstacle is placed in the path of light rays the rays do not change direction. Fig 13.26 shows a diagram, not drawn to scale, of a partial eclipse of the Moon with the Earth obstructing light from the Sun. Fig 13.27 is a photograph of such an eclipse. The blurring at the edge of the shadow of the Earth on the Moon results from the fact that the Sun is not a point source of light rather than from any bending of light around the Earth.

If these figures are compared with the situation shown in Fig 13.28 then a marked difference is apparent. In Fig 13.28(a) the scale is quite different.

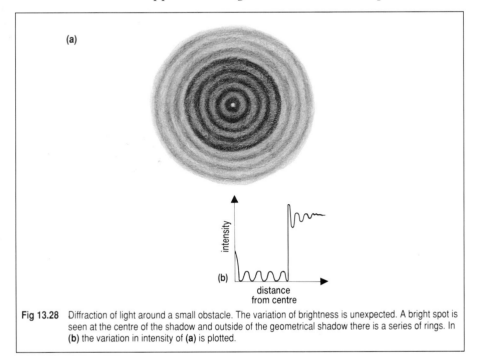

Fig 13.28 Diffraction of light around a small obstacle. The variation of brightness is unexpected. A bright spot is seen at the centre of the shadow and outside of the geometrical shadow there is a series of rings. In **(b)** the variation in intensity of **(a)** is plotted.

Fig 13.26 A partial eclipse of the Moon. The Moon is moving towards the reader.

Fig 13.27 An eclipse of the Moon. The Earth's shadow is blurred because it is caused by an extended source, the Sun – not because of diffraction.

The obstacle is a tiny ball bearing and the waves of light approaching it are appreciably deviated into the space which would be expected to be in shadow. Not only does this give illumination where it would not be expected, but the pattern of light surrounding the shadow is not constant and a series of rings appears as shown in Fig 13.28. This bending of waves near an obstacle is called **diffraction**.

The amount of diffraction which takes place around a solid object depends on the relative sizes of the obstacle and the wavelength. There is a corresponding effect when a wave passes through an opening. Fig 13.29

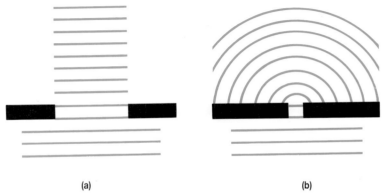

(a)   (b)

Fig 13.29 Waves are approaching an opening. In **(a)** the width of the opening is large compared with the wavelength and little diffraction occurs. In **(b)** the width of the opening is approximately the same size as the wavelength and there is considerable diffraction.

WAVE INTERACTION

shows this happening. In Fig 13.29(a) the width of the opening is about 8 wavelengths and the amount of diffraction occurring is small. In Fig 13.29(b) however the width of the opening is comparable with a single wavelength and now the amount of diffraction occurring is considerable. The actual sizes of the openings do not matter. It is a question of the relative size of the wavelength to the width of the opening which is all important.

For instance, Fig 13.29(a) could be used to illustrate why you cannot see objects round the corners of a door when looking from the inside. Fig 13.29(b) illustrates why, through the same door, you can hear someone outside speaking although you cannot see them.

If a photograph is taken of a screen placed in front of a slit through which light is passing, an unexpected result is found. This is shown diagrammatically in Fig 13.30. The light used is monochromatic laser light and the experimental arrangement is shown in Fig 13.31.

(a)

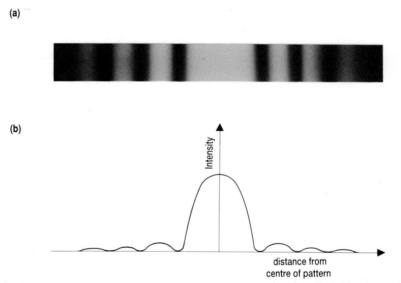

(b)

Fig 13.30 (a) The single slit diffraction pattern. (b) The variation of brightness across a single slit diffraction pattern.

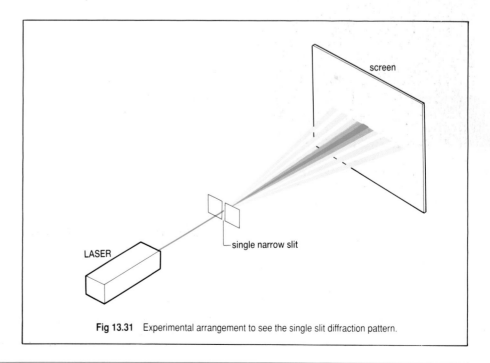

Fig 13.31 Experimental arrangement to see the single slit diffraction pattern.

A graph plotted to show how the intensity varies across the screen is shown in Fig 13.30(b). The central maximum is much brighter than any of the subsidiary maxima and it is also twice the width of the subsidiary maxima. The unexpected feature of this graph is the way that although the intensity falls off as it might be expected to, there is a dark region followed by positions where light is again seen. The angle θ, between the central maximum and the first minimum (in radians) and shown in Fig 13.30(a), is given by

$$\theta = \frac{\lambda}{d}$$

where d is the slit width, and λ is the wavelength of the light.

It can be seen from this that θ gets larger and the pattern spreads out as d is reduced. The pattern also gets dimmer as less light energy passes through a small slit.

This equation has important consequences. For example, it gives the minimum size possible for the image of a point object when using a pinhole camera. This by itself is of no great importance but the fact that it also applies to a real camera is of much greater significance. Cameras are required to show fine detail so the diffraction pattern produced by a camera must be as small as possible. The equation, with modification, can be used to show the theoretical limit for viewing small objects with a microscope, and this theoretical limit is larger than an atom so optical microscopes cannot be used to observe atoms. The ability of an instrument to observe small or distant objects is known as its resolving power.

The resolving power of some large telescopes enables them to pick out craters as small as 100 m diameter on the Moon, but stars appear as if they are point objects. The image which a telescope produces of a star on a piece of film is the diffraction pattern. It is a small blurred centre with surrounding rings.

Example 2
What is the diameter of the real image of a star formed by a telescope in which the objective lens has a focal length of 2.80 m and an aperture of 0.25 m? The average wavelength of visible light can be taken as 500 nm.

To solve this problem it is necessary to make certain assumptions. The first is that the image being considered is being formed, possibly on a photographic plate, at the principal focus of the objective. Secondly we assume that the aperture of the telescope behaves as a slit through which diffraction takes place. Finally we assume that the diameter of the image is equal to the distance between the first minima on either side of the central maximum.

This gives $\lambda = 5.00 \times 10^{-7}$ m, and $d = 0.25$ m

so $\quad \sin \theta = \dfrac{5.00 \times 10^{-7} \text{ m}}{0.25 \text{ m}} \approx \theta$ since for small angles $\sin \theta \approx \theta$

But $\quad \sin \theta = \dfrac{\frac{1}{2} \text{ width of image}}{\text{focal length}} \quad$ (see Fig 13.31)

so $\quad \dfrac{5.00 \times 10^{-7} \text{ m}}{0.25 \text{ m}} = \dfrac{\text{width of image}}{2 \times 2.80 \text{ m}}$

$\quad$ width of image $= \dfrac{2 \times 2.80 \times 5.00 \times 10^{-7} \text{ m}}{0.25 \text{ m}}$

$\qquad = 1.12 \times 10^{-5}$

$\qquad = 0.011$ mm (to 2 sig. figs)

A spot as small as this cannot be achieved in practice because the optics of the telescope will not be perfect. There is also a modification which strictly should be made to the theory because a circular aperture is being used rather than a rectangular aperture. The example nevertheless gives a guide to the size of images being formed by the telescope and also indicates that there is no possibility of getting separate images from two objects which have an angular separation less than about 2×10^{-6} rad.

13.5 DIFFRACTION GRATING

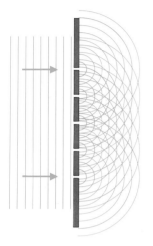

Fig 13.32 It is difficult to analyse the diffraction pattern from several sources.

If parallel, monochromatic light waves approach a series of narrow slits close to one another the resulting pattern of waves is as shown in Fig 13.32. This diagram was drawn showing only five slits for the light to go through, but even with as few as five slits the resulting pattern is too confusing to interpret easily.

The diagram is confusing because the light emerging from each slit is being spread over an angle of 180°, so too much information is being supplied for it to be able to be analysed. In Figs 13.33 and 13.34, however, just the light emerging over a limited angle is being considered. By

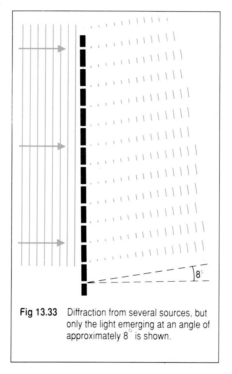

Fig 13.33 Diffraction from several sources, but only the light emerging at an angle of approximately 8° is shown.

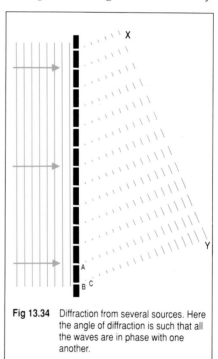

Fig 13.34 Diffraction from several sources. Here the angle of diffraction is such that all the waves are in phase with one another.

working with a limited angle in the first instance, the key features of the diagram can be found and subsequently the whole pattern can be described. To draw Fig 13.33 the angle at which the light emerges has been chosen at random and is about 8°. There is no pattern to the position of the wavefronts, and if all of these wavefronts were focused by a lens to a point the waves would arrive at the point with no pattern to their phase relationship. Some would be at a crest, some at a trough; some just starting to rise, some just starting to fall. If the number of slits through which the light starts is large enough, the total chaos of different phases produces almost *zero* intensity. This is why Fig 13.32 was so difficult to decipher. For each trough there will be somewhere among the waves a crest to cancel it out, and the same will be true for any wave. There will always be another wave in antiphase to it.

A large series of slits like this is called a **diffraction grating** and it is the ability of a diffraction grating to give a dark background, where the intensity is near zero, which makes it so useful for examining spectra. In contrast to Fig 13.33, now look at Fig 13.34. The angle here has not been

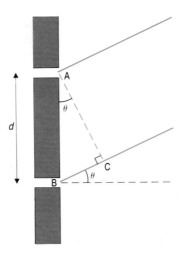

Fig 13.35

chosen at random but has been adjusted carefully so that the light from each slit, as you move down the pattern, travels exactly one wavelength further than the previous one. This means that all the waves arrive at the line XY in phase with one another. If these waves are focused by a lens on to a single point, they will all arrive at that point in phase with one another and will produce a bright image, a maximum intensity region. Because the angle needs to be so carefully chosen to get an image, the image is very sharp and since we are dealing with slits the image at this angle on a screen or on the retina of the eye will be a narrow line if the source itself was narrow.

Now, what will be visible at other angles? Over most of the field of view there will be nothing. If the angle is zero, XY will be parallel to the grating and again all the waves will be in phase at XY, and so again a maximum intensity will be found. The angle used in Fig 13.34 was $22\frac{1}{2}°$. A similar diagram could be drawn for $-22\frac{1}{2}°$ to give another maximum. There is another possibility for a maximum. If a larger angle is used the extra distance which light from one slit travels compared with light from the one next to it could be 2 wavelengths rather than one. This is said to give the second order spectrum. Note that it does not occur at twice the angle. If you try to find this angle by drawing you will find the second order occurs on Fig 13.34 at an angle of 50°.

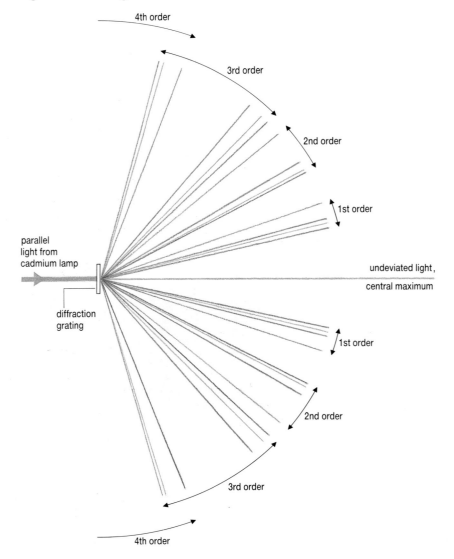

Fig 13.36   The diffraction pattern of cadmium light when viewed through a diffraction grating having 500 lines per millimetre.

WAVE INTERACTION

Once the principle of the diffraction grating is understood, the theory is easy.

Fig 13.35 is triangle ABC of Fig 13.34 drawn on a larger scale, and for a maximum BC must be a whole number of wavelengths.

Since

$$\sin \theta = \frac{BC}{AB}$$

we get

$$\sin \theta = \frac{n\lambda}{d} \quad \text{or}$$

$$n\lambda = d \sin \theta$$

where n is an integer and d is the grating spacing.

Example 3

When a cadmium light is viewed through a diffraction grating having 500.0 lines per millimetre, the following first order spectral lines were observed at the stated angles.

Angle/°	Colour
18·78	Red
14·74	Green
13·89	Light Blue
13·53	Dark Blue

Find the wavelength of each of these lines.

Find other angles at which spectral lines will be observed and draw a diagram to show the pattern.

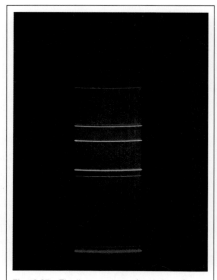

Fig 13.37 The line spectrum of cadmium. A photograph such as this is taken using an instrument called a spectrometer. The spectrum uses a diffraction grating to produce the spectrum and in this photograph the photographer has photographed just the first order spectrum.

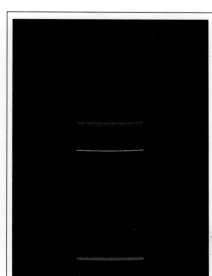

Fig 13.38 The line spectrum of hydrogen taken in the same way as Fig 13.37

First the value of d must be calculated.

Since there are 500 lines per millimetre:

$$d = \frac{0.001 \text{ m}}{500} = 2.00 \times 10^{-6} \text{ m}$$

(This is a typical value for a diffraction grating).

Then use $n\lambda = d \sin \theta$ for the 1st order red line giving

$$1 \times \lambda = 2.00 \times 10^{-6} \times \sin 18.78°$$
$$\lambda = 6.44 \times 10^{-7} \text{ m} = 644 \text{ mm}$$

This can be repeated for the other colours and the results presented in a table. See Table 13.1.

Table 13.1

Order		Red	Green	Light blue	Dark blue
1st	θ_1/degree	18.78	14.74	13.89	13.53
	$\sin \theta_1$	0.3220	0.2543	0.2400	0.2340
	$d \sin \theta_1 = \lambda$/nm	644	509	480	468
2nd	$\sin \theta_2 = 2\lambda/d = 2 \sin \theta_1$	0.6440	0.5086	0.4800	0.4680
	θ_2/degree	40.1	30.6	28.7	27.9
3rd	$\sin \theta_3 = 3 \sin \theta_1$	0.9660	0.7629	0.7200	0.7020
	θ_3/degree	75.0	49.7	46.1	44.6
4th	$\sin \theta_4 = 4 \sin \theta_1$			0.9600	0.9360
	θ_4/degree			73.8	69.4

A tricky step is in realising that although the angles do not increase by multiplying by 2, 3, 4 etc. the sines of the angles do. This is because

$$\sin \theta_1 = \frac{1 \times \lambda}{d}$$

$$\sin \theta_2 = \frac{2 \times \lambda}{d} \quad \text{etc.}$$

Note that there is a limit to the number of orders obtainable because $\sin \theta$ can never be above 1.

QUESTIONS

13.8 When passing monochromatic sodium light through a diffraction grating having $d = 1.8 \times 10^{-6}$ m the first order spectrum was obtained at an angle of 19.1° and the second order at 40.9°. What is the wavelength of sodium light?

13.9 Monochromatic light is observed through a diffraction grating at an angle of 12.2° for the first order. How many other orders can be found and at what angles will they be observed?

13.10 The visible spectrum extends from about 400 nm to 750 nm. If observed with a diffraction grating having 480 lines per millimetre find the angular width of the first, second and third order spectra. Give two reasons why such a spectrum will normally be viewed in the first order.

SUMMARY

- Waves may be added to one another. This is called superposition and three particular cases should be understood.
- Two waves of nearly equal frequency cause beats at a frequency equal to their difference in frequencies.
- Two waves of equal frequency travelling in opposite directions set up a stationary wave in which the node to node distance is half a wavelength.

WAVE INTERACTION

- An interference pattern is set up when two sources of the same frequency are close to one another. The fringe spacing is given by

$$x = \frac{D\lambda}{s}$$

- The spread of a wave by diffraction is large if the opening is small compared with the wavelength; it is small if the opening is large compared with the wavelength.

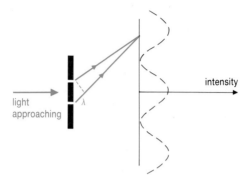

- Double slit interference: the first maximum occurs when the path difference equals the wavelength.

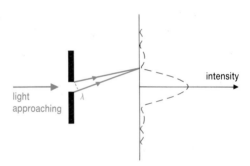

- Single slit diffraction: the first minimum occurs when the path difference for the two rays shown is one wavelength.

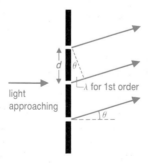

- Diffraction grating: maximum intensity occurs if $n\lambda = d \sin \theta$. Intensity everywhere else is virtually zero.

EXAMINATION QUESTIONS

13.11 (a) Distinguish between (i) the *pitch* and the *frequency* of a pure tone, (ii) the *loudness* and *intensity* of a sound.   (5)
(b) The figure shows the first overtone on a string stretched between points A and B.

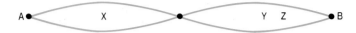

  (i) Explain why the mid-point of the string is stationary at all times.
  (ii) Compare and contrast the motions of the string at points X and Y, and at points Y and Z.   (6)
(c) The key on a piano corresponding to the note of frequency 880 Hz is depressed very gently. The hammer does not strike the string but the damper is lifted and the string is free to vibrate. When the key corresponding to the note of 440 Hz is struck firmly, it is found that the 880 Hz string also vibrates. Give a brief explanation of this observation.   (4)

(UCLES 1989)

13.12 (a) Explain the formation of beats.   (3)
(b) A microphone connected to a cathode-ray oscilloscope receives simultaneously two sound waves of frequencies 1000 Hz and 1100 Hz and equal amplitudes. Sketch and explain what is seen on the screen of the oscilloscope if the time-base repetition frequency is set at 50 Hz.   (5)

(OXFORD 1987)

13.13 (a) (i) How are standing (stationary) waves formed?   (2)
  (ii) State **four** features that distinguish a travelling (progressive) wave from a standing (stationary) wave.   (4)
(b) (i) With the aid of a diagram, describe a laboratory method for producing standing waves in air and for measuring their wavelength. You may assume that a source of sound of known frequency is available.   (6)
  (ii) How could your results be used to determine the speed of sound in air?   (2)
(c) The diagram shows a sinusoidal standing wave produced on a stretched cord by a vibrator oscillating at a frequency of 50 Hz. The mass per unit length of the cord is

0.080 kg m^{-1}. Using the data from the diagram, calculate:
  (i) the tension in the cord;   (4)
  (ii) the horizontal separation of a node and the nearest particle of cord whose amplitude of oscillation is 7.5 mm.   (5)

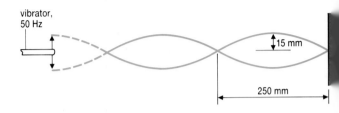

(d) Describe and explain what will be seen if the cord is viewed in light from a stroboscope set at a frequency of: (i) 50 Hz; (ii) 101 Hz.

(OXFORD 1987)

13.14 (a) Describe briefly how you would measure the wavelength of a monochromatic light source using a double slit. Your account should include
  (i) a labelled diagram showing the apparatus used, how it is arranged and giving rough values of the dimensions involved,
  (ii) a list of the measurements that need to be taken and the means by which each is measured,
  (iii) a diagram showing the appearance of the interference pattern.
The theory of the experiment is not required.   (7)
(b) A beam of microwaves of wavelength 3.1 cm is directed normally through a double slit in a metal screen and interference effects are detected in a plane parallel to the slits and at a distance of 40 cm from them. It is found that the distance between the centres of the first maxima on either side of the central maximum in the interference pattern is 70 cm. Calculate an approximate value for the slit separation. Why does the formula you have used only give an approximate value?   (4)
(c) Light from a point in a monochromatic source is emitted in pulses lasting about 10^{-9} s.
  (i) Two light beams derived from the same source can be recombined and produce stable interference patterns provided

the path difference introduced between the beams is not too great. Explain why this is so and calculate an approximate value for this path difference. (Speed of light is 3.0×10^8 m s^{-1}.)

(ii) Splitting the wavefront by means of a double slit is one method of using a single source to produce two beams. Describe with the aid of a diagram an alternative method.

(iii) Explain why light from two separate monochromatic sources of the same frequency will not, when recombined, result in a stable interference pattern.

(ULSEB 1986)

13.15 Sound travels in air as a longitudinal wave. Explain this statement. Discuss clearly how sound waves may be represented by a sinusoidal waveform relating
(i) displacement and position,
(ii) pressure and position. (6)
Show that, in a sound wave, displacement and pressure are out of phase by one quarter of a cycle. (2)
Explain what is meant by (i) frequency, (ii) node, in relation to stationary waves. (3)
Describe a method of measuring the frequency of a sound wave. (5)
Two loudspeakers, facing each other and about 50 m apart, are connected to the same oscillator, which gives a signal of frequency 165 Hz. Describe how the sound intensity varies along a line joining the loudspeakers. Calculate the separation of the nodes. (Take the speed of sound in air as 330 m s^{-1}.)

(UCLES 1987)

13.16 (a) A note of frequency 600 Hz is sounded continuously over the open upper end of a vertical tube filled with water. As the water is slowly run out of the bottom the air in the tube resonates, first when the water level is 130 mm below the top of the tube and next when the level is 413 mm below the top of the tube. Calculate
(i) the speed of sound in the air in the tube,
(ii) the position of the water level when the third resonance occurs. (5)
(b) Describe the motion of the air particles at various points along the axis of the tube when the air first resonates. (4)

(JMB 1986)

13.17 Two narrow parallel slits, distance d apart, act as coherent sources of monochromatic light of wavelength λ. Interference fringes are observed in a plane parallel to the plane of the slits and a distance D from it.

(a) Derive an expression for the distance apart of the centres of adjacent dark fringes near the centre of the pattern, if D is very much greater than d. (7)
(b) Describe an arrangement for producing fringes in this way, indicating appropriate dimensions for the apparatus. Explain how you would use the arrangement to determine the wavelength of light. (9)
(c) Explain how the fringe pattern would change if it were possible to immerse the whole arrangement in water.

(JMB 1986)

13.18 (a) Draw a labelled diagram of an experimental arrangement that would enable you to examine and to make measurements on the diffraction pattern produced when a parallel beam of light from a monochromatic source is incident on a single slit. Indicate approximate dimensions of the arrangement, including the width of the slit. (3)
(b) Derive an expression for the angular separation, θ, between the first minimum and the centre position of the diffraction pattern, in terms of the slit width, s, and the wavelength of the light, λ. (3)
(c) Sketch a graph of relative intensity of the diffraction pattern against angular separation from the centre of the pattern. Indicate on your graph the values of the angles which correspond to the first and second minimum for a slit of width 0.20 mm and light of wavelength 6.5×10^{-7} m. (3)
(d) If a white light source were used in place of the monochromatic source describe the pattern obtained and compare it with the original pattern. (2)
(e) Describe how the intensity variation of part (c) would alter if the single slit were replaced by two slits about 1 mm apart, both of the same width as the single slit. (2)

(JMB 1987)

13.19 Light from a cadmium discharge lamp can be used to determine the spacing of the lines on a plane diffraction grating. This is done by measuring the angle ϕ between the diffracted beams either side of the normal in the first order spectrum for light incident normally on the grating.

(a) If the measured value of ϕ is 46° 43′ and the red line used in the cadmium spectrum is of wavelength 644 nm, calculate the number of lines per metre on the grating.

(b) Make a suitable calculation to test whether the second order spectrum of this line will be visible. (7)

(ULSEB 1988)

13.20 (a) The spacing of the slits of a plane diffraction grating is d. It is illuminated normally by parallel light of wavelength λ. Light coming from the grating at an angle θ to the normal is observed using the telescope of a spectrometer. Show that maxima of intensity are observed at angles θ_n where $d \sin \theta_n = n\lambda$ and n is a whole number. (2)

(b) When $d = 1.3 \times 10^{-6}$ m and $\lambda = 500$ nm, it is found that the first and second order intensity maxima can be observed, but the third order cannot. Explain why this is the case. (3)

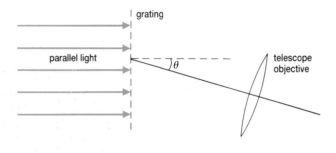

(OXFORD 1988)

13.21 (a) A parallel beam of light from an illuminated slit consists of one red wavelength and one blue wavelength. It is incident normally on a diffraction grating placed on a spectrometer turntable.

(i) Explain why the spectrometer telescope must be adjusted to receive parallel light in order to view a pure spectrum of the source.

(ii) If the telescope is set initially in the straight-through position and then turned in one direction only, give the colour sequence of the first and second spectral 'lines' seen in the first order, and explain why this is so.

(iii) At a diffracting angle of 54.8° a blue 'line' and a red 'line' coincide. If the grating has 600 rulings per mm and the blue wavelength is 454 nm, deduce the order numbers for both colours, and also the red wavelength. (8)

(b) The same diffraction grating is used to view a monochromatic light source placed 50.0 cm away from the grating. The observer holds the grating close to his eye, and observes an image on either side of the source. Explain, with the aid of a simple diagram, why this occurs and calculate the wavelength of the light if the images are 37.8 cm apart. (5)

(JMB 1986)

Theme 4

ELECTROMAGNETISM

Electricity is much associated with modern life. Yet for a large fraction of the total population it is totally mystifying and simply something that is plugged into. An understanding of the principles of electricity is perhaps the particular attribute expected of all students of physics. A study of these principles leads to the realisation that the fundamental work done by such scientists as Galvani, Volta, Ampere, Ohm, Faraday, Kirchhoff and Oersted was all the more remarkable because it was experimentation with an invisible quantity, the electric current.

Chapter **14**

ELECTRIC CURRENT

> **LEARNING OBJECTIVES**
>
> At the end of this chapter you should be able to:
>
> 1. define and use the terms charge, potential difference and resistance and the SI units for each of these quantities;
>
> 2. relate current flow to the speed and number of charge carriers;
>
> 3. relate electromotive force with potential difference and internal resistance;
>
> 4. define and use the terms resistivity and conductivity;
>
> 5. calculate the energy and power supplied to any electrical component.

14.1 ELECTRIC CHARGE

An electric current always causes a magnetic field. The strength of this magnetic field is used in the SI system's definition of the value of electric current. The definition of the ampere as the unit of electric current was given in Section 1.2 and more detail about the magnetic effect will be given in Chapter 18. In this chapter, use will be made of the fact that electric current is used as the base electrical quantity. This is not only the SI approach to the measurement of electrical quantities; it is also historically true that electric current was the first electrical quantity to be measured. For many years, determination of electric current was done by passing the current through a circular coil of wire and measuring the magnetic field which was caused at its centre. The whole instrument was called a tangent galvanometer. This emphasis on electric current was also necessary during the Nineteenth Century because at the time it was not known what was flowing in a wire carrying an electric current.

With the discovery of the electron in 1897 the nature of electric current in a wire as a flow of negative charge became apparent. It is possible that in the future some form of electron counter will be used as an ammeter, but even now it is the magnetic effect of the electric current which is used in ammeters to determine the value of the electric current. Experimentally it can be shown that an electric current of one ampere in a wire is a flow of 6.242×10^{18} electrons per second past any point in the wire.

Since the choice of base electrical quantity is electric current, this quantity must be used in the definition of all other electrical quantities. Charge is the first of these additional quantities which have been defined. The **charge** Q which flows past a point in time t if there is a constant current I, is given by the equation

$$\text{charge} = \text{current} \times \text{time}$$

$$Q = I \times t$$

The unit of charge is the coulomb. One **coulomb** is the charge that in one second crosses a section of a circuit in which there is a current of one ampere.

If the current is not constant in a circuit then the charge which flows can be found by using the area beneath the current–time graph. In calculus terms this can be written

$$Q = \int I \, dt$$

Example 1
A car battery supplies an electric current of 4.0 A for 2000 s; find the total charge which flows from the battery.

Charge which flows = $I \times t$ = 4.0 A × 2000 s = 8000 C.

Example 2
A more likely practical use of the car battery in Example 1 is for it to be used to supply varying currents. Find the total charge delivered from the battery if the variation in current supplied is given by the graph, Fig. 14.1.

In the first 2000 seconds; charge flowed = 3.0 A × 2000 s = 6000 C.
In the next 1000 seconds; charge flowed = 4.0 A × 1000 s = 4000 C.
In the 3000 seconds during which the battery is assumed to be supplying a current which is decreasing at a steady rate.

Charge flowed = average current × time = 3.0 A × 3000 s = 9000 C

Total charge flowed = (6000 + 4000 + 9000) C = 19 000 C.

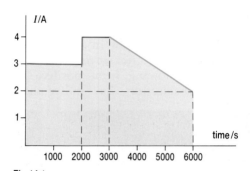

Fig 14.1

QUESTIONS

14.1 (a) What charge flows if a current of 6.5 A flows for a time of 800 s?
(b) What is the steady current if a charge of 750 C flows in a time of 30 000 s?
(c) How long will it take a current of 27 mA to supply a charge of 18.9 C?

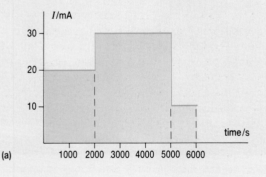

(a)

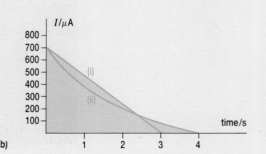

(b)

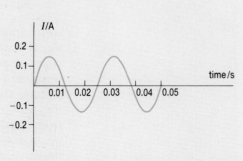

(c)

Fig 14.2

14.2 Using the current time graphs of Fig 14.2, find the charge which flows in each case.

14.2 TRANSPORT OF CHARGE

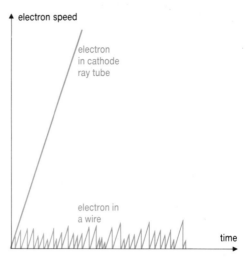

Fig 14.3 Graphs showing how the speed of an electron in a cathode ray tube varies with time in comparison with that of an electron in a wire. The graphs cannot be drawn to scale as the electron in a cathode ray tube is travelling immensely faster.

When there is an electric current in a wire, the electrons are moved along the wire by a force exerted on them by the electric field (see Chapter 16). If a calculation is made of the acceleration of the electron in an electric field, it is found that the electron reaches a very high speed in a small distance of travel. An electron moving in the vacuum inside a cathode ray tube will move through the tube at high speed.

In a wire, however, the influence of the fixed atoms on the movement of the electrons is considerable. The electrons are continually bumping into atoms so that they gain kinetic energy and then lose it in inelastic collisions with the fixed atoms as shown in Fig 14.3. This chaotic movement of the electrons in a wire results in a drift of negative charge through the wire. The drift speed is usually very low; typically only a fraction of a millimetre per second. Note that although the electrons drift through a wire at a low speed, this does not imply that it will take a long time for the current to start at any point in a circuit. When a current is switched on, the slow drift of electrons commences virtually instantaneously at all points around the circuit.

The fact that electrons have a negative charge and flow in a wire from negative to positive is an inconvenience in that it does not correspond with the historical convention of electric current flowing from positive to negative. Electronic engineers frequently use electron flow rather than

Fig 14.4 Nathan Road, Hong Kong at night. An electric current through a gas causes it to glow!

conventional current, but electrical engineers generally still use conventional current, regarding it as a flow in the external circuit of positively charged particles from positive to negative. There are situations when positive charge is flowing from positive to negative. These situations include the movement of positive ions in a conducting liquid or gas and the flow of positive charge carriers in semiconductor material. Because from a practical point of view a flow of negative charge from negative to positive is exactly the same as a flow of positive charge from positive to negative, there is now little pressure to change from using conventional current to electron flow.

The relationship between the current and the speed of the charged particles is illustrated by Example 3.

ELECTRIC CURRENT

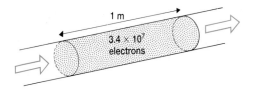

Fig 14.5 Illustrating individual electrons in a 1 m length of an electron beam. The number of 34 million used in example 3 is a more imaginable number than the millions of millions of electrons which flow every second past a point in a wire carrying a current.

Example 3

A cathode ray tube in a television tube has a beam of electrons which carries a current of 0.097 mA. Find the speed of the electrons if the beam contains 3.4×10^7 electrons per metre of its length.

Fig 14.5 shows the situation within a one metre length of the electron beam. Since the charge on a single electron is 1.6×10^{-19} C the number of electrons passing any point in one second is given by

$$\frac{0.097 \times 10^{-3} \text{ C s}^{-1}}{1.6 \times 10^{-19} \text{ C}}$$

$$= 6.06 \times 10^{14} \text{ s}^{-1}$$

Since the number of electrons in a metre of beam is 3.4×10^7, then all the

electrons in $\dfrac{6.06 \times 10^{14} \text{ s}^{-1}}{3.4 \times 10^7 \text{ m}^{-1}} = 1.8 \times 10^7$ metres of beam must pass a point in

one second to allow this current to flow.

The speed of the electrons in the beam must therefore be 1.8×10^7 m s^{-1}, as all the electrons in this length of beam must pass the reference point in a second.

Put into algebraic terms the above example becomes:

$$\text{Number of electrons passing a point per unit time} = \frac{I}{e}$$

where I is the current and e the electronic charge.

This gives the speed of the electrons, $v = \dfrac{I}{Ne}$

where N is the number of electrons in a metre length of the beam.

The same problem can be worked through for the flow of electrons in a wire, but it is then usual to quote the number of conduction electrons per unit volume n.

The relationship between n and N is

$$N = n \times A$$

where A is the area of cross-section of the wire.

Combining the two equations gives the speed of the electrons as

$$v = \frac{I}{nAe}$$

or the current in terms of the speed as

 $I = nAve$

It is worthwhile remembering this equation, but at the same time do realise that n is the number of charge carriers per unit volume and that e as the electronic charge would need to be changed to q, the charge on the charge carrier, if the current is a flow of particles other than electrons. In a liquid, for example, the flow might be of doubly charged ions.

The equation is sometimes written using the term current density. **Current density**, J, is the current per unit area of cross-section. This gives

$$J = \frac{I}{A} = nve$$

Example 4

A copper wire of diameter 1.4 mm connects to the filament of a light bulb of diameter 0.020 mm. A current of 0.42 A flows through both of the wires. Copper has 8.0×10^{28} electrons per cubic metre and tungsten can be assumed to have 3.4×10^{28} electrons per cubic metre. Find the speed of an electron in each of the materials and explain why it is that the copper wire stays cool although the tungsten filament reaches a high temperature.

Using the subscript t for tungsten and c for copper gives:

$$I = n_c A_c v_c e$$

$$0.42 = 8.0 \times 10^{28} \times \pi \times (0.70 \times 10^{-3})^2 \times v_c \times 1.6 \times 10^{-19}$$

$$v_c = \frac{0.42}{8.0 \times 10^{28} \times \pi \times (0.70 \times 10^{-3})^2 \times 1.6 \times 10^{-19}}$$

$$= 0.021 \times 10^{-3} \text{ m s}^{-1}$$

$$I = n_t A_t v_t e$$

$$0.42 = 3.4 \times 10^{28} \times \pi \times (0.010 \times 10^{-3})^2 \times v_t \times 1.6 \times 10^{-19}$$

$$v_t = \frac{0.42}{3.4 \times 10^{28} \times \pi \times (0.010 \times 10^{-3})^2 \times 1.6 \times 10^{-19}}$$

$$= 0.246 \text{ m s}^{-1}$$

The drift speed of an electron increases from 0.021 mm s^{-1} in the copper to a value of 250 mm s^{-1} as it enters the tungsten filament. The increased speed of the electrons in the tungsten means that they have a much greater kinetic energy in the tungsten and, as they collide with the fixed atoms, energy is lost to these atoms, resulting in their rise in temperature. It is interesting to note that as electrons move into the tungsten their speed increases. This is contrary to the popularly held belief that a resistance slows up electric current. Most of the speeding up effect is due to the constriction of the current into a smaller area of cross-section. There is a further effect however due to the fact that there are not as many charge carriers per unit volume in the tungsten as there are in copper.

QUESTIONS

14.3 The cable from a car battery can often be used to supply a current of 100 A. The diameter of the cable is 5.0 mm and it is made of copper with 8.0×10^{28} electrons per cubic metre. Find the speed of the electrons when the current is 100 A.

14.4 The density of copper is 8900 kg m^{-3}. 1 mole of copper atoms has a mass of 0.0635 kg. Find the number of moles in a cubic metre of copper and hence the number of atoms in a cubic metre of copper. Compare this value with the number of free electrons per cubic metre of copper, given in Question 14.3, and hence find the number of free electrons per atom of copper. ($N_A = 6.02 \times 10^{23}$ mol^{-1}.)

14.5 When a sufficiently high voltage is applied between two electrodes in a gas, the gas ionises. Electrons move towards the positive electrode, and positive ions move towards the negative electrode.
(a) What is the current in a hydrogen discharge if, in each second, 4.4×10^{15} electrons and 1.5×10^{15} protons move in opposite directions past a cross-section of the tube?
(b) What is the direction of the current?

14.3 POTENTIAL DIFFERENCE

Electrical circuits are basically used to transfer energy from one place to another. The energy may be very small in the case of a telecommunication signal or very large in the case of public energy supplies. A power station built near a coal mine can burn coal and transfer the chemical energy of the coal into electrical energy. The electrical energy can then be transferred very efficiently over long distances where it can be used in a city by the industrial, commercial and domestic customers of the electricity companies. A similar situation exists within a simple torch circuit. In this case the chemical energy supplied by the torch battery is converted into electrical energy and then supplied to the filament of the torch bulb where it is converted into heat and light energy.

A term is needed to indicate how much energy is transferred with each unit of electrical charge, and this quantity is called the potential difference.

 The **potential difference** between two points in a circuit is defined as the energy converted from electrical energy to other forms of energy when unit charge passes from one point to the other. In equation form this becomes

$$\text{potential difference} = \frac{\text{energy converted}}{\text{charge}}$$

The unit of potential difference is the joule per coulomb. This unit is given the name volt. So one **volt** (V) is the potential difference between two points in a circuit in which 1 joule of energy is converted when one coulomb passes from one point to the other.

Because the unit of potential difference is the volt, potential difference is frequently called the voltage. The abbreviation p.d. is also frequently used. Since the definition refers to two points in a circuit, the potential difference can only be used if the two points are clearly stated. For a single component in a circuit the two points are normally immediately before the component and immediately after the component. The expression 'the potential difference across the component' is therefore frequently used. It is correct to speak of the current through a component being caused by the potential difference across that component. A common mistake in solving electrical problems arises when there is more than one component in a circuit, as is usually the case. When several components are present it is essential to realise that each has its own particular potential difference. It is not possible to use a blanket term for the voltage, as if it were the same for every component.

Sometimes the term 'the potential at a point' in a circuit is used. This is normally the potential difference between the point and an arbitrary zero of potential. In many cases the arbitrary zero of potential is the potential of the metallic casing of the apparatus. For safety, the casing will be connected to the ground and so the zero of potential is called the electrical **earth**.

Potential difference is a quantity relating the energy supplied to a component with the charge delivered to it. A slightly different viewpoint can be obtained however if, instead of considering total quantities of energy and charge, the rate of supply of energy and charge are considered. An example of the two approaches should make the similarities and differences clear.

There is no need to memorise the definition of the unit of potential difference. Units can always be obtained from the definition of the quantity itself. Here the crucial fact to remember is that potential difference is energy converted per unit charge. Its unit **must** therefore be the joule per coulomb as that is the unit of energy divided by the unit of charge. You also need to know that *volt* is a shorthand form of *joule per coulomb*. Many problems connected with potential difference can be solved by re-writing volt as joule per coulomb.

Example 5

An immersion heater is rated at 3000 W and is switched on for 2000 s. During this time a charge of 25 000 C is supplied to the heater. Find the potential difference across the heater.

Energy supplied	= 3000 W × 2000 s	=	6 000 000 J
Charge supplied		=	25 000 C

$$\text{Potential difference} = \frac{\text{energy supplied}}{\text{charge supplied}} = \frac{6\ 000\ 000\ \text{J}}{25\ 000\ \text{C}} = 240\ \text{V}$$

This can equally well be determined in the following way:

$$\text{Energy supplied per unit time} \quad = 3000\ \text{J s}^{-1} \qquad = 3000\ \text{W}$$

$$\text{Charge supplied per unit time} \quad = \frac{25\ 000\ \text{C}}{2000\ \text{s}} \quad = 12.5\ \text{C s}^{-1} = 12.5\ \text{A}$$

$$\text{Potential difference} \qquad = \frac{\text{energy supplied per unit time}}{\text{charge supplied per unit time}}$$

$$= \frac{\text{power}}{\text{current}} = \frac{3000\ \text{W}}{12.5\ \text{A}} = 240\ \text{V}$$

The example shows that by working in unit time the potential difference is given not only as energy per unit charge but also as power per unit current. The volt is not only the joule per coulomb but also the watt per ampere. The following two equations are entirely equivalent.

$$\text{Potential difference} \quad = \frac{\text{energy converted}}{\text{charge}} = \frac{E}{Q}$$

$$\text{Potential difference} \quad = \frac{\text{power converted}}{\text{current}} = \frac{P}{I}$$

QUESTIONS

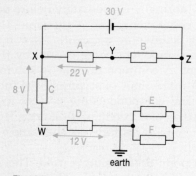

30 V

Fig 14.6

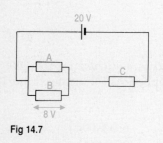

Fig 14.7

14.6 A light bulb is marked 240 V, 60 W. It is switched on for 10 000 s. Assuming that the bulb is being used correctly find:
(a) the total energy converted by the bulb from electrical energy;
(b) the total charge supplied to the bulb;
(c) the current.
Note that the answer to (c) can be obtained even if (a) and (b) had not been able to be determined as a result of the time not being known.

14.7 Fig 14.6 shows a circuit containing a 30 V battery and 6 resistors. The potential differences across A, C and D are 22 V, 8 V and 12 V respectively. Find the potential difference across each of the components B, E and F, and also the potential at the points W, X, Y and Z.

14.8 Fig 14.7 is a circuit diagram of a circuit containing a battery and 3 other components A, B and C. The circuit is switched on for 400 s. Table 14.1 gives some information about the charge supplied to, and the potential difference across, components in the circuit. Redraw the table and complete it. If you find difficulty in answering this question, more detail is given about the procedure required in Section 15.1.

Table 14.1

	Battery	A	B	C
Charge	800 C	100 C		
Power				
Energy				
Current				
Potential Difference	20 V		8 V	

As indicated in the previous section, when a potential difference is applied across a material which conducts electricity it causes an electric current in the material. Potential difference is the cause and electric current is the effect. Different currents are caused for a given potential difference depending on the nature of the material, on its shape and on its temperature. As is frequently the case with cause and effect problems, the ratio of the cause to the effect is a useful quantity. Resistance is the name given to the ratio here. Resistance is defined by the following equation.

$$\text{resistance } R \ = \ \frac{\text{potential difference } V}{\text{current } I}$$

The unit of resistance must be the volt per ampere. This is given the name **ohm** (Ω). Following the margin note in Section 14.3, the equation defining resistance can be used to give the definition of the unit of resistance, the ohm. A resistor has a resistance of one **ohm** if there is a current of one ampere through it when the potential difference across it is one volt. Notice that if a resistor in the laboratory is marked $16 \, \Omega$ it means that it requires 16 volts across it if it is to have a current of one ampere through it. Think of the resistance of any component as the voltage needed to cause a current of one ampere. A piece of thick copper wire has a low resistance because it only needs a low voltage across it to cause a current of one ampere to flow through it. A component with a high resistance is a poor conductor of electrical current as it needs a high potential difference to cause the one ampere of current. A $1 \, \text{M}\Omega$ resistor ($1 \, \text{M}\Omega = 1$ megohm = $1 \, 000 \, 000$ ohm) requires a million volts across it if one ampere is to flow through it. The fact that it would melt long before such a high voltage could actually be raised to this value is immaterial. The resistor is more likely to be used in a circuit in which 1 V causes a current of $1 \, \mu\text{A}$. (The same method of working is used with speeds; to say that you are travelling at $150 \, \text{km h}^{-1}$ does not mean that you necessarily have to travel for an hour. It is no defence against a prosecution for speeding to say that you could not possibly have been travelling at $150 \, \text{km h}^{-1}$ because the road was not 150 km long!)

INVESTIGATION

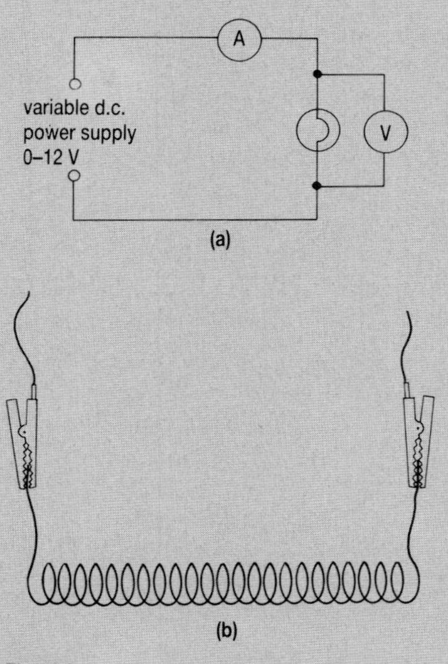

Find how the resistance of (a) a light bulb and (b) a coil of constantan wire vary with the potential difference across them.

A suitable light bulb would be a car bulb, say 12 V, 24 W. Connect it in the circuit shown in Fig 14.8(a) and measure the current flowing through it for a series of values of potential difference across it from 0 V to 12 V. Your voltmeter must be a d.c. voltmeter and must be capable of reading up to 12 V. The ammeter must also be a d.c. meter and you should be able to work out what the maximum current will be. Calculate the resistance for each reading taken and plot a graph of resistance against p.d. You may need to be reminded that $\frac{0}{0}$ does not equal 0. Here this implies that the graph does not go through the origin.

The specification of the coil can be changed to a considerable extent. You may have suitable coils already available even though the wire used may be of different length and diameter. One possible coil would be made from about 1.5 m of 28 s.w.g. constantan wire. Wind the coil neatly and closely around a pencil, remove the pencil and slightly stretch the coil so that separate turns are not touching one another, and attach leads to each end using crocodile clips, as shown in Fig 14.8(b). The coil will get hot in use so it must either rest on a heat-resistant mat or be supported in the air. Repeat the experiment with the coil replacing the light bulb.

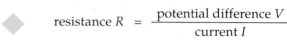

variable d.c.
power supply
0–12 V

(a)

(b)

Fig 14.8

Sometimes the term conductance is used to relate current with potential difference.

The **conductance** of a component is the reciprocal of its resistance and is measured in **siemens (S).**

$$\text{conductance} = \frac{\text{current}}{\text{potential difference}}$$

It is recommended that you do not use this term until you are thoroughly familiar with the term resistance.

The very marked difference in the behaviour of the two wires which is apparent from the investigation is that whereas the resistance of the light bulb varies, the resistance of the constantan coil remains nearly constant. The variation of resistance of the light bulb occurs mainly because of the large temperature rise which takes place. The temperature of the constantan varies much less but it was chosen because in any case the resistance of constantan does not change much with temperature. That is why it is called constantan and why it is used in standard resistors. It was experiments such as this which led Ohm to formulate his law as:

> The current through a conductor is proportional to the potential difference across it provided its temperature remains constant.

In Ohm's day, almost all of the conductors he used obeyed the law. Now there are many exceptions. A resistor which obeys Ohm's law has a variation of current with potential difference which gives a straight line graph which passes through the origin. The resistance can be marked on such a resistor. If the graph of current against potential difference is not a straight line, then the resistance varies with potential difference.

Example 6

A graph of current against potential difference for a component is given as line (a) in Fig 14.9.

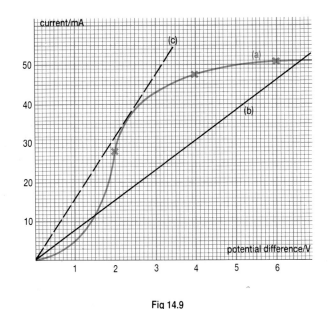

Fig 14.9

Find **(a)** the resistance when the p.d. across the component is (i) 2.0 V, (ii) 4.0 V, (iii) 6.0 V;
   (b) the value(s) of the voltage for which the resistance is 130 Ω;
   (c) the minimum resistance of the component.

ELECTRIC CURRENT

(a)

	i	ii	iii
p.d.	2.0 V	4.0 V	6.0 V
current	28 mA	48 mA	51 mA
resistance	$\dfrac{2\text{ V}}{0.028\text{ A}}$	$\dfrac{4\text{ V}}{0.048\text{ A}}$	$\dfrac{6\text{ V}}{0.051\text{ A}}$
	$=71\ \Omega$	$=83\ \Omega$	$=118\ \Omega$

(b) A resistance of 130 Ω means that 130 volts are required for a current of 1 ampere. This is the same as 0.13 V for 1 mA or 1.3 V for 10 mA, or 6.5 V for 50 mA. A straight line for this resistance is plotted as (b) on Fig 14.8 and shows that the diode has this resistance when the potential difference is either 1.5 V or 6.7 V.

(c) The minimum resistance occurs when the highest value of current/p.d. is obtained. This is shown by the line marked (c) on Fig 14.8. It is a tangent to the curve. There is no other point on the curve from which a line can be drawn to the origin which gives a higher current per unit potential difference. The value of resistance given is 3.2 V/0.050 A = 64 Ω.

QUESTIONS

14.9 A light-dependent resistor (LDR) has a p.d. of 6.0 V across it. When it is not illuminated a current of 0.87 mA flows through it, but when it is illuminated the current increases to 53 mA. Find the resistance of the LDR in both cases.

14.10 A graph of current against voltage for a diode is given in Fig 14.10. Note that the diode does not behave in the same way to forward potential difference as it does to reverse potential difference. The scales for the two parts of the graph are different.

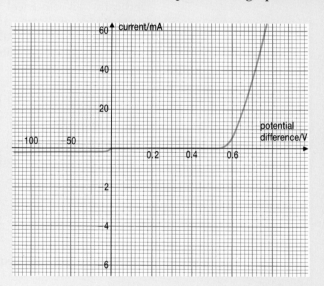

Fig 14.10 V–I characteristic for a diode.

Find **(a)** the resistance when the p.d. across the diode is (i) 0.40 V, (ii) 0.60 V, (iii) 0.70 V and (iv) – 50 V;

(b) the value of the potential difference for which the resistance is 15 Ω.

14.11 The equation relating I, the current, and V, the potential difference, for a semiconductor material is

$$I = aV^2 + bV$$

where $a = 0.023$ mA V^{-2} and $b = 3.3$ mA V^{-1}.

In Question 14.11(c) the gradient of the graph of V against I is required. This is sometimes called the **slope resistance** of a component. It is a quantity measured in ohms but it is not the resistance of the component. Resistance is not the gradient of a graph of V against I unless the graph is a straight line passing through the origin. That is, unless Ohm's law is obeyed. On a graph of I against V the gradient is $1/R$ if Ohm's law is obeyed. Be careful which way these graphs are plotted. Strictly, potential difference should be plotted on the x-axis as it is the independent variable. That is, it is the cause which can be determined independently: current is the effect and depends on the potential difference, so being the dependent variable it is plotted on the y-axis.

14.5 RESISTIVITY AND CONDUCTIVITY

Before dealing with these two terms separately a comment on word endings is appropriate. The suffix -or, on words such as resistor, inductor, indicates an electrical component. The suffix -ance, as in resistance, capacitance, indicates an electrical property which the component has. A resistor has the electrical property of resistance. The suffix -ivity, as is being introduced here, indicates an electrical property of a material. It is possible to state the resistance of a particular piece of copper wire but it is not possible to state the resistance of copper because it might be of any shape or size. Resistivity can be stated for copper however. The resistivity of a material enables comparisons to be made between the conducting ability of different substances.

The resistance R of a piece of material of uniform cross-section is found experimentally to be proportional to its length, l. It is also found that the resistance is inversely proportional to the area of cross-section, A.

$$R \quad \propto \quad l$$

$$R \quad \propto \quad \frac{1}{A}$$

Combining these two equations together gives

$$R \quad = \quad \frac{\rho l}{A}$$

where ρ is a constant of proportionality. ρ is called the **resistivity** of the material. ρ will be numerically equal to the resistance only in the unlikely event of having a wire of length one metre and of area of cross-section one square metre.

The equation gives

$$\rho \quad = \quad \frac{RA}{l}$$

and therefore the unit of resistivity is Ω m.

In the same way that conductance is the reciprocal of resistance, **conductivity** is the reciprocal of resistivity.

Fig 14.11 Air at normal atmospheric pressure will conduct if a high enough potential difference is used.

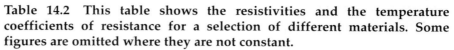

Table 14.2 This table shows the resistivities and the temperature coefficients of resistance for a selection of different materials. Some figures are omitted where they are not constant.

Substance	Resistivity at 25 °C /Ω m	Temperature coefficient of resistance /K^{-1}	Uses
Conductors			
metals			
copper	1.72×10^{-8}	0.004 0	connecting wires
gold	2.42×10^{-8}	0.003 6	microchip contacts
aluminium	2.82×10^{-8}	0.003 6	power cables
tungsten	5.51×10^{-8}	0.005 2	light bulb filaments
steel	20×10^{-8}	0.003 3	
alloys			
constantan	49×10^{-8}	0.000 010	standard resistors
nichrome	100×10^{-8}	0.000 40	heating elements
Semiconductors			
carbon	3.5×10^{-5}	−0.000 5	resistors
germanium	0.60		transistors
silicon	2300		transistors, chips
Insulators			
glass	$\sim 10^{13}$		power grid insulators
polythene	$\sim 10^{14}$		wire insulation

Table 14.2 shows the resistivities of several different conductors and insulators. Details shown about temperature coefficient of resistance will be dealt with shortly. Good conductors have very low resistivities whereas good insulators have very large resistivities. The vast difference in the conducting abilities of materials should be noticed. Measuring the resistivity of an insulator is a particularly difficult experiment to perform because it is difficult to measure any current at all flowing through it. When a current does flow, it often flows in the surface layer where there is likely to be contamination with impurity atoms. The resistance can also be highly voltage dependent. The current in many insulators will increase appreciably if the voltage rises above a particular value. This is of considerable practical importance in two situations. Atmospheric air is a most important insulator. Overhead power cables on the National Grid are insulated by the air surrounding them. If too high a voltage is used then the air ionises and becomes conducting, as has happened in the photograph, Fig 14.11. This also happens in a lightning flash, Fig 14.12. An interesting speculation is that if atmospheric air did conduct electricity then probably electricity itself would never have been discovered because, among other things, a battery would discharge itself through the surrounding air. The second important consideration for the breakdown of insulators so that they conduct is in the space between the plates of a capacitor. There is an advantage in a capacitor in having the plates very close together. (See Section 17.1.) If the insulator between the plates is only a few hundred molecules thick then a relatively small voltage can cause the dielectric, as this insulator is named, to conduct and so damage the capacitor. For this reason, capacitors are marked with the maximum voltage which can be applied to them.

Fig 14.12 A lightning flash occurs when the electric field is sufficient to ionise the air. The potential differences can be many billions of volts.

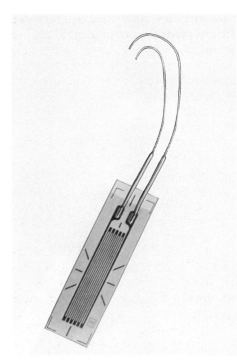

Fig 14.13 A strain gauge. Black shows a conducting region and white an insulator. When stretched the wire gets a small amount longer and thinner and so its electrical resistance increases.

Example 7

Using data from Table 14.2 find the resistance of a piece of copper wire whose length is 8.5 cm and whose diameter is 0.068 mm. The wire is then stretched so that its length increases by 2% and its diameter reduces by 1%. Find the value of its resistance when under strain.

$$R = \frac{\rho l}{A} = \frac{\rho l}{\pi r^2} = \frac{\rho l}{\pi (d/2)^2} = \frac{4\rho l}{\pi d^2}$$

$$R = \frac{4 \times 1.72 \times 10^{-8} \times 0.085}{\pi \times (0.068 \times 10^{-3})^2}$$

$$= 0.403 \ \Omega$$

There is no need to re-calculate the whole problem with the altered length and area of cross-section when under strain. If the equation for R is considered, then the new value for l will be $1.02l$, and the new value for d will be $0.99d$. This gives the new value for R to be

$$R' = 0.403 \times \frac{1.02}{(0.99)^2}$$

$$R' = 0.419 \ \Omega$$

This principle is used in a device called a strain gauge. Fig 14.13 shows a strain gauge. Strain gauges are used to measure distortion. Fig 14.14 shows an aircraft wing undergoing tests in which considerable loading is applied to the wing. Strain gauges stuck firmly onto the wing become stretched when a load is applied and their resistances are altered.

Fig 14.14 The wing of an aeroplane is here being stressed by applying downward forces to it. Strain gauges measure the resulting strain.

QUESTIONS

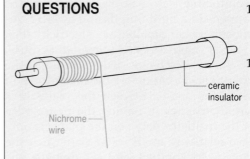

Fig 14.15 Winding the element of an electric fire.

14.12 A wire of length 4.6 m and diameter 0.83 mm is found to have a resistance of 0.24 Ω. What is the resistivity of the material of the wire? What substance might the wire be made of?

14.13 A manufacturer of electric fires makes elements for the fire out of nichrome wire. The wire is coiled on a spiral groove on a ceramic former as shown in Fig 14.15. The centre-to-centre separation of adjacent turns is 0.12 cm and the wire diameter is 0.080 cm, so turns do not touch. The length of the coil is 22.8 cm and the diameter of each turn is 1.3 cm. Find the length of wire used and hence the resistance of the element when it is at a temperature of 25 °C.

ELECTRIC CURRENT

14.6 TEMPERATURE COEFFICIENT OF RESISTANCE

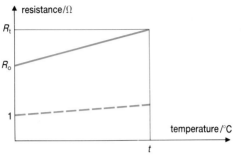

Fig 14.16 Resistance of most metals increases uniformly with temperature. The gradient of the graph depends on the material and on its resistance at 0°C. If the resistance at 0°C is 1Ω, as shown dotted, the gradient of this graph is the temperature coefficient of resistance.

The resistance of most materials varies with the temperature. For most metals the graph of resistance against temperature is a straight line as shown in Fig 14.16. The gradient of a graph such as this depends not only on the material but also on R_0, the resistance at 0 °C. The gradient of the graph when $R_0 = 1$ Ω is called the temperature coefficient of resistance, α. The **temperature coefficient of resistance** is defined by the equation

$$\alpha = \frac{R_t - R_0}{R_0 t}$$

where R_t is the resistance at the celsius temperature t.

Values of α are given in Table 14.2 for various conductors, and it should be noted that carbon has a negative temperature coefficient of resistance. That is, its resistance falls as the temperature rises, in contrast to most materials in which the resistance increases as the temperature rises. A negative temperature coefficient of resistance is a characteristic of semiconductors. The electrons in semiconductors are bound more firmly to individual atoms than electrons in a metal. In a metal an increase in the temperature causes a greater vibration of the atoms in the fixed framework of atoms, and hence there is more interaction between these vibrating atoms and the flowing electrons. This results in the resistance of the metal rising as the temperature increases. In a semiconductor this is a less important factor than the fact that the electrons are more likely to have sufficient energy to escape from a particular atom if the temperature is higher. They can then contribute towards the electric current and hence reduce the resistance of the semiconductor.

One other feature of temperature coefficient of resistance is that for many metals its value is close to 0.003 66 °C^{-1}. This figure is the reciprocal of 273 °C. The implication of this is that for many metallic conductors the resistance is proportional to the thermodynamic temperature.

Example 8

A standard resistor is made of constantan wire and has a value of 5.000 Ω at a temperature of 15 °C. In use, too much current is passed through it so its temperature rises to 158 °C. What will be the percentage error caused by overheating?

since $\qquad \alpha \quad = \dfrac{R_t - R_0}{R_0 t}$

$\qquad\qquad R_0 \alpha t \quad = R_t - R_0$

$\qquad\qquad\quad R_t \quad = R_0 + R_0 \alpha t = R_0 (1 + \alpha t)$

At 158 °C: $\quad R_{158} \ = R_0 (1 + 158\, \alpha)$

At 15 °C: $\quad R_{15} \ = R_0 (1 + 15\, \alpha)$

α for constantan is given in Table 14.2 as 0.000 010 °C^{-1}.

So $\qquad\qquad R_{158} \qquad = R_0 (1 + 0.001\ 58)$

$\qquad\qquad\quad R_{15} \qquad = R_0 (1 + 0.000\ 15)$

Divide these equations to get

$$\frac{R_{158}}{R_{15}} \ = \frac{1.001\ 58}{1.000\ 15} = 1.001\ 4 \ = \frac{R_{158}}{5.000}$$

$$R_{158} \quad = 5.000 \times 1.001\ 4 = 5.007\ \Omega$$

So the percentage error is

$$\frac{0.007}{5.00} \ \times \ 100 = 0.14\%$$

14.14 When an electric fire is first switched on, the current through it is 5.2 A, the potential difference across it is 240 V and its temperature is 0 °C. By the time it has reached its operating temperature the current has fallen to 3.7 A. If the element is made of nichrome wire, what is the operating temperature? (This illustrates the principle of one type of thermometer, the resistance thermometer. See Section 24.4.)

DATA ANALYSIS

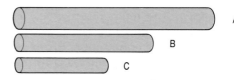

Fig 14.17

Light bulb construction

A light bulb manufacturer makes 240 V, 60 W bulbs. The filament of the bulb is at a temperature of 2600 °C and the bulb lasts for 1000 hours before the filament breaks. There are two schools of thought concerning these bulbs. One group of people want the manufacturer to raise the temperature at which the filament operates so that, for the same power, more light is emitted. The other group of people think that the manufacturer makes too much profit on his bulbs by making them so that they break after 1000 hours. They want him to make bulbs which last 2000 hours. The manufacturer can happily satisfy both of these requirements – but only by manufacturing two other bulbs alongside his original bulb.

Using the data given below, design the electrical characteristics of the three bulbs A, B and C, and hence decide which bulbs you would advise people to buy. (Some of the data have been modified here as in practice the coiling of filaments can make considerable differences to filament temperature, life, and radiated power.) Work to 3 significant figures in the exercise in order to keep 2 reliable significant figures at the end.

Material used for filament	tungsten
Resistivity of tungsten at 25 °C	5.5×10^{-8} Ω m
Temperature coefficient of resistance of tungsten	0.0052 K^{-1}

(a) Show that the resistivity of tungsten at 2600 °C is 7.9×10^{-7} Ω m.

(b) Find the current through, and hence the resistance of, a 240 V, 60 W bulb. All the bulbs have this resistance.

(c) What diameter of tungsten wire is needed for the filament of bulb B if it is 0.14 m long? This is the standard production bulb.

(d) The three filaments which could be used are illustrated in Fig.14.17. Explain how all can have the same resistance and state which is the long life filament and which is the high temperature filament.

(e) Complete the table, assuming that the resistivity of all the filaments is 7.9×10^{-7} Ω m.

	A	B	C
resistance			
diameter	0.0129 mm		0.0113 mm
length		0.14 m	
surface area			
power lost per unit area(P)			

(f) Using the graph of power lost per unit area (P) against temperature, Fig 14.19, find the temperature at which each filament operates.

(g) The bulb at the highest temperature gives out 12% of its power input as light but it only lasts 500 hours, whereas the one at the lowest temperature gives out only 9% of its power as light. If a kilowatt hour costs 8p and each bulb costs 50p find the cost of running bulbs A and C for 2000 hours and hence work out the cost of equal quantities of light energy.

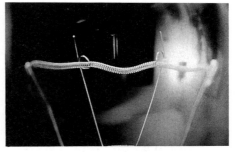

Fig 14.18 The coiled filament of an electric light bulb. It is made of tungsten so that it can operate at high temperatures without melting.

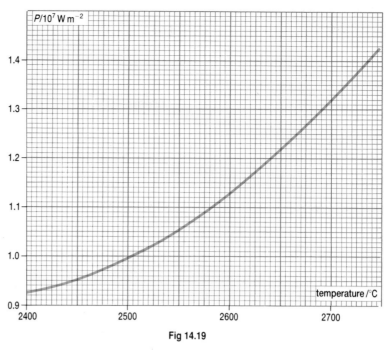

Fig 14.19

(h) Summarise your findings in the form of advice to a purchaser. Include not only financial considerations but also environmental considerations. Greenpeace, in 1988, suggested that this type of bulb should be used less and that fluorescent lamps should be used more because the efficiency of light output from filament bulbs is so low.

14.7 ELECTROMOTIVE FORCE

As explained earlier, in Section 14.3, electrical circuits are used to transfer energy from one place to another. Potential difference was the term used to state how much electrical energy was being converted into other forms of energy per unit charge. The term electromotive force (e.m.f.) is a complementary term to potential difference. It is **not** a force but, as for potential difference, it is a ratio of energy to charge and is therefore measured in volts. Whereas potential difference is used when the electrical energy is being changed into other forms of energy, electromotive force is used when electrical energy is being produced from other forms of energy. Within a battery, for instance, energy is given to the electrons which flow from it and the store of chemical energy within the battery is reduced.

A formal definition of electromotive force is:

 the **electromotive force** of any source of electrical energy is the energy converted into electrical energy per unit charge supplied.

In equation form this is

$$E = \frac{energy\ converted}{change\ supplied}$$

where E is the electromotive force.

Be careful with the symbol E. It is used in this chapter to mean only electromotive force, but it is often used for energy.

The units involved can, as always, be used as a check on any working done. Provided an e.m.f. of, say, 12 V is thought of as 12 joules per coulomb, then there will be no problem in getting the equation correct.

A torch battery usually has its electromotive force marked on its case. If 9 V is its stated e.m.f., this means that 9 joules of energy are converted into electrical energy with every coulomb of charge passing through the battery. Since an electron has a charge on it of 1.60×10^{-19} C, 6.25×10^{18} electrons are needed to supply one coulomb. 9 V actually means that the average energy converted is $9/6.25 \times 10^{18} = 1.44 \times 10^{-18}$ J per electron.

The law of conservation of energy can be applied to a situation such as this. An energy account for one coulomb of charge being supplied could be as follows:

Energy loss
Chemical energy in battery   24 J

Energy gain
Internal energy in battery   15 J
Electrical energy of charge   9 J

For a battery the loss of chemical energy is always much larger than the electrical energy produced. The conversion of chemical energy to electrical energy is not very efficient. A battery's temperature will rise when it is supplying charge, as a result of the increase in its internal energy. (See section 25.1 for further details on internal energy.) From this table we can see that the e.m.f. of the battery is 9 V.

A similar account can be made for what happens to the electrical energy on this 1 C of charge.

Energy loss
Electrical energy of charge   9 J

Energy gain
Internal energy in battery   1 J
Internal energy in light bulb 8 J

You should note that there is a further loss of energy, as internal energy, within the battery itself. Some of the electrical energy produced within the battery never leaves the battery but is immediately lost for further electrical use in the resistance of the battery itself. These values show that the potential difference across the light bulb is 8 V, and the potential difference across the internal resistance of the battery is 1 V. You should be able to deduce from this that the potential difference across the components in a circuit will always be smaller than the e.m.f., because the electrical energy lost within a battery can never be reduced to zero. It can however be made very small if the current supplied by the battery is small. If the current being supplied approaches zero then the e.m.f and the potential difference across the circuit components approach being equal. This gives rise to the statement that:

 the electromotive force of a source is the potential difference across its terminals as the current approaches zero.

From a practical point of view, if a good quality voltmeter is the only component connected across the terminals of a battery, then the reading on the voltmeter will be equal to the e.m.f. of the battery. The quality of the voltmeter needs to be considered here not only to ensure that it is giving an accurate reading but also because good voltmeters have a very high resistance and therefore take very little current. A typical modern digital voltmeter might have a resistance of 10 MΩ. It therefore takes an extremely small current and can give accurate values of e.m.f.

The simplest useful electrical circuit is shown in Fig 14.20. It consists of a battery, of electromotive force E and internal resistance r, connected to a resistor of resistance R. It is in many respects the most important electrical circuit of all, since it illustrates what all electrical circuits do in transferring energy from a source, which has internal resistance of its own, to an external consumer. The circuit could represent

• a microphone delivering energy to an amplifier or

• an amplifier delivering energy to a loudspeaker or

• a power station delivering energy to its customers or even

• an entire national generating company supplying all the electrical energy demand of a nation.

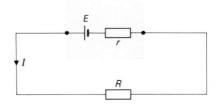
Fig 14.20   Arguably the most important electrical circuit.

     ELECTRIC CURRENT

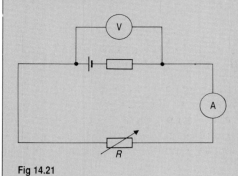

Fig 14.21

Set up the circuit given in Fig 14.21 using a torch battery as the source of electric current and with a variable external resistance R. Connect a suitable voltmeter across the terminals of the battery to measure the potential difference V. This gives the e.m.f. of the battery if there is zero current, and it is also the potential difference across R provided the ammeter has a low resistance, which is usually the case. Connect a suitable ammeter in the circuit to measure the current I. You may well need two ammeters (at different times) or an ammeter with more than one range to cope with the variation of current supplied. A suitable range of resistance is shown in the partly completed Table 14.3. Calculate the other quantities listed in the table and plot graphs to show how V, I, V_b and P vary with the external resistance, where V_b is the potential difference across the internal resistance and P is the power delivered to the external resistance.

Table 14.3

Resistance	Potential difference across R	Current	Potential difference lost across internal resistance	Power to R
R/Ω	V/V	I/A	V_b/V	P/W
infinity	*	0	0	0
50				
30				
20				
10				
8				
5				
4				
3				
2				
1				
0.5				
0.2				
0				

The * represents the e.m.f. of the battery. Do not leave the battery connected to the small or zero resistances for more than a minimum of time, because shorting out batteries in this way runs them flat very quickly.

Similar graphs to those plotted in the investigation may be obtained theoretically. Consider again Fig 14.20. If V is the p.d. across the resistance R, and V_b is the potential differences lost across the internal resistance then

$$E = V + V_b$$
but $$V = IR \text{ and } V_b = Ir$$
so $$E = IR + Ir = I(R + r)$$

This gives

$$I = \frac{E}{R + r}$$

$$V = \frac{ER}{R + r}$$

and

$$V_b = \frac{Er}{R + r}$$

The power to the external resistance, $V \times I$ is given by

$$P = V \times I = \frac{ER}{R + r} \times \frac{E}{R + r} = \frac{E^2 R}{(R + r)^2}$$

Since the total electrical power used is $E \times I$ this is

$$\frac{E^2}{R + r}$$

Electrically therefore the circuit has an efficiency given by

$$\text{efficiency} = \frac{\text{Power to external circuit}}{\text{total power used}} = \frac{\dfrac{E^2 R}{(R + r)^2}}{\dfrac{E^2}{R + r}}$$

$$\frac{R}{R + r}$$

The efficiency of an electrical circuit cannot be equal to 1 unless the internal resistance of the source is zero.

Example 9

A cell in a deaf aid supplies a current of 2.5 mA through a resistance of 400 Ω. When the wearer turns up the volume, the resistance is changed to 100 Ω and the current rises to 6.0 mA. What is the e.m.f. and the internal resistance of the cell?

When answering electrical problems, the following sequence is recommended. As you first read through a question, sketch out a circuit diagram freehand. As extra facts are given add the detail on to the diagram. If two sets of information are given, as in Example 9, use two diagrams. If you work as suggested, the diagrams take no time at all since you are simply using time when other people are thinking of what to do. Next put on to your diagram any further information which can immediately be calculated, such as the potential differences across the external resistors in Example 9. With the information now on the circuit diagram, find links between known facts. If you are stuck, put in an algebraic unknown and work out links between quantities using those unknowns. Continue to use the diagram. Add extra facts as they become available and at the end of the problem look at the diagram to see if what you have found is possible.

The two circuits are shown in Fig.14.22. (See marginal note.)

The currents are known in both cases and so the potential differences across the external resistors can be found and inserted on the diagram at once.

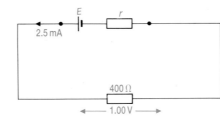

(a)

2.5 mA

400 Ω

1.00 V

(b)

6.0 mA

100 Ω

0.60 V

ELECTRIC CURRENT

Assume the e.m.f. to be E and the internal resistance to be r.
If I is the current then Ir is the potential difference lost across the internal resistance:

$$
\begin{aligned}
\text{in circuit (a)} \quad E - 0.0025r &= 1.00 \\
\text{in circuit (b)} \quad E - 0.0060r &= 0.60 \\
\text{Subtract} \quad 0.0060r - 0.0025r &= 0.40 \\
0.0035r &= 0.40 \\
r &= \frac{0.40}{0.0035} \\
&= 114\ \Omega
\end{aligned}
$$

$$
\begin{aligned}
\text{Substituting} \quad E - 0.0060 \times 114 &= 0.60 \\
E = 0.60 + 0.69 &= 1.29\ \text{V}
\end{aligned}
$$

(Check: $1.29 - 0.0025 \times 114 = 1.29 - 0.285 = 1.00$ to 3 sig. figs, thus satisfying the first equation.)

QUESTIONS

14.15 A 9.0 V battery has an internal resistance of 12.0 Ω. What is the potential difference across its terminals when it is supplying a current of 50 mA?
What is the maximum current which the battery could supply?
Draw a sketch graph to show how the terminal potential difference varies with the current supplied if the internal resistance remains constant. How could the internal resistance of the battery be obtained from this graph?

14.16 Explain why the headlights of a car go dim when the starter motor is used.

14.17 A battery is connected in series with a variable resistor and an ammeter. When the resistance of the resistor is 10 Ω the current is 2.0 A. When the resistance is 5 Ω the current is 3.8 A. Find the e.m.f. and the internal resistance of the battery.

14.18 A battery is connected to a variable resistor and a voltmeter is connected across its terminals. When the variable resistor has 6.0 Ω resistance the voltmeter reading is 4.0 V. When the resistance is 10 Ω the voltmeter reading is 4.4 V. Find the e.m.f. and the internal resistance of the battery.

14.8 ENERGY AND POWER

It was shown earlier that both energy and power can be expressed in terms of the potential difference. The expressions obtained, when combined with expressions for resistance, can be put into many different forms and there is a real danger of confusion if too much memorising of formulae is done. Work as much as possible from first principles, that is, by using the defining equations:

$$
\begin{aligned}
\text{for power } (P) \qquad & P = \frac{energy}{t} \\
\text{for charge } (Q) \qquad & Q = It \\
\text{for potential difference } (V) \quad & V = \frac{P}{I} \\
\text{for resistance } (R) \qquad & R = \frac{V}{I}
\end{aligned}
$$

where I is the current and t is the time.

The power supplied to a resistor therefore can be given by any of the following expressions:

$$P = V \times I = IR \times I = I^2R$$

$$P = V \times I = V \times \frac{V}{R} = \frac{V^2}{R}$$

and the energy supplied will be given by:

$$energy = VIt = I^2Rt = \frac{V^2t}{R}$$

Note that the energy supplied to a resistor has a 100% conversion into heat. This is always the effect of a resistor in a circuit, to convert electrical energy into heat energy. The process is irreversible.

The kilowatt-hour (kWh)

This is a large unit of energy. The joule is a comparatively small unit of energy, so for the commercial selling of energy by the area electricity boards the kilowatt-hour is used. (In contrast, area gas boards sell energy by the megajoule, MJ.)

> 1 kilowatt hour is the energy supplied by 1 kilowatt of power for an hour

So 1 kWh = 1000 W × 3600 s = 3 600 000 J.

Example 10

An iron is marked 240 V, 800 W. If it is used when the supply voltage is only 220 V, what power will be supplied to it?

At 240 V: current to iron $= \dfrac{800\ W}{240\ V} = \dfrac{10}{3}\ A = 3.33\ A$

resistance of iron $= \dfrac{240\ V}{3.33\ A} = 72\ \Omega$

Assuming that the resistance is the same when used on the lower voltage:

current to iron from 220 V supply $= \dfrac{220\ V}{72\ \Omega}$

power supplied to iron $= V \times I = 220 \times \dfrac{220}{72} = 670\ W$

(Note that this is not $\dfrac{220}{240} \times 800$ which is 730 W.)

QUESTIONS

14.19 An immersion heater has a resistance of 20.0 Ω and is used on 240 V mains. What current does it use and what is its power? How long will it take to raise the temperature of 80 kg of water from 18 °C to 60 °C if it takes 4200 J of energy to raise the temperature of 1 kg of water by 1 °C?

14.20 A storage battery, of the type used for emergency lighting in the event of a fire or a power cut, operates on 24 V. The battery has negligible internal resistance and is rated at 140 ampere-hours, which means that it can supply 1 ampere for 140 hours or two amperes for 70 hours, etc. What is the total energy which is stored by the battery? What external resistance would have to be connected to the battery if it were to be discharged in 14 hours? What power output is the battery giving at this rate of discharge?

14.21 A hot tap is accidentally left running for a long time. At first very hot water comes from the tap but after a while the temperature drops to 15 K above the cold water temperature. If the water is running at the rate of 0.044 kg s^{-1}, find the power of the immersion heater which is being used to heat the water. What assumption is being made in obtaining this answer? The specific heat capacity of water is 4200 J kg^{-1} K^{-1}.

14.22 In a particular household the average use per day of electrical appliances during a 100 day period was as shown in Table 14.4. If 1 kWh costs 6p, find how large a bill will be expected at the end of the 100 days. What would you recommend as an economy measure? A deep freeze and a refrigerator are normally left on all the time. Why is it therefore that the times given in the table are less than 24 hours?

Table 14.4

Appliance	Power	Time used
immersion heater	3 kW	2 hours
cooker ring	2 kW	3 hours
cooker oven	2 kW	1 hour
lights	60 W	30 hours
lights	100 W	20 hours
television	200 W	5 hours
refrigerator	120 W	10 hours
deep freeze	400 W	14 hours
central heating pump	50 W	16 hours
vacuum cleaner	400 W	15 minutes
washing machine	500 W	30 minutes
washing machine heater	3 kW	20 minutes
hair drier	1200 W	15 minutes
iron	800 W	15 minutes
stereo	40 W	1 hour 30 mins
various other items	2 kW	1 hour

SUMMARY

- The ampere is the electrical base unit in the SI system of units.

- Charge is defined as the product of current and time, $Q = I \times t$; 1 coulomb is 1 ampere for 1 second.

- Potential difference is the energy changed from electrical energy to other forms of energy per unit charge. This is equivalent to the power per unit current,

$$V = \frac{energy}{change} = \frac{power}{current} ;$$

1 volt is 1 joule per coulomb or 1 watt per ampere.

- Electromotive force (e.m.f.) is not a force. It is the energy changed from other forms of energy into electrical energy per unit charge. Its unit is the volt and it is the potential difference across a battery when it is not supplying current.

- Resistance is the potential difference per unit current, $R = V/I$; 1 ohm is 1 volt per ampere. For some conductors the current through the conductor is proportional to the potential difference across it, if the temperature is constant. For these conductors a graph of I against V is a straight line through the origin and the gradient is $1/R$.

$$\frac{V}{I} = R \qquad V \times I = P$$

- Resistivity ρ of a material is defined by the equation $R = \dfrac{\rho l}{A}$.

- Conductivity σ is the reciprocal of the resistivity, $\sigma = \dfrac{1}{\rho} = \dfrac{l}{RA}$.

- Conductance G is the reciprocal of the resistance $G = \dfrac{1}{R}$.

- Current I in terms of the charge q and the velocity v of charge carriers is $I = nAvq$ where n is the number of charge carriers per unit volume and A is the cross-section area of the conductor.
- The temperature coefficient of resistance α is given by

$$\alpha = \frac{R_t - R_0}{R_0 t}$$

- The terminal potential difference V for a cell of e.m.f. E and internal resistance r is

$$V = E - Ir = \frac{ER}{R + r}$$

$$I = \frac{E}{R + r} \text{ and useful power} = \frac{E^2 R}{(R + r)^2}$$

$$\text{Efficiency} = \frac{R}{R + r}$$

- Power to a resistor of resistance R:

$$P = VI = I^2 R = \frac{V^2}{R}$$

- Energy to a resistor of resistance R:

$$E = VIt = I^2 Rt = \frac{V^2 t}{R}$$

- 1 kilowatt hour (kWh) = 1000 W $\times$ 3600 s = 3 600 000 J.

EXAMINATION QUESTIONS

14.23 **(a)** The relative atomic mass of copper is 63.5 and its density is $8.93 \times 10^3 \, \text{kg m}^{-3}$. Its electrical properties show that in each cubic metre there are 8.34×10^{28} free electrons. Find, to three significant figures, the effective number of free electrons contributed by each atom of copper.

(b) (i) A copper wire of cross-sectional area $3.0 \, \text{mm}^2$ carries a current of 2.0 A. Find the drift velocity of free electrons in the wire.

(ii) How long would it take, on average, for a free electron to travel a distance of 0.50 m along this wire? (6)

(UCLES 1982)

14.24 Define electrical resistivity and state the unit in which it is measured.

Explain carefully why, in an experiment to measure the resistivity of polythene (a very poor electrical conductor), it would be advantageous to use **(a)** a large but thin sheet of polythene, **(b)** a high voltage supply. (6)

(ULSEB 1981)

14.25 An electrical heating element is to be designed so that the power dissipated will be 750 W when connected to the 240 V mains supply.

(a) Calculate the resistance of the wire needed

(b) The element is to be made from nichrome ribbon 1.0 mm wide and 0.050 mm thick. The resistivity of nichrome = $1.1 \times 10^{-6} \, \Omega \, \text{m}$. Calculate the length of ribbon required.

(c) Draw a circuit diagram to show how a second heating element would be connected to increase the power dissipated to 1.5 kW.

(d) State one important property of a conductor used to make heating elements. (9)

(AEB 1987)

14.26 **(a)** Sketch and label a graph to show how you would expect the current in a 12 V, 24 W tungsten filament lamp to change as the potential difference across the lamp increases slowly from, 0 V to 12 V.

(b) Describe briefly how you would check your graph experimentally.

(c) Explain whether or not you think the filament obeys Ohm's law. (10)

(ULSEB 1987)

14.27 A student is provided with a 2 V cell, a lamp, a switch and a thermistor with a negative temperature coefficient of resistance.

The lamp, which is in series with the cell as in diagram (a), lights immediately the switch is turned on.

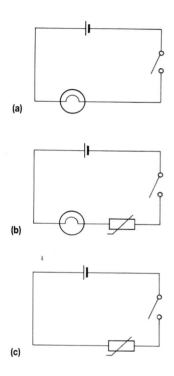

Explain why

(a) when the thermistor is connected in series with the lamp, as in diagram (b), and the switch is turned on, the lamp lights up slowly, and

(b) if the lamp is omitted, as in diagram (c), and the switch is turned on, the cell is soon destroyed by overheating. (5)

(ULSEB 1986)

14.28 Define the *resistance* of a resistor and the *resistivity* of a conductor. (2)

Describe how the resistivities of **(a)** a pure metal, and **(b)** a pure semiconductor, vary with temperature. In each case explain why such variations occur. (6)

One practical use of the variation is to measure temperature. Discuss the advantages and disadvantages of one type of resistance thermometer compared with the mercury-in-glass thermometer. (3)

Diagram (b), shows the relation between p.d. and current for a lamp bulb in a steady state. Calculate values for the resistance of the bulb when connected to (i) a 12 V supply, and (ii) a 1.5 V battery, both of negligible internal resistance. (2)

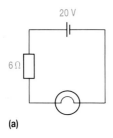

(a)

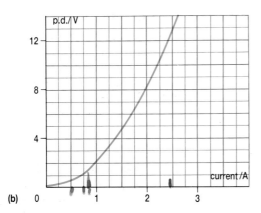

(b)

The circuit in diagram (a) contains a 20 V battery of negligible internal resistance, a 6 Ω resistance and the bulb whose characteristic is shown in the graph. The bulb obeys the relationship

(p.d., in V, across bulb) = $2 \times$ (current, in A, through bulb)^2

Calculate the current flowing in the circuit. (4) When a 10 V battery is used, deduce whether the current in the circuit is less than, greater than, or equal to half the current flowing when the 20 V battery is used. (3)

(O & C 1985)

Chapter 15

ELECTRICAL CIRCUITS

> **LEARNING OBJECTIVES**
>
> At the end of this chapter you should be able to:
>
> 1. apply Kirchhoff's laws to series and parallel circuits;
>
> 2. adapt ammeters and voltmeters and use them in making practical measurements;
>
> 3. understand the principles and uses of the potential divider, potentiometer and Wheatstone bridge circuits.

15.1 CIRCUIT LAWS

In the previous chapter a series of electrical terms, such as electromotive force and resistance, were introduced. These terms were used when dealing with circuits containing only the power source and a single component. In this chapter the same terms will be used but the circuits will be more complex. As is usually the case with increasing complexity, general rules are established which can be used repeatedly in analysing problems. Many practical electrical problems involve only a single circuit component but where more than one component exists it is essential to take care to find the relevant electrical quantities for that component.

The general rules in the case of electrical circuits are called Kirchhoff's laws. Kirchhoff's laws are central to the understanding of all electrical circuits, even the simplest. In the previous chapter no specific reference was made to the laws, but they were nevertheless being made use of. Essentially the two laws are conservation laws; the first law is a statement of conservation of charge and the second law is a statement, in electrical terms, of the law of conservation of energy.

Kirchhoff's first law

A formal statement of this law is:

⬥ the algebraic sum of the currents at a junction is zero

Put another way, this law says that charge cannot be created or destroyed. If within a complex circuit there is part of the circuit which is as shown in Fig 15.1, then at each junction the total current flowing into the junction equals the total current leaving the junction. At the left-hand junction of Fig 15.1 there is 3.0 A + 2.6 A coming into the junction and there must therefore be 5.6 A leaving the junction. Since a further 1.3 A joins with this current at the right-hand junction the total current becomes 6.9 A. Fig 15.2 shows this for four currents a, b, c and d. The first law emphasises the algebraic nature of the sum of the currents. Account must be taken of whether there is current into or out of the junction. Here a, b and d are currents into the junction while c moves out. The algebraic sum of the currents, which must be zero, is shown by the equation:

$$a + b - c + d = 0$$

Fig 15.1

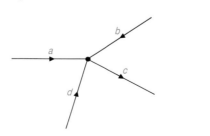

Fig 15.2

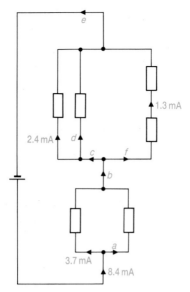

Fig 15.3

Example 1

Find the currents a, b, c, d, e and f as marked on Fig 15.3.

$$\text{At the bottom junction} \qquad 8.4 \text{ mA} \quad = 3.7 \text{ mA} + a$$
$$a \quad = 4.7 \text{ mA}$$

b must be equal to 3.7 mA + 4.7 mA $= 8.4$ mA again.

This current splits up and since the two series resistors have 1.3 mA through them, f must be 1.3 mA and so

$$c \quad = 8.4 \text{ mA} - 1.3 \text{ mA} = 7.1 \text{ mA}$$
$$\text{Since} \qquad 7.1 \text{ mA} \quad = d + 2.4 \text{ mA}$$
$$d \quad = 4.7 \text{ mA}$$

A check that 1.3 mA + 4.7 mA + 2.4 mA = e = 8.4 mA, confirms that Kirchhoff's first law has been obeyed.

This example illustrates that the use of Kirchhoff's first law is very straightforward. It seems to be common sense that charge cannot escape from a wire but has to flow round the circuit. It is surprising therefore to find that many mistaken solutions to problems are given in which the law has appeared to have been broken. Almost always these mistakes arise because no check has been made on the solution to the problem to ensure that the currents stated are possible. One quick check can always be made concerning a battery; it must **always** have the same current leaving the positive terminal as is entering the negative terminal. A battery does not produce electrical charge; it is a charge pump which pumps the charge taken in at one terminal out through the other terminal. In the same way a water pump is not a source of water; it can only pump out water which it takes in through its inlet. Kirchhoff's first law can be applied at any instant to all electrical circuits whether the current is a.c. or d.c., whatever the circuit components are, and whether or not there is an earth connection in the circuits.

Contrary to popular opinion, charge cannot flow to earth and somehow get lost. If charge is flowing to earth in some part of a circuit then an exactly equal charge must be flowing from earth in some other part of the circuit. There can be no current in the earth connection shown in Fig 15.4(a) because if some charge did flow to earth there would be more charge flowing from the battery's positive terminal than into its negative terminal. Current will exist in the circuit in Fig 15.4(b) because for every unit of charge which flows to one earth connection, a corresponding unit of charge can leave the earth through the other earth connection. This is a circuit used in cars. The metal bodywork of the car is used as the earth and one terminal of the battery is connected to it. A single wire then connects the other terminal of the battery to the component and the return current path is through the car's bodywork.

Kirchhoff's second law

A formal statement of this law is:

◆ Round any closed circuit the algebraic sum of the e.m.fs is equal to the algebraic sum of the potential differences.

Another way of putting this is to say that each and every point in a stable electrical circuit has a particular value of potential. If therefore a charge moves around the circuit but comes back to the original point, any gains in electrical energy which it might have had in its journey must be balanced by corresponding losses of energy. Provided the charge returns to its original place, the gains and losses of electrical energy are always equal no matter what route is taken by the charge. A mechanical system which involves similar principles concerns the movement of a body in a gravitational field. Each point on a mountain has a gravitational potential with respect to some arbitrary zero. Each kilogram of mass which moves around

(a)

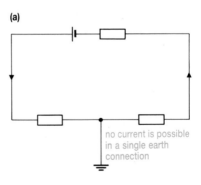

no current is possible in a single earth connection

(b)

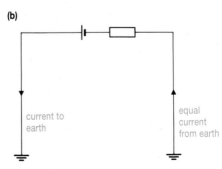

current to earth

equal current from earth

Fig 15.4 How a circuit is earthed determines whether or not there can be a current in the earth connection.

on the mountain will gain potential energy as it moves upwards and will lose potential energy as it moves down. If it finishes up at the point it started out from, then its potential energy gains must equal its potential energy losses no matter what route it takes. The independence of potential and potential energy on the route taken is an important feature of the law of conservation of energy as applied in both the mechanical and electrical cases.

To see how Kirchhoff's second law is applied to a circuit, work through Examples 2 and 3.

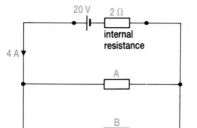

Fig 15.5

Example 2

A battery with an e.m.f. of 20 V and an internal resistance of 2.0 Ω supplies a total current of 4.0 A to two resistors A and B, as shown in Fig 15.5. B has a resistance of 12 Ω. Find the resistance of A and the power supplied to each circuit component.

Table 15.1(a) is drawn first and the known information is inserted. Once two electrical quantities are known for an individual component, then other quantities can be found for that component. This gradual build-up of known information can be done quite simply on a single table but to make the method clear, several tables have been drawn to show the sequence used. You are asked to follow the blue numbers through from 1 to 11.

Table 15.1 (a)

	Internal resistance	A	B	Battery	
Potential difference	1	2	3	20 V	e.m.f.
current	4 A				current
resistance	2 Ω		12 Ω		
power					

1. 4 A flowing through a 2 Ω resistance requires a potential difference of 8 V.

2. A charge of one coulomb flowing from the positive terminal of the battery is given 20 J of energy. A coulomb which flows through resistor A must have 8 J left to flow through the internal resistance of the battery and so will lose 12 J in heating up the resistor. This is applying Kirchhoff's second law. 20 V e.m.f. is the sum of the two potential differences 12 V + 8 V.

3. Using the same argument for a charge flowing through B must give 12 V also across B. You may have known before that the potential difference across two resistors in parallel is always the same. This is a direct consequence of Kirchhoff's second law. Table 15.1(b) now shows the following information and the sequence of numbers continues.

Table 15.1 (b)

	Internal resistance	A	B	Battery	
potential difference	8 V	12 V	12 V	20 V	e.m.f.
current	4 A	5	4	6	current
resistance	2 Ω	7	12 Ω		
power	8	9	10	11	power

4. If 12 V is the potential difference across a resistance of 12 Ω then the current will be 1 A.
5. If the battery supplies 4 A and 1 A is the current through B then 3 A must be the current through A. This is a direct example of Kirchhoff's first law.
6. The current through the battery is the same as the current through its internal resistance and the current from the battery is the same as the current to the battery.
7. If 12 V across A are causing a current of 3 A then the resistance of A must be 4 Ω.
8, 9, 10, 11. All of the powers can now be calculated using the fact that the potential difference across a component multiplied by the current through the component gives the power. The table now becomes as shown in Table 15.1(c).

Table 15.1 (c)

	Internal resistance	A	B	Battery	
potential difference	8 V	12 V	12 V	20 V	e.m.f.
current	4 A	3 A	1 A	4 A	current
resistance	2 Ω	4 Ω	12 Ω		
power	32 W	36 W	12 W	80 W	power

Many electrical problems can be solved using this table method. The method has the advantage that it forces you to concentrate on components one at a time, so that there is less danger of using incorrect values. In this problem, for instance, there are three different numerical values of the potential difference to consider. Fundamental mistakes are made if 20 V is used whenever a potential difference is required. Another advantage of a table such as this is that it provides a good check at the end of the problem. Here the power used in heating resistors is 32 W + 36 W + 12 W = 80 W. Independently, the power supplied by the battery is 20 V × 4 A = 80 W. This confirms that the answer is correct.

There is one obvious gap on the final table. It is not correct to speak of the resistance of anything apart from the resistance of a resistor. It is correct therefore to put into the table the internal resistance of the battery, but you cannot speak of the resistance of the battery's e.m.f. However dividing the e.m.f. of 20 V by the current supplied of 4 A does give a resistance of 5 Ω. This is the total circuit resistance. Note that you cannot get this figure by adding 2 Ω + 4 Ω + 12 Ω.

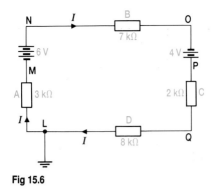

Fig 15.6

Example 3

The circuit shown in Fig 15.6 consists of two batteries connected with 4 resistors A, B, C and D. Point L on the circuit is earthed. Using the numerical data on the figure, find the current I and the potential of the points L, M, N, O, P and Q.

Using Kirchhoff's first law we know that there is the same current I through all the components in the circuit. There is no current through the earth connection.

Using Kirchhoff's second law we know that the sum of the potential differences equals the sum of the e.m.fs.

Potential difference across A = $I \times 3000 \ \Omega$

Potential difference across B = $I \times 7000 \ \Omega$

Potential difference across C = $I \times 2000 \ \Omega$

Potential difference across D = $I \times 8000 \ \Omega$

The sum of the potential differences is therefore $I \times 20\,000$.

Since both batteries are connected in such a way as to cause current in the same direction, the sum of the e.m.f.s = 6 V + 4 V = 10 V.

Therefore $20\,000\,I = 10$ V

$I = 10$ V / $20\,000 \ \Omega = 0.00050$ A = 0.50 mA.

This gives the potential difference across A = 1.5 V,

the potential difference across B = 3.5 V,

the potential difference across C = 1.0 V and

the potential difference across D = 4 V.

Positive charge flows from high potential to low potential. A current from L to M therefore is moving down an electrical potential gradient as it moves through resistor A. Since by convention L is the zero of potential, the potential of M must be –1.5 V. The e.m.f. of the cell then raises the potential by 6 V so the potential of N is +4.5 V.

The potential falls across B by 3.5 V to give the potential at O as +1 V and rises by 4 V as a result of the e.m.f. of the battery to a potential of 5 V at P. The potential falls by 1 V across C to give the potential at Q as +4 V and by a further 4 V across D to bring the potential back to zero. There is a built-in check here as, if the potential had not become zero back at L, then some mistake must have been made. Signs are often the cause of mistakes with this type of problem. The ability to find the potential at different points on a circuit such as this is essential when dealing with many electronic circuits, though usually they can be broken down into sections smaller than this even when the whole circuit is involved.

This particular circuit can be thought of as if it were an electrical roller coaster. Starting at ground level a charge drifts along to A where it falls and loses energy. The battery raises its energy and it coasts around to B where again it loses energy. Further energy is supplied by the second battery and C and D absorb energy before it gets back to where it started. Kirchhoff's second law can be applied to state that for unit charge, energy gains equal energy losses in a complete circuit.

QUESTIONS

15.1 Find the currents a, b, c and d in the part of a circuit shown in Fig 15.7

15.2 No knowledge of transistor function is needed to find the conventional currents in the circuit shown in Fig 15.8. Find the currents through R and R_L.

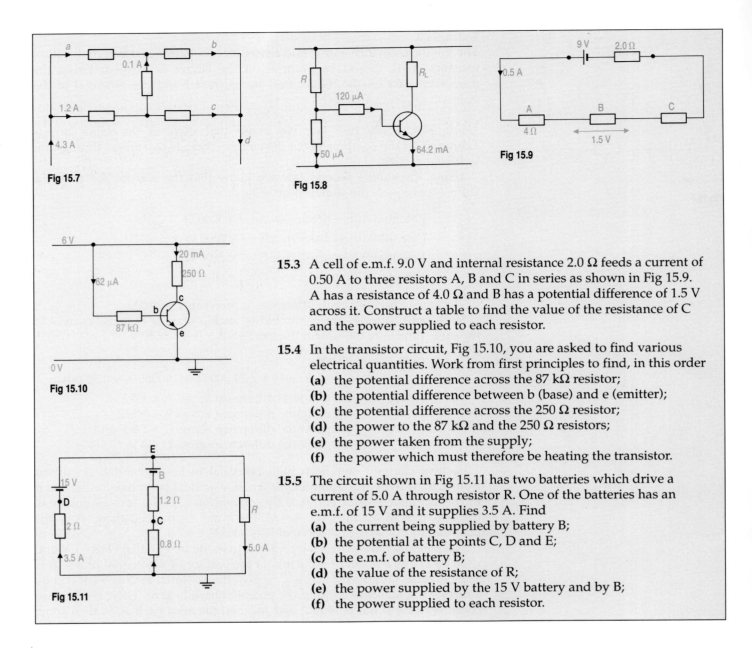

Fig 15.7

Fig 15.8

Fig 15.9

Fig 15.10

Fig 15.11

15.3 A cell of e.m.f. 9.0 V and internal resistance 2.0 Ω feeds a current of 0.50 A to three resistors A, B and C in series as shown in Fig 15.9. A has a resistance of 4.0 Ω and B has a potential difference of 1.5 V across it. Construct a table to find the value of the resistance of C and the power supplied to each resistor.

15.4 In the transistor circuit, Fig 15.10, you are asked to find various electrical quantities. Work from first principles to find, in this order
   (a) the potential difference across the 87 kΩ resistor;
   (b) the potential difference between b (base) and e (emitter);
   (c) the potential difference across the 250 Ω resistor;
   (d) the power to the 87 kΩ and the 250 Ω resistors;
   (e) the power taken from the supply;
   (f) the power which must therefore be heating the transistor.

15.5 The circuit shown in Fig 15.11 has two batteries which drive a current of 5.0 A through resistor R. One of the batteries has an e.m.f. of 15 V and it supplies 3.5 A. Find
   (a) the current being supplied by battery B;
   (b) the potential at the points C, D and E;
   (c) the e.m.f. of battery B;
   (d) the value of the resistance of R;
   (e) the power supplied by the 15 V battery and by B;
   (f) the power supplied to each resistor.

15.2 SERIES AND PARALLEL CIRCUITS

Kirchhoff's laws can be applied to all electrical circuits. When they are applied to circuits containing resistors in series and in parallel they give results which, because they occur frequently, are worth memorising.

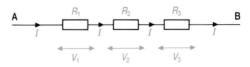

Fig 15.12 Resistors in series: the same current exists in each resistor.

Resistors in series

Fig 15.12 shows three resistors of resistance R_1, R_2 and R_3 connected to one another in series. Since there is only one possible path for the current through the three resistors, Kirchhoff's first law applied to this circuit shows that the current I through all of the resistors is the same. The potential differences across the resistances are therefore given by

$$V_1 = IR_1 \qquad V_2 = IR_2 \qquad V_3 = IR_3$$

Using Kirchhoff's second law gives the potential difference V, across all three resistances, as the sum of the individual potential differences, i.e.

$$V = V_1 + V_2 + V_3 = IR_1 + IR_2 + IR_3 = I(R_1 + R_2 + R_3)$$

But $\dfrac{V}{I} = R$, where R is the total resistance between A and B so

$$R = R_1 + R_2 + R_3$$

This equation can be extended to as many resistors as are present, provided there are no branches to the circuit and therefore each resistor has the same current. The result may seem to be just the application of common sense. It is, if you accept that the laws of conservation of energy and charge are also common sense. Note that the total resistance must be greater than any of the individual resistances, in contrast to the parallel case which follows.

Resistors in parallel

Fig 15.13 shows three resistors of resistance R_1, R_2 and R_3 connected to one another in parallel. Kirchhoff's first law applied to this circuit shows that the total current, I, is the sum of the individual currents, i.e.

$$I = I_1 + I_2 + I_3$$

Using Kirchhoff's second law gives the potential difference V, across all three resistances to be the same so

$$I_1 = \dfrac{V}{R_1} \qquad I_2 = \dfrac{V}{R_2} \qquad I_3 = \dfrac{V}{R_3}$$

Since also $\dfrac{V}{I}$ is the total resistance R between A and B this gives

$$I = \dfrac{V}{R} = \dfrac{V}{R_1} + \dfrac{V}{R_2} + \dfrac{V}{R_3}$$

Cancelling V throughout we get

$$\dfrac{1}{R} = \dfrac{1}{R_1} + \dfrac{1}{R_2} + \dfrac{1}{R_3}$$

This equation can be extended to as many resistors as are present, provided each resistor has the same potential difference across it.

For two resistors only this becomes

$$\dfrac{1}{R} = \dfrac{1}{R_1} + \dfrac{1}{R_2} = \dfrac{R_2 + R_1}{R_1 R_2}$$

and hence

$$R = \dfrac{R_1 R_2}{R_1 + R_2}$$

$$\text{Total resistance of 2 resistors in parallel} = \frac{\text{Product of two resistances}}{\text{Sum of two resistances}}$$

With parallel connections for resistors, the total resistance is always smaller than the smallest resistance. This must be so because each of the resistors has a current through it; each resistor therefore increases the total current and hence reduces the total resistance.

The circuits shown in Figs 15.14 and 15.15 need careful analysis. In Fig 15.14 only resistors R₂ and R₃ are in parallel. These two are in series with R₁. This circuit is basically the same as the circuit dealt with in Example 2. In Fig 15.15 the two resistors R₁ and R₂ cannot be considered to be in parallel because they do not have the same potential difference across them. The effect of E_1 in series with R₁ means that this type of circuit is best dealt with from basic principles using the table method shown in Example 2.

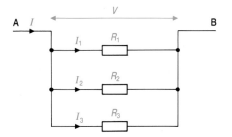

Fig 15.13 Resistors in parallel: the same potential difference exists across each resistor.

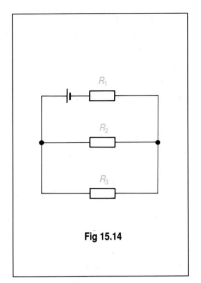

Fig 15.14

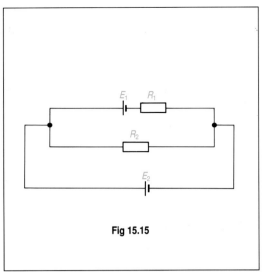

Fig 15.15

QUESTIONS

15.6 Find the total resistance of each of the circuits shown in Fig.15.16. Why do (c) and (d) give the same answers?

(a)
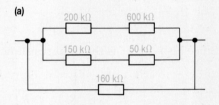

(d)
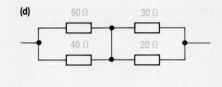

(b)
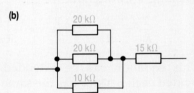

(c)
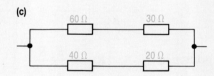

Fig 15.16

15.7 Resistors are manufactured in preferred values. One series of values, used for resistors with tolerance of ± 10% is
   1.0, 1.2, 1.5, 1.8, 2.2, 2.7, 3.3, 3.9, 4.7, 5.6, 6.8, 8.2, kΩ
then continuing
   10, 12, 15, 18, . . . 82 kΩ then
   100, 120, 150, 180 . . . 820 kΩ etc.
A particular resistor in a circuit is found to have a value of 2.10 kΩ and it is desired to reduce its value to 2.00 kΩ using resistors available from the preferred list. How can this be done?

15.8 Two variable resistors are often connected in series when it is desired to control closely the value of a current. Explain how the circuit in Fig 15.17 functions and give relative values for the two variable resistances.

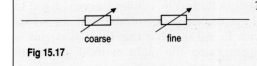

coarse   fine

Fig 15.17

ELECTRICAL CIRCUITS

15.3 THE POTENTIAL DIVIDER

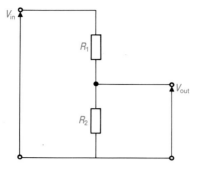

Fig 15.18   The potential divider circuit.

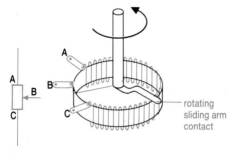

Fig 15.19   A potentiometer. This device, which enables a potential divider to be set up in practice, is frequently the electrical component behind the knobs on a radio or television which enable you to alter brightness, volume, etc. The circuit symbol for the device is also shown.

This is an extremely useful, simple circuit. The odd thing about it is how much difficulty it causes! The circuit in its simplest form is shown in Fig 15.18 which shows an input potential difference V_{in} being applied across two resistors having resistances of R_1 and R_2 and an output potential difference V_{out} being taken across R_2. If the output current is zero then the current I flowing through R_1 also flows through R_2 and the two resistors are in series. This gives

$$I = \frac{V_{in}}{R_1 + R_2}$$

or

$$V_{out} = IR_2 = \frac{V_{in}}{R_1 + R_2} \times R_2$$

$$V_{out} = \frac{R_2}{R_1 + R_2} \times V_{in}$$

This result is usefully thought of as being that the output p.d. is the same fraction of the input potential difference as R_2 is as a fraction of the total resistance $R_1 + R_2$. Example 4 shows how this can be used in practice. There is no need to calculate the current first.

Example 4

A resistance of 10 000 Ω in the form of a coil bent into a loop, as shown in Fig 15.19, is connected across a 12 V supply. A sliding arm can make contact with the coil and an output taken from one end of the coil and the sliding arm. This gives the potential divider circuit shown in Fig 15.20(a) where one part of the coil has a resistance of 3700 Ω leaving 6300 Ω for the other end. Find the output voltage. Find also the reading a voltmeter gives when connected across the output, as shown in Fig 15.20(b), if the resistance of the voltmeter is 9500 Ω.

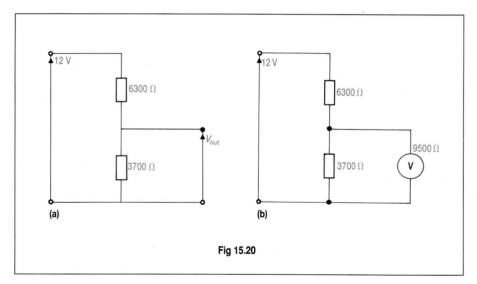

Fig 15.20

Output potential difference $= \dfrac{3700}{3700 + 6300} \times 12\text{ V} = \dfrac{3700}{10\ 000} \times 12\text{ V} = 4.4\text{ V}$

When the voltmeter is connected in parallel with the 3700 Ω resistor, the resistance of the two is

$$\frac{3700 \times 9500}{3700 + 9500} = \frac{35\ 150\ 000}{13\ 200} = 2660\ \Omega$$

The reading on the voltmeter is therefore given by

$$\frac{2660}{8960} \times 12\text{V} = 3.6\text{ V to 2 sig. figs}$$

Note in this case that the voltmeter is correctly reading the potential difference across itself and the 3700 Ω resistor. This however is not the same value of potential difference which existed across the 3700 Ω resistor before the voltmeter was connected in place. This is a characteristic of many instruments. They frequently affect the very readings which they are trying to measure. A thermometer, for instance, will cool down a hot liquid a little bit when it is placed in the liquid. The thermometer reads correctly, but does not record the value which would have existed if the thermometer had not been used.

15.9 A light-dependent resistor (LDR) is connected in series with a 12 kΩ resistor and a 9 V d.c. source as shown in Fig 15.21. Find the potential at the point P when the LDR is
(a) in the dark and has a resistance of 6.0 MΩ;
(b) in the light and has a resistance of 1.0 kΩ.
What is the value of the resistance of the LDR if the potential at P is 0.60 V?

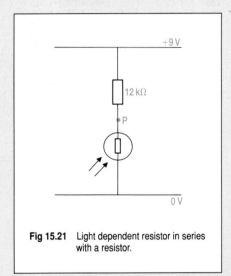

Fig 15.21   Light dependent resistor in series with a resistor.

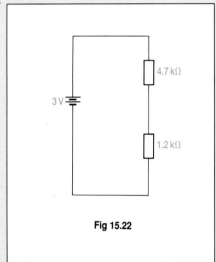

Fig 15.22

15.10 Find the potential difference across the 1.2 kΩ resistor in Fig 15.22. What is the smallest resistance a voltmeter can have if it is to measure the potential difference across the 1.2 kΩ resistor without introducing a systematic error of more than 1%?

15.4 ELECTRICAL INSTRUMENTS

In Section 15.3 it was pointed out that a voltmeter can give misleading readings because of the current which it takes from the circuit into which it is placed. How much current it takes depends on the resistance of the voltmeter. An ideal voltmeter will not take any current at all. An ammeter in a circuit causes a similar problem. When the ammeter is placed in the circuit to measure the current it increases the circuit resistance and hence reduces the current. An accurate ammeter then correctly reads this lower current. An ideal ammeter would have zero resistance. The principles behind the structure and calibration of an ammeter will be dealt with in Chapter 17 but here is a suitable place to deal with the effect of the resistance of an ammeter and the way the range of an ammeter can be altered by the use of resistors, called **shunts**, placed in parallel with it. At the heart of a traditional voltmeter or ammeter is a coil of wire which can rotate through about 90° in a magnetic field when it carries an electric current. A pointer attached to the coil moves over a scale. The basic instrument, with no shunts, is called a **galvanometer**.

Fig 15.23   A flight deck is a maze of electrical and electronic instrumentation and switching.

Ammeter

Example 5 shows how a galvanometer can be used as an ammeter to measure currents over the range from 0 to 3 A.

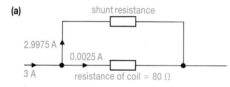
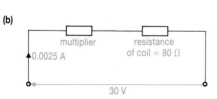

Fig 15.24   (a) Ammeter circuit. **(b)** Voltmeter circuit.

Example 5

The galvanometer in an ammeter has a resistance of 80 Ω and the pointer is deflected fully across the scale, full scale deflection (f.s.d.), when a current of 2.5 mA is passing through the coil. Find the value of the shunt resistance necessary to make it an ammeter reading up to 3 A.

The maximum current which can flow through the coil is 0.0025 A; the maximum current which needs to be measured is 3 A. If 2.9975 A pass through a by-pass resistance R then full scale deflection will occur when the current supplied is 3 A. This is shown in Fig 15.24(a).

The potential difference across the resistance of the coil is 0.0025 A $\times$ 80 Ω = 0.2 V.

Since the coil and the shunt resistor are in parallel, the potential difference across the shunt resistor is also 0.2 V.

The value of the shunt resistance needed is therefore

$$\frac{0.2 \text{ V}}{2.9975 \text{ A}} = 0.0667 \text{ } \Omega$$

If the current being measured is then reduced to 1 A, the shunt resistance and the coil resistance take one-third of their previous current and the pointer moves only one-third of the way across the scale. It can be quite tricky to read from the scale of a meter because there are often several ranges which can be used and these may be printed on the scale of the instrument. Be careful that you read from a scale which rises to the known full scale deflection and also be careful to get the decimal point in the right place when measuring A, mA or μA.

Commercial instruments are available in which the scale itself is changed as the shunt resistor is changed. Digital instruments do not have this difficulty but they may still have different ranges. It is standard practice when using an ammeter to start by setting it for high currents and then to work downwards. A small current can do no damage to an ammeter set for a high current but large currents through an ammeter set to measure low currents can damage the instrument irreparably.

An ammeter such as this does not behave ideally. As calculated earlier, the potential difference across the instrument when in use is 0.2 V. This indicates that when the ammeter is placed in series in a circuit and is measuring a full scale deflection current of 3 A, it has a low total resistance of 0.2 V/3 A = 0.67 Ω. If the other resistances in the circuit are high then this small resistance can be neglected.

Voltmeter

Example 6 shows how a galvanometer may be used as a voltmeter over the range from 0 – 30 V.

Example 6

How can the instrument used in Example 5 be adapted to measure a potential difference up to 30 V?

If a galvanometer is to be used to measure potential difference, then Ohm's law has to be assumed to hold. If a p.d. of 30 V is used to cause a current of 0.0025 A to flow, then the total resistance of the circuit must be

$$\frac{30 \text{ V}}{0.0025 \text{ A}} = 12\,000 \ \Omega$$

Since the meter resistance is 80 Ω, a series resistance of 11 920 Ω needs to be placed in the circuit in series with the coil. This series resistor is called a **multiplier**. The circuit is shown in Fig 15.24(b). In practice it is unlikely to be necessary to measure the series resistance to as many as 4 significant figures as the basic instrument is probably not that accurate.

A voltmeter is placed in parallel with the component across which the potential difference is required, and ideally has infinite resistance. If therefore the resistance of the voltmeter is too low, there will be an appreciable current through the voltmeter and this will affect the rest of the circuit. Here the resistance is 12 kΩ and there is only a small current of 2.5 mA. Provided other resistances are appreciably lower than the 12 kΩ then the voltmeter is suitable.

For testing many electronic circuits however, this type of instrument is not suitable because circuit currents are themselves small and circuit resistance values are high. It is now usual to use electronic voltmeters where the resistance of the meter can be of the order of 10 MΩ.

QUESTIONS

15.11 A galvanometer with full scale deflection 3.00 mA and resistance 75 Ω is to be used as an ammeter to read up to 6 A.
  (a) What value of shunt resistance will be required?
  (b) What will be the potential difference across the meter when it is being used to measure a current of 6 A?
  (c) Describe qualitatively what will happen if the electrical connection from the galvanometer to the shunt gradually works loose.

15.12 Using the same data as is given in Example 6, find the necessary series resistances for measuring potential differences up to 300 V, 3 V and 0.3 V as well as for 30 V. Draw a diagram to show how all of these resistors can be connected into a single circuit to give a multi-range voltmeter with different input terminals for the different ranges.

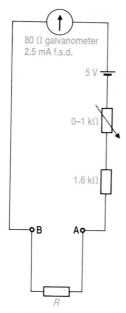

80 Ω galvanometer
2.5 mA f.s.d.

5 V

0–1 kΩ

1.6 kΩ

B   A

R

Fig 15.25 Ohmmeter circuit.

Ohmmeter

An ohmmeter circuit is shown in Fig 15.25. It is still basically an ammeter but it needs a source of e.m.f. to cause a current to flow through the resistance to be measured and is then calibrated so that the scale reads resistance directly. The current which flows depends on the total resistance, R_t, in the circuit.

This includes the resistance of the coil of the meter which is 80 Ω in the example, a fixed resistance of 1.6 kΩ which protects the meter from overload, a variable resistance of 0–1 kΩ which is used to set the zero of the instrument before use, and the resistance, R, to be measured. At first, terminals A and B are shorted out and the resistance of the variable resistor is adjusted so that full scale deflection is obtained. Then the unknown resistance is connected across A and B and the resistance is read from the

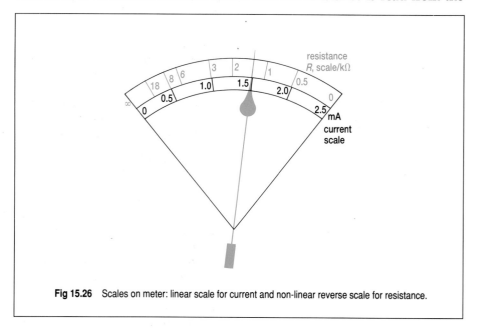

Fig 15.26 Scales on meter: linear scale for current and non-linear reverse scale for resistance.

scale. The scale is shown in Fig 15.26 and is non-linear. It can be calibrated from Table 15.2, which assumes that the e.m.f. of the supply is 5 V and that full scale deflection of the coil of the meter is 2.5 mA.

Table 15.2

Current /mA	R_t /Ω	R /Ω
2.50	2000	0
2.00	2500	500
1.67	3000	1000
1.25	4000	2000
1.00	5000	3000
0.625	8000	6000
0.500	10 000	8000
0.250	20 000	18 000
0	∞	∞

15.13 An ohmmeter uses a galvanometer which has a resistance of 5 kΩ and gives f.s.d. for 100 μA. It is used with a 10 V supply and a series resistance R. Show that R has to have the value of 95000 Ω if full scale deflection is to be obtained for zero external resistance. What external resistances correspond to currents of 0, 10, 20, 30, 40 . . . 100 μA? Hence draw a suitable scale for the ohmmeter.

15.14 For the ohmmeter in Figs 15.25 and 15.26 find
(a) the value of the variable resistance used;
(b) the reading the pointer is giving. (Note: it is **not** 1.5kΩ. You cannot interpolate directly on a non-linear scale. Use the known value of the current as the starting point.)
(c) Derive an equation connecting the current I with R.

15.5 THE WHEATSTONE BRIDGE

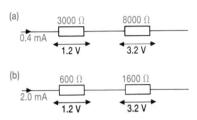

Fig 15.27

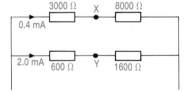

Fig 15.28

This electrical circuit is also used for measuring resistance. Strictly speaking, it only compares two resistances so if an unknown resistance is to be determined then a standard resistance is required. It gets its name from its inventor, Charles Wheatstone, and from the fact that in one arrangement of the circuit the two resistors to be compared are connected from one side to another across gaps in the apparatus.

The Wheatstone Bridge, and other similar bridge circuits used with alternating current, are still used a great deal and are rather different in their function from normal ammeter, voltmeter and ohmmeter circuits. They use the idea of a circuit balanced so that zero current passes through a meter. A reading taken in this way is called a **null deflection** reading and has the advantage that it is easier to detect a current rather than to measure it. In the Wheatstone Bridge circuit the aim is to adjust the circuit so that the potential difference across a meter is zero.

Consider part of a circuit, Fig 15.27(a):

the potential difference across the 3000 Ω resistor
= 0.0004 A × 3000 Ω = 1.2 V
the potential difference across the 8000 Ω resistor
= 0.0004 A × 8000 Ω = 3.2V

In the circuit shown in Fig 15.27(b), the current is different but the potential differences are the same. These two circuits can be joined without there being any effect on the currents in each branch as a result of the potential difference being the same across each. This gives the circuit shown in Fig 15.28. Not only are the potential differences across each branch the same but also the potential difference across the 3000 Ω resistor is the same as the potential difference across the 600 Ω resistor, so X and Y are at the same potential. If these points are connected with one another, no current will flow in the connecting wire. This is then said to be a balanced Wheatstone Bridge and is similar to the generalised circuit shown in Fig 15.29, where the four resistors have resistances P, Q, R and S.

The condition necessary for balance is that:
the potential difference across P = potential difference across R and
the potential difference across Q = potential difference across S.

If the current through P and Q is I_2 and the current through R and S is I_1 then

$$I_2P = I_1R$$
$$I_2Q = I_1S$$

Dividing gives $\dfrac{P}{Q} = \dfrac{R}{S}$

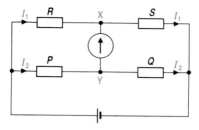

Fig 15.29 The Wheatstone bridge circuit.

In the example this checks, as

$$\frac{3000}{8000} = \frac{600}{1600}$$

Note that current I_2 in P will only equal I_2 in Q if no current flows through the galvanometer. It is quite a difficult electrical problem to find a current flowing through the meter of an unbalanced Wheatstone Bridge, but provided the bridge is balanced there is no need to find any current value at all.

There are many practical arrangements of such circuits. The most common school laboratory arrangement consists of a half-metre or metre resistance wire mounted on a wooden baseboard together with copper strips arranged in a way shown in Fig 15.30. The reason for using copper strip is that copper has a very low resistance. In use, the unknown resistance is connected across the left-hand gap and the standard resistance across the right-hand gap. A power supply is connected and a sensitive centre zero galvanomter is connected between X and Y as shown.

Contact at Y is made with a tapping key, which is able to be moved along the wire until a point is found at which zero current flows through the galvanometer. At this position we get

$$\frac{\text{Resistance of length } l_1 \text{ of wire}}{\text{Resistance of length } l_2 \text{ of wire}} = \frac{R}{S}$$

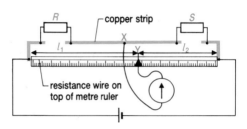

Fig 15.30 The Wheatstone metre bridge.

so $\quad \dfrac{l_1 \times \text{resistance of unit length of wire}}{l_2 \times \text{resistance of unit length of wire}} = \dfrac{R}{S}$

Since the resistance of unit length is the same at both ends of the wire

$$\frac{l_1}{l_2} = \frac{R}{S}$$

INVESTIGATION

Use a Wheatstone Bridge circuit to check the calibration of a resistance box. The investigation needs to be done in at least two parts.

1. Take several standard resistors of the same value and with any comparable resistor used as resistor S, insert the standard resistors one at a time into gap R and find a balance point. If the same balance point is found for 2 (or preferably 3) standard resistors then it is highly likely that these resistors really are standard at the value stated on them. Use one of these resistors as your standard resistor and place it in gap S.

2. With the standard resistor in gap S connect the resistance box in gap R using short connecting leads. Find the value of the resistance of each of the resistors in the box one at a time.

Check each reading by placing the standard resistor in gap R and the resistance box in gap S.

How can you improve the investigation if

(a) the length of l_1 or l_2 gets so small it is difficult to measure to more than 2 significant figures;

(b) the needle on the galvanometer does not remain steady;

(c) the needle on the galvanometer only moves a small distance either side of zero;

(d) the galvanometer needle is not on zero when the galvanometer is removed from the circuit altogether?

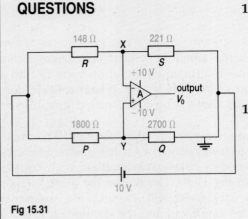

Fig 15.31

15.15 A Wheatstone Bridge circuit is used to find an unknown resistance R. (See Fig 15.30). When S is 4.00 Ω, l_1 is 28.3 cm and l_2 is 71.7 cm. Find the value of R.

  S is then changed to 10.00 Ω. What new value of l_1 is expected? If the new value is not the expected value, give three possible explanations.

15.16 A Wheatstone Bridge circuit is set up (Fig 15.31) with P = 1.8 kΩ, Q = 2.7 kΩ, R =148 Ω and S = 221 Ω. The circuit is not balanced, but there is negligible current between X and Y because circuit component A has a very high resistance.

(a) Find the potential at X and the potential at Y.

(b) A is an integrated circuit called an operational amplifier (see Section 21.6). The output V_0 from A is given by

$$V_0 = 10\ 000\ (V_y - V_x)$$

but V_0 cannot be outside the range of +10 V to –10 V. Find V_0 in this case.

(c) Explain why the output from the operational amplifier suddenly changes from +10 V to –10 V as R is decreased slowly to 147 Ω.

  Many control circuits use this principle. The sudden switch in output from the operational amplifier can be used to control a motor which itself is used to alter the value of R back so that balance is again achieved.

15.6 THE POTENTIOMETER

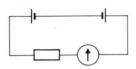

Fig 15.32

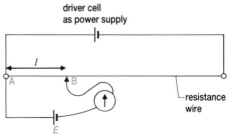

← current direction in *E* if *l* is too small for balance
→ current direction in *E* if *l* is too large for balance
  zero current in *E* and galvanometer at balance

Fig 15.33 The potentiometer circuit.

In some respects this is a similar instrument to the Wheatstone Bridge because it is also a null reading device and can be used with a tapping key on a long wire. It does however rely on a totally different principle and if the principle is understood there will be no danger of confusing the two circuits.

The principle used is illustrated in Fig 15.32 in which two cells of the same type are in a circuit back to back with one another. No matter what resistance there is in this circuit, no current will be recorded by the galvanometer. Put another way, if no current flows then the electromotive forces of the two cells are the same. This would be a very restricted instrument if it were limited just to verifying equal e.m.fs but it is possible to make one of these e.m.fs a variable e.m.f. whose value is adjusted until it is equal to the unknown e.m.f. This is done using a long resistance wire across which a stable potential drop must be maintained. The following would best be done with the apparatus. The same resistance wire as on the Wheatstone Bridge can be used.

A metre length of resistance wire is mounted on top of a metre rule set on a baseboard. A d.c. power supply (of about 2 V) is connected across the wire. This is the circuit shown in Fig 15.33 before the cell of e.m.f. E and the galvanometer are put in position. The power supply must remain constant throughout the experiment so a lab-pack or a well charged accumulator must be used, not a torch battery. A voltmeter is connected between A and B and the sliding contact B is now moved from the left-hand end of the wire to the right-hand end. A graph of voltmeter reading should be plotted against distance l to show that the potential difference across the wire is proportional to the length l. This varying potential difference can now be balanced against the unknown e.m.f., E as follows.

If l has a value less than that for balance, then the potential difference across l is less than E, the e.m.f. of the cell and the current will be one way through the galvanometer. If l is too large the current will flow the opposite way through the galvanometer. If there is zero current through the galvanometer the potential difference across l equals the e.m.f. of the cell. Note that this can only occur if the two cells are back to back with one another. They must therefore have both positive (or both negative) terminals connected together as shown. A balance point cannot be achieved if both cells are driving current in the same direction.

Fig 15.34 The instrumentation in the control room of a power station, such as this one at Hinkley Point B, is extensive and consists of computers and their monitors together with digital and ordinary meters, potentiometers, Wheatstone bridges and graph plotters.

When an e.m.f. is to be measured, it is usual first to do a calibration experiment with a standard cell. A Weston standard cell at 18 °C gives an e.m.f. E_s, of 1.0186 V. If this cell is used, and a balance point l_s is found, this calibrates the potential drop per unit length across the wire to be

$$\frac{E_s}{l_s}$$

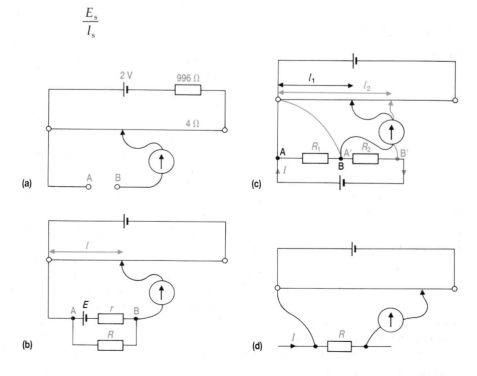

Fig 15.35

If the cell of unkown e.m.f. E, is placed in position instead of the standard cell and a new balance length l is obtained, the value of E is given by

$$E = \frac{E_s}{l_s} \times l_0$$

$$E = \frac{l_0}{l_s} \times E_s \quad \text{where } E_s = 1.0186 \text{ V}$$

Many modifications can be made to the basic potentiometer circuit but in all cases the instrument is comparing e.m.fs. These e.m.fs can then be used to find other quantities. Question 15.17 should be worked through as it gives guidance to the particular modifications shown in the four circuit diagrams of Fig 15.35.

QUESTIONS

Fig 15.36 A British Telecom engineer fault finding on a telephone line at Bolton Castle.

15.17 Fig 15.35 shows four modifications which can be made to potentiometer circuits. Answer the following questions, taking data from the diagrams where necessary.

(1) In circuit (a) a 996 Ω resistance is placed in series with the driver cell, which has negligible internal resistance. The resistance of the potentiometer wire is 4.0 Ω. What is the potential difference across the potentiometer wire?

(2) What is the maximum e.m.f. which can be measured between A and B for circuit (a)? How can this modification lead to an instrument which can measure temperature using a thermocouple? A thermocouple gives an e.m.f. of a few millivolts which is proportional to the temperature.

(3) In circuit (b) state an algebraic expression for the potential difference between A and B by considering only the cell, of e.m.f. E, its internal resistance r, and the known value of the external resistance R.

(4) If R is made to have infinite resistance, what will be the potential difference between A and B?

(5) If the balance length in part (3) is l and the balance length in part (4) is l_0 show that

$$\frac{l_0}{l} = \frac{R + r}{R} = 1 + \frac{r}{R}$$

Since l, l_0 and R are known or measured, r can be found.

(6) Using the above equation, a straight line graph can be plotted

of $\frac{l_0}{l}$ against $\frac{1}{R}$ if a series of values of l are obtained

for different values of R. How can the internal resistance be found from the graph?

(7) In circuit (c) the potential difference between A and B is measured first (black line connections) to get the balance length l_1. The connections to A and B are then moved to A' and B' (blue line connections) and a new balance length l_2 is obtained. If the same current l exists in both resistors, what are the potential differences across AB and A'B'.

(8) Find an expression connecting l_1 and l_2 to R_1 and R_2 and hence state the purpose of the circuit.

(9) How can circuit (d) be used to measure current?

- Kirchhoff's first law: the current into any junction equals the current leaving the junction. This is a law which is in principle a law of conservation of charge.
- Kirchhoff's second law: around any closed loop in a circuit the sum of the potential differences equals the sum of the e.m.fs. This is a law which is in principle a law of conservation of energy.
- Resistors in series: $R = R_1 + R_2 + \ldots$
- Resistors in parallel: $\dfrac{1}{R} = \dfrac{1}{R_1} + \dfrac{1}{R_2} + \ldots$
- For two resistors in parallel this becomes $R = \dfrac{R_1 R_2}{R_1 + R_2}$
- Ammeters have a low shunt resistance in parallel with the galvanometer. They have a low resistance and are placed in series in the circuit.

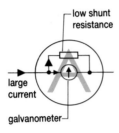

- Voltmeters have a high resistance in series with the galvanometer. They have a high resistance and are placed in parallel with the circuit.

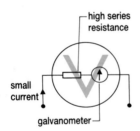

- The output from the potential divider circuit is $V_{out} = \dfrac{R_2}{R_1 + R_2} \, V_{in}$

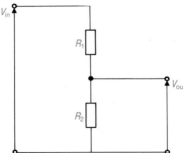

- In a potentiometer the driver cell maintains a constant potential difference across the wire. The value of this p.d. is determined by using a separate standard cell in place of E: then E_s/l_s = potential difference per unit length.

$$E = \frac{l_0}{l_s} \times E_s$$

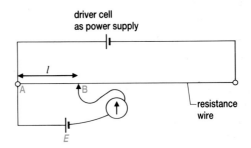

- When a Wheatstone Bridge is balanced then $\dfrac{R}{S} = \dfrac{P}{Q}$

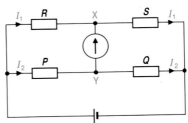

EXAMINATION QUESTIONS

15.18 A 12 V, 24 W lamp and a resistor of fixed value are connected in some way inside a box with two external terminals. In order to discover the circuit arrangement inside the box, a student connects a variable d.c. power source and an ammeter in series with the box and obtains the following results:

Applied potential difference	1.0 V	12 V
Current	1.17 A	4.00 A

(a) Draw a circuit diagram of the most likely arrangement inside the box, giving your reason.

(b) Use the 12 V, 4.00 A reading to deduce the value of the fixed resistor.

(c) What is the percentage increase in the resistance of the lamp as the applied potential difference changes from 1.0 V to 12 V?

(d) When the applied potential difference is increased to 24 V the current is again found to be 4.00 A. Explain this observation.

(You may neglect the internal resistance of the power supply and the resistance of the ammeter in all your calculations.) (10)

(ULSEB 1988)

15.19 State Ohm's law. Describe the experiment you would perform to show that a metal wire obeys this law. (6)

Write down expressions for the effective resistance of two resistors connected (a) in series, and (b) in parallel. (2)

The diagram illustrates a network of four resistors with a cell of e.m.f. 2 V and internal resistance $\frac{1}{3}$ Ω connected to A and C as shown. Calculate

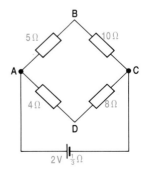

(i) the current in each resistor (5)
(ii) the potential difference between A and C
(iii) the potential difference between B and D (3)

(iv) the change in total current from the cell that occurs when a 20 Ω resistor is connected (a) across BD, (b) across AC. (4)

(UCLES 1985)

15.20 The voltmeter in the circuit shown below initially reads 12.0 V.

When A and B are connected by a piece of copper wire the voltmeter reading changes from 12.0 to 11.4 V.

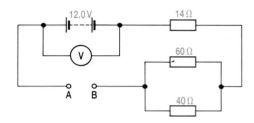

(i) Explain why this occurs.
(ii) Calculate the current flowing through the copper wire.
(iii) Hence calculate the internal resistance of the battery. (6)

The wire linking A and B is replaced by component X. The new voltmeter reading is 11.6 V. This reading does not alter when the connections to component X are reversed. Why does this suggest that X is a resistor? Caclulate the resistance of component X. (3)
X is now replaced by component Y. The voltmeter reading is 12.0 V but is considerably lower when the connections to Y are reversed. State, with reasons, what type of component Y is likely to be. (2)

(UCLES 1986)

15.21 The 4.00 V cell in the circuits shown has zero internal resistance.

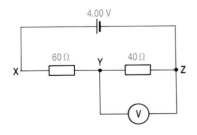

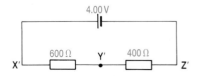

An accurately calibrated voltmeter connected across YZ records 1.50 V. Calculate (a) the resistance of the voltmeter, (b) the voltmeter reading when it is connected across Y'Z'. What do your results suggest concerning the use of voltmeters?

(ULSEB 1984)

15.22 State Ohm's law. (2)

Explain the advantage of using a voltmeter of extremely high resistance as opposed to one of low resistance. (2)

Voltage measurements are to be made in the circuits below using a voltmeter X of resistance 0.5 MΩ and full scale deflection 20 V. Calculate (i) the voltage across AB, (ii) the voltage across CD. (2)

What values would you obtain if you used voltmeter X to measure each of these voltages? (4)

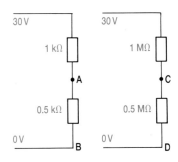

Explain how the behaviour of your voltmeter will be altered by adding a 0.5 MΩ resistor to it in series. What will its full scale deflection become?

The supply voltage is increased to 120 V. Will the modified voltmeter be able to read the voltages across AB and CD? If so, what will these readings be? (4)

Meter X can be described as having a sensitivity of 25 kΩ/V. Suggest what is meant by this statement. Calculate the sensitivity of the modified voltmeter. (3)

(UCLES 1987)

15.23 (a) Describe an experiment in which an unknown resistor may be compared with a standard resistor, where both resistances are less than one ohm. Include a suitable circuit diagram. (4)

(b) A galvanometer and several resistors are arranged to act as a multirange ammeter as shown in the diagram. The galvanometer has a resistance R_G of 10 Ω and is connected across four resistors in series of total resistance 90 Ω. The meter is to be used as an ammeter with full scale deflection

currents of 1A, 100 mA, 10 mA or 1 mA by using connections AB, AC, AD or AE respectively.

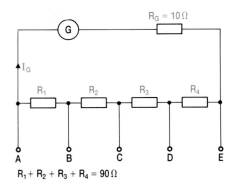

$R_1 + R_2 + R_3 + R_4 = 90\,\Omega$

(i) Find the full scale deflection current I_G of the galvanometer.

(ii) Calculate the values of the four resistors. (8)

(O & C 1984)

15.24 A 12 V battery of negligible internal resistance is connected as shown below. The resistance of the voltmeter is 100 Ω. What reading will the voltmeter show when the switch, S, is (a) open, and (b) closed? (4)

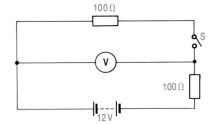

(ULSEB 1983)

15.25 Account for the increase in the resistance of a pure metal wire when
(a) the temperature of the wire is raised, and
(b) the wire is stretched at a constant temperature. (6)

A simple strain gauge consists of a straight metal wire of length 0.60 m, diameter 0.040 mm and resistivity 5.0×10^{-7} Ω m. Calculate the resistance of the gauge unstrained and when stretched by 1.00 mm. (Assume that the volume of the metal in the wire remains constant.) (6)

Four such identical gauges, R_1 to R_4 are mounted in the Wheatstone bridge circuit shown in the diagram. The internal resistance of the cell is negligible and the voltmeter has a very high resistance. Deduce the reading on the voltmeter with all four gauges unstrained, explaining your reasoning. (4)

ELECTRICAL CIRCUITS

Gauges R_1 and R_4 are now each stretched by 1.00 mm. Calculate the reading on the voltmeter. (4)

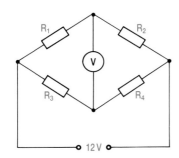

15.26 In the circuit below PQ is a uniform resistance wire. Explain how you would use the circuit to measure the e.m.f. of a dry cell assuming that a standard cell and any other apparatus you require are available. (8)

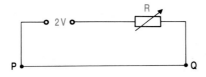

(AEB 1986)

15.27 In the circuit shown a 12 V battery of negligible internal resistance is connected to a potentiometer X of total resistance 100 Ω. Filament lamps L_1 and L_2 are each marked '6 V 18 W' and are connected as shown with the sliding contact positioned so that the resistance of X is equally divided.

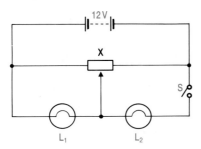

(a) What is the significance of the markings on the lamps?
(b) Calculate the resistance of each lamp when operating normally.
(c) Switch S is closed. Explain whether you would expect lamp L_1 to light to its normal brightness.
(d) Discuss the effect on the brightness of lamp L_1 of opening the switch. (9)

(AEB 1987)

15.28 Draw a labelled diagram of a potentiometer arrangement for obtaining a variable 0 V – 12 V d.c. supply from a fixed 12 V d.c. source. (3)

(ULSEB 1983)

Chapter 16

ELECTRIC FIELD

LEARNING OBJECTIVES

At the end of this chapter you should be able to:

1. calculate the force between point charges;

2. define and use the term electric field strength and state its unit;

3. relate electric field strength to potential gradient.

16.1 FORCE BETWEEN CHARGES

At the start of another chapter we come back again to the term force. The influence which one object has on another is of universal importance. In Section 5.5 it was stated that there are only two types of force outside the nucleus of an atom, the gravitational force and the electromagnetic force. In Chapter 9 details were given about the gravitational force. Here a further look will be taken at the electromagnetic force. An electromagnetic force can only exist between two objects if they have charge. If the charges are stationary the force will have a different value from its value if the charges are moving relative to one another.

The development of ideas and concepts within any science is often not logical because discoveries are not necessarily made in a logical order. In the case of charge, the knowledge that light objects could be attracted to a piece of amber was known by the Greeks. The very word electricity comes from the Greek word for amber, which is ελεχτροσ. If however the order of discovery had been different, Newton could have introduced the idea of gravitational attraction and he could have explained the movement of the planets and moons from his theories. Then someone else might have discovered the attraction of light objects to amber and Newton would have been at a total loss to explain how it took place. He would only have been able to say that it is impossible to explain the attraction of the light object by gravitation. A new property needs to be introduced to explain the concept; its mass is insufficient to explain the attraction. This is the property we call charge. Another fact which Newton would not have been able to explain is why solids can exist. A calculation of the value of gravitational attraction between neighbouring atoms results in an extremely small force, far too small to have any noticeable effect on the fast moving atoms in a solid. Solids are able to exist only because they have charged particles within them and have electromagnetic forces between their atoms. See Section 22.3.

The same development of ideas is still proceeding. A nucleus of an atom cannot be held together by the force acting because of the mass of the nuclear particles. The force is not big enough. The nucleus cannot be held together by the charge on the nuclear particles. The force due to the charge would scatter the nuclear particles because it is repulsive and not attractive. There must therefore be some other property of nuclear matter which results in strong nuclear binding forces. This property is still unnamed because it is not fully understood.

Fig 16.1 In a high voltage research laboratory a discharge at 7 500 000 V of 33 000 A is achieved to test distribution transformers.

       ELECTRIC FIELD

This discussion of charge in effect states that there is a property of matter which can result in there being a force between two charged objects. When the results of accurate quantitative experiments were first published by Coulomb in 1784 they showed that the force F between two charges Q_1 and Q_2 was proportional to each of the charges and inversely proportional to the square of the distance r between them. This is now known as Coulomb's law and can be written mathematically as

$$F \propto \frac{Q_1 Q_2}{r^2}$$

This formula, and the corresponding one for the gravitational force of attraction between two masses of mass M_1 and M_2, should be compared.

$$F \propto \frac{M_1 M_2}{r^2}$$

The similarity of these expressions was first realised by Cavendish who was working on both forces and measured the value of the constant of proportionality in the gravitational case as part of his determination of the mass of the Earth. He was also probably the first person to state Coulomb's law for the electrostatic force.

There is an important difference between these two effects. In the gravitational case the force is always an attractive one. Mass always attracts mass. In the case of electrostatic charge, however, the force can be one of attraction or repulsion. This is because there are two known types of charge which are called positive and negative. Positive charge repels positive charge, negative charge repels negative charge, but positive charge attracts negative charge.

In this chapter the particular concern is with the force between static charges. Experimental checking of Coulomb's law is difficult to do directly because it is difficult to keep the charges constant, but by using methods which will be dealt with later (Section 16.1) the constant of proportionality k for the equation

$$F = \frac{k Q_1 Q_2}{r^2}$$

can be found to be 8.99×10^9 N m^2 C^{-2} when the two charges are in a vacuum. The very high value for this constant means that very large forces exist even between small charges. This again indicates why solids exist. The interatomic forces binding all the atoms in a solid into a rigid structure exist as a direct result of the charge present within the atoms. To get some idea of the size of these forces, look at the values in Example 1 and then work through Questions 16.1 and 16.2.

Example 1

Find the force between:
(a) a proton and an electron in a hydrogen atom if their separation is 5.3×10^{-11} m;
(b) a charge of 7 nC and a charge of 20 nC placed 2 cm apart in a vacuum;
(c) the Earth, with a charge of 2 MC, at a distance of 1.5×10^{11} m from the Sun, with a charge of 6000 MC.

Substituting the numerical values into the Coulomb equation gives
(a)

$$F = \frac{9.0 \times 10^9 \times 1.6 \times 10^{-19} \times 1.60 \times 10^{-19}}{(5.3 \times 10^{-11})^2}$$

$$= 8.20 \times 10^{-8} \text{ N}$$

This is a very large force when it is considered that it acts on an electron of mass 9.1×10^{-31} kg. If you calculate the acceleration this force produces on the electron, it comes to the enormous value of 9.0×10^{22} m s^{-2}.

(b) This would be typical for the size of charge on an insulating rod charged by rubbing.

$$F = \frac{9 \times 10^9 \times 7 \times 10^{-9} \times 20 \times 10^{-9}}{(0.02)^2}$$

$$= 0.0032 \text{ N}$$

This force is small for most objects, but would be capable of lifting tiny scraps of paper, for instance.

(c) These charges are estimates. The Earth is charged and has a measurable electric field as a result. The value of the electric field does enable the charge to be calculated, as you will see later (Section 16.2). In such problems it is only sensible to give answers to one significant figure.

$$F = \frac{9 \times 10^9 \times 2 \times 10^6 \times 6 \times 10^9}{(1.5 \times 10^{11})^2}$$

$$= 5000 \text{ N}$$

This force is totally insignificant when compared with the size of the gravitational force acting on the Earth due to the Sun, which is of the order of 10^{22} N (Section 9.1).

QUESTIONS

16.1 Find the force of attraction between two isolated charges of 1 C each when placed 1 km apart. Suggest an object whose weight could be equal to this force. What does the answer to this question suggest about the practicality of obtaining an object with a charge on it of 1 C?

16.2 Find the force between the charge on two protons, a distance of 10^{-15} m apart in a nucleus? Your answer should have a huge value when considered as a force acting on the mass of the proton. What is the value of the acceleration of the protons? The mass of a proton is 1.67×10^{-27} kg.

The constant of proportionality in Coulomb's equation has the value of 8.99×10^9 N m^2 C^{-2} when the space between the charges is a vacuum. The numerical value of this fundamental constant is linked with the way in which the units of length, mass, time and electric current are defined, and on the SI system has this fixed value. However the constant is not normally written as k. By convention it is usual to write

$$k = \frac{1}{4 \pi \varepsilon_0}$$

This makes $\varepsilon_0 = \dfrac{1}{4 \pi k} = \dfrac{1}{4 \pi \times 8.99 \times 10^9} = 8.85 \times 10^{-12}$ C^2 N^{-1} m^{-2}

The equation for the force between two charges then becomes

$$F = \frac{Q_1 Q_2}{4 \pi \varepsilon_0 r^2}$$

It will be shown in Section 17.1 that F m^{-1} (farad/metre) is an equivalent unit to C^2 N^{-1} m^{-2}.

ε_0 is called the permittivity of free space. The reason for making this change is in part historical. The introduction of the 4π term, though, is a modern development and has a specific advantage. Putting a 4π into the constant does not alter the value of the constant in any way but it does rationalise certain equations. It has the effect of making any formula which applies to a system with spherical symmetry contain a 4π term. A formula for a system with cylindrical symmetry will contain a 2π term. A formula for a uniform system will not contain a π term at all. Examples of the use of these formulae will be pointed out when they occur.

QUESTION	**16.3** In a sodium chloride crystal a sodium ion has a charge of $+1.6 \times 10^{-19}$ C and a chlorine ion has a charge of -1.6×10^{-19} C. They are found by X-ray diffraction to be a distance apart of 0.2 nm. What force exists between them if they are in a vacuum? ($\varepsilon_0 = 8.9 \times 10^{-12}$ F m^{-1}.)

16.2 ELECTRIC FIELD STRENGTH

In Section 9.2 comparison was made between the idea of a magnetic field around a magnet with the gravitational field around the Earth. The fields were treated as regions of influence within which a force was exerted on either a magnet or an object having mass. The term gravitational field strength was introduced as the gravitational force per unit mass. We are now in a position to define electric field strength in a similar way. Whereas a gravitational field is a region in which a force is exerted on a mass, an electrical field is a region in which a force is exerted on a charge. The **electric field strength** at a point is defined as the force per unit positive charge placed at the point. Note that since there are two types of charge, positive charge has to be stated in order to get the direction of the field correct, and also that the force per unit charge phrase has to be treated as a division of a small force by a small charge and not as the force actually acting on a unit of charge. The experimental test for the existence of an electric field depends on placing a test charge in the field and measuring the force exerted on it. If this test charge has a large value it may distort the field which it is being used to measure, so it is essential to use tiny charges. The unit of charge, the coulomb, is a very large unit in this context so, in practice, test charges may well be measured in nanocoulombs or smaller units.

In equation form electric field strength E is given by

$$E = \frac{F}{q}$$

where F is the force acting on a small charge, q. Electric field strength is a vector and has the unit newton per coulomb, N C^{-1}.

Finding the electric field strength at any point can be done either by experimental or theoretical methods. Some of the mathematics in these calculations can be involved because of the vector nature of the quantity, but the principle of finding the electric field strength is straightforward.

Example 2
A charge of $+1.6 \times 10^{-19}$ C has a force of 8.7×10^{-15} N exerted on it when placed at a point in an electric field. Find the electric field strength at the point.

$$E = \frac{F}{q} = \frac{8.7 \times 10^{-15} \, \text{N}}{1.6 \times 10^{-19} \, \text{C}} = 5.4 \times 10^4 \, \text{N C}^{-1}$$

The direction of the electric field is in the direction of the force. In this example the force being exerted on a single fundamental charge is found. This is the force due to the electric field exerted on a hydrogen ion.

Example 3

The electric field strength between a pair of plates of length 4.0 cm in a cathode ray tube is 2.3×10^4 N C^{-1}. An electron enters the field at right angles to it with a velocity of 3.7×10^7 m s^{-1} as shown in Fig 16.2. Find the velocity of the electron when it leaves the electric field.

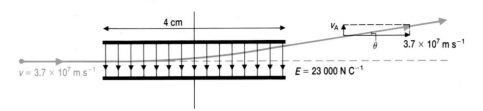

Fig 16.2 An electron passing through an electric field is deflected by the field. Its path is parabolic while in the field.

Force on an electron in the electric field

$$= qE = (-1.6 \times 10^{-19} \text{ C}) \times (2.3 \times 10^4 \text{ N C}^{-1})$$
$$= -3.68 \times 10^{-15} \text{ N}$$

The minus sign indicates that the force is in the opposite direction to the field, i.e. it is vertically upwards on the diagram.

Since there is no horizontal force on the electron its horizontal velocity will be unchanged while it travels through the field.

The time taken to travel through the field is therefore

$$\frac{\text{horizontal distance}}{\text{horizontal velocity}} = \frac{0.04 \text{ m}}{3.7 \times 10^7 \text{ ms}^{-1}} = 1.08 \times 10^{-9} \text{ s}$$

The acceleration of the electron will be upwards in the direction of the force and will have the value given by

$$\text{acceleration} = \frac{\text{force}}{\text{mass}} = \frac{3.68 \times 10^{-15} \text{ N}}{9.1 \times 10^{-31} \text{ kg}} = 4.04 \times 10^{15} \text{ ms}^{-2}$$

The vertical velocity v_A on leaving the field = vertical acceleration × time

$$= 4.04 \times 10^{15} \text{ ms}^{-2} \times 1.08 \times 10^{-9} \text{ s}$$
$$= 4.37 \times 10^6 \text{ ms}^{-1}$$

The resultant velocity is in a direction θ given by

$$\tan \theta = \frac{4.37 \times 10^6}{3.7 \times 10^7} = 0.118$$

$$\theta = 6.7°$$

and has magnitude

$$\frac{3.7 \times 10^7 \text{ m s}^{-1}}{\cos \theta} = \frac{3.7 \times 10^7 \text{ m s}^{-1}}{\cos 6.7°}$$
$$= 3.73 \times 10^7 \text{ m s}^{-1}$$

In this case the speed of the electron has increased by about 1% while undergoing a deflection of 6.7°.

The electric field strength near a point charge

A small test charge q is placed a distance r from a point charge Q (Fig 16.3). Coulomb's law is then used to find the force F acting on q:

$$F = \frac{Qq}{4 \pi \varepsilon_0 r^2}$$

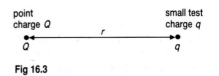

Fig 16.3

This gives the magnitude of the electric field strength

$$\frac{F}{q} = E = \frac{Q}{4 \pi \varepsilon_0 r^2}$$

A diagram showing the field near a positive point charge is shown in Fig 16.4. The field does of course exist in three dimensions. The field has spherical symmetry and because of rationalisation a factor of 4π occurs in the expression. The graph shown in Fig 16.5 shows how the electric field

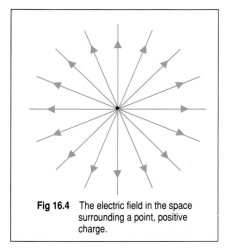

Fig 16.4 The electric field in the space surrounding a point, positive charge.

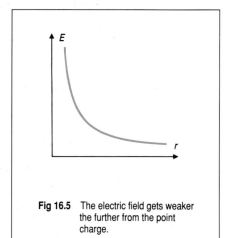

Fig 16.5 The electric field gets weaker the further from the point charge.

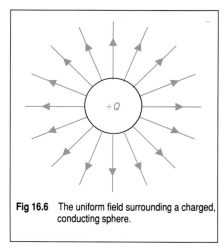

Fig 16.6 The uniform field surrounding a charged, conducting sphere.

strength falls off with distance according to an inverse square law. If the charge $+Q$ is not a point charge but is a charge $+Q$ uniformly distributed on the surface of a conducting sphere, then by symmetry the field is as shown in Fig 16.6. Its value is identical to that of the point charge provided the value of r is greater than that of the radius of the sphere.

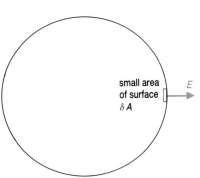

Fig 16.7

This principle is made use of in the explanation of the transmission of electrical impulses along a nerve. The nerve cell itself has long, thread-like axons spreading out from its nucleus, and the membrane of the wall of a nerve cell is shown in Fig 16.9. In its normal state, shown at A, the outer surface of the cell carries a positive charge and the inner layer a negative charge. This gives an electric field across the cell wall from the outside to the inside. At one place, B, the charges are reversed as a result of the movement of chlorine ions, so in this part of the axon the field will be from the inside towards the outside. A nerve impulse is the movement of the B region along the axon. The speed of nerve impulse depends on the diameter of the axon. Typically it is 50 m s⁻¹.

The electric field strength near a uniformly charged conducting sheet

The field outside a uniformly charged conducting sphere includes a situation as shown in Fig 16.7 where the field is to be found just outside a charged sphere of large radius. If the sphere carries a total charge Q and its radius is R, then the charge per unit area σ called the surface charge density, is given by

$$\sigma = \frac{\text{total charge}}{\text{surface area}} = \frac{Q}{4 \pi R^2}$$

The field just outside the sphere is given by

$$E = \frac{Q}{4 \pi \varepsilon_0 r^2}$$

so we can write

$$E = \frac{\sigma}{\varepsilon_0}$$

By making the sphere of large enough radius its curvature can be ignored, yet we still have the field near the surface to be

$$E = \frac{\sigma}{\varepsilon_0}$$

This therefore is the field near a uniformly charged plate with a surface density of charge σ, under conditions in which the field is in one direction only. Note that here the field is uniform and no 4π appears in the equation for the field because of the rationalisation of the constant in Coulomb's law.

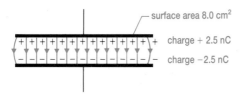

surface area 8.0 cm²

charge + 2.5 nC

charge −2.5 nC

Fig 16.8

This expression relating the electric field strength to the surface density of charge can be applied generally. It is one way of expressing a law called Gauss's law which is not included in advanced level syllabuses.

Example 4

Two plates each have an area of 8.0 cm² (= 0.000 80 m²) and are placed facing one another in a vacuum as shown in Fig 16.8. If the top plate carries a positive charge of 2.5 nC and the bottom one a charge of −2.5 nC, find the electric field strength in the space between them.

The attractive forces between the opposite charges pulls them so that all the charge on the top plate is spread on its lower surface and all the charge on the bottom plate is spread on its upper surface. Provided the edge of the plates is avoided, the charge is spread uniformly and the field is uniform throughout the space. The field will not therefore be dependent on the separation of the plates and we can find the value of the electric field strength by using the equation for the field near a uniformly charged plate.

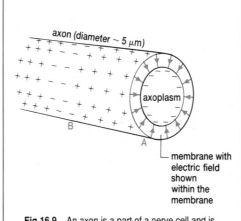

axon (diameter ~ 5 μm)

axoplasm

B

A

membrane with electric field shown within the membrane

Fig 16.9 An axon is a part of a nerve cell and is long and thread like. The axon transmits nerve impulses by reversing the electric field across the membrane of the wall.

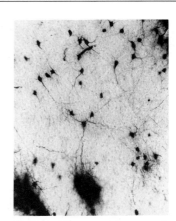

Fig 16.10 The axons of the nerve cells are clearly shown as thread like lines in this photograph.

The surface charge density of the top plate $\sigma = \varepsilon_0 E = \dfrac{Q}{A}$

$$E = \frac{Q}{\varepsilon_0 A}$$

$$= \frac{2.5 \times 10^{-9}}{8.9 \times 10^{-12} \times 0.000\ 80} = 3.5 \times 10^5 \text{ N C}^{-1}$$

The electric field strength near other charge distributions

Field pattern around an electric dipole

A dipole is simply a pair of charges of equal size but opposite sign. One such dipole is shown in Fig 16.11 and the electric field strength at a particular point has been found by scale drawing. First the electric field strength at a point due to the positive charge has been found and then that due to the negative charge at the same point. The two electric field strengths have then been added together using a parallelogram of vectors to find the resultant electric field strength. This procedure needs to be repeated at many points in the space around the dipole so that the total field pattern can be built up.

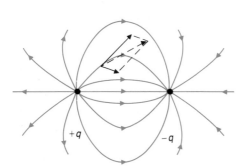

+q   −q

Fig 16.11 The electric field of a dipole. A dipole is a pair of opposite charges a small distance from one another.

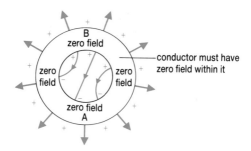

Fig 16.13 This diagram shows an impossible situation because uneven distribution of charge cannot occur on the inside surface of the sphere.

Fig 16.14 One of the first Van de Graaff generators built by Robert Van de Graaff in the 1930s.

Fig 16.15 The high voltage section of this computer is not only well sealed for safety but also to prevent stray electromagnetic waves escaping from any sparking which might occur. The waves will be absorbed by the metal container.

This procedure can be extended to any situation where any number of charges are involved.

Field pattern inside a conductor

If a current is flowing in a conductor it implies that there is a force exerted on the conduction electrons and hence there must be an electric field. If however there is no current flow within a conductor then there is no force causing the electrons within it to drift along and hence there can be no electric field. In electrostatic problems, therefore, there can be no field within the body of a conductor. A microwave oven is totally enclosed by a conducting case, Fig 16.12. Microwaves within the oven cannot be transmitted through the conductor because it absorbs any microwaves it does not reflect.

Fig 16.12 A Moulinex microwave cooker. The metal casing prevents any microwaves from escaping.

Field pattern in a hollow conductor

As was stated on page 287, the electric field near a surface depends on the surface density of charge. If an attempt is made to find the field in the space inside a hollow conductor, contradictions soon become apparent. For instance, if a hollow positively charged sphere could have its charge distributed as shown in Fig 16.13, then it would perhaps be possible to have its field as shown. This however is asymmetrical despite being a uniformly charged sphere. The only symmetrical solution to the problem is that there should be zero charge on the inside surface of the sphere and zero field within the hollow space. A more rigorous treatment using Gauss's law confirms this result. Faraday first showed there to be zero charge inside a hollow conductor and the result can be used to prove Coulomb's inverse square law.

The result has important practical applications. Charge placed inside a hollow conductor will immediately flow to the outside, leaving zero field inside. This is made use of in the design of the Van de Graaff generator where a very high voltage can be built up on the outside of a spherical dome while the inside has no charge or field present, see Fig 16.14. Another related example concerns the radio frequency part of a radio or television set which uses very high frequencies and needs to be unaffected by external electric fields. For this reason this part of the set is totally enclosed in a hollow aluminium box so that no external field can cause any field within it. These enclosures are sometimes called Faraday cages (Fig 16.15).

16.4 Charges of +60 nC and –60 nC are placed at the corners X and Y of an equilateral triangle XYZ of side length 4 cm. What is the electric field strength at Z?

16.5 The Earth is a conducting sphere of radius 6.4×10^6 m and it is uniformly charged. The Earth therefore has an electric field besides its gravitational and magnetic fields. Why cannot the Earth's electric field be used for navigation? Why does the electric field vary from time to time?

At a time when the intensity of the Earth's electric field is 300 N C^{-1} find
(a) the total charge on the Earth;
(b) the surface density of charge on the Earth.

16.3 ELECTRIC POTENTIAL

When field theory was first introduced in Chapter 9, the field being considered was the gravitational field. There it was stated that it was convenient to use a scalar property of the field at each point. The same is true with electric fields. Working with fields themselves is often difficult mathematically because of the vector nature of a field. If a scalar quantity is defined then addition of that scalar property can be done by normal arithmetic instead of by vector addition. The property used with electric fields is called the electric potential and it is defined in the same way as gravitational potential was defined. The **electric potential** of a point in an electric field is the work done in moving unit charge from infinity to the point. Infinity is a convenient way of saying a point far removed from all other electrical influence. If W is the work done in moving a small test charge q from infinity to the point, then the potential of the point V is given by

$$V = \frac{W}{q}$$

The unit of electric potential is the joule per coulomb, which is the volt.

Whereas gravitational potential can only be negative, electric potential can be positive or negative because the force between charges can be either one of attraction or of repulsion. In practice, differences of potential are more frequently used than potential itself. In Section 14.3, potential difference was introduced as the energy converted from electrical energy to other forms of energy per unit charge passing between the points. There is no conflict between that definition and the one just introduced for electric potential: there is an extension of the idea however. When dealing with potential difference in an electric circuit, the potential of a point was regarded as being a point somewhere within the confines of the wires of the circuit. Here the definition of potential and potential difference is extended to all points whether the point is in a conductor or in an insulator or in a vacuum.

To see how the idea of electric potential is used, work through Example 5.

Example 5

A small charge of +6.2 nC is moved a distance of 2.0 mm from one conducting plate to another against a field of 20 000 N C^{-1} (Fig 16.16). Find the work done on the charge and the potential difference between the plates.

Force exerted by the field on the charge = 20 000 N C$^{-1} \times 6.2 \times 10^{-9}$ C
$= 12.4 \times 10^{-5}$ N

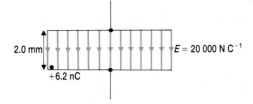

2.0 mm $E = 20\,000$ N C^{-1}

+6.2 nC

Fig 16.16

Work done in moving the charge at a constant velocity against the field
= force exerted on charge × distance moved
= 12.4 × 10^{-5} N × 0.0020 m
= 24.8 × 10^{-8} J

Potential difference = work done per unit charge

$$= \frac{2.48 \times 10^{-7}\,\text{J}}{6.2 \times 10^{-9}\,\text{C}} = 40\,\text{V}$$

It is not a coincidence that this cancels out so neatly. If any other charge had been specified, the answer for the potential difference would have been the same because the potential difference between the plates is dependent on the field and the separation of the plates but not on the value of the small charge.

The relationship between potential and field

If the calculation done in Example 5 is repeated using the algebraic values E for the uniform electric field, q for the charge, x for the separation of the plates and V for the potential difference, it becomes

Force exerted by the field on the charge = Eq

Work done in moving the charge at a constant velocity against the field
= force exerted on charge × distance moved
= Eqx

Potential difference = work done per unit charge

$$V = \frac{Eqx}{q} = Ex$$

This can alternatively be written as

$$E = \frac{V}{x}$$

It is always true for a uniform field that the potential difference per unit distance is the electric field.

The electric field is in a direction in which the potential falls. This also gives an alternative unit for electric field as volt per metre. V m^{-1} is an equivalent unit to N C^{-1}.

If the field is not uniform then calculus terminology can usefully be used, and the positive charge q is considered to move a small distance δx from A to B and to have a small quantity of work δW done on it by some external system. This is shown in Fig 16.17. In the calculation, the direction of the field is taken to be positive so the force exerted by the field on the charge = Eq.

The work done by some external system on the charge to move it from A to B must therefore be negative, i.e. work is being done on the external system.

$$\delta W = -Eq\delta x$$

Small change in potential δV = work done per unit charge

$$\delta V = \frac{-Eq\delta x}{q}$$

giving

$$E = \frac{-\delta V}{\delta x}$$

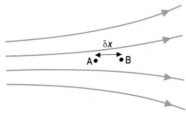

Fig 16.17

The electric potential gradient is the electric field; the direction of the field is the same as the direction of decreasing potential.

Another way of looking at this is to regard lines of equal electric potential as electrical contour lines. Force on a charge acts at right angles to these contour lines. If the contour lines are close together then there will be a large force acting on the charge; if the contour lines are well separated then the force will be small. A charge will always have a force on it in the direction of the field, and therefore its acceleration will be in the direction of the field but its velocity is not necessarily in the same direction as its acceleration.

One practical use of this topic is in producing focusing devices within an oscilloscope. If an electron beam is passed through two cylinders which are held at different potentials, then the electric field in the space inside the cylinders is as shown in Fig 16.18. The equipotential lines are also shown

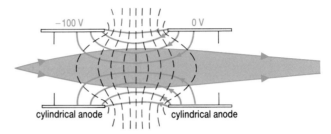

Fig 16.18 The focusing of a cathode ray tube is done by shaping an electric field within cylindrical anodes.

Fig 16.19 The surface of a record is shown with the stylus in the groove. The photograph was taken using a scanning electron microscope.

and they create a steep electrical valley. An electron which goes straight through the centre of the two cylinders is accelerated but is not deviated. One which deviates to the left has forces acting on it to the right to bring it back towards the centre, whereas one which deviates to the right will have forces acting on it to the left. Remember that an electron, since it has a negative charge, will have a force on it in the opposite direction to the electric field. The same principle is used in electron microscopes and television sets but the fields used are magnetic fields rather than electric fields, Fig 16.19.

Another example of the use of electric fields is shown by how some fish can detect their prey. The fish acts as a dipole with a positively charged head and a negatively charged tail as shown in Fig 16.20(a). When another

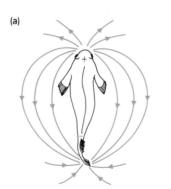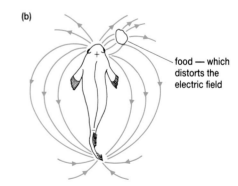

Fig 16.20 A fish using electric field to detect its prey. The distortion of the field caused by the presence of food can be sensed by the fish.

Fig 16.21 An electric eel – which has the Latin biological name *Electrophorus electricus.*

object is near the fish, the electric field is distorted and the fish feels the change in the field, Fig 16.20(b). In a similar way human beings can detect electric fields. If you hold the back of your hand near the screen of a television set, the movement of hairs in the electric field are able to be felt. Some people even claim to be able to forecast the weather by being able to detect changes in the atmospheric electric field.

ELECTRIC FIELD

The potential in a vacuum at a distance from a point charge

The potential difference between two points in a uniform electric field was easy to calculate because the force acting on the charge was constant. In general, the force which acts on a charge in an electric field is not constant and so calculation of the potential from the field usually requires calculus. To see why this is necessary, consider a small test charge q being brought from infinity to a point P a distance r from a point charge Q. The forces acting on it are shown in Fig 16.22(a). To move the charge from infinity requires the force P to be at least equal to force F. The work done by P will be at least equal to the area under the graph shown in Fig 16.22(b). If you can follow the following calculus through, well and good. If you cannot, do

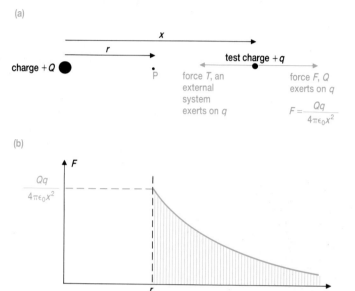

Fig 16.22

not worry. Integration is just a technique for finding the area under a graph by adding together the areas of a large number of narrow strips.

The small quantity of work δW done on q by some external system to move it a distance δx towards P was given above:

$$\delta W = -Eq\delta x$$

The total work done by this external system in moving the charge q from infinity to the point P is therefore

$$W = -\int_{\infty}^{r} Eq\delta x$$

Since the field E, a distance x from a point charge Q, is given by

$$E = \frac{Q}{4\pi\varepsilon_0 x^2}$$

$$W = \int_{\infty}^{r} -\frac{Qq}{4\pi\varepsilon_0 x^2}\, dx = -\frac{Qq}{4\pi\varepsilon_0}\int_{\infty}^{r} \frac{dx}{x^2}$$

$$= -\frac{Qq}{4\pi\varepsilon_0}\left[-\frac{1}{x}\right]_{\infty}^{r} = -\frac{Qq}{4\pi\varepsilon_0}\left[\left(-\frac{1}{r}\right) - \left(-\frac{1}{\infty}\right)\right]$$

$$= \frac{Qq}{4\pi\varepsilon_0 r}$$

The total work done on the charge q is therefore $\dfrac{Qq}{4\pi\varepsilon_0 r}$

and this gives the potential energy of charge q. To find the potential at P it is simply necessary to find the work done per unit charge. This gives

$$V_P = \frac{\text{Work done}}{q} = \frac{Q}{4\pi\varepsilon_0 r}$$

If the potential is required for a series of charges then, because potential is a scalar quantity, all that is necessary is to find the potential due to each charge separately and then add the results together as is shown in Example 6.

Example 6

Fig 16.23 shows four charges placed at the corners of a rectangle. Find the potential of the point P. This type of problem is becoming important in determining the shape and structure of complex molecules. The potential at different points in the space surrounding a group of atoms determines where additional atoms may be placed in order to have the lowest potential energy.

Potential at P due to 6 nC charge:

$$V_P = \frac{Q}{4\pi\varepsilon_0 r} = \frac{6\times10^{-9}}{4\pi\times8.9\times10^{-12}\times0.030} = 1790 \text{ V}$$

Potential at P due to –2 nC charge:

$$= \frac{-2\times10^{-9}}{4\pi\times8.9\times10^{-12}\times0.030} = -600 \text{ V}$$

Potential at P due to 8 nC charge:

$$= \frac{8\times10^{-9}}{4\pi\times8.9\times10^{-12}\times0.050} = 1433 \text{ V}$$

Potential at P due to –3 nC charge:

$$= \frac{-3\times10^{-9}}{4\pi\times8.9\times10^{-12}\times0.050} = -540 \text{ V}$$

Total potential at P = 1790 V – 600 V + 1430 V – 540 V = 2100 V to 2 sig. figs.

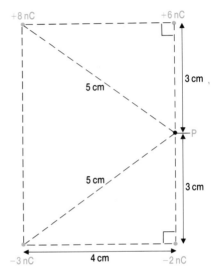

Fig 16.23

QUESTIONS

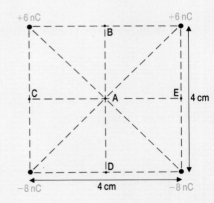

Fig 16.24

16.6 An electric dipole consists of charges $+e$ and $-e$ separated by a distance of 10^{-10} m. If the dipole is placed in an electric field of intensity 10^6 N C^{-1}, find the torque on the dipole when it is:
(a) parallel to the field;
(b) at right angles to the field;
(e is the fundamental charge and has the value 1.6×10^{-19} C.)

16.7 An electron in the vacuum of a cathode ray tube moves from rest at a point where the potential is –1400 V to a point where the potential is zero. Find
(a) the gain in potential of the electron;
(b) the loss in potential energy of the electron (why are these of opposite sign?);
(c) the gain in kinetic energy of the electron;
(d) the final speed of the electron.
(Mass of electron = 9.1×10^{-31} kg.)

16.8 Find the potential at each of the points A, B, C, D and E for the 4 cm square shown in Fig 16.24. Copy the diagram and sketch the field and equipotential lines.

- The force between two point charges in a vacuum

$$F = \frac{Q_1 Q_2}{4\pi\varepsilon_0 r^2}$$

- Electrical field strength (intensity) is the force acting on unit charge. It is a vector.

field inside a hollow charged conductor	$= 0$
field inside a conductor with no charge flowing	$= 0$
field near a charged plane surface	$= Q/\varepsilon_0 A$
field near a uniformly charged sphere or a point	$= Q/4\pi\varepsilon_0 r^2$

- Electric potential of a point is the work done per unit positive charge brought from infinity to the point. It is a scalar so additions of potential can be done by direct addition.

- Potential V, near a point charge $+Q$, is given by $V = Q/4\pi\varepsilon_0 r$.

- Electric field is the potential gradient $E = -\dfrac{dV}{dx}$

EXAMINATION QUESTIONS

16.9 **(a)** Write down an equation for the electrostatic force between two charges placed some distance apart in a vacuum. Name the physical quantities represented by each of the symbols used in the equation.

(b) In a simple model of the hydrogen atom, the electron is regarded as a point charge of magnitude -1.6×10^{-19} C moving in a circular path of radius 5.3×10^{-11} m with a point charge of magnitude $+1.6 \times 10^{-19}$ C at the centre.

(i) Calculate the magnitude of the electrostatic force acting on the electron.

(ii) Assuming that the electrostatic force provides the force necessary to maintain the circular motion of the electron, calculate the speed of the electron and the time it takes to complete one orbit.

(Permittivity of free space, $\varepsilon_0 = 8.8 \times 10^{-12}$ F m^{-1}

Mass of the electron = 9.1×10^{-31} kg)   (8)

(AEB 1986)

16.10 A gold nucleus has a radius of about 5×10^{-15} m and contains 79 protons. Assuming that the protons are uniformly distributed in the nucleus, find the electric field strength at the surface of the nucleus.   (3)

(UCLES 1982)

16.11 **(a)** Distinguish, in terms of their atomic structure, between solid materials that are electrical conductors and those that are insulators.

(b) Dry air is a good insulator yet it will conduct electricity well if it is subjected to an electrical field strength of greater than about 3 MV m^{-1}. Account for this.   (6)

(ULSEB 1982)

16.12 An oil drop of mass m carrying a charge q is maintained stationary between two plane, horizontal metal plates by the application of a suitable electric field. Write down an expression for the magnitude E of this field and state its direction.

State what would happen if (a) the drop acquired an additional charge of the same sign, (b) the plates were moved apart, the potential difference between them remaining constant.   (3)

(UCLES 1985)

16.13 Define *electric field strength, E* and *electric potential, V*.

Two equal positive point charges of magnitude Q are placed on the x-axis at positions $x = a$ and $x = -a$ as shown below. Make a sketch of the lines of force of the electric field in the x-y plane, assuming that the region is in vacuum and that no other charges are present.

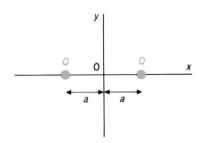

Sketch graphs showing how E and V vary with position along the x-axis. Calculate the values of E and V at the origin O.   (7)

(O & C 1985)

16.14 **(a)** A steel ball is projected horizontally from the top of a table.

An electron is projected horizontally into a uniform electric field acting vertically upwards in an evacuated space.

Explain why the ball and the electron follow similar paths.

(b) (i) Estimate the value of the electric field if the acceleration of the electron due to the field is to be 10^{15} times its acceleration due to gravity.

(ii) This electric field is to be set up between two parallel plates placed 0.1 m apart. Calculate the required potential difference between the plates.

(Charge on the electron = -1.6×10^{-19} C

Mass of the electron = 9.1×10^{-31} kg

Acceleration of free fall, $g = 10$ m s^{-2}.)   (10)

(AEB 1987)

16.15 Write down a general expression relating electric field strength to electric potential. Hence, explain why the electric potential on the surface of an isolated charged conductor is everywhere the same.   (4)

16.16 Fig (a) below shows a small area of the surface of the Earth, assumed to be flat and a conductor of electricity. There is a uniform electric field at the surface of 300 V m^{-1} directed away from the surface.

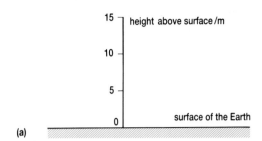

(a)

(a) On a similar sketch, draw three field lines and three equipotentials. Label the equipotentials with their potential relative to Earth. (3)

(b) Calculate the work done in moving a charge of +5.0 μC
   (i) a distance of 4.5 m parallel to the Earth's surface and 2.0 m above it, and
   (ii) from a height of 2.0 m down to the surface of the Earth. (3)

(c) Fig (b) shows a car parked on the surface and at Earth potential. Draw on a similar diagram four equipotentials near the surface with the car present. (2)

(b)

(O & C 1987)

16.17 The diagram shows a simple model of a water molecule where A and B are hydrogen atoms and C is an oxygen atom. The oxygen atom acts as a point charge of magnitude -1.1×10^{-19} C and each of the hydrogen atoms act as a point charge of magnitude $+0.55 \times 10^{-19}$ C. The distance x between the oxygen atom and each hydrogen atom is 1.0×10^{-10} m.

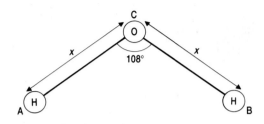

(a) Draw a diagram showing each of the forces acting on the oxygen atom due to the hydrogen atoms.

(b) Assuming that the model is correct, determine the magnitude and direction of the resultant force acting on the oxygen atom due to the hydrogen atoms.

(UCLES 1989)

Chapter 17

CAPACITANCE

> **LEARNING OBJECTIVES**
>
> At the end of this chapter you should be able to:
>
> 1. define and use the term capacitance and state its unit;
> 2. relate the capacitance of a parallel plate capacitor to its dimensions;
> 3. calculate the capacitance of capacitors in series and parallel;
> 4. calculate the energy stored in a capacitor;
> 5. describe how charge, current and potential difference vary as a capacitor is charged through a resistor.

17.1 CAPACITANCE

If a conductor has no movement of charge within it then it is all at the same potential. A copper wire at room temperature with a charge flowing through it must have a potential difference between its ends. If there is no potential difference then no charge flows. This is not true for an insulator. Some parts of the same insulator may be at different potentials from other parts. Indeed on an insulator the potential may have positive, zero or negative values at different places. This is why insulation is used around wires. The potential of the insulator near the wire will be different from the potential on the outside surface of the insulator.

The gravitational equivalent of this may help with understanding the idea. Mass is the property in gravitational terms which is equivalent to charge in electrical terms. If mass can flow, because it is in liquid form, then it will flow from higher to lower potential. Water flows down a river because of the potential gradient, so if the river surface becomes horizontal then flow ceases. In hot, dry countries many rivers do not flow at all during certain seasons of the year. However mass in solid form does not flow and it will not usually have a horizontal surface and so different potentials are possible even when the mass is not moving.

The value of the potential of a conductor depends, among other things, on the charge on it.

Since both the charge on the conductor and its potential have definite values, it is possible to define a quantity C, the **capacitance** of the object by the equation

$$C = \frac{Q}{V}$$

where Q is the charge on the conductor and V is its potential.

The unit of capacitance is the coulomb per volt and this unit is given the name **farad**. One farad, F, is the capacitance of a conductor which is at a potential of 1 volt when it carries a charge of 1 coulomb. The farad is a very large unit and therefore the microfarad, µF, and the picofarad, pF are frequently used.

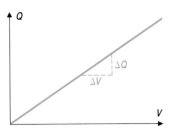

Fig 17.1 The charge stored by a capacitor is proportional to the potential difference across it.

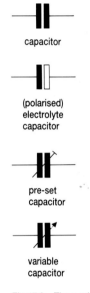

capacitor

(polarised) electrolyte capacitor

pre-set capacitor

variable capacitor

Fig 17.2 The symbols for various different types of capacitor.

$$1 \, \mu\text{F} = 10^{-6} \, \text{F}$$
$$1 \, \text{pF} = 10^{-12} \, \text{F}$$

Note that there is little meaning to the capacitance of an insulator since there is no single value of the potential for an insulator.

If more positive charge is added to the charge already on a conductor its potential will rise. If the conductor has a small capacitance its potential will rise by a large amount when a charge is added: in contrast, the potential of a large capacitor will rise by only a small amount for the same added charge. It is this ability to store charge which gave rise to the term capacity. It is not anything to do with the volume of the conductor. It is simply the charge stored per unit potential. It is found experimentally that the charge on an isolated conductor is proportional to its potential. A graph of charge Q, plotted against potential V, is shown in Fig 17.1 and is a straight line through the origin. The gradient of this graph is the capacitance C of the conductor and is given by

$$C = \frac{\text{total charge}}{\text{potential}} = \frac{\text{change in charge}}{\text{change in potential}}$$

$$C = \frac{Q}{V} = \frac{\Delta Q}{\Delta V}$$

So far, the term capacitance has been applied to a single conductor, but in practice the term is more usually applied to a pair of conductors called a capacitor. This has the practical advantage that a greater quantity of charge may be stored per unit potential difference on a pair of conductors, and the difficulty of using potential with reference to an arbitrary zero of potential is overcome by using the potential difference between the two conductors. The two conductors are usually a pair of parallel plates with an insulator, called the **dielectric**, between them. The electrical symbols for different types of capacitor are shown in Fig 17.2. The reason why an electrolytic capacitor requires to be connected into a circuit with one of its plates held at a positive potential will be explained later in this section.

In order to reduce the physical size of a capacitor, the plates are usually made of aluminium foil and are then coiled up with insulators placed so that there is no contact between the plates. A huge variety of different

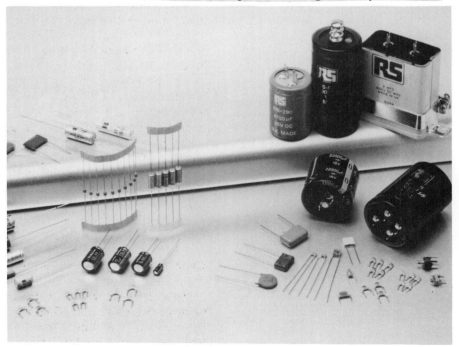

Fig 17.3 Some of the stock of capacitors held by R.S. Components Ltd.

CAPACITANCE

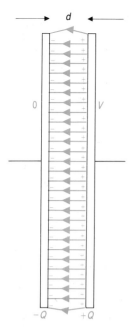

Fig 17.4 The electric field between the plates of a capacitor is very uniform.

values and types of capacitor are available commercially. A photograph of some of them is shown in Fig 17.3. One interesting point about the values of the capacitance of capacitors is that they are rather variable even when manufactured in the same way. For this reason they are sold with a large tolerance, up to 50%. This means that a capacitor marked 10 μF may have a capacitance anywhere between 5 μF and 15 μF. Because of this, manufacturers make a series of nominal values only. One series of values used for standard capacitors is

0.01, 0.022, 0.047, 0.1, 0.22, 0.47, 1, 2.2, 4.7, 10, 22, 47, 100 μF etc.

If values in between these values are specifically required, then either capacitors are joined together in series or parallel or several are measured until one of the correct value is obtained. Usually the value of a capacitor is not critical so the restriction on the values available is no great problem. If a greater variety of values is required, greater accuracy is needed in manufacture and then values in the series 1, 1.5, 2.2, 3.2, 4.7, 6.8, 10 are available – at greater cost. (Series such as these are geometrical series of numbers. In the first series each number is 2.16 times the previous number, $2.16^3 = 10$: in the second series the multiplying factor is the square root of 2.16.)

If a pair of plates are oppositely charged and arranged so that they are parallel to one another and only a small distance d apart, then the electric field between them will be uniform except at their edges as shown in Fig 17.4. If the edge effect is ignored the value of the electric field strength between the plates is

$$E = \frac{\sigma}{\varepsilon_0} = \frac{Q}{\varepsilon_0 A}$$

as was shown in Section 16.2. Since also the electric field strength was shown in Section 16.3 to be equal to the potential gradient and since here the field is uniform, we have

$$E = \frac{V}{d}$$

Equating these two values of E gives

$$\frac{Q}{\varepsilon_0 A} = \frac{V}{d}$$

So $$\frac{Q}{V} = \frac{\varepsilon_0 A}{d}$$

This gives the capacitance of the parallel plate capacitor, C, to be

$$C = \frac{\varepsilon_0 A}{d}$$

Note that a parallel-plate capacitor is said to have a charge Q when $+Q$ is the charge on one plate and $-Q$ is the charge on the other. You need not consider situations in which there are different charges on the two plates.

The assumption has been made in this calculation that the space between the plates is a vacuum. This will not in practice be the case. Some parallel-plate capacitors have air as their dielectric but most capacitors have a dielectric which is a sheet of insulating plastic or waxed paper.

The dielectric has three functions. First it acts to keep the plates a small fixed distance apart. Secondly it increases the voltage which would cause sparking. Any insulator will experience dielectric breakdown if the electric field gets too high. For dry air at atmospheric pressure, a field of about 3000 V mm^{-1} will allow the air to become ionised and a current to pass through it. Modern dielectric materials can withstand much higher fields before undergoing breakdown. It is however this problem which necessitates that capacitors have to be marked with the highest safe voltage which can be used across them. Thirdly the effect of the dielectric is to increase the capacitance. It is found experimentally that a capacitor with a dielectric present may have a capacitance several times larger than one which has a vacuum between the plates. This is because the effect of induced charge on the dielectric reduces the field in the dielectric, so a given charge can be stored with a smaller potential difference between the plates. The factor by which the capacitance is raised is called the **relative pemittivity** ε_r or the **dielectric constant**. In equation form this becomes

$$\varepsilon_r = \frac{\text{capacitance of capacitor with dielectric between the plates}}{\text{capacitance of capacitor with a vacuum between the plates}}$$

ε_r is a ratio of two capacitances and therefore does not have a unit.

The value of ε_r for air at atmospheric pressure is 1.0006, so in performing an experiment to measure ε_r it is usually acceptable to use air between the plates instead of a vacuum. Some values for ε_r are given in Table 17.1. The high value for water results in there being a much smaller force between ions when they are separated by water rather than by air. This is why many crystals dissolve in water. Water is not suitable for use as the dielectric in capacitors because it is a liquid and because impurities in it allow it to conduct electricity. High values of relative permittivity in ceramic materials containing strontium titanate are used to obtain high capacitance capacitors.

Table 17.1 Relative permittivity of some dielectrics

Dielectric	Relative Permittivity ε_r
vacuum	1 exactly
air	1.0006
teflon	2.1
polythene	2.3
waxed paper	2.7
polyvinyl chloride	3.18
neoprene rubber	6.7
glycerin	43
pure water	80
strontium titanate	310

Once the effect of the dielectric is taken into account, the full equation for the capacitance of a parallel-plate capacitor is

$$C = \frac{\varepsilon_0\,\varepsilon_r\,A}{d}$$

Here again the effect of rationalisation of units is apparent. There is a uniform electric field in a capacitor and consequently π does not appear in this equation. If the constant of $1/4\pi$ had not been introduced in Coulomb's law then a 4π would have appeared here. It can be seen from the equation that the capacitance of a parallel plate capacitor is proportional to the area of the plates and inversely proportional to their separation. A larger area of plates will therefore increase the capacitance but this also has the effect of making the capacitor larger in size. The miniaturisation of electronic circuits however requires capacitors of larger value and smaller size. This can be achieved by using high values of the relative permittivity or by having the plates closer together. The following example shows typical sizes for an electrolytic capacitor.

Example 1

A capacitor is to be constructed to have a capacitance of 100 μF. The area of the plates is 6.0 m × 0.030 m and the relative permittivity of the dielectric is 7.0. Find the necessary separation of the plates and the electric field strength in the dielectric if a potential difference of 12 V is applied across the capacitor.

The capacitance is given by $C = \dfrac{\varepsilon_0\, \varepsilon_r\, A}{d}$

$$100 \times 10^{-6}\ \text{F} = \frac{8.9 \times 10^{-12}\ \text{F m}^{-1} \times 7 \times (6\ \text{m} \times 0.030\ \text{m})}{d}$$

$$d = \frac{8.9 \times 10^{-12} \times 7 \times 6 \times 0.030\ \text{m}}{100 \times 10^{-6}}$$

$$= 1.12 \times 10^{-7}\ \text{m}$$

$$d = 1.1 \times 10^{-4}\ \text{mm (to 2 sig. figs)}$$

The electric field strength $= \dfrac{12\ \text{V}}{1.12 \times 10^{-7}\ \text{m}} = 1.1 \times 10^{7}\ \text{V m}^{-1}.$

This very small value of d is a layer less than 1000 atoms thick, so there is a real danger that the electric field will cause dielectric breakdown.

To obtain such thin layers of dielectric an electrolytic method of production is used. A pair of aluminium plates is separated by a piece of paper soaked in aluminium borate solution and a current is passed through the solution. Aluminium oxide is deposited on the positive plate, the anode, and this is an insulator which reduces the current and acts as the dielectric for the capacitor. It is essential to maintain the anode at a positive potential to prevent the electrolytic process being reversed, so an electrolytic capacitor always needs to be connected with its anode positive and its cathode negative.

17.2 CAPACITORS IN SERIES AND IN PARALLEL

The circuit laws, Kirchhoff's laws, which were applied to circuits containing resistors and in which there are direct currents, are laws which can be applied to any electrical circuit. Here we need to see how these laws can be applied to circuits containing capacitors.

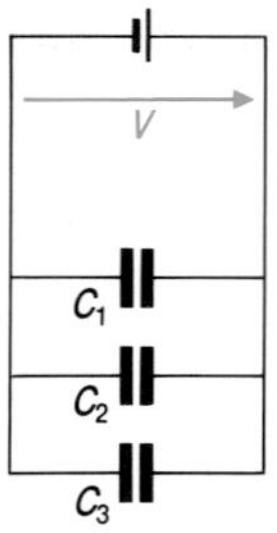

Capacitors in parallel

In Fig 17.5 three capacitors of capacitance C_1, C_2 and C_3 are shown connected in parallel and with a potential difference V across them. Because they are in parallel, the potential difference is the same for each capacitor and for the supply. When this circuit is switched on, a charge will flow from the cell for a short time. Some of the total charge Q which flows will flow to charge up capacitor C_1 with a charge Q_1.

Similarly C_2 will gain a charge Q_2 and C_3 will gain a charge Q_3.

Fig 17.5 Capacitors in parallel: each capacitor has the same potential difference across it.

Using Kirchhoff's laws $Q = Q_1 + Q_2 + Q_3$

but since $Q = CV$, $Q_1 = C_1 V$, $Q_2 = C_2 V$, $Q_3 = C_3 V$

$$CV = C_1 V + C_2 V + C_3 V$$

$$C = C_1 + C_2 + C_3$$

where C is the total circuit capacitance.

This result shows that when capacitors are in parallel the total capacitance is the sum of the capacitances of the individual capacitors.

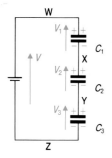

Fig 17.6 Capacitors in series: each capacitor stores the same charge.

Capacitors in series

In Fig 17.6 three capacitors of capacitance C_1, C_2 and C_3 are shown connected in series and with a potential difference V across them. When these capacitors are initially connected, a charge will flow in the circuit. Kirchhoff's laws state that if ammeters are placed at W, X, Y and Z, they will at all times show the same current values since the circuit is a series circuit. This being the case, the charges on C_1, C_2 and C_3 will all be the same as the charge Q, leaving the cell. (This frequently causes difficulty; if 6 mC leave the cell, people imagine that, if the charges on each capacitor are equal, 2 mC will be on each of the capacitors. This is not the case. If 6 mC leave the cell, 6 mC will be on each of the capacitors – even if there are 100 capacitors all in series and even if they all have different values. You should be able to see why this is so if you look at the piece of wire at X and the two plates to which it is connected. The two plates and the wire are isolated electrically from the rest of the circuit. If they start with zero charge on them they will end with a total of zero charge on them. This is the case. Any positive charge which moves to the right-hand plate leaves an equal and opposite charge on the left-hand plate. The total charge is zero.)

From Kirchhoff's second law we have

$$V = V_1 + V_2 + V_3$$

$$\frac{Q}{C} = \frac{Q}{C_1} + \frac{Q}{C_2} + \frac{Q}{C_3} \quad \text{giving}$$

$$\frac{1}{C} = \frac{1}{C_1} + \frac{1}{C_2} + \frac{1}{C_3}$$

Note that it is when capacitors are in series that their capacitances are added by this reciprocal addition, which gives the total capacitance to be smaller than any of the individual capacitances. When they are in parallel they are added by normal addition. This is the reverse situation of that for resistors. This arises because when resistors are in parallel more charge flows and therefore the resistance is reduced; with capacitors in parallel, more charge flows to be stored and therefore the capacitance is increased.

Tabulation of electrical quantities to solve numerical problems can be done with capacitors in the same way as with resistors, as is shown in Example 2.

Example 2

An 80 µF and a 20 µF capacitor are connected in series to a 10 V supply, as shown in Fig 17.7. Find the charge on each capacitor and the potential difference across each.

Start by constructing a table (Table 17.2a) and inserting the known information.

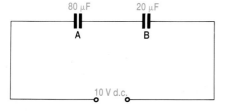

Fig 17.7

Table 17.2a

	Supply	Capacitor A	Capacitor B
charge/µC	Q	Q	Q
p.d./V	10		
capacitance/µF		80	20

This example illustrates one other aspect of tabulating answers to electrical problems. When the initial information is inserted it is found that for none of the components are two quantities known, so at first sight no progress seems to be possible. When this situation arises, use an algebraic term for something that needs to be known. In this table the charge delivered from the 10 V supply has been inserted as Q, assumed to be in microcoulombs. Since the capacitors are in series, both capacitors receive the same charge Q. Using Q, further quantities may now be found (Table 17.2b).

Table 17.2b

	Supply	Capacitor A	Capacitor B
charge/μC	Q	Q	Q
p.d./V	10	$\dfrac{Q}{80}$	$\dfrac{Q}{20}$
capacitance/μF		80	20

Now that all the potential differences have been found in terms of Q, some subsidiary working can be done to find Q. The potential difference across the supply must equal the sum of the potential differences across the capacitors, so:

$$10 = \frac{Q}{80} + \frac{Q}{20}$$
$$= \frac{Q + 4Q}{80} = \frac{5Q}{80}$$
$$5Q = 800$$
$$Q = 160\ \mu C$$

This value can then be inserted into the table and all other details found (Table 17.2c).

Table 17.2c

	Supply	Capacitor A	Capacitor B
charge/μC	$Q = 160$	$Q = 160$	$Q = 160$
p.d./V	10	$\dfrac{Q}{80} = 2$	$\dfrac{Q}{20} = 8$
capacitance/μF	$\dfrac{160\ \mu C}{10\ V} = 16$	80	20

As explained in Section 15.1, the gradual build up of information can be done on a single table although three tables have been used here to show the sequence. The tabulation method has the advantage of clarifying the need for electrical quantities for each component separately.

Two points should be noted about the answers to this question. First, the value entered in the space labelled for the capacitance of the supply is the total effective capacitance of the whole circuit. This value is lower than the capacitance of either of the capacitors separately. Two capacitors in series like this do indeed store less charge than either of them would separately. The second point to note is that the smaller capacitor has the higher potential difference across it. Because the charge on both capacitors is the same, the potential difference across each capacitor is inversely proportional to its capacitance.

CAPACITANCE

17.1 Find the capacitance of a parallel plate capacitor in which the plates are 2.0 m long and 1.0 cm wide and are separated by a dielectric which is 0.017 mm thick and of relative permittivity 2.3. (Be careful to put all lengths into metres.)

Capacitors of this construction are always coiled up. (2 m long capacitors are too clumsy!) When coiled, an extra layer of dielectric is added as shown in Fig 17.8. What is the effect of coiling on the capacitance?

17.2 Construct a table to find the charge on each of the capacitors shown in Fig 17.9 together with the charge supplied by the battery. What is the total capacitance of this circuit?

17.3 The 0.010 μF capacitor shown in Fig. 17.10 has a potential difference of 20 V across it. What charge is there on each capacitor and what is the e.m.f. of the battery? What is the total circuit capacitance?

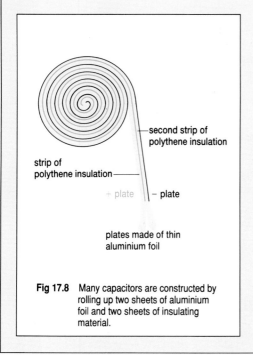

Fig 17.8 Many capacitors are constructed by rolling up two sheets of aluminium foil and two sheets of insulating material.

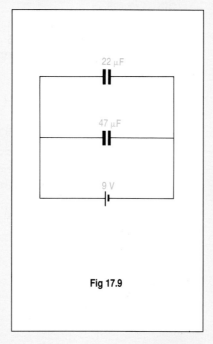

Fig 17.9

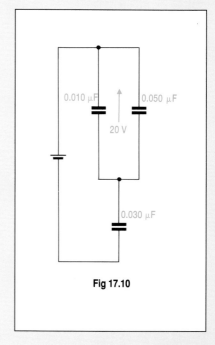

Fig 17.10

17.3 ENERGY STORED IN A CHARGED CAPACITOR

When a charge of 200 μC is stored on a capacitor with a potential difference across it of 4 V (4 joules per coulomb), it might seem reasonable to assume that the energy stored is 800 μJ. This is however **not** correct. To see why it is not correct, it is necessary to consider how the capacitor gained its charge in the first place. Initially the capacitor is uncharged. When the first small quantity of charge flows to the capacitor, the potential difference is very small and so little work is done. As the charge on the capacitor increases, so the potential difference increases and so more work has to be done to increase the charge on the capacitor. If a graph of potential difference is plotted against charge, it will be a straight line as shown in Fig 17.11. If, when the potential difference is v, a small charge δq is added to the capacitor, then the work done will be $v\,\delta q$. This is the area shown shaded on the graph. When the full charge Q has been added, the total work done to charge the capacitor, which will be equal to the energy stored by the capacitor, E, will be the total triangular area beneath the graph and is given by

$$E = \tfrac{1}{2} QV$$

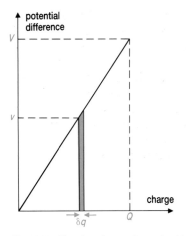

Fig 17.11 The triangular area beneath a V–Q graph is equivalent to the energy stored by a capacitor.

If the capacitor has been charged as a result of a charge Q flowing from a battery of e.m.f. V, then the work done by the battery is QV but the energy stored by the capacitor is only $\frac{1}{2}QV$. This implies that energy $\frac{1}{2}QV$ has been wasted within the resistance of the circuit as internal energy.

A mechanical analogy to this situation is a spring being extended a distance x by a force F. The force needs to increase from zero to F as the extension takes place. The work done by the force is the average force multiplied by the extension, i.e.

$$W = \tfrac{1}{2}Fx$$

If a fixed force F is applied to the spring, for instance by dropping a weight on it, then the spring will gain some kinetic energy, stretch beyond its equilibrium extension and oscillate until it loses energy as internal energy, so that it does store energy of $\frac{1}{2}Fx$. The $\frac{1}{2}$ in this equation and in the $\frac{1}{2}QV$ equation is often forgotten. So the energy stored when a capacitor with a potential difference across it of 4 V carries a charge of 200 μC is 400 μJ, not 800 μJ.

The expression

$$E = \tfrac{1}{2}QV$$

can be combined with the expression $C = Q/V$ to give the energy stored in terms of different quantities.

$$E = \tfrac{1}{2}QV = \tfrac{1}{2}Q\frac{Q}{C} = \tfrac{1}{2}\frac{Q^2}{C} \text{ or}$$

$$E = \tfrac{1}{2}QV = \tfrac{1}{2}CVV = \tfrac{1}{2}CV^2$$

QUESTIONS

17.4 Find the charge and the energy stored by a 1000 μF capacitor when the potential difference across it is **(a)** 6.0 V, **(b)** 12.0 V.

17.5 A 47 μF capacitor in a flash gun supplies an average power of 1.0×10^4 W for 2.0×10^{-5} s. What is the potential difference across the capacitor before it is discharged?

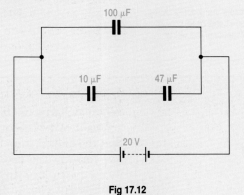

Fig 17.12

17.6 Fig 17.12 shows a 20 V supply connected to a 100 μF capacitor in parallel with capacitors of capacitance 10 μF and 47 μF in series. Calculate
(a) the charge on the 100 μF capacitor
(b) the charge on the 10 μF and 47 μF capacitors
(c) the energies stored by each capacitor
(d) the potential differences across each capacitor.

CAPACITANCE

A finite time is taken to charge a capacitor from any source of charge. This time has become of increasing importance with the development of computers and the increasing use of digital information. All computer circuits use capacitors to store energy and there is also stray capacitance between the various parts of any circuit which may have to be taken into account. A problem particularly arises when the frequency of transfer of information is high. If digital information is being handled by a circuit at a frequency of 1000 MHz then the time for one cycle is a billionth of a second. A capacitor may need to be charged and discharged during this time and, if its value is too high or if the resistance through which it is

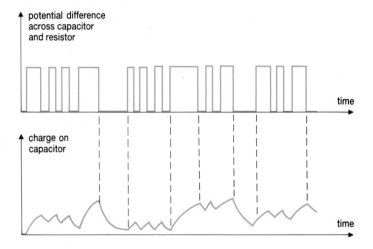

Fig 17.13 When a pulsed signal is applied across a capacitor and resistor in series, the charge on the capacitor may not have time to reach full charge. The resultant distortion of the signal can be a major problem at high frequency.

being charged is too high, it may not be able to gain or lose its charge quickly enough. Fig 17.13 shows how the potential difference varies across a circuit containing a capacitor and a resistance. The corresponding charge on the capacitor is also shown. If the intention is to maintain the shape of the original pulses, it has clearly not succeeded. A great deal of distortion has been introduced as a result of the inability of the capacitor to charge and discharge instantaneously.

INVESTIGATION

The numerical values given for components used in this investigation are for guidance only. If you cannot find precisely these values, use the nearest available ones.

1. Connect a square-wave generator with a frequency of 1 kHz to a 0.47 µF capacitor in series with a 220 Ω resistor. Connect an oscilloscope across the capacitor and adjust the Y-gain and/or the amplitude of the square-wave generator output so that the vertical deflection occupies most of the screen. Then adjust the time base of the oscilloscope so that approximately two complete cycles of the square wave are shown stationary on the screen. If your trace is not stationary it will probably be because the synchronisation (synch) needs to be adjusted. Pay particular attention to the earth connections of the instruments you are using. These must be connected together as shown in Fig 17.14.
   (a) What is the length of time for one complete cycle?
   (b) Approximately how long does it take for the capacitor to become fully charged?

CAPACITANCE

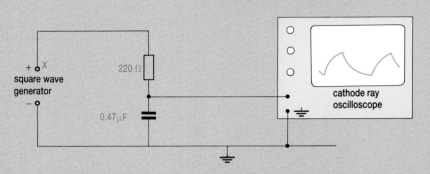

Fig 17.14 A cathode ray oscilloscope measuring the potential difference across a capacitor as it is charged and discharged.

(c) Approximately how long does it take for the capacitor to discharge fully?

(d) Draw a sketch graph showing how the potential difference across the capacitor varies with time.

2. Repeat the experiment with different values of the resistance. First reduce the resistance in convenient steps to about 20 Ω and then increase the resistance in steps to more than 10 000 Ω. For each resistance, make a sketch of the shape of the trace seen on the screen.

3. Reverse the positions of the capacitor and the resistor. Use the 220 Ω resistor again.

(a) Sketch the pattern shown for the potential difference across the resistor and compare it with the potential difference across the capacitor obtained previously. Do this by drawing graphs of the two potential differences against time, using the same scale on both time axes.

(b) Find the sum of the two potential differences by adding the values shown on the two sketch graphs together.

4. Disconnect the oscilloscope from the resistor and connect it to X.

(a) The sum of V_C and V_R should equal the supply voltage at all times. Why?

A full mathematical analysis of the charge and discharge of a capacitor of capacitance C in series with a resistor of resistance R requires the use of calculus. An understanding of why the curves have the shape they do can be obtained by working through the following exercise. It can be programmed on a computer quite easily if you have one available. Do not let the number of significant figures which you use increase beyond four.

DATA ANALYSIS

Charging a capacitor through a resistor

A 400 µF capacitor is to be charged from a 10.0 V battery through a 200 000 Ω resistor as shown in Fig 17.15. The switch is turned on at time $t = 0$ so that the sum of the potential differences across the capacitor and the resistor is fixed at 10.0 V.

The capacitor starts in an uncharged state so the potential difference across it is zero. The potential difference across the resistor must therefore be 10 V. This will only be the case if a current of 50 µA flows through it. The assumption is made that for the next 10 seconds the current remains constant, so a charge of 500 µC will flow from the battery to charge up the

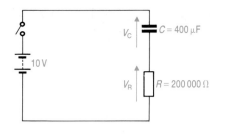

Fig 17.15

capacitor. Using the notes which are given, Table 17.3 can be completed by working across and down it line by line.

Table 17.3

Time/s	V_C / V	V_R/ V	I / μA	Charge flowing in 10 s /μC	Charge on capacitor 10 s later /μC
0	0	10	50	500	500
10	1.25 (1)	8.75 (2)	43.8 (3)	438	938. (4)
20	2.35				
					3926. (5)
400	9.82	0.18	0.90	9.0	3935
410	9.84	0.16	0.8	8.0	3943
420	9.86	0.14	0.7	7.0	3950
430					
440					
∞ (6)					

Notes to table

1. 500 μC on a capacitor of value 400 μF necessitates 1.25 V across the capacitor.

2. If there is now 1.25 V across the capacitor, the remainder of the 10 V of the supply must be across the resistor, i.e. 8.75 V.

3. A potential difference of 8.75 V across a 200 000 Ω resistor causes a current of 43.8 μA.

4. 438 μC arrive on the capacitor to be added to the 500 μC already there.

5. No one is expecting you to have worked through this far manually. It is hoped that you can see the way in which as the potential difference across the capacitor rises, the potential difference across the resistor falls and the current falls correspondingly. Now proceed a little further from here to see that there is a limit to the quantities involved.

6. What are these limits at infinite time?

Plot a graph of the potential difference across the capacitor against time.

The assumption made to do the working in the data analysis exercise is that the current remains constant for a time after each current calculation is made. This assumption is only accurate if the time interval is small compared with a constant called the time constant of the circuit. In this case the time constant is 80 s and we have used a time interval of 10 s. It would have been better to have used 1 s instead of 10 s but then there would have been much more arithmetic to do. This problem can be overcome if you have a computer available and can program it to do the calculation for you. The process of needing to use smaller and smaller time intervals is exactly the one which calculus provides. Using calculus, small time intervals are used and the time intervals are then made vanishingly small to get a precise theoretical answer rather than an approximate answer.

The general equation for charging a capacitor through a resistor can be obtained in the following way. Fig 17.16 shows a capacitor of capacitance C in series with a resistor of resistance R. The capacitor is initially uncharged and will start to charge at time $t = 0$ when the switch is closed, from a source of e.m.f. E. Let Q be the charge on the capacitor when the time is t.

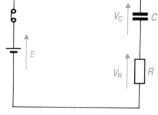

Fig 17.16

Using Kirchhoff's law that the e.m.f. is the sum of the potential differences gives

$$E = V_C + V_R$$

$$E = \frac{Q}{C} + IR = \frac{Q}{C} + \frac{dQ}{dt} R$$

The solution to this differential equation is

$$Q = EC\,(1 - e^{-t/CR})$$

E here is the e.m.f. of the battery used to charge the capacitor. It is therefore equal to the potential difference across the capacitor at the end of the charging process because when the current is zero V_R is also zero. EC therefore is the final charge Q_0 on the capacitor, so the equation can also be written:

$$Q = Q_0\,(1 - e^{-t/CR})$$

Several mathematical steps have been omitted in this deduction. The final equation however is one which you should recognise and be able to use.

QUESTIONS

17.7 Use the equation for the charge on a 10 µF capacitor when it is being charged through a 2000 Ω resistor from a source of e.m.f. 9.0 V to plot a graph of charge against time. You are recommended to do this question by tabulating the values needed for the graph as shown in Table 17.4.

Table 17.4

t second	t/CR no unit	$e^{-t/CR}$ no unit	$1 - e^{-t/CR}$ no unit	Q coulomb
0	0	1	0	0
0.005	0.25	0.779	0.221	
0.010				
0.015				
.				
.				
.				
.				
.				
0.10				

17.8 Using the graph which you have drawn from Question 17.7, draw a graph of current against time for the same charging process. (Use the fact that the gradient of a charge–time graph is the current $I = dQ/dt$.)

17.9 Using the same numerical values as in Questions 17.7 and 17.8, find the following:
 (a) the current the moment the switch is connected at $t = 0$;
 (b) the charge on the capacitor when it is fully charged;
 (c) the length of time it would take the capacitor to charge up fully if this initial rate of flow of charge were continued (this time is called the time constant of the circuit).
 (d) the value of C multiplied by R (the answers to (c) and (d) should be the same).
 (e) the unit of CR;
 (f) the fraction of the total charge which a capacitor has after one time constant and after five time constants.

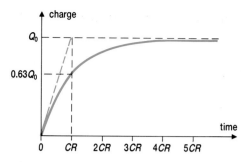

Fig 17.17 After a time equal to one time constant, $t = CR$, the charge on the capacitor is 63% of the maximum charge. If the original rate of charging had been maintained then the capacitor would have been fully charged by this time.

The term time constant was mentioned in the last question. The **time constant** of a resistor–capacitor circuit is the product CR. It is also the time the capacitor would take to become fully charged if the initial rate of charging is maintained. This is shown in Fig 17.17.

If a charged capacitor is discharged through a resistor, the charge on the capacitor falls exponentially. The equation for discharge, using the same symbols as previously, is:

$$Q = Q_0\, e^{-t/CR}$$

If the capacitor is discharged for a time equal to its time constant, the equation becomes

$$Q = Q_0\, e^{-CR/CR} = Q_0\, e^{-1} = \frac{Q_0}{e} = \frac{Q_0}{2.718}$$

$$Q = 0.368\, Q_0$$

After 5 time constants the charge is given by

$$Q = Q_0\, e^{-5CR/CR} = Q_0\, e^{-5} = \frac{Q_0}{(2.718)^5}$$

$$Q = 0.0067\, Q_0$$

(handwritten) $E = \dfrac{Q}{C} =$

(handwritten) $Q = 1.5 \times 10^{-4}$

An exponential decay such as this occurs in many branches of physics. The cooling curve for a cup of tea follows approximately the same pattern and so does the rate of decay of radioactive materials. It is unusual, but possible, to calculate the half-life of a capacitor–resistor circuit or even of the excess temperature of a cup of tea above its surroundings.

If $Q = \tfrac{1}{2} Q_0$ then

$$\tfrac{1}{2} Q_0 = Q_0\, e^{-t/CR}$$

(handwritten) $0.3675\, Q_0 = Q$

Q_0 can be cancelled and taking natural logarithms gives

$$-\ln 2 = -\frac{t}{CR} = -0.693$$

$$t = 0.693\, CR$$

(handwritten) $\ln 0.3675 = \dfrac{-t}{CR}$

(handwritten) $71.00 = \dfrac{t}{t}$

In other words, a capacitor will discharge by half after 0.693 time constants.

When a capacitor discharge occurs, the current flows in the opposite direction to the charging current. The potential difference across the resistor is therefore in the opposite direction for discharge to that for charge. This is shown in the graphs of Figs 17.18(a) and (b) and their equations which summarise this work on the charging and discharging of capacitors through resistors.

(handwritten margin notes, left side)

2.9H

$E = \dfrac{Q}{C}$

$V = 8.$
$Q = 2.4 \times 10^{-5}$

$V = 2.94$
$Q = 8.82 \times 10^{-6}$

Charge

$$Q = EC\,(1 - e^{-t/CR}) = Q_0\,(1 - e^{-t/CR})$$

$$V_C = E\,(1 - e^{-t/CR})$$

$$V_R = E\, e^{-t/CR}$$

$$I = \frac{E}{R}\, e^{-t/CR}$$

Discharge

$$Q = Q_0 e^{-t/CR}$$

$$V_C = E\, e^{-t/CR}$$

$$V_R = E\, e^{-t/CR}$$

$$I = -\frac{E}{R}\, e^{-t/CR}$$

(handwritten) $2.94 = 8e^{-t/1.2}$

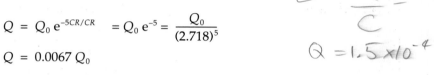

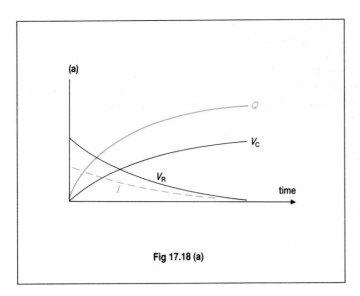

Fig 17.18 (a)

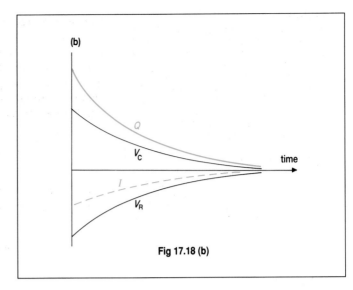

Fig 17.18 (b)

When a series of equations such as this is given, it is important not to use them mechanically but to see how they are related to one another. Here the following points should be noted:

- All the expressions contain an exponential term $e^{-t/CR}$ which governs the rate of rise or fall with time.
- A quantity whose numerical value is rising has a $1 - e^{-t/CR}$ term whereas one which is falling has just a $e^{-t/CR}$ term.
- V_C and Q are always related by the expression $Q = CV_C$.
- V_R and I are always related by the expression $V_R = IR$.
- Current must always tend to zero as the time increases since a capacitor can only have a current flowing to or from it when it is charging or discharging. If it is left long enough no current will flow. The current can be either positive, on charging, or negative, on discharging.
- $V_C + V_R$ equals E during charge and equals zero during discharge.

Example 3

A 100 pF capacitor is charged to a potential difference of 50 V and then discharged through a 2.2 K resistor.

Find the initial discharge current and how long it will take for the potential difference across the capacitor to fall to 1% of its original value.

At the start the potential difference across the capacitor is +50 V and since the only other component in the circuit is the resistor, the potential difference across it must be –50 V. The current is therefore given by

$$I = \frac{V_R}{R} = \frac{-50\ V}{2200\ \Omega} = -0.023\ A = -23\ mA$$

The equation will of course give the same answer:

$$I = -\frac{E}{R}\ e^{-t/CR} = -\frac{E}{R}\ e^{-0}$$

$$= \frac{E}{R} = -\frac{-50\ V}{2200\ \Omega} = -23\ mA$$

Using the equation for the potential difference across the capacitor gives

$$V_C = E\ e^{-t/CR}$$

$$\frac{E}{100} = E\ e^{-t/CR}$$

(handwritten notes in left margin:)

$8 \times 3 \times 10^{-5}$

$Q = C V_C$

$Q = 1.5 \times 10^{-4}$

(handwritten notes in right margin:)

$V = \dfrac{E}{100}$

312

CAPACITANCE

A note on logarithms.
Since $10^3 = 1000$ we write $\log_{10} 1000 = 3$, i.e. the logarithm of 1000 to the base of 10 is 3. The logarithm of a number is just the power to which the base has to be raised to get the number.
Similarly since $2.718^3 = 20.08$ we write $\log_{2.718} 20.08 = 3$ and since $e^3 = 20.08$ we write $\log_e 20.08 = 3$.
$\log_e x$ is written $\ln x$ and is called a natural logarithm. (e is a number and is approximately 2.718). Scientific calculators have the facility to find the natural logarithm of a number and the reverse of this process, to find the term e^x. You are advised to become familiar with what these keys and the lg and 10^x keys do by practising using them.

Cancel E to get

$$= e^{-t/CR}$$

At this stage, take natural logs of both sides of the equation. If you are not familiar with this procedure, work carefully through the margin note as this is a technique which will be used frequently and needs to be understood.

$$-4.605 = -\frac{t}{CR}$$

$$t = CR \times 4.605$$
$$= (100 \times 10^{-12}\text{F}) \times (2200\ \Omega) \times 4.605$$
$$= 1.01 \times 10^{-6}\text{s}$$
$$t = 1.0\ \mu\text{s}$$

QUESTIONS

17.10 What percentage of the original charge remains on a capacitor when it is discharging, after 1, 2, 3, 4 and 5 time constants?

17.11 A square wave signal which is either at zero volts or 500 mV and of frequency 100 kHz, see Fig 17.19, is applied to a capacitor of capacitance 0.010 μF in series with a resistor of resistance 100 Ω. Copy Fig 17.19 and sketch, on the same time axis, graphs of V_C, the potential difference across the capacitor, and V_R, the potential difference across the resistor. This is called a short CR circuit because the time constant of the circuit is short compared with the period of oscillation.

17.12 What would the graphs look like in Question 17.11 if the resistor used had been 20 000 Ω instead of 100 Ω? Working through this question fully is difficult. You are only expected to show a rough sketch of what happens. This is called a long CR circuit because the time constant of the circuit is long compared with the period of oscillation.

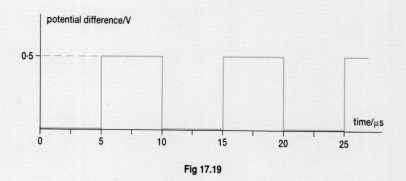

Fig 17.19

17.13 A 100 μF capacitor carries a charge of 20 mC. It is discharged through an 86 kΩ resistance. Find the time it takes to lose half of its charge (its half-life). Find the time taken for its charge to fall to 1.25 mC. What is the time constant for this circuit? What is the relation between the time constant and the half-life?

- Capacitance is charge stored per unit potential difference.

$$C = \frac{Q}{V}$$

- 1 farad is the capacitance of a capacitor which stores 1 coulomb of charge when a potential difference of 1 volt is across it. μF (10^{-6} F) and pF (10^{-12} F) are more usually used.

- The capacitance of a parallel plate capacitor is given by $\dfrac{\varepsilon_0 \, \varepsilon_r \, A}{d}$ where ε_r is 1 for a vacuum.

- Capacitors in series $\dfrac{1}{C} = \dfrac{1}{C_1} + \dfrac{1}{C_2} + \dfrac{1}{C_3}$

- Capacitors in parallel $C = C_1 + C_2 + C_3$

- Energy stored in a capacitor $= \frac{1}{2} QV = \frac{1}{2} CV^2 = \frac{1}{2} \dfrac{Q^2}{C}$

- Charging a capacitor $Q = Q_0 \, (1 - e^{-t/CR})$

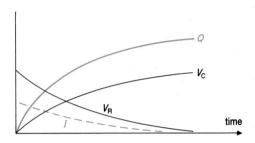

- Discharging a capacitor $Q = Q_0 \, e^{-t/CR}$

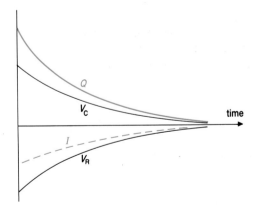

EXAMINATION QUESTIONS

17.14 (a) Diagram (a) shows a typical capacitor. What do the markings on the capacitor indicate?

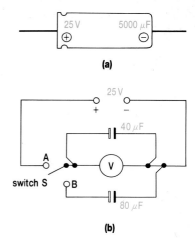

(a)

(b)

(b) In the circuit shown in (b) V is a very high resistance voltmeter and the two capacitors are initially uncharged.
   (i) When switch S is moved to position A, what will be the final steady reading on the voltmeter, and how much charge will then be stored on the 40 μF capacitor?
   (ii) Switch S is then moved to position B. State and explain whether the voltmeter reading will increase or decrease. Calculate the new steady voltmeter reading. (10)

(AEB 1987)

17.15 Given a number of capacitors each with a capacitance of 2 μF and a maximum safe working potential difference of 10 V, how would you construct capacitors of
   (a) 1 μF capacitance, suitable for use up to 20 V?
   (b) 2 μF capacitance, suitable for use up to 20 V?

(UCLES 1987)

17.16 Examine the diagram below and calculate

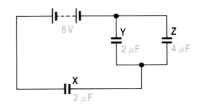

(a) the potential difference across capacitor X,
(b) the charge on the plates of capacitor Y,
(c) The energy associated with the charge stored in capacitor Z. (6)

(ULSEB 1984)

17.17 A resistor is connected across a charged capacitor, the time constant of the circuit being 2.0 s. Calculate the fraction of the initial stored energy which remains 1.0 s after connection. (4)

(UCLES 1986)

17.18 (a) Define each of the following terms
   (i) the capacitance of a capacitor,
   (ii) the time constant of a circuit containing capacitance and resistance. (4)
(b) You are to determine the capacitance of an electrolytic capacitor (known to be about 5000 μF) by discharging it through a resistor.
   (i) Draw a diagram of the circuit you would use, describing the procedure you would employ, and stating the measurements which you would make. (7)
   (ii) Specify a suitable size for each of the circuit components and a suitable range for any meter you use, justifying your choice with calculations where appropriate. (6)
   (iii) Explain the theoretical ideas on which the experiment is based and how the capacitance would be calculated. (5)
   (iv) Identify the principal sources of inaccuracy in the experiment, and suggest what, if anything, could be done to reduce these inaccuracies. (3)

(AEB 1987)

17.19 (a) (i) Define *capacitance*.
   (ii) Describe, with the aid of a circuit diagram, an experiment to determine the ratio of two capacitances each roughly equal to 0.1 μF. (8)
(b) A parallel plate capacitor is charged by connecting it to the terminals of a battery. Explain how the energy stored in the capacitor is changed when the distance between the plates is increased if (i) the battery is left connected, (ii) the battery is first disconnected. (8)
(c) A capacitor and a resistor are joined in series and connected across the terminals of a battery of negligible internal resistance.

Show that, when the capacitor is fully charged, the energy stored in the capacitor is equal to the energy dissipated in the resistor during the charging process. (4)

(JMB 1986)

17.20 (a) A 2000 µF capacitor is fully charged using a 9 V battery and then discharged through a 200 kΩ resistor.

Calculate the time constant for the circuit. Draw a labelled diagram of the circuit arrangement, including a means for measuring the p.d., V, across the resistor. Describe what measurements you would take to obtain a graph showing how V varies during the discharge.

Equation (1) shows how V varies with time, t.

$$V = V_0 \, e^{-t/RC} \qquad (1)$$

What does V_0 represent?

(b) The diagram shows the circuit for a photographic flash unit.

The capacitor is initially charged by the battery and when the switch is closed, the capacitor is discharged through the flash bulb. The flash bulb can only be used once and its electrical resistance is much less than 5 kΩ.

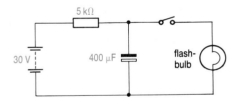

Explain why the average current in the flash bulb during the discharge is much higher than the average current in the resistor during the charging process. Why does the current only flow for a very short time in the flash bulb? (12)

(ULSEB 1988)

17.21 A capacitor is connected to a d.c. supply through a resistor. The graph shows how the charge on the capacitor increases with time.

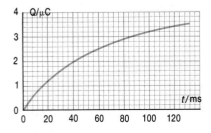

Calculate a value for the average current in the charging circuit during the interval $t = 40$ ms to $t = 100$ ms.

Find the instantaneous value for the current at time $t = 40$ ms.

With reference to these two results, explain how instantaneous current is described as 'the limiting value of an average current'. (8)

(ULSEB 1987)

17.22 (a) Explain the term *capacitance* in relation to two neighbouring but separated conductors.

Write down a relation for the capacitance of a pair of conducting plates of area A and separation t in air.

Describe a laboratory experiment, based on the relation you have just written down, for measuring the permittivity of air. Your account should include a circuit diagram, the basic method and how you would calculate the result. (10)

(b) The diagram shows a condenser (i.e. capacitor) microphone DB and part of an associated circuit. The two plates of the capacitor are the conducting flexible plate D, which oscillates when a sound wave arrives, and the rigid perforated backing plate B. The flexible plate D has an area 0.80 cm² and the microphone has capacitance 25 pF.

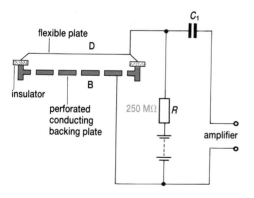

(i) Estimate the separation of the plates B and D.

(ii) Why does the potential difference across the capacitor oscillate when a sound wave arrives?

(iii) Suggest a reason for the perforations in the backing plate B.

(iv) Explain why the value chosen for the resistor R is so high.

(v) What is the function of the capacitor C_1?
(Permittivity of air = 8.9×10^{-12} F m⁻¹.) (8)

(ULSEB 1986)

CAPACITANCE

17.23 A capacitor is made from two parallel metal sheets each of area A. The sheets are connected into the circuit shown at C.

S is a switch which vibrates f times per second between contacts X and Y, fully charging and discharging the capacitor.

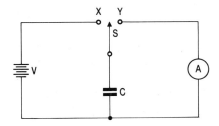

(i) The distance d between the sheets is varied and the discharge current is recorded by the microammeter A. Explain how the capacitance C of the plates is related to the ammeter reading. Draw a graph to show the expected relationship between the current and the plate separation. (4)

(ii) Suggest modifications to this experiment to investigate how the capacitance of the capacitor depends on quantities other than the distance between the plates. (2)

(O & C 1984)

Chapter 18

MAGNETIC FIELD

> **LEARNING OBJECTIVES**
>
> At the end of this chapter you should be able to:
>
> 1. define and use the term magnetic flux density and state its unit;
>
> 2. calculate the force on a moving charged particle;
>
> 3. define and use the term magnetic flux and state its unit;
>
> 4. calculate the torque on a rectangular coil carrying a current in a magnetic field;
>
> 5. calculate the force between parallel wires carrying a current;
>
> 6. calculate the value of the permeability of free space from the definition of the ampere;
>
> 7. use expressions to calculate the magnetic flux density near wires carrying currents;

18.1 MAGNETIC PHENOMENA

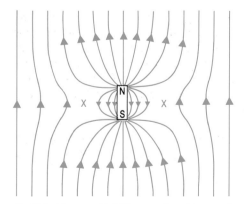

Fig 18.1 The magnetic field surrounding a bar magnet freely suspended in the Earth's magnetic field.

Magnets have been discovered from archaeological sites more than two thousand years old, and have certainly been used for navigation since the Eleventh Century. Early study of magnetism depended on the properties of the poles of magnets and their behaviour in magnetic fields, such as that of the Earth. If a magnetic field is strong enough its shape can be found by scattering iron filings in the field. The pattern of weaker fields can be found by using compass needles. Fig 18.1 shows the magnetic field around a magnet when it is freely suspended in the magnetic field of the Earth. The direction of the magnetic field is taken as the direction in which a force is exerted on the North pole of a magnet.

With the definition of **gravitational field strength** as the force exerted on unit mass, and **electric field strength** as the force exerted on unit charge, it appears reasonable to define **magnetic field strength** as the force exerted on unit magnetic pole. This was the way in which magnetic field strength was originally defined but there is a problem with this definition which does not exist with the other two. Whereas it is possible to use experimentally a unit of mass or a single quantity of charge, it is not possible to use a single magnetic pole. Because of the way magnetic field is generated, magnetic poles always come in pairs. In order to avoid this difficulty, the SI definition and unit of magnetic field strength are approached in a slightly different way from the definitions of gravitational or electrical fields. Instead of considering the force on a magnetic pole, the force on an electric current in a wire is used. The discovery that there is a force acting on a wire carrying a current in a magnetic field was made by Oersted in 1819, but it was not until the 1950s that the effect became the standard way of measuring magnetic field.

18.2 MAGNETIC FLUX DENSITY

A region in which there is a force acting on an electric current in a wire is called a magnetic field. As is always the case with fields in physics, their presence is detected by the force which they exert. The value of the force per unit mass in gravitation is the gravitational field strength in both magnitude and direction. The value of the force per unit charge in electrostatics is the electric field strength in both magnitude and direction. Because of the use of electric current to define and measure magnetic field strength, an exactly corresponding statement is not made for magnetic field. To begin with, the force acting on a current-carrying wire is not in the same direction as the field. Fig 18.2(a) shows a wire XY carrying an electric

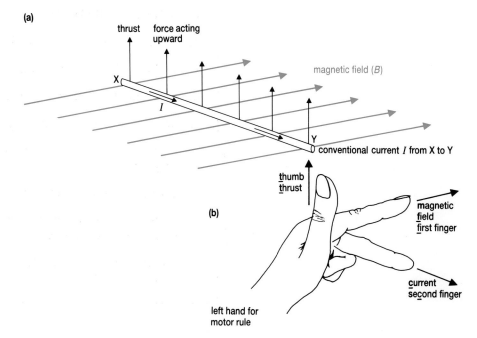

Fig 18.3 An experimental arrangement for measuring the force acting on a wire carrying a current in a magnetic field.

current I in a direction at right angles to a magnetic field B. The force acting on the wire is at all places vertically upwards and depends not only on the current but also on the length of the wire; the longer the wire, the greater the total upward force. Since this is the way magnetic field strengths are measured however, this is the arrangement which is used for its definition.

The quantitative value of magnetic field strength is called the magnetic flux density and it is given the symbol B.

> **Magnetic flux density** is defined as the force acting per unit current in a wire of unit length at right angles to the field. Put in equation form, B is given by
> $$B = \frac{F}{Il}$$

where F is the total force acting on the wire and l is its length.

The direction of the field is at right angles to both the force and the current and is given by Fleming's left-hand rule, as shown in Fig 18.2(b).

The unit of magnetic flux density is the **tesla (T)**. A tesla is the magnetic flux density if a wire of length $1\,m$ carrying a current of $1\,A$ has a force exerted on it of $1\,N$ in a direction at right angles to both the flux and the current. The tesla is named after Nikola Tesla, an American electrical engineer who designed the power generating station at Niagara Falls and was responsible for many features of high voltage electrical transmission.

If the angle between the current and the field is not 90° then the force is smaller than that given by the equation $F = BIl$. By resolving I perpendicularly to B, the force becomes

$$F = BIl \sin \theta$$

where θ is the angle between the field and the current.

INVESTIGATION

Find the value of the magnetic flux density inside a solenoid. A possible experimental arrangement is shown in Fig 18.3 though several others are available commercially. It is suggested that the solenoid is wound on a cardboard tube of greater than 5 cm diameter and about 40 cm long.

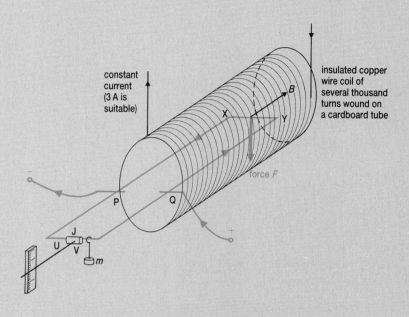

Fig 18.3 An experimental arrangement for measuring the force acting on a wire carrying a current in a magnetic field.

The wire used should be insulated copper wire of about 24 s.w.g. and the coil should have at least 500 turns in a single layer. A double layer containing 1000 turns is better if you have the patience to do the winding! (s.w.g. stands for standard wire gauge and is a frequently used measure of wire diameter.) The wire on which the force is to be measured is XY in the diagram. A stiff copper wire rectangular frame UPXYQV is needed, small enough for one end of it to go into the solenoid, and with an insulator, J, used to join its ends. The frame is supported on knife edges at P and Q through which current can pass from a d.c. source. It helps if the wire frame is notched at P and Q so that it can easily be returned to its position if it gets displaced. A pointer and a scale showing zero need to be positioned as shown and a rider, of known mass m, placed on UV. An alternative arrangement would be to measure the force on UV by attaching it to the pan of a top-pan balance.

First the frame is balanced by adding Plasticine. Then a steady current is passed through the solenoid and current I is passed through XY. In operation, the currents must be in suitable directions for the force which is to be measured to be acting downwards. Fig 18.4 shows the arrangement in elevation and by applying the principle of moments to the system we get

$$mgd_1 = Fd_2$$

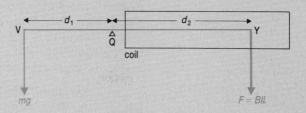

Fig 18.4

This enables the force F on the wire to be determined if m, d_1 and d_2 are measured. Since $F = BIl$, B can be determined once l, the distance XY, has been found. Note that any force which is exerted on PX will be cancelled out by the force on YQ because the currents in these two sections are in opposite directions. In practice there is little force on either of these sections of the rectangular frame as the current in these sections is parallel with the magnetic field of the solenoid.

Use the apparatus to plot a graph of magnetic flux density on the axis of the solenoid against distance along the solenoid. It will almost certainly be easier to move the solenoid rather than the rectangular frame when you do this. Do not stop when XY is outside the solenoid. Take another few readings until the field is too small to be measured. You should also record the number of turns per unit length on the solenoid and the current passing through it. The value of the Earth's magnetic flux density is about 2×10^{-5} T in a horizontal direction, so it should not affect your result significantly.

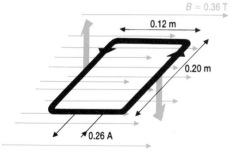

Fig 18.5

Example 1

In an electric motor a rectangular coil of wire has 150 turns and is 0.20 m long and 0.12 m wide. The coil has a current of 0.26 A through it and is parallel to a field of magnetic flux density 0.36 T, as shown in Fig 18.5. Find the torque which is exerted on the coil.

Force F on 0.20 m of wire with 0.26 A through it is given by

$$
\begin{aligned}
F &= BIl \\
&= 0.36 \text{ T} \times 0.26 \text{ A} \times 0.20 \text{ m} \\
&= 0.0187 \text{ N}
\end{aligned}
$$

Turning effect of this force	=	$0.0187 \text{ N} \times 0.60 \text{ m}$
	=	$1.12 \times 10^{-3} \text{ N m}$
Torque on a single coil	=	total turning effect of both forces
	=	$1.12 \times 10^{-3} \text{ N m} \times 2$
	=	$2.24 \times 10^{-3} \text{ N m}$
Torque on 150 turns	=	$2.24 \times 10^{-3} \text{ N m} \times 150$
	=	0.337 N m
	=	0.34 N m to 2 sig. figs.

QUESTIONS

18.1 A wire of superconducting niobium of radius 0.100 cm can carry a current of 1500 A. The density of niobium is 8600 kg m^{-3}. What field must the wire be placed in so that it levitates, that is, it floats with no visible means of support? (The length of the wire is not needed to solve the problem.)

18.2 A straight wire is placed in a uniform magnetic field of magnetic flux density 0.023 T at an angle of 30° with the magnetic field. The wire carries a current of 8.6 A. Calculate the force on a 3.4 cm length of the wire. Show the direction of the force in a diagram.

18.3 MAGNETIC FLUX

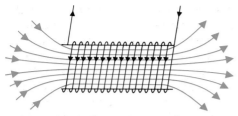

Fig 18.6 The field within a solenoid is uniform apart from near each end.

In the investigation above you were asked to find how the magnetic flux density varied along the axis of a solenoid. At the end of the solenoid the flux density falls. Diagrammatically the field can be drawn as shown in Fig 18.6. The field lines get further apart as the magnetic flux density falls but a term is needed to show that the total number of field lines is constant. The term used is magnetic flux is given the symbol Φ (phi).

 Magnetic flux is the product of the flux density, B, and the area, A, when the flux is at right angles to the area. In equation form

$$\Phi = BA$$

The unit of magnetic flux is the **weber**. 1 weber is the magnetic flux if a field of flux density 1 T exists at right angles to an area of 1 m². Another way of putting this is to say that 1 weber of magnetic flux per square metre is a magnetic flux density of 1 tesla. At first sight there seem to be a lot of terms and units here, but they tend to be used with particular types of problem and do not usually cause much difficulty. Magnetic flux is a term which is used frequently in the next chapter.

These are the terms introduced so far this chapter:

- **Magnetic field** A name for a region in which a force is exerted on a current-carrying conductor. The pattern of the field can be shown by a field diagram.
- **Magnetic flux density** When it is obvious that the magnetic effect is required, this is often simply called the flux density. The flux density is a vector, it is measured in tesla, and it has a numerical value and direction for every point in the magnetic field.
- **Magnetic flux** The flux density is not a flux per unit volume but a flux per unit area. Flux is the product of flux density and area. It is measured in weber. In field diagrams a large value of magnetic flux density is shown by the lines of magnetic flux being drawn close together.

18.4 FORCE ON A MOVING CHARGED PARTICLE IN A MAGNETIC FIELD

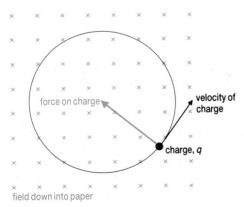

Fig 18.7 A charge moving at right angles to a magnetic field experiences a force at right angles to its direction of motion and will travel along a circular path.

If a charge moves through a magnetic field at right angles to the field, it experiences a force on it at right angles both to its direction of travel and to the direction of the magnetic field. This is because a moving charge is a current and so Fleming's left-hand motor rule can be used to find the direction of the force.

Consider a charge q travelling at a constant speed v at right angles to a magnetic field of flux density B. Assume the particle travels a distance x in time t, so $v = x/t$.

The moving charge constitutes a current q/t.

The force on this current is $BIl = \dfrac{Bqx}{t} = Bqv$.

The direction of the force acting on a positively charged particle moving in a uniform magnetic field is shown in Fig 18.7. Because the force is of constant magnitude and is always at right angles to the velocity, the conditions are met for circular motion. Using Newton's second law gives therefore

$$\text{Force} = \text{mass} \times \text{acceleration}$$

$$Bqv = m \times \frac{v^2}{r}$$

where m is the mass of the charged particle and r is the radius of the circle in which it travels. Reorganising this equation to find the radius of the orbit gives

$$r = \frac{mv}{Bq}$$

showing that the radius is proportional to the velocity of the particle if the charge and mass are constant.

Another feature of the effect of a magnetic field on a moving charged particle is that, since the force acting on the particle is at right angles to the direction of travel, the work done on the particle is always zero. This must be so since work done is defined as the product of the force and the distance moved in the direction of the force. Here the distance moved in the direction of the force must always be zero, so the work done on the particle is zero. This implies that there can be no increase or decrease in the kinetic energy of the particle, and the motion of a charged particle under the action of a magnetic field must always be at constant speed. This is different from the situation in an electric field. The action of an electric field on a charged particle is to change its kinetic energy.

Example 2

An electron travels with a constant velocity through a magnetic field of flux density 0.0076 T and an electric field of electric field strength 56 000 V m^{-1}. Show on a diagram how this is possible and find the speed of the electron. If the electron then emerges from the electric field, what will be the radius of curvature of the path of the electron in the magnetic field?

Since the electron travels with a constant velocity, the resultant force on it must be zero. This can happen if the force which is exerted on it by the magnetic field is equal and opposite to the force exerted on it by the electric field. This is shown in Fig 18.8. The force due to the electric field is in the

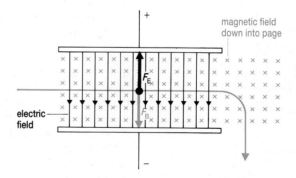

Fig 18.8   A charged particle moving in a straight line through electric and magnetic fields. The force on the charge due to the electric field is equal and opposite to the force due to the magnetic field. When the charged particle leaves the electric field it travels in a circular path in the magnetic field alone.

opposite direction to the electric field because the electron is negatively charged. In order to find the direction required for the magnetic field, Fleming's left-hand rule is used. The current is in the direction from right to left because of the electron's negative charge, and in order to obtain a force opposing that due to the electric field the magnetic field must be directed away from the reader. Since

$$F_E = F_B$$
$$Ee = Bev$$

$$v = \frac{E}{B} = \frac{56\,000 \text{ V m}^{-1}}{0.0076 \text{ T}}$$

$$= 7.4 \times 10^6 \text{ m s}^{-1}$$

$$Bev = m\frac{v^2}{r}$$

$$r = \frac{mv}{Be}$$

$$= \frac{(9.1 \times 10^{-31}\ \text{kg}) \times (7.4 \times 10^{6}\ \text{m s}^{-1})}{(0.0076\ \text{T}) \times (1.60 \times 10^{-19}\ \text{C})}$$

$$= 5.5 \times 10^{-3}\ \text{m}$$

$$= 5.5\ \text{mm}$$

QUESTIONS

18.3 An electron travelling through a cathode ray tube with a velocity of 3.2×10^{7} m s^{-1} enters a magnetic field of flux density 0.47 mT at a right angle. What will be the radius of curvature of the electron's path while in the field?

18.4 In a bubble chamber, high-energy particles leave a track of tiny bubbles of gas as they travel through liquid hydrogen at just above its boiling point (see Fig 18.9). A positive nuclear particle travels at right angles to a magnetic field of flux density 2.0 T in a bubble chamber. The speed of the particle is 1.24×10^{5} m s^{-1} and the radius of curvature of its track is 1.3 mm. Find the ratio of the charge to

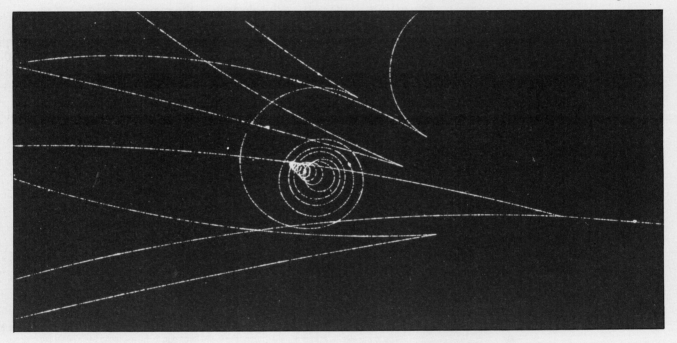

Fig 18.9 Particle tracks in a bubble chamber. The tracks of positive particles curve anticlockwise and negative particles clockwise in the magnetic field.

the mass of the particle. Draw a diagram showing the relative directions of velocity, field and force acting, and suggest a possible particle to cause this track.

18.5 A copper wire of area of cross-section 1.0 mm^2 carries a current of 5.0 A. The wire is placed at right angles to a magnetic field of 0.048 T. Copper contains 7.8×10^{28} conduction electrons per cubic metre. Find:

(a) the drift velocity of the conduction electrons;

(b) the average force exerted by the magnetic field on a single electron as it drifts through the wire;

(c) the number of free electrons per metre of the wire's length;

(d) the force on the wire due to the magnetic field.

Check that the answer which you get for (d) is in agreement with that predicted by the use of the formula $F = BIl$.

18.5 THE HALL EFFECT

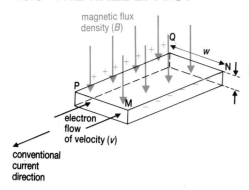

magnetic flux density (B)

electron flow of velocity (v)

conventional current direction

Fig 18.10 The Hall Effect. A potential difference is set up across a slice of semiconductor when it carries a current at right angles to a magnetic field.

Question 18.5 asked you to find the force exerted by a magnetic field on an electron flowing through a wire. Because the drift velocity of electrons flowing through wires is small, the force is correspondingly small. The total force exerted may be large because of the huge number of electrons which flow. The Hall effect is an effect which arises because of the force which a magnetic field exerts on a moving charge. It is simpler if a current of electrons is imagined to be flowing in a wire of rectangular cross-section placed in a magnetic field as shown in Fig 18.10.

An individual electron, travelling with velocity v through the strip, experiences a force F_B given by $F_B = Bqv$. If this force were the only force exerted on the electron then it would accelerate towards the side MN of the strip, but this would have the effect of giving MN an ever-increasing negative potential and PQ an ever-increasing positive potential as the charge continues to flow. In practice, an equilibrium is set up in which the negative potential on side MN and the positive potential on side PQ set up an electric field to exert a force on the moving electrons which is equal and opposite to that exerted by the magnetic field. The electrons will then drift through the strip without being deflected to one side.

The force which the magnetic field B exerts on the electron = Bev.

The force which the electric field E exerts on the electron = $Ee = V_H e/w$.

$$Bev = Ee = \frac{V_H e}{w}$$

$$V_H = Bvw$$

Since also $I = nAve$ where A = area of cross-section = tw, it is possible to combine these two equations to find the value of V_H.

$$V_H = B \frac{I}{nAe} w$$

$$= \frac{BIw}{nwte} = \frac{BI}{nte}$$

$$V_H = \frac{BI}{ten}$$

V_H is called the Hall potential difference and is a potential difference set up across the width of a current-carrying conductor in a magnetic field.

Example 3

A strip of copper 10 mm wide and 2.0 mm thick carries a current of 6.3 A. It is placed so that a magnetic field of flux density 0.083 T passes through it at right angles to its surface. Find the potential difference set up across its width if the copper has 7.8×10^{28} conduction electrons per cubic metre.

The current I is related to the drift velocity v of the electrons by the equation

$$I = nAve$$

where n is the number of conduction electrons per unit volume, A is the area of cross-section and e is the fundamental charge.

$$6.3 \text{ A} = (7.8 \times 10^{28}) \times (0.010 \text{ m} \times 0.0020 \text{ m}) \times v \times (1.6 \times 10^{-19} \text{ C})$$

$$v = \frac{6.3}{7.8 \times 10^{28} \times 0.010 \times 0.0020 \times 1.6 \times 10^{-19}}$$

$$= 2.52 \times 10^{-5} \text{ m s}^{-1}$$

MAGNETIC FIELD

The force on the moving charge is given by:

$$F = Bev - eE = 0$$

$$Bv = E = \frac{V_H}{w}$$

$$(0.083 \text{ T}) \times (2.52 \times 10^{-5} \text{ m s}^{-1}) \times (0.010 \text{ m})$$

$$V_H = Bvw$$

$$= 0.083 \times 2.52 \times 10^{-5} \times 0.010 \text{ V}$$

$$= 2.1 \times 10^{-8} \text{ V}$$

This potential difference is too small to be measured easily but as can be seen from the equation for the Hall potential difference

$$V_H = \frac{BI}{ten}$$

the potential difference obtained for a given current is inversely proportional to n, the number of charge carriers per unit volume. Copper has a huge number of charge carriers per unit volume and so each one of them travels slowly, giving a small Hall potential difference. If however a piece of semiconductor material is used in place of the copper, then the value of n might well be reduced by a factor of 10^9. The current is likely to be less in the poorer conductor, but nevertheless a Hall potential difference of millivolts is readily attainable. The Hall effect is often used in instruments which, after calibration, can be used to measure magnetic flux density. The principle is illustrated by question 18.6.

In practice the instrument, called a Hall probe, needs careful adjustment before the Hall potential difference can be measured. Fig 18.11 shows a suitable circuit. The variable resistance S is adjusted to give a convenient current from the battery and the milliammeter is used to measure the current through the germanium. The potential divider R is adjusted so that when there is zero magnetic field a millivoltmeter connected across XY reads zero. The slice of germanium is then placed at right angles to the field to be measured and the millivoltmeter will then record the Hall potential difference.

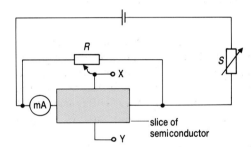

Fig 18.11 An experimental arrangement for measuring a Hall effect potential difference.

QUESTIONS

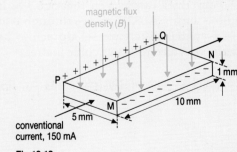

magnetic flux density (B)

conventional current, 150 mA

Fig 18.12

18.6 A piece of germanium has dimensions of 10.0 mm × 5.0 mm × 1.00 mm, and has 4.3×10^{21} charge carriers per m³. A current of 150 mA enters one of its smallest faces and the germanium is placed so that a magnetic field is at right angles to its largest faces as shown in Fig 18.12. It is found that the side at PQ is 57 mV positive with respect to the side at MN. Find
(a) the velocity of the elementary charge carriers through the germanium;
(b) the value of the magnetic flux density;
(c) whether the charge carriers have a positive or a negative sign.

18.7 In an experiment on the Hall effect a current of 0.25 A is passed through a metal strip having thickness 0.20 mm and width 5.0 mm. The Hall voltage is measured to be 0.15 mV when the magnetic field is 0.20 T. What is the number of elementary charge carriers per unit volume and what is the drift speed of these carriers? (The reason for stating *elementary* charge carriers here is to guard against the possibility that the charge carriers can be holes with a + charge or electrons with a − charge in a semiconductor.

MAGNETIC FIELD

18.6 THE TORQUE ON A COIL CARRYING A CURRENT IN A MAGNETIC FIELD

This topic has already been considered arithmetically in example 1 earlier in the chapter but, as it is a topic of practical importance, it is necessary to find an algebraic expression for the torque. It is worth while pointing out that the use of algebra is of particular importance if the same process has to be done many times. A general rule, the algebraic equation, can then be established and used. A problem which is done only once does not have the same need for an algebraic expression. Whichever approach is required, the solving of a problem for the first time should always be done from first principles. Finding an equation into which numbers can be put is *not* solving a problem, it is using a solution someone else has found. Spending time looking for an equation which exactly fits the problem you have to solve is time badly spent. It is far better to establish the basic facts.

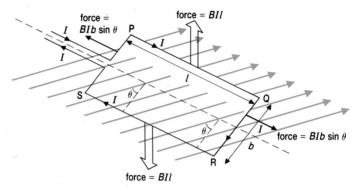

Fig 18.13 A flat coil carrying a current *I*. The plane of the coil is at an angle θ to a magnetic field.

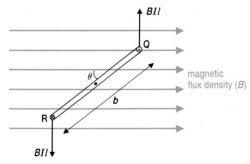

Fig 18.14 A diagram showing the end-on view (elevation) of Fig 18.13.

If a rectangular coil PQRS, carrying current *I* is placed in a uniform magnetic field so that the plane of the coil is at an angle θ to the magnetic field then the coil will have forces acting on it which will tend to cause it to rotate to make $\theta = 90°$. It will tend to rotate so that as much flux as possible passes through it. To see why this is so, consider Fig 18.13 which shows the coil in the field. Fig 18.14 shows the same situation but is a two-dimensional sketch from the right-hand side.

The force acting on PQ is upwards and is of magnitude *BIl*. The force on RS is of the same magnitude but is downwards. There are forces on QR and PS of *BIb* sin θ in magnitude, but since these are in the same straight line and in opposite directions to one another they have no turning effect on the coil. The torque on the coil is provided by the forces on PQ and RS.

$$\text{torque} = \text{force} \times \text{perpendicular distance from axis}$$

$$\text{total torque} = 2 \times BIl \times \frac{b}{2} \cos \theta$$

$$= BIlb \cos \theta$$

If the coil has *N* turns of wire instead of one turn then the torque (*T*) becomes

$$T = BINA \cos \theta$$

where *A* is the area of the coil *lb*.

The total force on the coil is zero. There is only a turning effect. This is a good example of a couple.

If the coil is lying parallel to the magnetic field then θ has the value of zero and the torque has its maximum value.

$$T_{\text{max}} = BINA$$

This equation also holds even if the coil is not rectangular.

Use is made of this principle in an electric motor where the torque acting on a rectangular coil is the torque causing motion. Once the coil has reached the position at which $\theta = 90°$ the current is reversed through the coil. This has the effect of reversing the forces so that a torque is maintained on the rotating coil and this torque can always be in the same direction. More detail will be given on the electric motor in Section 19.7.

The moving coil galvanometer

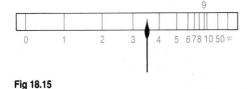

Fig 18.15

A galvanometer is an instrument for measuring small electric currents. It may be modified by the use of additional resistors to make it suitable for measuring larger currents or potential differences or resistances. As with all instruments it needs to be calibrated. The instrument is essentially a rectangular coil of many turns of fine, insulated copper wire which is able to rotate in a magnetic field as a result of the torque which is exerted on it. As shown above, the torque on a rectangular coil is $BINA \cos \theta$. This implies that the torque reduces as the angle of deflection increases. If this were to be used directly in constructing a galvanometer then the scale of the galvanometer would not be a linear scale. This does not imply that the scale cannot be curved on the instrument but that the size of the divisions on the scale are of unequal size. A non-linear scale on an instrument makes it difficult to read. The scale shown in Fig 18.15 is an example of a non-linear scale. The scale is not only non-linear on its major divisions but also non-linear between them. The pointer on this scale is half-way between the 3 and the 4 but the reading is **not** 3.5. The mark for 3.5 must be to the right of the pointer, so this reading is probably 3.4.

It is to avoid these problems that in a galvanometer the angle between the field and the plane of the coil is kept at zero by the use of a radial field. This is shown in Fig 18.16. An iron cylinder, held in position by non-magnetic brass supports, is fixed centrally between semi-circular pole

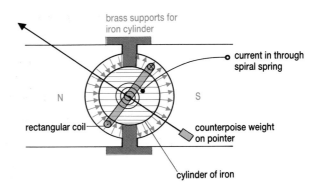

Fig 18.16 A moving coil galvanometer.

pieces of a permanent magnet. The coil has a pointer attached to it and is pivoted on jewelled bearings which are placed so that the coil is also centrally placed between the pole pieces. The coil moves in the annular space between the iron cylinder and the magnet. In order to be able to measure the current which is flowing it is necessary to provide a restraining torque. This is supplied by springs placed above and below the coil. The springs also act as electrical leads to pass the current into and out of the instrument.

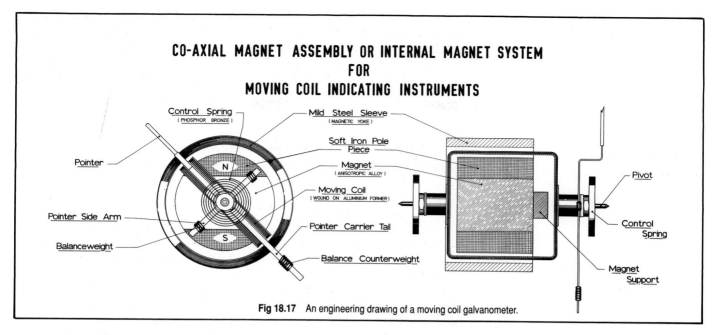

CO-AXIAL MAGNET ASSEMBLY OR INTERNAL MAGNET SYSTEM
FOR
MOVING COIL INDICATING INSTRUMENTS

Fig 18.17 An engineering drawing of a moving coil galvanometer.

An engineering diagram of a moving coil meter is shown in Fig 18.17. You should be able to explain the principles behind the way the instrument functions. Diagrams such as Fig 18.16 are adequate for such explanations. Another type of moving coil galvanometer is the light beam galvanometer, Fig 18.18. The principle of this galvanometer is the same but it is usually more sensitive as a result of using a fine suspension ribbon instead of a spring, and being able to use a larger scale with a light beam pointer. The lens is used to illuminate the fine wire so that the concave mirror on the coil can focus a sharp image of the wire on the scale. The wire is called a cross wire, and its image enables readings to be taken accurately.

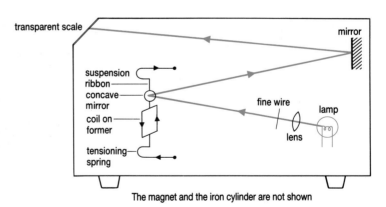

The magnet and the iron cylinder are not shown

Fig 18.18 A mirror galvanometer. The magnetic and iron cylinder are not shown in the diagram.

For both of these instruments the torque provided by the current in the coil equals the restraining torque provided by the suspension when the coil is in equilibrium. The manufacturers of these instruments have to ensure that the restraining torque provided by the suspension is proportional to the angular deflection ϕ. This gives

torque provided by current in coil in radial field $= BINA$
restraining torque supplied by suspension $= k\phi$
where k is a constant called the **torsion constant**.

Since the two torques are equal and opposite when the coil is in equilibrium

$$BINA = k\phi$$

or

$$\frac{\phi}{I} = \frac{BNA}{k}$$

ϕ/I is called the **current sensitivity** of the meter. It is the angular deflection per unit current. For some moving-coil meters it will be about 1 radian per milliamp, whereas for light beam meters it can be a thousand times greater – 1 radian per microamp. The current sensitivity can be increased by

- increasing B, the magnetic flux density
- increasing N, the number of turns of wire on the coil
- increasing A, the area of the coil
- decreasing k, the torsion constant of the suspension.

As is often the case with sensitivity of instrumentation, some of these requirements are mutually contradictory. If the area of the coil is increased it is more difficult to provide as strong a magnetic field; if the number of turns of wire on the coil are increased then the coil becomes heavier and the suspension needs to be stronger to support it. In practice, a compromise has to be reached to achieve sufficient sensitivity while maintaining good accuracy. Good quality engineering of the meter is essential for accuracy. You should now do the data analysis exercise to see what effect the number of turns on the coil has on the current sensitivity and also on another similar term called the voltage sensitivity of the meter.

DATA ANALYSIS

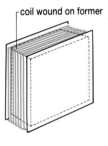

coil wound on former

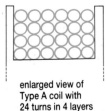

enlarged view of Type A coil with 24 turns in 4 layers of 6 turns

Fig 18.19

Meter sensitivity

A manufacturer of light beam meters of the type shown in Fig 18.18 will supply the meters with different characteristics. The basic meter is the same in every case so that from the outside the different types look the same. The same magnet is used in each type and the moving coil is wound on the same rectangular former which fits in the radial field of the magnet. The difference between the four types A, B, C and D is that different diameter insulated copper wire is used to make the moving coil. The thickness of the insulation on the wire does make an appreciable difference to the number of turns which can be packed on to the former, but for the sake of simplicity in this exercise assume that the insulation is of negligible thickness.

The data given for the Type A galvanometer shows that the former is large enough to have 24 turns of wire of diameter 0.200 mm. The turns are arranged in a rectangular pattern of 4 layers of 6 turns as shown in Fig 18.19. In use, this coil gives a deflection of the coil which is sufficient to cause a deflection of the light spot of 50 mm for 1 mA passing through the coil. This is another way of writing the current sensitivity of the instrument and is quoted as 50 mm mA^{-1}.

The resistivity of copper is 1.7×10^{-8} Ω m. Taking the length of one turn to be, on average, 0.060 m, find the resistance of the Type A coil. Use this value to find how much deflection of the spot occurs for 1 volt being applied across the coil. This is called the voltage sensitivity. You can assume that the deflection is proportional to the current through the coil. To avoid unrealistic numbers, a voltage sensitivity of 100 000 mm V^{-1} can be quoted as 100 mm mV^{-1}.

Complete Table 18.1 to find the voltage and current sensitivities of Type B, Type C and Type D meters. It will save you some time if you look for a pattern to the numbers which you are finding rather than working each one out independently.

Table 18.1

Type	Wire diameter /mm	Approx. s.w.g. size	Number of turns	Resistance /Ω	Current sensitivity /mm mA^{-1}	Voltage sensitivity /mm mV^{-1}
A	0.200	36	24		50	
B	0.100	42				
C	0.050	47				
D	0.025	50				

QUESTIONS

You may well be familiar with this type of instrument. School laboratories are not well known for the accuracy of their meters because the meters are not always handled with sufficient care. You can overcome some of the practical difficulties of using meters by having two ammeters in series or two voltmeters in parallel. If they read the same value it is unlikely that they both have the same systematic error, so it is probable that you do not have an appreciable systematic error. If they do not read the same value then at least you know you have a problem! You can also help to prevent damage to meters by choosing suitable meters and by watching them carefully. If a meter needle moves violently to the maximum value, it is likely that it is being overloaded. An a.c. current passed through a d.c. meter can do a lot of damage to the meter even though the needle does not appear to be responding.

18.8 What would be the effect on the current sensitivity of a meter in each of the following cases, taken separately? Say what would be the effect on your readings if you were using a calibrated instrument which was affected in each of these ways:
  (a) as a result of ageing, and being dropped, the magnetic flux density is reduced;
  (b) as a result of ageing, and being overheated, the torsion constant is reduced
  (c) as a result of too much current being allowed to pass through the coil, the insulation around some of the turns of wire on the coil is burnt away.

18.9 The construction of a moving coil galvanometer gives the following information:
  the area of the coil is 1.2 cm^2 (= 0.000 12 m^2)
  the number of turns on the coil is 20
  the flux density of the radial field is 0.43 T
  the torsional constant of the suspension is 2.8×10^{-6} N m rad^{-1}
  the resistance is 72 Ω
  (a) Calculate the deflection of the coil when the current through the galvanometer is 5.0 mA.
  (b) What potential difference will there be across the galvanometer when the current is 5.0 mA?
  (c) What potential difference is required to cause a deflection of 1 radian?

18.7 THE MAGNETIC FIELD NEAR CURRENT-CARRYING CONDUCTORS

So far in this chapter the measurement of magnetic flux density has been achieved by considering the force which acts on a current in the field. Nothing has been written about the magnitude of the magnetic flux density caused by an electric current or about the factors which affect the magnitude of this flux density.

Fig. 18.20 shows a wire carrying a current I. One small length of the wire, XY, has been shaded. The current in XY creates magnetic field in the space all around it. If we choose a point P at random then we would like to know what field is produced just by this small section. Unfortunately, in practice, we must have some leads connected to carry the current to and from XY, but for the moment we will ignore the field created by the current in the leads. The field at P is found to be

inversely proportional to r^2
proportional to $\sin \theta$
proportional to l, the small length XY
proportional to I
dependent on the material in the space between XY and P

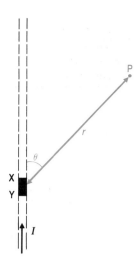

Fig 18.20

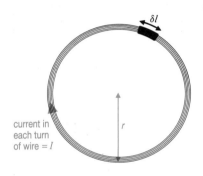

current in each turn of wire = I

Fig 18.21 A coil in which there is a current I in each of N turns of insulated wire.

Combining these statements gives

◆   magnetic flux density at P $\propto \dfrac{Il \sin \theta}{r^2}$

While experiments can be carried out to check this expression they are difficult and frustrating to do. The real test of such expressions lies in their ability to predict magnetic flux densities in many situations over many years.

The expression can be written in the equation form

$$B = \frac{kIl \sin \theta}{r^2}$$

where k is a constant dependent on the material in the space between XY and P.

For a vacuum in this space k, the constant of proportionality is written as $\dfrac{\mu_0}{4\pi}$ where μ_0 is called the permeability of free space. This gives

$$B = \frac{\mu_0 Il \sin \theta}{4\pi r^2}$$

The reason for writing k this way is to rationalise the equations for magnetic field strength in the same way as was done for electric field strength. The formula for the flux density of any magnetic field showing spherical symmetry will contain a 4π term: one showing cylindrical symmetry will contain a 2π term, and any uniform magnetic field will not contain a π term at all.

Put in calculus terms the small flux density δB, caused by the current in the small length of wire δl, is

$$\delta B = \frac{\mu_0 I\delta l \sin \theta}{4\pi r^2}$$

The numerical value of μ_0 is found to be $4\pi \times 10^{-7}$. Why this is so will be explained in section 18.8. Using this equation to find the flux densities near different configurations of wires involves calculus in most cases, but an expression for the flux density at the centre of a circular coil can be found without using calculus, and illustrates the principle of the use of the equation. Expressions for the flux densities for other situations are given in Fig 18.22, without proof, so that they may be calculated when required. Do not try to memorise these equations; they will be given to you if and when they are required.

The flux density at the centre of a flat, circular coil of N turns of radius r and carrying a current I can be found easily because θ is always 90° and r is fixed, and the direction of the field caused is the same for every piece of the wire, Fig 18.21. The total flux density at the centre B is produced from the total length of the wire, which is $2\pi rN$. Since

$$\delta B = \frac{\mu_0 I\delta l \sin \theta}{4\pi r^2}$$

for a piece of the wire of length δl, for the whole length of wire this becomes

$$B = \frac{\mu_0 I \times 2\pi rN \times \sin 90°}{4\pi r^2}$$

$$= \frac{\mu_0 IN}{2r}$$

MAGNETIC FIELD

Wire arrangement	Field pattern	Formula for field
(a) Flat circular coil	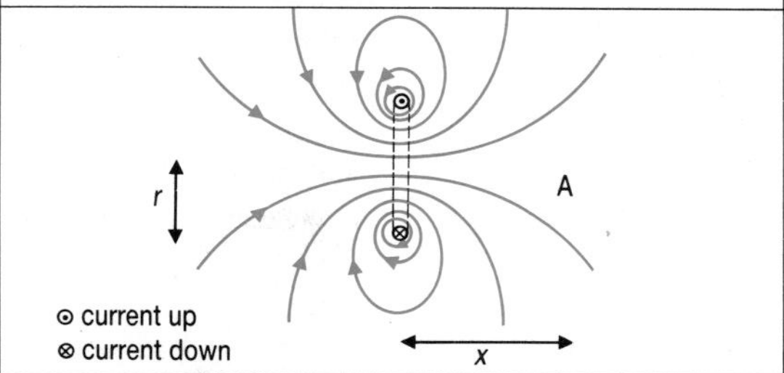 ⊙ current up ⊗ current down	Field at centre $B = \dfrac{\mu_0 IN}{2r}$ Field at A $B = \dfrac{\mu_0 IN}{2} \dfrac{r^2}{(r^2 + x^2)^{3/2}}$
(b) Long straight wire	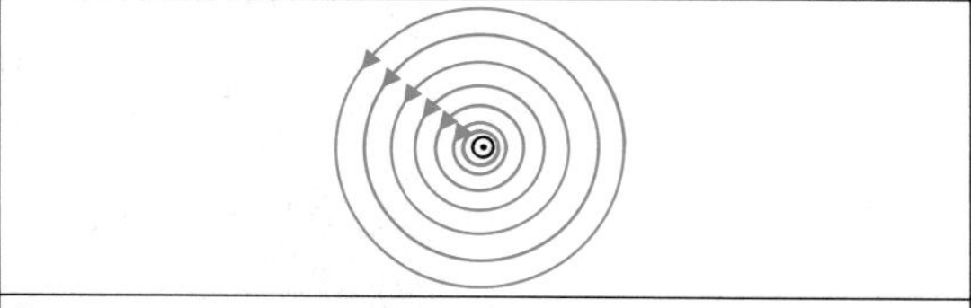	Field close to wire a distance r from it $B = \dfrac{\mu_0 I}{2\pi r}$
(c) Solenoid — a long helical coil	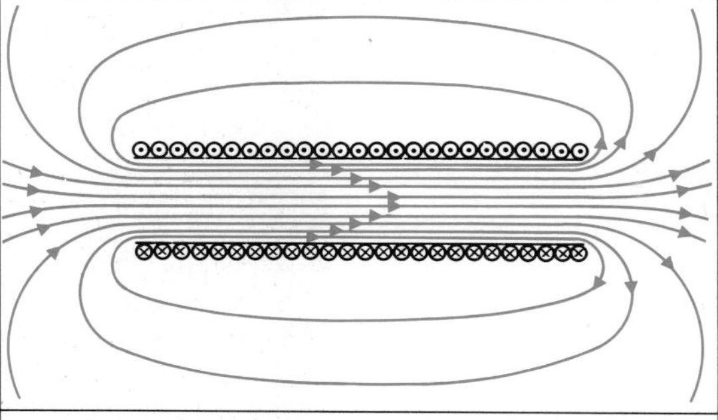	Uniform field within solenoid $B = \mu_0 nI$ n = number of turns per unit length
(d) Helmholz arrangement — radius of coils equals separation of coils (r)	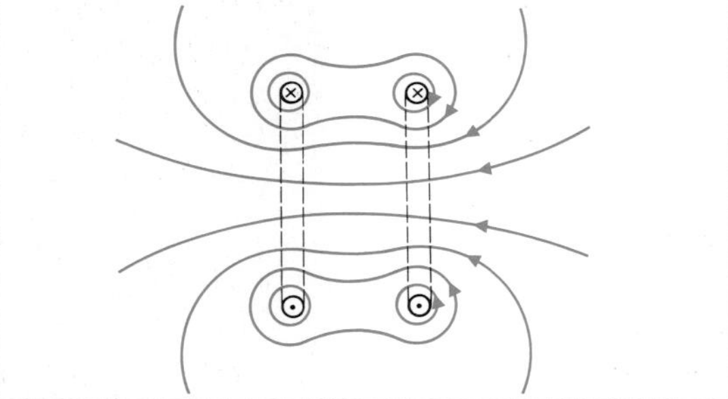	Nearly uniform field at the centre $B = \dfrac{8\mu_0 NI}{5\sqrt{5}r}$

Fig 18.22 The magnetic field patterns and strengths for some commonly used arrangements of current-carrying wires.

In the above expressions it has been assumed that the space around any current-carrying wire is a vacuum. If this is not so then the field will be different from the value stated by a factor μ_r. Note that μ_r is called the **relative permeability** of the material and is a plain number. The product $\mu_r \mu_0$ is called the **permeability** μ of the material. It is possible for the relative permeability to be slightly less than one; it is 0.999 99 for copper, for instance. Often it is slightly greater than one; it is 1.000 002 for oxygen. For a few materials, called ferromagnetic materials, it is very large. Iron has a value of μ_r which varies with temperature but is around a value of 1000, and alloys containing iron, cobalt and nickel can have relative permeabilities up to 10 000. This is why placing an iron bar in the centre of a solenoid increases the magnetic flux density to a high value.

Example 4
The submarine power cable between England and France uses d.c. current. The cables used for the link are some of the longest, continuous lengths of cable in the world. Each cable consists of a central copper conductor, 10 cm in diameter, insulated with oil-impregnated paper and protected by sheaths of lead and polyethylene plastic. The outside is 6 mm thick steel

When the cross channel power link was first established it was the intention to have a two way exchange of energy. Peak demand is not at the same time in both countries because of the time difference and differing national habits. In practice it has not worked out this way. France generates a great deal of electrical energy from its nuclear power stations and always has a surplus so it imports very little energy from Britain. Britain has saved itself the need to build an extra power station by taking electrical energy from France on a large scale at a competitive price.

armouring and the whole cable has a mass of 1700 tonnes. Eight of these cables were buried in the sea bed at a depth of about 1.5 metres by an embedding machine which was able to work at a rate of around 100 metres per hour. Each cable can carry a current of 1850 A and the system is designed to operate at 270 kV. Fig 18.23 shows the project under construction.

Fig 18.23 One of the cables which now link England and France laid out on the beach at Bakers Gap.

(a) What power is transmitted by each cable?

(b) What power is transmitted by the whole system of 8 cables?

(c) At a place where the cable is 60 m below the surface of the sea what magnetic field is caused by a cable when working at its design power?

(d) Bearing in mind that the Earth's magnetic field in the Straits of Dover is 1.9×10^{-5} T, how would you arrange the cables in the sea bed to avoid affecting ships' compasses?

Power transmitted by each cable = 1850 A × 270 000 V = 500 MW.
   2 cables are required for each circuit
therefore total power transmitted = 2000 MW.
   Magnetic flux density at a distance r from a long straight wire = $\dfrac{\mu_0 I}{2\pi r}$

$$B \quad = \quad \frac{4\pi \times 10^{-7} \times 1850}{2\pi \times 60} \quad = \quad \frac{2 \times 10^{-7} \times 1850}{60}$$

$$= \quad 6.2 \times 10^{-6} \text{ T}$$

Magnetic flux density caused by a cable = 6.2×10^{-6} T.

MAGNETIC FIELD

This is about a third of the Earth's natural magnetic flux density and since it is not in the same direction as the Earth's field it will significantly alter the direction in which a compass needle will point. At maximum, for this depth of water, it would alter the compass direction by an angle whose tangent is 6.2/19. This is an angle of 18°. In shallower water the deflection could be greater. In order to avoid causing such a large field, the cables are laid in their trenches in pairs. Since the current in one wire of each pair is in the opposite direction to the current in the other wire in the same trench, the magnetic fields cancel one another out.

QUESTIONS

18.10 What is the magnetic flux density 20 cm from the centre of one of the cables linking England and France and carrying a current of 1850 A as in example 4?

18.11 A solenoid gives a very uniform field within it. One particular solenoid is constructed out of a single layer of turns of wire of 1.0 mm diameter with a current of 10 A through it. What is the magnetic flux density inside the solenoid?
If a stronger field is required then more layers of turns can be added to the solenoid. What problem is likely to arise as more layers are added?

18.8 FORCES BETWEEN CURRENT-CARRYING CONDUCTORS

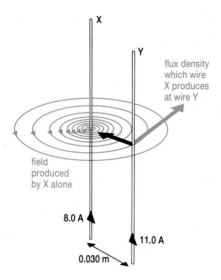

field produced by X alone

flux density which wire X produces at wire Y

8.0 A

11.0 A

0.030 m

Fig 18.24

If two current-carrying conductors are placed near to one another, then each of the conductors is in the magnetic field which the current in the other creates. This will cause a force to be exerted on each of the conductors and, by Newton's third law, the forces will be equal and opposite. The simplest situation to consider is when the two conductors are parallel to one another and carry their currents in the same direction. The example 5 shows how this can be done.

Example 5
Find the force per unit length acting on a long straight wire carrying a current of 11.0 A when placed 3 cm away from a similar wire carrying a current of 8.0 A. See Fig 18.24.

The flux density which wire X produces at wire Y is given by

$$B = \frac{\mu_0 I_X}{2\pi r} = \frac{\mu_0 \times 8\,A}{2\pi \times 0.030} = \frac{4\pi \times 10^{-7} \times 8\,A}{2\pi \times 0.030}$$
$$= 5.3 \times 10^{-5}\,\text{T}$$

The force which X exerts on Y is given by

$$F = BI_Y l = 5.3 \times 10^{-5}\,\text{T} \times 11\,\text{A} \times l$$

The force per unit length on Y = 5.9×10^{-4} N m^{-1}.
The force per unit length on X is of the same magnitude but in the opposite direction:

$$\frac{F}{l} = \frac{\mu_0 \times 11.0\,\text{A}}{2\pi \times 0.030\,\text{m}} \times 8.0\,\text{A}$$
$$= \frac{4\pi \times 10^{-7} \times 11.0 \times 8.0}{2\pi \times 0.030} = 5.9 \times 10^{-4}\,\text{N m}^{-1}$$

The wires are pulling each other together.
Example 5 uses the numerical value of μ_0 quoted earlier but so far nothing has been written about how μ_0 is obtained or about the unit in which it is measured. The definition of the ampere involves a very similar arrangement to that given in the example. In Table 1.2 the ampere was defined as

the constant current which, if maintained in two straight parallel conductors of infinite length placed 1 metre apart in vacuum, would produce between these conductors a force of 2×10^{-7} newton per metre of length. If we use this definition then we can calculate the value of μ_0.

In general the flux density, B, created by wire X at wire Y is given by

$$B = \frac{\mu_0 I_X}{2\pi r}$$

$$\text{Force on Y} = BI_Y l$$

$$= \frac{\mu_0 I_X}{2\pi r} \times I_Y \times l$$

$$\text{Force per unit length on Y} = \frac{\mu_0 I_X I_Y}{2\pi r}$$

Substituting the numerical values from the definition of the ampere gives

$$\frac{2 \times 10^{-7}\,\text{N}}{\text{m}} = \frac{\mu_0 \times 1\,\text{A} \times 1\,\text{A}}{2\pi \times 1\,\text{m}}$$

$$\mu_0 = 2 \times 10^{-7} \times 2\pi \quad \frac{\text{N}}{\text{A}^2\,\text{m}}$$

$$= 4\pi \times 10^{-7}\,\text{N A}^{-2}$$

This value of μ_0 comes directly from the arbitrary definition of the ampere. The way the ampere is defined is, in effect, fixing the value of μ_0 at the exact value of $4\pi \times 10^{-7}\,\text{N A}^{-2}$.

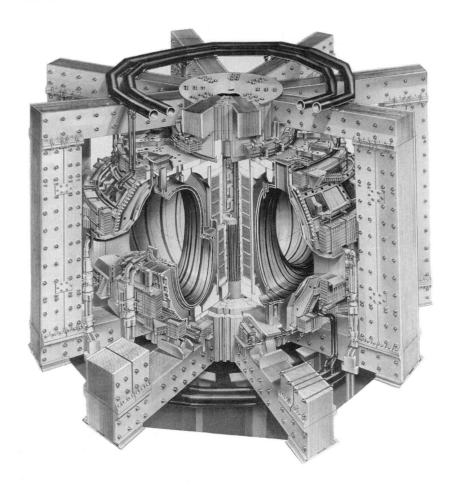

Fig 18.25   The Joint European Torus at Culham in which very strong magnetic fields are used to contain extremely high temperature plasma in the doughnut shaped ring, the torus.

MAGNETIC FIELD

The force pulling two parallel wires, carrying currents in the same direction, towards one another is usually quite a small force, but it can be important when a current is passing through an ionised gas or plasma. Under these circumstances there is an effect called the pinch effect which squeezes the current into a very small area of cross-section. This effect is used in a tokamak nuclear fusion reactor, Fig 18.25, to achieve temperatures in excess of 50 000 000 K so that fusion of lighter atoms into heavier atoms may take place.

QUESTIONS

18.12 Two long parallel wires separated by a distance of 1.0 cm carry opposite currents of 20 A. Find the magnetic flux density at their mid-point and at a point 1.0 cm from one of the wires and 2.0 cm from the other.

18.13 Find the force per unit length between two adjacent turns in a transformer if they are separated by a distance of 0.000 50 m and carry a current of 0.46 A. You may assume that the wires behave as long straight wires because their separation is much smaller than their length. Use your answer to explain why a transformer using mains alternating current can often be heard to buzz. What is the frequency of the buzz?

18.9 PRACTICAL APPLICATIONS

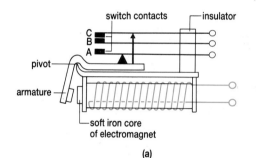

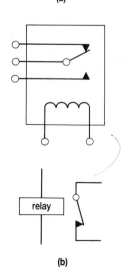

Fig 18.26 A relay.

There are many domestic and engineering applications of electromagnets and the motor effect. These range from the tokamak, mentioned above, and high energy particle accelerators which are used to probe into the structure of the nucleus, right down to simple electromagnets for lifting scrap iron. Electric motors of a bewildering variety are manufactured by the million every year.

At one time the entire telephone system relied on electromagnetic switches, for making connections. Now much of this linking is done using electronic switching systems but electromagnetic switches are still used in many items of domestic equipment such as central heating boilers and washing machines.

The relay

Many electronic control systems, even very complex ones, at the final stage in the system make use of a relay to switch on any mains-operated equipment. A relay is an electromagnetic switch which has the advantage that it isolates the equipment being switched on from the control system itself. It also has the advantage that it can operate with a small current on the control side and a large current on the mains side. When the controlling current is passing through the electromagnet, the soft iron armature is attracted to the electromagnet and as it rocks on the pivot the lever attached to it pushes the switches open or closed. It is possible to purchase many different types of relays working on different sizes of control current and with an assortment of switches, some normally off, some normally on and some changeover. Fig 18.26 shows a relay with two switches. One of them will switch on, and the other will switch off, when a current passes through the electromagnet.

The moving-coil loudspeaker

In Section 18.6 the value of the torque on a coil in a magnetic field was determined. It was then stated that the resultant force on the coil was zero although the torque was not zero. Sometimes it is desirable to arrange that a coil carrying a current in a magnetic field has a resultant force on it but zero torque. This is achieved in loudspeakers by carefully arranging the shape of the magnetic field. Fig 18.27 shows a cylindrical S pole of a permanent magnet surrounded by N poles. The magnetic field is therefore a radial field which is directed inwards towards the centre. If you use the left-hand motor rule for any part of the wire carrying the current, you should find that the force on the wire is down into the page. The entire coil

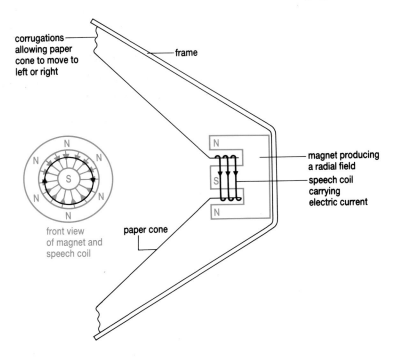

corrugations allowing paper cone to move to left or right

frame

N

magnet producing a radial field

speech coil carrying electric current

front view of magnet and speech coil

paper cone

Fig 18.27 A moving coil loudspeaker.

therefore will move into the body of the magnet and will pull the paper cone with it. If the direction of the current is now reversed, then the coil will move in the opposite direction and will push the paper cone outwards. If a.c. is passed through the coil then the coil will oscillate. By varying the frequency, amplitude and pattern of the alternating current, the whole vast range of responses of the cone can be achieved to produce any desired sound.

SUMMARY

- Magnetic flux density is the force acting on unit current in unit length of wire at right angles to the field:

$$B = \frac{F}{Il}$$

- One tesla is the magnetic flux density if a force of 1 newton acts on a current of 1 ampere in a wire of length 1 metre.
- If θ is the angle between a magnetic field and a wire carrying a current then the force on the wire is given by

$$F = BIl \sin \theta$$

- Magnetic flux = magnetic flux density × area: $\Phi = BA$

- The weber (Wb) is the flux if a flux density of 1 tesla exists over an area of 1 m^2.

- Magnetic flux density near long straight wire $B = \dfrac{\mu_0 I}{2\pi r}$

- Magnetic flux density at the centre of a flat circular coil $B = \dfrac{\mu_0 NI}{2r}$

- Magnetic flux density in a long solenoid $B = \mu_0 nI$.

- The force between two long parallel wires $F = \dfrac{\mu_0 I_X I_Y l}{2\pi r}$

- The permeability of free space $\mu_0 = 4\pi \times 10^{-7}$ N A^{-2}. This figure is exact and comes directly from the arbitrary definition of the ampere. The relative permeability is the factor by which the magnetic flux density is changed from the value it would have in a vacuum, by the presence of some material.

- The torque T on a rectangular coil whose plane is at an angle θ to a magnetic field of flux density B is given by

$$T = BINA \cos \theta$$

EXAMINATION QUESTIONS

18.14 Define *magnetic flux density*. (2)
With the aid of a labelled diagram, state and explain the rule determining the direction of the force on a current-carrying conductor situated in and at right angles to a magnetic field. (4)

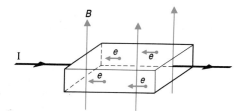

In the diagram electrons are shown moving from right to left in a semiconducting block. A magnetic field B is applied as indicated and a concentration of electrons occurs at the front face of the block.
Explain why this happens. What would be the effect if the current were due to positive charges moving from left to right? (3)
Describe an experiment using a Hall probe to compare the magnetic flux densities at two points in a magnetic field. (6)
The Hall voltage V_H is given by the expression

$$V_H = \frac{BI}{nqt}$$

where B is magnetic flux density, I is current, n the number of charge carriers per unit volume, q the carrier charge, and t the probe thickness. Calculate the magnitude of the Hall voltage for a copper sample of thickness 1 mm in a field of 1.0 T. Take the current to be 1.0 A, n to be 1.0×10^{29} m^{-3}, and the carrier charge to be 1.6×10^{-19} C. Comment on the order of magnitude of your answer. (3)
Explain (i) why semiconductors rather than metals are often used as Hall probes,
(ii) how the sign of the charge carrier can be determined. (2)

(UCLES 1987)

18.15 Describe, by means of diagrams and mathematical equations, the forces due to
(i) a gravitational field,
(ii) an electric field,
(iii) a magnetic field
which would act on a stationary electron.
Show how each of your answers would change, if at all, when the electron moves at right angles to each field. (9)

(UCLES 1988)

18.16 **(a)** A particle, of mass m, carrying positive charge q and travelling at speed v, enters a region of uniform magnetic field of flux density B directed at right angles to the motion of the particle.
(i) Write an expression for the force acting on the particle, and draw a diagram which shows the direction of the force relative to the direction of the field as it enters the field.
(ii) Explain carefully why the particle follows a circular path, and derive an expression for the radius of the path. (5)

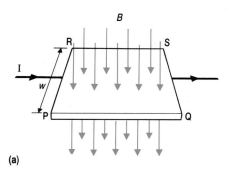

(a)

(b) Diagram (a) above shows a rectangular slice of a semiconducting material which is connected to a battery so that a current I flows from left to right as shown. The current is due to electrons of charge $-e$ moving with speed v. A uniform magnetic field of flux density B acts vertically downwards through the plane of the slice.
(i) Explain the production of a Hall voltage between the edges PQ and RS of the slice.
(ii) Write down an equation for the force on an electron in the slice due to the Hall voltage V_H, and show that $V_H = Bvw$, where w is the width of the slice.
(iii) Describe and explain the effect on the value of the Hall voltage of first increasing the current in the slice and then replacing the semiconductor slice by one made of copper of the same dimensions and carrying the same current as originally. (10)

(c) The Hall effect can be used to measure the rate of flow of an ionised liquid through a pipe. Diagram (b) shows the principle of the method. An electromagnet produces a magnetic field vertically downwards perpendicular to the direction of flow, and

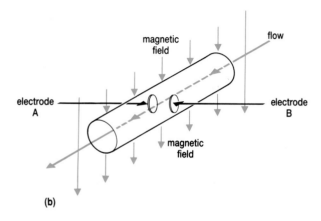

(b)

the Hall voltage is measured between electrodes A and B placed in each side of the pipe as shown.

(i) Explain why a Hall voltage is produced in this case.

(ii) In one experiment an ionised liquid is flowing at a rate of 9.0×10^{-6} m^3 s^{-1} through a plastic pipe of diameter 7.5 mm. A magnetic field of flux density 8.0×10^{-3} T acts at right angles to the flow. Calculate the average speed of flow of the liquid, and determine the potential difference which is developed across the electrodes, assuming them to be 7.5 mm apart.   (6)

(AEB 1986)

18.17 (a) Explain qualitatively the origin of the Hall effect in a single carrier semiconductor.   (3) Show how the effect can be used in a Hall probe to measure magnetic field. Assume a probe consists of width a and thickness b carrying a current I. The magnetic field is perpendicular to the direction of current as shown below.

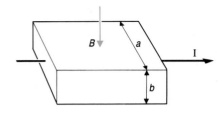

Obtain an expression for the Hall voltage in terms of the current I, the magnetic field B, the dimensions of the sample, the carrier concentration n in the sample and the charge e of the carriers.   (4) The probe consists of a germanium slab with b equal to 0.20 mm and containing 1.0×10^{23} conducting electrons per cubic metre. Calculate the Hall voltage when B is 80 mT and I is 100 mA.   (2)

(b) A square coil of side 2.0 cm and of 100 turns is pivoted between the poles of a magnet as shown. The magnetic field is uniform and of magnitude 80 mT. A current of 1.5 A passes through the coil

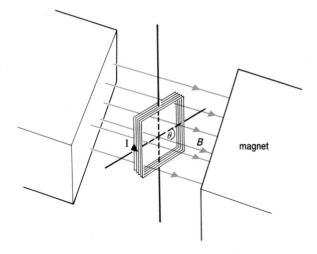

which is pivoted about an axis of rotation perpendicular to the magnetic field. Obtain an algebraic expression for the couple exerted on the coil when its plane is at an angle θ to the magnetic field. State the angle θ at which the maximum couple occurs and calculate the maximum couple.   (5) The arrangement of the suspension of the coil is modified to produce a d.c. motor. Explain

(i) the arrangement necessary to obtain continuous rotation from a d.c. supply;   (2)

(ii) why energy losses occur such that all the electrical power is not converted to mechanical power;   (2)

(iii) what limits the angular speed of the motor if it is not driving any load and the bearings are frictionless.   (2)

(O & C 1986)

18.18 Diagram (a) shows a rectangular coil PQRS of N turns, with sides of length a and b and with sides QR and SP vertical. A current I flows through the coil in the direction shown. Diagram (b) shows a view from above of the coil in a uniform horizontal magnetic flux density B, with the plane of the coil at an angle θ to the direction of the flux density.

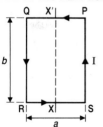

(a)

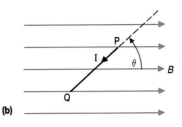

(b)

(a) State the magnitude and direction of the forces acting on each of the sides PQ, QR, RS and SP.   (4)

(b) (i) Derive an expression for the moment of the couple due to the magnetic flux density acting on the coil in the position shown in diagram (b).

(ii) If the coil is free to rotate about the vertical axis XX' indicate, with the aid of a diagram, the positions of the coil as viewed from above when the moment of the couple is a minimum.   (4)

(JMB 1987)

18.19 (a) The force F experienced by a wire of length l carrying a current I at right angles to a magnetic field of flux density B is given by the expression $F = BIl$. Use this expression to derive the torque T acting on a rectangular coil of n turns of mean cross-section area A carrying a current I and placed with its plane at an angle q with a similar field, as shown.   (4)

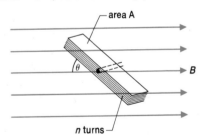

(b) With the aid of a clear diagram, describe the construction of a moving coil milliammeter. (6)
Explain clearly how the following features of the meter are obtained:
(i) sensitivity;   (2)
(ii) linearity;   (2)
(iii) damping;   (2)

(c) Describe a potentiometer method for checking the accuracy of a milliammeter whose full-scale reading is 1.0 mA d.c. You may assume that a standard cell of known e.m.f. and a selection of standard resistors are available.   (7)

(d) Describe:
(i) how a 1.0 mA d.c. meter of resistance 1000 Ω could be converted to give full-scale deflections of 10 V d.c.   (3)
(ii) a method for converting a d.c. voltmeter to record a.c. readings.   (4)

(OXFORD 1987)

18.20 (a) Draw a diagram showing two parallel wires in which the currents are flowing in *opposite* directions. Show on the diagram
(i) the direction of the *resultant* B-field at a point midway between the wires (ignoring the Earth's magnetic field),
(ii) the direction of the electromagnetic force on each wire.

(b) The ampere is defined as 'the constant current which, flowing in two infinitely long, straight parallel conductors of negligible circular cross-section, placed in a vacuum 1 m apart, produces between them a force per unit length of 2×10^{-7} N m^{-1}. Using this definition and the equation given on page 336 for the force per unit length between two long parallel conductors, deduce a value for μ_0 and for the unit in which it is measured.   (6)

(ULSEB 1988)

18.21 (a) A rectangular coil PQRS has N turns of area A. It carries constant current I and is mounted in a uniform magnetic field of strength B whose direction is parallel to the sides PQ and RS.
Sketch a diagram of the arrangement and show on your diagram the directions of the current I, the magnetic field B and the forces, if any, on each of the four sides. Show that the coil is acted upon by a couple $NAIB$.
What design features of a moving coil galvanometer allow the magnitude of this couple to be constant while the coil deflection increases?   (8)

(b) A moving coil galvanometer of resistance 200 Ω has full-scale deflection for a current of 100 μA. Explain with diagrams and relevant calculations how this galvanometer can be converted to an ammeter giving full-scale deflection at 5.0 A.   (4)

(ULSEB 1986)

Chapter 19

ELECTROMAGNETIC INDUCTION

LEARNING OBJECTIVES

At the end of this chapter you should be able to:

1. state the conditions necessary for electromagnetic induction;

2. deduce the value of an induced electromotive force;

3. state Faraday's law and Lenz's law;

4. define and use the term inductance and state its unit;

5. describe how the current and potential difference vary as a current is established in an inductor;

6. describe the action of a generator;

7. understand the importance of the induced electromotive force in the operation of a motor.

19.1 INDUCTION PHENOMENA

Until 1831 electricity was not much more than a novelty. Various scientists had discovered some effects of electric currents and there were suggestions that perhaps electrical effects were of fundamental significance but a major restraint on further development was that sources of electric current were batteries, and batteries were at the time very inefficient. Even today, batteries are totally unsuitable for large power supplies. A link between electric current and magnetic field had been established by Oersted in 1819. He had discovered that a magnetic field was produced whenever there was an electric current. In 1822 Ampère showed that there was a force acting between two current carrying conductors because of their magnetic field. The 1831 publications by Henry in the United States and Faraday in England announced that an electric current could, in certain circumstances, be induced from a magnetic field. This reverse process is called electromagnetic induction and is of enormous practical importance. Virtually all electrical current is now produced using electromagnetic induction rather than from batteries, and even where batteries are used for large-scale electric current production the batteries are recharged using the effect of electromagnetic induction. It is not difficult nowadays to demonstrate electromagnetic induction because sensitive galvanometers are available to detect even the small current produced in a single wire when a bar magnet is moved near it. You are recommended to perform the demonstration illustrated in Fig 19.1. The aim is to use the magnet to make the electrons flow in the wire. The light-beam galvanometer needs to be on its most sensitive setting. If a brushing action is used to sweep the electrons along the wire, the demonstration is very unconvincing. There is only the occasional jerky movement of the spot on the scale. If however the magnet is moved sideways across the wire, then there is a surge of current in one direction when the magnet passes over the wire. The magnet makes the

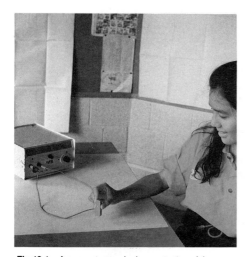

Fig 19.1 A suggested basic demonstration of the use of a sensitive galvanometer to detect electromagnetic induction.

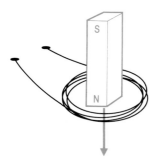

Fig 19.2

electrons move in a direction at right angles to its own direction of movement. If the magnet is moved sideways in the opposite direction, then the current will be induced in the opposite direction. Reversing the polarity of the magnet also causes the current direction to be reversed.

If the wire is now shaped into a circle, then the magnet can move past all of it sideways if it is pushed through the circle. This can be extended, as shown in Fig 19.2, by using a coil of many turns in place of a single straight wire. The sensitivity on the galvanometer may have to be reduced for this demonstration. The essential requirements for a current to be produced are that there is a complete circuit and that the wire cuts magnetic flux. Here it has been suggested that the wires are kept stationary and that the magnet moves, but it is equally possible for the magnet to be stationary and for the wires to be moved. Relative motion is needed. By using larger, stronger magnets, and by using more turns of wire, it is possible to generate as much electrical power as is required. The large generators used by power stations have fixed coils and the magnets are moved past them. The output of electrical power is huge but the principle is exactly the same as the demonstration with the light-beam galvanometer.

19.2 INDUCED ELECTROMOTIVE FORCE

Before analysing electromagnetic induction and contrasting it with the motor effect, it is worth while comparing a similar pair of mechanical processes, lifting and dropping. In lifting, work is done on the object and it gains potential energy: in dropping, potential energy is lost and kinetic energy is gained. However, consider an object either (a) being lifted on the end of a rope with a constant velocity or (b) being let down with a constant velocity. The force diagrams, Figs 19.3(a) and 19.3(b), must be the same in both cases because, since there is no acceleration, the resultant force on the object must be zero.

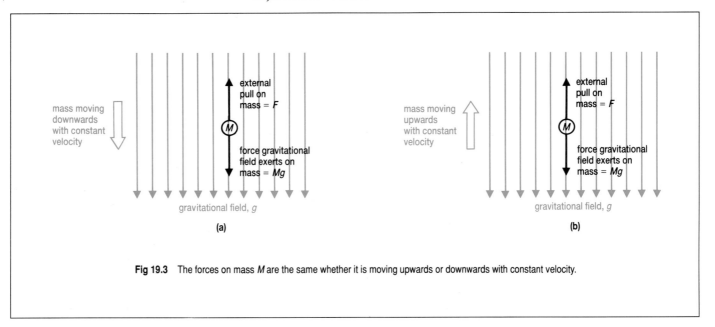

Fig 19.3 The forces on mass M are the same whether it is moving upwards or downwards with constant velocity.

The pull on the mass is provided by some external agent. There may be a rope attached to it with a crane lifting it up or letting it down. Here the important point is that if its velocity is constant, whether it is upwards or downwards movement, the pull of the rope F is equal and opposite to Mg, the weight of the mass. (If you do not believe this is possible – try it. Use something attached to the end of a piece of string and then lift it or lower it with a constant velocity. While getting it moving requires that the forces are different, keeping it moving with a constant velocity simply requires that the upward force equals the downward force.)

Since the two forces are equal and opposite we can state the following :

Downward motion of mass		Upward motion of mass	
constant velocity	$= v$	constant velocity	$= v$
time considered	$= t$	time considered	$= t$
distance travelled	$= x$	distance travelled	$= x$
external force	$= F$	external force	$= F$
force due to field	$= Mg$	force due to field	$= Mg$
work done by field		work done by	
force on mass	$= Mgx$	external agent $= Fx$	$= Mgx$
energy gained by		energy gained by	
external agent $= Fx$	$= Mgx$	mass	$= Mgx$

Or, going up the external agent does work Mgx which results in the mass gaining potential energy of Mgx, and going down the gravitational force on the mass does work Mgx which results in a gain of energy Mgx by the external agent.

Now consider the corresponding situation when the field is a magnetic field and the mass is replaced by a wire carrying an electric current, I, see Figs 19.4(a) and 19.4(b). Apart from the forces acting at right angles to the field, the situation is very similar to the one analysed above.

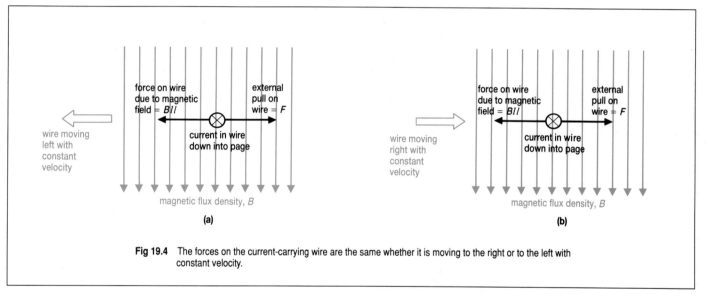

Fig 19.4   The forces on the current-carrying wire are the same whether it is moving to the right or to the left with constant velocity.

Again, at constant velocity the two forces in each situation must be equal and opposite. This time the corresponding facts are:

Motion of wire to the left		Motion of wire to the right	
constant velocity	$= v$	constant velocity	$= v$
time considered	$= t$	time considered	$= t$
distance travelled	$= x$	distance travelled	$= x$
external force	$= F$	external force	$= F$
force due to field	$= BIl$	force due to field	$= BIl$
work done by field		work done by	
force on wire	$= BIlx$	external agent $= Fx$	$= BIlx$
energy supplied to		energy supplied to	
external agent $= Fx$	$= BIlx$	current in wire	$= BIlx$

Motion to the left is a motor effect. The wire pulls the external agent and hence supplies energy to it as it moves along with constant velocity. Motion to the right is a dynamo effect. Here the external agent pulls the wire through the field and does work on it. As a result, energy is supplied to the current in the wire in the same way as when a mass is lifted energy is supplied to it.

The energy supplied to the current in the wire was shown to be equal to $BIlx$, but the definition of electromotive force E, as the energy changed from other forms of energy to electrical energy per unit charge Q, gives the energy change to electrical energy $= EQ = EIt$.

Here the energy changed to electrical energy $= BIlx$

so we get $EIt = BIlx$, or

$$E = \frac{BIx}{t}$$

The electromotive force generated by electromagnetic induction when a wire cuts magnetic flux is an **induced electromotive force**.

If Fig 19.4(b) is redrawn to show the wire moving from its initial position, a distance x to the right, it appears in plan view as in Fig 19.5. Connecting wires to the piece of wire considered are not shown. Here it can be seen that lx is the area swept out by the wire, so that Blx is the flux Φ which the wire cuts. This gives the induced electromotive force, E as

$$E = \frac{\Phi}{t}$$

This quantitative way of finding the induced electromotive force supplied by electromagnetic induction as the flux cut per unit time also shows that an equivalent unit to the weber is the volt second. The direction in which the induced current flows can be obtained from Fleming's right-hand rule. The left-hand rule applies to Fig 19.4(a) and so a mirror image is necessary for Fig 19.4(b) as the velocity is in the opposite direction. If you use both hands together on the two diagrams in Fig 19.4 (and a bit of contortion), you can see how the motion is given by the thumb in both cases.

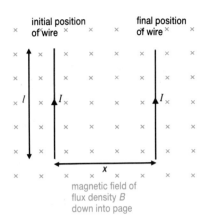

initial position of wire   final position of wire

x

magnetic field of flux density B down into page

Fig 19.5

19.3 THE LAWS OF ELECTROMAGNETIC INDUCTION

Faraday's law of electromagnetic induction is a formal statement of the deduction made in the previous section.

◆ The induced electromotive force across a conductor is equal to the rate at which magnetic flux is cut by the conductor.

(In Faraday's original statement of the law, when the units used were not SI units, the equality was not established and the statement was that the induced e.m.f. is proportional to the rate at which flux is cut.)

The direction in which the e.m.f. is set up is important, as has been shown already. Any current which flows as a result of the e.m.f. must be in such a direction that it opposes the motion. This can be seen by looking at Fig 19.4(b). The current induced in the wire is down into the page. As a result of this current, the wire has a force acting on it because it is in the magnetic field. This force opposes the pull on the wire from the external agent. If the force assisted the external agent, then the wire would be accelerating and it would be supplying electrical energy and gaining kinetic energy. This would be breaking the law of conservation of energy. If the force assisted motion, it would become possible to use a bicycle dynamo to run a motor on the bicycle, which would eliminate the need for pedalling! This impossibility was formally stated in **Lenz's law** which is an electrical statement of the law of conservation of energy. It states:

◆ The direction of any induced current is such as to oppose the flux change which causes it.

Put another way, this says that any electrical energy which can be supplied by a dynamo has to be obtained by doing at least an equivalent amount of mechanical work on the dynamo. To indicate this opposition of the direction of the induced current, the induced e.m.f. which causes it is given a negative sign in the equation which summarises both Faraday's and Lenz's law. The equation is

$$E = -\frac{\Phi}{t}$$

or more accurately as a rate of change of flux using calculus notation

$$E = -\frac{d\Phi}{dt}$$

Example 1

A coil of 1000 turns and resistance 23 Ω enclosing an area of 4.0 cm² is rotated from a position where its plane is perpendicular to the Earth's magnetic field to one where it is parallel to the field, in 0.0050 s.

(a) What is the average induced e.m.f. if the magnetic flux density of the Earth's magnetic field at the point is 6.0×10^{-5} T?

(b) What is the average current flowing in the coil if its ends are short-circuited?

(c) What charge will flow past any point in the coil in the time it is moving?

flux through one turn of the coil	= flux density × area
	$= 6.0 \times 10^{-5}$ T $\times 0.0004$ m² $= 2.4 \times 10^{-8}$ Wb
flux through 1000 turns	$= 2.4 \times 10^{-5}$ Wb
average flux cut per second	$= 2.4 \times 10^{-5}$ Wb$/0.0050$ s
	$= 4.8 \times 10^{-3}$ V
	= average induced e.m.f.
average current caused	= e.m.f/resistance
	$= 4.8 \times 10^{-3}$ V$/ 23$ Ω $= 2.1 \times 10^{-4}$ A
charge = current × time	$= 2.1 \times 10^{-4}$ A $\times 0.0050$ s $= 1.05 \times 10^{-6}$ C
	$= 1.0$ μC to two significant figures

This example, with its average currents and e.m.fs, seems to be rather impracticable. In fact, it is possible to show that although the currents and e.m.fs are averages, the charge is not. If the coil is connected up to a charge-measuring instrument, the charge can be measured and the charge is independent of the time taken to turn the coil.

Example 1 is the basis of one way of measuring magnetic flux density. A small coil with a large number of turns, N, of very fine wire, called a search coil, is connected to a meter which will measure charge, and placed so that it is at right angles to the magnetic field to be measured. The coil is then suddenly removed from the field and the induced e.m.f. causes a charge Q to flow through the meter. If the area of one turn of the coil is A, then the total flux cut is BNA. This gives

$$\text{induced e.m.f.} = BNA/t = \text{current} \times \text{total resistance}$$

or
$$BNA = t \times I \times R$$

Since $Q = It$ we get
$$BNA = QR$$

so
$$B = \frac{QR}{NA}$$

19.1 In order to measure the movement of sea-water, an oceanographer immerses two electrodes into the water at a distance apart of 500 m. When she measures the e.m.f. between the two electrodes, she finds that it is 8.3×10^{-3} V. If the vertical component of the Earth's magnetic flux density is 3.7×10^{-5} T, find the velocity of the water in a direction at right angles to the line joining the electrodes.

(**Note:** The conductor moving in this case is the sea-water. The problem can easily be solved if the area of sea-water which passes the line joining the electrodes per unit time is found, as the flux through this area is the flux cut per unit time. Note also that different values are often given for the flux density of the Earth's magnetic field. This is because it varies from place to place and also because sometimes its horizontal component is needed and sometimes, as here, its vertical component is needed.)

19.2 An aeroplane has a distance between its wingtips of 36 m and it is travelling horizontally with a velocity of 280 m s^{-1}. If the vertical component of the Earth's magnetic flux density is 3.7×10^{-5} T find the induced e.m.f. between the plane's wingtips. Why would a voltmeter in the plane connected between the wingtips always read zero?

DATA ANALYSIS

Electromagnetic flow metering

If a conducting liquid flows through a magnetic field, the conditions exist for an e.m.f. to be set up across the liquid. This principle is used in electromagnetic flow meters to measure the rate of flow of liquid along a pipe. A diagram of this type of flowmeter is given in Fig 19.6. As the liquid flows through the tube it cuts the magnetic field set up by the field winding coils, causing an e.m.f. E_1 to be induced. This e.m.f. is sensed by two electrodes X and Y which are opposite each other and in contact with the liquid at right angle to the axis of the magnetic field.

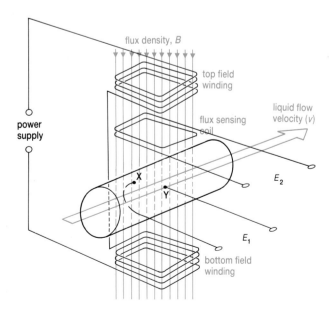

Fig 19.6 The principle of electromagnetic flow metering can be explained using this diagram.

ELECTROMAGNETIC INDUCTION

In addition, there is a flux sensing coil which gives an output E_2, proportional to the magnetic flux density.

(a) Why does the pipe, in the vicinity of the electrodes, have to be non-conducting?

(b) Show that $E_1 = kvB$
and that the velocity $v = k' \dfrac{E_1}{E_2}$
where k and k' are constants.

(c) Show that for a liquid travelling along a pipe of internal diameter 65 mm with velocity 0.73 m s^{-1}, when $B = 0.35$ T, the value of E_1 is 16.6 mV.

(d) Only a very small conductivity is necessary for the liquid used in such a flowmeter. Only one of the following liquids is unsuitable. Flow of all of the others has been measured with one of these meters. Make a reasoned guess as to which of these liquids does not conduct sufficiently.

beer	molasses
blood	nitric acid
drinking water	sewage
fruit juice	soap slurry
hydrocarbons	sugar syrup
ice cream mixture	

(e) What advantage is there in using platinum electrodes?

(f) Why does E_2 become zero if the power supply is d.c.?

(g) How does using a.c. overcome this problem? What complication occurs because a.c. is used?

(h) The accuracy of these instruments when calibrated is typically 0.5% at full flow, i.e. to 1 cm s^{-1} in 200 cm s^{-1}. Why is the accuracy less for low flow rates?

(i) What liquid speed would you expect a flowmeter to read on a 450 mm bore pipe if liquid passes through the pipe at a rate of 200 m^3 h^{-1} (cubic metres per hour). How can this type of calculation enable a flowmeter to be calibrated? What assumption are you making?

(j) Fig 19.7 shows a graph of usable cable length from electrodes to measuring instrument plotted against conductivity of the liquid in use. What type of scales are used on the graph and why are they necessary?

(k) What is the maximum cable length which can be used with a liquid of conductivity 4 μS cm^{-1}?

(l) Why does the usable cable length depend on liquid conductivity?

(m) What effect, if any, could the following have on the system and its calibration?
   (i) air getting into the pipe. Why is it preferable to mount the meter in a vertical rather than a horizontal pipe?
   (ii) magnetic particles in the liquid
   (iii) bends in the pipe near the instrument
   (iv) rise in temperature
   (v) solids transported in the moving liquid.

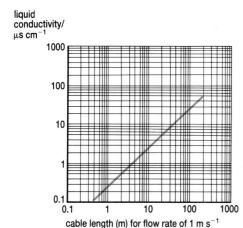

liquid conductivity/ μS cm^{-1}

cable length (m) for flow rate of 1 m s^{-1}

Fig 19.7

19.4 INDUCTANCE

One method which is frequently used to produce an induced e.m.f. uses two coils arranged so that the field produced by one of the coils passes through the other. If this field then changes, the conditions are such that an induced e.m.f. is set up in the second coil. This arrangement is shown in Fig 19.8. A current I_p is passing through the primary coil and this sets up magnetic flux through itself and through the secondary coil. If the current changes then the magnetic flux density through the secondary coil also

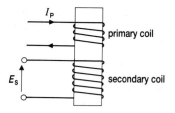

Fig 19.8 A pair of coils with mutual inductance.

changes and an induced e.m.f. is produced across the secondary coil. It is possible, by knowing the dimensions of the coils and the material on which they are wound, to calculate the value of the induced e.m.f. for a known rate of change of current in the primary coil. In practice, however, it is often more useful to be able to relate the induced e.m.f. to the rate of change of current without considering the actual structure of the two coils.

Clearly, a coil manufacturer needs to know how the output is affected by the structure of the coils, but once manufacture has taken place there needs to be a quantity, which can be quoted for the pair of coils, which links the rate of change of current input to the value of the induced e.m.f. output. This quantity is called the mutual inductance.

Mutual inductance is defined by the equation

$$\text{Mutual inductance} = \frac{\text{induced e.m.f. across secondary}}{\text{rate of change of current in primary}}$$

The unit of mutual inductance is the **henry**. One henry (H) is the mutual inductance if an e.m.f. of 1 volt is produced across the secondary when the rate of change of current in the primary is 1 A s^{-1}. A mutual inductance of a few millihenries between two coils is common unless they have an iron core, in which case the mutual inductance is increased hundreds of times.

The defining equation above did not state in what direction the e.m.f. is induced. Lenz's law can be applied to the generation of this e.m.f. as it can to any other induced e.m.f. The e.m.f. is therefore produced in such a direction so that any current which it produces will be in such a direction as to oppose the flux change which causes it. This opposition is again taken into account by inserting a minus sign into the mathematical equation, hence

$$M = -\frac{E_s}{\left(\dfrac{\mathrm{d}I}{\mathrm{d}t}\right)_P} \quad \text{or} \quad E_s = -M\left(\frac{\mathrm{d}I}{\mathrm{d}t}\right)_P$$

where M is the mutual inductance, E_s is the induced e.m.f. across the secondary and $\left(\dfrac{\mathrm{d}I}{\mathrm{d}t}\right)_P$ is the rate of change of current in the primary coil.

Example 2

A pair of coils have a mutual inductance of 0.0073 H. Find the output e.m.f. across the secondary coil if a current of 2.8 A in the primary coil is reduced to zero at a uniform rate in 0.0018 s.

$$\text{The rate of change of current} = \frac{2.8 \text{ A}}{0.0018 \text{ s}} = 1560 \text{ A s}^{-1}$$

Note that high rates of change of current are possible even though the current itself may not have a high value. This is likely to be the case as currents are switched off or switched on.

$$\text{Mutual inductance} = \frac{\text{induced e.m.f. across secondary}}{\text{rate of change of current in primary}}$$

$$0.0073 \text{ H} = \frac{\text{induced e.m.f. across secondary}}{1560 \text{ A s}^{-1}}$$

Induced e.m.f. across secondary $= 0.0073 \text{ H} \times 1560 \text{ A s}^{-1} = 11.4 \text{ V}$

In the above discussion about mutual inductance, emphasis was laid on the fact that when the flux cutting the secondary coil changed, then an e.m.f. was set up across the secondary coil. The flux however also changes through the primary coil and so there must be an e.m.f. set up across the

ELECTROMAGNETIC INDUCTION

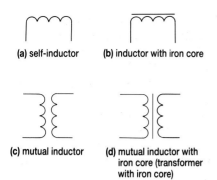

(a) self-inductor   (b) inductor with iron core

(c) mutual inductor   (d) mutual inductor with iron core (transformer with iron core)

Fig 19.9 Circuit symbols for inductors.

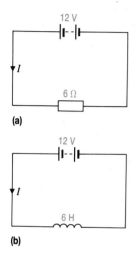

(a)

(b)

Fig 19.10 (a) A resistive circuit (b) an inductive circuit.

primary coil. Any coil with a changing current through it must have an induced e.m.f. set up across itself in a direction which opposes the change of flux causing it. Using similar terminology to that used for mutual inductance, the **self-inductance** of a coil can be defined by

$$\text{Self-inductance} = \frac{\text{induced e.m.f. across coil}}{\text{rate of change of current in coil}}$$

giving in symbols $\quad L = -\dfrac{E_s}{\left(\dfrac{dI}{dt}\right)} \quad$ or $\quad E = -L\dfrac{dI}{dt}$

where L is the self-inductance. The henry is also the unit of self-inductance.

The symbols used for mutual and self inductors are given in Fig 19.9.

To see how to use the term, and to see how it is necessary to be careful with the signs, work through example 3 which contrasts an e.m.f. applied to a resistor with the same e.m.f. applied to a self-inductance.

Example 3

Fig 19.10(a) is a circuit diagram showing a $6\,\Omega$ resistor connected to a battery of negligible internal resistance and an e.m.f. of 12 V. Fig 19.10(b) is the circuit in which the resistor has been replaced by an inductor of self inductance 6 H and of negligible resistance. Apply Kirchhoff's second law to the circuit and hence plot graphs to show how the current in each circuit varies with time, see Figs 19.11(a) and (b).

E.m.f. of supply battery	= 12 V	E.m.f. of supply battery	= 12 V
Potential difference across resistor	= 12 V	Potential difference across coil	= 12 V
Current	= 2 A		

This is a potential difference across the coil as 12 J of energy supplied with each coulomb is being changed into another form of energy, magnetic energy so

$$E = L\frac{dI}{dt}$$

$$12 = 6 \times \frac{dI}{dt}$$

$$\frac{dI}{dt} = 2\ \text{A s}^{-1}$$

QUESTIONS

19.3 In example 3, after the current has been produced in the coil, the circuit is suddenly switched off and the coil is shorted out. What will happen? Suddenly stopping a d.c. current in this way is rather idealised, but the circuit has important practical application in induction coils and some heart pacemakers. Induction coils are used to provide, at the correct instant, a high voltage pulse to a sparking plug to ignite the fuel in a petrol engine. In a pacemaker a sudden drop in the current through a coil outside the body causes a high voltage pulse in another coil, implanted in the body and wired to the heart muscle.

19.4 A solenoid of inductance 2.3×10^{-2} H and negligible resistance is suddenly connected across a 12 V battery. What is the rate of increase of the current in the solenoid?

ELECTROMAGNETIC INDUCTION   **351**

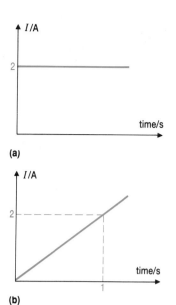

(a)

(b)

Fig 19.11 **(a)** The current–time graph for the resistive circuit of Fig 19.10(a).
(b) The current–time graph for the inductive circuit of Fig 19.10(b). The area beneath a current–time graph gives the charge.

Fig 19.12 A selection of coils made by Salford Electrical Instruments. The smallest coil is less than a centimetre diameter, the largest about 10 cm diameter.

Example 3 can also be used to find the **energy stored in the magnetic field** of a coil. In that example a constant 12 V induced e.m.f. was maintained while the current rose at a fixed rate of 2 A s^{-1}. Consider the current to have a value I. The energy supplied to the magnetic field is given by the e.m.f. multiplied by the charge which has flowed, therefore

$$\text{energy supplied} = E \times Q = E \times \frac{It}{2}$$

The factor of 2 must be included because the current is not constant and the charge is the area beneath the current–time graph (Fig 19.11(b)).

Since the induced e.m.f. is the inductance multiplied by the rate of change of current, we have for a constant current growth

$$\text{energy supplied} = L \; \frac{I}{t} \times \frac{It}{2}$$

$$\text{energy stored} = \tfrac{1}{2} L I^2$$

ELECTROMAGNETIC INDUCTION

This energy is energy stored in the magnetic field of the inductor. A fixed current flowing through an inductor does not increase the energy stored by the magnetic field, but if the current increases then there is an increase in the energy stored. This equation has been determined for a specific case of a current increasing linearly with time. It can be shown that the equation is true however the current increases.

19.5 GROWTH AND DECAY OF CURRENT INDUCTIVE CIRCUITS

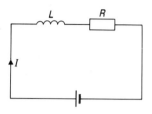

Fig 19.13

Practical circuits are more complex than the idealised circuits examined in example 3. At room temperature it is impossible to create a circuit which has inductance but not resistance, and a normal resistive circuit must have some inductance even if it is usually ignored. Any circuit which has leads which start from and return to a battery must behave as a coil having at least one turn. A practical inductive circuit will therefore have both inductance L and resistance R in series as is shown in Fig 19.13. The resistance may be the resistance of the coil itself. Applying Kirchhoff's second law to this circuit

$$E = L\,\frac{dI}{dt} + IR$$

After a long time, conditions will become steady and so dI/dt will become zero. The current will then have its maximum value I_0 given by

$$I_0 = \frac{E}{R}$$

This gives

$$I_0\,R = L\,\frac{dI}{dt} + IR$$

or

$$(I_0 - I)\,R = L\,\frac{dI}{dt}$$

This is a similar differential equation to the one given in Chapter 17 for charging capacitors through a resistance. (section 17.4). The solution to this equation is

$$I = I_0\,(1 - e^{-(R/L)t})$$

The pattern of current growth in this circuit is similar to the pattern of charge growth in a capacitance–resistance circuit. The **time constant** for this circuit is L/R, and is the time it would take the full current to be established in the circuit if the initial rate of increase were maintained. Growth and decay of current in inductive circuits are shown graphically in Fig 19.14. The equations for these graphs are:

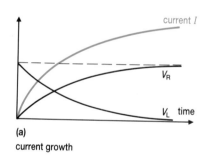

(a)
current growth

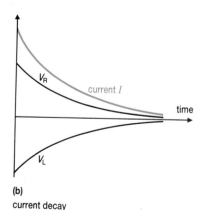

(b)
current decay

Fig 19.14 Current and potential difference growths and decays in inductive-resistive series circuits.

growth of a current in an inductive circuit	**decay of a current in an inductive circuit**
$I = I_0\,(1 - e^{-(R/L)t})$	$I = I_0\,e^{-(R/L)t}$
$V_R = V_0\,(1 - e^{-(R/L)t})$	$V_R = V_0\,e^{-(R/L)t}$
$V_L = V_0\,e^{-(R/L)t}$	$V_L = -V_0\,e^{-(R/L)t}$

This work should be compared with the corresponding section for charging capacitors, section 17.4.

Example 4

An inductor has an inductance of 0.030 H and a resistance of 2.3 Ω. If a potential difference of 6.0 V is applied across, it how long will it take before the current is **(a)** 1 A **(b)** 99% of maximum current?

The current growth equation is

$$I = I_0 (1 - e^{-(R/L)t})$$

$$I_0 = \text{maximum current} = \frac{6.0 \text{ V}}{2.3 \text{ Ω}} = 2.61 \text{ A}$$

$$1 \text{ A} = 2.61 \text{ A} (1 - e^{-(2.3\,t/0.030)t})$$

$$1 = 2.61 - 2.61 \, e^{-76.7\,t}$$

$$2.61 \, e^{-76.7\,t} = 1.61$$

$$e^{-76.7\,t} = \frac{1.61}{2.61} = 0.617$$

Now take natural logarithms:

$$-76.7t = -0.483$$

$$t = \frac{0.483}{76.7} = 0.0063 \text{ s} = 6.3 \text{ ms}$$

For 99% of maximum current:

$$\frac{99}{100} \times 2.61 = 2.61 \, (1 - e^{-(2.3\,t/0.030)t})$$

$$e^{-76.7\,t} = 1 - \frac{99}{100} = \frac{1}{100}$$

Taking natural logarithms gives

$$-76.7t = -4.61$$

$$t = \frac{4.61}{76.7}$$

$$= 0.060 \text{ s} = 60 \text{ ms}$$

Be careful with the arithmetic of these problems. There are many different approaches possible but all of them necessitate care. Realising that negative indices imply reciprocals is one important point. e^{-2} means $1/e^2$.

The natural logarithm of e^{-2} is −2 because the natural logarithm of a number *is* the power to which e has to be raised to give the number.

Example 5

A coil of inductance 40 mH and resistance 200 Ω carries a current of 5.0 A. If it is suddenly shorted out how long will it take for the current to fall to 0.010 A?

Shorting the coil means that a low resistance wire is connected across its ends. The current I through the coil drops away according to the formula

$$I = I_0 \, e^{-(R/L)t}$$

where $I_0 = 5 \text{ A}$

this gives $0.010 = 5e^{-(200/0.040)t}$

or $0.0020 = e^{-(5000)t}$

taking natural logs $-6.21 = -5000t$

so $t = \frac{6.21}{5000} = 1.24 \times 10^{-3} \text{s}$

$$= 1.24 \text{ ms}$$

ELECTROMAGNETIC INDUCTION

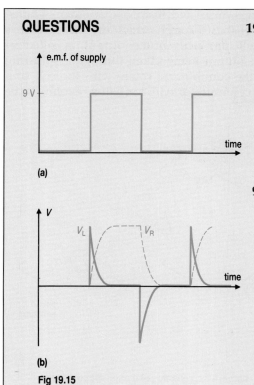

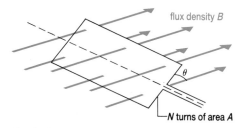

(a)

(b)

Fig 19.15

19.5 A current of 25 A passes through a coil of inductance 0.68 H and resistance 0.50 Ω. Find:

(a) the power being used by the resistance of the coil;

(b) the power being used by the inductance of the coil;

(c) the magnetic energy stored by the inductance of the coil;

(d) the electrical energy stored by the resistance of the coil.

The coil is then short-circuited. What will be the current when the time is 1.36 s? This time is called the time constant of the circuit and is L/R.

How long will it take for the current to fall to 0.0025 A?

9.6 Fig 19.15 shows how V_L and V_R vary when a square wave of 9 V maximum and zero minimum is applied to an inductor and resistor in series. If $R = 10.0$ Ω and $L = 0.48$ H, find the following if the frequency of the square wave is 2 Hz:

(a) the period of the square wave;

(b) the time for a positive pulse of the supply;

(c) the time constant of the circuit;

(d) the time taken for V_R to reach 99% of 9 V during the current growth period;

(e) V_L, 50 ms after the supply voltage has been switched either from +9 V to 0 V or from 0 V to +9 V;

(f) the value of the rate of change of current at both of the instants mentioned in **(e)**.

19.6 GENERATORS

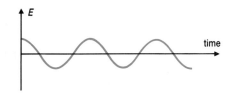

Fig 19.16

If a flat coil of N turns of area A is placed in a magnetic field of flux density B, then the flux through each turn is given by $BA \sin \theta$ where θ is the angle between the plane of the coil and the direction of the magnetic field, see Fig 19.16. The total flux cut by all the turns of the coil is therefore given by

$$\theta = BAN \sin \theta$$

If now the coil is rotated with an angular velocity ω, the angle $\theta = \omega t$ and the flux cut by the coil will vary continually and so induce an e.m.f. across the ends of the coil. The induced e.m.f. can be found by differentiating

$$E = \frac{d}{dt} = \frac{d(BAN \sin \theta)}{dt}$$

$$= BAN \frac{d(\sin \theta)}{dt}$$

$$= BAN \frac{d(\sin \omega t)}{dt}$$

$$E = BAN\omega \cos \omega t$$

It can be seen from this equation that the output e.m.f. is directly proportional to the flux density, the area of the coil, the number of turns on the coil and the angular velocity. It will also vary sinusoidally as shown in Fig 19.17 reaching a maximum value of

$$E_0 = BAN\omega$$

when $\theta = 0$. That is, the maximum output occurs when the plane of the coil is parallel to the magnetic field. This is where flux is cut fastest.

This theory forms the basis of all generators. Generators vary enormously in their construction depending on the requirements of their user and on their manufacturer. The output from a generator can be taken through slip rings or it may be taken through a split ring commutator. Fig 19.18(a) shows a generator using slip rings. The leads from the ends of

Fig 19.17

the coil are connected to two rings, insulated from one another. Against these rings are pressed carbon brushes through which the a.c. output is taken. A generator like this, which produces an a.c. output, is called an alternator. Fig 19.18(b) shows the output being taken through a split-ring commutator. The two halves of the commutator rotate with the coil so a carbon brush pressed against the commutator will, for half a revolution, be

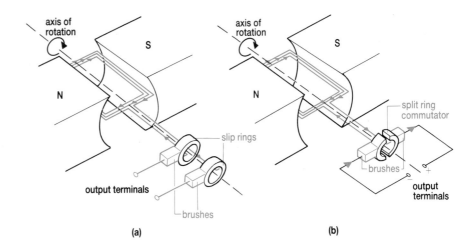

(a)   (b)

Fig 19.18   (a) An a.c. generator using slip rings (b) a d.c. generator using a split ring commutator.

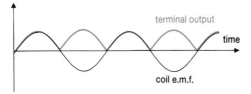

Fig 19.19

pressed against one half of the commutator, and for the other half revolution be connected to the other half of the commutator. In this way, one of the carbon brushes will always be positive and the other carbon brush will be negative. The output is shown in Fig 19.19. It is not a smooth output but it is d.c. in the sense that the current only ever flows in one direction. The output can be smoothed with smoothing capacitors.

Fig 19.20   Generators under construction by G.E.C.

ELECTROMAGNETIC INDUCTION

Another way of producing direct current is to put the a.c. output of the alternator through a rectifier. This is standard now for supplying the d.c. power required for a car although old cars used the system shown in Fig 19.18(a) for their dynamos. For the largest a.c. generators in power stations, the coils remain fixed and the magnet, which is an electromagnet, rotates. This does away with the necessity for slip rings or commutators, except for the connection to the electromagnet under construction.

A photograph of 600 MW generators is shown in Fig 19.20. The arrangement of the coils in such a generator is shown in Fig 19.21(a). There are normally three separate a.c. outputs, called phases, from such a generator. These come from three coils, each of which is divided into two parts. The two parts are opposite to one another on either side of the stator, and the rotor rotates between them. The outputs from the three coils are shown in Fig 19.21(b) and are out of phase with one another by a third of a revolution. The advantage of this system lies in the smoothness of running of the generator and in economy of operation.

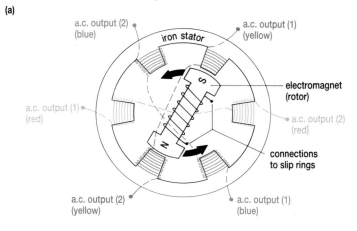

(a)

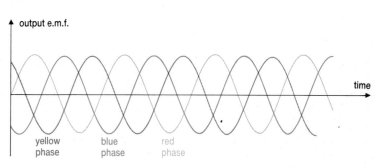

(b)

Fig 19.21 (a) A generator in which a single rotating magnet generates an a.c. output in three separate fixed coils. This is the system used for large generators at power stations (b) shows the output e.m.f. from each of the three coils. These are called three phases. They are a third of a cycle out of phase with one another.

19.7 THE MOTOR

The variety in the size, type and construction of electric motors is even greater than the variety of construction of generators, and there is no intention here to cover more than the basic principles of the motor. One construction for a simple d.c. motor is shown in Fig 19.18(b). When that particular diagram was discussed earlier it was described as a d.c. generator, but if the generator, instead of being turned to produce an electric current, has a d.c. current passed through it, then it behaves as a motor. A more practical motor is a motor on a washing machine or vacuum cleaner. These motors, however, need to work on a.c. mains and their rotation is caused by clever manipulation of the magnetic field within the motor. They are called induction motors. A simpler d.c. motor is the starter motor on a car. It uses brushes to feed current into a series of coils. By using

many coils there is no possibility that the coils will not rotate in the required direction immediately the current is switched on.

A photograph of a car starter motor is shown in Fig 19.22 and a diagram, showing the construction of the coils and commutator is given in Fig 19.23. This diagram shows a commutator of 8 segments, 2 for each of the 4 coils. In practice, there are more coils than are shown and there is inter-connection between them so that each coil has current through it for a large fraction of every cycle, rather than just when that particular coil is touching the pair of brushes.

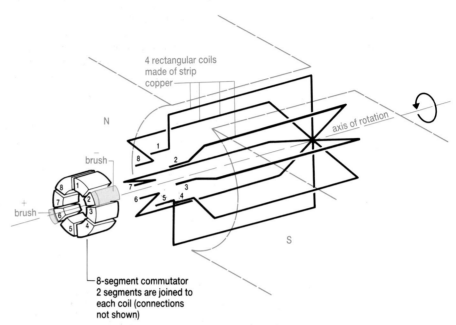

Fig 19.23 A car starter motor.

The coils themselves are fixed on to a soft iron cylindrical core which rotates and comprises the armature of the motor. There is also an electro-magnet rather than a permanent magnet. One of the coils for this is shown in the photograph. It is a rectangular coil and is wrapped around the armature coils. It does not rotate.

It may seem odd to delay discussion of the electric motor until the end of a chapter on electromagnetic induction rather than putting the explanation where it seems to fit, namely in a chapter dealing with the motor effect. The reason for doing this is that when a motor is in operation it is a current-carrying coil rotating in a magnetic field. It therefore generates an induced e.m.f. which affects the current through it. The interaction of the input e.m.f. and the induced e.m.f. are fundamental to the operation of any motor. The torque exerted on a current-carrying coil in a magnetic field was dealt with in section 18.6. When a motor is switched on, a current passes through the armature and produces a torque on it which causes rotation. At the same time an induced e.m.f. is produced which, according to Lenz's law, must be in a direction such as to oppose motion. Also opposing the motion will be the system on which the motor is doing work. The dynamic balance between these various torques controls the speed of the motor. In many practical situations an electric motor does not need a gearbox to control its speed. The speed is controlled automatically by the motor itself and its response to the system to which it is connected.

When an electric drill is switched on, it soon reaches a constant rate of rotation even if it is running freely. It is the induced e.m.f. set up in the

ELECTROMAGNETIC INDUCTION

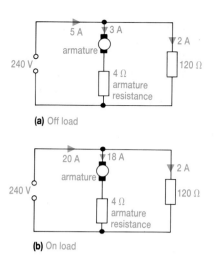

5 A   3 A
armature   2 A
240 V   120 Ω
4 Ω
armature
resistance

(a) Off load

20 A   18 A
armature   2 A
240 V   120 Ω
4 Ω
armature
resistance

(b) On load

Fig 19.24

armature which limits its maximum rate of rotation. If the drill is then made to do some work, so that there is a mechanical torque applied to resist its rotation, the motor will slow down until it reaches a new lower rate of rotation at which, again, the reduced induced e.m.f. is such as to keep the rate of rotation constant. The following example illustrates this balance and should also provide an insight into the meaning of potential difference and electromotive force as applied to energy conversion.

Example 6

A shunt-wound motor is a motor in which the magnetic field is supplied by an electromagnet connected in parallel with the motor's armature. One such motor has an armature whose resistance is 4.0 Ω and field windings whose resistance is 120 Ω. It runs off load at a rate of rotation of 2800 revolutions per minute, at which speed it takes a current of 5.0 A from 240 V mains. Find the speed that the motor runs at when the supply current is 20 A, and find how much useful power is then being supplied by the motor.

The circuit diagram for the motor is shown in Fig 19.24. The resistance of the armature is within the armature, but on the circuit diagram it is shown outside and in series with the armature. This is to emphasise that the resistance which the armature has causes a quite different effect from the effect caused by it being a coil rotating in a magnetic field. The diagram drawn this way also ought to prevent you from considering the 4 Ω and the 120 Ω as being in parallel with one another. They are not in parallel because, as a result of the induced e.m.f. they do not have the same potential difference across them.

Fig 19.25 Electric motors are used here to tow people up the ski slopes of Aviemore in Scotland. The versatility of an electric motor stems from its ability to adjust its speed as the load varies without needing a gearbox.

Off load: Fig 19.24(a).

Potential difference across 120 Ω field resistance = 240 V

Current through field resistance = 240 V/120 Ω = 2.0 A

Since supply current is 5 A, the armature current = 5.0 A – 2.0 A = 3.0 A

Potential difference across 4 Ω armature resistance = 3.0 A × 4.0 Ω = 12 V

Induced e.m.f. = 240 V – 12 V = 228 V

This last line needs more comment. Across the armature is an e.m.f. of 240 V. That is, 240 J are supplied with every coulomb of electric charge which passes through the armature. We have established that of these 240 J, 12 J are wasted as heat in the resistance of the armature. This leaves a further 228 J able to do work (against friction and air resistance in this off-load case). If we use our definition of potential difference as the energy per unit charge converted from electrical energy into other forms of energy, then there is a potential difference of 12 V across the resistor as 12 joules per coulomb are converted from electrical energy into heat energy, and a potential difference of 228 V across the armature as 228 joules per coulomb are converted from electrical energy into mechanical energy. This is the induced e.m.f. It is directly proportional to the rate of rotation of the armature.

Power wasted as heat in field windings = 2 A × 240 V = 480 W

Power wasted as heat in armature = 3 A × 12 V = 36 W

Mechanical power output doing work against friction = 3 A × 228 V = 684 W

Total power input = 5 A × 240 V = 1200 W

The term **back e.m.f.** is often used instead of induced e.m.f. when dealing with motors, but this is a confusing term for two reasons. First, back e.m.f. seems to imply opposition, something not wanted, yet in a motor it is the induced e.m.f. which is responsible for the mechanical work output of the motor. Secondly a back e.m.f. is a potential difference. This is indeed the case here. Energy is being converted from electrical energy into other forms of energy so perhaps the quantity should be placed on the other side of any energy equation and given a positive sign. One thing is certain: you will need to be particularly careful with signs in these problems so you are recommended to check at the end that energy supplied equals energy used. A term used in the United States for induced e.m.f. is motional e.m.f. It has the advantage that it clearly indicates how the e.m.f. is produced, and it would be recognised if you wish to use it.

On load: Fig 19.24(b)

The problem can be repeated when on load. The figures have been inserted on to the circuit diagram. Circuit diagrams are intended to be used: they are useful working diagrams and an extra diagram should be sketched out when changes are made.

Since 18 A is now the current through the armature, the induced e.m.f. must be 168 V. The speed drops in the same proportion so

$$\text{speed} = \frac{168}{228} \times 2800 = 2063 \text{ revolutions per minute}$$

The new values of the powers become

power wasted as heat in field windings = 2 A × 240 V = 480 W

power wasted as heat in armature = 18 A × 72 V = 1296 W

mechanical power output doing work = 18 A × 168 V = 3024 W

Total power input = 20 A × 240 V = 4800 W

This large motor has a mechanical power output of 3024 W. The work done against friction when off load was 684 W. It is unlikely that this has dropped very much at the lower speed, so the useful power output when on load will be about 2400 W. The answer is not reliable to any more than 2 significant figures.

ELECTROMAGNETIC INDUCTION

19.7 A motor in a toy runs off a 6 V battery of negligible internal resistance, and uses a permanent magnet to provide the magnetic field. The armature has a resistance of 12 Ω and the motor uses a current of 0.087 A when it is off load and is rotating at 2400 revolutions per minute. Find:
  (a) the current through the motor when it is jammed and the power then wasted as heat;
  (b) the power used to overcome friction and the power wasted as heat when it is off load;
  (c) its speed when on load and drawing a current of 0.18 A.

19.8 A 12 V d.c. motor uses a permanent magnet to provide its magnetic field. It has an armature with a resistance of 2 Ω, and when it is off load it takes a current of 0.50 A and turns at 3000 r.p.m. Plot a sketch graph to show how the speed of the motor on load varies with the current.

INVESTIGATION

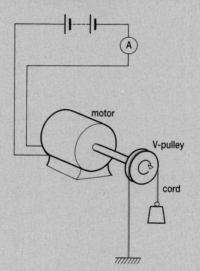

Fig 19.26 An experimental arrangement for finding the power output of a motor.

Carry out question 19.8 experimentally. That is, plot a graph of rate of rotation against current for a motor. A low voltage d.c. motor is required together with a suitable power supply. A small, low-power motor as used in toys is suitable. There is no problem in measuring the current to the motor with a d.c. meter but it is quite tricky to keep the speed constant when the motor is on load. You will need to clamp the motor securely and, depending on its physical size and power, fix a brake to it. This can be done in the way shown in Fig 19.26 with a cord under tension rubbing on a V-belt pulley wheel. Adjusting the weight changes the load on the motor. (This is the way many engines are tested for output power. The power output is often called the brake horse power: it is the power output when under these test conditions.) A rev. counter or a stroboscopic lamp has to be used to measure the speed of the motor.

  Be careful of the following, particularly if your motor is quite powerful.
  • The brake getting too hot. Industrial brakes can be very large and are often water cooled.
  • The motor getting too hot as it is made to do more work. Keep it running only long enough to take a reading.
  • The ammeter having too much current through it. If the motor gets jammed for any reason, not only might the motor burn out but also the ammeter may be damaged.

SUMMARY

  • Faraday's law of electromagnetic induction states that the e.m.f. produced when a wire cuts a magnetic field is equal to the rate of cutting of magnetic flux. Lenz's law states that any induced current is always in a direction to oppose any flux change which causes it.

$$E = -\frac{d\Phi}{dt}$$

  • Self-inductance is the e.m.f. produced in a coil per unit rate of change of current in the coil. Mutual inductance is the e.m.f. produced in a secondary coil per unit rate of change of current in a primary coil. One henry is the inductance of a circuit if an e.m.f. of 1 volt is produced when the current changes at a rate of 1 ampere per second.

$$E = -L\frac{dI}{dt} \qquad E = -M\left(\frac{dI}{dt}\right)_P$$

- Energy stored in the magnetic field of an inductor $= \frac{1}{2}LI^2$.

- growth of a current   decay of a current
  in an inductive circuit   in an inductive circuit

$$I = I_0 \left(1 - e^{-(R/L)t}\right)$$

$$I = I_0\, e^{-(R/L)t}$$

$$V_R = V_0 \left(1 - e^{-(R/L)t}\right)$$

$$V_R = V_0\, e^{-(R/L)t}$$

$$V_L = V_0\, e^{-(R/L)t}$$

$$V_L = -V_0\, e^{-(R/L)t}$$

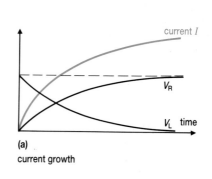

(a)
current growth

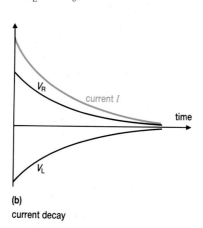

(b)
current decay

- The work done against the induced e.m.f. in a motor is the useful work output of the motor. Work done against the resistance of the wires is wasted as thermal energy.

- If the e.m.f. generated by a generator causes a current, then work must be done on the generator. Some of this work is changed into electrical energy, its useful output, and the rest is changed into thermal energy in the resistance of the wires of the generator.

- Output e.m.f. of a generator $E = BAN\omega \cos \omega t$

EXAMINATION QUESTIONS

19.9 Draw a labelled diagram showing clearly the direction of the force on a straight conductor carrying current in a magnetic field at right angles to it. (1)

Explain why two long parallel wires attract each other when they both carry a steady current. (4)

Diagram (a) shows two turns of a long solenoid wound with a single layer of wire. The solenoid has a diameter of 0.15 m, is wound with wire spaced 2.0 mm apart and carries a current of 2.5 A. By regarding two adjacent turns of the coil as two isolated parallel wires, calculate the force of attraction between the two turns. Explain how this force changes when the effect of the other turns of the coil is considered for (a) the turn at the centre of the coil, and (b) the end turn of the coil. (8)

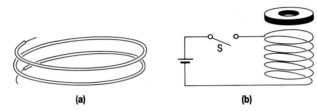

(a)   (b)

$$\frac{BI}{nqt}$$

Diagram (b) shows a conducting ring resting over the end of a coil whose axis is vertical. When switch S is closed the ring is repelled momentarily and then falls back. Explain this observation. Your explanation is **not** expected to be mathematical. It should contain
(i)   a diagram of the magnetic field set up by the coil,
(ii) an explanation of the relative directions of the currents in the coil and the ring.
(iii) sketch graphs of how the field and current vary with time for the coil and the ring. (7)

(O & C 1987)

19.10 (a) Define *magnetic flux density* and *magnetic flux*.
(b) A large flat coil is connected in series with an ammeter and a 50 Hz sinusoidal alternating supply whose r.m.s. output can be varied. At the centre of this coil is situated a much smaller coil which is connected to the Y-plates of a cathode ray oscilloscope (c.r.o.). The planes of the two coils are coincident (see diagram).

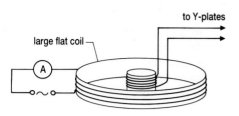

(i)   Draw sketch graphs, one in each case, to show the variation with time of (1) the magnetic flux, and (2) the induced e.m.f. in the small coil. Give physical explanations for the shapes of your graphs. (6)
Hence, describe how you could use this apparatus to demonstrate how the magnetic flux density at the centre of a large flat coil varies with the number of turns on the coil. (6)
(ii) Explain how the trace on the screen of the cathode ray oscilloscope would be affected if the angle between the planes of the two coils were slowly to increase from zero to 90° whilst maintaining a constant r.m.s. current in the large coil. (5)

(UCLES 1988)

19.11 (a) What is meant by the *inductance* of a coil?
(b) In the diagram is a battery of e.m.f. 9 V and negligible internal resistance, R is a resistor, S is a coil of wire wound on an iron core

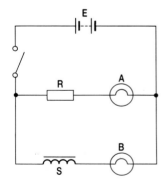

and A and B are lamps, each rated at 6 V 3 W. After the switch is closed each lamp operates at its rated power. A lights up very quickly; B takes more time.
(i)   Why does the current through B rise more slowly than the current through A?
(ii) Immediately after the switch is closed, the current through B rises at a rate of 0.6 A s^{-1}. Calculate the inductance and resistance of the coil S. (8)

(AEB 1986)

19.12 The coils of a certain electromagnet have an inductance of 2.5 H and a maximum current rating of 3.0 A. In order to prevent arcing when the current in the coils is switched off, a capacitor of capacitance 1000 μF is connected in parallel with the switch contacts. Calculate the energy stored in the coils when the electromagnet is carrying the maximum current. What will be the maximum potential difference across the capacitor if the current is suddenly switched off? (6)

(UCLES 1988)

19.13 (a) Describe a method of measuring the mean magnetic flux density in the region between the poles of an electromagnet. (8)

(b) A coil with n turns of area A is mounted in a uniform magnetic field of flux density B that rotates around the coil at a uniform angular speed ω, as shown below.

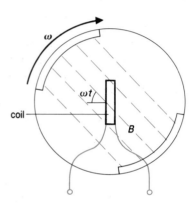

(i) Draw a graph on labelled axes of the waveform of the e.m.f. induced in the coil, showing clearly the time-scale involved. (4)

(ii) Show that the peak voltage E_0 induced in the coil is given by $E_0 = \omega BAn$ (4)

(iii) Derive an expression in terms of E_0 and R for the mean power developed when the coil is connected in a circuit of total resistance R. (4)

(c) A hospital emergency electrical supply consists of a bank of 240 V storage batteries charged periodically by a d.c. generator. The generator, which is 90% efficient, has a mean power output of 25 kW and is driven by a diesel engine at a constant speed of 50 revolutions per second.

Calculate:

(i) the charge supplied to the storage batteries during one hour's charging; (4)

(ii) the mean torque T required to drive the generator. (6)

(OXFORD 1986)

19.14 (a) Electromagnetic induction is the effect in which a changing magnetic field gives rise to an e.m.f. in a conductor.

State the laws which define (i) the magnitude of the induced e.m.f. and (ii) the direction of the induced e.m.f. (3)

(b) Diagram (a) shows a conducting disc of diameter 0.10 m with its plane perpendicular to a uniform magnetic field

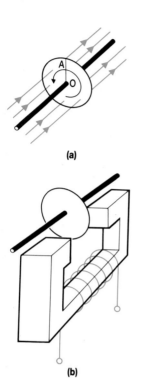

(a)

(b)

of flux density 0.25 T. The disc is rotated at 15 revolutions per second about an axis through its centre and perpendicular to its plane.

(i) Calculate the quantity of magnetic flux cut by a radial line such as OA as the disc completes one revolution.

(ii) Use your answer to (i) to calculate the value of the e.m.f. generated between the centre of the disc and any point on its rim.

(iii) State and explain whether a continuous current will flow in the disc as a result of this e.m.f.

(iv) Diagram (b) shows the disc placed above an electromagnet so that only part of the disc is between the poles. It is found that the disc spins freely with the electromagnet switched off, but quickly comes to rest when the electromagnet is switched on. Explain these observations. (12)

(c) The arrangement shown in diagram (b) could be used as an emergency braking system for machinery. The disc would be mounted on an axle which rotates as the machine operates. In an emergency, the electromagnet would be switched on and the machine would be decelerated.

(i) At a particular speed, the value of the deceleration of a given machine could be increased by increasing the flux density of the field. Explain one other way in which the effectiveness of the brake could be improved.

(ii) During a test of the brake, current was applied to the electromagnet when the machine was rotating freely at a speed of 1600 revolutions per minute, and the speed was reduced to 800 revolutions per minute in 5 s. Assuming that the deceleration is proportional to the speed of rotation, sketch and label a graph which shows how the speed of the machine will vary with time over the next 15 s of the test.

(iii) Discuss the effectiveness of the system as an emergency brake, and as a means of preventing rotation of a stationary machine. (10)

(AEB 1987)

19.15 Define *self inductance*.

A 12 V battery of negligible internal resistance is connected in series with a coil of resistance 1.0 Ω and inductance L. When switched on the current in the circuit grows from zero. When the current is 10 A the rate of growth of the current is 500 A s^{-1}. What is the value of L? (5)

(ULSEB 1983)

19.16 A shunt wound motor with armature resistance of 0.50 Ω and field coil resistance of 50 Ω is connected to a 25 V supply of negligible internal resistance, as shown below. When the motor is running on no load, the supply

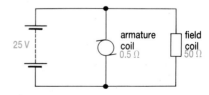

delivers 4.5 A and the motor turns at 2800 revolutions per minute. Determine

(i) the powers dissipated in the armature coil and the field coil. (4)

(ii) the power required to overcome frictional forces in the motor. (2)

When the motor is running on a certain load, the current supplied by the d.c. source rises to 22.5 A. Assuming that the frictional power loss stays constant, determine

(iii) the speed of the motor. (4)

(iv) the useful power supplied. (2)

Increasing the load causes the current to rise still further.

(v) Investigate how the speed of the motor depends on current. Plot your results graphically and determine the current when the speed of the motor would be zero. (4)

(vi) By considering the variation in output power, describe in practice what would happen to the motor as the load is increased. (4)

(UCLES 1987)

19.17 (a) State Lenz's law of electromagnetic induction. Describe the apparatus you would use and the observations you would make to verify your statement. (8)

(b) The diagram shows a copper disc held between the poles of an electromagnet. An *alternating* current flows through the coils of the electromagnet.

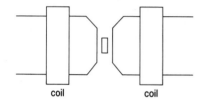

coil   coil

(i) Explain why heat is developed in the disc.

(ii) State and explain **two** ways in which, without moving the specimen, the rate of production of heat could be increased.

(iii) Explain how the rate of production of heat would change if the disc were replaced by another of identical dimensions, but made of a material of lower electric conductivity. (12)

(JMB 1986)

Chapter 20

A.C. THEORY

LEARNING OBJECTIVES

At the end of this chapter you should be able to:

1. describe sinusoidally varying currents and voltages in terms of their amplitude, frequency and phase;

2. use a phasor diagram to analyse an a.c. circuit;

3. define and use the terms reactance and impedance;

4. calculate the power supplied to an a.c. circuit;

5. define and use the terms root mean square current and root mean square potential difference.

20.1 CHARACTERISTICS OF ALTERNATING CURRENT

A.C. is an abbreviation for alternating current but it is often used in a descriptive way just to indicate that the electrical quantities being considered are oscillating. For this reason the term a.c. current is often heard in laboratory discussions despite the fact that strictly this term repeats the word current. Similarly, a.c. voltage is heard. This is even more of a nonsense if analysed carefully, as you cannot have a current voltage. These terms will not be used here but the meaning behind them is usually clear and they do highlight the fact that currents and potential differences do not always have constant values.

An alternating current is one in which the charge carriers, usually the electrons, oscillate about a fixed point. That is, there is zero drift velocity. The way in which the oscillation takes place may be very complex. The purest oscillation is one of simple harmonic motion in which there is a sinusoidal variation in the current. This oscillation can be caused by an alternator in which the coil rotates in a uniform magnetic field, see section 19.6.

Some examples of alternating currents and their practical uses are shown in Fig 20.2. These may be added to direct currents to make matters even more complex, but to start with we will consider the simplest, sinusoidal alternating current and progress to the more complicated situations later.

Once the simple situation has been thoroughly understood, there will be no problem with the more complicated one.

A pure alternating current I is shown in Fig 20.2(a) where the value of the current at any time t is plotted. It has the equation

$$I = I_0 \sin 2\pi ft$$

I_0 is the peak value of the current and f is the frequency of the oscillation. A term called the angular frequency ω is often used and is defined by the expression

$$\omega = 2\pi f$$

Fig 20.1 A sound engineer looking at an oscilloscope screen on which is displayed the pattern of the word 'HELLO'.

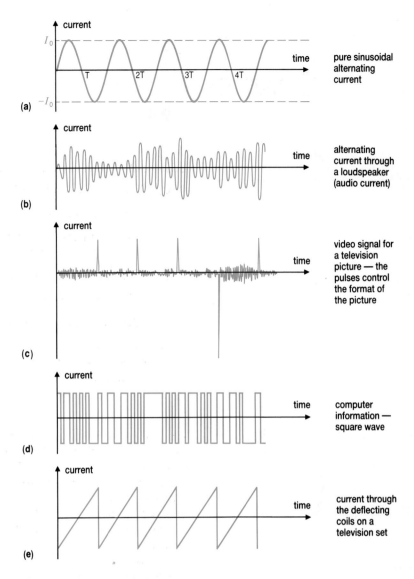

Fig 20.2 Some commonly occurring a.c. signals. Only the first one is a pure sine wave oscillation.

As with simple harmonic motion and waves the frequency is related to the period T by the expression

$$f = \frac{1}{T}$$

Note that the angle $2\pi ft$ has the unit radian. If you are unfamiliar with sine waves you are advised to plot a few to see the effect the different terms have. These are suggested in question 20.1 which follows.

QUESTION	20.1 Tabulate values for the following graphs and plot them on graph paper. You will need to choose small values of t if you are not to get $I = 0$ too frequently, and your calculator must be set for angles in radians.

        (a) $I = I_0 \sin 2\pi ft$ where $I_0 = 2.0$ A and $f = 3$ Hz
        (b) $I = I_0 \sin 2\pi ft$ where $I_0 = 2.0$ A and $f = 12$ Hz
        (c) $I = I_0 \sin 2\pi ft$ where $I_0 = 6.0$ mA and $f = 12\,000$ Hz
        (d) $I = I_0 \cos 2\pi ft$ where $I_0 = 6.0$ mA and $f = 12\,000$ Hz

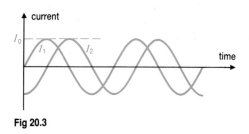

Fig 20.3

Question 20.1(d) asks you to plot a current against time when the equation is a cosine variation rather than a sine variation. You should know or have seen that the shape of this curve is the same as the shape of the sine wave (question 20.1(c)) but that it has been displaced by a quarter of a cycle. This is called a **phase difference** of a quarter of a cycle. Since one cycle corresponds to an angle of 2π radians, there is a phase difference of $\frac{1}{2}\pi$ between the sine and the cosine curves.

Often, when two pure alternating currents of the same frequency are compared, it is found that they are out of phase with one another. This is illustrated by the graph, Fig 20.3, where the two currents have the same peak value I_0. Current I_2 reaches its maximum at a later time than I_1 and is said to **lag** behind I_1. I_1 is said to **lead** I_2. Be careful here. It might look as though I_2 is in front of I_1, but I_2 reaches its maximum later than I_1 and it is definitely I_2 which lags behind I_1. Since these two currents have the same frequency, the phase difference between them is constant and its value may be stated – in radians. If the two currents do not have the same frequency then the phase difference between them is continually varying, so there is little point in stating it. The value of the phase difference as an angle in radians can be found by realising that one cycle corresponds to 2π radians so in this case, where I_2 lags behind I_1 by a quarter of a cycle, it has a phase difference of $\frac{1}{2}\pi$ with I_1.

The equations for I_1 and I_2 will be

$$I_1 = I_0 \sin (2\pi f t)$$

$$I_2 = I_0 \sin \left(2\pi f t - \frac{\pi}{2}\right) \text{ or } I_2 = -I_0 \cos (2\pi f t)$$

20.2 PHASOR DIAGRAMS

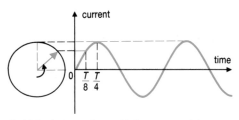

Fig 20.4 A phasor shown with the current variation it represents.

Fig 20.5

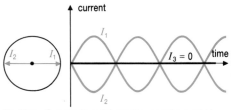

Fig 20.6 Currents in antiphase give a total current of zero, as shown by the phasor diagram or the graph.

Drawing sine waves is not easy and in any case if several of them have to be compared with one another the diagrams become hopelessly confused. A phasor diagram is a way of representing a pure oscillation with a single straight line. The phasor is a rotating vector. In Fig 20.4 the phasor is shown on the left-hand side of the graph and is assumed to be rotating in an anticlockwise direction, so that it completes one revolution in the period, T, of the oscillation of the current.

The phasor is shown after it has completed an eighth of a revolution. As it rotates, the height of the tip of the phasor marks the value of the current at that instant and hence the height of the sine wave graph. Apart from the ease of drawing phasors compared with drawing the sine wave, they have the advantage that they can be treated as vectors when the addition of alternating currents is necessary. Consider the part of a circuit shown in Fig 20.5 in which two alternating currents I_1 and I_2 combine at a junction to produce a total current I_3. Kirchhoff's first law can be applied to this situation so that at any instant $I_3 = I_1 + I_2$. If however I_1 and I_2 are not in phase with one another, some surprising results can occur. For instance, if I_1 and I_2 are in **antiphase**, that is exactly out of phase with one another or out of phase by a phase angle π, then I_3 will be zero because the two currents will cancel one another out. This is shown in Fig 20.6.

Also on this figure are shown the phasors which represent each of the currents. If the phasors are added together as if they were vectors, then they give a resultant of zero. This procedure can be carried out with as many different currents as are required provided that all those plotted on one phasor diagram have the same frequency. The example which follows illustrate this, together with questions 20.2 and 20.3.

A.C. THEORY

Example 1

An alternating current I_1, of peak value 1.6 A, meets at a junction with an alternating current I_2 of the same frequency and peak value 2.2 A. The two currents are out of phase by a quarter of a cycle. I_2 leads I_1. Find the peak value of the total current, I_3 and find its phase relationship to I_1.

The first thing to notice about this problem is that the peak current of I_3 is **not** $1.6 + 2.2 = 3.8$ A. As an example, the full diagram of the sine wave variation of the currents will be drawn, (Fig 20.7). Once you are certain you understand what the phasor diagram is doing, the full sine wave diagrams can be omitted. On the wave diagram the addition of I_1 and I_2 is done here by adding the two currents together point by point at many different times.

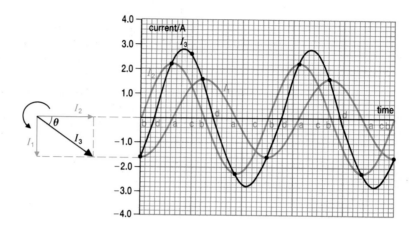

Fig 20.7 The phasor diagram is much easier to draw and to work from than the sine wave graphs it represents.

The following are useful points to start with. At times marked

- (a) I_1 is zero so I_3 must equal I_2
- (b) I_2 is zero so I_3 must equal I_1
- (c) $I_1 = I_2$ so $I_3 = 2I_1$
- (d) $I_1 = -I_2$ so I_3 is zero

Note that Kirchhoff's law is never broken. At all times $I_1 + I_2 = I_3$. When I_3 is found it has a peak value of 2.8 A and has a phase angle so that it lags behind I_2 by a bit less than an eighth of a cycle, that is, $<\frac{1}{4}\pi$ rad.

The phasor diagram enables this value to be found quickly and accurately. I_1 and I_2 can be added together using the Pythagoras' theorem to get

$$I_3{}^2 = I_1{}^2 + I_2{}^2$$

$$= (1.6 \text{ A})^2 + (2.2 \text{ A})^2 = 7.4 \text{ A}^2$$

$$I_3 = 2.7 \text{ A}$$

I_3 lags behind I_2 by angle θ where

$$\tan \theta = \frac{I_1}{I_2} = \frac{1.6 \text{ A}}{2.2 \text{ A}} = 0.727$$

$$\theta = 36°$$

$$= 0.63 \text{ rad}$$

20.2 Two pure sine wave alternating currents of the same frequency have the same peak values of 2.8 mA. Use a phasor diagram to find their sum if they are out of phase by an angle of $\frac{1}{2}\pi$. What is the phase relationship of each of the currents to their sum?

20.3 The mains supply uses three phases called the red phase, the blue phase and the yellow phase. These ideally are equal currents out of phase with each other by a third of a cycle as shown in Fig 19.21(b). Draw a phasor diagram to represent these currents and find their sum (no answer given).

Use your answer to explain why, on power lines, the current for customers is carried along thick cables but that the neutral cable shown in Fig 3.15 at the very top of the pylon can be thin.

As stated earlier, a pure alternating current can be produced by an alternator in which the coil rotates in a uniform magnetic field. The alternating current in a resistor connected to the alternator is being caused by an output e.m.f. from the alternator which oscillates in the same way as the current. Alternating potential differences can be shown on phasor diagrams in the same way as alternating currents can be. Fig 20.8 shows how the current I and the potential difference V across a resistance R vary with time. When plotting such variations the time scale must be the same for both quantities, but of course the vertical scales will be different since for one graph the y-axis will be calibrated in volts and for the other in amperes.

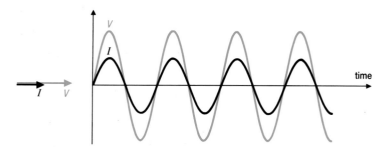

Fig 20.8

For a resistance R, the current is in phase with the potential difference as Ohm's law applies.

If the p.d. applied is given by

$$V = V_0 \sin 2\pi ft$$

then the current I, flowing at any instant is

$$I = \frac{V}{R} = \frac{V_0 \sin 2\pi ft}{R} = I_0 \sin 2\pi ft$$

20.3 POWER IN A RESISTIVE CIRCUIT

When an alternating current exists in a resistance there is a heating effect as there would be when a direct current flows. In the case of alternating current, however, the current is sometimes in one direction and sometimes in the other. The direction makes no difference to the heating effect. If the frequency of the alternating current is f, the power has a frequency $2f$ because maximum current occurs twice per cycle.

Since the frequency of the power is not the same as that of the current or potential difference, power cannot be shown on a phasor diagram. A graph is drawn in Fig 20.9 which shows how the power does vary with time.

Every point on the power curve is obtained by multiplying the current at that moment by the potential difference. Note that when the potential difference is negative, the current is also negative and so the power is positive.

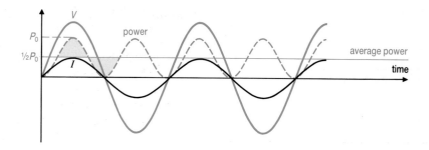

Fig 20.9 The variation of potential difference, current and power when the circuit component in use is a resistor.

Power cannot be supplied from a resistor back to the supply, so the power cannot be negative. The relative heights of the three graphs are not significant since these will depend on the scales used on the axes. The shape of the power graph is important though. It is of twice the frequency, and the top half is the same shape as the bottom half so the two areas shaded are equal. The average power is therefore half of the peak power, P_0.

We therefore can write

$$P = VI = V_0 \sin 2\pi ft \times I_0 \sin 2\pi ft$$
$$= V_0 I_0 (\sin 2\pi ft)^2$$

and the maximum value of the power is given by

$$P_0 = V_0 I_0$$

since the maximum value of the sine of any angle is 1.

$$\text{average power} = \tfrac{1}{2} P_0 = \tfrac{1}{2} V_0 I_0$$

Since $V_0 = I_0 R$ this gives

$$\text{average power} = \tfrac{1}{2} I_0^2 R$$

The $\tfrac{1}{2}$ is awkward. In electrical work with direct current the power is given by $P = I^2 R$ and there would be many complications in having one rule for power with a.c. and a different rule with d.c. so, to overcome the problem when working with alternating current, it is not usual to use peak currents. Instead a current called the **root mean square** current is used. This is written $I_{\text{r.m.s.}}$ and it is defined as the *direct current* which produces the same heating effect as the alternating current. This is shown in Fig 20.10.

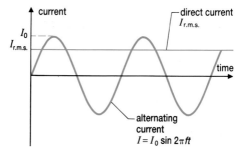

Fig 20.10

The average power supplied by the alternating current $= \tfrac{1}{2} I_0^2 R$

Power supplied by the direct r.m.s. current $= I^2_{\text{r.m.s}} R$

If these are to be the same then

$$(I_{\text{r.m.s}})^2 R = \tfrac{1}{2} I_0^2 R$$
$$(I_{\text{r.m.s}})^2 = \frac{I_0^2}{2}$$
$$(I_{\text{r.m.s}})^2 = \frac{I_0}{\sqrt{2}} = \frac{I_0}{1.414} = 0.707 \, I.$$

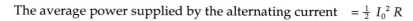

A.C. THEORY

Since R is constant

$$V_{\text{r.m.s.}} = 0.707\, V_0$$

Where $V_{\text{r.m.s.}}$ is the direct potential difference which would cause the same heating effect as the alternating potential difference:

$$V = V_0 \sin 2\pi ft.$$

When a mains voltage is said to be 240 V, the value being quoted is the r.m.s. value. The peak value of a 240 V mains is given by

$$V_0 = \sqrt{2 \times V_{\text{r.m.s.}}}$$

$$= 1.414 \times 240 \text{ V}$$

$$= 339.4 \text{ V}$$

Its minimum value is –339.4 V so it has an overall maximum change in value of 678.8 V. This value can be important in some instances. A diode in mains circuit can be subjected to this peak inverse voltage. That is, it can have the full 678.8 V across it in the opposite direction to that in which it will conduct, and it is important that it is not damaged by it. The peak inverse voltage which can be applied across a diode is normally quoted by the manufacturer.

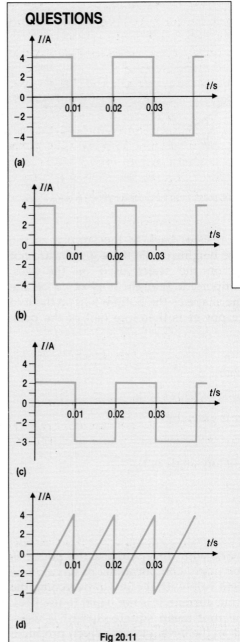

QUESTIONS

(a)

(b)

(c)

(d)

Fig 20.11

20.4 An electric kettle is marked 240 V 1440 W. Although not stated, these values may be taken to be r.m.s. values. Find:
   (a) the r.m.s. current
   (b) the peak current
   (c) the peak potential difference
   (d) the peak power
   (e) the average power.

20.5 The r.m.s. current is the d.c. current with the same heating effect as the a.c. current used. For a sinusoidal a.c., the peak current is 1.41 × r.m.s. current, but the value 1.41 only applies for sinusoidal a.c. What is the r.m.s. value for each of the currents shown in Fig 20.11. You should be able to give answers for (a) and (b) easily but (c) and (d) are more difficult. The answer to (c) is *not* 2.5 A. Find the average power in a resistor R for the negative part of the cycle, and the average power for the positive part of the cycle. Use these to find the average power for the whole cycle. To calculate (d) requires the use of calculus but if the question is set as a multiple choice question with answers you should be able to eliminate all but one of the answers as being impossible. Explain how to eliminate all but 2.3 A, which is the correct answer.

   (i) 4.0 A   (ii) 2.82 A   (iii) 2.3 A   (iv) 2.0 A   (v) 1.41 A

20.4 REACTANCE

If a capacitor of capacitance C has a potential difference V across it, then using the definition of capacitance, the charge Q on the capacitor is given by

$$Q = CV$$

This will apply even when the potential difference is changing, so if the potential difference is alternating and has the value $V = V_0 \sin 2\pi ft$, the charge on the capacitor at time t will be

$$Q = CV_0 \sin 2\pi ft$$

In graphical form the charge will therefore vary in the way shown in Fig 20.12. The charge on the capacitor increases as long as the current is

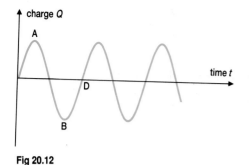

charge Q

A

D

B

time t

Fig 20.12

positive. At the peak of the graph, A, there is no increase in charge so the current flowing to or from the capacitor is zero. At B also the current will be zero. On the other hand, at D, the charge on the capacitor is changing rapidly. This can only occur if there is a flow of charge: that is, a current in the circuit. The current which is necessary to alter the charge stored in the capacitor is given by the rate of change of charge on the capacitor. This current is the gradient of the charge–time graph, or in calculus terms is given by

$$I = \frac{dQ}{dt}$$

The graph of current–time is shown in Fig 20.13 where numerical values have been supplied for both the charge and the corresponding current.

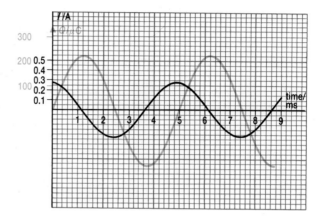

Fig 20.13   Graphs showing how the charge on, and the current to, a capacitor varies with time.

Although the maximum charge is only 220 μC, the maximum current is found from the gradient to be 0.28 A. The calculation of the current using calculus gives

$$Q = CV_0 \sin 2\pi ft$$

$$\frac{dQ}{dt} = I = CV_0\,2\pi f \cos 2\pi ft$$

As can be seen from the graph, the current is a cosine curve if the charge, and hence the potential difference, is a sine curve. The current is leading the potential difference by a quarter of a cycle, $\frac{1}{2}\pi$ rad.

The maximum value of the current (I_0) must be $CV_0\,2\pi f$ since $\cos 2\pi ft$ has a maximum possible value of 1.

As with resistors, it is important to know what the alternating current will be for a given alternating potential difference so the ratio of

$$\frac{\text{maximum potential difference}}{\text{maximum current}}$$

is called the **reactance** of the capacitor, and is given the symbol X_C.

Reactance is measured in ohms because it is a potential difference in volts divided by a current in amperes and for a capacitor has the value

$$X_C = \frac{V_0}{I_0} = \frac{V_0}{C\,V_0\,2\pi f}$$

$$X_C = \frac{1}{2\pi fC}$$

Note that $\dfrac{V_0}{I_0} = \dfrac{1.41\,V_{r.m.s.}}{1.41\,I_{r.m.s.}} = \dfrac{V_{r.m.s.}}{I_{r.m.s.}}$

and therefore we also get the ratio

$$X_C = \frac{\text{r.m.s. potential difference}}{\text{r.m.s. current}} = \frac{1}{2\pi f C}$$

Note that the reactance of a capacitor depends not only on its capacitance but also on the frequency being used. The reactance cannot therefore be marked on a capacitor. The higher the frequency, the lower the reactance.

The phasor diagram for a capacitor is shown in Fig 20.14. The current leads the potential difference by $\frac{1}{2}\pi$ rad.

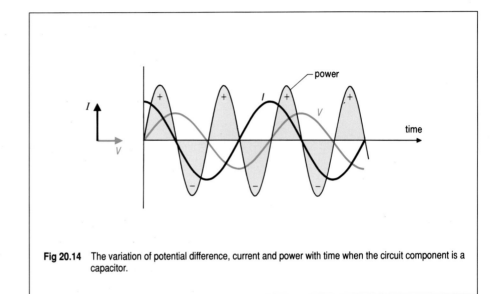

Fig 20.14 The variation of potential difference, current and power with time when the circuit component is a capacitor.

The full graph for this phasor diagram is also shown. It shows too the interesting feature that if the potential difference and current graphs are multiplied together at each instant to find the instantaneous power, the power is sometimes negative. If the current and the potential difference are in the same direction, then the capacitor is being charged up and extra energy is being stored. There are times however when the potential difference and the current exist in opposite directions. Here energy is being supplied from the capacitor back to the a.c. source.

These two possibilities give the power graph as shown with alternating positive and negative loops of equal size. The power supplied to a capacitor over a long period of time is therefore zero; and energy supplied to it in one quarter of a cycle is given back to the source during the next quarter of a cycle. It is for this reason that the word reactance is used for the peak potential difference divided by the peak current. To use the word resistance would be misleading because a circuit having resistance receives power from the source and changes it into thermal energy. A reactive circuit, that is, one with a reactance but not resistance, cannot change electrical energy into thermal energy.

The power changed from electrical energy into thermal energy by a resistive circuit was given in section 20.3 as

$$P = V_{\text{r.m.s.}} \times I_{\text{r.m.s.}}$$

In a reactive circuit where the potential difference and the current are a right angle out of phase, the average power is zero.

The equation which will satisfy all the situations is

$$P = V_{\text{r.m.s.}} \times I_{\text{r.m.s.}} \times \cos \phi$$

where ϕ is the angle between the phasors of current and potential difference.

Example 2

An r.m.s. alternating current of 5.7 mA passes to a capacitor when the applied r.m.s. alternating potential difference is 13.2 V. Find the reactance of the capacitor and the value of capacitor if the frequency of the supply is 250 Hz.

$$\text{reactance} = \frac{\text{peak potential difference}}{\text{peak current}}$$

$$= \frac{V_{r.m.s.}}{I_{r.m.s.}} = \frac{13.2 \text{ V}}{5.7 \text{ mA}}$$

$$= 2316 \ \Omega = 2300 \ \Omega \text{ to 2 sig. figs}$$

$$\text{reactance} = \frac{1}{2\pi f \, C}$$

$$2316 \ \Omega = \frac{1}{2\pi \times 250 C}$$

$$C = \frac{1}{2\pi \times 250 \times 2316} = 2.75 \times 10^{-7} \text{ F}$$

$$C = 0.275 \ \mu\text{F}$$

QUESTIONS

20.6 Use Fig 20.13 to find the current from the graph of charge against time, and check your values on the current–time graph. Make the calculations on tangents to the graph at points when $t = 0, 2$ ms, 2.5 ms and 4 ms.

20.7 Plot a graph to show how the reactance of a 1.0 μF capacitor varies as the frequency changes from 100 Hz to 3000 Hz. Explain why the reactance is smaller with higher frequency.

20.8 A capacitor in a fluorescent lamp is needed to pass an r.m.s. current of $\frac{1}{2}$ A when the r.m.s. potential difference across it is 240 V and the frequency of the supply is 50 Hz. Find the reactance of the capacitor and its capacitance.

20.9 A 0.47 μF capacitor has an r.m.s. potential difference of 20.0 V across it at a frequency of 4000 Hz. What is its reactance at this frequency and what current will flow through it?

20.5 CAPACITOR/RESISTOR CIRCUITS

A capacitor is often in a circuit in series or in parallel with a resistor. When a circuit contains both resistive and reactive parts, a phasor diagram is needed to deduce the current in each component. Consider the series circuit shown in Fig 20.15.

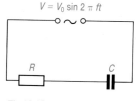

$V = V_0 \sin 2\pi ft$

Fig 20.15

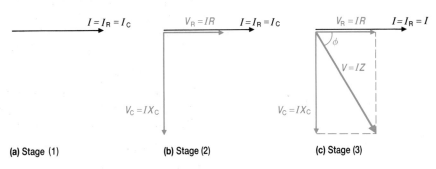

(a) Stage (1)   (b) Stage (2)   (c) Stage (3)

Fig 20.16

In this circuit the capacitor C and the resistor R are in series with the a.c. supply, so the current from the supply and through C and R must be the same at all instants. This step is important. It tells us where to start drawing the phasor diagram, which is shown in stages of construction in Fig 20.16.

Stage 1
The supply current I, the current through the capacitor I_C and the current through the resistor I_R, are identical at all times.

Stage 2
This now enables the potential differences across the resistor and the capacitor to be determined. $V_R = IR$ and V_R is in phase with the current through the resistor. The current and potential difference are always in phase with one another for a resistive component.

For any capacitor the current leads the potential difference by a right angle and so the potential difference phasor is behind the current phasor and has a value of IX_C.

Stage 3
In the third stage the total potential difference V is found by adding V_R and V_C to one another vectorially. The value of the total potential difference can be calculated using the Pythagoras' theorem.
The phasor diagram gives

$$V^2 = V_R{}^2 + V_C{}^2$$
$$= (IR)^2 + (IX_C)^2$$
$$= I^2(R^2 + X_C{}^2)$$
$$= I^2 \left\{ R^2 + \left(\frac{1}{2\pi f C} \right)^2 \right\}$$

$$\frac{V^2}{I^2} = \left\{ R^2 + \left(\frac{1}{2\pi f C} \right)^2 \right\}$$

$$\frac{V}{I} = \sqrt{ \left\{ R^2 + \left(\frac{1}{2\pi f C} \right)^2 \right\} }$$

Because this circuit contains both resistive and reactive components, the ratio V/I is not a pure resistance or a pure reactance. Another term is used to indicate this. The ratio is called the **impedance (Z)** and it too is measured in ohms because it is a potential difference in volts divided by a current in amperes.

We have therefore

$$Z = \frac{V_{max}}{I_{max}} = \frac{V_0}{I_0} = \frac{V_{r.m.s.}}{I_{r.m.s.}}$$

$$= \sqrt{ \left\{ R^2 + \left(\frac{1}{2\pi f C} \right)^2 \right\} }$$

this can also be written

$$Z = \sqrt{ (R^2 + X_C{}^2) }$$

The phasor diagram here contains 4 phasors and each of these represents a sine wave variation. The full sine wave diagram is given, not to scale, in Fig 20.17. You can see how difficult it becomes to complete sine wave diagrams for more complex circuits.

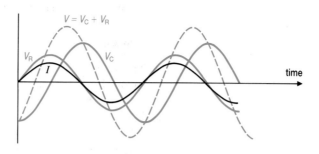

$$V = V_C + V_R$$

V_R V_C

I

time

Fig 20.17

In a circuit with both reactance and resistance, it is only the resistance which can continually absorb energy. The power P supplied to the resistance is, as before, given by

$$P = V_R I \text{ where } V_R \text{ and } I \text{ are r.m.s. quantities}$$

This gives $P = VI \cos \phi$ where V is the r.m.s. value of the supply potential difference.

Example 3

Find the current through each component of Fig 20.18 and the power to the resistor.

$$\text{Reactance of capacitor} = \frac{1}{2\pi f C} = \frac{1}{2\pi \times 1000 \text{ Hz} \times 0.10 \times 10^{-6} \text{ F}} = 1590 \ \Omega$$

The phasor diagram is shown in Fig 20.19. Make a habit of using a phasor diagram to record useful information as you go along. It is not a diagram to be drawn and then ignored. There is one point to be aware of in their use which has not been mentioned so far. It is that, strictly, phasor diagrams show peak values of current and potential difference. In practice, this means that the diagram has all its vectors 1.41 times longer than the values of the r.m.s. values, and a diagram which used r.m.s. values throughout would be of the same shape. No difficulty is therefore likely to arise provided that you are consistent and always use either r.m.s. values or always use peak values. In this example, peak values will be used so that you can see how the 1.41 term cancels out.

From the diagram we get

$$(1590 \ I)^2 + (2200 \ I)^2 = (2 \times 1.41)^2$$

where I is the peak current.
Solving for I gives

$$(2.53 \times 10^6) \ I^2 + (4.84 \times 10^6) \ I^2 = 8$$
$$(7.37 \times 10^6) \ I^2 = 8$$

$$I^2 = \frac{8}{7.37 \times 10^6}$$

$$I = 1.04 \times 10^{-3} \text{ A}$$

The r.m.s. value of I is therefore

$$\frac{I}{\sqrt{2}} = \frac{1.04 \times 10^{-3} \text{ A}}{\sqrt{2}} = 0.74 \text{ mA}$$

Note an unexpected feature of a.c. circuits.

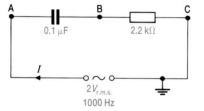

A   B   C

0.1 µF   2.2 kΩ

I

$2V_{\text{r.m.s.}}$
1000 Hz

Fig 20.18

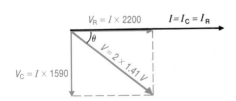

$V_R = I \times 2200$ $I = I_C = I_R$

θ

$V = 2 \times 1.41 \ V$

$V_C = I \times 1590$

Fig 20.19

If a good voltmeter (one with a high resistance) is placed across the supply terminals it should read $2\,V_{\text{r.m.s.}}$ If it is placed across the capacitor it will read

$$IX_C = \frac{0.74\,\text{A}}{1000} \times 1592\,\Omega = 1.18\,\text{V}$$

If it is then transferred to measure the p.d. across the resistor, it will record

$$IR = \frac{0.74\,\text{A}}{1000} \times 2200\,\Omega = 1.63\,\text{V}$$

At first sight it seems as if $1.18\,\text{V} + 1.63\,\text{V} = 2.0\,\text{V}$, but you need to remember that V_C and V_R are not in phase so there is no reason why they should add arithmetically. If you check:

$$1.18^2 + 1.63^2 = 2^2 \text{ to 2 sig. figs}$$

The power supplied to the resistor is

$$V_R I = 1.63\,\text{V} \times 0.74 \times 10^{-3}\,\text{A} = 1.21\,\text{mW}$$

and this is the power from the supply.

The power from the supply is **not** $2 \times 0.74 \times 10^{-3}\,\text{W}$ because there is a phase difference between the current and the supply voltage, and because the power to the capacitor is zero.

INVESTIGATION

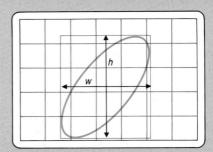

Fig 20.20

This investigation is to find how the phase angle in a C–R series circuit varies with the frequency. The phase angle being considered is the angle θ between V_R and V, the supply voltage.

Set up the circuit shown in Fig 20.18. The values suggested for the components are not critical, but given for guidance only. You will require a signal generator which can give an output which can be varied from around 200 Hz to 5000 Hz, but you do not need to know the potential difference of the output.

Some calibration of the X and Y deflection is necessary first. Switch off the time base and connect a 2 V accumulator (or a 1.5 V dry cell) across the Y-plates. Adjust the Y-gain so that on connection the spot on the screen moves 2 cm (or 1.5 cm with the dry cell). This gives a Y-plate calibration of $1\,\text{V cm}^{-1}$. Repeat on the X-plates to give the same $1\,\text{V cm}^{-1}$. It is important to have the *same* calibration on the two sets of plates. Check with a cell as suggested rather than relying on any calibration given on the instrument.

Connect point B to the X input and check that point C is at earth potential for both the oscilloscope and the signal generator. Connect point A to the Y input and you should have an ellipse on the screen. This is a Lissajou figure, as referred to in section 11.6. Adjust the input voltage so that the whole ellipse is on the screen. Do not adjust the oscilloscope controls as this would alter the calibration you did at the start. Measure the horizontal width w, and the vertical height h, of the ellipse, Fig 20.20. It can be shown that

$$\cos \theta = \frac{w}{h}$$

Repeat for different supply frequencies and plot a graph of θ against f. For the values suggested, a range of frequencies from 2000 Hz to 400 Hz is suitable. **Note:** if your ellipse is almost a straight line, it implies that the capacitor is having a negligible effect; its capacitance is therefore too high.

A.C. THEORY

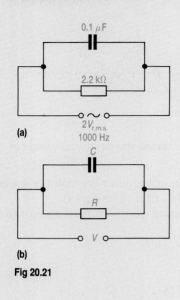

0.1 μF

2.2 kΩ

2 V$_{r.m.s.}$
1000 Hz

(a)

C

R

V

(b)

Fig 20.21

20.10 A 0.022 μF capacitor and a 10 kΩ resistor are connected in series with a power supply of frequency 1 kHz and an r.m.s. potential difference of 6.0 V. What is the current through the circuit and what is the potential difference across the resistor and the capacitor? Without making any further detailed calculations, find the approximate current through the circuit if the capacitor is replaced by a 2.2 μF capacitor.

20.11 A 120 V 60 W bulb is used on 240 V, 50 Hz mains. A capacitor is put in series with the bulb in order to limit the current to the value needed for normal bulb operation. What capacitance is required for the capacitor?

20.12 Find the current through each component and the power to the resistor if the two components used in Example 3 are connected in parallel as shown in Fig 20.21(a). (You will need a different phasor diagram here. Start with the quantity which is common to all three components in the circuit.)

20.13 Show that the impedance Z of the circuit shown in Fig 20.21(b) is given by

$$Z = \cfrac{1}{\sqrt{\cfrac{1}{R^2} + (2\pi fC)^2}}$$

What is the impedance when the frequency is zero (that is, when d.c. is used) and when the frequency is very high.

20.6 INDUCTOR/RESISTOR CIRCUITS

An inductor behaves in a complementary way in an a.c. circuit to a capacitor. If a current is required which leads the voltage, then a capacitor is used. If the current is required to lag behind the voltage, then an inductor is used. The reactance of a capacitor decreases as the frequency is raised: the reactance of an inductor increases with increasing frequency. In these respects the two components perform opposite functions but they also have similarities. Neither a capacitor nor an inductor loses any energy as heat. Energy is stored by a capacitor in its electric field; energy is stored by an inductor in its magnetic field but any energy which they store for a time can be returned to the circuit later. Both components have a phase angle of $\frac{1}{2}\pi$ between the current and the potential difference. Because of the contrasts in their behaviour, it is sensible to compare the theory for the two components.

If an inductor of inductance L has a current through it which is changing at a rate dI/dt then the potential difference V_L across it is given by

$$V_L = L\,\frac{dI}{dt}$$

If therefore the current through it is alternating and given by

$$I = I_0 \sin 2\pi ft$$

the potential difference will be

$$V = LI_0\, 2\pi f \cos 2\pi ft$$

This is shown in Fig 20.22 where the basic phasor diagram for an inductor is also given. You should notice carefully that here the voltage leads the current by a right angle. A useful way of remembering that for a

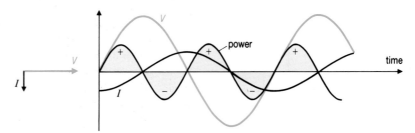

Fig 20.22 The variation of potential difference, current and power with time when the circuit component is an inductor.

capacitor the current leads the voltage but for an inductor the voltage leads the current is the word $\underline{CIVIL}$ (for C, I before V, for L, V before I).

Working in a similar way to that for a capacitor gives the reactance of the inductor:

$$X_L = \frac{V_{max}}{I_{max}} = \frac{V_0}{I_0} = \frac{L I_0 \, 2\pi f}{I_0}$$

$$X_L = 2\pi f L$$

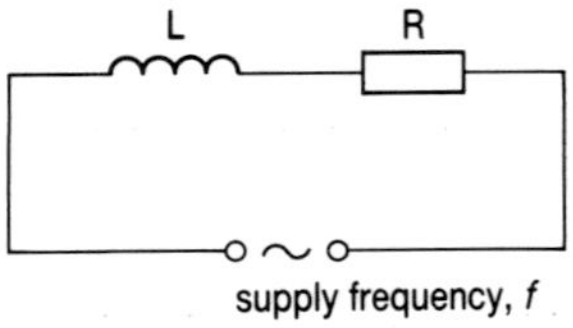

At room temperature, a circuit which only has inductance is unobtainable. Any coil has inductance, but it also has resistance and since the same current passes through the resistance and the inductance, a coil can be considered as a resistance and an inductance in series. This is shown in Fig 20.23 and you are asked in question 20.14 to show that the impedance of such a circuit is given by

$$Z = \sqrt{X_L^2 + R^2} = \sqrt{(2\pi f L)^2 + R^2}$$

Fig 20.23

QUESTION

20.14 An inductor of inductance 0.63 H is in series with a resistance of 100 Ω. A 50 Hz a.c. supply of r.m.s. potential difference 12 V is connected across the components.

(a) Calculate the reactance of the inductance using the formula $X_L = 2\pi f L$.

(b) Start the construction of the phasor diagram by drawing, not to any scale, a current phasor. This is the current through all components.

(c) Add phasors to represent V_R, the voltage across the resistor, and V_L, the voltage across the inductor. Label these in terms of the current through them.

(d) Show the supply phasor, which is the vector sum of V_L and V_R.

(e) Write down an equation linking V, V_L and V_R.

(f) Show that

$$12^2 = (198I)^2 + (100I)^2$$

(g) Find the current and hence the impedance of the circuit.

(h) How much power is supplied to the resistance?

(i) Find $\cos \phi$, where ϕ is the angle between the phasor representing V and that representing V_R. $\cos \phi$ is called the power factor.

(j) By consideration of the working you have already done, show that the impedance of an L–R series circuit is given by

$$Z = \sqrt{(2\pi f L)^2 + R^2}$$

A.C. THEORY

Example 4

A pure inductor of inductance 4.0 mH is placed in parallel with a resistance of 1200 Ω. A supply of r.m.s. potential difference 2.0 V is connected across both components. Sketch graphs to show how the impedance and the current from the supply vary with the frequency of the supply.

The phasor diagram for this circuit must be started by using the voltage phasor, as voltage is common to both components and to the supply. The phasor diagram is drawn in Fig 20.24(a). Some of the calculation has been done on the phasor diagram. Table 20.1 gives a few values of I_{total} and Z to assist with plotting the sketch.

Hints:
- It is useful to start off with $f = 0$ and $f =$ infinity in order to get the start and finish of the sketch graph in the right position.
- This circuit is unlikely to be used at low frequency because then the current through the coil is large and the current through the resistor is negligible.
- The total current is found by using Pythagoras' theorem to get

$$I^2 = \left(\frac{1}{600}\right)^2 + \left(\frac{20}{0.0251 \, f}\right)$$

$$= \frac{1}{600^2} + \left(\frac{79.6}{f}\right)^2$$

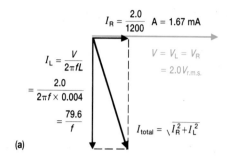

$$I_R = \frac{2.0}{1200} \, A = 1.67 \, mA$$

$$I_L = \frac{V}{2\pi f L}$$

$$= \frac{2.0}{2\pi f \times 0.004}$$

$$= \frac{79.6}{f}$$

$$V = V_L = V_R = 2.0 V_{r.m.s.}$$

$$I_{total} = \sqrt{I_R^2 + I_L^2}$$

(a)

(b)

Fig 20.24

Table 20.1

Frequency /Hz	I_L /mA	I_{total} /mA	Z /Ω
0	∞	∞	0
∞	0	1.67	1200
100 000	0.796	1.85	1080
50 000	1.59	2.30	870
30 000	2.65	3.13	640
10 000	7.96	8.13	250

QUESTION

20.15 A fluorescent tube is marked 240 V$_{r.m.s.}$, 60 W and is for use on 50 Hz mains. In operation it has a resistance of only 50 Ω and so a coil, sometimes called a choke, is placed in series with it to limit the current.
  (a) Find the current through a 50 Ω resistor if the power used in it is 60 W.
  (b) Find the potential difference across the 50 Ω resistance.
  (c) Using a phasor diagram, find the potential difference across the coil.
  (d) Find the reactance of the coil.
  (e) Find the inductance of the coil.
  (f) (difficult) The current supplied from the mains to such a fluorescent tube is out of phase with the mains voltage. How could this be corrected for? (This is necessary since public electricity suppliers insist on their consumers providing resistive loads, that is, keeping the voltage and the current in phase.)

So far in this chapter, inductors and capacitors have been considered in circuits by themselves or with resistors. An important development takes place, however, if a capacitor is put in a circuit with an inductor. Before the theory of this circuit is looked at, answer question 20.16. If you understand how to use phasor diagrams, you will have no difficulty with it.

QUESTION

20.16 An inductor of inductance 0.023 H is placed in parallel with a capacitor of capacitance 4.7 µF. A 500 Hz a.c. supply of r.m.s. potential difference 0.180 V is placed across the components.
  (a) Calculate the reactance of the inductor and of the capacitor.
  (b) Calculate the current through each component.
  (c) Draw a phasor diagram for the circuit.
  (d) Calculate from the phasor diagram the current from the supply.
  (e) Describe the effect on your results so far of reducing the frequency by a small amount.

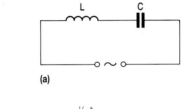

(a)

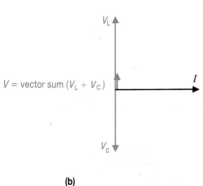

V = vector sum ($V_L + V_C$)

(b)

Fig 20.25

A circuit with similar properties to the one you were asked to analyse in question 20.16 is shown in Fig 20.25(a). It is called a series resonance circuit. The phasor diagram for the circuit is shown in Fig 20.25(b). The current through each component is the same, so the first stage in drawing the diagram is to draw a phasor for the current. The potential difference across the capacitor V_C is $\frac{1}{2}\pi$ behind the current and the potential difference across the inductor V_L is $\frac{1}{2}\pi$ in front of the current. The **sum** of these two potential differences, the supply potential difference V is therefore given numerically by $V_L - V_C$ since they are in antiphase. The effect of the minus sign in this expression is of particular importance if V_L and V_C have similar values. It is then possible to have a large current through the circuit, and to have high potential differences across each component even though the input potential difference is very small.

Circuits such as this are used on tuned circuits. A television aerial supplies only a very small voltage to the tuned circuit to which it is connected, but it can cause much higher voltages to be established across other components in the circuit. Not only is this amplifying effect caused but also the circuit is designed to receive only the frequency required. It is only for a particular frequency that V_L equals V_C, so it is only for that frequency that the high voltages are obtained. When you turn the tuning control knob on a radio, you are altering the value of a capacitor and hence changing the frequency at which the circuit gives its high output. A circuit used in this way is said to be a resonant circuit. A small input of energy sustains a large oscillation. It is the electrical equivalent of mechanical resonance, which was dealt with in Chapter 10.

A graph of V_L against the frequency, for constant L, C and V, is shown in Fig 20.26 and it is apparent from the graph that the frequency at which resonance occurs needs to be known in terms of L and C. This can be found easily since it is the frequency at which $X_C = X_L$. Therefore

$$\frac{1}{2\pi fC} = 2\pi fL$$

$$4\pi^2 f^2 LC = 1$$

$$f^2 = \frac{1}{4\pi^2 LC}$$

$$f = \frac{1}{2\pi\sqrt{LC}}$$

Fig 20.26 A characteristic series electrical resonance graph. The potential difference across each component is much larger than the supply potential difference when the circuit is resonating.

A.C. THEORY

This is the same frequency which is obtained for resonance in the parallel resonant circuit for which you were asked to draw the phasor diagram earlier. The difference between the series and parallel cases is that whereas in the series case a small supply voltage causes a large current (low overall impedance), in the parallel case a large current can exist in the capacitor and the inductor, with only a very small current being drawn from the supply (high overall impedance). The circuit diagram for the parallel resonant circuit is shown in Fig 20.27 together with its phasor diagram when at resonance. Fig 20.28 shows how the impedance and the current supplied from a constant voltage source vary with the frequency.

Practical resonance circuits always have some resistance in them besides inductance and capacitance. The effect of resistance, which is mostly in the inductor circuit, is much the same as the damping effect mechanical resistance has on mechanical oscillations. Electrical resistance changes

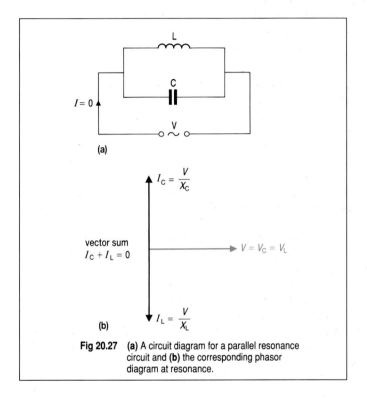

Fig 20.27 **(a)** A circuit diagram for a parallel resonance circuit and **(b)** the corresponding phasor diagram at resonance.

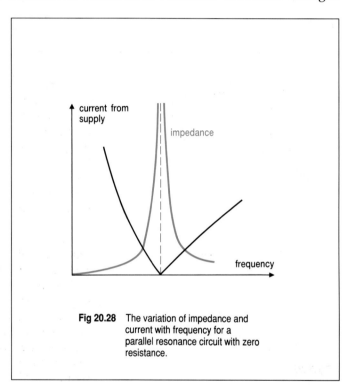

Fig 20.28 The variation of impedance and current with frequency for a parallel resonance circuit with zero resistance.

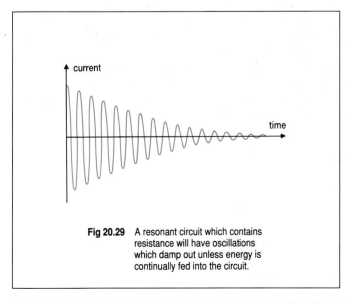

Fig 20.29 A resonant circuit which contains resistance will have oscillations which damp out unless energy is continually fed into the circuit.

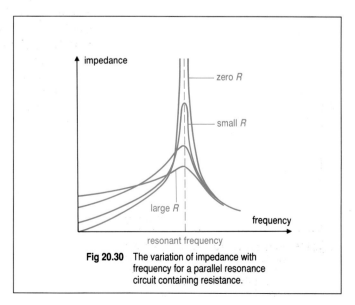

Fig 20.30 The variation of impedance with frequency for a parallel resonance circuit containing resistance.

electrical energy to thermal energy, and hence results in the amplitude of the current falling unless that energy is replaced. If a charged capacitor is connected to a coil, an ammeter in the circuit will show the oscillating current being damped out as is shown in Fig 20.29. At resonance, the impedance of a resonant circuit containing resistance is always purely resistive. Its value is such that the peak on the resonance graphs is less pronounced. There is also a change in the resonant frequency, but for small values of resistance the effect is small as can be seen from Fig 20.30.

QUESTIONS

20.17 Find the resonant frequency of an *LC* circuit in which the capacitor has a value of 10.0 pF and the coil an inductance of 4 µH. Higher resonant frequencies can only be obtained in such a circuit by reducing the inductance or the capacitance. What problems arise in trying to reduce these values?

20.18 A series circuit consisting of a capacitor of value 20 µF and an inductor of inductance 0.30 mH and negligible resistance is connected to a 0.40 V source of frequency 2500 Hz. What is the maximum value of the current and the maximum value of the potential difference across the capacitor and across the inductor?

INVESTIGATION

This investigation is to examine the behaviour of a parallel resonant circuit.

1. Set up the circuit shown in Fig 20.27 using a capacitor of capacitance 1.0 µF and an inductor of inductance 0.10 H. The coil used to have an inductance of 0.10 H will probably have a resistance of about 30 to 100 Ω. All of these values are not critical but are for guidance only. A signal generator is needed as a supply and it is preferable to use digital meters. An a.c. voltmeter and an a.c. 0–100 mA milliammeter are required.
2. Measure the resistance of the coil using a multimeter
3. Connect the voltmeter across the supply and the milliammeter in series to measure the current from the supply to the resonant circuit.
4. Calculate the impedance of the circuit from your meter readings over a range of frequencies, and plot graphs of circuit impedance and current against frequency. A range of frequencies from 100 Hz to 2000 Hz is suitable for the values suggested.
5. The investigation can be repeated to find how resistance affects resonance. It is preferable to add a resistance to the coil branch of the circuit, which doubles the resistance already present.

20.8 THE TRANSFORMER

Alternating current is used for the transmission of electrical power throughout the world for two reasons.

The first is that, compared with d.c., it is much easier to switch a.c. on and off. The reason for this is that a.c. is zero many times per second as a result of the variation in current, so a switch in an a.c. circuit can interrupt the current at one of these instants. Switching large d.c. currents off is surprisingly difficult, because if a larger current is to be reduced to zero in a short time then the rate of change of current is high; this sets up a large e.m.f. by electromagnetic induction which can cause sparking across the terminals of the switch. Switching equipment therefore has to be robust and carefully designed if the current is to be switched off safely.

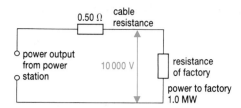

Fig 20.31

The other reason why a.c. is used for the commercial transmission of electrical power is because of the need for high voltage transmission. The following example illustrates this essential requirement.

Example 5

A power station supplies a factory with 1.0 MW of electrical power at a p.d. of 10 000V. See Fig 20.31. The total resistance of both of the cables between the power station and the factory is 0.50 Ω. Find the percentage of the power station's output power which is delivered to the factory. How does this figure change if the supply potential difference is only 250V?

$$\text{current to factory} = \frac{1\ 000\ 000\ \text{W}}{10\ 000\ \text{V}} = 100\ \text{A}$$

$$\begin{aligned}\text{potential difference across cables} &= 100\ \text{A} \times 0.50\ \Omega \\ &= 50\ \text{V}\end{aligned}$$

$$\begin{aligned}\text{power wasted in cables} &= 50\ \text{V} \times 100\ \text{A} \\ &= 5000\ \text{W}\end{aligned}$$

$$\text{power output from power station} = 1\ 005\ 000\ \text{W}$$

$$\begin{aligned}\text{\% of power station's output supplied} &= \frac{1\ 000\ 000}{1\ 005\ 000} \times 100 \\ &= 99.5\ \%\end{aligned}$$

If the supply p.d. is reduced to 250 V the current supplied to the factory, for the same power is

$$\frac{1\ 000\ 000}{250\ \text{V}} = 4000\ \text{A}$$

The p.d. across the cables is

$$\begin{aligned}&4000\ \text{A} \times 0.50\ \Omega \\ &= 2000\ \text{V}\end{aligned}$$

The power wasted in the cables becomes

$$2000\ \text{V} \times 4000\ \text{A} = 8\ 000\ 000\ \text{W}$$

so the % of the power station's output which is supplied to the factory is only

$$\frac{1\ 000\ 000}{9\ 000\ 000} \times 100$$

that is, only 11%, a totally unacceptable value.

There is therefore always commercial pressure on the electrical supply industry to use higher and higher potential differences. Local transmission at 240 V can only be through thick cables for distances up to about 500 m. For larger distances potential differences of 11 000 V, 25 000 V, 125 000 V, 275 000 V and 400 000 V are used. Experimentation is taking place to use even higher values of potential difference for long-distance transmission.

Since these high potential differences are extremely dangerous, it is essential to be able to change the potential difference from one value to another with high efficiency, reliability and safety. Because transformers meet these requirements, the supply must be a.c. since a transformer is only able to function on a.c. This is the main reason why a.c. is used for the transmission of electrical power. The photograph of Fig 20.32 shows insulators under test.

A transformer consists of two coils, a primary coil and a secondary coil, wound on the same soft iron core. The design of a transformer and its

Fig 20.32 This insulator is designed for a 275 kV system. Flashover occurred under test at 1500 kV. Three separate flashovers are recorded on the single negative of the photograph.

Fig 20.33 A large transformer.

A.C. THEORY

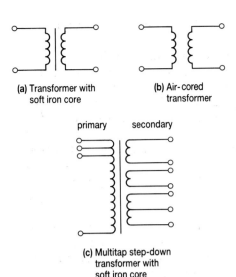

(a) Transformer with soft iron core

(b) Air-cored transformer

primary   secondary

(c) Multitap step-down transformer with soft iron core

Fig 20.34 Circuit symbols for transformers.

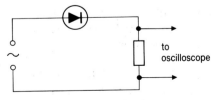

Fig 20.35 The cooling system of a transformer can be seen on this photograph. Oil within the transformer becomes heated and as the oil rises, it sets up a convection current. The oil cools as it falls through the external tubes.

construction depend on the use to which it is to be put. A Transformer is shown in Fig 20.33 but it is difficult to illustrate the vast range of different types and powers of commercial transformers. The smallest are for high frequency work and might use power less than a microwatt. The largest are mighty constructions capable of handling many megawatts at high voltage. The symbol for a transformer is given in Fig 20.34(a) together with some variations. Note that there must be at least 4 connections to it: 2 primary and 2 secondary.

Assume that an alternating current flows through the primary. This will set up a varying magnetic field in the core which links with the secondary coil and by electromagnetic induction sets up an e.m.f. across each turn of wire in the secondary coil. The e.m.f. across the secondary (E_s) is therefore dependent directly on the number of turns in the secondary coil (n_s) and the rate of change of the magnetic flux in the primary coil. This gives

$$\frac{E_s}{E_p} = \frac{n_s}{n_p}$$

If n_s is greater than n_p, then E_s is greater than E_p and the transformer is called a step-up transformer. A step down transformer is when E_s is less than E_p.

Transformers are usually very efficient. The largest ones can be over 99% efficient. The lost 1% of power can be a problem with high power transformers as, for example, 1% of the power of a transformer handling 1 MW of power is 10 000 W. This power causes considerable heating of the transformer, so on large transformers some form of cooling system is needed. This is frequently a convected flow of oil, up through the heart of the transformer and down in cooling tubes on the exterior. These are clearly visible in Fig 20.35.

An ideal transformer does not lose any power so in that case

electrical power input   =   electrical power output

$$E_p\, I_p = E_s\, I_s$$

and the turns ratio $\dfrac{n_s}{n_p} = \dfrac{E_s}{E_p} = \dfrac{I_p}{I_s}$

where I_p and I_s are the primary and secondary r.m.s. currents respectively. A transformer which steps up the voltage steps down the current.

QUESTIONS

20.19 A transformer used to step up 240 V to 600 V has a primary coil of 100 turns. What must be the number of turns on the secondary coil?

20.20 One of the largest transformers ever made handles a power of 1.50×10^9 W. This transformer is used to step down 765 kV to 345 kV. What is the r.m.s. current in the primary? What is the r.m.s. current in the secondary? What is the turns ratio? How much power is wasted by the transformer if it is 99.8% efficient?

20.9 RECTIFICATION

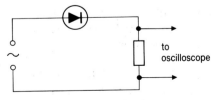

to oscilloscope

Fig 20.36 Basic rectifier circuit.

If a diode is placed in a circuit with an a.c. power supply in series with a resistor, then the current can be in one direction only. This circuit is shown in Fig 20.36 where the p.d. across the resistor V_r is being monitored with an oscilloscope.

The trace on the oscilloscope will be as in Fig 20.37 and shows only positive-going peaks. This circuit is called a **half-wave rectifier**. If the potential difference across the diode V_d, is monitored, it will have a potential difference across it when it is not conducting, so will show the negative-going peaks. If these voltages are compared, the three traces

A.C. THEORY

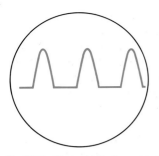

Fig 20.37 Characteristic shape of half wave rectified power supply.

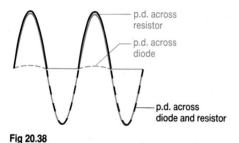

p.d. across resistor

p.d. across diode

p.d. across diode and resistor

Fig 20.38

shown in Fig 20.38 are obtained. The potential difference of the supply, V_s, equals $V_d + V_r$ at all instants.

There is a small forward potential difference across the diode when it conducts, but if the value of the resistor's resistance is high and the forward resistance of the diode is low, almost all the forward p.d. will be across the resistor. When the negative potential difference is applied, very little current flows and so there is very little potential difference across the resistance.

Although the potential difference across the resistor is in one direction only, it is far from being a smooth d.c. potential difference. In order to smooth the potential difference, a capacitor can be added in parallel with the resistor. This is called a reservoir capacitor and it performs a similar function to a reservoir for a water supply.

A reservoir fills up with water when it rains so that it can supply water at a smooth rate all the time. A capacitor charges up when charge can flow to it and can discharge smoothly all the time. This circuit is shown in Fig 20.39(a) and the corresponding potential differences and currents in Fig 20.39(b). The symbol used for the reservoir capacitor in Fig 20.39(a) is that for a polarised capacitor. These are usually electrolytic capacitors and they must be connected into the circuit the correct way round. They are used because it is possible to make electrolytic capacitors with high values of capacitance see section 17.1).

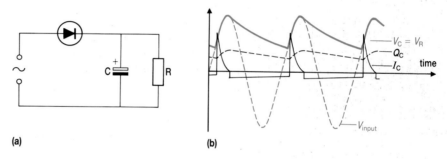

$V_C = V_R$

Q_C

I_C

time

V_{input}

(a)   (b)

Fig 20.39 A capacitor placed across a load resistor in a rectifier circuit has the effect of smoothing the output voltage, V_R.

Fig 20.39(b) contains a great deal of information but it is worth while analysing it properly. The input potential difference to the circuit is sinusoidal, V_{input}. The potential difference across the capacitor, V_C, is equal to the potential difference across the resistance, V_R, as they are in parallel. The V_C and V_R graph enables Q_C, the charge on the capacitor, to be plotted, and the gradient of this graph is the current to the capacitor, I_C. What will the graph of I_R look like?

The time when current flows to the capacitor is shown as a positive current, and the time it is flowing from the capacitor is shown as negative. The following example shows how the value required for a reservoir capacitor can be estimated.

Example 6
A half-wave-rectified mains supply is used to obtain a p.d. of 340 V d.c. which must have a ripple of less than 2 volts when supplying a current of 40 mA. Find the value of the smoothing capacitor required.

Mains frequency is 50 Hz and the peak value of the mains 340 V. Fig 20.40 shows the problem involved. In the time between two peak values of potential difference, the potential difference must fall by less than 1V.

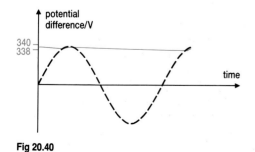

potential difference/V

340
338

time

Fig 20.40

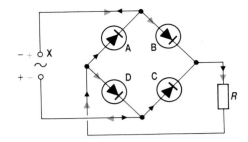

Fig 20.41 A bridge rectifier circuit.

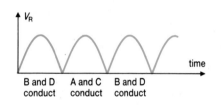

B and D   A and C   B and D
conduct   conduct   conduct

Fig 20.42 The output voltage from a full wave rectifier.

Time between peaks	= 0.020s
Current during this time	= 40 mA and is assumed to be constant.
Charge lost from reservoir capacitor	= 40 mA × 0.020
	= 0.8 mC = 800 µC

This causes a drop in potential difference across the capacitor of only 2 volts, so since the graph of charge against potential difference for a capacitor is a straight line through the origin

$$C = \frac{Q}{V} = \frac{\text{change in } Q}{\text{change in } V}$$

$$= \frac{800 \text{ µC}}{2V} = 400 \text{ µF}$$

This means the charge stored when the capacitor is fully charged is

$$400 \text{ µF} \times 340 \text{ V} = 136\,000 \text{ µC}$$

and this falls to 400 µF × 338 V = 135 200 µC during the 0.020 s – a fall of 800 µC as required.

Other points to note about reservoir capacitors are that they do have a large charge on them which is gained while the first few cycles after switching on take place. The surge of current at switch on can damage the rectifier, and when switching off care must be taken not to touch them as they may still be charged. For this reason many television sets have labels on them warning people not to open the back of the set even when it is switched off. It can take several minutes for the unwanted charge to leak away, usually through the resistance of the dielectric of the capacitor.

One way of reducing the problem of ripple is to use both halves of the a.c. cycle instead of just the positive half of the cycle. A bridge rectifier is often used: this is 4 diodes connected in the way shown in Fig 20.41. Whether terminal X is positive or negative, current will pass through R in the same direction. The output is as shown in Fig 20.42. The output can be smoothed, as with half wave rectification, but here the time for which the smoothing capacitor supplies the charge is only half the previous time; for mains this is therefore a reduction from 0.02 s to 0.01 s and the capacitor can be reduced to half its previously calculated value.

Fig 20.43 A bridge rectifier.

A.C. THEORY

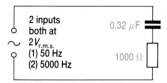

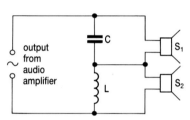

Fig 20.44

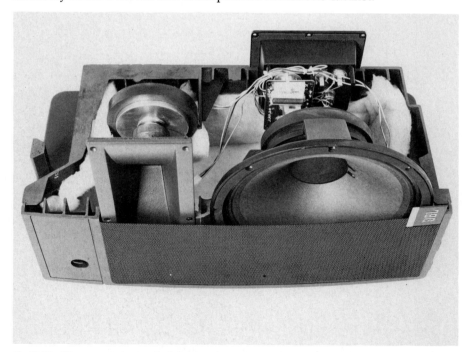

Fig 20.45 A cross-over filter circuit.

Filter circuits

Modern electronics makes a great deal of use of filter circuits, which are used to allow different frequencies to be passed to different parts of a circuit. To see the principle behind filter circuits answer questions (a) to (d) below which refer to the circuit shown in Fig 20.44 in which two inputs are being used.

(a) What is the reactance of the 0.32 µF capacitor at both 50 Hz and 5000 Hz?

(b) Draw a phasor diagram. What is the impedance of the circuit at 50 Hz?

(c) What is V_C and V_R when the frequency is 50 Hz?

(d) What is V_C and V_R when the frequency is 5000 Hz?

If you have done this correctly (check the answers) you will see that for the 50 Hz frequency most of the output voltage is across the capacitor, whereas for the 5000 Hz frequency most of the output voltage is across the resistor. An amplifier connected across the resistor will therefore pick up mostly high frequencies, and one across the capacitor will pick up mostly low frequencies. This is a rather unsophisticated filter. In practice, filters can be tuned to different bands of frequencies. One commonly used circuit is with loudspeakers. The output from an amplifier covers a range of frequencies to cover the audible range from around 40 Hz to 12 000 Hz, depending on the quality of the amplifier. Since it is difficult to construct good loudspeakers for the whole of this range, it is common practice to use a woofer for the low frequencies and a tweeter for the high frequencies. Fig 20.45 shows one way of connecting these loudspeakers. This circuit is normally fitted with the two loudspeakers inside one cabinet.

Fig 20.46 This cut away photograph of a loudspeaker enclosure shows the two loudspeakers and the electronic circuitry needed to direct the low frequencies to the large loudspeaker and the high frequencies to the small loudspeaker.

(e) What is the function of L?

(f) What is the function of C?

(g) In one loudspeaker cabinet C was found to have a value of 3.3 µF. What is its reactance at 3000 Hz?

(h) What must be the value of the inductance of L if it is to have the same reactance as C at 3000 Hz?

(i) Which of S_1 and S_2 is the tweeter?

(j) Without working out numerical values, draw phasor diagrams for this circuit at both low and high frequencies. The speakers may be considered to have resistance only.

SUMMARY

- A pure a.c. current has the equation $I = I_0 \sin 2\pi ft$.
- I is the instantaneous current, I_0 is the peak current and f is the frequency. This equation assumes that $I = 0$ when the time, $t = 0$.
- The root mean square current, $I_{\text{r.m.s.}}$, is the direct current which gives the same heating effect as the alternating current. For sinusoidal currents its value is given by

$$I_{\text{r.m.s.}} = \frac{I_0}{\sqrt 2} = 0.707\, I_0$$

- Instantaneous power has a frequency twice that of the current and has a maximum value which is twice the mean power in a resistive circuit.

Resistor	Capacitor	Inductor
V and I in phase	I leads V by 90°	V leads I by 90°
resistance $= \dfrac{V_0}{I_0} = R$	reactance $= \dfrac{V_0}{I_0} = \dfrac{1}{2\pi fC}$	reactance $= \dfrac{V_0}{I_0} = 2\pi ftL$
power $= V_{\text{r.m.s.}} \times I_{\text{r.m.s.}}$	power $= 0$	power $= 0$

- A phasor is a rotating vector and represents a sinusoidal variation of voltage or current. Impedance can be found from a phasor diagram as the vector sum of resistance and reactance.
- In a resonant circuit, the reactance of the capacitor equals the reactance of the inductor. The circuit is purely resistive. Increased resistance in a resonant circuit reduces the height of the resonance curve; the circuit is damped.
- The resonant frequency in series and parallel L–C circuits is given by

$$f_0 = \frac{1}{2\pi \sqrt{LC}}$$

A.C. THEORY

EXAMINATION QUESTIONS

20.21 The diagram shows (in part) the variation with time of a periodic current.

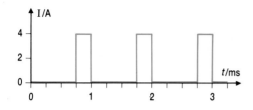

(a) What is the average value of the current?
(b) Find the root-mean-square current.
The periodic current passes through a resistor, producing heat at a certain rate.
(c) What steady current, passing through the same resistor, would have an identical heating effect? (3)

(UCLES 1983)

20.22 (a) Explain the terms (i) *frequency*, (ii) *phase difference*, as applied to alternating current. (3)
A generator is connected in the three circuits shown below.

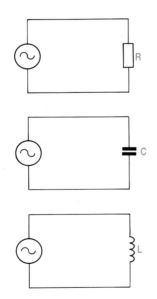

For each circuit element R, C, L, draw a graph to show the phase relationship between current and potential difference. Account for each phase relationship you have shown. (6)
(b) Account for the relationship between the potential difference across the coils and the number of turns in the coils of a transformer. How would you test this relationship experimentally? (6)

In an experiment, a 24 V 12 W lamp is connected satisfactorily across the secondary of a transformer with its primary connected to a 240 V a.c. supply. Given that the transformer has 100 turns in its secondary, calculate (i) the number of turns in the primary, (ii) the current in the primary. (5)

(UCLES 1982)

20.23 (a) A capacitor is marked '470 μF 25 V'.
(i) What is meant by a *capacitance of 470 μF*?
(ii) What is the significance of the '25 V'? (3)
(b) Explain the features in the structure of a capacitor which ensure that it has a high capacitance. (4)
(c) Explain in qualitative terms why a capacitor
(i) blocks the flow of direct current in a circuit but allows the flow of alternating current,
(ii) impedes the flow of a low frequency alternating current more than the flow of a high frequency alternating current. (4)
(d) A student has designed two circuits, shown below, which enable a 240 V lamp to operate at 150 V when a switch is opened. At this lower voltage the power dissipated by the lamp is 60 W. A 240 V 50 Hz supply is used.

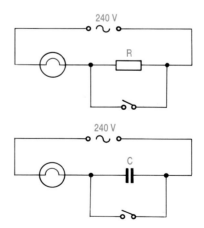

Calculate each of the following quantities showing clearly how each answer has been obtained.
(i) The current flowing in the lamp when it dissipates 60 W.
(ii) The resistance of the lamp at 150 V.
(iii) The resistance of the resistor R.

(iv) The power developed in R.
(v) The reactance of the capacitor C.
(vi) The capacitance of C.
State and explain which design you consider to be the better. (10)

(AEB 1986)

20.24 (a) Define the term *self-inductance*. (2)
Describe **two** simple experiments to show that a coil of wire has a self-inductance. (6)
(b) An alternating potential difference $V = V_0 \sin \omega t$ from a source of negligible impedance is applied to (i) a pure inductor of inductance L, (ii) a pure capacitor of capacitance C. Deduce the amplitude of the current in each case. Deduce the value of the phase-shift between the current and the applied potential difference in each case. Explain how these results arise. (7)
(c) A parallel plate capacitor with air between the plates is connected in series with an inductor of inductance 10 mH and a source of alternating potential difference of negligible impedance and constant amplitude. The frequency of the alternating supply is varied and the current shows a resonance maximum at 1.6×10^5 Hz. The capacitor is then immersed in oil so that the space between the plates is completely filled. The frequency of the current maximum is reduced to 7.8×10^4 Hz. Calculate
(i) the capacitance of the capacitor with air between the plates,
(ii) the relative permittivity of the oil. (5)

(O & C 1986)

20.25 (a) Distinguish between the *peak* value and the *r.m.s.* value of an alternating sinusoidal current. What is the numerical relationship between peak and r.m.s. values? (3)
In many filament lamp dimming circuits, the waveform of the current is altered by the components in the device from sinusoidal to a form like that shown (full line).

Explain how this waveform, while still having the same peak value as before, reduces the brightness of the filament. (4)

(b) An alternating current source of constant r.m.s. voltage 3.0 V is connected in series with an a.c. ammeter of negligible impedance, a coil having both inductance and resistance and a variable capacitor. When the capacitor is adjusted so that maximum current is recorded by the ammeter, r.m.s. voltages of 5.0 V and 4.0 V are developed across the coil and the capacitor respectively. If it were possible to separate the inductive and resistive elements of the coil what would be the voltages across these parts? (4)
Explain the effect on the voltage developed across the capacitor of
(i) reducing the frequency of the source to zero,
(ii) increasing the frequency of the source. (3)
The capacitance of the capacitor is kept at the value giving maximum current in the circuit, and a resistor of resistance equal to that of the coil is connected in series with the other components in the circuit. What will be the new values of the voltages across the coil and the capacitor? (4)

(ULSEB 1983)

20.26 A full-wave rectifier provides an output potential difference of 6.0 V (r.m.s.) when on open circuit.
A capacitor of capacitance 25 μF is connected between the output terminals of the rectifier. What is the maximum energy stored in the capacitor?

(UCLES 1988)

20.27 In the circuit shown in, the signal generator gives a sinusoidal output maintained at 10 V r.m.s. The capacitor C has capacitance 0.10 μF, the inductor L has inductance 20 mH and negligible resistance and the resistor R has resistance 500 Ω.

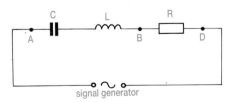

(a) Explain why, when the frequency of the generator output is varied from a low to a high value, the r.m.s. current in the circuit rises to a maximum and then falls. (6)

A.C. THEORY

(b) For the condition when *the current in the circuit is a maximum* what will be
  (i) the frequency of the generator output,
  (ii) the r.m.s. potential difference between A and B,
  (iii) the r.m.s. potential difference between B and D
  (iv) the r.m.s. current
  (v) the phase difference between the generator voltage and the current? (8)
(c) Describe how you would use a cathode ray oscilloscope to determine
  (i) the phase difference between the generator voltage and the current,
  (ii) the phase difference between the generator voltage and the potential difference across C, when the current in the circuit has its maximum value. For both (i) and (ii) state what would be observed on the C.R.O. (6)

(JMB 1987)

20.28 Explain what is meant by *resonance* in an electrical circuit. Why do you expect a circuit containing an inductor and a capacitor in series to exhibit resonant properties? (6)
Write down expressions for the reactances of an inductance L, and of a capacitance C, at a frequency f. (2)
At what frequency are the reactances of the two components equal in magnitude? What is the significance of this condition? (2)
An ideal inductor of 10 mH and an ideal capacitor of 4.0 μF are connected in series to an a.c. supply of zero internal resistance. On the same set of axes draw graphs of the reactance of the inductor and the reactance of the capacitor against frequency, for the frequency range 0 to 2.0 kHz. Taking due account of the phase relationship between the potential differences across the inductor and the capacitor, draw a graph of the total reactance of the inductor and capacitor in series. (6)
Predict a value for the current at resonance from your values of the reactance at resonance with ideal components. Comment on this. (4)

(O & C 1982)

20.29 The data from experiments to study the variation of the reactance of an ideal capacitor C and an ideal inductor L with frequency are plotted in Fig Q20.29.

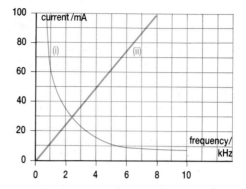

(a) Draw the circuit containing suitable measuring instruments and a signal generator to perform each experiment. The output of the signal generator is adjusted to maintain a potential difference of 2.0 V across C or L at all frequencies. Which graph corresponds to C and which to L? (4)
(b) Use the graphs to find the values of C and L. (6)
(c) The experiment is repeated with C and L in parallel. The p.d. across the combination is maintained at 2.0 V. Copy Fig Q20.29 onto graph paper and add to it a sketch of the variation of current with frequency for this combination. (4)
(d) By considering the relative phase differences between the current in and the p.d. across each component, explain why the circuit containing L and C together has a resonant frequency. (6)

(O & C 1988)

20.30 A power station feeds 1.0×10^8 W of electrical power at 760 kV into a transmission line. 10% of this power is lost heating the resistance of the transmission line itself. What percentage of the power would be lost if the power station were to feed 340 kV into the transmission line instead of 760 kV?

Chapter 21

ELECTRONICS

LEARNING OBJECTIVES

At the end of this chapter you should be able to:

1. distinguish between digital and analogue systems;

2. use a variety of different sensors to provide inputs to digital and analogue systems;

3. use a variety of different output systems and appreciate the importance of linking the output of one system to the input of another;

4. understand how a transistor can function as a switch;

5. construct truth tables for NOT, AND, NAND, OR, NOR, EX-OR and EX-NOR gates;

6. combine gates to perform required functions;

7. understand the use of an operational amplifier with feedback.

8. describe the operational amplifier astable circuit.

21.1 SYSTEMS APPROACH

The subject of electronics is vast. It is almost entirely a Twentieth Century subject, the electron only having been discovered in 1897. Whole degree and further degree courses are planned and taught to give students an insight into its basic principles and to point them in possible directions for further study. Many volumes of books are written on the subject and even more on just some aspect of electronics. In writing a single chapter on the subject, to be studied in only a few weeks, the problem is to know what to omit and what to include. The same problem faces the designers of advanced level physics syllabuses and they have all accepted that some electronics should be taught, that detailed understanding of electronic components and circuits themselves is too time-consuming and demanding, and that therefore the systems approach should be used. In this approach, many of the circuits are regarded as black boxes.

The man shown in Fig 21.1 is holding an electronic component called a silicon chip. This particular chip can store over a million bits of information and was, in 1988, among the densest and fastest of any megabit chips in existence. The circuit of the chip is etched in silicon and is known as an integrated circuit (IC). The construction of an integrated circuit need not concern you, but you will often need to know what a particular integrated circuit does.

In studying this chapter, you should be prepared to experiment frequently. The electronic components required are cheap and simple to set up on a breadboard. A battery as a power supply together with occasional use of a multimeter, an oscillator and an oscilloscope complete the apparatus requirements.

Fig 21.1 An IBM engineering manager holds a computer memory chip that can store over a million bits of information on a piece of silicon about 5 mm × 10 mm. Behind the chip is one of its storage cells – magnified 20 000 times on the screen of a scanning electron microscope.

The systems approach to electronics depends very largely on the ability, which can be built in, to link systems together. To give a simple example: the output of a microphone can be connected to the input of an amplifier; the output of the amplifier can be connected to the input of a loudspeaker. What happens in each of the three black boxes, the microphone, the amplifier and the loudspeaker, need not concern us. What does matter is that we need to know that we can inter-link the systems and that, for instance, if we want to use a record player rather than a microphone we can disconnect the microphone and connect the output from the record player instead. That this can be done with electronic systems is dependent on the input and output impedances of the circuits used. (Often the voltage and current in the input and in the output are in phase with one another, so the impedance is then equal to the resistance.) A connection between two components is shown in Fig 21.2(a). The circuit is very similar to the circuit in Fig 21.2(b) which was a circuit emphasised as being important in section 14.7. Since all electric currents are concerned with the distribution of energy from one place to another, it is essential to see how the circuit in Fig 21.2(b) behaves. If R is large compared with r, most of the energy

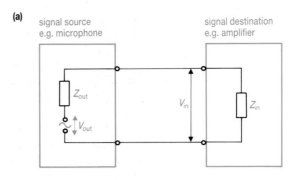

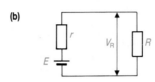

Fig 21.2 The transfer of energy from a source to a destination takes place in all electrical circuits.

supplied by the cell will be transferred from the cell to the resistance R. That is, V_R will be very nearly equal to E. The value of V_R is given by

$$V_R = \frac{R}{r + R} \times E$$

as this is a potential divider circuit. For the circuit shown in Fig 21.2(a), the same equation holds true with both d.c. and a.c. sources provided the impedances are resistive.

The input voltage to the amplifier V_{in} is given by

$$V_{in} = \frac{Z_{in}}{Z_{in} + Z_{out}} \times V_{out}$$

and this will have its maximum value if Z_{out} is small compared with Z_{in}. For most of the components which you will use, this will be so. For integrated circuits Z_{in} could be many megohms, whereas Z_{out} may be only a few kilohms. With high input impedances and low output impedances, it is possible to chain together thousands of circuit components one after the other.

21.2 INPUTS AND OUTPUTS

Both inputs to and outputs from electronic circuits can be a.c. or d.c. or a combination of the two as shown in Fig 21.3. The behaviour of a circuit can be very different for a steady current as opposed to a varying one, so it is necessary to be quite clear which type of current is being used.

Inputs

The devices used to provide inputs to circuits are very varied. Some of them produce an electromotive force but many do not. For example, microphones are of no fewer than four basic types. In a moving coil microphone the changes in air pressure resulting from a sound wave move a small coil within a strong magnetic field. By electromagnetic induction a small, varying e.m.f. is created which can be connected across the input of an amplifier. A crystal microphone also creates an e.m.f. When some crystals are squeezed, an e.m.f. is created across them and this effect is made use of in a crystal microphone where the sound wave vibrations are used to force a diaphragm against the crystal.

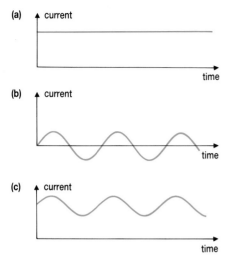

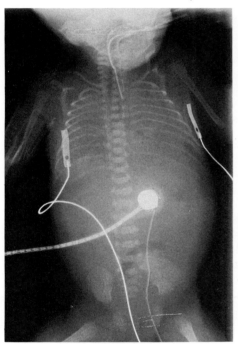

Fig 21.3 (a) the graph shows a smooth d.c. current (b) shows a sinusoidal a.c. current and (c) shows an actual current which contains both d.c and a.c. components.

Fig 21.4 An X-ray photograph of an eleven week premature baby in intensive care. Various life support devices are visible. The round white object over the baby's abdomen is a probe to monitor carbon dioxide levels through the skin. On either side of the chest are electrodes connecting to an electrocardio-gram, used to monitor the function of the baby's heart. A feeding tube is visible passing down the throat.

The other two types of microphone require some external source of potential difference. In a capacitor microphone, the sound waves move one plate of a capacitor and hence adjust its capacitance. If a battery is connected across the capacitor then a change in the charge residing on the plates will occur when the capacitance changes. The fluctuation of charge results in a current in a resistor and hence a potential difference across it which varies with the sound oscillations. The carbon microphone also requires an external source of power. Carbon powder is compressed and relaxed by the sound waves and the current through it fluctuates. The circuits into which these components are connected have to take account of their different characteristics.

Inputs to electronic circuits are by no means always microphones. Circuits can be used to monitor almost anything from the vital to the trivial. Fig 21.4 shows a picture of a baby in an intensive care unit. The baby is wired up to a series of sensors which monitor such features as heart beat, breathing and temperature. Devices are now available for use in the home so that babies can be monitored while they are asleep to check that they are still breathing. If they stop breathing an alarm sounds so that the parents can be alerted. It is being found that this is an encouraging way of reducing the relatively large number of distressing cot-deaths which happen each year.

Sensors which function on temperature, light intensity and mechanical force are clearly needed, besides straightforward switching and resistive devices. These are incorporated into potential divider circuits as shown in Fig 21.5. The sensor can be connected either to switch on, or to switch off, a circuit when a particular low or high value is reached.

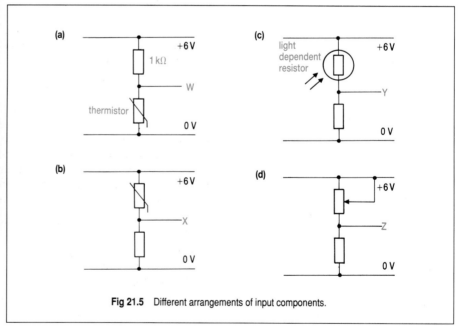

Fig 21.5 Different arrangements of input components.

Several of these sensors can be used in conjunction with one another or with other sensors which may respond to any of magnetic field, sound, infra-red, ultra-violet, radiation, humidity, speed, etc.

Thermistor

The thermistor is one particularly frequently used input device. It is made of a semiconductor material, such as nickel oxide, and its resistance falls markedly when the temperature is raised. The extra kinetic energy of the atoms at the higher temperature enables more electrons to move from atom to atom. Table 21.1 is a guide to how the resistance of some common thermistors varies with temperature. The resistance of a length of copper wire which happens to be 1 Ω at 20 °C and a constantan wire with the same dimensions are shown for comparison.

Table 21.1

| | Temperature | |
	20 °C	100 °C
a wire made of copper	1.0 Ω	1.3 Ω
a similar wire made of constantan	31 Ω	31.1 Ω
disc thermistor	25 Ω	1 Ω
rod thermistor	380 Ω	28 Ω
bead thermistor	5000 Ω	80 Ω

Example 1
In Fig 21.5(a) the thermistor has a resistance of 80 Ω when hot and 5000 Ω when cold. What is the potential at W in each case?

When hot: potential difference across thermistor is:

$$\frac{80}{80 + 1000} \times 6 \text{ V} = 0.44 \text{ V}$$

so potential at W = 0.44 V

When cold: potential difference across thermistor is:

$$\frac{5000}{5000 + 1000} \times 6 \text{ V} = 5.0 \text{ V}$$

so potential at W = 5.0 V.

Note that as the temperature rises, the potential at W decreases. This change can be used to switch off a circuit when a pre-determined temperature is reached. This could be used in a central heating control system to switch the heating off when a room is warm enough.

QUESTIONS

21.1 If the same thermistor is used in circuit 21.5(b) (see example 1), how does the potential at X change when the temperature of the thermistor is altered from cold to hot?

21.2 The same thermistor as used in example 1 has a resistance of 80 Ω at 100 °C. What value resistance should be used in place of the 1 kΩ resistor if it is required that the potential at W (Fig 21.5(a)) is to be 0.6 V at this temperature?

21.3 The light dependent resistor (LDR) in Fig 21.5(c) has a resistance of 1200 Ω when illuminated and a resistance of 10 MΩ when in the dark. Find the value of a suitable resistor which can be placed in series with it so that the potential at Y changes from near zero to 2 V if the LDR is illuminated.

21.4 In Fig 21.2(a) the output impedance, Z_{out}, is resistive and has a value of 600 Ω, and the input impedance, Z_{in}, is also resistive and has a value of 50 kΩ. Find V_{in} as a fraction of the output voltage, V_{out}.

ELECTRONICS

Outputs

Outputs from electronic circuits can be used to operate almost anything, but there are a few details which always need to be taken care of. For example, there is no point feeding a d.c. output into a loudspeaker. A lamp, on the other hand, can function on either d.c. or on a.c. provided the current to it is not too high and not too low. A light emitting diode (LED) is a common indicator for microelectronic circuits but two points have to be watched. Firstly, LEDs are diodes; they only allow current through them in one direction so they must be connected the correct way round. Secondly, LEDs made of gallium phosphide are easily damaged if the current through them is too large. In order to limit the current to around 10 mA, a resistor having a resistance of about 600 Ω is placed in series with the LED. For a.c. operation a diode needs to be connected in parallel with the LED and to allow the current to pass in the opposite direction to the LED.

While LEDs and lamps can be connected directly to an output, there are very many devices which need to be controlled by microelectronic circuits but which require a much greater current than the circuit can supply. In these cases, relays are used.

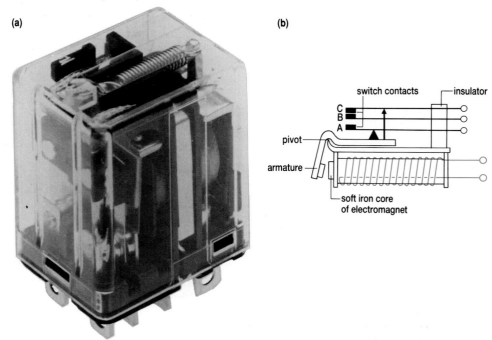

Fig 21.6 **(a)** A relay which is about 3 cm high and is used for switching and **(b)** a diagram showing how it operates.

Relays are electromagnetic switches which can be used to allow small currents to switch on or off large currents. Fig 21.6 shows a relay (see also Fig 18.26). For the relay shown in Fig 21.6, when there is a current through the coil it magnetises the core of the electromagnet and this attracts the armature. The armature pivots and the other end of the armature which has two insulating rods attached to it moves upwards and pushes through holes against the springy metal supporting the switch contacts. The probe will move contact A so that it touches contact B, and at the same time the other probe will push contact C away from contact B. This type of relay is called a change-over relay. A diode should always be connected across a relay in a microelectronics circuit because when the current through the coil of the relay is switched off there is a surge in potential across the coil as a result of electromagnetic induction. The diode will allow the consequent current to flow through itself for the brief time the surge exists, so preventing any damage to the circuit.

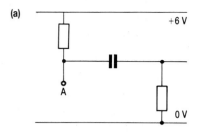

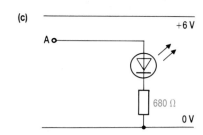

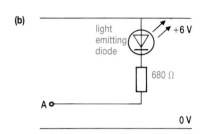

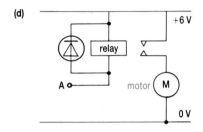

Fig 21.7 Different arrangements of output components.

Fig 21.7 shows various configurations which can be used for the output from a point A in the circuit.

(a) is used with a.c. The capacitor blocks d.c. from A but allows a.c. to pass to the next stage in the circuit.
(b) shows a light emitting diode which will be on only when A is at low potential.
(c) shows the rearrangement necessary for the LED to be on only when A is at a high potential.
(d) shows a relay, with its protective diode, operating a 6 V motor on its switch terminals.

QUESTION	**21.5** For the circuits of Figs 21.7(b) and 21.7(c), the current through the LED is 5 mA. If the potential difference across the LED under these operating conditions is 2.2 V, find the potential at A in each case.

21.3 THE TRANSISTOR AS A SWITCH

Having considered inputs and outputs for electronic circuits, it is now time to consider the component at the heart of modern electronics, the **transistor**.

The name transistor is a contraction of transfer resistor, and transistors were first constructed in 1948 in the Bell Telephone Laboratories in the United States by Shockley, Bardeen and Brattain who were awarded the Nobel Prize for Physics in 1956.

A glance at an electronics catalogue shows how many thousands of types of transistor are manufactured for different purposes, and at this stage it is necessary to consider only the basic function of a common transistor and see why it might be called a transfer resistor.

INVESTIGATION

It is recommended that circuits are set up on a circuit board. There are several different types available and the solderless ones are easier to experiment with as components can be interchanged easily. Do not be tempted to use crocodile clips and long trailing wires. They are a source of much frustration. Firm connections are essential.

One of the main problems in setting up electronic circuits is having components which are compatible with one another. Because of the number of different values of components, it is possible to have a circuit functioning perfectly but not to know that it is doing so, simply because a current is not large enough to make the particular bulb you are using light up. Please therefore experiment. Try different components, logically, when things seem not to work – you will not be the first person to damage a transistor by over-loading it!

Start with the simplest possible transistor circuit.

Apart from the transistor itself all you will require is a bulb, a 6 V battery and some connecting wire. A suitable transistor is a BC 108 and the bulb should be 6 V 60 mA. The transistor is an n–p–n silicon transistor and it has three connections to it: emitter, base and collector. The arrangement of these and the circuit diagram is given in Fig 21.8.

Keep X and Y apart and nothing happens. No current is possible through the transistor if the base current is zero. Now hold X in one hand and Y in the other hand. If your resistance is low enough there will be a current of up to 100 microamperes around the circuit from the battery, through you to the base, emitter and back to the battery.

As soon as this base current flows, the lamp should light although it will probably be dim. The important feature of the transistor's function is that a current of about 30 milliamperes from the collector to the emitter is possible as a result of a current of 100 microamperes in the base. In other words, a small current in a circuit of high resistance (you) becomes transferred into a larger current in a circuit of low resistance (the bulb) – a transfer resistor. This is the basis of transistor action so it is worth while making sure that you have seen it clearly. Any difficulty here will probably be because of one of three things:

1. You let X touch Y and thereby over-heated the transistor. Remedy – use a new transistor and do not do the same thing again.
2. Your resistance is too high to allow sufficient base current. Remedy – moisten your hands or hold X and Y, still separated from one another, between the same two fingers, or replace yourself by a 100 kΩ or 33 kΩ or 10 kΩ resistor (or by all three, one at a time).
3. Your bulb has too small a current to make it glow. Remedy – put a milliammeter in series with the bulb to show the collector current. Then try again – at least you should be able to see the milliammeter needle fall as your hands dry out. This is the basis of some lie detectors.

Interplay between the resistance connected between X and Y and the current through the bulb should be observed. The higher the resistance between X and Y, the lower the current through the base. A small base current results in a small collector current.

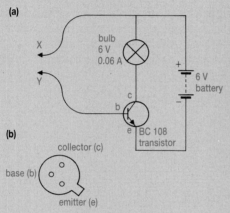

(a)

bulb
6 V
0.06 A

6 V
battery

c
b
e BC 108
transistor

(b)

collector (c)

base (b)

emitter (e)

Fig 21.8

The base current in a transistor controls the collector current. The current through the collector, I_c, is directly proportional to the current through the base, I_b so with zero base current there will be zero collector current as seen in the investigation.

 The ratio $\dfrac{I_c}{I_b}$ is called the **dc current gain**, h_{FE}.

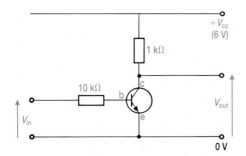

Fig 21.9

Typically h_{FE} is of the order of 100.

From now on, a resistor will be put into the lead from the base. This resistor is used to limit the base current and hence to limit the collector current to a safe value.

Having seen the basic action of a transistor, the next stage is to see how a load resistor in the collector lead enables the transistor to act as a switch, Fig 21.9.

If the input potential difference V_{in} is zero then there will be zero base current. Zero base current implies zero collector current, and so there will be no current through the 1 kΩ load resistor and consequently no potential difference across it. Since one end of this resistor is connected to the 6 V supply lead, the other end, connected to the collector, must also be at 6 V and the output is 6 V.

If now the input potential difference is made positive (in practice, with a silicon transistor it has to be more than 1.4 V) a base current is caused and this will in turn cause a collector current. This results in there being a potential difference across the load resistor V_L making the potential of the collector less. The transistor is therefore behaving as a kind of inside-out switch. When no input voltage is supplied, there is maximum output. When an input voltage is supplied the output falls nearly to zero.

Table 21.2 summarises this and Example 2 shows how a calculation can be done.

Table 21.2

	Input zero	Input positive
V_{in}	zero	few volts
I_b	zero	few tens of μA
I_c	zero	few mA
V_L	zero	few volts
V_{out}	6 V	near zero

Example 2

Using Fig 21.9, find the output voltage if the input voltage causes a base current of 30 μA and the d.c. current gain of the transistor is 150.

$$\frac{I_c}{I_b} = 150$$

$$I_c = 150 \times I_b = 150 \times 30 \text{ μA} = 4500 \text{ μA}$$

The potential difference, V_L across the load resistor is therefore

$$V_L = IR = 4500 \text{ μA} \times 1000 \text{ Ω}$$

$$= 4\,500\,000 \text{ μV}$$

$$= 4.5 \text{ V}$$

Therefore the output potential difference, V_{out}, is

$$V_{out} = 6 - V_L = (6 - 4.5) \text{ V}$$

$$= 1.5 \text{ V}$$

ELECTRONICS

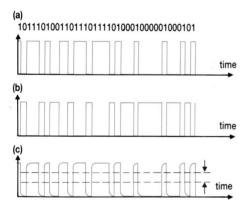

Fig 21.10

21.6 For a transistor with a current gain of 200, find the base current necessary to make the output 1.2 V if the transistor is used with a 2.0 kΩ load resistor and V_{cc}, the supply voltage, is 6.0 V. What is the maximum collector current possible for such a circuit?

21.7 For the circuit in Fig 21.9, the potential difference between the base and the emitter, V_{BE} is 0.6 V. (This is a fixed standard figure for silicon transistors). If V_{in} varies between 0.6 V and 2.0 V and the d.c. current gain of the transistor is 150, plot a sketch graph to show how V_{out} changes with V_{in}.

21.8 Fig 21.10 shows a circuit combining an input, a transistor as a switch, and an output. Identify, and suggest suitable values for each component and explain the function of the circuit. Design similar circuits to:
(a) Switch on a fan when it gets hot.
(b) Switch on a light when it gets dark.
(c) Switch on a pump when a water lever gets low.
(d) Switch off a heater when it gets hot.
(e) Switch off a refrigerator when it is cold enough in the food compartment.

21.4 LOGIC GATES AND TRUTH TABLES

(a)
10111010011011101111010001000001000101

time

(b)

time

(c)

time

Fig 21.11 **(a)** digitally coded information, **(b)** is the reverse of **(a)**. Such signals are often drawn without graph axes which are then assumed to be potential difference on the y-axis and time on the x-axis. **(c)** How the shape of a distorted signal can be improved by clipping and then amplifying the clipped signal.

	input	output
NOT gate	0	1
	1	0

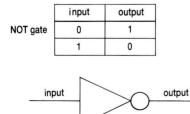

Fig 21.12 The truth table and the symbol of a NOT gate.

Table 21.2 could be simplified to

input on results in output off
input off results in output on

This is regarding the transistor circuit described as a circuit for which the input is either off or on, and whose output is the opposite to its input. Circuits which handle pulses rather than continuously varying voltages are said to be using digital signals, and this type of signal is used almost exclusively in computers, calculators and electronic watches. Digital information is also being used increasingly in fields where continuously varying voltages, analogue signals, have formerly been used. Many telephone conversations are now translated into digital information before being transmitted, and then converted back into analogue information to operate the earpiece of the telephone of the receiver.

When dealing with digital information, the signal is a sequence of high and low voltages with a frequency which can be over 10^9 Hz. A high voltage is said to be logic state 1 and a low voltage is logic state 0. This is shown in Fig 21.11(a). If such a signal is passed through the transistor circuit of Fig 21.9, the output will be the reverse of the input as shown in Fig 21.11(b), and the circuit is called a **gate**. This particular gate is a **NOT** gate or **inverter**, and its function can be shown in a **truth table**, Fig 21.12, where the symbol for the NOT gate is also given. Fig 21.11(c) shows a digital signal which has become distorted as a result of continued amplification or transmission. One of the important virtues of digital signals is that when this happens the signal can be clipped top and bottom before being amplified back to its original shape.

A NOT gate is a single input gate. The circuit for a transistor as a switch can be modified to work from two inputs rather than one. If this is done, then there are 4 possibilities for digital signals to the two inputs. These are shown on Fig 21.13 together with a series of different outputs which are used. Each of these outputs can be obtained with suitable circuits. The names and symbols are also given in Fig 21.13. There is a lot of information in this diagram but it is logical, once some of the conventions which are used are clear.

ELECTRONICS

INPUTS		AND	NAND	OR	NOR	EX-OR	EX-NOR
A	B	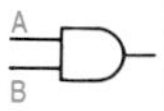	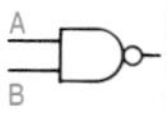	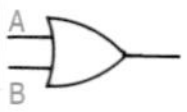	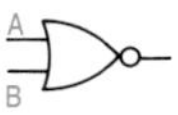		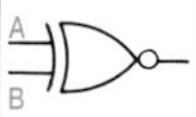
0	0	0	1	0	1	0	1
0	1	0	1	1	0	1	0
1	0	0	1	1	0	1	0
1	1	1	0	1	0	0	1

Fig 21.13 Truth tables and circuit symbols for 6 gates. The output of each gate is shown for the 4 possible input combinations.

- Each of the gates only has one output.
- For an AND gate, both A AND B have to be logic 1 for the output to be logic 1 (The symbol is a like a D).
- For an OR gate, either A OR B has to be logic 1 for the output to be logic 1 (The symbol is shaped like an OAR).
- A small circle before the output reverses the gate's output. So a NAND gate is the reverse of an AND gate, and a NOR gate is the reverse of an OR gate.
- The EX-OR and EX-NOR gates are shown by an additional line on the symbol. For an OR gate, either A or B is logic 1 for output logic 1. An Exclusive-OR gate is similar: A or B must be logic 1 **but not both**. For this reason an EX-OR gate is called a difference gate. The reverse of the EX-OR gate is the EX-NOR gate, and this is a parity gate: an output is obtained only when the inputs are the same.
- Although not shown here, three and more input gates are available.

Individually these gates have limited use, but it is possible to link them together in ways which enable a multitude of different functions to be performed. A computer uses thousands of linked gates. The output of one gate is connected to the input of another, and the process can be continued indefinitely. The following investigation into the behaviour of gates should be carried out in parallel with the theoretical analysis of them.

INVESTIGATION

The basic apparatus you will require for this investigation is:

a breadboard similar to the one shown in Fig 21.14
a smooth d.c. power supply within the range of 5–9 V
several LEDs, each with a 680 Ω resistor
connecting wires
certain specified integrated circuits (ICs)

The ICs suggested are all CMOS devices in the 4000B series. ICs are often covered with manufacturers' reference numbers and letters besides their series numbers. Usually you can ignore most of these numbers and letters.
Each IC in use must be inserted into the breadboard so that it straddles the centre of the breadboard.

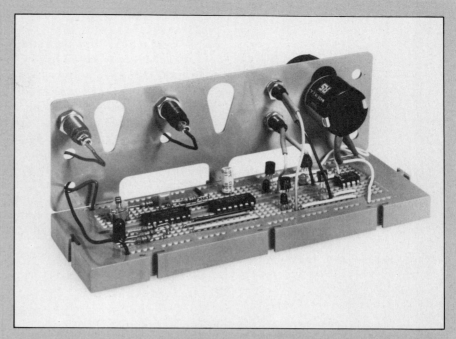

The top rail on the breadboard must be connected to the positive pole of the battery (say + 6 V) and the bottom rail must be connected to the negative terminal, 0 V. Fig 21.15 shows how the connections are made on a small section of breadboard. You must make sure that both ends of any component are never connected in the same vertical line of 5 holes because if it were, the component would be short-circuited by the connection beneath the five holes, shown by the red lines.

(a) To examine the behaviour of a single gate

A 4011 IC contains 4 NAND gates, of which 3 will not be used initially. The connections to its 14 terminals are shown in Fig 21.16 together with the other necessary connections and an LED. Connect terminal 14 to

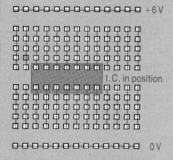

Fig 21.15 The blue lines on this section of blockboard indicate where internal connections are made. An integrated circuit is always placed across the central gap.

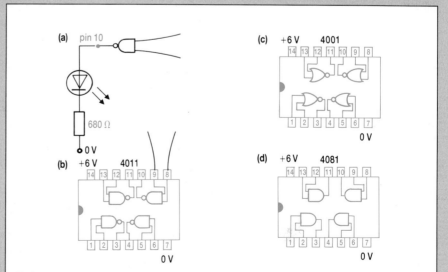

Fig 21.16 (a) The connections to a NAND gate in use. The other diagrams give the pin configurations for 3 common ICs. (b) is for the 4011 QUAND NAND gate, (c) is for the 4001 QUAND NOR gate and (d) is for the 4081 QUAND AND.

+6 V and terminal 7 to 0 V. The NAND gate being used has inputs on pins 8 and 9 and an output on pin 10. This output is connected through an LED and a 680 Ω resistor to the 0 V rail as shown in Fig 21.16(a). Free wire leads are shown used for the inputs, but these can be fed through switches if they are readily available.

The free leads should first be connected to 0 V. The truth table for a NAND gate indicates that this should give an output which is logic 1, so the LED should be lit. If it does not light, it is probably in the circuit the wrong way round and should be reversed. If it still does not light, check that the numbering on the pins of the IC is correct before you assume that your IC or LED is not working. If then one or other of the leads is transferred from 0 V to 6 V, the LED will still light up; but if both leads are connected to the +6 V rail, the LED will go out. Before completing this check, transfer the output connection through the LED to + 6 V instead of 0 V, and reverse the LED. How can you now connect the inputs to get the LED alight?

This simple check on a gate can be done on an NOR gate on IC 4001 or on an AND gate on IC 4081. Their pin connections are shown in Figs 21.16(c) and Fig 21.16(d).

(b)   Combinations of gates

You should now be able to use any combination of gates, but to start it is suggested that you use all four gates on a 4011 IC in the combination shown in Fig 21.17(a). Before you start this experimentally, work out what you expect in Fig 21.17(b). To do this, work through the table in the sequence shown. To find C, you should use A and B as inputs; to find D, you should use C and A as inputs, etc. F is the output from the circuit which can be connected to an LED but you could, if you so desired, connect LEDs at C and D and E also. Do not forget that every LED must have its own series resistor.

(c)   The half-adder

Binary arithmetic for the addition of two figures gives only the following possibilities:

$$0 + 0 = 00 \quad 0 + 1 = 01 \quad 1 + 0 = 01 \quad 1 + 1 = 10.$$

Fig 21.18 is a circuit which will perform this simple piece of arithmetic. The digits to be added are the inputs A and B, and the second digit of the answer is obtained from the output marked *sum*. The first digit of the answer is obtained from the output marked *carry*. This circuit can be constructed using 2 NOR gates on the 4001 IC and an AND gate on the 4081 IC.

(a)

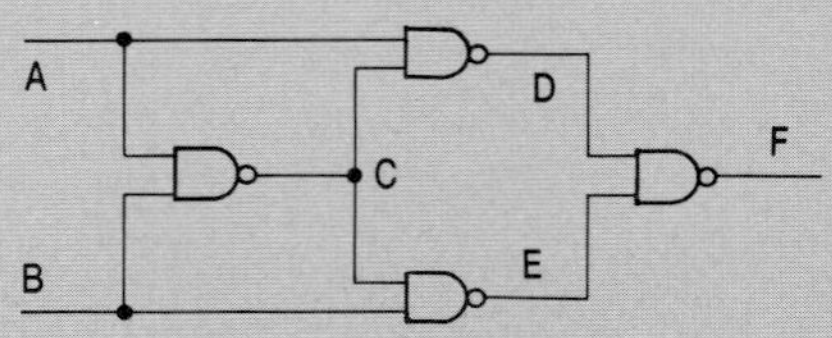

(b)

A	B	C	D	E	F
0	0				
0	1				
1	0				
1	1				

Fig 21.17

21.5 THE USE OF LOGIC GATES

Addition

A half-adder, as shown in Fig 21.18 and used in the investigation, is a useful point to start at when considering the use of logic gates. If 2 half-adders are put together, as shown in Fig 21.19, then a full-adder is made capable of adding 2 digits and a *carry* from a previous calculation. A truth table for a full-adder is given in Fig 21.19(c).

Chains of full adders can be used when the arithmetic to be done is more complex, and a computer also has to accept numbers in decimal form, change them into binary form, perform the calculation and then convert back into decimal form and display them. All of these operations are done by combination of different gates.

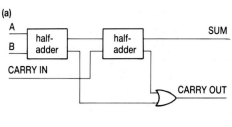

(a)

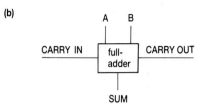

(b)

(c)

INPUTS			OUTPUTS	
A	B	CARRY IN	SUM	CARRY OUT
0	0	0	0	0
0	1	0	1	0
1	0	0	1	0
1	1	0	0	1
0	0	1	1	0
0	1	1	0	1
1	0	1	0	1
1	1	1	1	1

Fig 21.19 The full-adder.

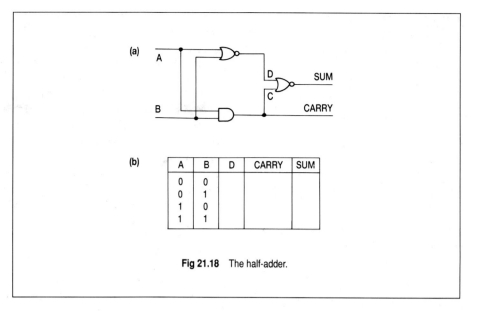

(a)

(b)

A	B	D	CARRY	SUM
0	0			
0	1			
1	0			
1	1			

Fig 21.18 The half-adder.

Multiplication

Multiplication of two digits

$$0 \times 0 = 0 \quad 0 \times 1 = 0 \quad 1 \times 0 = 0 \quad 1 \times 1 = 1$$

can be done directly with a single AND gate. The multiplication of other numbers involves this operation together with a shift register. This does the same thing in binary arithmetic as is done in decimal arithmetic when, say, multiplying 200×3. The answer 6 is the result of 2×3. The answer 600 is obtained by shifting the 6 two places to the left.

Counting

Up to this point, the output from any gate or combinations of gates has, at least in theory, been easy to find. In circuits like that shown in Fig 21.20, this is not the case.

Completing the truth table for the circuit involves an element of guesswork. You cannot find output D until you know output C, but you cannot find output C until you know output D. In this chicken-and-egg situation you need to try something and see if it works. For the last three lines of the truth table there is no problem in doing this, but if the inputs A and B are both zero then there are two possibilities. C will always be the reverse of D but can be either 1 or 0. A circuit such as this, with two stable states, is said to be **bistable**. Which of the two stable states it is in depends on how it reached that state. If the inputs were A = 0, B = 1 giving C = 1, D = 0 and then the inputs were changed to A = 0, B = 0, then C will remain at 1 and D at 0. If however the inputs started at A = 1, B = 0, before going to A = 0, B = 0 then C will be 0 and D will be 1. It is worth while setting this circuit up on the breadboard to see its behaviour in practice. This particular bistable circuit is called a set–reset (RS) flip-flop and illustrates **memory**. The behaviour of the circuit depends not only on its present condition but also on what has preceded it.

One particular use of an RS flip-flop is given in sequence in Fig 21.21. It is not essential to work out truth tables for this series of circuits but you should now be able to do so if necessary (this is question 21.21). The basic RS flip-flop is modified in Fig 21.21(a) by the addition of a clock pulse (CP) to make a so-called clocked RS flip-flop. The clock pulse can be supplied from a square wave oscillator changing regularly from logic 0 to logic 1 and back (shown as the input in Fig 21.22). The inputs to the clocked RS flip-flop are shown as R, S and CP, and the outputs are marked Q and $\overline{Q}$. $\overline{Q}$ is

(a)

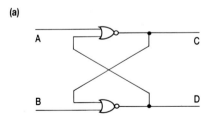

A

B

C

D

(b)

INPUTS		OUTPUTS	
A	B	C	D
0	0	or 1 / 0	0 / 1
0	1	1	0
1	0	0	1
1	1	0	0

Fig 21.20

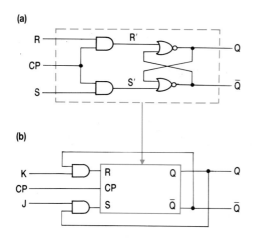

(a)

(b)

Fig 21.21 **(a)** a clocked set-reset gate and **(b)** a J-K flip flop.

read as not Q. It is always the opposite of Q, as the output $Q = 0$ and $\overline{Q} = 0$ is always avoided with this circuit. The clocked RS flip-flop is again modified in Fig 21.21(b) by the addition of two more AND gates. This circuit is called a JK flip-flop. Note how Fig 21.20 is inside Fig 21.21(a) and Fig 21.21(a) is inside Fig 21.21(b).

The clocked RS flip-flop only allows Q and $\overline{Q}$ to swap over if the clock pulse is logic 1. The JK flip-flop gives rise to a toggle condition if both J and K are logic 1. The outputs then reverse their state every time the clock goes to logic 1. Nothing happens when the clock goes to logic 0. This results in a halving of the clock frequency. If a whole series of JK flip-flops are used to produce a counter, a modification with a NOT gate is needed so that the outputs reverse their state every time the clock goes to logic 0 rather than logic 1. This is used in the circuit shown in Fig 21.22 where four modified

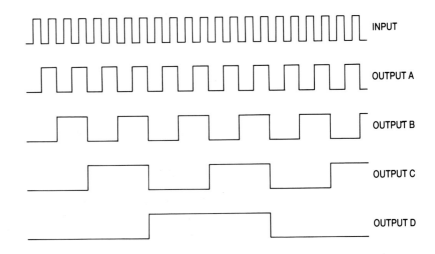

Fig 21.22 The counter circuit.

JK flip-flops are used with the J and K inputs fixed at logic 1. The clock input and the four outputs are shown in Fig 21.23.

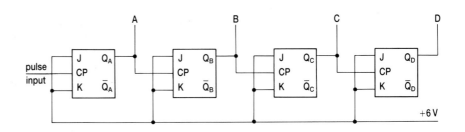

Fig 21.23 Pulses being counted. Each output flips from logic 0 to logic 1, or *vice versa*, only on negative going changes to its input. That is the voltage is going from more positive to more negative. The frequency is halved by each circuit unit.

There are a huge number of uses for logic gates in the whole field of microelectronics, and almost a different language is used to describe them. However, once you have established the principles on which they are based, you should have no difficulty in analysing any particular use, be it a memory register or a cascading counter or a synchronous counter or decoder or a Schmitt trigger or . . . etc.

ELECTRONICS

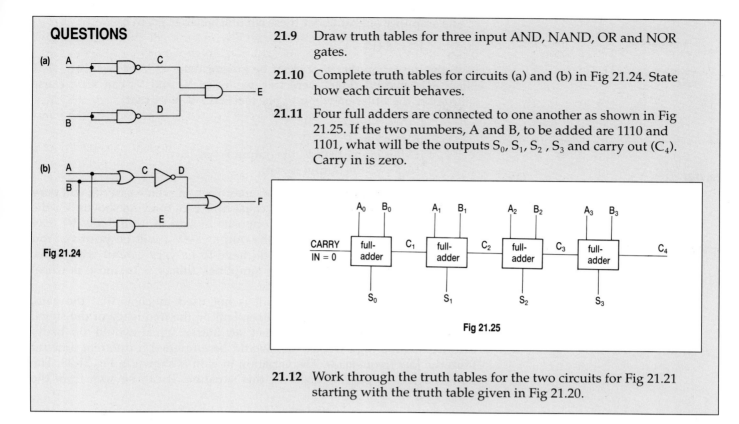

21.9 Draw truth tables for three input AND, NAND, OR and NOR gates.

21.10 Complete truth tables for circuits (a) and (b) in Fig 21.24. State how each circuit behaves.

21.11 Four full adders are connected to one another as shown in Fig 21.25. If the two numbers, A and B, to be added are 1110 and 1101, what will be the outputs S_0, S_1, S_2, S_3 and carry out (C_4). Carry in is zero.

Fig 21.24

Fig 21.25

21.12 Work through the truth tables for the two circuits for Fig 21.21 starting with the truth table given in Fig 21.20.

21.6 THE OPERATIONAL AMPLIFIER (OP-AMP)

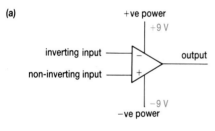

(b)

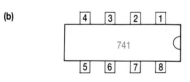

1 Offset null
2 Inverting input
3 Non-inverting input
4 −ve power
5 Offset null
6 Output
7 +ve power
8 No connection

Fig 21.26 (a) The op amp circuit symbol and (b) the pin connections for the 741 op amp.

The last three sections in this chapter concern amplification of analogue information using a particular circuit item called an operational amplifier (usually abbreviated to op amp). An op amp is an integrated circuit amplifier containing many components and having the following characteristics. It has

- a high gain (A) of the order of 10^5
- a high input impedance of the order of 2 MΩ
- a low output impedance of the order of 100 Ω
- an ability to accept positive, negative or alternating inputs
- two inputs, one of which gives an output of the opposite sign to the input

The symbol for an op amp is given in Fig 21.26 together with a plan of the pin configuration of the 741 op amp. The positive and negative power supplies must be equal and opposite. +9 V and −9 V are convenient values but any value between 5 V and 15 V can be used. These essential connections are always assumed, but not usually put on to circuit diagrams. The output voltage can never be higher than the voltage of the power supply. Two other connections need to be mentioned. They are marked as 'offset null' on pins 1 and 5, and enable the output to be adjusted to zero when the input is zero. This is only important when high gain is required for a d.c. input, and will not concern us here.

When a basic op amp circuit is used – and it must be admitted that the basic circuit is almost never used – inputs V_1 and V_2 can be applied to the inverting and the non-inverting inputs respectively as shown in Fig 21.27, and an output voltage will be produced. It is necessary to be particularly careful with the direction of an applied voltage. The convention used is for the head of the arrow showing the applied voltage to be taken to be positive, and for the voltage to be taken relative to the zero voltage of the power supply. It is usually convenient on a breadboard to have the bottom rail connected to the mid-voltage point between the positive and negative power supplies, here assumed to be +9 V and −9 V.

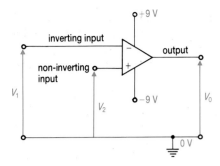

Fig 21.27

The amplifier output under these circumstances is given by

$$V_0 = A (V_2 - V_1)$$

with the limitation that V_0 cannot be greater than +9 V or less than –9 V. Since A is about 100 000, very tiny values of V_2 and V_1 can soon cause saturation, the situation when V_0 has reached 9 V. Note that

if V_2 is zero $V_0 = -AV$
if V_1 is zero $V_0 = +AV_2$
if V_1 is larger than V_2 the output is negative
if $V_2 = V_1$ the output is zero

Of course, the values of V_1 and V_2 may vary, in which case V_0 will vary correspondingly. An a.c. input will cause an a.c. output. As shown, V_1, the inverting input, causes an output opposite in sign to itself. Now V_1 may itself be negative, in which case the output, $-AV_1$, will be positive. You should gather from this that you do have to be very careful with signs, although there is one fact which simplifies things – in most practical circuits only one of the inputs is used.

The reason why this basic circuit is not used much is that the gain, quoted as about 100 000, is very dependent on the frequency of the signal being used. This would mean that an audio signal would be badly distorted as the high frequencies would be amplified a different amount from the low frequencies. The variation in gain is shown in Fig 21.28. The term 'open loop gain' is used for this situation. It can be seen from the

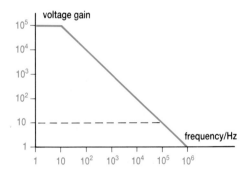

Fig 21.28 The open loop voltage gain for the 741 op amp plotted as a function of frequency. The meaning of the dotted line will be explained later in the chapter.

graph that the op amp amplified 1 kHz signals 1000 times, but 10 kHz signals only 100 times. The way in which an op amp can be used so that it does not produce distorted signals is by using a technique called negative feedback. This technique drastically reduces the gain of the op amp but it also enables good quality amplification to be obtained.

21.7 NEGATIVE FEEDBACK

This is not only used with op amps. Negative feedback is an extremely important control technique. The correct function of the human body is impossible without it. Standing erect puts you in unstable equilibrium. When a minor disturbance occurs, you start to fall over and set in train a series of actions – which you do not even notice unless you have your eyes closed, because the system of feedback is so good. As you move, your eyes detect small changes in the image on the retina, so a signal is sent to the brain. The fluids in the semicircular canals in your ears start to move. That too is detected and messages are sent to your brain. Your brain controls muscles in your feet which restore you to equilibrium. The foot muscles exert forces to move you back if you fall forward, or to move you forward if you move back. They exert large forces if you have moved a long way, small forces if you have only moved a short way. This is feedback in action.

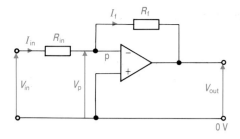

Fig 21.29 The basic negative feedback circuit.

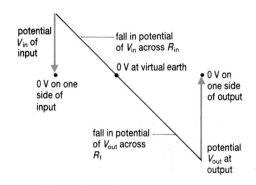

Fig 21.30 A chart showing how the potential changes around the circuit. Three points at which the potential is almost zero make the theory of the op amp very much simpler than it would otherwise be.

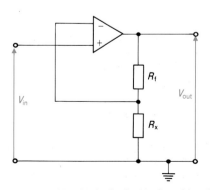

Fig 21.31 Negative feedback when the non-inverting input is used.

The same principle is involved in steering a car or bicycle. If you find yourself going too far to the right when on a straight course you move the steering wheel or handlebars to the left. If you have not corrected enough, then you move them a bit further. It is a sensitive system and is carried out continually and almost imperceptibly.

Feedback in an op amp is done by connecting a feedback resistor, R_f, from the output back to the inverting input as shown in Fig 21.29. The feedback is in antiphase to the input because the inverting input of the op amp is being used.

The gain in a circuit using negative feedback is called the **closed loop gain** and to calculate it some approximations need to be made. The approximations are good provided the resistances of circuit components are low compared with the input impedance of the op amp. The approximations made are:

(a) The current through R_{in} equals the current through R_f. At first sight this seems strange, since perhaps a feedback current is expected in the opposite direction, but remember, the input impedance of the op amp is of the order of 2 MΩ whereas R_{in} and R_f are usually in the range of 1 kΩ–100 kΩ. Only a small fraction of I_{in} can therefore go into the op amp inverting input.

(b) The potential at P is near to zero. P is called a **virtual earth**. This arises because of the very large open loop gain A of the op amp. The value of A could well be 100 000 and since

$$\frac{V_{out}}{V_P} = A$$

this makes V_P extremely small since the maximum possible value of V_{out} is only 9 V or so.

These two approximations now make the theory extremely easy – especially if a chart of potential around the circuit is used (Fig 21.30). Because of the virtual earth, any rise in potential caused by the input is balanced by a similar loss in potential across R_{in}. Similarly the fall in potential across R_f must be equal to the rise in potential across the output. This gives

$$V_{in} = I_{in} R_{in}$$

and $-V_{out} = I_f R_f$

But by using the approximation $I_{in} = I_f$ we get

$$\frac{V_{out}}{V_{in}} = -\frac{I_f R_f}{I_f R_{in}}.$$

Closed loop gain for inverting input $= \dfrac{V_{out}}{V_{in}} = -\dfrac{R_f}{R_{in}}.$

This shows that the closed loop gain does not depend upon the open loop gain, A, at all. It is determined only by the values of R_{in} and R_f. The gain can be altered at will by choosing suitable values for R_{in} and R_f. If the gain required is 10, then R_f could be 10 kΩ and R_{in} 1 kΩ. This constant gain can be achieved over a range of frequencies from zero to 10^5 Hz. The dotted line on Fig 21.28 shows the frequency range for this gain.

This circuit should be set up on a breadboard and with an input fixed at, say, 0.5 V, the output should be measured for different values of R_{in} and R_f.

If the input signal is to be used with the non-inverting input, the feedback signal is still fed into the inverting input. The circuit is shown in Fig 21.31 and the closed loop gain in this case can be shown to be given by

Closed loop gain for non-inverting input $\dfrac{V_{out}}{V_{in}} = 1 + \dfrac{R_f}{R_{in}}$

21.13 Plot a sketch graph to show how the output of an op amp with open loop gain of 100 000 varies with the inverting input if the non-inverting input is zero. (See Fig 21.27.)

21.14 An op amp using negative feedback (Fig 21.29) has an input voltage of 26 mV and an input resistor (R_1) of 1.2 kΩ. Find the value of the output potential difference if the feedback resistor R_f has a value of 8.6 kΩ. Find also the values of input current I_{in} and feedback current I_f.

21.15 A non-inverting amplifier, Fig 21.31, has a 50 Hz a.c. input of 250 mV r.m.s. Its supply potentials are +5 V and −5 V. Sketch a graph showing how the output potential varies with time if $R_f = 19$ kΩ and $R_x = 1.0$ kΩ.

21.8 PRACTICAL CIRCUITS USING THE OP AMP

The obvious example for use of an op amp is for producing amplification in an amplifier. The op amp is however much more than just an amplifier. It is better to think of it as a manipulator of electrical signals. It can take any signal and multiply it or add a constant to it, or subtract it from another signal, or differentiate it or integrate it – to give a few examples.

Signals can be combined with one another in a bewildering variety of circuits. The complexity of outputs which are possible from electronic synthesisers, for example, are due to the use of many op amps. The photograph of a keyboard synthesiser is shown in Fig 21.32, and it can be seen from the variety of different control switches how different sounds can be produced although the same notes are played. Mixers, for use with amplifying systems, use op amps, and even the sound output of a record player can be adjusted for personal preference by using op amps which control different frequencies.

Fig 21.32 A keyboard synthesiser. Some of the controls and options are marked, pitch, modulation, volume, tempo, memory, edit, rhythm, oscillator. These enable sounds to be manipulated and memorised to give a wide variety of musical output.

In analogue computers, op amps can be used to solve differential equations. If the displacement x of an object is measured electrically, its velocity can be found by passing the electrical signal x through a differentiating op amp to find the velocity

$$\frac{\mathrm{d}x}{\mathrm{d}t}$$

If this output is then passed through another differentiating op amp, the output will be the acceleration

$$\frac{\mathrm{d}x^2}{\mathrm{d}t^2}$$

All of these outputs could then be put into an adding and multiplying op amp to give say

$$\frac{dx^2}{dt^2} + 4\,\frac{dx}{dt} + 3x$$

as an output. If values for this to be zero are required, they can be found from the final output.

You will not at this stage be required to analyse op amp circuits of any great complexity, and there is certainly no need to memorise any of the circuits given in Fig 21.33. They are given to illustrate the variety of circuit possible and to see if you understand the principles of their operation and can work out some details. All of them are straightforward to construct on a breadboard with a 741 op amp.

Fig 21.33 Six common op amp circuits – see question 21.17.

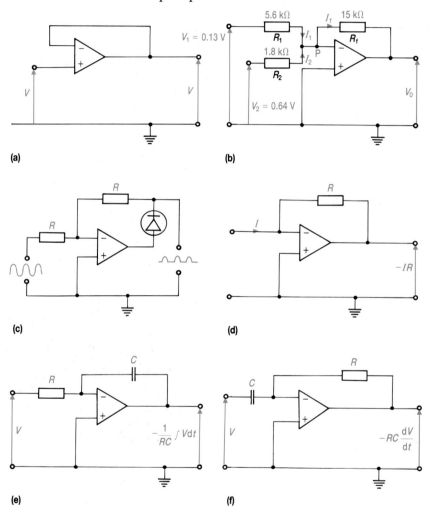

(a)

(b)

(c)

(d)

(e)

(f)

Example 3

Find the output from the circuit in Fig 21.33(b) using the details given on the figure.

Since the input impedance of the op amp inverting input is high, a negligible current enters it at P. Therefore

$$I_f = I_1 + I_2$$

Since the potential at P is a virtual earth:

the p.d. across $R_1 = V_1 = I_1 R_1$
the p.d. across $R_2 = V_2 = I_2 R_2$
the p.d. across $R_f = -V_0 = I_f R_f$

$$\text{so} \qquad -\frac{V_0}{R_f} = \frac{V_1}{R_1} + \frac{V_2}{R_2}$$

$$-\frac{V_0}{15 \text{ k}\Omega} = \frac{0.13 \text{ V}}{5.6 \text{ k}\Omega} + \frac{0.64 \text{ V}}{1.8 \text{ k}\Omega}$$

$$= 0.0232 \text{ mA} + 0.3556 \text{ mA}$$

$$V_0 = -15 \text{ k}\Omega \times 0.3788 \text{ mA}$$

$$= -5.68 \text{ V}$$

In this example, the output is given algebraically by

$$V_0 = -\left(\frac{R_f V_1}{R_1} + \frac{R_f V_2}{R_2} \right)$$

It should be noted that the output is therefore the sum of the outputs which each input would have given on its own. This is an adder circuit. If the circuit is to be used for direct addition then R_1, R_2 and R_f are made equal, in which case $V_0 = (V_1 + V_2)$.

This can be extended as required by using an R_3, R_4, R_5 etc. . . . If subtraction is required, then one of the inputs can be connected into the circuit the opposite way round. An adder circuit such as this is used as a mixer and it would not be unusual to have the inputs from thirty microphones all connected into the same op amp.

QUESTIONS

21.16 Sketch circuit diagrams showing circuits which will give the following outputs. V_1, V_2, V_3 etc. are inputs.
(a) $-(V_1 + V_2 + V_3)$
(b) $-(V_1 + V_2 - V_3)$
(c) $-(V_1 + \frac{1}{2}V_2)$
(d) $-(V_1 + 3V_2)$
(e) $-(2V_1 + 3V_2 + 4V_3)$
(f) $-6(V_2 - V_1)$
(g) $-3(V_1 - \frac{1}{2}V_2 + \frac{1}{3}V_3 + \frac{1}{4}V_4)$

21.17 All these questions refer to the circuit diagrams of Fig 21.33. Remember that you are not expected to know these circuits (apart from (b)), but you should be able to see the principle behind their operation.

The circuits are, not in this order, a voltage adder, an integrator, a current to voltage converter, a voltage follower, a differentiator and a precision rectifier.
(i) Which circuit is which?
(ii) What is the purpose of circuit (a) where the input and the output potential difference are the same?
(iii) Why are the two resistors in (c) the same?
(iv) Show that the output in (d) is $-IR$.
(v) If V has a constant value of 6 mV in circuit (e), $C = 2.0$ μF and $R = 120$ kΩ, how long will it take for the output to saturate if the supply voltages are +9 V and –9 V?
(vi) Sketch a graph of the output from circuit (f) if the input is a square wave of frequency 200 Hz varying between +2 V and +5 V. $R = 120$ kΩ and $C = 2.0$ μF as in (v). In what respect is your sketch inaccurate?
(vii) Why would the values of R and C need to be changed if the frequency were 200 kHz instead of 200 Hz?
(viii) Explain the action of circuit (c). Why do you think it is given this name?

The last practical circuit which will be mentioned in this chapter is that of an astable oscillator. The op amp has been described as a manipulator of electrical signals. In the case of the astable oscillator, however, it is a creator of an electrical signal. A simple two transistor circuit can also give an astable output, as can a digital integrated circuit. To create oscillations in any of these circuits, feedback is again needed. This time, however, the feedback must be positive feedback. Positive feedback gives rise to instability in circuits. The gain rises until the output potential swings wildly from supply potential + to supply potential −. An example of positive feedback creating oscillations, which you are probably familiar with, is the effect you get if you put a microphone of a public address system too near to its loudspeaker. Tiny vibrations in the microphone are amplified, sent to the loudspeaker and then picked up again by the microphone with increased amplitude. The process repeats itself rapidly so it takes very little time before there is an awful screeching sound coming from the loudspeaker.

This effect is made use of in the circuit called an astable multivibrator shown in Fig 21.34. Both the inverting and the non-inverting inputs are used. Analysis of oscillator circuits is more difficult than other circuits in that it is not sensible to begin at the beginning. This is because initially the oscillation process depends on transient currents in the circuit – as with the screeching loudspeaker. What needs to be shown is that once the circuit is oscillating it will continue to oscillate. Assume that for some reason the output has just become saturated positive – say to $+V_s$. We then need to show that a cycle of events takes place which ends with the output going to $+V_s$ again so the cycle is ready to start once more.

With the output potential $+V_s$, the potential at P is

$$\frac{R_3 V_s}{R_2 + R_3}$$

since R_2 and R_3 make a potential divider. This assumes that R_2 and R_3 are small compared with the input impedance of the non-inverting input, so that I_2 is approximately zero. The potential at V_2 equals the potential at P and has this constant value of

$$\frac{R_3 V_s}{R_2 + R_3}$$

provided V_0 is constant.

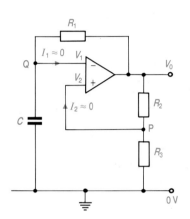

Fig 21.34 An op amp astable multivibrator circuit.

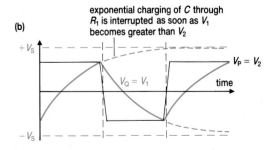

Fig 21.35 Graphs showing how the potentials different parts of the circuit vary with time in a multivibrator. V_P is the potential at P and V_Q is the potential at Q.

Also connected between the output and earth is R_1 and C in series. C will charge up exponentially through R_1, since I_1 is also approximately zero, and the rate of charging depends on the time constant CR_1 (see Fig 21.35). Whatever the state of charge of C initially, it would eventually have a potential difference across it of $+V_s$ were it not for the fact that while charging up the potential at Q, which is V_1, will become larger than V_2. Since the output from the op amp is $A(V_2 - V_1)$, the switch from V_1 being smaller than V_2 to being greater than V_2 will suddenly cause the output V_0 to change from $+V_s$ to $-V_s$.

V_2 becomes

$$-\frac{R_3 V_s}{R_2 + R_3}$$

and the capacitor which had been charging up through R_1 now discharges through R_1 and V_1 falls. When it has fallen far enough, V_1 becomes less than

$$-\frac{R_3 V_s}{R_2 + R_3}$$

and V_0 suddenly changes back from $-V_s$ to $+V_s$ to complete one cycle. With suitable values of C and R_1, the cycle can be completed in as little as a few microseconds, giving a square wave output with a frequency up to hundreds of kilohertz. If a triangular waveform is required instead of a square wave, then R_3 can be made smaller so that less of the exponential charge and discharge is used, and the potential at Q is nearly triangular in form.

The circuit is called a multivibrator because a square wave has many harmonics of the fundamental frequency present. A filter circuit connected to the output can therefore be suitably adjusted to supply a wide range of frequencies.

SUMMARY

- The base current for a transistor is small and it controls the collector current. The current gain, $h_{fe} = I_c/I_b$. When a transistor is used as a switch, an input gives no output and no input gives an output.
- Truth tables and symbols for 7 gates:

1 input gate

	input	output
NOT gate	0	1
	1	0

input ⟶ ⊳○ ⟶ output

2 input gates

INPUTS		AND	NAND	OR	NOR	EX-OR	EX-NOR
A	B	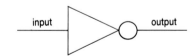					
0	0	0	1	0	1	0	1
0	1	0	1	1	0	1	0
1	0	0	1	1	0	1	0
1	1	1	0	1	0	0	1

- A bistable circuit has two stable states. A trigger can be used to make it change from one state to the other. This is used in counting circuits.
- An astable circuit has no stable states. It continually changes from one state to another generating square waves (a multivibrator).
- An operational amplifier (op amp) has very high input impedances on both its inverting input and on its non-inverting input, and a very high voltage gain, called the open loop gain, A, which is dependent on the frequency. The output is given by $V_0 = A (V_2 - V_1)$.
- By the use of negative feedback, the op amp's gain can be made to depend on the value of the resistances of the resistors in the circuit. Since the inputs are virtual earths

$$\text{using the inverting input} \quad \frac{V_{out}}{V_{in}} = - \frac{R_f}{R_{in}}$$

$$\text{using the non-inverting input} \quad \frac{V_{out}}{V_{in}} = 1 + \frac{R_f}{R_{in}}$$

EXAMINATION QUESTIONS

21.18 Write down the truth table for a NAND gate and hence show how it may be made to function as a NOT (inverting) gate. (4)

(UCLES 1988)

21.19 (a) Construct a truth table for each part of the circuit outlined by a dotted box in the diagram. For values of the inputs at A and B give the corresponding values at each labelled point in the circuit and hence identify the logic gate equivalent of each box.

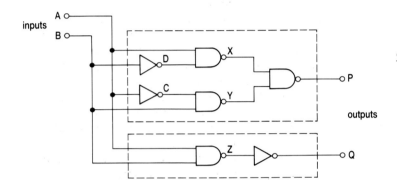

State the function of the complete circuit and distinguish between the two outputs. (5)

(JMB 1987)

21.20 (a) Give the truth tables for simple (i) AND, (ii) NOR (iii) NOT logic gates. Describe in words the action of each type of gate. (6)

(b) Draw up a truth table for the logic system shown in (a) below, including the logic states for C and D. (4)

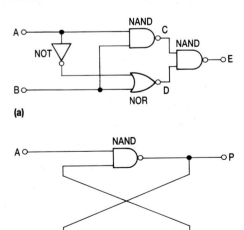

(c) A set of three warning lights is used in the control panel for an industrial process. An alarm sounds when any one of the following states occurs:
(i) all lamps are lit
(ii) lamps A and B are lit
(iii) lamps A and C are lit.
Complete a truth table showing the alarm response to all possible states of the lights. Draw a circuit diagram of a system to operate the alarm. (6)

(d) Write down the truth table for the system shown in diagram (b) and comment on your result. (4)

(UCLES 1986)

21.21 (a) Explain what is meant by the terms *astable* and *bistable*.
The diagrams illustrate a circuit which has inputs P_1 and P_2 and outputs S_1 and S_2.
(i) Identify the three types of logic gate shown in (a) and write out their respective truth tables. (3)

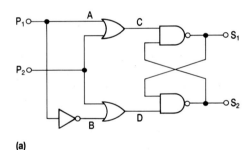

(a)

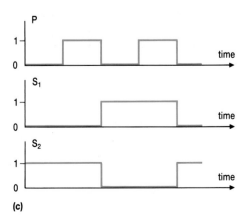

(ii)

	P_1	P_2	A	B	C	D	S_1	S_2
1st state	1	0						
2nd state	1	1						
3rd state	0	1						
4th state	0	0						
5th state	0	1						

Copy out the logic table given. P_1 is initially set to 1 and P_2 to 0 (1st state). Complete the first line of the logic table. P_1 and P_2 are then changes to the states shown in order down the table; complete the rest of the table.   (5)

(iii) By reference to the table, explain how the circuit has exhibited the property of memory.   (2)

(b) (i) Diagram (b) represents a circuit with a single input and outputs S_1 and S_2. Diagram (c) shows how the states of S_1 and S_2 vary as logic pulses are applied at P.
Describe in words the action of this circuit.   (2)

(ii) Draw a block diagram to illustrate how four of these circuits may be connected together to produce a simple binary counter. Show at which points logic indicators would be connected to enable the counter to be read.   (3)

(UCLES 1986)

21.22 Draw an amplifying circuit for which the voltage gain is 8.
The power output signal may carry more power than the input signal. What is the source of this extra power?
Explain, with reference to your circuit, the term *negative feedback*.   (8)

(ULSEB 1987)

21.23 (a) Explain what is meant by (i) the open loop gain of an operational amplifier, (ii) saturation of the output of the amplifier.

(b) The open loop gain of an operational amplifier is 10^5 and the supply voltage is ±15 V. Calculate the minimum potential difference which may be applied between the input terminals to produce saturation of the amplifier.

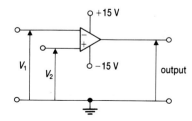

(c) In the circuit shown in the diagram, V_1 and V_2 are input voltages to an operational amplifier of large open loop gain. V_1 is a sinusoidal signal of peak value 10.0 V and V_2 is a 5 V d.c. signal.
On the same axes, sketch graphs of voltage against time for each of the two input voltages and for the output voltage.   (10)

(AEB 1987)

21.24 Explain the relationship between the r.m.s. current in a resistor in an a.c. circuit and the power dissipated in it. On one set of labelled axes, sketch graphs showing the instantaneous and mean power dissipated in a resistor carrying a current $I = I_0 \sin \omega t$.   (4)
A transformer can be arranged to give a larger output voltage than the input voltage applied to the primary, yet it is not usually regarded as an amplifier. Explain the distinction between the step-up transformer and the amplifier.   (3)
Explain what is meant by *feedback* in operational amplifier circuits. Discuss the part played by positive and negative feedback in operational amplifier circuits. Give one example of each. Draw suitable circuit diagrams and identify the feedback components in each case.   (7)
The diagram shows an ideal operational amplifier and its associated components. The input voltage V_{in} is a sinusoidal signal of amplitude 0.14 V.

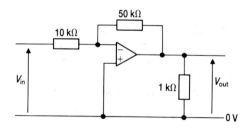

(a) Calculate values for (i) the voltage gain, and (ii) the power gain of the circuit.
The power gain is defined as the ratio of the power dissipated in the load resistor (here 1.0 kΩ) to the power absorbed by the input circuit.

(b) On one set of axes, sketch graphs of the input and output voltages against time. Put a scale on the voltage axis.   (6)

(O & C 1987)

21.25 Resistors of the values shown in the diagram are connected to an operational amplifier A. If the two d.c. input signals are $V_1 = 0.1$ V and $V_2 = 0.6$ V, what is the output V_0?   (3)

(UCLES 1987)

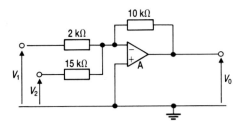

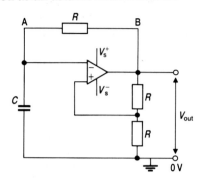

21.26 The diagram shows an operational amplifier used as an astable multivibrator.

(a) Assuming that initially the capacitor is uncharged and that V_{out} is positive, explain how a square wave output is produced.   (5)

(b) Explain how the form of the output would change if the resistance of the resistor between the points A and B were increased.   (2)

(JMB 1986)

21.27 (a) Draw a circuit diagram of an operational amplifier used in the inverting mode as a summing amplifier with two inputs. Show that, if the input resistors of resistance R_1 and R_2 respectively are of equal magnitude and the feedback resistor has resistance R_f, the output voltage V_{out} is given by

$$V_{out} = \frac{R_f}{R_1} \; (V_1 + V_2)$$

where V_1 and V_2 are the voltages applied to the two inputs. Show how you would modify the summing amplifier so that it would accept positive input voltages but ignore negative input voltages. Explain your modification.   (7)

(b) Taking the values of R_1 and R_2 to be 5 kΩ draw another circuit diagram, giving component values which would *multiply* the sum of positive input voltages by –2 but would *divide* the sum of negative input voltages by –2. Explain your modification. Calculate V_{out} when
(i) $V_1 = +2$ V and $V_2 = +3$ V,
(ii) $V_1 = +2$ V and $V_2 = -8$ V.
Under which circumstances would the output voltage not have the expected value?   (6)

(JMB 1987)

Theme 5

MATTER

The structure of matter has intrigued people from our very earliest history until the present day. In this century many answers have been found to age-old problems, yet new questions are continually being asked. Much remains to be discovered about the solids, liquids, gases and plasmas which together provide a universe of which we occupy such a tiny part.

Chapter 22

PHASES OF MATTER

LEARNING OBJECTIVES

At the end of this chapter you should be able to:

1. define and use the terms density and relative density;

2. describe the arrangement of molecules within solids, liquids and gases;

3. use the mole and the Avogadro constant correctly;

4. explain the significance of both kinetic and potential energies in determining the phase of matter;

5. define and use the term specific latent heat;

6. define and use the term pressure;

7. explain how the effect of pressure in a fluid gives rise to a buoyancy force on an object within the fluid, and why bodies float.

22.1 DENSITY

Density is defined as mass per unit volume. An object with a mass m and volume V therefore has a density ρ given by the equation

$$\rho = \frac{m}{V}$$

The SI unit of density is the kilogram per cubic metre, $kg\,m^{-3}$. The densities of various materials are given in Table 22.1. Some of the values given are only estimates because either the density can vary within the material or there is uncertainty in making the measurements. Other values are known more accurately. The density of copper, for example, has the value of $8900\ kg\,m^{-3}$, and this value varies only a very small amount with temperature and pressure, and does not vary within the structure of different pieces of copper. The density of a substance such as wood, however, does vary from place to place within any particular tree. Different parts of the ring system within each growing tree have different densities, depending on the time of year and on whether the growing season was wet or dry. Substances, like copper, which have a uniform density throughout are said to be **homogeneous**. Wood, on the other hand, is non-homogeneous. Average densities can be quoted for non-homogeneous substances, but they are often not very reliable.

In dealing with the density of gases, the temperature and pressure must be quoted as these have a large effect on the density. For liquids, the density varies with the temperature and to a lesser extent with the pressure. Pressure variation makes very little difference to the density of a solid, and temperature has only a small effect.

Table 22.1 Densities of various materials. Where given, pressures are quoted in atmospheres. 1 atmosphere $\approx 10^5$ Pa

Material	Density /kg m^{-3}
interstellar space	~10^{-20}
air in 'vacuum' in TV tube	~10^{-7}
air at 0 °C and 1 atmos.	1.29
air at 100 °C and 1 atmos.	0.95
air at 0 °C and 50 atmos.	65
balsa wood	150
ethanol	810
ice at 0 °C and 1 atmos.	920
water at 100 °C and 1 atmos.	958
water at 0 °C and 1 atmos.	1000
water at 0 °C and 50 atmos.	1002
aluminium	2700
iron	7800
copper	8900
lead	11 300
mercury	13 600
gold	19 300
osmium (densest element)	22 500
centre of Sun	160 000
white dwarf stars	~10^9
atomic nuclei	~10^{16}

The term **relative density** is sometimes used. This is a dimensionless quantity defined by the equation

$$\text{relative density} = \frac{\text{density of substance}}{\text{density of water}}$$

QUESTIONS

A common mistake in problems dealing with density is to substitute relative density into formulae instead of density. An error of 1000 is consequently introduced.

22.1 Using values given in Table 22.1, find the relative density of mercury. Find the mass of 1 cm^3 of mercury. (These two values are numerically the same but their units are different.)

22.2 The density of antifreeze for a car's radiator determines the temperature at which it will freeze. Antifreeze is a mixture of water and ethylene glycol. If the density of the antifreeze in the radiator of a particular car is 1045 kg m^{-3}, find the proportion by volume of ethylene glycol in the antifreeze mixture. The density of ethylene glycol is 1120 kg m^{-3}. What is the proportion of ethylene glycol by mass?

22.2 PHASE

Mention was made in section 22.1 of the different properties of solids, liquids and gases. **Phase** is the general term used to indicate these three properties. (The word phase as used here has no connection with the use of the word in waves and oscillations.) It is not only in density measurements that the differences between the three phases of matter are important. In making measurements of elasticity, heat conductivity, electrical conductivity, heat capacity and many more, there is often a startling difference between a property of the solid and the corresponding property of the liquid. The difference is even greater when the substance changes to a gas. In this section, the phase of a substance will be related to the pattern of molecular movement. Three phases of matter are usually easily recognised as **solid**, **liquid** and **gas**. There are materials, however, which do not fit well into any of these categories, but usually these materials are those with a wide variety of different molecules in their constitution. Butter, for example, does not melt at one particular temperature. It gradually gets softer and more liquid as the temperature rises, so it is difficult to know whether to classify soft butter as a solid or a liquid. But then, butter is not a simple substance chemically. It is its complex molecular composition which gives rise to its lack of a unique melting point. To solid, liquid and gas should be added a fourth phase at extremely high temperatures. This high temperature phase is called **plasma** and consists of atomic nuclei and electrons all moving with very high speed but not connected to one another as would be the case at lower temperatures. Since stars are at high temperatures and most of the mass of the universe is concentrated in the stars, it has been estimated that 90% of all the matter in the universe is plasma.

Not all substances can exist in all phases, though by adjusting the temperature and the pressure many can. Many biological substances, for example, would undergo chemical changes at much lower temperatures than those at which they would be able to melt. Heating sugar will give a sticky black mess before liquid sugar is produced. This is because, on heating, the carbohydrate molecules of the sugar lose water, leaving behind black carbon.

22.3 THE MOLECULAR STRUCTURE OF MATTER

Many scientists have contributed to the idea of molecular structure. Early scientists had no direct evidence of molecules but were working from indirect evidence. Dalton first proposed that an element was composed of many identical atoms. Avogadro and Gay-Lussac's experiments on gases led them to believe that chemical reactions took place between elementary units of matter. Mendeleyev laid down the basis for the modern periodic table of elements.

Certain terms used when dealing with the molecular structure of matter need to be given at this stage.

The **mole**, which was defined in section 1.2 as one of the seven base units of the SI system of units, is a measure of the amount of substance. It is the amount of substance which contains as many entities as there are atoms in 0.012 kg of carbon 12. Experimental determination shows that the number of atoms in 0.012 kg of carbon 12 is $(6.022\ 17 \pm 0.000\ 02) \times 10^{23}$. A mole of hydrogen molecules therefore contains $6.022\ 17 \times 10^{23}$ molecules. It will be two moles of hydrogen atoms because each hydrogen molecule contains two hydrogen atoms. These figures enable the mass of a single atom of carbon 12 to be calculated as $0.012\ kg/(6.022\ 17 \times 10^{23}) = 1.992\ 64 \times 10^{-26}$ kg. The mass of a single atom of carbon 12 is also, by definition, twelve times the **unified atomic mass constant**.

The unified atomic mass constant is therefore

$$(1.992\ 64 \times 10^{-26}\ kg)/12$$
$$= 1.6611 \times 10^{-27}\ kg$$

This mass is sometimes simply called the unit u, of atomic mass. It is approximately the mass of a hydrogen atom and also approximately the mass of a proton or neutron.

$$1\ u = 1.6611 \times 10^{-27}\ kg$$

There may be a time in the future when the unit of mass will be based, say, on the mass of a proton, but at present the link between the mass of sub-atomic particles and the mass of any piece of matter has to be found experimentally.

The **molar mass** M_m is the mass per mole of the substance and has the unit kg mol⁻¹.

The **molar volume** V_m is the volume per mole of a substance and has the unit m³ mol⁻¹.

The **Avogadro constant** N_A is numerically equal to the number of atoms in 0.012 kg of carbon 12.

Its value is $N_A = 6.022\ 17 \times 10^{23}$ mol⁻¹

The **relative molecular mass** M_r is defined by the equation

$$M_r = \frac{\text{mass of a molecule}}{\text{a twelfth the mass of a carbon 12 atom}}$$

$$= \frac{\text{mass of a molecule}}{1.6611 \times 10^{-27}\ \text{kg}}$$

M_r has no unit.

Direct evidence for molecules and for their movement came with the discovery of the tiny, jerky motion of minute grains of pollen when suspended in water. The motion is called Brownian motion after Robert Brown who first recorded it in 1827. Einstein first analysed the motion mathematically and was able to deduce a value for the Avogadro constant from his deduction. Agreement between the value so deduced and other experimental values gave strong evidence for the theory. Brownian motion can be seen clearly in the motion of smoke particles in air and gives direct evidence of collisions between individual molecules and the tiny smoke particles.

INVESTIGATION

Brownian motion

If you have not previously seen Brownian motion, then you should try to use a microscope to see the effect now. Brownian motion gives direct evidence for perpetual molecular movement. High power is not necessary for the magnification of the microscope, but good illumination of the particles is required.

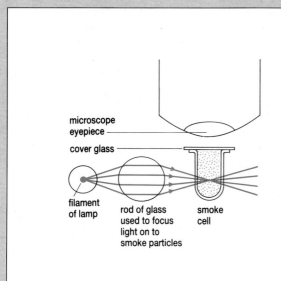

microscope eyepiece

cover glass

filament of lamp

rod of glass used to focus light on to smoke particles

smoke cell

Fig 22.1 The experimental arrangement for viewing Brownian motion. The rod of glass is to focus from the lamp on to the smoke particles.

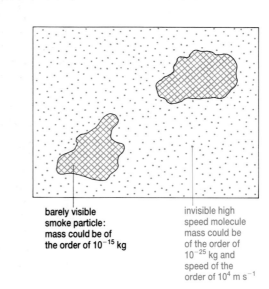

barely visible smoke particle: mass could be of the order of 10⁻¹⁵ kg

invisible high speed molecule mass could be of the order of 10⁻²⁵ kg and speed of the order of 10⁴ m s⁻¹

Fig 22.2 On a large scale the barely visible smoke particles appear as large patches. They might have a mass of about 10⁻¹⁵ kg. The dots represent invisible molecules whose mass may be about 10⁻²⁵ kg and whose speed is of the order of 10³ m s⁻¹.

There are several commercial cells available for viewing the motion of smoke particles in air and one of these is illustrated in Fig 22.1. If you use a cell such as this, it is important to focus the microscope on the region in which the light is also being focused. The smoke particles are barely visible; they appear as minute spots of light which are probably much smaller than you would expect. You do not really see a particle and certainly could not describe its shape. What you can just see is a small amount of light scattered from the surface of a particle whose mass is probably about 10^{-15} kg. A particle such as this is illustrated in Fig. 22.2. It is surrounded by a vast number of invisible molecules which continually strike it. The random nature of these impacts causes just enough irregularity for the movement of the smoke particles to be visible. If you do not have one of these cells available, then simply place a spot of water on to a microscope slide, add a trace of some fine powder of intense colour (carmine is suitable), put a cover glass on top of the drop and view the powder. Watch a single smoke or carmine particle for a time while it remains in focus, and then draw a sketch showing a typical pattern of its movement.

QUESTION

22.3 The kinetic energy of a smoke particle of mass 10^{-15} kg is the same as that of a nitrogen molecule of mass 5×10^{-26} kg and speed 1×10^3 m s^{-1}. What is the speed of the smoke particle?

Modern experiments have reinforced earlier ideas that matter consists of particles and that these particles are in rapid motion. The photographs, Figs 22.3 and 22.4, taken using a field ion microscope and an electron microscope respectively, both show the pattern of molecules within a crystal.

Fig 22.3 A field ion micrograph of atoms of iridium. The tiny dots are the locations of individual atoms.

Fig 22.4 A high resolution electron micrograph of the atomic lattice of a thin gold crystal. At 16 000 000 times magnification each patch represents an individual gold atom.

Any theory of the structure of matter must be able to explain why some substances are solid, some are liquid and some are gaseous. The theory that the particles of matter are in rapid motion can explain why gases are possible, but seems to make the explanation of the solid phase difficult. The

PHASES OF MATTER

molecules photographed in Figs 22.3 and 22.4 do not appear to be in rapid motion. They do not appear to be particularly blurred, as photographs of fast-moving objects often are.

The key to understanding this paradox lies in realising that there are two conflicting influences on the molecules in any substance. On the one hand there is the fact that they are moving and therefore have kinetic energy: on the other hand they clearly attract one another. This is why the molecules in Fig 22.3 are so close to one another and so uniformly packed. Because they attract one another, they would need to gain potential energy to become separated. It is the balance between the kinetic energy and the potential energy of the molecules that is all important.

The appreciable force acting between atoms, molecules and ions is an electrical force. The gravitational force between atomic particles exists, but is far too small to have any important influence on their individual behaviour. In the complex variety of materials there are many different ways in which the electrical force exists. The force of attraction between atoms of sodium and chlorine in crystalline sodium chloride is said to be an ionic bond. Each sodium atom loses one electron to become a positive sodium ion. Each chlorine atom gains one electron to become a negative chlorine ion. The sodium and chlorine ions attract one another to give the ionic bonding in crystalline sodium chloride. A different type of bonding occurs within a hydrogen molecule. This bond is called a covalent bond and it is characterised by electrons being shared between atoms. Another way atoms can be bonded together is with a metal bond. Here electrons are not attached to particular atoms but are free to move through a network of positive metallic ions within the crystalline structure of a metal.

Interatomic forces frequently lead to the formation of molecules. The whole of chemical science is concerned with such interatomic bonding and the way in which different molecules can be formed as a result of interatomic forces. Here it is useful to be able to discuss interatomic forces for a particular example without having the complication of the formation of molecules. For this reason an inert gas, argon, will be taken as an example. Argon forms about 1% of the Earth's atmosphere and it is called a monatomic gas because the atoms of argon move around singly and do not form molecules. There are nevertheless weak forces of attraction which exist between argon atoms. These forces are called Van der Waals forces. Besides forces of attraction, forces of repulsion exist between argon atoms if they are close enough to one another. In Fig 22.5 both the force of repulsion and the force of attraction between two atoms are plotted, together with their sum, against the separation of the atoms d. In the absence of any other influence on the atoms, they would come to be in equilibrium a distance d_0 apart where the force of attraction equals the force of repulsion. If some other influence then moves them together they will repel one another, and if they move apart from one another then they will attract each other. Notice also how steep the graph is for low values of d. This makes it extremely difficult to squash argon atoms much closer than a separation of d_0, and effectively sets a limit to the maximum density of argon.

This graph can also be used to find the potential energy of a pair of atoms. The total force graph is redrawn in Fig 22.6(a) where it is drawn to scale. If the potential energy of the two atoms is taken to be zero when they are separated by a large distance, then the work done on one of the atoms in moving it to a distance d from the other is given by the (negative) area beneath the force graph as shown. This has been done and the graph of potential energy against d has also been plotted in Fig 22.6(b). This graph shows that the atoms have minimum potential energy when in their equilibrium position; they are said to be at the bottom of a potential well.

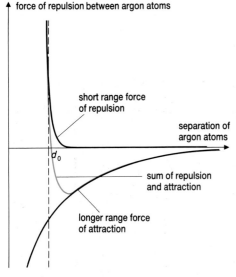

force of repulsion between argon atoms

short range force of repulsion

separation of argon atoms

d_0

sum of repulsion and attraction

longer range force of attraction

Fig 22.5   The force between two atoms can be one of either attraction or repulsion depending on their separation.

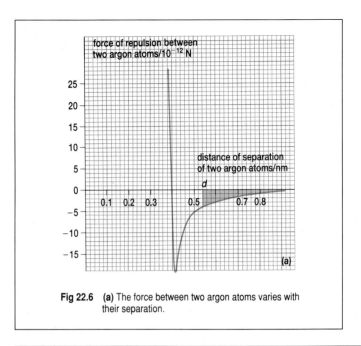

Fig 22.6 **(a)** The force between two argon atoms varies with their separation.

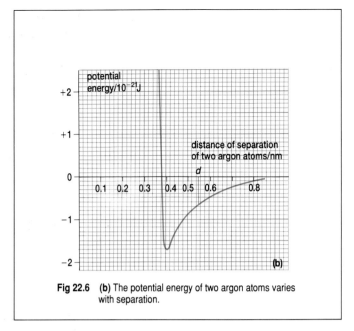

Fig 22.6 **(b)** The potential energy of two argon atoms varies with separation.

QUESTION

22.4 Use the graphs of Fig 22.6 to find:
  (a) what force exists between the two atoms if they are at a separation of 0.39 nm – is this force one of attraction or repulsion?
  (b) how much energy is needed to remove an argon atom from a distance of 0.50 nm from a neighbouring argon atom to a large distance (infinity)
  (c) the maximum kinetic energy gained by an argon atom if it is released a large distance from another argon atom and then falls in towards it;
  (d) where the atom in **(b)** will first stop;
  (e) what the gradient of the energy–distance graph is when the separation of the molecules is 0.50 nm. (There is a quick way of finding this.)

If some argon atoms are cooled down to near absolute zero, then the argon will be a solid and all of the atoms will have very low total energy. The kinetic energy will be nearly zero and the atoms will have minimum potential energy. An argon atom at the bottom of the potential well would require 1.7×10^{-21} J to escape from one other argon atom. In practice, there are going to be many argon atoms surrounding it and not just one. To escape from a solid mass of other atoms actually requires about 6 times this energy. If some heat is then supplied to the argon atoms, the molecules gain kinetic energy. At first, even the fastest molecules have nothing like enough kinetic energy to escape from the potential well that they are in, but eventually there are enough molecules escaping to break down the structure of the solid material. This happens at a particular temperature, the melting point of the argon. The graph enables an approximate value of the melting point to be obtained, and the calculation is done in the following example.

Example 1
The mean kinetic energy E_k of an atom at a temperature T is given by $E_k = \frac{3}{2} kT$ where k is the Boltzmann constant, 1.38×10^{-23} J K^{-1}. (See section 25.4.)

PHASES OF MATTER

A significant number of atoms have 10 times the mean kinetic energy. Find the melting point of argon using data from the graphs in Fig 22.6.

Gain in potential energy required to release an argon atom from one other argon atom $= 1.7 \times 10^{-21}$ J

Gain in potential energy required to release an argon atom from solid argon $= 6 \times 1.7 \times 10^{-21}$ J

Kinetic energy of high speed molecules
$$= 10kT = 10 \times (1.38 \times 10^{-23} \text{ J K}^{-1}) \times T$$

Enough high speed molecules break away from the crystal when this kinetic energy is equal to the potential energy gain required, so

$$6 \times 1.7 \times 10^{-21}\text{J} = 10 \times (1.38 \times 10^{-23} \text{ J K}^{-1}) \times T$$
$$T = 74 \text{ K}$$

Despite the approximations made in making this calculation, it gives a figure surprisingly close to the actual melting point of argon, which is 84 K.

This example shows how the attraction between atoms is binding them together whereas the kinetic energy of the atoms is responsible for separating them. At one particular temperature, the melting point, enough bonds are broken so that the atoms do not remain in a fixed pattern. At a higher temperature, the boiling point, the higher value of the kinetic energy of the atoms is sufficient not only for atoms to be removed from their potential well but also to be able to do work against the air molecules surrounding them and to escape from the surface into vapour form. For argon, this occurs at 88 K.

22.4 SOLIDS

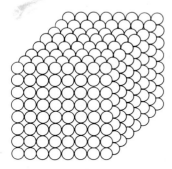

Fig 22.7 Molecules considered as spheres packed in a cubic pattern.

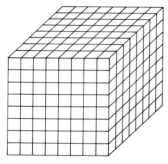

Fig 22.8 Molecules assumed to be cubes packed in a cubic pattern.

A solid normally has a fixed shape; it remains in its shape because there is a fixed pattern of the molecules within it. The molecules vibrate about their mean position but the amplitude of the vibration is small compared with their separation. The arrangement of the molecules in a mass of a solid is controlled by several factors. One of these factors is the shape of the molecule, another factor is the strength of the attractive force between the molecules. Temperature has an influence on the pattern. Many solids have different molecular arrangements at different temperatures. The stresses which have been applied to the solid also affect its molecular pattern, so the way that the solid has been treated in the past has an effect on its pattern of molecules and hence on its strength. The strength of solids is a complex matter and is part of the study of a materials scientist. More detail will be given in Chapter 23 about the effects of applying stress to a solid, but here it is necessary to consider how the pattern of molecules within a solid is related to the density of the solid. To simplify the problem at the start, assume that the molecules behave as solid spheres and they are packed in a cubic structure as shown in Fig 22.7. For ease of calculation it is often simpler to imagine that each molecule occupies the volume of a cubical box as shown in Fig 22.8. If we use the data for argon from the graph in Fig 22.6, we can estimate the density of solid argon as shown in the following example.

Example 2
Find the density of solid argon using data from the potential energy graph, Fig 22.6. Explain why the value you obtain is less than the true value. The relative atomic mass of argon is 40.

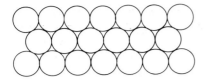

Fig 22.9 Closer packing can be achieved by using a hexagonal pattern. This is called close packing.

The atomic mass unit is 1.67×10^{-27} kg.

Mass of an atom of argon = $40 \times 1.67 \times 10^{-27}$ kg = 6.68×10^{-26} kg.

Separation of the centres of two argon atoms in equilibrium = 0.40 nm
$$= 4.0 \times 10^{-10} \text{ m.}$$

Volume of cubical box in which there is one argon atom = $(4.0 \times 10^{-10})^3$.

Density of argon

$$= \frac{\text{mass of one atom}}{\text{volume occupied by one atom}}$$

$$= \frac{6.68 \times 10^{-26} \text{ kg}}{(4.0 \times 10^{-10} \text{ m})^3}$$

$$= 1040 \text{ kg m}^{-3}$$

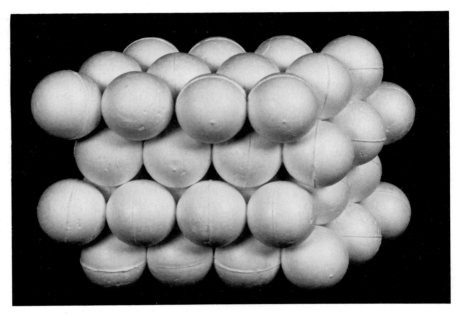

Fig 22.10 A model showing hexagonal close packed atoms.

Fig 22.11 A model showing the cubic closed packed structure.

PHASES OF MATTER

This value is lower than the value found by experiment. The reason for this is that argon atoms can and do pack in more tightly than in a simple cubic structure. Fig 22.9 shows how closer packing of spheres can be achieved. When spherical molecules are packed in three dimensions, they can be close packed in two ways. Photographs of models, Figs 22.10 and 22.11, show these two arrangements, called hexagonal close-packed and cubic close packed (sometimes called face centred cubic).

QUESTIONS

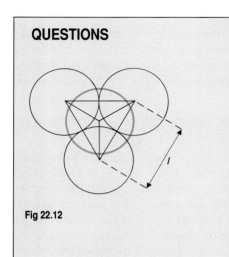

Fig 22.12

22.5 What is the mean separation of the atoms in argon gas of density 1.4 kg m^{-3}?

22.6 What is the spacing of the atoms in copper of density 8900 kg m^{-3}?

The relative atomic mass of copper is 63.5. Assume that each atom occupies a cubical box to begin with and then, if your geometry is good enough, find the distance between layers of atoms. Fig 22.12 shows three atoms in a layer with a single atom from the next layer, shown coloured blue, resting on them. Join together the centres of the four atoms to make a tetrahedron of length of side *l*. The height of the tetrahedron is the separation of the layers of atoms.

22.5 FLUIDS

In a fluid the pattern of the arrangement of a few molecules is, at any instant, very similar to the pattern in a solid. For this reason, the density of liquids is about the same as the density of the corresponding solid. In most cases the liquid has a lower density than its solid, but water is an exception. The density of ice, as given in Table 22.1, is only 920 kg m^{-3}, as compared with water of density 1000 kg m^{-3}. At the molecular level therefore, molecules in water are packed in more closely than they would be in ice.

The essential difference between molecular behaviour in a solid and in a liquid is that in the liquid the pattern of molecules is not fixed, and where pattern does exist it does not extend far within the liquid. Individual molecules do not maintain contact with the same adjacent molecules for any appreciable time and the liquid is therefore able to change its shape. The ability of molecules to flow past one another in a liquid means that a liquid is unable to retain its shape when stretching forces act on it. Neither can it retain its shape when shear forces act on it. Shear forces cause an object to be twisted. A liquid can be put under compressive forces, however, in which case its volume will be little changed. Fig 22.13 is a sketch illustrating some of these differences, and can be useful to get a visual

(a) Solid   (b) Liquid   (c) Gas

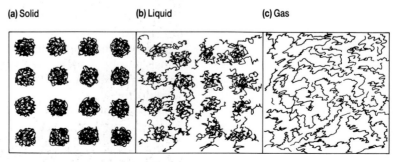

Fig 22.13 **(a)** The atoms in a solid move around but stay in the same mean position. The atoms are in a regular pattern. **(b)** In a liquid there is some pattern but atoms are able to drift from place to place. **(c)** In a gas no pattern of movement is discernible.

impression of the differences between molecular behaviour in solids, liquids and gases. Sketches and models of atoms must not be taken too literally, however, and the complexity of the actual substance must not be lost sight of. The huge number of molecules, and the three-dimensional nature of matter, cannot adequately be sketched.

22.6 CHANGE OF PHASE

In order to change phase, energy must be supplied. This energy is called **latent heat**.

The **specific latent heat of fusion** is the quantity of energy required to change unit mass of the substance from the solid phase to the liquid phase.

The **specific latent heat of vaporisation** is the quantity of energy required to change unit mass of the substance from the liquid phase to the gas phase.

$$L = \frac{Q}{m}$$

where L is the specific latent heat of fusion or of vaporisation, m is the mass and Q is the quantity of heat energy supplied. If the change in phase takes place at a fixed temperature then the energy supplied does not increase the kinetic energy of the molecules of the substance. The average kinetic energy of water molecules at its boiling point of 100 °C is the same whether the water is in the form of liquid water or steam. The energy supplied, as indicated in section 22.3, is used in two ways. Firstly it increases the potential energy of the system by increasing the distance between the molecules. Secondly it does work by pushing molecules in the atmosphere away.

Table 22.2 Latent heats of fusion and vaporisation; melting and boiling points.

Substance	Normal melting point /°C	Specific latent heat of fusion /kJ kg^{-1}	Normal boiling point /°C	Specific latent heat of vaporisation /MJ kg^{-1}
helium	−270	5.23	−269	0.021
hydrogen	−259	58.6	−253	0.452
oxygen	−219	13.8	−183	0.213
nitrogen	−210	25.5	−196	0.201
ethanol	−114	104	78	0.854
mercury	−39	11.8	357	0.272
water	0	334	100	2.26
sulphur	119	38.1	445	0.326
silver	961	88.3	2190	2.34
copper	1083	134	1187	5.07

Example 3

The specific latent heat of vaporisation of water is 2.26×10^6 J kg^{-1} at a temperature of 100 °C. Steam at a temperature of 100 °C and a pressure of 1.01×10^5 Pa has a density of 0.59 kg m^{-3}. Find the work done against the atmosphere by 1 kg of water molecules escaping from a water surface to form 1 kg of steam, and hence find the increase in the potential energy of the molecules.

Volume occupied by 1 kg of steam at 100 °C $= \dfrac{1 \text{ kg}}{0.59 \text{ kg m}^{-3}} = 1.695 \text{ m}^3$

Volume occupied by 1 kg of water at 100 °C $= \dfrac{1 \text{ kg}}{1000 \text{ kg m}^{-3}} = 0.001 \text{ m}^3$

Change in volume, $\Delta V = 1.695 \text{ m}^3 - 0.001 \text{ m}^3 = 1.694 \text{ m}^3$

Work done against the atmosphere $= p\,\Delta V$ (see section 7.2)

$$= 1.01 \times 10^5 \text{ Pa} \times 1.694 \text{ m}^3 = 0.171 \times 10^6 \text{ J}$$

Total energy supplied $= 2.26 \times 10^6 \text{ J}$

Increase in the potential energy of the molecules $= (2.26 - 0.171) \times 10^6 \text{ J}$

$$= 2.1 \times 10^6 \text{ J to 2 sig. figs.}$$

Note from this example that even in the case of vaporisation, most of the work is done in increasing the potential energy of the molecules. The work needed to be done against the atmosphere to produce the increase in volume is very low, and in the case of a solid melting is negligible.

The numerical values of the specific latent heats vary with the temperature at which the change in phase occurs. To change water into water vapour by evaporation does not require quite as much heat as to boil the same quantity of water at the boiling point. This is partly because the density of the vapour depends on the pressure.

To measure the value of the specific latent heat of vaporisation at the boiling point of the liquid a continuous flow method can be used. Energy is supplied electrically at a constant rate to the liquid and the rate at which the liquid boils is found. A suitable apparatus is shown in Fig 22.14 where vapour from the boiling liquid provides a jacket surrounding the liquid itself. Outside this another insulating jacket is placed and at first the heater is left switched on until the temperatures throughout the whole apparatus are steady. At this stage the mass M of liquid which is boiled off in time t is measured. If the potential difference across the heater is V and the current is I then:

$$\text{the power supplied to the heater} = V \times I$$

$$\text{the power supplied to the liquid} = \frac{M}{t} \times L_v$$

where L_v is the specific latent heat of vaporisation

These two expressions are the same if no heat is lost so

$$VI = \frac{ML_v}{t}$$

$$L_v = \frac{VIt}{M}$$

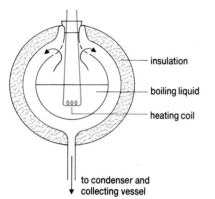

insulation

boiling liquid

heating coil

to condenser and collecting vessel

Fig 22.14 Apparatus for measuring the specific latent heat of a liquid.

INVESTIGATION

An electric kettle, of the older type which does not automatically switch off when it boils, can be used to give a good estimate of the value of the specific latent heat of vaporisation for water. The power of the kettle will be written on its base. Some water is put in the kettle so that the element is well covered. The full kettle is weighed and the water is then boiled. Once the water has started to boil, the time is noted and the kettle is left switched on until a reasonable amount of water has been boiled away. The kettle is then switched off, the time noted and the energy supplied during the measured time is calculated. If the kettle and remaining water is now reweighed, then the mass of water boiled away by the energy supplied can be found.

Find what fraction of the energy supplied is wasted by comparing your value with the accurately measured value of the specific latent heat of vaporisation of water, which is given in Table 22.2.

What factors determined the length of time you used for boiling water away?

electrical
input 150 W

energy to surroundings
at a rate of 250 W

refrigerator

thermal energy
from water at 0° C

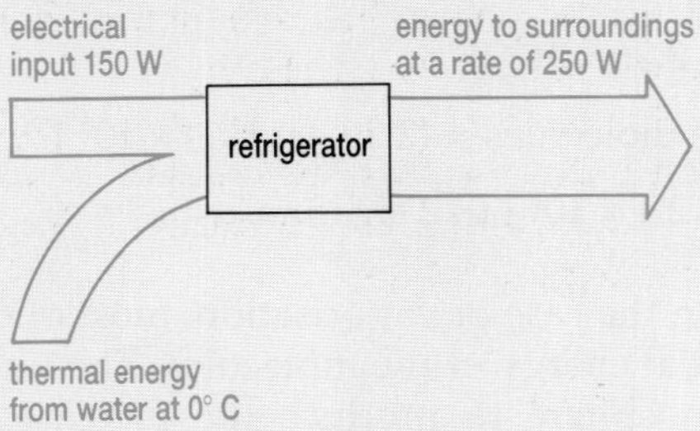

Fig 22.15

22.7 A refrigerator uses 150 W of electrical power. It extracts thermal energy from water at 0 °C and pumps 250 W of thermal energy into its surroundings, as shown in Fig 22.15. Using data from Table 22.2, find the maximum rate at which it can produce ice.

22.8 The Mediterranean Sea loses a substantial quantity of water by evaporation. This water is replaced by water from the Black Sea, from the Atlantic Ocean and from rivers flowing into it. Using the following data, estimate the rate of evaporation on a fine summer day.

Area of the Mediterranean Sea	$= 2.9 \times 10^6 \text{ km}^2$
Latent heat of vaporisation of water at sea temperature	$= 2.45 \text{ MJ kg}^{-1}$
Power of sunlight used for evaporation	$= 0.9 \text{ kW m}^{-2}$

22.7 PRESSURE

Pressure

Pressure is defined as the normal force per unit area. Strictly the **average pressure** on a surface is the total force divided by the total area. The **pressure at a point** p is given by the limit

$$p = \frac{\delta F}{\delta A} \text{ as } \delta A \text{ tends to zero}$$

where δF is the small normal force applied to a small area δA.

The SI unit of pressure is the **pascal**. A pascal is a pressure of one newton per square metre. Although pressure is defined in terms of a vector quantity, force, it is not itself a vector. This is because the area on which the force is acting is a vector having an associated direction. In dividing force by area, a vector is being divided by a vector and the result is a scalar quantity.

Example 4

Find the pressure exerted on the foundations by a wall which is 2.0 m high, 20 m long and 0.30 m wide, if the concrete blocks from which it is constructed have a density of 2500 kg m^{-3}.

Volume of wall = 2 m × 20 m × 0.30 m = 12 m^3

Mass of wall = 12 m^3 × 2500 $\dfrac{\text{kg}}{\text{m}^3}$ = 30 000 kg

Weight of wall = 30 000 kg × 9.8 $\dfrac{\text{N}}{\text{kg}}$ = 294 000 N

Area of base of wall = 20 m × 0.30 m = 6.0 m^2

Pressure on foundations = $\dfrac{\text{Total force}}{\text{Total area}}$ = $\dfrac{294\,000 \text{ N}}{6 \text{ m}^2}$ = 49 000 Pa

Note that some unnecessary information has been given in this example. The width of the wall and its length are not needed. Work through the problem again using algebraic symbols for width and length and see that they cancel out.

QUESTION

22.9 What is the maximum mass which a house can have if it is built on clay foundations which can have a maximum pressure on them of 220 000 Pa. The area of the foundations is 20 m^2.

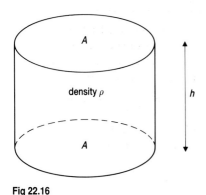

Fig 22.16

Fluid pressure

The word **fluid** means capable of flowing. When a material is said to be a fluid it must therefore be in its liquid or gas phase. In other words, both liquids and gases are fluids. The pressure exerted by a column of fluid on its base can be found by reference to Fig 22.16 where a column of height h and area of cross-section A is formed from a fluid of density ρ. The column is shown without any support. In practice, any column of liquid requires support at the sides, possibly from glass tubing.

Volume of fluid in column	$= hA$
Mass of fluid	$= hA\rho$
Weight of fluid	$= hA\rho g$

$$\text{Pressure} = \frac{\text{weight}}{\text{area}} = \frac{hA\rho g}{A} = h\rho g$$

The pressure due to a column of liquid is independent of its area of cross-section. Since the pressure is dependent on the density, care has to be taken when the fluid has a variable density. In finding atmospheric pressure, for instance, the density of the atmosphere must not be taken as having a uniform value. The standard of **atmospheric pressure** is taken as being equivalent to that of a column of mercury 0.760 m high. This gives standard atmospheric pressure p_a as

$$p_a = (0.760 \text{ m}) \times (13\ 600 \text{ kg m}^{-3}) \times (9.8 \text{ N kg}^{-1}) = 101\ 000 \text{ Pa}$$

Because of variations of g from place to place, this value varies slightly. To overcome this problem when high accuracy is required, a standard atmosphere is now defined as 101 325 Pa.

DATA ANALYSIS

Blood pressure

A doctor measures the blood pressure of a patient with an instrument called a sphygmomanometer. The doctor measures the maximum, systolic, and the minimum, diastolic, pressures and records the systolic/diastolic ratio. A typical, healthy adult at rest will have readings about 120/80. These readings are the differences in levels in a mercury manometer measured in millimetres and give the pressure excess above atmospheric pressure. Readings above 140/90 are usually regarded as high blood pressure (hypertension). To take the readings, the doctor forces air into the air sack wrapped around the upper arm at the level of the heart, while at

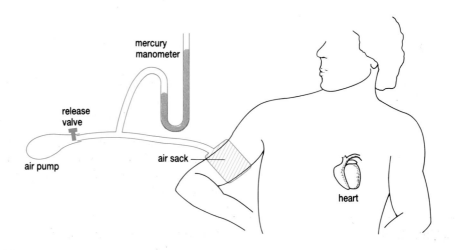

Fig 22.17 A doctor measures the blood pressure of a patient at the level of the heart.

the same time listening to the blood flow through the main artery in the arm, Fig 22.17. When the pressure in the air sack is just sufficient to prevent all blood flow, the pressure is the systolic pressure and the difference in the levels of the mercury is measured. A doctor will normally raise the pressure in the air sack to a value above this pressure and then let it fall gently until blood just starts to flow immediately after the heart has pumped and the pressure is at a maximum. The diastolic pressure can be obtained by lowering the pressure in the sack still further so that blood is only just being stopped from flowing for a brief interval at the end of the pumping cycle.

A person has blood pressure readings of 120/80.

(a) What would these readings be when measured in pascals?

(b) What are the maximum and minimum pressures actually present in the blood of the person's arm if atmospheric pressure is 100 000 Pa?

(c) Show that it is reasonable that the maximum blood pressure in the foot of the person is 130 000 Pa.

(d) Find by how much the pressure in a person's brain changes when she bends so that her brain, instead of being 0.4 metres above her heart, is 0.3 m below her heart.

(e) (A difficult concept is involved here.) Show that the systolic blood pressure in an astronaut's brain will be zero if he is accelerating in a direction straight upwards towards his head with an acceleration of about 28 m s^{-2}. Assume that his brain is 0.4 m above his heart and that no body mechanisms operate to compensate for such conditions. You may be familiar with a similar effect to this. If you stand up suddenly you can sometimes feel light-headed as a result of the drop in blood pressure in your brain.

Relative density of mercury = 13.6. Relative density of blood = 1.06.

Buoyancy and Archimedes' Principle

The equation given in section 22.7 for the pressure beneath the surface of a fluid shows that the pressure increases with the depth of the fluid. If therefore an object is submerged in a fluid, the pressure at the bottom of the object is greater than the pressure at the top.

Fig 22.18 shows a submerged submarine. On the submarine are drawn forces which the water pressure causes, and they can be seen to be acting at right angles to the surface of the submarine. Sideways forces cancel out, but because there are larger upward forces on the bottom than there are downwards forces on the top, the resultant of all of these forces is an upward force. The resultant force due to the pressure of the fluid surrounding it is called the **buoyancy force** or **upthrust**. The same reasoning would show that as a result of the air surrounding a person, everybody has a buoyancy force acting **upwards** on them due to the pressure of the air. The force due to air pressure acting downwards on your head is less than the upward force on your feet because of the higher air pressure near the ground. The buoyancy force on a person is small. Someone with a weight of 800 N usually has a buoyancy force of about 1 N.

To calculate the value of the buoyancy force, consider an object with uniform cross-section area A, and which has a horizontal top and bottom. It has a length l and is placed so that its top is a distance d beneath the surface of a liquid of uniform density ρ, Fig 22.19.

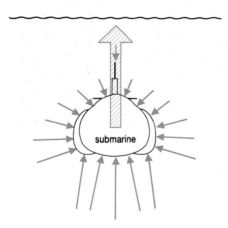

Fig 22.18   The forces acting on a submarine due to water pressure result in a resultant upward force on the submarine. This is the buoyancy force, or upthrust.

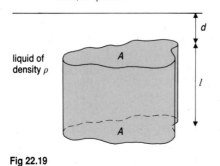

liquid of density ρ

Fig 22.19

pressure on top	$= d\rho g$
force downwards on top	$= d\rho g A$
pressure on bottom	$= (d + l)\rho g$
force upwards on bottom	$= (d + l)\rho g A$
buoyancy force	$= (d + l)\rho g A - d\rho g A = l\rho g A$

PHASES OF MATTER

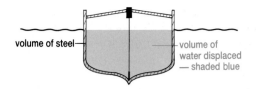

volume of steel ——

—— volume of
water displaced
— shaded blue

Fig 22.20 The volume of water a ship displaces is very much larger than the volume of steel from which it is constructed. In equilibrium the upthrust on the ship equals its weight.

Since lA is the volume of the object, $lA\rho$ is the mass of the fluid which the object displaces and $l\rho gA$ is the weight of fluid displaced. This is a deduction of **Archimedes' Principle** which states that for an object immersed in a fluid the buoyancy force is equal to the weight of the fluid displaced. It can be applied generally and not just when the object has a regular shape.

If an object floats, it does so because the buoyancy force acting on it is equal and opposite to its weight. An object which sinks has a buoyancy force acting on it which is less than its weight. The shape of any piece of material is bound to have an effect on the weight of fluid which it displaces. A steel boat can float because the steel is shaped to displace a large volume of water; the volume displaced is clearly much larger than the volume of the steel itself, Fig 22.20.

Fig 22.21 The oil tanker Norrisia has a total mass of approximately 150 000 tonnes when fully laden. It therefore needs to displace 150 000 tonnes of water in order to float.

QUESTIONS

22.10 What is the buoyancy force on a human body of volume 7.4×10^{-2} m³ when totally immersed in
  (a) air of density 1.3 kg m⁻³
  (b) sea water of density 1030 kg m⁻³

22.11 24 m of a supertanker is below the surface when it is in sea-water of density 1030 kg m⁻³. What depth will be below the surface when it enters fresh water of density 1000 kg m⁻³? Assume the supertanker has vertical sides.

22.12 The Arctic ice cap is a floating mass of ice. By how much would the sea level in the oceans rise if the Arctic ice cap melted as a result of an increased greenhouse effect? Why is this answer different from the answer which would be given if the question had referred to the Antarctic ice cap?

- Density is mass per unit volume $\rho = \dfrac{m}{V}$

- Relative density $= \dfrac{\text{density of substance}}{\text{density of water}}$

- The mole is the amount of substance which contains as many entities as there are atoms in 0.012 kg of carbon 12.

- The unified atomic mass constant, u, is one twelfth of the mass of a single atom of carbon 12.

$$1\ u = 1.6611 \times 10^{-27}\ kg.$$

- The Avogadro constant, N_A is numerically equal to the number of atoms in 0.012 kg of carbon 12.

$$N_A = 6.022\ 17 \times 10^{23}\ mol^{-1}.$$

	Solid	**Liquid**	**Gas**
Pattern of atoms	Usually ordered over relatively large distances.	Ordered over relatively short distances.	No order.
Mobility of atoms	Atoms remain in fixed positions relative to one another.	Atoms can slide over one another. No fixed structure.	Atoms move at random.
Density	Atoms closely packed. High density.	Atoms closely packed. Similar high density to solid.	Atoms widely scattered. Density of the order of a thousandth the density of the liquid.

- Pressure is force acting perpendicularly per unit area. 1 pascal is $1\ N\ m^{-2}$.
- Pressure at a depth in a fluid $p = h\rho g$.
- Atmospheric pressure is 0.760 m of mercury
$$= (0.760\ m) \times (13\ 600\ kg\ m^{-3}) \times (9.8\ N\ kg^{-1})$$
$$= 101\ 000\ Pa$$
- Archimedes' Principle: A body immersed in a fluid has an upthrust on it equal to the weight of the fluid displaced. For a floating body this is equal and opposite to the weight of the body.

EXAMINATION QUESTIONS

22.13 The nucleus of an iron atom is spherical and has a radius of 4.6×10^{-15} m; the mass of the nucleus is 9.5×10^{-26} kg. What is the density of the nuclear material?

22.14 The nuclei of all atoms have approximately the same density. The nucleus of a copper atom has a mass of 1.06×10^{-25} kg and has a radius of 4.8×10^{-15} m. The nucleus of a lead atom has a mass of 3.5×10^{-25} kg; what is its radius? The nucleus of an oxygen atom has a mass of 2.7×10^{-26} kg; what is its radius? Assume that all the nuclei are spherical.

22.15 **(a)** Find the volume occupied by one mole of an ideal gas at a temperature of 290 K and a pressure of 1.0×10^5 Pa.
(b) The diameter of a molecule of this gas is 2.5×10^{-10} m. What fraction of the volume you have calculated in (a) above do the molecules occupy?   (4)

(UCLES 1987)

22.16 In an experiment to find the length of a molecule of olive oil a droplet of olive oil, which was estimated to have a volume of $\frac{1}{8}$ mm^3, was placed on a water surface. The olive oil droplet spread out across the surface to form a circular patch of diameter 250 mm. Assuming that the patch has a thickness which is equal to the length of the molecule find the molecule's length.

22.17 **(a)** Explain what is meant by each of the following statements:
(i) The molar heat capacity of oxygen at constant volume is 20.8 J mol^{-1} K^{-1}.
(ii) The specific latent heat of fusion of ice is 3.34×10^5 J kg^{-1}.   (3)
(b) The first law of thermodynamics may be written as

$$\Delta Q = \Delta U + \Delta W$$

(i) Explain the meaning of each term in the equation as it applies to the heating of a fixed mass of gas at constant pressure thereby increasing the temperature of the gas.
(ii) Explain the effect on each term in the equation of producing an equal temperature change in the same mass of gas at constant volume.
(iii) Hence explain why it is necessary to specify the conditions when a value for the molar heat capacity of a gas is quoted.   (8)

(c) The diagram shows an experimental arrangement designed to measure the specific latent heat of vaporisation of water.

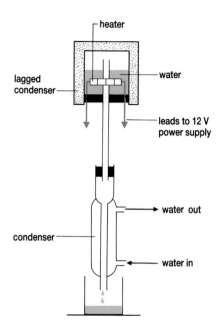

The heater is rated at 12 V/50 W. Once the water is boiling, the steam produced passes down the vertical tube and into the condenser. Here it is condensed and collected in the beaker. The experimenter waits until all the conditions are steady and then records that 2.0 g of condensed water are collected in 100 s.
(i) What is meant by 'rated at 12 V/ 50 W'?
(ii) Why is it necessary for the experimenter to wait until all conditions are steady?
(iii) Calculate the specific latent heat of vaporisation of water as indicated by these experimental results. Is the value obtained likely to be too high or too low? Explain your answer.
(iv) In an attempt to obtain a more accurate value for the specific latent heat, the experiment is repeated using a lower power. The voltage is reduced to 10.5 V and the current flowing in the heater is 3.7 A. When conditions are steady, 1.5 g of water are collected in 100 s.
Use both sets of experimental results to obtain a more accurate value for the specific latent heat of vaporisation.   (10)

22.18 A steel ball falls vertically into a flask filled with thick oil. The ball hits the oil surface at speed and slows down almost to a halt after falling 10 cm within the oil.
Show on a free body diagram the three principal forces which act on the steel ball when it is about 5 cm below the oil surface. Explain the origin of each of these forces. (6)
(ULSEB 1987)

22.19 Within a tornado the air pressure is much lower than normal – about 80 kPa compared to the normal value of 100 kPa. Suppose that such a tornado suddenly envelopes a house; the air pressure inside the house is normal with the abnormally low pressure outside. This will cause the house to burst explosively. What is the net outward force on a 12 m × 3 m wall of this house? Is the house likely to suffer less damage if all the windows and doors are left open?

22.20 Commercial jetliners have pressurised cabins enabling them to carry passengers at a cruising altitude of 10 000 m. The air pressure at this altitude is 210 mm Hg. If the air pressure inside the jetliner is 760 mm Hg, what is the net outward force on a door of area 2.0 m^2 in the wall of the cabin. (1 mm Hg is the pressure caused by one millimetre depth of mercury; density of mercury = 13 600 kg m^{-3}.)

22.21 An oil tanker is full of oil of density 880 kg m^{-3}. The flat bottom of the hull is at a depth of 26 m below the surface of the surrounding water. Inside the hull, oil is stored with a depth of 30 m. What is the pressure of the water on the bottom of the hull? What is the pressure of the oil on the hull? What is the net vertical pressure on the bottom?

22.22 Icebergs commonly found floating in the North Atlantic are 30 m high (above the water) and 400 m × 400 m across. The density of ice is 920 kg m^{-3} and the density of sea water can be taken to be 1000 kg m^{-3}.
What is the total volume of such an iceberg (including the volume below the water)? What is the total mass of the iceberg?

22.23 When an object is weighed on an analytical balance, a correction must be made for the buoyancy force exerted by the air. If the object being weighed has a density of 589 kg m^{-3} and the balance weights are made of brass of density 8.70×10^3 kg m^{-3} what percentage difference to the balance reading for the mass of the object does the buoyancy force make? Should this difference be added to or subtracted from the balance reading to obtain the true mass of the object. The density of air is 1.3 kg m^{-3}.

Chapter 23

DEFORMATION OF SOLIDS

LEARNING OBJECTIVES

At the end of this chapter you should be able to:

1. define and use the terms stress and strain;

2. distinguish between elastic and plastic behaviour;

3. explain the behaviour of solids in terms of molecular structure;

4. describe the properties of substances which are ductile, brittle, glassy and polymeric.

23.1 STRESS AND STRAIN

Rigidity

Newton's laws state that when a resultant force is applied to an object, the object will accelerate. On the other hand, a piece of railway track does not seem to accelerate if suddenly a train travels over it. The track seems to have a force applied to it and yet not to accelerate. The track is held rigidly in position and so does not appear to move. The key word here is rigidly. What is meant by a rigid substance? Rigid substances are all solids. Solids have the ability to retain their shape when forces are applied to them in a way that fluids cannot. As explained in section 22.4, this is because if the individual atoms are moved closer to one another they will repel, whereas if they are pulled further apart they will attract. Note, however, that a small movement is necessary to cause a change in the force which exists between the atoms.

This explains the paradox posed at the opening to this section. The force applied to the track by the train **does** cause downwards acceleration of the track, but the additional large forces set up within the parts of the track by atoms being pushed closer together, limit the distance within which acceleration occurs. The vertical distance moved downwards by the track is small but is visible if you look closely at a railway track as a train passes over it. This effect, and similar effects, are normally ignored in everyday life, except in such cases as when a person lies on a mattress and the distortion is clearly visible. When we think of a rigid object we assume that it does not distort when forces are applied to it, but from the point of view of a study of physics it is essential to realise that all real substances do distort when under the influence of an applied force. All substances are to some extent elastic. Elastic distortion of a tennis ball can be seen in Fig 23.1. When you drop a piece of steel on to a concrete floor there might not be much distortion occurring, but there is some. The concrete and the steel both change shape, and large inter-atomic forces are set up which will stop the movement and, after some bouncing, result in an equilibrium position.

Fig 23.2 illustrates a similar effect when a mass M is allowed to drop on to a spring. Distortion of the spring occurs, the mass stops after some potential energy has been stored by the spring, and oscillation takes place until damped out by friction. The spring and mass reach a different equilibrium position after the mass is added. Putting a book down on

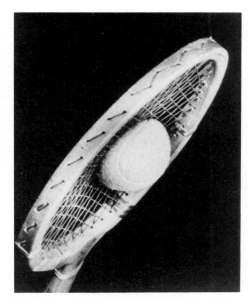

Fig 23.1 The large acceleration of the ball is caused by the large force acting on it. This large force causes considerable elastic distortion.

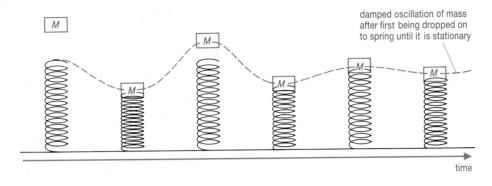

damped oscillation of mass after first being dropped on to spring until it is stationary

time

Fig 23.2 Distortion of a spring when a mass falls on to it.

table is physically similar to this, with M being replaced by the book and the spring representing the table. The table is equivalent to a very stiff spring, so the new equilibrium position is established very quickly and the amount of distortion is very little.

Tension

Another important feature of solid behaviour is the way in which solids can be stretched as well as being squashed. When a stretching force is applied to a solid, the solid is said to be in a state of **tension**. There are some tension effects in liquids but because of the ease of moving adjacent atoms in liquids, the effect is small and is normally only noticeable as a surface tension. In a taut horizontal wire being pulled by equal forces, T, at both ends, each small section of the wire could be considered separately. This is shown in Fig 23.3 where a few of the sections have been labelled A,B,C,D,E. . . . Free body diagrams for three of the sections have been

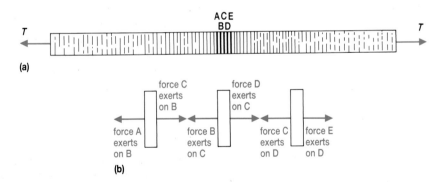

Fig 23.3 A horizontal wire considered as a series of slices, for some of which free body force diagrams are drawn.

drawn to show the forces on individual sections. Since all the sections are in equilibrium, the resultant force on each section must be zero. Using Newton's third law gives the force C exerts on B to be equal and opposite to the force B exerts on C. This can be continued along the length of the wire, so it follows that throughout the wire the two forces shown acting on each section in Fig 23.3(b) must be equal to T. T is called the tension in the wire. Note that a tension T is caused by two opposite forces both of magnitude T acting on the ends of the wire. If a wire is held vertically then the tension in the wire will vary because of the weight of the wire. The tension will be greater nearer the top. A wire undergoing an acceleration must have a different force applied at one end from the other. Very often, however, the assumption is made that the mass of the wire is negligible, and under these circumstances the difference between the forces at each end is negligible.

Stress

The **stress** applied to a wire to stretch it is defined as the force applied per unit area of cross-section. This is sometimes called the tensile stress, as forces can be applied in different ways to objects. Under compression, for instance, a compressive stress is applied. All stresses are defined as a force per unit area and therefore they all have the same unit as pressure, the pascal, Pa. If a force T is applied to a wire of area of cross-section A, then the stress σ is given by the equation

$$\sigma = \frac{T}{A}$$

Example 1

The high tensile steel used to support the roadway of a suspension bridge breaks under a stress of 1.61×10^9 Pa. Find the minimum area for a cable if it is to support a load of mass 2.0×10^5 kg (200 tonnes).

$$\text{Weight of load} = 2.0 \times 10^5 \text{ kg} \times 9.8 \, \frac{\text{N}}{\text{kg}} = 1.96 \times 10^6 \text{ N}$$

$$= \text{tension in cable, } T$$

$$\sigma = 1.61 \times 10^9 \text{ Pa} = \frac{T}{A}$$

$$A = \frac{1.96 \times 10^6 \text{ N}}{1.61 \times 10^9 \text{ Pa}} = 1.22 \times 10^{-3} \text{ m}^2 = 12.2 \text{ cm}^2$$

The cable must therefore have an area which is greater than 12.2 cm^2.

QUESTION

23.1 The calf muscle exerts a tensile force of 1500 N on the Achilles' tendon in the lower part of the leg. If the area of cross-section at the centre of the muscle is 70 cm^2 ($= 0.0070 \text{ m}^2$), and the area of cross-section of the tendon is 1.1 cm^2, find the tensile stress in each of these cross-sections.

Strain

As a result of applying a tensile stress to a wire a tensile strain is set up within the wire. Stress is the cause and strain is the effect. The tensile **strain**, ε, is defined as the extension per unit length. Strain is a ratio of two lengths and therefore does not have a unit. Using the symbols from Fig 23.4 this can be written in equation form as

$$\text{tensile strain } \varepsilon = \frac{\text{extension}}{\text{original length}} = \frac{x}{l}$$

Strain is related to stress, and for most materials the strain is proportional to the stress provided the stress is small enough. When strain is proportional to stress a material is said to be obeying **Hooke's law**, named after Robert Hooke who in the Seventeenth Century first formulated the law. A graph of stress plotted against strain is straight up to a certain limit. More detail will be given about stresses beyond this limit in section 23.2. Two stress–strain graphs are shown in Fig 23.5. A more rigid material, such as steel, has a steeper slope than a less rigid material, such as aluminium.

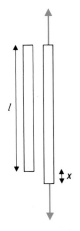

Fig 23.4

Graph axes are normally chosen to have the cause of any change plotted on the x-axis and the effect plotted on the y-axis. In this particular case, however, the axes are, by convention, reversed so that stress is plotted on the y-axis.

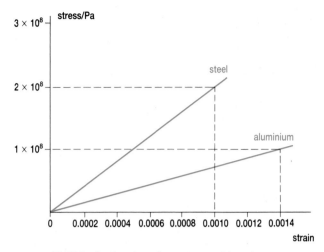

Fig 23.5 Steel requires a larger stress on it for a given strain than does aluminium.

The ratio of stress to strain is called a modulus of elasticity. If the ratio of stress to strain is large, then the material distorts only a little under the influence of the applied stress and the material is stiff. For tensile stress and strain, the ratio is called the **Young modulus, E.** This gives

$$\text{Young modulus} = \frac{\text{tensile stress}}{\text{tensile strain}}$$

$$E = \frac{\sigma}{\varepsilon} = \frac{\dfrac{T}{A}}{\dfrac{x}{l}}$$

$$E = \frac{Tl}{Ax}$$

Hint: Note that E has a very high numerical value. In order to obtain this, the two numerical values which are large occur on the top of this expression and the two normally small values appear on the bottom.

Example 2

Use the graphs in Fig 23.5 to find the Young moduli of steel and aluminium.

Since the graphs are straight lines through the origin, the gradient of each graph is the Young modulus:

$$\text{Young modulus of steel} = \frac{2.0 \times 10^8 \, \text{Pa}}{0.0010} = 2.0 \times 10^{11} \, \text{Pa}$$

$$\text{Young modulus of aluminium} = \frac{1.0 \times 10^8 \, \text{Pa}}{0.0014} = 7.1 \times 10^{10} \, \text{Pa}$$

DEFORMATION OF SOLIDS

Fig 23.6   The stressing rods in the concrete shielding of a nuclear reactor are tested for strength.

QUESTIONS

23.2   A wire of diameter 1.0 mm and length 2.3 m is made of copper whose Young modulus is 1.1×10^{11} Pa. Find:
(a) the radius and hence the area of cross section of the wire in m^2;
(b) the strain if the wire is stretched by 0.85 mm;
(c) the stress;
(d) the force necessary to cause this stress.

23.3   The steel cable on a crane has a diameter of 13 mm. What force is necessary to produce a strain of 0.1% in the wire?
    What is the maximum acceleration with which it can lift a load of mass 1000 kg if the strain in the cable is not to exceed 0.1%? The Young modulus of the high tension steel used is 2.4×10^{11} Pa. (Be certain to draw a free body diagram for the load.)

23.4   A copper wire 4.00 m long and 0.96 mm in diameter was used in an experiment to measure the Young modulus of copper. One end of the wire was clamped in a fixed position. The following readings were obtained from the scale of a travelling microscope positioned at the other end of the wire for each load on the wire.

Load /N	Scale reading /cm
20	3.021
30	3.070
40	3.118
50	3.167
60	3.215
70	3.264
80	3.314
90	4.270

(a) Plot a graph of scale reading against load and use the graph to find
   (i) the Young modulus of copper
   (ii) the stress at the limit of proportionality.

(b) This is not a very accurate way of using these data. Why not?
(c) How can a more reliable value for the Young modulus of copper be found from the readings?

23.2 ELASTIC AND PLASTIC BEHAVIOUR

In section 23.1 the stress applied to any piece of material was assumed to be sufficiently small so that the resultant strain was proportional to the applied stress. If larger stresses are applied, then this ceases to be the case. A graph of tensile stress against tensile strain is shown in Fig 23.7 for copper. The actual values given on the graph can be considered only as typical values because they vary from one piece of copper wire to another, depending on the crystalline structure of the copper in the wire and on its previous treatment. Temperature also affects the strain. The end of the straight line portion of the graph is called the **limit of proportionality**, and is marked as point A. At a point near to the limit of proportionality the copper ceases to behave elastically and starts to behave plastically. This

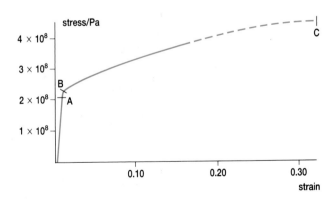

Fig 23.7 A graph of tensile stress plotted against tensile strain for copper.

point is B and is called the **elastic limit** or yield point. The distinction between elastic and plastic behaviour is that when a material is elastic it returns to its original shape on removal of the distorting force, whereas if plastic it does not. Plastic behaviour means that the applied force has caused permanent deformation: elastic behaviour means temporary distortion. Most materials are elastic for low stresses and plastic for high stresses. It is important in civil engineering that constructional materials are subjected only to loads which cause elastic deformation. A bridge, distorted by a heavy load moving across it, must spring back into its original shape after the load has passed.

Once a material is plastic, then a small increase in the stress can cause a very large increase in the strain. A material which behaves in this way is said to be **ductile**. It has a large plastic region. Within the plastic region it is more difficult to measure the strain at a particular stress. There are several reasons for this difficulty. In the first case, as the wire is stretched its area of cross-section reduces. This can be taken into account to begin with, but once the force is sufficiently large then localised narrowing, necking, takes place at weak points and the wire will eventually break at one of these points. Another problem is that plastic distortion is time dependent. For an applied stress, the initial strain will have a particular value; but if the strain is measured later, it is often found to have increased. This is known as **creep**. A third problem is that even if two pieces of wire are taken from the same reel, when stresses are applied it is often found that the strain they reach before breaking can be remarkably different. The microstructure of the two wires cannot be expected to be identical and neither can the wires be clamped identically where the force is applied, so whereas one wire might snap with a strain of, say, 10%, the other may not break until the strain is 40%. The latter part of the graph is therefore uncertain and has been shown dotted. C is the **breaking point** for the specimen and also shows the **breaking stress**. The breaking stress is the maximum tensile strength of the material. See Table 23.1.

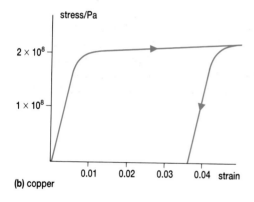

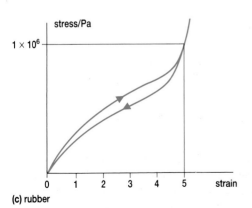

Table 23.1 The Young modulus and the breaking stress for different materials

Material	Young modulus / $\times 10^{11}$ Pa	Breaking stress / $\times 10^8$ Pa
aluminium	0.70	2.2
copper	1.1	4.9
brass	0.91	4.7
iron	1.9	3.0
mild steel	2.1	11.0
glass	0.55	10
tungsten	4.1	20

Approximate values only are given for breaking stress since it varies markedly from one specimen to another and from alloy to alloy.

Certain other features of different materials can be shown by stress–strain graphs. Fig 23.8(a) is a stress–strain graph for glass. It shows no plastic distortion at all. A material such as glass, which does not behave plastically, is said to be **brittle**. Concrete is another substance which is brittle as it cannot be permanently distorted. If it does not break under an imposed load, then it will always return to its original shape after the load is removed. Figs 23.8(b) and 23.8(c) show that the graph can also depend on whether the applied stress is increasing or decreasing. Fig 23.8(b) is the stress–strain graph for copper again, but in this case once the strain has reached a value of 0.050, the stress has been reduced to zero and the consequent permanent set in the strain can be seen to be 0.035. A similar feature is apparent when rubber is stretched. The first thing to notice about this graph, Fig 23.8(c), is that the strain reaches the high value of 5. This means that the extension of the rubber is 5 times its original length, whereas in the example given for copper the elastic extension was only one twentieth of its original length. Quite different values for the strain are obtained for increasing stress as compared with decreasing stress.

Fig 23.8 **(a)** Glass shows a characteristic stress/strain graph for a brittle substance. **(b)** The corresponding graph for a ductile substance (copper) **(c)** The graph for a polymeric substance (rubber).

Load–extension graphs

All of the graphs so far discussed in this section have been stress–strain graphs plotted for the material in a particular specimen. A similar graph will be obtained if extension is plotted against load. The extension is proportional to the strain but the stress is only proportional to the load if the area of cross-section of the wire is constant. In practice, this will affect the shape of the graph only a little for low stress, but more as the load reaches the breaking stress. The area beneath a load–extension graph is the work done in stretching the material (see section 7.2). For the straight line portion of the graph this will give:

$$\text{work done} = \tfrac{1}{2} (\text{load} \times \text{extension})$$

This is the potential energy stored by the wire provided the graph for decreasing loads is the same as that for increasing loads.

The load must have the unit newton and the extension must be measured in metres to give the joule as the unit of work.

QUESTION

23.5 A copper wire of length 2.30 m and area of cross-section 8.6×10^{-8} m^2 is stretched by a tension of 17.3 N. Taking the Young modulus for copper to be 1.1×10^{11} Pa find:
(a) the extension of the wire
(b) the potential energy stored in the wire as a result of doing work on it.
If a force of 173 N were used instead, it is likely that the work done by the force will be much greater than 10 times the value given in (b), but the potential energy stored will be less than 10 times that value. Why is this?

INVESTIGATION

The creep of copper wire

Wear eye protection for this investigation in case the wire snaps.

Set up the arrangement shown in Fig.23.9, using bare copper wire of about 28 s.w.g. Pay particular attention to the way that the wire is clamped. It must not slip at the clamp. Holding it firmly between two

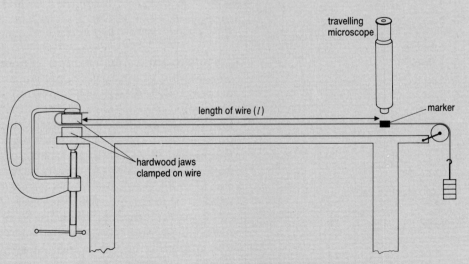

Fig 23.9   Apparatus for measuring the Young modulus of the material of a wire.

---

blocks of hardwood used as jaws is one way of fixing it, and it is then easy to measure the distance between the marker and the jaws to find the original length of the wire. The length of wire used needs to be at least 2 metres between the wire clamp and the marker if the extension is to be measured with a ruler, but smaller lengths of wire can be used if a travelling microscope can be used for the measurement of the extension.

If you have not done an experiment such as this before, it is worth measuring the Young modulus for copper first. Plot a graph of load against extension, and for the straight line portion of the graph find the gradient. Use this gradient, together with measured values of the length and area of cross-section of the wire, to find the Young modulus.

The load must be in newtons, the length of the wire in metres and the area of cross-section, which is πr^2, in square metres, r is the radius of the wire, not its diameter. Measurement of the diameter must be taken at several places along the wire and the zero reading of the micrometer must be checked. All readings taken must be recorded. It is standard practice to record values as you go and not to do any arithmetical working or averaging before recording readings.

Now apply just sufficient load to the wire to make it plastic. Record values of the extension of the wire at intervals over a period of a few minutes. Add a small additional load and repeat the experiment. Carry on adding load until eventually the wire breaks. You will probably break a few wires too quickly to begin with. Plot a series of graphs, one for each load, of extension against time. Summarise your findings.

The investigation can, if desired, be modified to measure breaking stress and to find how breaking stress varies with the diameter of the wire.

Note: Particular care must be applied in this experiment to get the units correct.

23.3 MICROSTRUCTURE

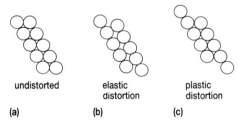

undistorted   elastic   plastic
            distortion   distortion

(a)   (b)   (c)

Fig 23.10 Elastic distortion results in no permanent change in shape when the distorting force is removed. With plastic distortion a permanent change in shape occurs.

The pattern of molecular arrangement in a solid, discussed in sections 22.3 and 22.4, needs to be able to explain the strength, elasticity and plasticity of materials. A full explanation cannot be given here as the topic of materials science is a huge subject in its own right. The complexity of material structure also makes it difficult to do more than give a few, simplified generalisations.

Elastic behaviour does not result in a permanent change in the shape of the material. This implies that the molecules within the material stay in the same basic position in relation to their nearest neighbours. Since elastic strain is usually less than about 1%, it is reasonable to suggest that the individual molecules move a small distance from their equilibrium position on the application of a force and then settle back into their original equilibrium position once the force is removed. Fig 23.10(a) shows a few molecules in their undistorted arrangement and Fig 23.10(b) shows the pattern after a distorting force has been applied. If the distorting force is now removed, they will return to their original position. If, however, the force is increased then **slip** occurs and the molecules move into a new position, Fig 23.10(c). The effect of this occurring is shown on a larger scale in Fig 23.11. Layers of molecules sliding across one another like this results

Fig 23.11

in visible lines being formed on the surface of the material. See the photograph of a metallic surface after slip has occurred. Fig 23.12.

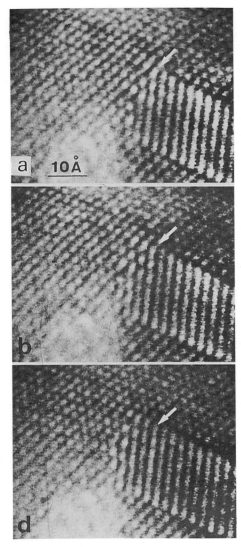

Fig 23.12 These high resolution electron micrographs show slip in an atomic lattice of gold. The arrow shows the movement of whole blocks of atoms along planes of weakness, called slip planes.

This simple explanation cannot explain how a substance such as rubber can undergo a strain of 500% or more. To have any understanding of the stretching of rubber it is necessary to know that rubber is a **polymer** with very long chain molecules. The molecules do have an attraction for one another and therefore solid rubber is possible, but they also tangle their molecules to give what can perhaps best be thought of as being like a plateful of cooked spaghetti having occasional cross linkage between the strands. If the rubber is then pulled, the strands can be pulled out to lie more nearly parallel to one another. See Fig 23.13 where (a) shows the tangled molecules before stretching and (b) shows them stretched. Some reorganisation of the cross links does take place and therefore there can be a permanent distortion of the rubber.

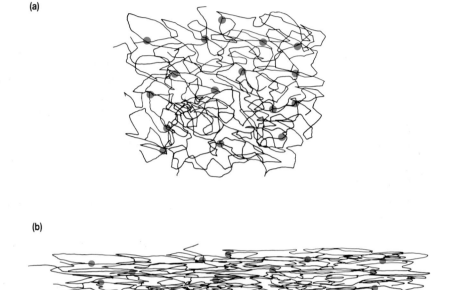

Fig 23.13 A polymeric material such as rubber can undergo a large strain. Its long molecules become stretched to lie parallel to one another.

Besides crystalline and polymeric solids there are also solids which are said to be **amorphous**. The word itself means lacking in shape. Glass and soot are amorphous solids and the shape which is lacking in these materials is the crystalline arrangement of the molecules. When pattern does occur it takes place over only very short distances, of the order of a few molecular diameters.

The force which exists between atoms in a solid should give a guide to the strength of the solid. In Fig 22.6(a) the maximum force between two argon atoms was given on the graph as approximately 2×10^{-11} N. The force between the more strongly bonded atoms in metals is greater. In Example 3, the strength of a copper wire is calculated after taking the maximum force of attraction between two copper atoms to be 5×10^{-11} N.

Example 3

Find the maximum load which can be put on a wire of copper of area of cross-section 1 mm^2, if the maximum force between two copper atoms is 5×10^{-11}N. The density of copper is 7900 kg m^{-3} and the relative atomic mass of copper is 63.

The Avogadro constant gives 6.02×10^{23} atoms in 0.063 kg of copper.

1 kg of copper has a volume of $\dfrac{1 \text{ kg}}{7900 \text{ kg m}^{-3}} = 0.000\,127 \text{ m}^3$

0.063 kg of copper has a volume of $0.063 \times 0.000\,127\ m^3 = 8.00 \times 10^{-7}\ m^3$

The volume occupied by 1 atom of copper $= \dfrac{8.00 \times 10^{-7}\ m^{-3}}{6.02 \times 10^{23}}$

$$= 1.33 \times 10^{-29}\ m^3$$

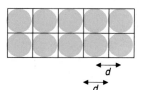

Fig 23.14

If the atoms are considered to be in a cubical pattern, then the volume of the cube occupied by each atom is the value given above, and the cube root of this value will be the diameter of a copper atom. This is shown in Fig 23.14. It should be noted that the distance between the centres of two atoms is the diameter of the atoms.

$$\text{Separation of molecules } = \sqrt[3]{(1.33 \times 10^{-29}\ m^3)}$$

$$= 2.37 \times 10^{-10}\ m$$

Now consider a layer of copper atoms across the wire at one place. It is easier to imagine that the wire has a square cross-section as shown in Fig 23.15 in which the layer being considered has been imagined to be separated from the next layer up and is shaded blue.

The number of atoms in this layer = (the number along a side of 1 mm)^2

$$\text{number in a layer } = \left(\dfrac{0.001\ m}{2.37 \times 10^{-10}\ m} \right)^2$$

$$= 1.78 \times 10^{13}$$

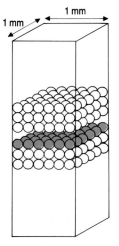

Fig 23.15

Since each of the atoms in this layer pulls on a corresponding atom in the next layer:

$$\text{total tensile force} = (1.78 \times 10^{13}) \times (5 \times 10^{-11}\ N) = 890\ N$$

Many approximations have been made in making this calculation, so the only valid deduction which we can make from the calculation is that the maximum tension in a copper wire of area of cross-section 1 mm^2 should be approximately 900 N.

The value found in example 3 for the tension in a piece of copper wire would imply that we ought to be able to place a mass of approaching 100 kg on the end of the wire and be able to lift it. When this is tried in practice, the mass is found to be much too large. Something has gone wrong with the theory. If instead of using a normal piece of copper with all its imperfections we use a piece of copper of greater crystalline purity, we find that instead of getting stronger, as might be expected, the pure crystal of copper is in fact weaker still. The deduction we are forced to make is that the strength of a piece of material, while having a maximum value limited by the size of the inter-molecular attractions, will in practice break at lower stresses. The fundamental flaw in the argument given in the above example is making the assumption that all the molecules need to be pulled apart simultaneously. They can equally well be un-zipped from one another a few at a time. Whether or not this un-zipping can take place depends crucially on how regularly the atoms are arranged. It also depends on the number of impurity atoms present.

The structure and strength of crystalline materials cannot be predicted as easily as might be expected. The photograph Fig 23.16 shows that, in practice, many crystals are present in most metal specimens. The irregularity of this multitude of individual crystals very much affects the strength of any specimen. The boundary between adjacent crystals, the grain boundary, can hinder a specimen from breaking. Metalworkers often change the crystal pattern in a piece of metal to increase its hardness. This process is called work hardening and it can be achieved by hammering the metal. Rapid changes in the temperature can also be used to alter the

Fig 23.16 The etched surface of a metal, seen here under a microscope, shows the polycrystalline nature of metals.

strength of a metal. A fault in the crystal pattern is known as a **dislocation** and the number and type of dislocations within any material has a marked effect on its strength. Dislocations can be deliberately introduced by using impurity atoms. The strength of steel is much greater than the strength of pure iron. Steel is manufactured by introducing carbon into iron. Much more detail on the behaviour of materials is given in *Physics of Materials* by Brian Cooke and David Sang.

SUMMARY

- A solid behaves elastically if no permanent deformation occurs after a distorting force is removed.
- A solid behaves plastically when permanent distortion occurs.
- All solids are elastic when small forces are applied to them; as the force applied to them increases, plastic behaviour may occur before breaking.
- Stress is force per unit area of cross section. It is measured in pascals.
- Strain is the increase in length per unit length. It has no unit.

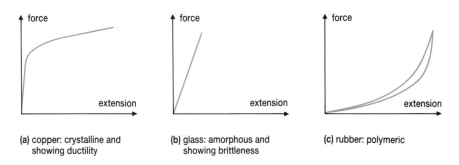

(a) copper: crystalline and showing ductility

(b) glass: amorphous and showing brittleness

(c) rubber: polymeric

- For the straight line portion of these graphs (up to the limit of proportionality) stress is proportional to strain and Hooke's law is said to apply.

- Young modulus $= \dfrac{\text{Stress}}{\text{Strain}} = \dfrac{\text{Force per unit area}}{\text{Extension per unit length}} = \dfrac{Fl}{Ax}$

- Work done in stretching a solid is the area beneath the force–extension graph. If Hooke's law applies, the work done is stored as strain energy. This energy is $\frac{1}{2}Fx$.

- Crystalline solids have long range order in the pattern of molecules.

- Amorphous solids have little ordered arrangement of molecules.

- Polymeric solids have long chain molecules which can cause stiff or flexible solids according to the degree of cross linking between the chains.

EXAMINATION QUESTIONS

23.6 **(a)** The Young modulus for copper is 1.2×10^{11} Pa and its yield stress is 7.5×10^7 Pa.

   (i) Explain what is meant by each of the statements in italics.

   (ii) A tensile load of 50 N is applied to a 3.0 m length of copper wire of cross-sectional area 1.0 mm^2.

   Explain whether such a load can be applied without the wire becoming permanently extended, and calculate the extension produced by the load. (8)

   (b) You have been asked to perform an experiment to determine the Young modulus of copper using copper wire of diameter 1.2 mm.

   (i) Suggest a suitable range of loads to be applied to the wire. (2)

   (ii) Describe, with the aid of a diagram, the apparatus which you would use, paying particular attention to the method of measuring the extension. (4)

   (iii) Describe the important steps in the procedure you would adopt and state the measurements you would make. (4)

   (iv) Show how you would use the results of your experiment to determine the Young modulus. (3)

   (v) Identify the principal sources of inaccuracy in your experiment, and suggest what could be done to reduce these inaccuracies. (4)

   (AEB 1987)

23.7 A large tensile force is needed to increase the length of a steel wire by about 0.1% but a modest tensile force doubles the length of a rubber band. Explain how the difference in behaviour is accounted for by the different molecular structures of steel and rubber. Explain why, if a steel wire is formed into a helical spring, the amount of elastic potential energy it can store increases enormously. (6)

   (ULSEB)

23.8 The diagram shows the compression produced in a human tibia (shin bone) when it is subjected to an increasing compressive load.
Assuming that the bone is 0.40 m long and has an effective area of 4.0×10^{-4} m^2, calculate

   (a) the compressive stress produced by a force of 2000 N, and obtain a value for the corresponding strain

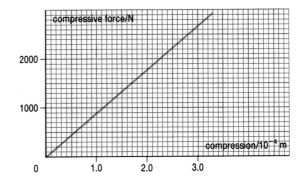

   (b) the modulus of elasticity of bone under compressive stress

   (c) the energy stored in the tibia when a compressive load of 1000 N is applied. (8)

   (AEB 1987)

23.9 The diagram shows a force/extension graph for a rubber strip of unstretched length 0.10 m and area of cross-section 6.0×10^{-6} m^2.

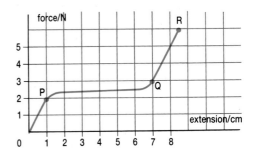

   (a) Account for the change in shape of the graph between PQ and QR in terms of the polymeric structure of rubber. (2)

   (b) Calculate

   (i) the stress in the strip at point P, (1)

   (ii) the strain in the strip at point P, (1)

   (iii) the Young modulus for the strip for small stresses. (1)

   (c) Estimate the work done in stretching the strip up to the point Q. Explain how you worked out your estimate. (3)

   (O & C 1988)

23.10 In the model of a crystalline solid the particles are assumed to exert both attractive and repulsive forces on each other. Sketch a graph of the potential energy between two particles as a function of the separation of the particles. Explain how the shape of the graph is related to the assumed properties of the particles. (5)
The force F, in N, of attraction between two particles in a given solid varies with their separation d, in m, according to the relation

$$F = \frac{7.8 \times 10^{-20}}{d^2} - \frac{3.0 \times 10^{-96}}{d^{10}}$$

State, giving a reason, the resultant force between the two particles at their equilibrium separation. Calculate a value for this equilibrium separation. (3)

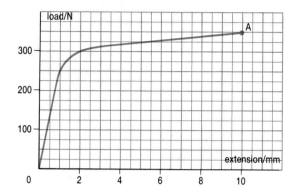

The graph displays a load against extension plot for a metal wire of diameter 1.5 mm and original length 1.0 m. When the load reached the value at A the wire broke. From the graph deduce values of **(a)** the stress in the wire when it broke, **(b)** the work done in breaking the wire, **(c)** the Young modulus for the metal of the wire. (8)

Define *elastic* deformation. A wire of the same metal as the above is required to support a load of 1.0 kN without exceeding its elastic limit. Calculate the minimum diameter of such a wire. (4)

(O & C 1983)

23.11 The graph represents the relationship between the potential energy E and the separation r of two adjacent atoms of a metal.

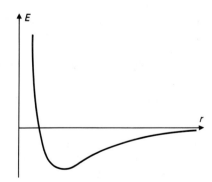

(a) Copy the graph and mark with
(i) a P any point at which the force between the atoms is attractive;
(ii) a Q the separation at which the force between the atoms is zero.

(b) Use the graph to explain why
(i) metals are much easier to stretch by a large amount than they are to compress by a large amount;
(ii) at room temperature the atoms move within a range of possible separations;
(iii) thermal expansion occurs in metals. (7)

(AEB 1986)

23.12 The scale of a certain spring balance reads from 0 to 100 N and is 0.15 m long. Find the strain energy of the spring when the balance reads 40 N. (Assume that the spring obeys Hooke's law.) (3)

(UCLES 1986)

23.13 The greatest instantaneous acceleration a person can survive is 25 g, where g is the acceleration of free fall. A climber should therefore use a rope such that, if he or she falls when the rope is attached to a fixed point on a vertical rock, the fall will be survived.

A climber of mass m is attached to a rope which passes through a ring attached firmly to a rock face at B. The end of the rope is held by a companion as shown. When the climber is at point A, a distance L above the ring, he falls; the companion holds the rope so that it binds at the ring at the instant the climber falls.

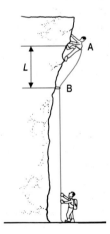

Assuming that the rope obeys Hooke's law up to breaking, use the principle of conservation of energy and the condition for greatest acceleration to show that the part AB of the rope (of length L) must be able to stretch by more than $L/6$ without breaking for the climber to survive. (10)

A particular rope has a breaking strength of 25 times the weight of a climber and obey's Hooke's law until breaking when it has stretched by 20% of its length. Determine whether this rope is suitable. (6)

Discuss how the companion could react to minimise injuries to the falling climber if the ring at B did not bind. (4)

(UCLES 1986)

23.14 **(a)** Using the same horizontal scale, sketch graphs showing how (i) the force between two atoms, and (ii) the potential energy of the two-atom system, vary with atomic separation. Mark on your graphs the equilibrium separation of the atoms.
Use your graphs to explain the thermal expansion of a crystalline solid.
Point out the relationship between a feature of your graphs and the Young modulus of a crystalline solid and hence suggest how the Young modulus might vary with temperature. (9)

(b) A man weighing 700 N has a leg-bone whose minimum diameter is 3.0 cm. Estimate the maximum stress in this bone when the man stands on one leg.
Consider a giant, each of whose external linear dimensions is a factor 2 greater than that of the man. Write down the ratio of the mass of the giant to the mass of the man. Assume now that the stress in the giant's *leg-bone*, when he stands on one leg, is to be the same as that of the corresponding *bone* of the man. Calculate the minimum diameter of the leg-bone of the giant. (Ignore any effect which a greater increase in the diameter of the giant's leg-bone has upon his mass.)

If d_L and d_T are the diameter of the leg-bone and torso (upper body) respectively, show that the ratio d_L/d_T is greater by a factor $\sqrt{2}$ for the giant than it is for the man, if their leg-bones are to undergo equal stresses. The drawings show two animals, X and Y, not drawn to the same scale. State which animal you believe to be the bigger, explaining the reasons for your choice in terms of your earlier analysis. (9)

(ULSEB 1988)

Animal X

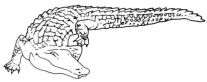

Animal Y

Chapter 24

TEMPERATURE

LEARNING OBJECTIVES

At the end of this chapter you should be able to:

1. define the ideal gas temperature and know that it is the same as the thermodynamic temperature;

2. use the Kelvin and Celsius temperature scales;

3. use empirical temperature scales;

4. suggest suitable thermometers for any use.

24.1 THERMAL EQUILIBRIUM

Temperature is a word which will be familiar to all readers, but it is also a word which poses scientific difficulty. We know that if it is a hot day the temperature is high and that on a cold day the temperature is low. We know what it feels like to touch a cold or a hot object. Children quickly learn that hot objects can cause pain or burning. Temperature is directly associated with the physiological sense of hotness, but one cannot measure temperature using a physiological effect because different people are differently sensitive to hotness and because how hot or cold an object feels depends on what materials it is made from. The air inside an oven is at approximately the same temperature as the shelves in the oven, yet it is possible to put your hands into the hot oven provided you are careful not to touch the shelves. The shelves seem to be much hotter because they conduct heat well and much more energy can flow from them into your hands than can flow from the hot air, air being a bad conductor of heat. Hot and cold are therefore imprecise terms on which it is difficult to base a scientific definition. Robert Boyle in the Seventeenth Century realised this difficulty when he wrote 'We are greatly at a loss for a standard whereby to measure cold.'

Various attempts have been made to overcome this problem. All of the early solutions involved attaching certain values to particular temperatures and then fixing a scale between them. A familiar proposal was suggested by Fahrenheit who gave 0 °F as the lowest temperature he could obtain with an ice–salt mixture and 96 °F as the temperature of the human body. Neither of these temperatures is sufficiently precise to be used as a standard. Newton also set up a temperature scale. He divided the temperature difference between body temperature and that of melting snow into twelve equal parts. In another way of tackling the problem the Royal Society thermometer was taken as the standard in the years from 1663 to 1730.

Temperature is a quantity which cannot be defined in terms of other quantities. Like time and mass it is one of the fundamental quantities. An important statement concerning temperature is the zeroth law of thermodynamics – so called because by the time it was seen to be necessary to have such a basic law, the first and second laws of thermodynamics had already been stated. The **zeroth law of thermodynamics** concerns thermal equilibrium and states that if two bodies are in thermal equilibrium with a

---

Fig 24.1 While everyone is familiar with the effect of being cold, measurement of temperature is not as straightforward as might be expected.

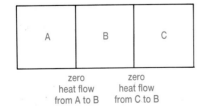

Fig 24.2 A is in thermal equilibrium with B; C is in thermal equilibrium with B. The zeroth law of thermodynamics states that A must be in thermal equilibrium with C. The property which is the same for A, B and C is their temperature .

third body, then they must be in thermal equilibrium with each other. Put in less formal language, if two objects are placed in contact with one another and allowed to reach thermal equilibrium, then no heat will flow from one to the other. If one of these objects is also in thermal equilibrium with a third object, then all three objects will be in thermal equilibrium with one another. The property which they then all have in common is said to be their temperature. **Temperature** is the property of an object which determines which way heat will flow from it to another object. If you put a hand into some water and heat energy flows from your hand to the water, then the water is at a lower temperature than your hand. If heat energy flows from the water to your hand, then the water is at a higher temperature than your hand. If no heat energy flows, then there is thermal equilibrium between your hand and the water and they must be at the same temperature. This is further illustrated in Fig 24.2. The relationship between heat flow and temperature is used to define the thermodynamic temperature.

TEMPERATURE

24.2 THERMODYNAMIC TEMPERATURE

The full definition of thermodynamic temperature depends on the theoretical efficiency of a perfectly reversible heat engine. It is a theoretical scale which has the important advantage that it does not depend on the way a physical property of any particular substance changes with temperature. The theory of reversible heat engines is not dealt with at A-level but it is essential to realise that there exists an absolute thermodynamic scale of temperature. Its symbol is T, and since it was proposed by Lord Kelvin, the unit of thermodynamic temperature has been named the **kelvin** (K). The kelvin is the fraction 1/273.16 of the thermodynamic temperature of the triple point of water. The **triple point** of any substance is the single temperature at which the solid, liquid and vapour phases of the substance co-exist. Note that referring to the triple point of water in this way simply attaches a numerical value to the scale of temperature. It does not contradict the statement made above about the scale being independent of the way a property changes with temperature. We are not concerned with the way any property of water changes with temperature, but a single reference temperature is needed. The choice of the exact number 273.16 seems unusual at first sight. In the absence of historical precedence no doubt a number such as 1000 would have been chosen, but the size of a unit of temperature would then have been different from units customarily used. An International Conference on units made the choice of 273.16 in order to make the temperature interval of one kelvin very nearly equal to the old interval of one centigrade degree.

It is still difficult to measure thermodynamic temperature to a high degree of accuracy. The measurement of the thermodynamic temperature of well defined fixed temperatures is carried out by Standards Laboratories around the world and a list of some of their results is given in Table 24.1 to the nearest 0.01 K, although the uncertainty in the readings is often greater than 0.01 K.

Table 24.1 Temperature reference points

Defining fixed point	Thermodynamic temperature/K
absolute zero	0
triple point of hydrogen	13.81
boiling point of hydrogen	20.28
boiling point of neon	27.10
triple point of oxygen	54.36
boiling point of oxygen	90.19
freezing point of water	273.15
triple point of water	273.16 exactly
boiling point of water	373.15
triple point of benzoic acid	395.52
freezing point of lead	600.65
boiling point of sulphur	717.82
freezing point of aluminium	933.52
freezing point of silver	1235.08
freezing point of gold	1337.58

24.3 OTHER TEMPERATURE SCALES

The Celsius scale

The Celsius scale of temperature is merely an arithmetical adjustment to the thermodynamic scale measured in kelvin. Its value is given by subtracting 273.15 from the thermodynamic temperature measured in kelvin. The symbol t is used for Celsius temperatures. We have therefore the defining equation for Celsius thermodynamic temperature:

 $$t/°C = T/K - 273.15$$

Since the interval of one kelvin is very nearly the same as the old centigrade degree, the effect of subtracting 273.15 is to make the Celsius scale very nearly the same as the old centigrade scale. The change to thermodynamic temperatures has made the centigrade scale redundant, but the international conference which set up the new scales accepted that since changing the custom of people is difficult, they would make the new scale agree with the old scale if no great accuracy is required. Table 24.2 should make clear that there was a fundamental change in the basis on which temperature is measured with the introduction of thermodynamic temperatures. In the table, temperature values shown in blue are exact by definition. The other temperatures are experimental values and improvements in experimental techniques in the future may make additional significant figures possible. When more significant figures are able to be given, it will be apparent that centigrade and Celsius temperatures are not the same.

Table 24.2   A few temperatures shown on different scales. The temperatures which are exact by definition are shown in blue

	Kelvin	Celsius	Centigrade
absolute zero	0	–273.15	–273.15
freezing point of water	273.15	0.00	0
triple point of water	273.16	0.01	0.01
boiling point of water	373.15	100.00	100.00
freezing point of silver	1235.08	961.93	961.93

Temperature scales

You have decided that the value 273.16 is too awkward to use in a basic definition and that the International Conference which set up the kelvin scale should have been bold and set up an entirely new thermodynamic temperature scale. You would like the Conference to use the number 500 for the value of the triple point of water on the new scale instead of 273.16 (and you would not object if it decided to call the new scale of temperature after you).

Prepare a table showing some temperatures, taken from Table 24.1, on your new scale.

Someone else at the Conference, whose name began with the letter F, suggests that you subtract 468 from your scale of values. He says that this makes things familiar. Do you think this helps? What is it similiar to? What would be the temperature on his scale on a nice warm summer day?

The ideal gas scale

The meaning of the term ideal gas will be dealt with more fully in Chapter 25. At this stage an ideal gas can be stated to obey the gas law $pV \propto T$ exactly, where p is the pressure of the gas, V its volume and T its temperature on the ideal gas scale. Since we have already defined pressure and volume, this equation effectively defines an unknown temperature from the equation

$$\frac{pV}{T} = \text{constant}$$

$$\frac{(pV) \text{ at unknown temperature}}{\text{unknown temperature}} = \frac{(pV) \text{ at triple point}}{273.16}$$

In symbols this is written:

$$T = \frac{(pV)_T}{(pV)_{tr}} \times 273.16 \text{ K}$$

where T is the unknown temperature on the ideal gas scale. It can be shown theoretically that the value of temperature given by the ideal gas scale is identical with the thermodynamic temperature. This is why the symbol T has been used for temperature on the ideal gas scale. Not only does this agreement reinforce the decision to use a scale which does not depend on the properties of any substance, but it also increases the possibility of measuring a thermodynamic temperature. The main difficulty with the reversible heat engine definition of the thermodynamic temperature is that it is at present impossible to use it directly to get accurate measurements of temperature. The ideal gas scale provides a way around this difficulty.

A real gas at low pressure behaves in a way similar to an ideal gas. As the pressure is reduced the agreement between the two gets progressively better. By extrapolating results to zero pressure it is possible to measure the ideal gas temperature of a system to within a few millikelvins. This can be written in the following way:

$$\frac{(pV) \text{ at unknown temperature}}{\text{unknown temperature}} = \frac{(pV) \text{ at triple point}}{273.16}$$

when p in each bracket approaches zero.

It is by the use of such methods that the values quoted in Table 24.1 were obtained. Before considering the practical arrangements for such measurements, the following example should be worked through to show how the extrapolation to zero pressure can be done.

Example 1

Using a gas thermometer, the pressures p_{tr} and p_T of different quantities of different gases were obtained at the triple point of water and at an unknown temperature T. For any pair of pressure readings, the volume and mass of gas remains constant.`

Since for an ideal gas

$$T = \frac{(pV)_T}{(pV)_{tr}} \times 273.16 \text{ K}$$

$$T = \frac{p_T}{p_{tr}} \times 273.16 \text{ K}$$

if, as in this case, the volume is kept constant.

The pressure readings measured and the temperatures calculated, assuming the gas to be ideal, are given in Table 24.3.

Table 24.3

Hydrogen			Nitrogen			Oxygen		
p_T /Pa	p_{tr} /Pa	T /K	p_T /Pa	p_{tr} /Pa	T /K	p_T /Pa	p_{tr} /Pa	T /K
219 180	83 460	717.36	274 260	104 360	717.87	231 440	87 880	719.39
164 400	62 590	717.49	205 690	78 270	717.85	173 490	65 910	719.02
109 540	41 700	717.55	137 120	52 180	717.82	115 620	43 950	718.61
54 780	20 850	717.68	68 560	26 090	717.82	57 760	21 970	718.15

You should note that the temperature as calculated in this way fluctuates from one gas and from one pressure to another. The last stage in this very lengthy process, when all the readings have to be obtained to high accuracy, is to plot a graph of calculated temperature against the pressure at the triple point for that reading. The graph obtained is reproduced in Fig 24.3 and shows an apparent variation in temperature which follows a pattern. It also shows that as the pressure of the gas falls, the calculated temperatures get closer together. The gas is approaching the ideal gas. If the graphs are extrapolated to zero pressure, that is, continued backwards

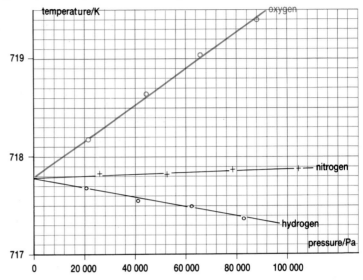

Fig 24.3 As the pressure of a gas falls it becomes more nearly an ideal gas. When the graphs are extrapolated to zero pressure the gases give a common temperature of 717.78 ± 0.05 K. This is the ideal gas scale temperature.

until each one crosses the temperature axis, they all meet at a point. This point gives the ideal gas temperature. Here the value is (717.78 ± 0.05) K. It can be seen here that which gas is used does not matter. All the graphs cross at the same point, reinforcing the idea that the ideal gas scale of temperature and the thermodynamic scale of temperature do not depend on the property of any particular substance. All the temperatures calculated in Table 24.3 are different from one another because the gases are real gases. By extrapolating to zero pressure the gases become ideal, and the ideal gas scale temperature is found.

Empirical scales

Many other properties of substances vary with temperature besides the pressure or volume of a gas. Any property which varies with temperature may be used as a basis for constructing a thermometer. All of the following properties vary with temperature and provide the principle on which a thermometer is based:

volume of a liquid, such as mercury or ethanol
resistance of a metal, such as platinum
thermocouple effect, between two metals such as copper and nickel
length of a metal rod, such as brass
resistance of a semiconductor, such as silicon
magnetic properties of a crystal, such as chromium potassium alum
vapour pressure of a gas, such as helium
radiation from a hot body, such as a lamp filament

Different properties are used over different ranges of temperatures, as will be explained in the next section. Here it is necessary to consider what constitutes a good property to use in making a thermometer.

First, the change in the proprerty must be large enough to measure accurately or the sensitivity of the thermometer will be low. This can be done with a normal mercury in glass thermometer by having a very narrow capillary tube to exaggerate the expansion of the mercury, and by using a glass which itself expands very little. What would happen if a thermometer were constructed out of a glass with large expansion containing a liquid with small expansion? Some semiconductors change their resistance by a very large factor, 100 times or more, for comparatively small changes in temperature. A device which makes use of this property is called a **thermistor**.

Secondly, the value of the temperature recorded must be reproducible. The thermometric property is useless if sometimes one value is given for, say, the melting point of tin, yet on other occasions the value is different. Some thermometric properties are particularly good on this account. The resistance of a platinum resistance thermometer can, if well used, give reproducible results in temperature measurement up to six significant figures. This creates a difficulty because often the thermodynamic temperature itself is not known to that degree of accuracy.

Thirdly the property being used must be suitable over the temperature range being measured. The magnetic property of crystals is used for measuring temperatures right down to near absolute zero, but is of no use at high temperatures. The radiation of light from a filament can only be used at high temperatures. Some properties do not change appreciably over a range of temperature. The volume of water cannot be used as a thermometric property for this reason. Between $0\,°C$ and $4\,°C$ water contracts, from $4\,°C$ upwards it expands as shown in Fig 24.4. This implies that if a water in glass thermometer were to be constructed then its scale would be so bunched up, and doubled up on itself, that it would be useless in the range of $0\,°C$ to $8\,°C$, Fig 24.5. Many thermocouples have ranges of temperature within which they cannot be used, as the thermoelectric e.m.f. does not vary sufficiently.

Finally the thermometers must be able to be calibrated without too much difficulty. If the property changes uniformly with thermodynamic temperature, then once two fixed points are calibrated a linear scale can easily be added. If the property varies irregularly, then calibration is more involved. In the case of a thermistor, which is a piece of suitably prepared semiconductor material, the resistance varies in a far from linear manner. This necessitates many calibration points or an equation relating temperature with resistance. Calibration equations made up to fit observed facts are called **empirical equations** and are much used in practical thermometry. An equation you may be familiar with can be considered as an empirical equation. It is the one which used to be used for defining the centigrade scale of temperature, but since the Celsius scale is so similar to the centigrade scale it is useful as an empirical scale for the Celsius scale. As an empirical scale it makes the assumption that the length of the mercury column in a mercury in glass thermometer increases linearly with the thermodynamic temperature. At the freezing point of pure water, the length of column (l_0) is measured. At the boiling point, the length of column is also measured (l_{100}) and a straight line graph plotted of length against temperature. The graph, or the equation

$$\frac{t}{100} = \frac{l_t - l_0}{l_{100} - l_0}$$

which is derived directly from the graph, can then be used to find an unknown temperature t if the length of the mercury thread l_t is found at t. (See Fig 24.6).

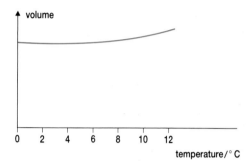

Fig 24.4 A graph showing qualitatively how the volume of water varies with the temperature. It shows that water has a maximum density at $4.0\,°C$.

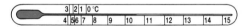

Fig 24.5 This unlikely thermometer uses coloured water as its thermometric liquid. It is shown recording a temperature of $10.5\,°C$ but when the temperature falls, two possible temperatures are given for a single reading. What would happen when the temperature is $-1\,°C$?

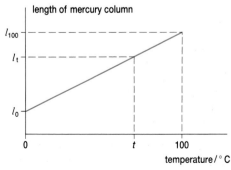

Fig 24.6

24.1 Find the Celsius temperature t given by a mercury in glass thermometer if the length of the mercury column is 40 mm at the ice point, 240 mm at the boiling point of water, and 178 mm at t. Answer this question first by using common sense, and then by using the equation above. You should, needless to say, get the same answer in both cases.

24.2 Find the Celsius temperature t on the platinum resistance thermometer if an empirical temperature scale for the thermometer is

$$\frac{t}{100} = \frac{R_t - R_0}{R_{100} - R_0}$$

The resistance at the ice point is 8.452 Ω, the resistance at the steam point is 11.550 Ω, and the resistance at t is 6.707 Ω.

The following example shows how empirical scales can be used with specific thermometers to obtain consistency in measuring temperature.

Example 2
The empirical equation which gives the thermoelectric e.m.f. E of a standard rhodium–platinum alloy and platinum thermocouple, when one junction is at a temperature of 0 °C, is

$$E = a + bt + ct^2$$

where a, b and c are constants. These constants are first calculated by using the following information:

at a temperature of 630.74 °C the e.m.f. is 5534 mV
at a temperature of 961.93 °C the e.m.f. is 9117 mV
at a temperature of 1064.43 °C the e.m.f. is 10 300 mV

These calibration values, when put into the equation, give three equations with three unknowns. Solving the three equations is tedious. It gives the values:

$$a = -278.4 \text{ mV}$$
$$b = 8.163\,9 \text{ mV } °C^{-1}$$
$$c = 0.001\,666\,8 \text{ mV } °C^{-2}$$

Once this has been done, then, in effect a parabolic calibration curve has been fitted around the facts concerning the variation of thermoelectric e.m.f. with Celsius temperature. If now an e.m.f. is measured at some other temperature, then that temperature can be calculated. This is the procedure with all empirical equations.

For instance, if the e.m.f. on the thermocouple is found to be 7121 mV then the Celsius temperature is given by solving the equation

$$7121 = -278.4 + 8.1639\,t + 0.001\,666\,8\,t^2 \text{ (all terms in mV)}$$

Solution of this quadratic equation gives $t = 781.62$ °C.

The above example shows that working on a curved empirical scale can be difficult, but it does have the advantage that reproducible temperature measurements can be made. Table 24.4 shows how certain empirical temperature values differ from the thermodynamic temperature if it is assumed that there is a straight line variation of the thermometric property with temperature between 0 °C and 100 °C. The variation in values should make it quite clear why it is necessary to use more complex equations as in example 2.

Table 24.4 Showing how near different thermometers approach the thermodynamic temperature.

Thermo-dynamic temperature	Celsius temperature	Chromel-alumel thermocouple	Platinum resistance thermometer	Constant volume hydrogen thermometer
/K	/°C	/°C	/°C	/°C
273.15	0	0.00	0.00	0.00
293.15	20	19.5	20.25	19.97
313.15	40	39.3	40.35	39.94
333.15	60	59.5	60.35	59.94
353.15	80	79.8	80.20	79.97
373.15	100	100.00	100.00	100.00

24.4 THERMOMETERS

The choice of thermometer to be used for a particular temperature measurement depends on many factors such as accuracy, sensitivity, convenience, size, nature of read-out required, cost, range of temperatures being measured, speed of response and availability. Some of these factors will be dealt with here but, since there is a huge variety of different types of thermometer, the list of thermometers will by no means be complete. Before dealing with individual thermometers, it is worth while making some general points about choice of thermometer.

Perhaps the first point to make is the obvious one that the thermometer must be capable of recording temperatures in the range required. Standard laboratory liquid in glass thermometers have clearly defined ranges. A common temperature range is –10 °C to +110 °C. Unless the thermometer is longer than 300 mm this limits the distance for each degree to not more than about 2 mm. These thermometers cannot be relied on therefore to measure to much greater sensitivity than the nearest half degree. If the thermometer is shorter, many are only 150 mm in length, then the sensitivity is bound to fall. The sensitivity will also fall if the range is increased. A mercury in glass thermometer of 150 mm length, which reads from –10 °C to +350 °C, will probably have a distance of only about 3 mm for each 10 °C rise in temperature. A suitable length of scale should therefore be chosen for the range of temperatures to be measured. Wasted scale range increases the uncertainty of measurement.

Another important factor is to be clear what reading is required. In using a clinical thermometer, the temperature required is the maximum temperature reached when the thermometer is placed under the patient's tongue. The thermometer's construction allows this maximum temperature to be recorded. A greenhouse thermometer is normally used to record maximum and minimum temperatures. A meteorological thermometer normally is required to produce a graphical record of the variation in temperature over a period of time. Many thermometers are used to control temperature. That is, they need to be able to operate electrical switches at pre-set temperatures. These switching thermometers are called **thermostats** and may use expansion of metals for their operation, as in an oven, or may use electronic circuitry involving thermistors. In other situations it may be important to have remote sensing. The thermometers in a nuclear reactor, for example, cannot be read directly so an electrical output is required from them. Hundreds of thermocouples are used throughout the reactor and the electrical output each gives enables the temperature at all points to be monitored continuously.

The heat required to raise the temperature of the thermometer itself is negligible for the thermocouples within a nuclear reactor. The thermocouple is permanently in a high temperature environment and the temperature does not change very much, nor very quickly. In other situations this may not be the case. If a temperature is changing rapidly then a thermometer can only respond quickly to the change in temperature if it has a low heat capacity. A low heat capacity is also needed when the temperature of a small object is required. To take an extreme example, if a fly lands on a mercury thermometer, the thermometer will continue to record its own temperature, not the temperature of the fly, which could be above room temperature as a result of muscular activity.

Thermometers do, of course, normally record their own temperature. When in use they are placed in thermal contact with the body whose temperature is required (manufacturers give details of the depth of immersion required and whether the thermometer is to be used vertically or horizontally). If given enough time to come into thermal equilibrium with their surroundings, they have the same temperature as their surroundings, as stated by the zeroth law of thermodynamics. Usually the thermal energy required to heat up the thermometer itself is negligible, but it must be considered that the final temperature recorded for a system is the temperature it acquires after its temperature has been changed by the insertion of the thermometer.

In calibration, a method must be used to obtain the triple point of water. This can be done using a triple point cell. A diagram of a **triple point cell** for water is shown in Fig. 24.7. It consists of a double walled Pyrex container with high purity water nearly filling the space between the walls. All the air is extracted from the water enclosure. In use, the triple point cell is cooled until ice is present. So long as ice, water and water vapour are present then the temperature in the cell is 273.16 K exactly. The reason that this temperature is not the same as the freezing point of water is that the pressure condition for the normal freezing point is that the pressure must be one atmosphere (101.3 kPa). In the triple point cell, the pressure is that due to the water vapour alone and has a value of 0.61 kPa.

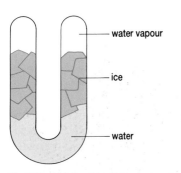

Fig 24.7 A triple point cell. The temperature in the cell is 273.16 K.

Constant volume gas thermometer

This thermometer is used to obtain temperature standards on the thermodynamic scale. Basically it consists of a flask containing gas whose volume can be maintained at a fixed value by controlling the pressure. The pressure of the gas is measured using a manometer and barometer. The barometer can be incorporated into the thermometer as shown in Fig 24.8, or it may be necessary to use a separate instrument to find atmospheric pressure. To find an unknown temperature, say, for example, the triple point of carbon dioxide, the flask is placed in a carbon dioxide triple point cell and after thermal equilibrium has been reached, the level of the reservoir is adjusted until the level of the mercury reaches the fixed mark A on the left-hand limb of the manometer. Height h is then measured to give the pressure p_T exerted by the gas at the unknown temperature. The flask is then placed in a water triple point cell and the procedure repeated to find p_{tr} at the triple point of water.

For accurate measurements using this thermometer, the heights h would be measured with a travelling microscope fitted with a vernier scale and corrections would be made for thermal expansion of the flask, surface tension effects on the mercury, the gas being at room temperature in the dead space of the thermometer, the vapour pressure of the mercury and for the gas not being an ideal gas.

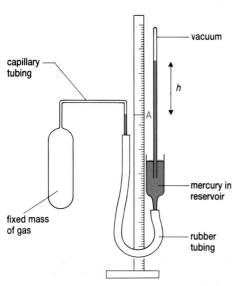

Fig 24.8 A constant volume gas thermometer.

TEMPERATURE

24.3 The readings taken in carrying out an experiment like that illustrated in Fig 24.8 were as follows:

h at the triple point of carbon dioxide   = 621.1 mm
h at the triple point of water   = 557.7 mm

(a) What is the temperature of the triple point of carbon dioxide? Give your answer in kelvin and in degrees Celsius.

(b) Explain why this is not the thermodynamic temperature. What procedure would be necessary to obtain the thermodynamic temperature with this apparatus?

Platinum resistance thermometer

The resistance of platinum varies with the temperature and so can be used as a thermometric property. The particular advantage of using platinum in this way is that its resistance may be measured electrically to a high degree of accuracy and over a wide range of temperature. The readings so obtained are difficult to relate to the thermodynamic scale of temperature, although tables are available to help with this conversion, but the readings

Fig 24.9 The platinum coil is wound on an insulating strip of mica and is wound half clockwise and half anticlockwise. This gives the coil zero inductance. Two of the terminals are connected to the coil; the other two are connected to dummy leads.

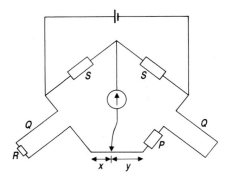

Fig 24.10 The circuit for use with a platinum resistance thermometer.

themselves are extremely reliable and reproducible. This makes comparison of temperature possible even if the exact thermodynamic temperature is not known. The thermometer itself, Fig 24.9, is usually a coil of pure platinum wire of diameter about 0.1 mm and of resistance about 25 Ω. To compensate for any change in the resistance of the leads, a pair of dummy leads are connected in the measuring circuit, Fig 24.10. The resistance is measured with a Wheatstone bridge using a current not exceeding 2 mA. If a larger current is used, the wire heats up appreciably on account of the current through it, and so gives a false temperature reading. In the circuit the two standard resistances, S, are accurately equal to one another, and the Wheatstone bridge equation (section 15.5) gives at balance

$$\frac{S}{S} = 1 = \frac{Q + R + xr}{Q + P + yr}$$

$$Q + R + xr = Q + P + yr$$

where r is the resistance per unit length of the bridge wire.

The resistances of the two pairs of leads, Q, may change as the temperature changes, but since they will change in the same way they will always be equal and will therefore cancel out from both sides of the equation. P is a resistance approximately the same size as the platinum resistance R. This gives

$$R + xr = P + yr$$
$$R = P + yr - xr$$
$$= P + r(y - x)$$

Once the resistance of the thermometer has been obtained at fixed points, the unknown temperature can be found by measuring the resistance at that temperature and using an empirical equation. If the straight line empirical equation

$$\frac{t}{100} = \frac{R_t - R_0}{R_{100} - R_0}$$

is used, as it was in question 24.2, we get

$$\frac{t}{100} = \frac{\{P + r(y-x)_t\} - \{P + r(y-x)_0\}}{\{P + r(y-x)_{100}\} - \{P + r(y-x)_0\}}$$

$$= \frac{(y-x)_t - (y-x)_0}{(y-x)_{100} - (y-x)_0}$$

where $(y - x)_t$ means the difference between y and x at temperature t etc. This calculation is complex when first seen, but is quite easy to follow through with numbers. Try it by answering question 24.4.

The temperature is obtained to better than half a degree, as shown in Table 24.4. Greater accuracy can be obtained by using more complex empirical equations involving squared or cubic functions.

QUESTION

24.4 A platinum resistance thermometer is used in a Wheatstone bridge circuit, Fig 24.10. When the temperature of the thermometer is 90.19 K, the value of x is 906.1 mm and the value of y on a metre wire is therefore 93.9 mm; when the temperature is 273.15 K, x is 287.0 mm. Use a straight line empirical equation to find the temperature when x is 51.3 mm and show that the values of P and r are not needed to make this calculation.

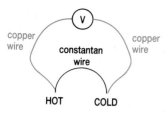

Fig 24.11 Circuit to show the Seebeck effect used in thermocouples.

Thermocouple thermometers

When a junction is made between two different metals, a small e.m.f. is set up between them. If any attempt is made to measure this e.m.f., then a continuous circuit has to be made and there will necessarily be other junctions to consider as well. These other junctions usually nullify the effect at the first junction. If however two junctions are used and they are at different temperatures, then a measureable e.m.f. is set up which depends on the temperature difference between the two junctions. This effect is called the **Seebeck effect** and it is made use of in all thermocouple thermometers. It can be demonstrated quite simply by connecting up the circuit shown in Fig 24.11 using copper and constantan wires twisted firmly together at the two junctions. The e.m.f. generated is of the order of a millivolt so a very sensitive meter or a suitably adapted potentiometer must be used, see section 15.6. In practice the thermometer can be used without an obvious cold junction. The cold junction is kept away from the hot junction and is simply at room temperature. One particular use of thermocouple thermometers is in the measurement of rapidly fluctuating temperatures. The junction itself can be made very small so that it has very low heat capacity. It will therefore quickly reach thermal equilibrium with its surroundings.

The e.m.f. of common thermocouples varies in a non-linear way with temperature, so again an empirical equation needs to be used for accuracy. Over small changes of temperature, however, the variation can often be regarded as linear.

Fig 24.12 Thermocouples being manufactured by arc welding. The wire for the thermocouples is above the arc welding unit.

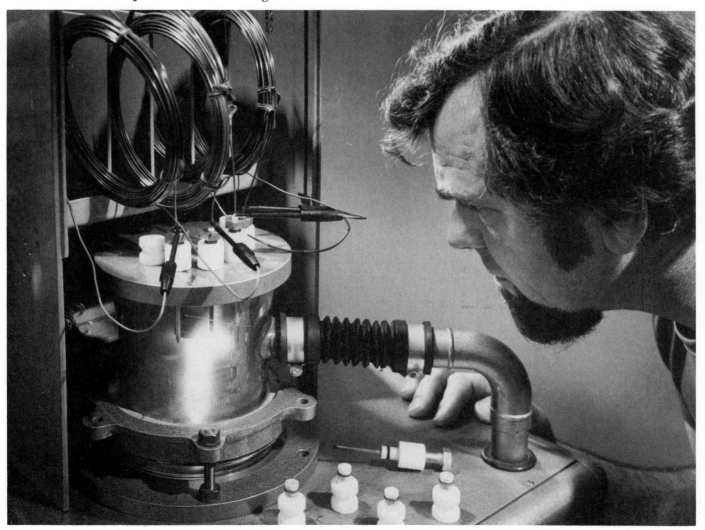

Table 24.5 lists some commonly used thermocouples and states their approximate sensitivities.

Table 24.5. Thermocouples and their sensitivities

Thermocouple	Sensitivity /μV K^{-1}
copper–iron	5
platinum–platinum rhodium	6
copper–constantan	40
iron–constantan	50

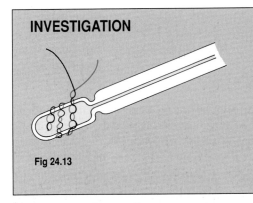
Radiation thermometers

Above temperatures of about 1700 K all of the thermometers mentioned so far become useless. Even with careful choice of metal they will be about at their melting point. Above this temperature, therefore, a totally different method has to be used. In practice, above the temperature of the melting point of gold, the concentration of radiant energy emitted by the hot object is used to measure its temperature. Detailed analysis of how this is done will not be given here but it depends on the brilliance and colour of the hot object. Table 24.6 merely gives a guide to the temperature and comparative colours of a few commonly observed hot bodies.

Table 24.6

Example	Colour	Approximate temperature/K
simmering hot plate (just visible)	dull red	900
fully on hot plate	red	1000
element of electric fire	orange	1100
filament of electic lamp	yellow	2500
surface of red giant star (Betelgeux)	yellow	3300
surface of the Sun	white	6000
surface of white dwarf star (Sirius)	white	11 000
surface of white giant star (Rigel)	blue/white	25 000

24.5 THERMAL EXPANSION

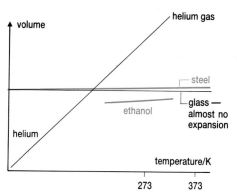

Fig 24.14 Compared with gases, solids and liquids expand only a little when heated.

Use has been made of the fact that materials change in size as they are heated. In almost all cases the volume of an object increases as its temperature increases, and the increase is nearly linear with thermodynamic temperature. For ideal gases the volume, at constant pressure, is directly proportional to the thermodynamic temperature, but for solids and liquids the volume is not proportional to the thermodynamic temperature. In graphical terms, a straight line graph only shows proportion if it passes through the origin. Graphs showing the expansion of a solid and a liquid are nearly straight line graphs not passing through the origin, Fig 24.14.

If only the change in the length of a solid is required then the equation

$$l_t = l_0 (1 + \alpha t)$$

can be used. l_t is the length at Celsius temperature t, l_0 is the length at 0 °C and α is called the linear expansivity.

A glass container at almost absolute zero looks the same as it does at room temperature. Between these two temperatures the amount its volume increases by is very small. Similarly a room radiator does not appreciably grow in size as hot water is first passed through it. Increase in the volume of a substance results in a decrease in its density. This is demonstrated very clearly by a hot air balloon, Fig 24.15. The air inside the balloon is hotter

Fig 24.15 Because hot air is less dense than cold air it rises. A hot air balloon depends on this effect for lift.

than the air outside and so is less dense. Because of its lower density, it rises (see section 27.3). The fall in the density of a fluid when it is heated is responsible for convection. The less dense, hot fluid rises while the denser colder fluid falls.

QUESTION

24.5 A hot air balloon has an average temperature inside the balloon of 60 °C when the outside temperature is –3 °C. Find the density of the air inside the balloon if the density of air outside the balloon is 1.3 kg m⁻³. Estimate the volume of the balloon if it has a total mass of balloon, fuel and passengers of 300 kg.

SUMMARY

- The **zeroth law of thermodynamics** states that if two bodies are in thermal equilibrium with a third body, then they must be in thermal equilibrium with each other. It introduces temperature as a useful concept.
- A temperature T on the ideal gas scale is defined by the equation

$$T = \frac{(pV) \text{ at unknown temperature}}{(pV) \text{ at triple point}} \times 273.16$$

when the pressure in each bracket approaches zero. It can be proved that this temperature is identical to the absolute thermodynamic temperature.
- The Celsius temperature is based on the thermodynamic temperature but with a numerical shift of 273.15, so if t is the Celsius temperature

$$t \, / \, °C = T \, / K - 273.15$$

- Empirical equations are often used with thermometers. Frequently they assume a uniform increase in a property between two temperatures. The two temperatures are often 0 °C and 100 °C. In which case

$$\frac{t}{100} = \frac{X_t - X_0}{X_{100} - X_0}$$

where X is the property being used.

Thermometer	Range/°C	Accuracy	Sensitivity	Other comment
mercury in glass	–39 to 500	poor	moderate	direct reading
constant volume gas	–270 to 1500	excellent	good	slow response: used as standard
platinum resistance	–200 to 1200	good	poor	large heat capacity
thermocouple	–250 to 1500	fair	fair	good for varying temperatures
thermistor	–50 to 200	poor	very high	low heat capacity, good for varying temperatures
Radiation (pyrometer)	over 1000	poor	poor	usable over 1500 °C

- For the expansion of a solid

$$l_t = l_0 \, (1 + \alpha t)$$

where l_t is the length at Celsius temperature t, l_0 is the length at 0 °C and α is called the linear expansivity.

TEMPERATURE

EXAMINATION QUESTIONS

24.6 **(a)** Draw an electrical circuit for use with a metal resistance thermometer. (3)
(b) Explain how the measurements are converted to temperature readings. (3)
(c) State **two** advantages (in each case) of using a resistance thermometer instead of:
(i) a mercury in glass thermometer; (2)
(ii) a thermocouple thermometer. (2)

(OXFORD 1989)

24.7 Explain what is meant by a *scale of temperature*, quoting **two** examples of materials that may be used to establish a scale in the range –10 °C to +110 °C. (4)
Why are scales of temperature based on different materials likely to be different?
What is the relationship between a temperature in degrees Celsius (°C) and in kelvin (K)?
What properties are required of a satisfactory thermometric material? (8)
What is a *thermocouple*? Describe how one may be calibrated as a thermometer and discuss its particular advantages. (5)
When one junction X of a thermocouple is placed in melting ice and the other junction Y in steam at 373 K, the e.m.f. recorded is 8.2 mV. Junction X is removed from the ice and placed in a liquid bath, while junction Y remains in the steam. The e.m.f. is now 12.3 mV in the same direction as before. Calculate a value for the temperature of the liquid bath. (3)

(UCLES 1984)

24.8 The resistance of the element in a platinum resistance thermometer is 6.750 Ω at the triple point of water and 7.166 Ω at room temperature. What is the temperature of the room on the scale of the resistance thermometer? The triple point of water is 273.16 K. State one assumption you have made. (4)

(ULSEB 1986)

24.9 Give simple explanations to account for the typical variation of electrical resistance with temperature in metals and semiconductors. The electrical resistance R of a device varies with thermodynamic temperature T according to the relation

$$R = ae^{b/T}$$

where a and b are constants. Values of R and Celsius temperature t are given in the following table.

R/Ω	3480	2910	2410	2000	1630
$t/°C$	2.0	7.0	12.0	18.0	24.0

By plotting a suitable graph, determine the value of b. (7)

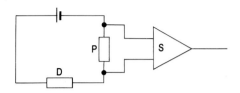

The diagram shows a circuit in which a device is used as a temperature sensor. The device, D, is connected in series with a resistor, P, and a cell of e.m.f. 1.50 V and negligible internal resistance. S has a very high input impedance and acts as a switch, changing the logic state of its output when the potential difference across P passes through 0.60 V. What must be the value of P so that the logic state changes at 16 °C? Suggest one reason why it would not be practicable to replace the sensor D by a wire-wound resistor. (5)

(UCLES 1985)

24.10 List the desirable properties of a material which enable it to be used as a thermometric substance. (2)
Two thermometers are constructed to measure the temperature θ, in degrees Celsius on the Gas Scale, as follows:
(a) a constant volume gas thermometer using helium at a pressure of 1.000×10^5 Pa at 0 °C. (Helium is assumed to behave as an ideal gas.)
(b) a small thermistor resistance thermometer of mass 0.1 g and resistance 5.0 kΩ at 0 °C. The thermistor resistance R_θ in ohms obeys the relation $R_\theta = 5000\,e^{-0.0400\theta}$.
Calculate the values of the pressure in (a) and the resistance in (b) for these thermometers at 50 °C and at 51 °C. Around 50 °C the error in measuring the gas pressure is ±100 Pa and in measuring the resistance of the thermistor is ±5 Ω. From this calculate the errors in the temperatures measured by the two thermometers, and deduce which is the more accurate at 50 °C. (6)
Draw a circuit diagram, with appropriate component values, suitable for use with the thermistor thermometer. (3)

You are provided with these two thermometers and also with an ordinary mercury in glass laboratory thermometer. Discuss the relative merits of each of these thermometers for each of the following measurements.

(i) The temperature of a water tank of volume 1 m^3 maintained at 50 °C.

(ii) The variation in skin temperature over the surface of your body.

(iii) The temperature of liquid oxygen (−183 °C) in a vacuum flask.   (9)

(O & C 1983)

24.11 The resistance of the element of a platinum resistance thermometer is 2.00 Ω at the ice point and 2.73 Ω at the steam point. What temperature on the platinum resistance scale would correspond to a resistance value of 8.43 Ω?

Measured on the gas scale, the same temperature corresponded to a value of 1020 °C. Explain the discrepancy.

(ULSEB 1983)

24.12 The e.m.f. of a certain thermocouple with one junction X in melting pure ice and the other Y in steam from water boiling at standard pressure is 4.1 mV. With Y still in the steam and X in a certain boiling liquid, the e.m.f. is 11.6 mV in the same direction as before. Deduce the boiling point of the liquid on the centigrade scale of the thermo-electric thermometer.   (3)

(UCLES 1985)

Chapter 25

GASES

LEARNING OBJECTIVES

At the end of this chapter you should be able to:

1. state and use Boyle's law;

2. use the universal gas law equation for an ideal gas;

3. describe the kinetic theory of an ideal gas;

4. relate the kinetic energy of molecules to the temperature;

5. describe qualitatively the distribution of molecular speeds within a gas.

25.1 BOYLE'S LAW

The behaviour of gases was referred to in Chapter 24 where the variation in the pressure and the volume of an ideal gas was used as a basis for measuring the thermodynamic temperature. In this chapter the properties of a gas will again be considered, but this time the properties will also be related to the behaviour of the molecules within the gas. When the properties under consideration are large scale properties which can be felt or measured with instruments, such as the pressure of a gas, they are called **macroscopic** quantities. Properties which cannot be sensed directly, such as the velocities of the individual molecules, are called **microscopic** quantities.

Early experiments on gases were done using macroscopic quantities only. In the Seventeenth Century Robert Boyle measured the volume of a fixed mass of gas at different pressures, while keeping the temperature constant. He found that the volume was inversely proportional to the pressure. The apparatus shown in Fig 25.2 can be used to repeat this experiment. The volume V of trapped gas is measured for different pressures p. Care must be taken to leave enough time in between taking one reading and the next to allow the gas to be in thermal equilibrium with its surroundings. If this is not done then the rise in temperature, which takes place when the

Fig 25.1 The atmosphere seems more real when it can be photographed. In fact light is being reflected from dust or water droplets in the atmosphere.

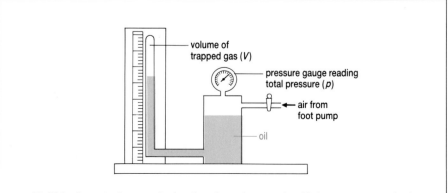

Fig 25.2 Apparatus for measuring how the volume of a gas varies with the pressure exerted on it.

pressure of the gas is increased, will affect the readings of the volume. If a graph is plotted of p against V then it is difficult to establish whether or not there is inverse proportion between the two quantities, because the graph is not a straight line. In order to establish inverse proportion either plot a graph of p against $1/V$ or plot pV against V. A formal statement of Boyle's law is:

 The volume of a fixed mass of gas is inversely proportional to the pressure applied to it if the temperature is kept constant.

Expressed mathematically this is:

$$p \propto \frac{1}{V} \ \text{ or } \ pV = \text{constant}$$

The law is an experimental law and has been found to be reliable provided the gas pressure is not too high.

25.2 THE IDEAL GAS

The theoretical model of an ideal gas was introduced in Chapter 24. An ideal gas must obey Boyle's law.

Since the equation used to define the ideal gas scale of temperature is

$$T = \frac{(pV)_T}{(pV)_{tr}} \times 273.16 \text{ K}$$

it follows that since 273.16 and p and V at the triple point of water are all constants

$$T \propto pV \text{ for an ideal gas.}$$

In other words because temperature is defined in this way it follows that the volume of a fixed mass of an ideal gas at constant pressure **must** be proportional to the temperature. If the measurements of an experiment show that the volume of a particular gas is proportional to the ideal gas temperature, the conclusion which can be drawn from the experiment is that the gas is behaving ideally.

Since for the ideal gas $pV \propto T$ it follows that

$$\frac{p_1 V_1}{T_1} = \frac{p_2 V_2}{T_2}$$

where 1 refers to the initial conditions in the state of a gas and 2 refers to the final conditions. Note that in all gas law problems the temperature is the absolute temperature measured in kelvin and not the Celsius temperature.

Additionally an ideal gas must not cool down when allowed to undergo a free expansion into a vacuum; this implies that the molecules in an ideal gas do not attract one another, so an ideal gas would never be able to be turned into a liquid. More detail will be given about this when considering the internal energy of gases in section 25.3.

Example 1

A fixed mass of gas in passing through a jet engine has its pressure increased from 3.0×10^5 Pa to 1.3×10^6 Pa, while its temperature rises from 80 °C to 1500 °C. By what factor does the volume of the gas change?

Hint for solution of gas law problems. Start with a table of information given.

	Initial conditions	Final conditions
pressure	3.0×10^5 Pa	13×10^5 Pa
volume	V_1	V_2
temperature	80 °C	1500 °C
	= 80 + 273	= 1500 + 273
	= 353 K	= 1773 K

$$\frac{p_1 V_1}{T_1} = \frac{p_2 V_2}{T_2}$$

$$\frac{(3.0 \times 10^5 \text{ Pa}) \times V_1}{353 \text{ K}} = \frac{(13.0 \times 10^5 \text{ Pa}) \times V_2}{1773 \text{ K}}$$

$$\frac{V_2}{V_1} = \frac{1773 \times 3.0}{353 \times 13} = 1.16$$

The volume of the gas has increased by a factor of 1.2 (to 2 sig. figs) in passing through the engine. The assumption has been made that the gas behaves as an ideal gas. For the information which is given here to two significant figures, this is a reasonable assumption since two significant figures clearly does not give a value capable of high accuracy.

Whether or not gases can be considered as having ideal behaviour depends on the pressures being used and on the accuracy required. With nitrogen for example, at atmospheric pressure and below, calculations using the ideal gas law give results which are within 0.1% of the experimental value. If the pressure is raised to 100 atmospheres, then the error introduced by using the ideal gas law is still only of the order of 5% provided the nitrogen is not cooled down to near its boiling point.

QUESTIONS

25.1 During the compression stroke of an internal combustion engine, the volume of the gas in the cylinder is reduced from 4.0×10^{-4} m^3 to 5.0×10^{-5} m^3. During the stroke, the pressure changes from 1.0×10^5 Pa to 18×10^5 Pa. If the temperature of the gas at the start of the stroke is 22 °C, what will be its temperature at the end of the stroke?

25.2 Standard atmospheric pressure and temperature (**stp**) are 1.01×10^5 Pa and 0 °C. Find the volume at stp of a mass of air which has a volume of 3.6×10^{-5} m^3 at a pressure of 5.7×10^4 Pa and a temperature of 400 °C.

25.3 The pressure in a car tyre when at a temperature of 10 °C is 1.9×10^5 Pa above atmospheric pressure. If the volume of the tyre does not change, what will be the pressure in the tyre at a temperature of 33 °C? (Atmospheric pressure is 1.0×10^5 Pa.)

The value of the term $\frac{pV}{T}$ is directly proportional to the amount of gas, n in moles.

This therefore gives an equation relating p, the pressure of the gas, V, the volume of the gas, T, the temperature of the gas and n, the amount of gas:

$$\frac{pV}{T} \propto n$$

or $$\frac{pV}{T} = nR$$

where R is a constant called the **molar gas constant**.

If 1 mol of an ideal gas is considered, then its volume at standard temperature and pressure is 0.0224 m³. Substituting in the numerical values gives:

$$(1.014 \times 10^5 \, \text{Pa}) \times (0.0224 \, \text{m}^3) = 1 \, \text{mol} \times R \times 273.15$$
$$R = 8.31 \, \text{J K}^{-1} \, \text{mol}^{-1}$$

The volume which a mole of gas occupies is called its **molar volume** and is given the symbol V_m. $V_m = V/n$. The gas equation, which is called the **equation of state** of the gas, can therefore be written either as

$$pV = nRT \quad \text{or as} \quad pV_m = RT$$

The equation is known as the **universal gas law equation**.

25.3 THE KINETIC THEORY OF GASES

The behaviour of a gas on a macroscopic scale concerns the variation of its temperature, volume and pressure, but to obtain a clearer understanding of gas behaviour the microscopic movement of the individual gas molecules must be able to be related to these macroscopic quantities. It is the kinetic theory which enables this to be done. As usual, any theory must be able to be checked by experiment and must provide greater understanding of the physical problem. The kinetic theory of gases applies the laws of mechanics to molecular movement so that expressions for the pressure and temperature of a gas are obtained in terms of molecular speed, mass and number.

Brownian motion, which was discussed in section 22.3, gives clear evidence that particles of matter are in perpetual movement. The kinetic theory uses the idea that movement of gas molecules is responsible for the pressure of a gas. Certain assumptions are made about the properties of gas molecules:

1. Any gas consists of a very large number of molecules.
2. The molecules of the gas are in rapid, random motion.

Since the gravitational force on each molecule acts downwards, there is not quite random motion. Within a small volume, however, this effect is very small; the pressure on the top of a flask is very nearly the same as the pressure on the bottom. It is only when dealing with, say, the Earth's atmosphere that the lack of uniformity of the pressure would need to be taken into account.

3. Collisions between gas molecules are perfectly elastic.
4. Collisions between gas molecules and the walls of the container are elastic.
5. There are no intermolecular attractive forces.
6. Intermolecular forces of repulsion act only during collisions between molecules. The duration of collisions is negligible compared with the time interval between collisions.
7. The volume of the gas molecules themselves is negligible compared with the volume of the container. That is, almost all the gas is empty space.

Several of these assumptions have the effect of defining the microscopic properties of an ideal gas. Assumption 7, for example, implies that the gas must not be at high pressure, because if the pressure is high then the concentration of molecules within the available space will be high. Densities of gases and liquids give a guide to how much space is actually occupied by the molecules. The density of a liquid is roughly 1000 times the density of the corresponding gas at atmospheric pressure, so of the total volume of the container, the volume of the molecules themselves is only 0.1% of the total volume.

The model of a gas which we are using therefore is one of a huge number of point masses, moving around in completely random zig-zag

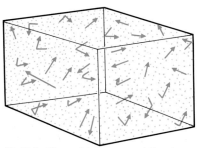

Fig 25.3 The model of a gas consisting of a very large number of molecules in rapid, haphazard motion.

GASES

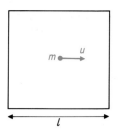

Fig 25.4

fashion as they bounce perfectly off one another and off the walls of the container, Fig 25.3.

To find the pressure caused by the continual bombardment of the walls of the container by the molecules, consider first a single molecule in a cubical box with sides of length l. If the mass of the molecule is m and it is travelling with velocity u directly towards the right-hand wall of the box, Fig 25.4, then it has:

momentum to the right before collision with wall $= mu$
momentum immediately after a perfect collision $= -mu$
change in momentum $= 2mu$ to the left.

The time interval before it makes another collision with the same wall will be the time taken to travel across the box and back:

$$\text{time between collisions with same wall} = \frac{2l}{u}$$

$$\text{number of collisions with this wall per unit time} = \frac{u}{2l}$$

$$\text{rate of change of momentum of molecule} = 2\,mu \times \frac{u}{2l} = \frac{mu^2}{l}$$

By Newton's second law, the rate of change of momentum of the molecule caused by one wall is the force the wall exerts on the molecule. This is equal in size to the force on the wall. So

$$\text{the force on the wall} = \frac{mu^2}{l}$$

The force here is an average force over a period of time, but if instead of one molecule we consider N molecules, all going backwards and forwards, Fig 25.5, then the total force on a wall will be

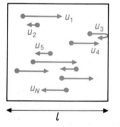

Fig 25.5

$$\frac{mu_1^2}{l} + \frac{mu_2^2}{l} + \frac{mu_3^2}{l} + \dots \frac{mu_N^2}{l}$$

$$= \frac{m}{l}(u_1^2 + u_2^2 + u_3^2 + \dots u_N^2)$$

where u_1 is the velocity of the first molecule, u_2 the velocity of the second, etc. If the average value of u^2 is written as $\overline{u^2}$, then this expression can be written more simply as

$$\frac{m}{l}(N \times \overline{u^2}) = \frac{mN\overline{u^2}}{l}$$

This gives the pressure on the wall

$$= \frac{\text{force on wall}}{\text{area of wall}} = \frac{m}{l}\frac{N\overline{u^2}}{l^2} = \frac{mN\overline{u^2}}{V}$$

where V is the volume of the box. Also mN is the total mass of the gas molecules, so mN/V is the density ρ of the gas. The pressure, p, exerted by all these molecules moving back and forth is given by

$$p = \rho u^2$$

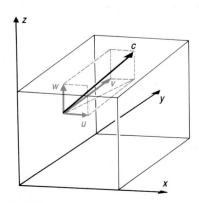

Fig 25.6

In an ideal gas, however, the molecules will not just move backwards and forwards in one dimension. They will move in three dimensions as shown in Fig 25.6. Using Pythagoras' theorem for the velocity c_1 of the first molecule gives

$$c_1^2 = u_1^2 + v_1^2 + w_1^2$$

If you imagine doing this for all the molecules of the gas and then adding all the equations together you would get

$$N\overline{c^2} = N\overline{u^2} + N\overline{v^2} + N\overline{w^2}$$

or $\qquad \overline{c^2} = \overline{u^2} + \overline{v^2} + \overline{w^2}$

where $\overline{c^2}$ is the average value of c^2 and is called the average or **mean square velocity**.

But since there is no tendency for the gas to exert a greater pressure in one direction rather than any other it follows that

$$\overline{u^2} = \overline{v^2} = \overline{w^2}$$

so $\qquad \overline{c^2} = \overline{u^2} + \overline{u^2} + \overline{u^2} = 3\overline{u^2}$

and $\qquad \overline{u^2} = \frac{1}{3}\overline{c^2}$

The pressure exerted on the wall was shown to be given by $p = \rho u^2$. This can therefore be written

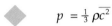

$$p = \tfrac{1}{3}\rho\overline{c^2}$$

This deduction has not involved collisions between molecules at all. In practice they will collide with one another, but since both momentum and kinetic energy are being conserved, any gain in these quantities by one molecule will result in another molecule losing an equal amount of that quantity. There is therefore no overall effect on the pressure as a result of collisions between molecules.

DATA ANALYSIS

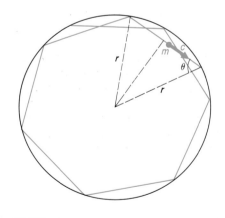

Fig 25.7

Kinetic theory

The pressure exerted by the gas does not depend on the shape of the container. This exercise enables you to work out the equation for the pressure exerted by a gas, $p = \frac{1}{3}\rho c^2$, for a spherical container of radius r. You are asked to complete each line with an algebraic expression, using the terms given on Fig 25.7. A single molecule, of mass m, and travelling with velocity c, hits the wall of the container at an angle of incidence θ.

Component of velocity in a direction towards the wall	=
Momentum towards wall before collision	=
Momentum away from wall after collision	=
Change in momentum at a collision	=
Distance travelled between collisions with wall	=
Time between collisions with wall	=
Number of collisions per unit time	=
Rate of change of momentum	=

(θ should cancel out at this stage; many glancing collisions have the same effect on the wall as fewer direct collisions)

Surface area of a sphere	= $4\pi r^2$
Total force on wall	=
Pressure on wall due to one molecule	=
Total pressure on wall	=

Volume of sphere	$= \frac{4}{3}\pi r^3$
Total pressure in terms of volume, N, m and mean square velocity	$=$
Total pressure in terms of mean square velocity and density	$= \frac{1}{3}\rho c^2$

Example 2

The density of nitrogen at stp is 1.25 kg m^{-3}. Find the mean square speed of a nitrogen molecule in air at stp and hence deduce the mean kinetic energy of a nitrogen molecule. (Relative molecular mass of nitrogen = 28.)

Standard atmospheric pressure = 1.01×10^5 Pa

$$
\begin{aligned}
\text{Substituting into } p &= \tfrac{1}{3}\rho c^2 \text{ gives} \\
1.01 \times 10^5 \text{ Pa} &= \tfrac{1}{3} \times 1.25 \text{ kg m}^{-3} \times c^2 \\
c^2 &= 3.03 \times 10^5 / 1.25 = 242\,400 \text{ m}^2 \text{ s}^{-2}
\end{aligned}
$$

1 mole of nitrogen molecules has a mass of 0.028 kg so using the Avogadro constant, N_A, gives the mass of a molecule of nitrogen as

$$\frac{0.028 \text{ kg}}{6.02 \times 10^{23}} = 4.65 \times 10^{-26} \text{ kg}$$

The mean kinetic energy of a nitrogen molecule = $\frac{1}{2}mc^2$

$$= \tfrac{1}{2} \times 4.65 \times 10^{-26} \text{ kg} \times 242\,400 \text{ m}^2 \text{ s}^{-2} = 5.64 \times 10^{-21} \text{ J}$$

QUESTIONS

25.4 Repeat example 2 but for oxygen instead of nitrogen. The density of oxygen at stp is 1.43 kg m^{-3} and the relative molecular mass of oxygen is 32. Compare your answer with example 2. Can you explain why they should be the same?

25.5 Seven molecules have speeds of 200, 300, 400, 500, 600, 700 and 800 m s^{-1} respectively. Find:
   (a) their mean speed;
   (b) their mean square speed (this is different from their mean speed squared);
   (c) the square root of their mean square speed (r.m.s. speed).

When using the kinetic theory, the term which keeps arising is the mean square speed of all the molecules. In the above example it was indicated that this term cannot be obtained by squaring the mean speed of the molecules. To overcome this difficulty, the term root mean square speed of the molecules is used. Be careful, when using this term, to work in the correct order. The **root mean square speed** of molecules is the square root of the mean (average) of the squared speeds of the molecules. It is **not** the square root of the mean speed squared.

$$c_{\text{r.m.s.}} = \sqrt{\overline{c^2}}$$

In solving problems using the kinetic theory of gases, extra care must be taken with algebraic symbols and units.
Temperature will always be in kelvin.
Relative molecular mass, M_r, has no units.
Molar mass is the mass of a mole and will be quoted in kg mol^{-1}.

For example, for oxygen:
an oxygen atom has a relative atomic mass of 16. An oxygen molecule, O_2, has a relative molecular mass of 32. Since the molar mass of oxygen atoms is 0.016 kg mol^{-1} = 16 g mol^{-1}, the molar mass of oxygen molecules is 0.032 kg mol^{-1}= 32 g mol^{-1} then the mass of an oxygen atom = $(0.016/6.02 \times 10^{23})$ kg.

A frequent cause of mistakes is to use 16 instead of 0.016 here. Another common error is not to distinguish carefully between a mole and a molecule. n is used to represent the number of moles of gas. N is likely to be a huge number as it is the number of molecules.

Always state exactly what amount of gas is being used and then, as always, look at answers critically to see if they make sense. An answer such as 'Number of molecules = 4.1×10^{-12}', is nonsensical.

25.4 APPLICATIONS OF THE KINETIC THEORY OF GASES

If $p = \frac{1}{3}\rho c^2$ is compared with the ideal gas law, we find that whereas at a microscopic level

$$p = \frac{1}{3} \frac{Nm}{V} \overline{c^2}$$

at a macroscopic level

$$p = \frac{n}{V} RT$$

where n is the amount of gas in moles.
This gives

$$\frac{1}{3} Nm\overline{c^2} = nRT$$

The total kinetic energy E_k of all the molecules $= \frac{1}{2} Nm\overline{c^2}$

so $\qquad Nm\overline{c^2} = 3nRT = 2E_k$

This gives $E_k = \frac{3}{2}nRT$

The kinetic energy of the molecules is therefore directly proportional to the temperature.

For an ideal gas the total kinetic energy of the molecules as they move around in their container is the **internal energy** U of the gas. This is given by

 $\qquad U = \frac{3}{2}nRT$

Note that this depends on the amount of ideal gas n. It is also directly proportional to temperature. Internal energy will be dealt with in more detail in section 26.3.

This expression not only gives the total kinetic energy of the gas molecules but also enables the average kinetic energy of one molecule to be obtained. Since a mole of molecules contains N_A molecules, where N_A is the Avogadro constant, n moles contain nN_A molecules. The average kinetic energy of one molecule is therefore

$$\frac{3nRT}{2nN_A} = \frac{3RT}{2N_A} = \frac{3kT}{2}$$

k is a constant called the **Boltzmann constant** and is the gas constant for an individual molecule. R is the molar gas constant, $R/N_A = k$.

The numerical value of k is 1.380×10^{-23} J K^{-1}.

The equation of state of the ideal gas can be expressed in terms of k rather than R giving:

$$pV = NkT, \text{ where } N \text{ is the total number of molecules.}$$

QUESTIONS

25.6 Find the average kinetic energy of a molecule in air at 300 K.

25.7 Sketch two graphs on the same temperature axis, to show how

(a) the average kinetic energy and

(b) the root mean square speed of a gas molecule in air vary with the temperature.

Avogadro's law

If we consider two gases A and B then

$$\text{for the first gas } p_A V_A = \tfrac{1}{3} N_A m_A \overline{c_A}^2$$

$$\text{for the second gas } p_B V_B = \tfrac{1}{3} N_B m_B \overline{c_B}^2$$

If the gases have the same pressure and volume then

$$N_A m_A \overline{c_A}^2 = N_B m_B \overline{c_B}^2$$

but if they are also at the same temperature then

$$\tfrac{1}{2} m_A \overline{c_A}^2 = \tfrac{1}{2} m_B \overline{c_B}^2$$

so $\qquad N_A = N_B$

This is Avogadro's law which states that

◆ Equal volumes of gas under the same conditions of temperature and pressure have equal number of molecules.

Dalton's law of partial pressures

If two gases A and B are mixed with one another in a single container of volume V, then for each gas separately

$$p_A V = \tfrac{1}{3} N_A m_A \overline{c_A}^2$$

$$p_B V = \tfrac{1}{3} N_B m_B \overline{c_B}^2$$

Since the gases must be at the same temperature as one another it follows that

$$\tfrac{1}{2} m_A \overline{c_A}^2 = \tfrac{1}{2} m_B \overline{c_B}^2$$

and dividing gives

$$\frac{p_A}{p_B} = \frac{N_A}{N_B}$$

This shows that the pressure which each gas exerts is proportional to the number of molecules of that gas present, and therefore that the total pressure is proportional to the total number of molecules. This is Dalton's law of partial pressures which states that when two or more gases which do not react with one another are present in the same container, then the total pressure is the sum of the pressures which each gas would exert by itself.

Mean free path

A spherical molecule, of radius r, when moving in a gas sweeps out a cylindrical volume in each unit of time, see Fig 25.8. If it does not hit any other molecule in this time then the centres of other molecules cannot be any closer to this volume than their radii. The volume of space within which the centres of other molecules must not be if a collision is to be avoided must therefore be

$$\pi(2r)^2 l$$

where r is the radius of the molecule and l is the distance the molecule travels in unit time. This volume would normally contain $n \times \pi(2r)^2 l$ molecules if n is the number of molecules per unit volume, their **number density**.

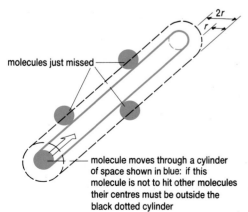

molecules just missed

molecule moves through a cylinder of space shown in blue: if this molecule is not to hit other molecules their centres must be outside the black dotted cylinder

Fig 25.8 If a molecule is to travel for unit time without hitting any other molecule it must be able to travel freely in the space shown by the cylindrical volume.

The **mean free path** λ of molecules is defined as the average distance a molecule travels between collisions. It is therefore given by

$$\lambda = \frac{\text{distance travelled}}{\text{number of collisions in this distance}} = \frac{l}{n\pi(2r)^2 l} = \frac{l}{4\pi r^2 l}$$

Note that the mean free path does not depend on the speed of the molecule, it therefore does not depend directly on the temperature, only on the number of molecules per unit volume.

In air at sea level the mean free path is about 7×10^{-8} m. At an altitude of 300 km, the air pressure has fallen to 10^{-10} of its value at sea level and the mean free path is then about 3 km.

QUESTION

25.8 The pressure and temperature at the top of Mount Everest are 3.26×10^4 Pa and 250 K respectively, compared with values of 1.0×10^5 Pa and 300 K at sea level. If the mean free path of a molecule of nitrogen at sea level is 1.0×10^7 m, what will be the mean free path at the top of Mount Everest?

25.5 DISTRIBUTION OF MOLECULAR SPEEDS

In any gas, the molecules are not all travelling with the same speed. As a result of collisions taking place, some molecules gain energy and other molecules lose energy. In between collisions, the ideal gas molecules have a constant speed. A graph, plotted to show the assortment of different speeds which exist, is shown in Fig 25.9 for nitrogen gas at a temperature of 1000 K. The graph shows the variation from molecule to molecule of the constant speeds between collisions. It does not deal with speeds during the collisions; these are assumed to take zero time in an ideal gas.

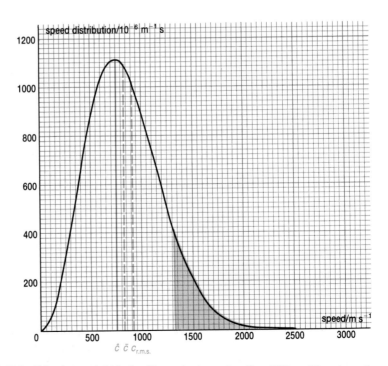

Fig 25.9 Molecular speed distribution. More molecules are travelling at 750 m s⁻¹ than at any other speed.

The vertical axis of the graph needs some explanation. Since the number of molecules in a gas is likely to be very high, the unit used on the vertical axis has been adjusted to make the total area beneath the graph equal to one. That is, the whole of the gas is represented by the area under the graph. A tenth of the total number of molecules making up the gas has been coloured blue on the graph to illustrate this; the area covered by the colouring is a tenth of the total area beneath the graph. The molecules in this 10% of the gas are the fastest 10% of all the molecules. The graph is not symmetrical and so the mean speed $\bar{c}$ of the molecules is slightly higher than the most probable speed, $\hat{c}$. The root mean square speed, $c_{r.m.s.}$, is different from the mean speed and the most probable speed. For the molecules in a gas the approximate relation between them is:

$$\hat{c} \approx 0.8c_{r.m.s.}$$

$$\bar{c} \approx 0.9c_{r.m.s.}$$

The following example should be worked through to see how the graph may be used.

Example 3

Use the graph, Fig 25.9, to answer the following questions:
(a) What is the most probable speed?
   The peak of the graph occurs at a speed of 740 m s^{-1}. More molecules have this speed than any other speed.
(b) What fraction of the total number of molecules travel with speeds between 700 m s^{-1} and 800 m s^{-1}?
   The height of the graph between 700 and 800 m s^{-1} has a nearly constant value of 0.001 15 m^{-1} s. The area beneath the graph between 700 and 800 m s^{-1} is

$$0.001\ 15\ \text{m}^{-1}\ \text{s} \times 100\ \text{m s}^{-1} = 0.115$$

   The fraction of the total number of molecules travelling with speeds between 700 m s^{-1} and 800 m s^{-1} is therefore 0.115.
(c) What fraction of the molecules of the gas travel at more than twice the most probable speed?
   Twice the most probable speed is $2 \times 740 = 1480$ m s^{-1}.
   The fraction required is the area under the graph for values of v above 1480 m s^{-1}. This is difficult to find accurately from the graph but there are very few molecules travelling at more than 2000 m s^{-1} so they can be neglected. The area found by counting squares is approximately 0.05. In other words, only 1 in 20 molecules travels at more than twice the average speed.

The figure that 1 in 20 molecules travels at more than twice the average speed is worth remembering. It can be used to give rough values for even higher speeds, for example

1 molecule in 20 travels at twice the average speed
1 molecule in 20^2 (1 in 400) travels at four times the average speed
1 molecule in 20^3 (1 in 8000) travels at 8 times the average speed etc.

Extending this scale it is possible to find that about

1 molecule in 20 000 travels at 10 times the average speed but that only
1 molecule in 400 000 000 travels at more than 100 times the average speed

The reason that these values are important is that in many situations it is the high speed molecules which matter. For instance, evaporation takes place as a result of molecules leaving the surface of a liquid. The faster molecules are more able to leave than the slower molecules. The rate at which a chemical reaction proceeds is often dependent on the temperature. Since high speed molecules are effectively at a higher temperature than slow molecules, the high speed molecules are important in activating the reaction. The molecular composition of the atmosphere of a star or planet depends on its escape velocity. The escape velocity of the Earth is

11 000 m s^{-1}; the average velocity of a hydrogen molecule in the Earth's atmosphere is about 1700 m s^{-1}. These values seem to indicate that hydrogen gas will not be able to escape from the Earth, but, using the figures from above, since 1 in 8000 hydrogen molecules travels at 8 times the average speed, there is 1 in 8000 hydrogen molecules travelling at more than 13 600 m s^{-1}. This is greater than the escape velocity and so some hydrogen molecules can escape from the Earth. Given a long enough time, virtually all of them will escape and so the atmosphere will contain very little hydrogen. The corresponding values for oxygen and nitrogen would show that very few of these gas molecules can escape and that is why our atmosphere contains them.

The graph showing variation of molecular speed within a gas is very temperature dependent. As the temperature rises, the number of low speed molecules decreases and the number of high speed molecules increases. The total number of molecules remains the same and so the area under the graph remains constant. The following data analysis exercise shows how different graphs are obtained at different temperatures, and gives some of the data used to plot the graphs. The area beneath all of the graphs is 1 (indicating all of the molecules in the gas).

Molecular speeds

Table 25.1 gives the height of the speed distribution graph (multiplied by a factor of 1 000 000 for convenience so that all the numbers used are greater than 1) for various temperatures of nitrogen gas. The graphs showing these data are plotted in Fig 25.10. Use either the graphs or the table to answer the questions which follow.

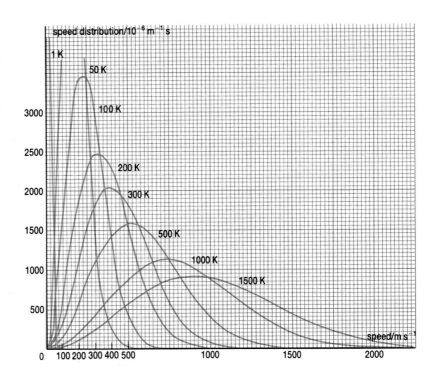

Fig 25.10 As the temperature of a gas is increased the average speed of its molecules increases. It has a higher internal energy.

GASES

Table 25.1

Speed /m s⁻¹	Height of distribution graph at different temperatures/10^{-6} m⁻¹s							
	1 K	50 K	100 K	200 K	300 K	500 K	1000 K	1500 K
0	0	0	0	0	0	0	0	0
20	33900	200	70	25	19			
40	15200	750	270	98	52			
60	900	1570	591	216	140			
80	10	2510	1000	375	205			
100	0	3450	1460	570	320	150	54	30
150		4920	2620	1240	660	325		
200		4620	3390	1720	1060	540	206	115
300		1690	3070	2460	1760	1070	420	240
400		230	1520	2310	2040	1400	660	400
500		14	460	1590	1850	1580	880	560
600		0	90	842	1370	1520	1040	700
700			11	350	840	1290	1110	820
800			1	117	440	980	1110	890
900			0	32	200	670	1030	920
1000				7	78	410	900	900
1100				1	26	232	740	840
1200				0	8	120	580	760
1400					0	24	310	550
1600						4	134	350
1800						0	49	192
2000							15	94
2200							4	41
2400							1	16
2600							0	6
2800								2
3000								0

(a) What is the most probable speed of the molecules at a temperature of 300 K?

(b) What is the speed of sound in nitrogen at a temperature of 300 K? (Calculate it using the equation given in Chapter 11, Table 11.1.) Why do you think these values are similar?

(c) Why are there so many molecules travelling at 20 m s⁻¹ in nitrogen at 1 K? (Note that nitrogen will not be gaseous at this temperature but the gaseous theory does give approximately the correct values for the average speed of the molecules even in a solid.)

(d) Why is there an insignificant number of molecules travelling at more than 100 m s⁻¹ when the temperature is 1 K?

(e) Of the total number of molecules, what fraction have speeds:
   (i) between 400 and 500 m s⁻¹ when the temperature is 300 K
   (ii) between 800 and 900 m s⁻¹ when the temperature is 300 K
   (iii) between 800 and 900 m s⁻¹ when the temperature is 900 K
   (iv) greater than 800 m s⁻¹ when the temperature is 300 K
   (v) greater than 800 m s⁻¹ when the temperature is 1000 K?

(f) Plot the graph for a temperature of 200 K and mark on it:
   (i) the most probable speed
   (ii) the mean speed
   (iii) the rms speed.
   Shade in the area of the graph showing:
   (i) the slower half of all the molecules
   (ii) the fastest 10% of all the molecules.

(g) Why is the fastest molecule at a temperature of 1500 K **not** travelling at about 3000 m s⁻¹?

- Boyle's law: at a constant temperature the pressure of a gas is inversely proportional to its volume.
- From the definition of thermodynamic temperature $pV \propto T$ for an ideal gas.
- Equation of state of a gas: $pV = nRT$ or $pV_m = RT$ for an ideal gas.
- Kinetic theory of gases:

  Basic assumptions: a gas consists of a large number of molecules in rapid, random motion, colliding with each other and with the walls of the container.

  Simplifying assumptions: intermolecular forces are negligible except during a collision; the volume of the molecules themselves is negligible compared with the volume of the container; gravitational forces are negligible; all collisions are elastic.

  The pressure exerted by a gas is given by $p = \frac{1}{3}\rho c^2$.

  k, the Boltzmann constant, is defined as R/N_A hence $pV = NkT$.
- Avogadro's law: equal volumes of gas under the same conditions of temperature and pressure have equal numbers of molecules.
- Dalton's law of partial pressures: in a mixture of gases the total pressure is the sum of the partial pressures of each constituent gas.

EXAMINATION QUESTIONS

25.9 **(a)** Explain what is meant by an ideal gas. What properties are assumed for the model of an ideal gas molecule in deriving the expression

$$p = \tfrac{1}{3}\rho\overline{c^2} \text{ [equation 1]}$$

where the symbols have their usual meanings? (5)

(b) How is pressure explained in terms of the kinetic theory of gases? Describe carefully, using diagrams where necessary, but without detailed mathematical analysis, the steps in the argument used to derive equation 1 (8)

(c) Show that for a fixed mass of ideal gas at constant temperature, equation 1 can be written $pV = A$, where A is a constant.

For some real gases, the pressure can be described in terms of the equation

$$p(V + B) = A \text{ [equation 2]}$$

where B is also a constant for a fixed mass of the gas at a particular temperature.
Show that equation 2 implies a pressure less than the value predicted for an ideal gas. Suggest a reason for this in molecular terms. (5)

(ULSEB 1983)

25.10 **(a)** **(i)** List **four** assumptions concerning the *motion* of molecules which are made in the kinetic theory of gases.

(ii) A molecule of mass m travelling at velocity c approaches the wall of its container along a line making an angle θ with the wall, as shown below.

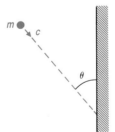

If it makes a perfectly elastic collision with the wall, calculate the magnitude and direction of its change of momentum in the collision. (7)

(b) According to the kinetic theory, the pressure p of an ideal gas comprising molecules each of mass m is given by $p = \tfrac{1}{3}mn\overline{c^2}$ where n is the number of molecules per unit volume.

What is the physical significance of the product mn? Explain the meaning of the term $\overline{c^2}$ and show how it would be calculated for four molecules travelling at speeds $v_1, v_2, v_3,$ and v_4. Also write down an expression for the mean speed of these four molecules. (3)

(c) Derive an expression for the speed of the molecules of an ideal gas at temperature T, and use it to calculate the speed of argon molecules at 27 °C. Hence, or otherwise, find the total internal energy of 2.00×10^{-2} kg of argon at 327 °C. (The internal energy may be assumed to be entirely translational kinetic energy.)
Mass of one molecule of argon $= 6.63 \times 10^{-26}$ kg; the Boltzmann constant $= 1.38 \times 10^{-23}$ J K^{-1}; the Avogadro constant $= 6.02 \times 10^{23}$ mol^{-1}. (7)

(JMB 1987)

25.11 A cubical container of volume 0.10 m^3 contains uranium hexafluoride gas at a pressure of 1.0×10^6 Pa and a temperature of 300 K.

(a) Assuming that the gas is ideal determine

(i) the number of moles of gas present, given that the universal gas constant $R = 8.3$ J K^{-1} mol^{-1};

(ii) the mass of gas present, given that its relative molecular mass is 352;

(iii) the density of the gas;

(iv) the r.m.s. speed of the molecules.

(b) A student suggests that since the molecules are so massive, the density of the gas at the bottom of the container would be significantly greater than the density at the top. Explain whether you agree or disagree. (8)

(AEB 1986)

25.12 Two closed vessels A and B, each of volume 3.0×10^{-3} m^3, are connected through a tap. Initially the tap is closed and both vessels are at 300 K. Vessel A contains helium (^{4}He) gas at a pressure of 1.0×10^5 Pa and B contains neon (^{20}Ne) gas at a pressure of 0.5×10^5 Pa. Both gases are monatomic.

(a) Find the amount, in moles, of each gas. (2)

(b) **(i)** Define the r.m.s. speed of atoms in a gas. (1)

(ii) Find the ratio of the r.m.s. speed of the helium atoms to that of the neon atoms. Explain your reasoning. (2)

GASES

(c) When the tap has been opened and equilibrium is eventually reached (again at 300 K), find the amount in vessel A of (i) helium and (ii) neon. Explain your reasoning. (3)

(O & C 1987)

25.13 Gas is contained in a fixed inverted cylinder having a frictionless and leak-tight piston as illustrated. The piston is attached to a load by a steel wire of natural length 2.0 m and of diameter 0.50 mm. The steel has Young modulus 2.0×10^{11} Pa and breaking stress 9.0×10^{8} Pa. Before a significant load is applied, the pressure, P, of the gas equals that of the surrounding atmosphere (1.0×10^{5} Pa) and the length, l, occupied by the gas in the cylinder is 0.1 m. The cylinder diameter is 50 mm. You may assume that the gas behaves as an ideal gas and that all the changes take place at constant temperature.

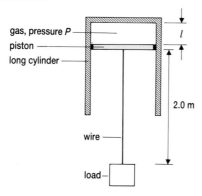

The suspended load is gradually increased from 0 to 12 kg.
(a) Calculate the extension of the wire. (2)
(b) Calculate the new value of the length l occupied by the gas in the cylinder. (3)
As the load is further increased, the suspension will eventually fail.
(c) Calculate the load at which the wire would snap. (2)
(d) Calculate the load at which the piston would be pulled out of the cylinder, whatever the length of the cylinder. (2)
(e) What therefore would you expect to happen as the load is increased from 0 towards 25 kg? (2)
(f) How would your answer to part (e) be affected if the wire were replaced by one of diameter 1.0 mm?

(O & C 1987)

25.14 A container holds a mixture of hydrogen and oxygen in thermal equilibrium at a temperature of 500 K. Find the mean translational kinetic energies of both types of molecule. Given that

the mass of a hydrogen molecule is 3.34×10^{-27} kg and the oxygen molecule is 16 times more massive, find the r.m.s. speed of both types of molecule. (7)

(UCLES 1988)

25.15 A gas molecule collides 100 times each second with the wall of its container. The graph shows how the force F exerted on the wall by the molecule might vary with time t over 20 ms.

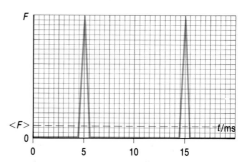

(a) If the momentum change during *one* collision is 6.0×10^{-23} Ns, calculate the average force acting on the wall during one collision.
(b) The broken line on the graph represents the force $<F>$ acting on the wall averaged out over the time spent in collision and the time spent between collisions. Explain how, using the graph, its value is obtained.
(c) Which of these two average values, from (a) or from (b), is effectively used in the kinetic theory calculation of the pressure exerted on the walls of a container by the molecules of the enclosed gas? (7)

(ULSEB 1988)

25.16 (a) Explain what is meant by
  (i) the mean free path; (1)
  (ii) the root mean square speed of molecules in a gas. (2)
(b) Derive a relationship between the pressure p of an ideal gas, its density ρ and the root mean square speed $c_{\text{r.m.s.}}$ of its molecules. (6)
(c) The diagram shows a Pirani gauge, a device for measuring very low gas pressures of the order of 1 Pa.
  In this device the temperature of the heated filament depends on how rapidly heat is conducted away to the wall by the low pressure gas G. This in turn depends on the pressure of G.
  (Consider G to be an ideal gas of molar mass 0.029 kg. Take the gas constant R to be 8.3 J K^{-1} mol^{-1} and the Avogadro constant N_{A} to be 6.0×10^{23} mol^{-1}.)

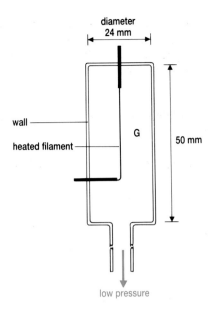

diameter
24 mm

wall

heated filament

G

50 mm

low pressure

(i) What mass of gas is present in the gauge at a pressure of 1.3 Pa and a mean temperature of 350 K?   (4)

(ii) How many molecules of gas are present?   (3)

(iii) What is the root mean square speed of the molecules?   (4)

(iv) At a pressure of 1.3 Pa the molecules of G have a mean free path equal to the distance from the filament to the wall of the gauge. Calculate the diameter of a molecule of G.   (6)

(v) Explain how and why the rate of heat conduction through G to the wall of the gauge (assumed at room temperature) would change if the pressure of G was steadily reduced from 1.3 Pa to zero.   (4)

(OXFORD 1987)

25.17 The average energy of an atom of a solid at a kelvin temperature T is $3kT$, where k is the Boltzmann constant. Calculate (a) the average energy of a copper atom in a mole of copper at a temperature of 300 K, and (b) the total energy of all the atoms in a mole of copper at the same temperature.

If this total energy could be transformed into linear kinetic energy of the whole mass of copper, at what speed would the mass be travelling?

(The Boltzmann constant = 1.38×10^{-23} J K^{-1}; the Avogadro constant = 6.02×10^{23} mol^{-1}; mass of 1 mole of copper atoms = 0.064 kg.)   (5)

(ULSEB 1982)

Chapter 26

THE FIRST LAW OF THERMODYNAMICS

LEARNING OBJECTIVES

At the end of this chapter you should be able to:

1. define and use the term internal energy;

2. state a general form of the first law of thermodynamics;

3. apply the first law of thermodynamics to a gas;

4. relate the specific heat capacity at constant volume for a gas to its specific heat capacity at constant pressure;

5. use the terms isothermal, isobaric and adiabatic correctly;

6. use an indicator diagram for a heat engine;

7. find the efficiency of a heat engine;

8. see the need for the second law of thermodynamics.

26.1 INTERNAL ENERGY

In any substance the molecules have kinetic energy because they are moving. They may also have potential energy because of the attraction between them. This has been referred to in section 25.4 where the term internal energy was introduced. The **internal energy** U of any object is defined as the sum of all the microscopic kinetic and potential energies of the molecules within the object. It is measured in joules. The word microscopic is introduced here to indicate that it is the kinetic and potential energies of the random movement of the molecules which is the internal energy, and not any large scale movement of the whole object. For instance, the internal energy of a cup of tea might be 40 000 J. An identical cup of tea at the same temperature but travelling on Concorde at 500 m s^{-1} still has an internal energy of 40 000 J. It might, in addition, have a further 10 000 J of energy due to its movement with the aeroplane. One other problem associated with internal energy is the need to choose an arbitrary zero of potential energy. This problem is always present when potential energy is being used. In most circumstances, however, it is changes in potential energy which are required, in which case the position taken to have zero potential energy does not matter. All that is needed is consistency; use a particular value for zero and then use that zero always.

In the particular case of an ideal gas, this difficulty does not arise. An ideal gas obeys the gas laws precisely and there is no inter-molecular attraction. The molecules in an ideal gas cannot therefore change their potential energy so the reference is chosen to give them zero potential energy. This means that the internal energy of an ideal gas is just the kinetic energy of the molecules and it can only be changed if the kinetic energy changes. Since a change in the kinetic energy of the molecules implies a change in temperature, it follows that the internal energy of an ideal gas is constant at a constant temperature. Its pressure and volume may be changed but the internal energy remains fixed. In many cases only a small error is made by assuming that a real gas is an ideal gas.

26.2 THE FIRST LAW OF THERMODYNAMICS

The laws of thermodynamics are concerned with energy transfer, and they may be applied to all systems. They can be applied to the energy changes taking place within a wire being stretched or to a battery supplying an electric current. They can be applied to the tension in the surface of a liquid or to a chemical reaction. In particular, they enable theoretical calculations to be made concerning the efficiency of heat engines. **Heat engines** are devices which convert thermal energy into work. Steam engines were the earliest types of heat engine, see Fig 26.1.

Fig 26.1   Boulton and Watt's rotative beam engine. The efficiency of such early steam engines was less than 1%.

In this chapter the laws of thermodynamics will be applied only to the behaviour of gases. The first law of thermodynamics is, in part, a statement of the law of conservation of energy (section 7.7). It goes beyond the law of conservation of energy however, by stating also that the internal energy of a system depends only on the state of the system. If we regard all the oxygen in an oxygen cylinder to be the system, then the first law states for the oxygen in the cylinder that its internal energy depends on factors such as the amount of oxygen in the cylinder, the pressure of the oxygen in the cylinder and its temperature. It does not depend on how the oxygen came to be in the cylinder or on the previous history of those particular oxygen molecules. Another cylinderful of oxygen under the same conditions, that is, in the same state, will have the same internal energy. This may seem to be obvious but note that one could not make the same statement about the work done on the oxygen molecules to put them into the cylinder. That can be done in different ways involving different amounts of work. It is a sensible question to ask how much internal energy there is in the oxygen in the cylinder; it is not sensible to ask how much work there is in it.

A general statement of the first law of thermodynamics is:

 The internal energy of a system depends only on its state; the increase in the internal energy of a system is the sum of the work done on the system and the heat supplied to the system.

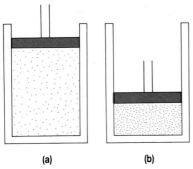

Fig 26.2 Work being done by a piston on a gas.

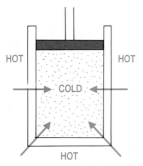

Fig 26.3 Heating a gas by having the surroundings of a cylinder at a higher temperature than the gas within the cylinder.

As usually happens when a basic law is stated in formal language, its meaning and uses are not at first clear, so some examples need to be given. As indicated above, the examples used in this chapter will be concerned with gas behaviour. This will enable you to see how to apply the law to the gases in heat engines such as a car engine, or a refrigerator or a steam turbine in a power station.

In all of these cases it is a gas on which, or by which, work is being done. The gas is known as the working substance. The two ways referred to for increasing the internal energy of a gas are illustrated in Figs 26.2 and 26.3.

Working on the gas is shown in Fig 26.2. By moving the piston downwards from the position shown in Fig 26.2(a), molecules are struck by the moving piston and hence gain internal energy in the same way that a tennis ball gains kinetic energy when struck by a moving racquet.

Heating the gas is shown in Fig 26.3. Here the space surrounding the cylinder is at a higher temperature than the gas within the cylinder, and so the molecules gain internal energy as a result of heat flow through the walls. Heating also takes place if a fuel is burnt within the cylinder.

Both heating and working are processes which can change the internal energy of the gas, so the first law can be written as an equation in the convenient shorthand form:

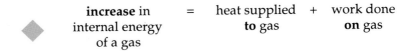

You are advised to write the equation using words and in three columns rather than using algebraic symbols.

Note: It is important when applying the first law to get the signs correct. In this equation you should note the emphasis on the words **increase, to** and **on**. Different books use different algebraic notations when stating the law. Some use the work done by the gas, some use the work done on the gas. This introduces a minus sign or the term appearing on the other side of the equation. This problem can be overcome by a clear understanding of the law. The law is stating that the internal energy of a gas may be increased either by heating it or by working on it.

Example 1

A fixed mass of an ideal gas has its state changed from state A to state B as given in Table 26.1.

Table 26.1

	Pressure /Pa	Volume /m³	Temperature /K	Internal Energy /J
State A	200 000	0.0010	100	300
State B	300 000	0.0030	450	1350

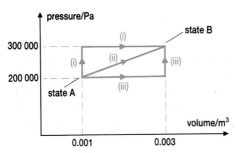

Fig 26.4 An indicator diagram showing change in pressure with volume for three different ways of changing the state of a gas.

Find how much heat has been supplied to the gas, and how much work has been done on it, if the change of state is carried out in the three different ways as shown on the p – V graph, Fig 26.4.

Note from the data that the internal energy can be given for each state of the gas, in accordance with the part of the statement of the first law that the internal energy of the gas depends only on its state.

THE FIRST LAW OF THERMODYNAMICS

The three ways being considered for changing the state of the gas from state A to state B are:

- a change of pressure at constant volume followed by a change of volume at constant pressure. In both of these sections there will be an increase in temperature.
- a straight line change of pressure and volume. This would be difficult to carry out experimentally.
- a change in volume at constant pressure followed by a change of pressure at constant volume. Again both of these sections will involve a rise in temperature.

The area beneath a pressure–volume graph is the work done. See section 7.2 on page 85. Here we must be careful about whether work is being done **on** the gas to compress it or **by** the gas in an expansion. Vertical lines on the graph imply no change in volume and therefore no work being done. In all three cases, expansion of the gas is taking place, work is being done by the gas. The value we expect to get for the work done **on** the gas is therefore negative.

Many problems on the first law of thermodynamics can conveniently be done by completing a table, with the first law statement making the table headings.

	increase in internal energy of a gas	=	heat supplied **to** gas	+	work done **on** gas
1	1350 − 300 = 1050 J				− 300 000 Pa × 0.0020 m³ = − 600 J
2	1050 J				− 250 000 Pa × 0.0020 m³ = − 500 J
3	1050 J				− 200 000 Pa × 0.0020 m³ = − 400 J

So far the table has included in it data given in the question and calculations made of the area beneath the graphs. We can now complete the table by using the first law statement to find how much heat needs to be supplied in each case.

	increase in internal energy of a gas	=	heat supplied **to** gas	+	work done **on** gas
1	1050 J		1650 J		− 600 J
2	1050 J		1550 J		− 500 J
3	1050 J		1450 J		− 400 J

A remarkable result which can be seen from these calculations is that the heat supplied to a gas to change its temperature from 100 K to 450 K is different for the three cases. The importance of the first law can be seen from this. The heat supplied to a system and the work done on a system can have different values and are dependent on how the changes are carried out. The increase in the internal energy does not depend on how the changes are carried out, only on the initial and final states of the gas.

It has been assumed up to this stage that the term *heat energy* is familiar to the reader but it is a term which needs some explanation. In everyday speech the word *heat* is used to mean a variety of different things. It is used as a noun in *heat rises*, as a verb in *heat the kettle* and as an adjective in *a heat pump*. It is very often used incorrectly in such statements as *the heat of the oven is* 220 °C. You can probably remember at some time in your physics lessons being warned to distinguish carefully between *heat* and *temperature*, heat being measured in joules and temperature being measured in °C. A similar warning is now being given to be careful to distinguish between internal energy and heat. Both of these quantities are measured in joules and probably in the past you would have used the word *heat* to cover both of them. In future be careful to use the words in this way:

internal energy is the sum of all the microscopic kinetic and potential energies of the molecules in an object
heat is energy transferred as a result of a temperature gradient.

It is therefore correct to say that heat escapes by conduction through a window. It is warmer inside than out so there is a temperature gradient across the window and this causes heat to flow from the room to outside where it will increase the internal energy of the atmosphere. Heating is a process in which energy is transferred from a region of high temperature to a region of low temperature. As shown in example 1, it is possible to answer the question 'How much internal energy is possessed by an object?' It is not possible or sensible to try to answer the question 'How much heat is possessed by an object?'

QUESTIONS

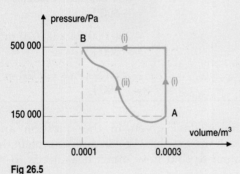

Fig 26.5

26.1 A gas has its state changed from A to B by two different paths as shown on the graph, Fig 26.5. If path (i) requires the input of 60 J of heat find:
(a) the work done on the gas in changing its state along path (i);
(b) the increase in internal energy between A and B.
What can be deduced about the work done on the gas, the heat supplied to the gas and the increase in the internal energy of the gas if path (ii) is used?

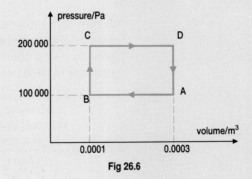

Fig 26.6

26.2 An ideal gas undergoes a cycle of changes A→B→C→D as shown in the graph, Fig 26.6. Complete the following table for the cycle.

	increase in internal energy of gas /J	heat supplied to gas /J	work done on gas /J
A→B	−50		
B→C	25		
C→D			140
D→A			

THE FIRST LAW OF THERMODYNAMICS

26.3 HEAT CAPACITY

The **heat capacity**, C, of an object is the quantity of heat which must be supplied to it to give it unit rise in temperature. The SI unit for heat capacity is the joule per kelvin, $J\ K^{-1}$. In equation form this definition becomes

$$C = \frac{\Delta Q}{\Delta T} \quad \text{or} \quad \Delta Q = C\,\Delta T$$

where ΔQ is the quantity of heat energy supplied and ΔT is the temperature rise.

The term specific heat capacity is also used. In physics the word *specific* means *per unit mass*. The **specific heat capacity**, c, therefore is the heat capacity per unit mass. It has the SI unit $J\ kg^{-1}\ K^{-1}$ and in equation form is

$$c = \frac{C}{m}$$

where m is the mass of the object. Combining the last two equations gives

$$\Delta Q = mc\Delta T,$$

an equation you will need frequently.

The specific heat capacity is given for a material, whereas the heat capacity is given for a particular object. This is illustrated in the following example.

Example 2

A hot water tank for a house contains 120 kg of water at 15 °C. The tank itself has a heat capacity of 6000 J K^{-1}. Find how long it will take an immersion heater to raise the temperature of the water to 50 °C if the tank is well insulated and the power of the heater is 2500 W.

(Specific heat capacity of water = 4200 J $kg^{-1}\ K^{-1}$.)

Heat supplied to the water
$= m\ c\ \Delta T$
$= 120\ kg \times 4200\ J\ kg^{-1}\ K^{-1} \times (50 - 15)\ K$
$= 17\ 640\ 000\ J$

Note that a difference in temperature of 35 °C is by definition the same as a difference in temperature of 35 K.

Heat supplied to tank $= C\,\Delta T = 6000\ J\ K^{-1} \times 35\ K = 210\ 000\ J$

assuming that the tank reaches the same final temperature as the water. This is a reasonable assumption since the tank is well insulated and is a good conductor of heat. Note that the quantity of heat required to heat the tank is very small compared with that required to heat the water.

Total quantity of heat required = 17 640 000 J + 210 000 J = 17 850 000 J
Power supplied = 2500 W = 2500 J s^{-1}

$$\text{Time required} = \frac{17\ 850\ 000\ J}{2500\ Js^{-1}} = 7140\ s$$

The assumptions made in answering this question make it unreliable to quote 3 or more significant figures so, having quoted the values which the data give, it is sensible to say that the time required is approaching 2 hours.

A problem arises with the heat capacity of gases. If you look again at the answers in example 1 you will see that the heat required to change the temperature of the gas in that example was

1650 J for the first way of changing the temperature from 100 K to 450 K

1550 J for the second way of changing the temperature from 100 K to 450 K and

1450 J for the third way of obtaining the same temperature change.

This means that the heat capacity of the gas involved can be 4.71 J K^{-1}, 4.43 J K^{-1} or 4.14 J K^{-1} depending on how the change in temperature takes place. This lack of constancy of heat capacity means that whenever the heat capacity of a gas is given it is essential also to say under what conditions the value was obtained. For example

the specific heat capacity of air at constant pressure, c_p is 1010 J kg^{-1} K^{-1} but

the specific heat capacity of air at constant volume, c_V is 722 J kg^{-1} K^{-1}

Specific heat capacities can have any value – positive, zero or negative – so the term is meaningless unless the conditions are stated. Values of heat capacity for gases are usually given per mole rather than per unit mass, so it is essential to be very careful and know exactly what value is being used, and to make it agree with the quantity of gas actually present. The molar heat capacity of a gas at constant pressure is given the symbol $C_{p,\,m}$ and the molar heat capacity of a gas at constant volume is $C_{V,\,m}$. Some values for these constants are given in Table 26.2.

Table 26.2

	$C_{p,\,m}$ /J mol^{-1} K^{-1}	$C_{V,\,m}$ /J mol^{-1} K^{-1}
helium	20.8	12.5
air	29.1	20.8
oxygen	29.4	21.1
carbon monoxide	29.2	20.9
sulphur dioxide	40.4	31.4

If you look at the figures given in Table 26.2, you may notice that there is a nearly constant difference between the two lines of figures. The difference has the value of 8.3 J mol^{-1} K^{-1} and it is not coincidence that this is the same numerical value and the same unit as the molar gas constant. $C_{p,\,m} - C_{V,\,m} = R$ for an ideal gas. The table shows that sulphur dioxide does not behave as an ideal gas. You may just need to remember this equation. The following example explains why it is true.

Example 3

Show that for an ideal gas $C_{p,\,m} - C_{V,\,m} = R$

Consider one mole of gas in state X at pressure p, volume V and temperature T, as shown on the graph, Fig 26.7.

If the gas is changed to being in state Z, where it has the same pressure but a volume $V + \Delta V$ and a temperature $T + \Delta T$, it can reach the new state by path (i) or path (ii) as shown. Path (ii) involves first changing to state Y at constant volume where the temperature is $T + \Delta T$, and then altering the volume at constant temperature until the pressure is back down to p. Now apply the first law of thermodynamics to the changes.

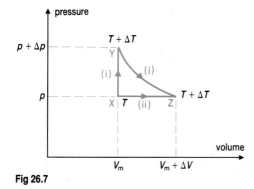

Fig 26.7

THE FIRST LAW OF THERMODYNAMICS

	Increase in internal energy of a gas	Heat supplied to gas	Work done on gas
X→Y		$C_{V,m}\,\Delta T$	0
X→Z		$C_{p,m}\,\Delta T$	$-p\Delta V$

The two blank spaces **must be the same** since the temperature at Y is the same as the temperature at Z, and for an ideal gas the internal energy depends on the mass of gas and the temperature but not on the pressure and volume.

The sum of the terms given in the top line must therefore be equal to the sum of the terms in the bottom line

$$C_{V,m}\,\Delta T \; + \; 0 \; = \; C_{p,m}\,\Delta T - p\,\Delta V$$

Rearranging gives

$$C_{p,m}\,\Delta T \; - \; C_{V,m}\,\Delta T = p\,\Delta V$$

and dividing by ΔT we get

$$C_{p,m} \; - \; C_{V,m} \; = \; \frac{p\,\Delta V}{\Delta T}$$

Since from X to Z the pressure is constant, we can use the fact that $\dfrac{V}{T}$ is constant and therefore:

$$\frac{V_x}{T_x} = \frac{V_z}{T_z}$$

$$\frac{V_m}{T} = \frac{V_m + \Delta V}{T + \Delta T}$$

$$V_m T + V_m \Delta T = V_m T + T\Delta V$$

$$V_m \Delta T = T\Delta V$$

$$\frac{\Delta V}{\Delta T} = \frac{V_m}{T}$$

so $$C_{p,m} - C_{V,m} = \frac{pV_m}{T} = R$$

26.4 INDICATOR DIAGRAMS

When the pressure, volume and temperature of a gas change, it is useful to be able to show those changes on a graph. A single two-dimensional graph cannot display three variables however, so it is usual to plot a whole series of pressure against volume graphs, each at a different temperature, and then to superimpose on these graphs a line to show how the state of a gas varies when certain changes are made. The lines on these graphs are called **indicator diagrams** and they have been used already in Figs 26.5, 26.6 and 26.7. Some features of indicator diagrams need to be understood in order to make use of them when considering different types of heat engine.

1. An indicator diagram needs to be drawn for a fixed mass of gas. This implies that for an ideal gas, pV/T will be constant for all points drawn on the diagram.
2. The area beneath the graph is a measure of work done. If the gas is expanding then work is being done by the gas; if the gas is being compressed then work is being done on the gas, usually by a piston compressing it in a cylinder.
3. A vertical line on an indicator diagram indicates a change in pressure at constant volume. For an ideal gas the pressure will be proportional to the temperature. No work is being done since there is no movement

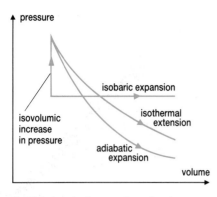

Fig 26.8 Isovolumic – no change in volume
isobaric – no change in pressure
isothermal – no change in temperature
adiabatic – no heat supplied.

pressure

isobaric expansion

isothermal
extension

isovolumic
increase
in pressure

adiabatic
expansion

volume

if the volume is constant. This change is called an **isovolumic** change in the state of the gas.

4. A horizontal line indicates a change in volume at constant pressure. For an ideal gas the volume will be proportional to the temperature. The area beneath the graph is the work done. This change is called an **isobaric** change in the state of the gas.

5. An **isothermal** change in the state of the gas is one in which there is no change in the temperature of a gas. For an ideal gas the volume will be inversely proportional to the pressure and there will be no change in the internal energy of the gas.

6. A change which takes place in the state of a gas without any heat being supplied to it, or taken from it, is called an **adiabatic** change. For an ideal gas an adiabatic expansion always results in a fall in the temperature as the work which the gas does can only be done by using some of its internal energy. This results in a curved line of steeper gradient than the isothermal on the indicator diagram. These changes are illustrated in Fig 26.8 and are summarised in a table at the end of the chapter.

The adiabatic change is always one in which no heat is supplied to the system. Mathematically the changes which take place in the pressure, volume and temperature are rather complex and are not dealt with in this book. You may, however, find it necessary to use the expression relating p, the pressure of an ideal gas with V, its volume while undergoing an adiabatic change. The relationship is

$$pV^\gamma = k$$

where k is a constant and γ is the ratio $C_{p,\,m}/C_{V,\,m}$. For air, γ has the value 1.4.

QUESTION

26.3 The graphs in Fig 26.9 show isotherms for 1 mole of an ideal gas. At all points on the graphs $pV/T = R$ where R is the molar gas constant having a value of 8.31 J mol^{-1} K^{-1}. Heat capacity at constant volume $C_{V,m} = 20.8$ J K^{-1} mol^{-1}. Heat capacity at constant pressure $C_{p,m} = 29.1$ J K^{-1} mol^{-1}.

(a) Show, by considering at least three points on the graphs, that pV/T does have a constant value of 8.3 (to 2 sig. figs) and show that its unit is J K^{-1} mol^{-1}.

(b) What gas law is being obeyed if the state of the gas changes along
   (i)   the isotherm AB
   (ii) a straight line AC
   (iii) a straight line AD?

(c) How much heat needs to be supplied to the gas to change its state from
   (i)   A to C in a straight line
   (ii) A to E in a straight line
   (iii) A to B along the isotherm?

(d) How much work needs to be done on the gas to change its state from
   (i)   A to E in a straight line
   (ii) A to C in a straight line (be careful with the sign)?

(e) Estimate, from the area under the graph, the work done on the gas for the state of the gas to change along the isotherm from A to B.

THE FIRST LAW OF THERMODYNAMICS

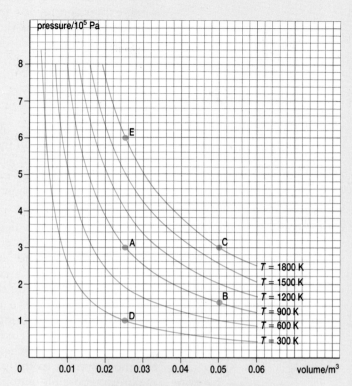

pressure/10^5 Pa

$T = 1800$ K
$T = 1500$ K
$T = 1200$ K
$T = 900$ K
$T = 600$ K
$T = 300$ K

volume/m^3

Fig 26.9 Isotherms for one mole of an ideal gas.

(f) Use the first law of thermodynamics to find the increase in internal energy if the state of the gas changes from
  (i) A to C
  (ii) A to E.
  Why are these two values the same?(Note that whereas the path taken for finding the heat supplied and the work done always needs to be stated, this is not necessary for the change in internal energy. This is because internal energy of a gas is a function of its state.)

(g) Apply the first law of thermodynamics to the isothermal change AB. Check your answer to (c) (iii). It should not be zero. How is it possible for a gas to be supplied with heat and yet for its temperature not to rise?

(h) What is the heat capacity of a gas at constant temperature???

(i) (Difficult) Assume a gas starts at point A and is then allowed to expand to a volume of 0.50 m^3 without any heat being supplied to it. Its temperature will fall. Find a possible final temperature so that the work done by the gas during the expansion equals its loss in internal energy. State what assumptions you make about the path taken for the change in state.

In section 26.3 it was stated that the equation for ΔQ, the heat supplied to a substance in terms of its specific heat capacity, would be needed frequently. The equation

$$\Delta Q = m\, c\, \Delta T$$

becomes $$\Delta Q = n\, C_{V,\mathrm{m}}\, \Delta T$$

when dealing with n mole of a gas whose molar heat capacity at constant volume is $C_{V,m}$ and which is undergoing a rise in temperature ΔT at constant volume. Using the first law on a change at constant volume gives this same expression for the increase in the internal energy of the gas since at constant volume, no work is done on or by the gas. That is

Increase in internal energy of gas	=	Heat supplied **to** gas	+	Work done **on** gas
$n\,C_{V,m}\,\Delta T$	=	$n\,C_{V,m}\,\Delta T$	+	0

The internal energy of an ideal gas however depends only on its temperature and not on the pressure or volume of the gas, so the change in the internal energy of a gas is **always** $n\,C_{V,m}\,\Delta T$ whatever the conditions for the change. If a change in state takes place at constant pressure, then the change in the internal energy of the gas is still $n\,C_{V,m}\,\Delta T$.

26.5 HEAT ENGINES

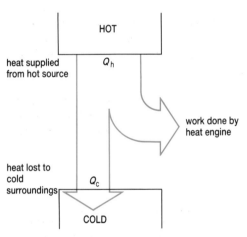

Fig 26.10 A generalised diagram showing the function of a heat engine.

As explained earlier, heat engines are devices which convert thermal energy into work. Heat engines absorb heat from a source, perform some mechanical work and lose some heat to an exhaust. The internal combustion engine in a car does just this. As a result of burning a small quantity of petrol inside the cylinder, a quantity of heat is supplied to the gas in the cylinder. The pressure in the cylinder rises and the piston is pushed downwards doing work in the process. The hot gases remaining then have to be ejected from the car through the exhaust pipe. A mechanical engineer, when designing an engine, needs to apply the laws of thermodynamics to the processes taking place in order to try to get the maximum possible amount of work done from a minimum quantity of petrol used. The **thermal efficiency** of a heat engine is defined by the equation:

$$\text{Thermal efficiency, } \varepsilon = \frac{\text{Work done by the engine}}{\text{Heat supplied to the engine}}$$

This is illustrated diagrammatically by Fig 26.10 where the heat supplied to the engine from the hot source, Q_h, is converted into work W, and an amount of heat Q_c is lost to the cold surroundings. Applying the law of conservation of energy to this process gives $W = Q_h - Q_c$, so the efficiency can also be written

$$\varepsilon = \frac{(Q_h - Q_c)}{Q_h}$$

As can be seen from the diagram, or from the equation, the efficiency of the heat engine can be increased if more of the energy supplied from the hot source can be changed into work so that less is wasted to the cold surroundings. There are both practical and theoretical reasons why Q_c cannot be zero. In practice, efficiencies are typically 40%.

QUESTION

26.4 The thermal energy supplied by a kilogram of fuel is 48 000 kJ. Find the efficiency of a car which is producing mechanical energy at a rate of 40 kW if it is using petrol at a rate of 8 kg per hour.

Any heat engine needs to be able to do work continuously. After expansion of the burning gases takes place in a petrol engine the waste gases have to be ejected through the exhaust and more petrol vapour and air have to be supplied to get the engine ready for the next power stroke. This cyclical pattern of changes can conveniently be shown on an indicator diagram. The Otto cycle is an idealised sequence of changes which take

THE FIRST LAW OF THERMODYNAMICS

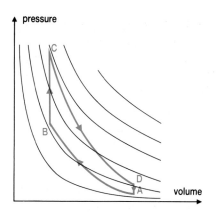

pressure

C

B

D

A

volume

Fig 26.11 Indicator diagram for an idealised petrol engine.

place on a fixed amount of gas in a petrol engine, Fig 26.11. It consists of the following separate parts.

1. During an intake stroke, air and petrol vapour are transferred from outside the cylinder to inside the cylinder, ideally without change in volume, temperature or pressure. This all occurs at point A on the indicator diagram.

2. From A to B the piston squashes the gas to a small volume. No heat is supplied to the gas in this stage. This is called the compression stroke and it is an adiabatic compression.

3. From B to C as a result of a spark from the sparking plug igniting the petrol, heat is supplied to the gas in a brief moment of time and this causes a large rise in the pressure at constant volume.

4. From C to D is the power stroke of the engine. Work is done by the expanding gases on the piston and their temperature and pressure fall. No heat is being supplied during this stage so the expansion is an adiabatic expansion.

5. The stage from D back to A, ready to start again, is complex. It can ideally be considered as a cooling down to the starting temperature, during which time heat is lost and then an exhaust stroke at constant pressure, volume and temperature just to move the waste gases from the cylinder to the outside.

Note that work is done by the gases in the cylinder on the piston during the power stroke, and that work has to be done on the gases by the cylinder during the compression stroke. The area within the loop is therefore the net work output of the engine in a cycle.

Example 4
A car engine has a cylinder whose maximum volume is 0.000 50 m³ and minimum volume 0.000 050 m³. It goes through 1 Otto cycle. Atmospheric pressure is 1.0×10^5 Pa and atmospheric temperature is 300 K. The engine has 0.0125 g of petrol injected during each stroke which gives 600 J of heat on burning. Calculate the maximum possible efficiency of the engine. The molar heat capacity $C_{V,m}$ for the gas in the engine is 21 J K⁻¹ mol⁻¹.

The procedure used in solving such problems is to assume that the engine behaves ideally. The efficiency will then be the maximum possible efficiency.
The first stage in solving the problem is to find the amount of air being used in each cycle.

At the start $pV = nRT$ so

$$1.0 \times 10^5 \text{ Pa} \times 0.000\ 50 \text{ m}^3 = n \times 8.3 \text{ J K}^{-1} \text{ mol}^{-1} \times 300 \text{ K}$$
$$n = 0.020 \text{ mol}$$

The second stage is to find the temperature rise which takes place when the petrol is burnt. This is given by

$$\Delta T = \frac{Q}{n\ C_{V,m}} = \frac{600 \text{ J}}{0.20 \text{ mol} \times 21 \text{ J K}^{-1} \text{ mol}^{-1}} = 1430 \text{ K}$$

The third stage involves applying the gas equation to each of the four sections of the cycle. Some of this information can be put in directly and is shown in normal type. Coloured type is used to show subsequent working with bold type showing the more difficult steps in the adiabatic expansions. Letters refer to Fig 26.11.

	p / kPa	V / m^3	T / K
start A	100	0.000 50	300
after compression B	**2510**	0.000 050	753
after ignition C	7270	0.000 050	2180
after power stroke D	**290**	0.000 50	870
after exhaust A	100	0.000 50	300

At all stages in the cycle the value of pV/T has the constant numerical value of 0.020×8.3 which is 0.166. Note that the temperature rise of 1430 K has taken place between B and C.

The fourth stage concerns the first law of thermodynamics particularly. All the information in the above table is information which, here, is obtained by using the gas equation. It could equally well have been obtained by experiment. It is interesting to note that the temperature reached by the gases in an engine is above the melting point of the metal used in the construction of the engine. This is one reason why the engine always needs to be well cooled. In applying the first law to the changes in state taking place, it is worth while again to use a table.

For all the changes the change in internal energy is $n\, C_{V,m}\, \Delta T$
$$= 0.020 \text{ mol} \times 21 \text{ J K}^{-1} \text{ mol}^{-1} \times \Delta T$$
$$= 0.42 \text{ J K}^{-1}\, \Delta T$$

	Increase in internal energy of a gas/J	=	Heat supplied to gas /J	+	Work done on gas /J
A to B	190		0		190
B to C	600		600		0
C to D	−550		0		−550
D to A	−240		−240		0

Note that there is no overall change in the internal energy of the gas.

The final stage is to examine this data to see that the efficiency of the engine is

$$\varepsilon = \frac{\text{Net work done by gas in engine}}{\text{Heat supplied to gas in engine}} = \frac{550 \text{ J} - 190 \text{ J}}{600 \text{ J}}$$
$$= 0.60 = 60\%$$

This value of 60% gives the maximum possible efficiency which an engine such as this could have even under such idealised conditions as have been assumed. In practice, there are many factors which will reduce the actual output to well below this figure. These factors include friction, the gas not behaving ideally, ignition not taking place instantly, heat being lost during the adiabatic compression, and losses of energy on transferring the gas into and out of the cylinder. The indicator diagram for a real engine is more likely to be as shown in Fig 26.12. Deviations from idealised behaviour do always reduce the efficiency but the important point for the operation of a heat engine is that even if it were working ideally it would still not be 100% efficient. A final point to note about this problem on a car engine is that the wasted 40% of energy is partly made use of in the car's heater. Some of the wasted heat is lost too through the exhaust pipe but that which is wasted in heating up the engine itself is used to heat the cooling water up to a temperature of around 80°C. It is this hot water which is used in the heater.

The following questions refer to different types of heat engine. Much of the tedious calculation of pressures and volumes has already been done for you so that the questions themselves can concentrate on the thermodynamic properties of the engine.

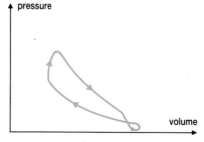

Fig 26.12 Indicator diagram for a practical petrol engine.

THE FIRST LAW OF THERMODYNAMICS

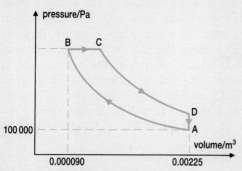

Fig 26.13 An indicator diagram for a large diesel engine.

26.5 In a large diesel engine air is adiabatically compressed (A to B, Fig 26.13) by the piston in a cylinder until the air reaches a very high pressure and temperature. A small amount of diesel oil is then injected into the cylinder and it burns in the hot air (B to C). Expansion continues adiabatically until the piston is pushed down its full amount (C to D) and then cooling and exhaust take place (D to A) before the cycle recommences. The compression ratio of this engine is 25 to 1. That is, the volume occupied by the gas in the engine has a value at maximum volume which is 25 times its minimum volume. Compression ratios have a considerable influence on efficiencies and are typically much higher in diesel engines than in petrol engines.

Use the data given on the graph and in the table to answer the questions which follow.

	P / kPa	V / cm^3	T / K
start A	100	2250	300
after compression B	9060	90	1090
after injection C	9060	173	2090
after power stroke D	250	2250	750
after exhaust A	100	2250	300

$C_{V,m} = 20.8$ J K^{-1} mol^{-1}: $C_{p,m} = 29.1$ J K^{-1} mol^{-1}

(a) Use the gas law to find how many moles of gas are used in each cycle.

(b) Complete the following table using the first law of thermodynamics. It is suggested that you start by putting in terms known to be zero and then work vertically down the increase in internal energy column.

	Increase in internal energy of a gas/J	Heat supplied to gas /J	Work done on gas /J
A to B			
B to C			
C to D			
D to A			

(c) Use the table to show that the efficiency of this diesel engine is 68%.

(d) What is the net power output of the engine if it completes 6000 power strokes each minute?

(e) Suggest a possible use for such an engine.

(f) If the fuel supplies 45 000 kJ of heat energy per kilogram, how fast is fuel being used when running at 6000 power strokes per minute?

The Behaviour of gases in a Turbofan jet engine (Rolls Royce Type RB211)

The Rolls Royce Turbofan jet engine is shown in Fig 26.14. Four of these engines are fitted to a Boeing 747 airliner. The thermodynamics of the engine is illustrated by Fig 26.15. Air enters the engine at a rate of 420 m^3 s^{-1} when the aircraft is cruising at 250 m s^{-1} at an altitude of 9000 m.

Fig 26.14 A Rolls Royce Turbofan engine, RB 211.

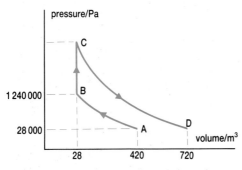

Fig 26.15 An indicator diagram for a turbofan engine.

Fig 26.16 An early steam engine such as this 1897 Fowler traction engine could be expected to have an efficiency of as little as 1%.

The air at this altitude has a pressure of 28 000 Pa and a temperature of 225 K. The air first passes through a compressor, which is a fan rotating with a high angular speed. This reduces the volume of gas to one-fifteenth of its original volume. The compressed air then has fuel continuously sprayed into it and the fuel burns creating a very high pressure, high temperature gas which expands rapidly out of the rear of the engine. In doing so it does work on a turbine, which is used to operate the compressor, and on the aircraft itself to provide thrust. The data used in this exercise are for one particular model of the RB211 engine which provides a thrust of 220 000 N. The indicator diagram is started in Fig 26.15 and shows A→B compression, B→C combustion at constant volume, C→D power and exhaust.

Throughout this exercise consider the operation of the engine for 1 second. Many approximations are made in the calculations so the answers are only a guide to the actual values. In reality the engine is less efficient than expected. It is worth while making intermediate calculations to 3 significant figures so that 2 significant figures are reliable for the idealised engine. If you get stuck with a part of this problem, just go on to the next part. The problem has been arranged so that later parts can be done even if earlier parts cannot. Information required can be found in the Answers section at the end of the book.

(a) Find the number of moles of air which enter the engine in one second using $pV = nRT$.
(b) A→B is an adiabatic compression. No heat is supplied to the air during the compression. For the compression, $pV^{1.4}$ is constant. Use the equation
$$p_A V_A^{1.4} = p_B V_B^{1.4}$$
to show that the pressure after compression is 1.24×10^6 Pa.
(c) What is the temperature at B?
(d) Fuel is burnt to raise the temperature of the air at C to 1400 K. What is the pressure at C? Assume that there is a constant amount of air throughout the cycle, that is, ignore the chemical processes which can alter the number of moles of gas present.
(e) How much heat is supplied by the fuel per second? The molar heat capacity of the air at constant volume, $C_{V,m}$ is 20.8 J mol^{-1} K^{-1}.
(f) How much aviation fuel is burnt per second? Aviation fuel gives 53 000 kJ kg^{-1}.
(g) How much fuel is used while cruising for 6 hours during a flight? (4 engines)
(h) The exhaust gases then expand adiabatically to atmospheric pressure. What is their temperature at D if they have then expanded to 720 m^3?
(i) Why do large jets cause vapour trails?
(j) It appears at first sight as though this indicator diagram is not a complete loop. However, while a continuous supply of new air is available at A, the gas supplied to the atmosphere at D occupies a larger volume at D than it originally did at A. Work has to be done against the atmosphere to create this space. How much work has to be done? This is equivalent to completing the indicator diagram with a horizontal line from D to A.
(k) **Table 26.3**

	Increase in internal energy of a gas/J	=	Heat supplied + **to** gas /J	Work done **on** gas /J
A to B				
B to C				
C to D				
D to A				

(i) Use the molar heat capacity at constant volume to find the increase in internal energy for each of the four sections of the cycle.

(ii) Insert zeros in Table 26.3 for the sections in which no work is done on the air, or no heat is supplied to, or taken from, the air.

(iii) Complete Table 26.3.

(l) Use the table to find the theoretical efficiency of the engine.

(m) The thrust of the engine was stated to be 220 000 N and the speed of the aircraft is 250 m s⁻¹. Find the actual power of the engine.

(n) What is the efficiency of the engine in practice?

(o) Why is the efficiency of the engine less at low altitude?

Fig 26.17   A high efficiency modern steam turbine. The laws of thermodynamics do not permit 100% efficiency. Turbines of this type are usually about 40% efficient.

26.6 THE SECOND LAW OF THERMODYNAMICS

All the examples given in section 26.5 on heat engines have been used to find the efficiency of engines. Even when these engines are operating under idealised conditions their efficiencies are not 100%. The question immediately arises therefore as to what factors control the maximum possible efficiency of a heat engine. Early steam engines were extremely inefficient, less than 1%. The most efficient steam engine which British Rail ever had was only about 11% efficient. The heat engines in power stations, steam

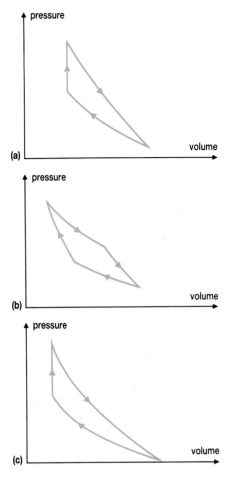

(a)

(b)

(c)

Fig 26.18 **(a)** This indicator diagram, in which no heat is lost, breaks the second law of thermodynamics. **(b)** The indicator diagram of an engine with the maximum possible efficiency for the temperatures used. **(c)** An indicator diagram in which the efficiency is 100%, but this is only possible with the surroundings at a temperature of 0 K.

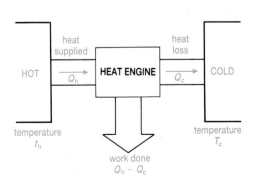

Fig 26.19 The efficiency of any heat engine is limited by the temperature between which it is operating. All heat engines give up heat to a low temperature reservoir, often the atmosphere.

turbines, are usually not more than 40% efficient. Car engines typically are about 30% efficient. The second law of thermodynamics deals with the question of heat engine efficiency and states that it is impossible, in principle, for any heat engine to have a thermal efficiency of 100%. The problem stems from the fact that heat engines have to work in a cycle and that the end of one expansion stroke **cannot** take the gas back to a state at which it is ready to start a new cycle. A 100% efficient engine would have an indicator diagram as shown in Fig 26.18(a) but the second law says that this is impossible. Fig 26.18(b) shows a cycle which can be shown to be the most efficient cycle operating between any two given temperatures, and Fig. 26.18(c) shows the indicator diagram of a 100% efficient engine which could only be achieved if the gas could start and finish at a temperature of absolute zero! Since we have to run engines in a world where the outside temperature is in the approximate range of 250–300 K, we have no possibility of achieving 100% efficient heat engines. The second law is also related to the definition of energy as the **stored** ability to do work. The internal energy of the atmosphere includes the kinetic energy of all the molecules in the atmosphere: energy which is stored in their movement. This energy is however extremely difficult to make use of to do work. The energy is present, stored in the movement of the molecules, but the second law results in it being impossible to get this energy out of store to do work unless there is some lower temperature system available. A temperature difference is necessary before work can be done. The Earth's atmosphere can be considered as a huge heat engine using hot air near the Equator as the source of heat and cold air at the poles as the low temperature reservoir. Work is done in creating wind movement. One statement of the Second law of Thermodynamics is that it is impossible for any process to have as its sole result the transfer of heat from a colder to a hotter body.

It can be shown that when an ideal heat engine takes a quantity of heat Q_h from a hot source at a temperature T_h and wastes a quantity of heat Q_c to its cold surroundings at a temperature T_c (Fig 26.19), then these quantities are related by the equation

$$\frac{Q_c}{Q_h} = \frac{T_c}{T_h}$$

This equation is, in fact, the fundamental equation for the definition of thermodynamic temperature, but here it is needed to calculate the maximum possible efficiency ε_{max} of any ideal engine.

Using Fig 26.19 we get:

$$\varepsilon_{max} = \frac{\text{Work done by ideal engine}}{\text{Heat supplied from hot source}} = \frac{Q_h - Q_c}{Q_h}$$

$$= 1 - \frac{Q_c}{Q_h}$$

$$= 1 - \frac{T_c}{T_h}$$

Since the value of T_c cannot be below the temperature of the outside air for an engine, the only practical way of improving the efficiency of heat engines is by raising the temperature T_h. Problems associated with this are the fact that the materials used in the engine may start to melt or become less strong. The high pressures involved with using steam at high temperatures also cause difficulties.

The implications of the formula for the efficiency of heat engine are enormous, particularly in the electrical supply industry. At present if a power station operates at 40% efficiency, the remaining 60% of the heat energy supplied by the fuel is wasted, probably in heating up a river or the atmosphere. The laws of thermodynamics always put upper limits on the efficiency of power stations as they do on all heat engines. If more people

THE FIRST LAW OF THERMODYNAMICS

lived near power stations they could be supplied with warm water from the cooling system of the power station just as a car heater uses the cooling water of the engine to heat the inside of the car. Such systems do exist, and they will probably become more popular as fuel costs rise. They are called CHP schemes; the initials stand for combined heat and power.

QUESTIONS

26.6 What is the maximum possible efficiency for a steam engine using steam at a temperature of 100 °C on a day when the temperature is 24 °C? (The answer is *not* 76%.)

26.7 The Magnox nuclear reactors which operate at a temperature of 370 °C use uranium metal as a fuel and this is contained in magnesium alloy cans. The Advanced Gas Cooled reactors use a ceramic uranium oxide as a fuel contained in stainless steel cans. These materials allow the higher temperature of 645 °C to be used. What improvement can theoretically be made in the efficiency of a nuclear power station if the output temperature of the reactor is raised from 370 °C to 645 °C, and if the cooling water for the reactor is at a temperature of 15 °C?

26.7 HEAT PUMPS

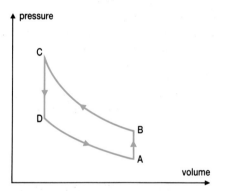

Fig 26.20 An indicator diagram of a heat pump.

An interesting application of the laws of thermodynamics is in the operation of a refrigerator. A refrigerator can be considered as a heat engine working in reverse. One particular refrigerator uses a gas which undergoes a cycle of changes of state as shown by the indicator diagram Fig 26.20. Notice that the arrows are in the opposite direction to those in indicator diagrams for heat engines. More details of the cycle are shown in Table 26.4.

Table 26.4

	Increase in internal energy of a gas	Heat supplied **to** gas	Work done **on** gas
A to B	47 J gas temperature increasing	47 J heat taken from food	0 no volume change
B to C	42 J gas temperature still increasing	0 adiabatic compression	42 J area under line BC
C to D	–63 J gas temperature falls	–63 J heat lost to atmosphere	0 no volume change
D to A	–26 J gas temperature falls to starting temperature	0 adiabatic expansion	–26 J area under line DA

The net work done to run this refrigerator for one cycle is 42 J – 26 J = 16 J. This will allow 47 J to be extracted from the food in the refrigerator and 63 J to be released into the atmosphere. This is why the back of a refrigerator is warm. Heat is being extracted from the food and being added to the work done to supply the heat to the atmosphere. The refrigerator is acting as a heat pump to move heat from inside the refrigerator to the outside as is shown in Fig 26.21. This principle can be used for heating systems and is particularly effective if there is something readily available outside a house which can be cooled down. The

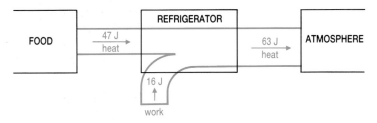

Fig 26.21   A refrigerator considered as the reverse of a heat engine, a heat pump.

atmosphere is a possibility but a stream would be much better. A heat pump uses electrical energy to operate and extracts heat from the stream and pumps it into the house as illustrated diagrammatically in Fig 26.22.

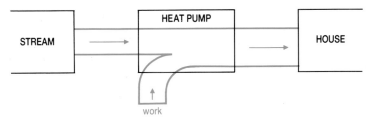

Fig 26.22   A heat pump being used for central heating.

Fig 26.23 shows a picture of a commercially available heat pump which uses the atmosphere as its source of heat. A system such as this can supply thermal energy at a rate of 6.2 kW for an electrical power output of only 3.3 kW if the outside temperature is 8 °C. When the outside temperature falls to –1 °C the output falls to 4.5 kW for the same electrical power input.

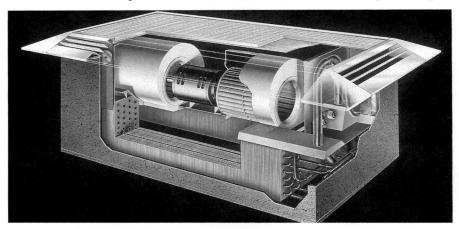

Fig 26.23   A modern heat pump. Heat is pumped from the atmosphere into the building to be heated. Such pumps depend for their usefulness on the fact that for every kWh of electrical energy used to operate the pump several kWh can be pumped in. The law of conservation of energy is not broken. The atmosphere is cooled down in the process.

Inevitably there are some snags with heat pumps. They operate much better in the summer than in the winter because it is easier to cool down the outside source of heat if it is at a high temperature. A heating system has severe limitations however if it works less and less well as the outside temperature drops. As mentioned earlier, they do not work particularly well if only the atmosphere is available for cooling down, because the air passed through the heat pump has too little mass. Running water is much better but not many houses have running water flowing past them. The capital cost of installing heat pumps can also be large because of all the pipe work necessary within and outside the pump.

- The internal energy of a gas is the sum of all the microscopic kinetic and potential energies of the molecules within the gas.
- The first law of thermodynamics states that the internal energy of a system depends only on its present state; the internal energy may be increased either by doing work on the system or by heating it. That is

| **Increase** in internal energy of a gas | = | Heat supplied **to** gas | + | Work done **on** gas |

- Indicator diagrams for a gas are graphs of pressure against volume and show how the state of a gas changes.

	Isothermal	Adiabatic	Isobaric	Isovolumic
definition	no change in temperature	no heat supplied	no change in pressure	no change in volume
work done on gas	area under graph	area under graph	$p\Delta V$	zero
heat supplied to gas	NOT zero	zero	$C_p\Delta T$	$C_V\Delta T$
increase in internal energy	zero	$C_V\Delta T$	$C_V\Delta T$	$C_V\Delta T$
law applying	$pV = k$	$pV^{\gamma} = k$	$\dfrac{V}{T} = k$	$\dfrac{p}{T} = k$

$pV = nRT$ applies additionally in all cases

$$C_{p,\,m} - C_{V,\,m} = R$$

- Since the internal energy of a fixed mass of an ideal gas depends only on its temperature, the change in internal energy can always be calculated from $C_V\Delta T$ no matter how the pressure and volume change.
- The second law of thermodynamics makes the maximum possible efficiency of a heat engine to be

$$\varepsilon = 1 - \frac{T_c}{T_h}$$

EXAMINATION QUESTIONS

26.8 The diagram shows the apparatus used in a determination of the specific heat capacity of a liquid. The total thermal capacity of the calorimeter, the heater, the stirrer and the thermometer was 107 J K^{-1}. The mass of liquid in the calorimeter was 0.241 kg, and the ammeter and voltmeter readings were 3.40 A and 12.2 V respectively.

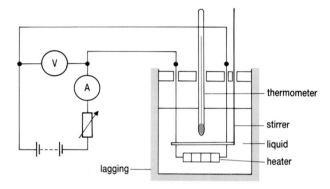

When the temperature of the liquid was equal to that of the surroundings, its value was found to be increasing at the rate of 46.3 × 10^{-3} K s^{-1}.

(a) (i) What is the advantage of measuring the rate of rise of temperature of the liquid when its temperature is equal to that of the surroundings rather than measuring the rise of temperature taking place in a given length of time?

(ii) Suggest a suitable procedure for making this measurement.

(b) Determine a value for the specific heat capacity of the liquid   (8)

(AEB 1986)

26.9 The diagram shows a light belt attached to two force meters and wrapped around a copper cylinder of mass 0.50 kg and radius 0.030 m.

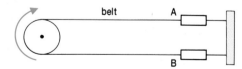

When the cylinder is rotated clockwise as shown, at three revolutions per second, the readings on the meters A and B are 100 N and 250 N respectively. At what rate is energy being dissipated because of friction between the belt and the cylinder?

If the specific heat capacity of copper is 400 J kg^{-1} K^{-1} at what rate will the temperature of the cylinder start to rise when first rotated?

If the rotation is continued why will the rate of rise of temperature probably decrease?   (6)

(ULSEB 1983)

26.10 State the law of conservation of energy. The braking system on all four wheels of a car consists of metal discs attached to the wheels and pads attached to the framework of the car. The pads, which are made from a poor thermal conducting material, grip the discs and slow down the car. A disc has a mass 2.8 kg and specific heat capacity 460 J kg^{-1} K^{-1}. Calculate

(i) the heat capacity of a disc, and

(ii) an approximate value for the temperature rise of a disc when the car of mass 900 kg brakes from 30 m s^{-1} to rest.   (8)

A single disc is removed from a car. It has a diameter of about 20 cm. Describe a simple experiment by which you could check that your calculated value for the heat capacity of the disc is of the right order of magnitude.   (6)

(ULSEB 1987)

26.11 (a) The specific heat capacities of air are 1040 J kg^{-1} K^{-1} measured at constant pressure and 740 J kg^{-1} K^{-1} measured at constant volume. Explain briefly why the values are different.

(b) A room of volume 180 m^3 contains air at a temperature of 16 °C having a density of 1.13 kg m^{-3}. During the course of the day the temperature rises to 21 °C. Calculate an approximate value for the amount of energy transferred to the air during the day. Assume that air can escape from the room but no fresh air enters. Explain your reasoning   (5)

(ULSEB 1984)

26.12 (a) (i) Write down the equation which defines a temperature on the Kelvin scale in terms of the properties of an ideal gas. Explain the symbols you use.

(ii) A simple form of gas thermometer consists of a capillary tube sealed at one end and containing a thread of mercury which traps a mass of dry air. Describe how you would calibrate it on the gas scale and use it to determine the boiling point of a liquid known to be about 350 K. Explain how the temperature is calculated from the readings and state any assumptions you make.   (10)

(b) A cylinder fitted with a piston which can move without friction contains 0.050 mol of a monatomic ideal gas at a temperature of 27 °C and a pressure of 1.0×10^5 Pa. Calculate (i) the volume, (ii) the internal energy of the gas.   (4)

(c) The temperature of the gas in (b) is raised to 77 °C, the pressure remaining constant. Calculate
(i)   the change in internal energy,
(ii) the external work done,
(iii) the total heat energy supplied.   (6)
(Molar gas constant = 8.3 J mol^{-1} K^{-1})
(JMB 1986)

26.13 Explain the meaning of the terms *heat*, *work* and *internal energy* as used in the first law of thermodynamics.   (3)
An amount, 0.20 mol, of an ideal gas is trapped in a closed cylinder sealed by a piston. Moving the piston varies the volume occupied by the gas from 1.0×10^{-3} m^3 to 5.0×10^{-3} m^3. When the gas is maintained at a temperature of 300 K, calculate the pressures exerted by the gas when it occupies volumes of 1.0, 2.0, 3.0, 4.0 and 5.0×10^{-3} m^3. The gas can also be maintained at 500 K. Calculate the pressures exerted by the gas maintained at 500 K for the same values of the volume. On one set of axes, plot accurate graphs of pressure against volume for this sample of gas maintained at 300 K and at 500 K.   (7)
The following operations are carried out on the gas in the cylinder.
(a) At 300 K the gas is compressed isothermally from a volume of 5.0×10^{-3} m^3 to a volume of 1.0×10^{-3} m^3. From your graph, or otherwise, find the work done on the gas during the compression. Explain the energy transfers which take place in this operation.   (4)
(b) The compressed gas is heated at constant volume from 300 K to 500 K. Calculate the heat energy supplied.   (2)
(c) The hot gas expands isothermally at 500 K from a volume of 1.0×10^{-3} m^3 to a volume of 5.0×10^{-3} m^3. From your graph, or otherwise, find the work done by the gas. Explain the energy transfers which take place during the expansion.   (4)
(O & C 1988)

26.14 (a) (i) Explain what is meant by the terms *internal energy of a gas*, and *ideal gas*.   (4)
(ii) State, in words, the relation between the increase in the internal energy of a gas, the work done on the gas, and the heat supplied to the gas.   (1)

(b) The gas in the cylinder of a diesel engine can be considered to undergo a cycle of changes of pressure, volume and temperature. One such cycle, for an ideal gas, is shown on the graph.

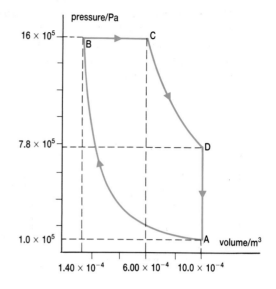

(i) The temperatures of the gas at A and B are 300 K and 660 K respectively. Use the ideal gas equation and data from the graph to find the temperatures at C and D.   (4)
(ii) During each of the four sections of the cycle, changes are being made to the internal energy of the gas. Some of the factors affecting these changes are given in the table.

Section of cycle	Heat supplied to gas / J	Work done on gas / J	Increase in internal energy of gas / J
A to B	0	300	
B to C	2580	−740	
C to D	0	−440	
D to A	−1700		

Explain why the work done on the gas is sometimes negative and find the work done on the gas in section D to A. Deduce the values of the 'increase in internal energy of gas' for each section and list them.   (7)
(iii) Explain why the total change in the internal energy of the gas during a complete cycle must be zero.   (2)
(iv) What is the net work output during a complete cycle?   (2)
(v) Assuming that the efficiency of a heat engine is defined as the ratio of work output to heat input, calculate the efficiency of this engine.   (2)
(UCLES 1988)

26.15 (a) State the first law of thermodynamics. (4)
Explain clearly what is meant by the *internal energy* of a system. (2)

(b) Describe a method for measuring the specific latent heat of vaporisation of a liquid. (9)

(c) The specific latent heat of vaporisation of water boiling at 100 °C is 2280 kJ kg^{-1}, and when 1 kg of water vaporises its volume increases from 1.0×10^{-3} m^3 to 1.7 m^3. (Take atmospheric pressure to be 100 kPa, the molar mass of water to be 0.018 kg, and the Avogadro constant to be 6.0×10^{23} mol^{-1}.)
Determine:
  (i) the external work done when 1 kg of water vaporises at 100 °C; (4)
  (ii) the change of internal energy that occurs; (2)
  (iii) the mean binding energy per molecule for the bonds between water molecules at 100 °C. (5)

(d) Draw a sketch graph relating mutual potential energy of molecules to their separation. Indicate the features of the graph that relate to (i) equilibrium separation, (ii) latent heat. (4)
(OXFORD 1986)

26.16 (a) The first law of thermodynamics is usually stated in the form of an equation

$$\delta Q = \delta U + \delta W.$$

Explain fully the meaning of each term in the equation. (6)

(b) Explain why an ideal gas is, from its definition, a substance in which the internal energy is wholly kinetic. (3)

(c) One mole of liquid water occupies 1.87×10^{-5} m^3 at 100 °C. At one atmosphere pressure (1.03×10^5 Pa) and 100 °C one mole of water vapour occupies 2.96×10^{-2} m^3. The heat transfer required to vaporise one mole of water at one atmosphere pressure and 100 °C (the molar latent heat capacity) is 4.05×10^4 J. Calculate the change of internal energy in this process. (5)

(d) Use the results of the kinetic theory to calculate the internal energy of one mole of helium gas at 300 K, assuming that helium behaves as an ideal monatomic gas. Calculate also the root mean square speed of helium atoms at 300 K. (3,3)
(Mass of one mole of helium = 4.0×10^{-3} kg.)

26.17 The diagram shows steam from a power station boiler passing through a turbine which is used to turn an electric generator.

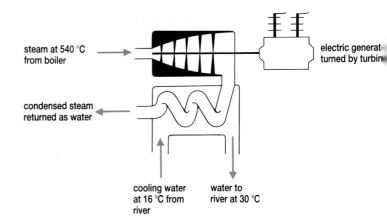

steam at 540 °C from boiler

electric generator turned by turbine

condensed steam returned as water

cooling water at 16 °C from river

water to river at 30 °C

Once steam has passed through the turbine, it is cooled by river water and condensed before being pumped back to the boiler for re-use.

(a) The cooling water is removed from the river at 16 °C and returned to the river at 30 °C. The water flow used is 15 tonnes per second (15×10^3 kg s^{-1}).
Calculate the power used to heat this water.
(Specific heat capacity of water = 4200 J kg^1 K^{-1}.) (1)

(b) The generator turned by the turbine produces 560 MW of electrical power. Calculate the efficiency with which energy in the steam is actually converted into electrical energy. (2)

(c) If the temperature of the water taken from the river is much lower, suggest what effect this may have on the efficiency of the system. (1)

Chapter 27

THE TRANSFER OF THERMAL ENERGY

LEARNING OBJECTIVES

At the end of this chapter you should be able to:

1. explain the two mechanisms of thermal conduction;

2. use the term thermal conductivity correctly and use it in calculations;

3. explain why convection occurs;

4. explain how radiation is important in establishing the temperature of a body.

27.1 THERMAL CONDUCTION

As discussed in Chapter 24, a difference in temperature between two bodies results in a flow of heat energy from the one at a higher temperature to the one at a lower temperature. Similarly if there is a temperature gradient within any object, then heat will flow from the region of higher temperature to the region of lower temperature. If a saucepan is placed on a ring on a cooker, the surface of the bottom of the saucepan in contact with the ring reaches a higher temperature than the the surface in contact with the food. Heat energy will therefore flow through the metal from the ring into the food. Heat ceases to flow only when there is no temperature gradient. This process of thermal conduction takes place in materials as a result of two quite different ways of energy transfer.

Any solid metal has a crystalline structure in which the positive metallic ions form a three-dimensional lattice. Within the spaces provided by the lattice there exists a sea of free electrons. Most of the electrons in an atom of a metal are associated with a particular atom but the outer electrons are not attached to particular atoms and are therefore able to drift from place to place within the lattice. These are the electrons which are responsible for electrical conduction, and are called the conduction electrons. When there is a temperature gradient within a metal, the conduction electrons gain energy at the hotter part of the metal and can diffuse to the colder part of the metal, resulting in a transfer of thermal energy. This process can only take place in electrical conductors because only electrical conductors have conduction electrons. Thermal energy can however be transferred through electrical insulators, so there must be some other process which allows heat to flow besides that using conduction electrons.

The second mechanism responsible for heat flow depends on coupling between adjacent molecules. The atoms in a solid have vibrational kinetic energy. At a hot region within the solid, the vibrational energy is high. If there is another region where the vibrational energy is lower, then heat transfer can take place. As a result of these two effects we can state that any good conductor of electricity will be a good conductor of heat. The reverse statement is not true. Some good electrical insulators are good conductors of heat. Quartz, for example, has an electrical resistance about 10^{20} times that of copper, but it conducts heat only a little less well than copper.

27.2 THERMAL CONDUCTIVITY

If a bar of metal is placed so that one end of it is in contact with an object kept at a high temperature and the other end kept in contact with an object at a low temperature, then heat will flow through the rod, Fig 27.1. How much heat flows depends on additional factors. If the bar is cold when it is placed in position then to begin with the temperature of the bar will rise at the hot end. Heat will flow towards the cold end and this will gradually raise the temperature of the bar, but it is never possible for the whole of the bar to reach the temperature of the hot object. When the temperature of each part of the bar has stopped rising, the bar is said to be in the steady state. No heat will subsequently be used to raise the temperature of the bar itself. The final temperature achieved at different points along the bar will depend on how the bar is lagged, as shown in Fig 27.1(a) and Fig 27.1(b).

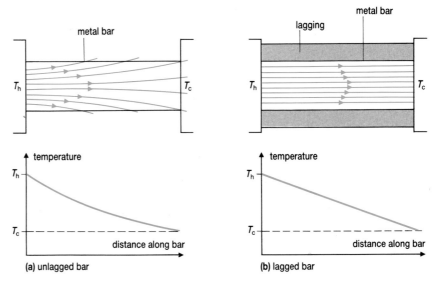

Fig 27.1 Heat flow through a lagged and an unlagged bar. The temperature gradient is constant along a bar from which heat does not escape.

If heat escapes through the sides of the bar, then there is a greater flow of heat through the bar near the hot end than there is through the cold end. This means that there must be a larger temperature gradient near the hot end than near the cold end. The variation of temperature with distance is not linear. If perfect heat insulation is imagined around the bar, then no heat escapes through the sides and the rate of heat flow must be the same at all points along the bar. This parallel flow of heat in the steady state implies that the temperature gradient must be constant.

In the steady state it is found experimentally that the rate of flow of heat, Q/t, depends on the material of the bar and is proportional to

the cross-section area of the material, A
the temperature gradient

For parallel flow, the temperature gradient is the difference in temperature between the ends of the bar divided by x, the length of the bar. Here it does not matter whether the temperatures are measured in kelvin or degree Celsius, as it is the difference in temperature which is required and the size of the interval of one kelvin is exactly the same as the interval of one degree Celsius. A temperature difference is normally quoted in kelvin even though it is likely that when measurements are made of the temperatures at the ends of the bar the readings are taken in degrees

Celsius. If T_h is the kelvin temperature at the hot end of the bar and T_c the temperature at the cold end then

$$\blacklozenge \qquad \frac{Q}{t} = \frac{\lambda A (T_h - T_c)}{x}$$

where λ is a constant called the **thermal conductivity** of the material.

λ is a constant which has a unit. To find the SI unit, λ can be made the subject of the equation giving

$$\lambda = \frac{Q}{t} \frac{x}{A (T_h - T_c)}$$

and hence the unit of λ is $W\ m^{-1}\ K^{-1}$, as watt is the unit of rate of flow of heat.

Some numerical values of λ are given in Table 27.1 for some common substances.

Table 27.1 Thermal conductivities of selected materials

Material	Thermal conductivity (λ) /$W\ m^{-1}\ K^{-1}$
Metals	
aluminium	205
brass	109
copper	385
silver	406
steel	50
Typical domestic materials	
cotton poplin	0.048
interlock wool	0.040
velour overcoating	0.040
cellular cotton	0.050
woollen blanketing	0.037
carpet	0.050
Typical building materials	
brick	0.6
insulating brick	0.15
concrete	0.8
fibreglass	0.04
glass	0.8
insulating Foam	0.01
soft wood	0.1
Gases	
air	0.024
hydrogen	0.14
oxygen	0.023

A useful analogy can be made between thermal conductivity and electrical conductivity. For thermal conductivity the rate of flow of heat:

$$\frac{dQ}{dt} = \lambda A \frac{\Delta T}{x}$$

where ΔT is the temperature difference across the length of the bar.

For electrical conductivity the rate of flow of charge, in a specimen having the same dimensions:

$$\frac{dQ}{dt} = I = \frac{\Delta V}{R}$$

where ΔV is the potential difference across the length of the bar.
Since

$$R = \frac{\rho x}{A}$$

where ρ is the resistivity:

$$\frac{dQ}{dt} = I = \frac{A\Delta V}{\rho x}$$

By comparison of the two rate of flow equations it can be seen that temperature gradient causes heat flow in the same way that potential gradient causes charge flow.

Example 1

Estimate the amount of heat flow through a window if the temperature indoors is maintained at 20 °C and the temperature outdoors is 5 °C. The area of the window is 2.0 m² and it is made of 6 mm glass.

$$\text{temperature difference} = 20\ °C - 5\ °C = 15\ K$$

$$\text{temperature gradient} = \frac{(20-5)\ K}{0.006\ m} = 2500\ K\ m^{-1}$$

$$\frac{Q}{t} = \frac{\lambda A (T_h - T_c)}{x}$$

$$= (0.8\ W\ m^{-1}\ K^{-1}) \times (2\ m^2) \times (2500\ K\ m^{-1})$$

$$= 4000\ W$$

In example 1 the rate at which heat is estimated to be lost through a single, not particularly large, window, for a fairly modest temperature difference, is not in agreement with common sense. The estimate implies that just to cope with heat losses from the window would require 4 single bar electric fires to be switched on in the room. By the time heat losses through the walls, ceiling, floor and doors are included the estimate might well be 20 electric fires, a totally unreasonable number.

The reason that this discrepancy occurs is because of the assumption that the temperature of the inside surface of the window is the same as the temperature of the room, and that the temperature of the outside surface of the window is the same as the outside temperature. In practice, a layer of still air in the room near the window is cooled by the window. The thermal conductivity of the air itself therefore also has an influence on the rate of heat flow. Fig 27.2(a) shows how the temperature has been assumed to vary with distance across the glass. Fig 27.2(b) shows a more likely practical variation in the temperature. This second graph has a much smaller temperature gradient and would give a heat flow which is correspondingly smaller, probably about a tenth of the value originally calculated. 400 W is a possible heat loss through a window; 4000 W is improbable and would be very expensive.

In order to overcome these practical problems, heating engineers use a less theoretical approach to heat losses. They measure heat losses through typical parts of building constructions and quote U-values which take the practical situation into account, and also include the thickness of the material in the U-value. The following example shows how this can be used.

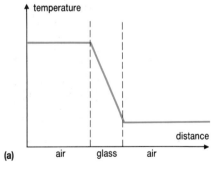

(a)

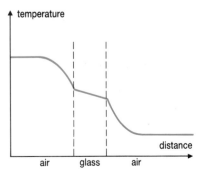

(b)

Fig 27.2 Idealised and practical temperature changes near the glass of a window.

THE TRANSFER OF THERMAL ENERGY

Fig 27.3 In the construction of modern houses a heat insulating material is placed between the inside and outside walls.

Example 2

A wall of a room has an area of 8.0 m² and is made of 2 thicknesses of brickwork separated by insulation, as shown in Fig 27.3. Find the rate at which heat will escape through the wall if the design temperature inside is 23 °C and outside –5 °C and the wall has a U-value of 0.51 W m⁻² K⁻¹.

$$\text{Rate of heat loss} = (0.51 \text{ W m}^{-2}\text{ K}^{-1}) \times (8 \text{ m}^2) \times (28 \text{ K})$$
$$= 110 \text{ W (to 2 sig. figs)}$$

The unit of U-value gives the information that the U-value is the rate of heat flow through unit area per unit difference in temperature.

$$U\text{-value} = \frac{\text{rate of flow of heat}}{\text{area} \times \text{temperature difference}}$$

If care is taken with the units of any equation, they can be very helpful in avoiding careless working and nonsensical answers.

QUESTIONS

27.1 What is the temperature difference across the base of an aluminium saucepan if it is 3 mm thick, has an area of 0.018 m^2 and heat flows through it at a rate of 1200 W. Why is your answer so small?

27.2 A steam boiler is made of steel 1.0 cm thick and has a surface area of 5.0 m^2. It contains steam at a temperature of 210 °C. It is surrounded by lagging 3 cm thick and the outside temperature of the lagging is 40 °C.

Assuming that the lagging has the same area as the boiler find
(a) the rate of heat loss from the boiler (this must be the same as the rate of heat loss through the lagging)
(b) the temperature at the interface between the boiler and the lagging. Thermal conductivity of steel = 50 W m^{-1} K^{-1}; thermal conductivity of the lagging = 0.04 W m^{-1} K^{-1}.

DATA ANALYSIS

Heat Losses from a house

The following data are taken from a book of building regulations:

Material	U-value /W m^{-2} K^{-1}
insulated external cavity wall	0.51
internal wall	3.5
door	5.2
window	5.7
floor	0.60
insulated ceiling	0.35

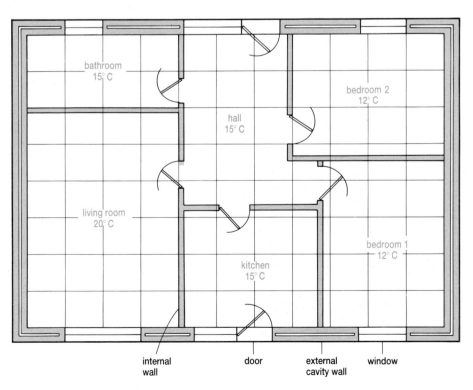

Scale: 1 cm : 1 m
The background squares are 1 m^2

Fig 27.4 Plan of a one storey house with required temperatures for different rooms.

THE TRANSFER OF THERMAL ENERGY

Fig 27.4 is a plan of a single-storey house drawn on squared paper with a scale so that the side of each square is 1 metre. The temperature required for each room is shown on the plan. Find the power necessary for a heater for each room to replace heat lost by conduction if the outside temperature is –5 °C. You are advised to tabulate all your calculations.

Additional necessary information: height of a door 2 m
height of all windows 1.2 m
height of walls 2.5 m

In making this type of standard calculation, a heating engineer uses the inside dimensions of each room and assumes that the ground underneath the house is at the same temperature as the outside, that is, –5 °C in this case.

What would be the output power requirement for the central heating system for the house?

In addition to the heat lost by conduction, the system is required to allow for a complete change of air in the house every hour. Make an approximate calculation to find whether this makes an appreciable extra load on the system. The heat capacity of a cubic metre of air is approximately $1000 \, J \, K^{-1}$.

List six different ways in which the energy bill for heating the house could be appreciably reduced and indicate the relative importance of each of your suggestions.

27.3 CONVECTION

Most fluids when heated expand and so have a lower density when hot than when cold. The hot fluid floats upwards with the colder fluid falling as a result of the buoyancy force exerted on the hot fluid by the surrounding cold fluid being greater than its weight. This movement of the hot fluid upwards and the cold fluid downwards is called a convection current. Convection currents can be set up in any fluid so are possible in gases as well as in liquids.

Convection currents are responsible for the familiar land and sea breezes. They are also responsible for the less familiar katabatic winds illustrated in Fig 27.5. In mountain areas, air over high land cools and flows down through valleys, sometimes causing damaging frosts. The Mistral in Southern France is a wind of this type. Fig 27.6 is a photograph of a hang glider rising in a convection current of warm air. Glider pilots rely on rising air to gain height and they become very skilled at finding it. They look for indications of rising air. These include corn fields, over which air tends to

Fig 27.6 In order to rise a hang glider must find rising air. This can occur when wind blows towards a ridge or when hot air rises in a convection current. The tiny droplets of water which make up a cloud also rise in upward convection currents.

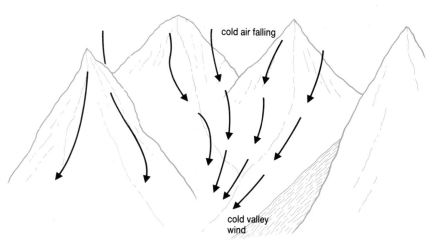

Fig 27.5 Cold air falling into a valley by convection and possibly giving rise to damaging frosts in the valley.

be hotter than over grass fields, bird movement, cumulus clouds, which are rising water droplets, and other gliders.

The rate of heat transfer by convection from a hot object depends on such factors as whether there is a wind blowing or not, the shape and the surface area of the object. These factors are difficult to quantify, so that although it is possible to state that the rate of loss of heat depends on the temperature difference between the object and its surroundings, it is difficult to state a mathematical relationship for that dependency.

27.4 RADIATION

All objects emit some electromagnetic radiation. The quantity and wavelength of this radiation depend on several factors. In some cases it is clear that electromagnetic radiation is being emitted because it is possible to see and feel the effect of the radiation whereas in other cases this is not so. The Sun, for example, emits a great deal of radiation. The light which we see is only a small part of the total emitted radiation which includes ultra-violet, infra-red and microwave radiation. If a graph is plotted of intensity of radiation against wavelength for the Sun then a graph similar to that in Fig 27.7 is obtained. Only about 20% of the Sun's radiation is emitted as visible light.

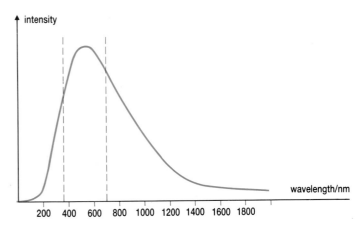

Fig 27.7 Intensity of radiation plotted against wavelength for a hot body.

The most important factor controlling radiation is temperature. The power radiated is proportional to the fourth power of its temperature so an object at a temperature of 2000 K radiates 16 times as much power per unit area as the same object would at 1000 K.

Example 3
The total energy arriving from the Sun and falling on the upper atmosphere of the Earth is 1400 W m^{-2}. The distance of the Earth from the Sun is 1.5×10^{11} m. Find the power output of the Sun.

The Sun cannot lose energy by conduction or convection because of the vacuum of space; it must therefore lose all its energy by radiation.

If a sphere of radius 1.5×10^{11} m is imagined around the Sun, then each m^2 of its surface must receive a power of 1400 W (Fig 27.8).

Area of surface of this imaginary sphere
$$= 4\pi r^2$$
$$= 4\pi \times (1.5 \times 10^{11} \text{ m})^2$$
$$= 2.8 \times 10^{23} \text{ m}^2$$

Power output of Sun
$$= 2.8 \times 10^{23} \text{ m}^2 \times 1400 \text{ W m}^{-2}$$
$$= 4.0 \times 10^{26} \text{ W}$$

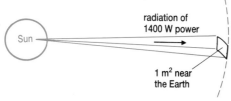

Fig 27.8

THE TRANSFER OF THERMAL ENERGY

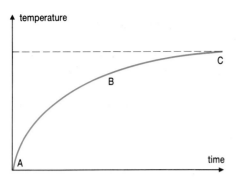

Fig 27.9 The establishment of a fixed temperature by the element of an electric fire.

The temperature of the surface of the Sun is controlled by this figure. Because any changes in the power output of the Sun take place only slowly, the Sun can be considered to be in a dynamic power equilibrium. This is not the same as temperature equilibrium referred to in Chapter 24 where no heat flowed from one body to another at the same temperature. This is certainly not true for the Sun because it is losing heat to its surroundings which are at a lower temperature than itself. The power equilibrium implies that a balance has been reached between the power output of the nuclear reactions going on inside the Sun and the power being radiated away from the Sun. When these two quantities are equal then there is no net gain or loss of energy by the Sun and its temperature will therefore remain constant. This happens for the Sun at its surface temperature of about 6000 K.

It is not only for astronomical bodies that this balance is achieved by radiation. When an electric fire is switched on the element is cold: point A on Fig 27.9. Electrical energy supplied to the element therefore has the effect of raising its temperature. At B the temperature is at a higher value and the element can lose heat to its surroundings. This it will do by both convection and radiation, so less of the supplied electrical energy will be used to raise the temperature of the heater and its temperature will rise more slowly. Heating continues until a final fixed temperature is reached at C at which the rate at which energy is being lost by convection or by radiation equals the rate at which electrical energy is being supplied. If something should happen to disturb the rate at which energy is being lost by convection, by putting the fire in a draughty position for instance, then a new temperature will be established at which this dynamic balance can be maintained.

INVESTIGATION

Examine how the balance of convection and radiation losses from a filament affect its temperature.

Using three different thicknesses of bare nichrome wire, for example, 24 s.w.g., 26 s.w.g. and 30 s.w.g., make three *close wound* coils, each of the same resistance of 3.0 Ω at room temperature, by wrapping the wires around a pencil. The lengths required for these particular wires are approximately 70 cm of 24 s.w.g. 50 cm of 26 s.w.g. and 30 cm of 30 s.w.g. The resistances should be checked experimentally, using an ohmeter or an ammeter and voltmeter. Adjustments can be made to the resistances by making connections to the wires at different places, so make sure that at the start your wires are too long rather than too short.

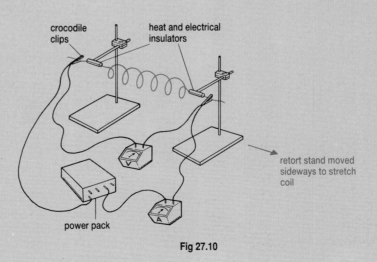

Fig 27.10

Mount the coils, one at a time, between two supports, which can withstand being heated and provide electrical insulation, Fig 27.10, and connect to a variable low voltage power supply. Adjust the power supply until the coil glows red hot and measure the current supply and the potential difference across the coil. Repeat the experiment as the coil is gradually stretched. What do you notice about the temperature of the coil as it is stretched?

It is possible to find the approximate temperature of the coil as each reading is taken by using the calibration graph given in Fig 27.11, provided the resistance of each coil is 3.0 Ω at room temperature.

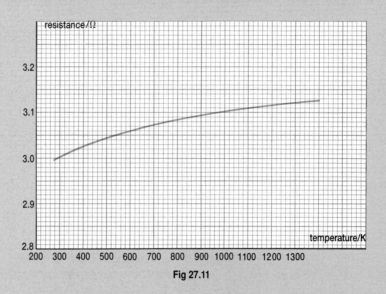

Fig 27.11

Repeat the experiment using the other two coils. Why do you think that the thinner wires require a smaller current to reach a temperature at which they are red hot? Why is it that a filament of a lamp bulb is made of very fine wire of short length, a radiant electric fire element is made of longer, thicker wire, and that an electrical convector heater element is made of wire longer and thicker again?

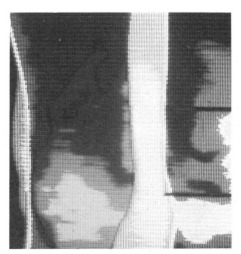

Fig 27.12 Thermographs of a back view of a person's knees. The left knee shows evidence of arthritis and is up to 5 degrees hotter than the right knee.

The radiation loss from an electric fire is something which can be felt readily on your skin. Your skin is a good detector of infra-red radiation in that it will experience a temperature rise which you can interpret as being due to the radiation. You can use your hand to detect the infra-red radiation from a radiator containing hot water. If you place your hand about 3 cm from a domestic radiator you will be away from the convection currents around the radiator but you can still feel the heat energy reaching you as a result of the radiation. Clearly there is not now as much radiation as from a radiant electric fire, but there is some radiation although it is not visible. The radiator does not glow in the dark; it is not giving out light, but it is emitting infra red radiation and by using infra-red film and special detectors it is possible to take photograhs using infra-red radiation. This principle is made use of medically. A person, like everything else, emits infra-red radiation at a rate dependent on temperature. A patient with joint inflammation caused by rheumatoid arthritis has a higher temperature at the joint than on surrounding skin. This is shown in the photograph in Fig 27.12 where the extra radiation emitted by the knee is shown as a black patch. Early diagnosis of this condition can help in delaying the onset of the arthritis.

Not only do objects emit radiation, they also absorb it from their surroundings. Where the surroundings are at a higher temperature than the body then the body will have a net gain in energy, whereas there will be a net loss of energy from the body if it has its surroundings at a lower temperature. This is a direct consequence of the zeroth law of thermodynamics (section 24.1) which applies equally well to transfers of energy by radiation as it does to conduction.

The ability to radiate and absorb radiation is a complex matter, but the two are related. Any good absorber of a particular wavelength is also a good radiator of that wavelength. The colour and texture of a surface are also important. A lemon is yellow because it absorbs blue and violet light well but reflects the red/yellow/green region of the spectrum.

DATA ANALYSIS

Insulation provided by clothing

The textile trade make great efforts to ensure that clothing and bedding provide good heat insulation. They are more interested in heat insulation than in heat conduction and they use a quantity of thermal resistance which is the reciprocal of the U-value. The SI unit of thermal resistance would be $m^2\,K\,W^{-1}$ but the textile trade use the *tog*. One tog is the practical unit of thermal resistance and is defined by the relationship that

$$10\ \text{tog} = 1\ m^2\ K\ W^{-1}$$

A medium quality blanket has a thermal resistance of approximately one tog

To measure the thermal resistance of a textile material, the Shirley Togmeter is used. The instrument works on the Ohm's law principle, namely that when current (in this case heat) flows through conductors in series, the ratio of the potential (temperature) difference across each conductor equals the ratio of their resistances.

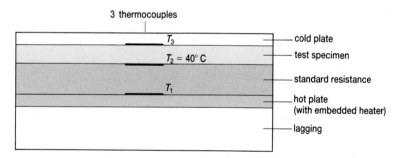

Fig 27.13   The structure of a togmeter.

The test specimen is laid over a material of known thermal resistance which is heated electrically by a hot plate to a surface temperature of 40 °C. Thermocouples are embedded in each surface of the 'standard resistance'. On top of the specimen is a lightweight cold plate of expanded polystyrene with a thermocouple embedded in its surface. This gives a pressure on the textile material of 6.9 Pa. A diagram of the instrument is shown in Fig 27.13 and a photograph is given in Fig 27.14.

Use the information given in the preceding section and in Table 27.1 to answer the following questions.

(a) Thermal conductivity for a textile material is in fact a combination of heat transfer by conduction, convection and radiation. This is taken into account in Table 27.1. Why is it that heat in a textile can flow by all three methods of heat transfer but for a metal only conduction is possible?

(b) Why are all the values for λ given for domestic materials in Table 27.1 so similar?

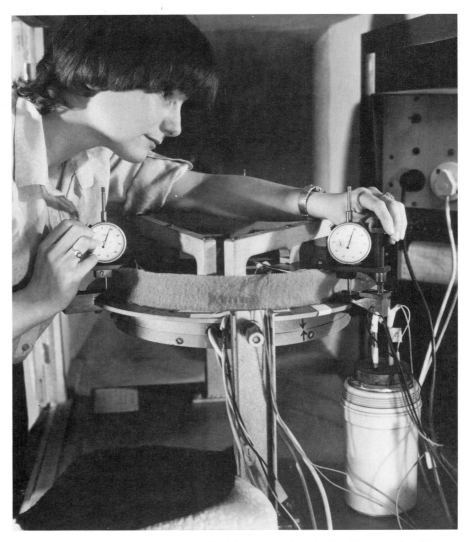

Fig 27.14 Setting up the Shirley togmeter for an insulation test on a piece of quilting. The temperature difference across the quilting is measured for standardised conditions when the pressure on the quilting is 6.9 Pa.

(c) Define thermal resistance.
(d) What is the thermal resistance of cotton poplin 1 mm thick in
(i) m^2 K W^{-1} (ii) tog?
(e) When using a togmeter with a particular piece of quilting, T_1 was 41.7 °C, T_2 was 40 °C and T_3 was 21.5 °C. If the thermal resistance of the standard resistance is 1.20 tog, show that the thermal resistance of the quilting is 13 tog. Explain your reasoning.
(f) Why is it necessary to state the pressure which the cold plate exerts on the specimen?
(g) In use, the togmeter needs to be switched on for up to 3 hours before readings can be taken. Why is such a long time necessary?
(h) In what respects is thermal resistance similar to electrical resistance? In what respects is it different?
(Data for this exercise were provided by the British Textile Technology Group, Shirley Towers, Manchester, England.)

THE TRANSFER OF THERMAL ENERGY

SUMMARY

- Thermal conduction takes place in electrical conductors and insulators by lattice vibration. In electrical conductors it can take place additionally by electron diffusion.
- The rate of flow of heat is proportional to temperature gradient and area

$$\frac{dQ}{dt} = \lambda A \frac{\Delta T}{x}$$

- Compared with the equation for resistivity, the rate of heat flow corresponds with the rate of charge flow; the temperature gradient corresponds with the potential gradient
- Convection takes place as a result of hotter, less dense, fluid rising and colder, denser, fluid falling.
- All objects emit electromagnetic radiation. The quantity of power radiated by an object depends on the nature of the surface of the object and on its temperature.

EXAMINATION QUESTIONS

27.3 Outline two processes by which heat may be conducted through a solid. In general, metals conduct both heat and electricity well; why is this? (3)

(UCLES 1986)

27.4 **(a)** A sheet of glass has an area of 2.0 m² and a thickness 8.0×10^{-3} m. The glass has a thermal conductivity of 0.80 W m⁻¹ K⁻¹. Calculate the rate of heat transfer through the glass when there is a temperature difference of 20 K between its faces.

(b) A room in a house is heated to a temperature 20 K above that outside. The room has 2 m² of windows of glass similar to the type used in (a). Suggest why the rate of heat transfer through the glass is much less than the value calculated above.

(c) Explain why two sheets of similar glass insulate much more effectively when separated by a thin layer of air than when they are in contact. (8)

(AEB 1986)

27.5 Two metal bars P and Q of the same length and cross-sectional area, but of different thermal conductivities, are joined end to end as shown. The sides of the bar are well

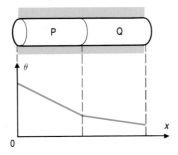

lagged. The left-hand end of P is maintained at a higher temperature than the right-hand end of Q. When steady state conditions are reached, the temperature θ varies with distance x along the bar as shown in the graph. State whether

(a) the magnitude of the temperature gradient in P is less than, equal to, or greater than that in Q;

(b) the heat flux in P is less than, equal to, or greater than that in Q;

(c) the thermal conductivity of P is less than or greater than that of Q (3)

(UCLES 1985)

27.6 **(a)** Draw an electrical circuit suitable for use with a metal resistance thermometer. Explain how the measurements are converted to temperature readings. (7) State two advantages (in each case) of using a resistance thermometer instead of:
(i) a mercury in glass thermometer; (2)
(ii) a thermocouple thermometer. (2)

(b) Define the *thermal conductivity* of a material. (3) Distinguish between the parts played by *lattice vibrations* and by *free electrons* in the conduction of heat through solids. (3, 3)

(c) A copper rod, an iron rod and a copper rod, each of diameter 8.0 mm and length 120 mm, are joined with the iron rod in the middle to make a composite rod 360 mm long. The free ends are maintained at 100 °C and 0 °C respectively, and the rods are insulated to maintain a linear temperature gradient.

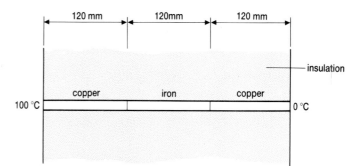

(Take the thermal conductivity of copper to be 400 W m⁻¹ K⁻¹ and of iron to be 60 W m⁻¹ K⁻¹.) Calculate for steady conditions:
(i) the temperatures of the ends of the iron rod; (7)
(ii) the rate of flow of heat through the composite rod. (3)

(OXFORD 1987)

27.7 Define the term *thermal conductivity* (2) Why are different methods required to measure the thermal conductivities of good and poor conductors? Illustrate your answer by brief descriptions of actual methods used. (8) The ends of a uniform metal bar of length l are maintained at constant temperatures θ_1 and θ_2 which are both greater than the temperature of the surroundings. θ_1 is greater than θ_2. Sketch graphs showing the variation of temperature along the length of the bar when (a) the bar is lagged to minimise heat losses, (b) the bar is

unlagged. Variations of temperature over any cross-section of the bar may be neglected. Explain the shape of the graphs that you draw. (5)
Such a bar, AB, made of copper of thermal conductivity 4.0×10^2 W m^{-1} K^{-1} has an area of cross-section of 1.5×10^{-3} m^2. The temperature of end A is greater than that of B. The temperature gradient near end A is -38 K m^{-1} and that near end B is -25 K m^{-1}. Calculate the rate of heat loss from the sides of the bar due to imperfect lagging. (5)

(O & C 1987)

27.8 A well-lagged solid copper cylinder of cross-section 5.0 cm^2 has one of its ends maintained at a constant temperature well above room temperature. Graph P shows how the temperature, θ, of the cylinder varies along part of its length when steady state conditions are reached. Graph Q shows how the temperature changes in an identical unlagged cylinder in the same conditions.

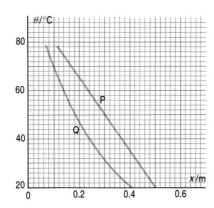

(i) Explain why the temperature gradient is constant along the well lagged cylinder. Draw a diagram, suitably labelled, showing the lines of heat flow in this cylinder.
(ii) Use graph P to calculate the temperature gradient along the well lagged cylinder. Assuming the thermal conductivity of copper to be 390 W m^{-1} K^{-1}, calculate the rate of heat flow along the cylinder.
(iii) Describe how the temperature gradient changes along the unlagged cylinder by referring to graph Q. Account for the change.
Where, along the unlagged cylinder, is the temperature gradient equal to that along the lagged cylinder? (11)

(ULSEB 1986)

27.9 Explain what is meant by a *temperature gradient*. Write down an equation relating the rate of flow of heat through a thin slice of a solid to the temperature gradient across it. State the meaning of any other symbol which appears in the equation. (2)
Explain what is meant by a *potential gradient* in electricity. Write down an equation relating the current in a conductor to the potential gradient across it. State the meaning of any other symbol which appears in the equation. (2)
These equations are similar in form. Comparing them, which thermal quantity is analogous to electrical potential? Which electrical quantity is analogous to heat? (2)
In experiments to measure thermal conductivity, it is necessary to wait for some time to ensure steady conditions prevail. Why is this? (1)
A theory suggests that the thermal conductivity λ and the electrical conductivity σ of metals at temperature T are related by

$$\lambda = \frac{\pi^2 \, k^2 \, \sigma \, T}{3 \, e^2}$$

where e is the electron charge and k is the Boltzmann constant. The table give values of λ, σ and σT for a number of metals at temperature T.

Metal	T/K	λ/W m^{-1} K^{-1}	$\sigma/10^6$ Ω^{-1} m^{-1}	$\sigma T/10^9$ K Ω^{-1} m^{-1}
lead	80	40	21	1.7
sodium	273	135	21	5.7
rhodium	273	152	23	6.4
tin	373	60	6	2.2
gold	373	310	35	13.1
silver	373	417	47	17.5
aluminium	573	226	17	9.7
copper	573	380	28	16.0

(a) Plot a graph of λ against σT. Draw the best straight line through the points and find its gradient. (4)
(b) Discuss the extent to which your graph supports the theory. (3)
If measurements of thermal and electrical conductivity are made on a given metal over a range of temperature, it is generally found that both quantities decrease with increasing temperature. Making reference to the conduction mechanisms involved, suggest an explanation. (3)

(UCLES 1987)

Chapter **28**

CHARGED PARTICLES

LEARNING OBJECTIVES

At the end of this chapter you should be able to:

1. understand the significance of Thomson's and Millikan's experiments in finding the elementary charge;

2. calculate the paths of beams of charged particles in both electric and magnetic fields;

3. find the specific charge of an electron or ion;

4. use and describe the principles of a cathode ray oscilloscope;

5. describe a mass spectrometer.

28.1 ELECTRONS

During the Nineteenth Century, sources of electrical energy became more powerful and reliable. The development of dynamos and induction coils in place of batteries meant that higher voltages could be used. At the same time, vacuum pumps using electric motors were becoming more efficient. By combining these two developments, the conduction of electricity through gases could be investigated. It was discovered that if a high voltage was applied across a gas at low pressure then the gas glowed with a colour characteristic of the gas at that pressure. The glowing gas conducts electricity and indicates that within the gas must be some charged particles. It is now known that any light being emitted from a gas shows that the gas is ionised. This phenomenon is used in many modern examples of lighting technology. Street lights frequently use sodium vapour discharge lamps which emit a characteristic orange glow. Higher pressure sodium vapour lamps are now increasingly being used because they produce additional colours besides orange and this gives a more pleasing, whiter light. These lamps have high efficiency of conversion of electrical energy into light energy in comparison with ordinary filament lamps. A common fluorescent tube is also an example of a discharge tube. This contains a mixture of argon and mercury vapour. Examples of emission of light from ionised gases are shown in the photographs of Fig 28.1 and Fig 28.2.

If the pressure in a discharge tube is reduced it is found that even though little visible light comes from the gas itself a glow can still arise from the glass of the tube, and that the position of any glow occurring can be affected by the presence of a magnetic field. It is also found that there is a small electric current through the tube. These findings suggest that a flow of charged particles through the tube is the cause of the current and the glow, and that the particles come from the material of the cathode itself. It was soon realised that the properties of these particles were the same no matter what metal was used for the cathode, so the particles seem to be present in all metals. These beams were originally called cathode rays and are now known to be beams of electrons.

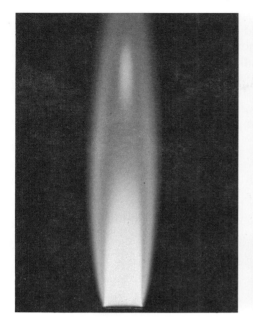

Fig 28.1 If a gas is visible then ionisation is occurring as in this propane gas flame from a bunsen burner.

Fig 28.2 The atmosphere glows when it is bombarded by charged particles from the Sun. The particles collect in the strong magnetic field near the North Pole causing the *aurora borealis* or northern lights. This picture was taken in Alaska. The exposure was several seconds. Long enough to expose the film with starlight but not long enough to notice the rotation of the Earth. Unusually northern lights were visible in the south of England during the winter of 1989.

INVESTIGATION

crocodile clips

heat and electrical insulators

3 mm

E.H.T power pack

μA

Fig 28.3

The Resistance of a bunsen flame

Use two steel rod probes about 3 mm thick and place them in retort stands so that they are about 3 mm apart, Fig 28.3. Use crocodile clips on the probes to connect them to an E.H.T. d.c. power supply through a 100 μA ammeter. E.H.T. stands for Extra High Tension. There are several E.H.T. power units available commercially capable of supplying a voltage of a few kilovolts. For safety, all of them have very high value resistors in series with their output and they often have built-in voltmeters. Some preliminary experimentation will probably be necessary to obtain a suitable potential difference across the probes for the range of meter being used. 2000 V has been found to be suitable when using a 100 μA ammeter.

No current will flow when the bunsen flame is not present. By supporting the bunsen at different heights, find the current through the flame at different places within and around the flame. Use the value of the potential difference given on the E.H.T. meter or scale to calculate the resistance of the flame. The values obtained for resistance will not be very accurate because of the internal resistance in the power supply; however the variation of resistance from place to place can be found. Draw a plan of the flame and mark on it the resistance of the flame at different places. Another plan can be drawn showing conductance at different places within the flame. Conductance is dependent on ion concentration.

28.2 THE SPECIFIC CHARGE OF AN ELECTRON

At the end of the Nineteenth Century, several fundamental experiments were done. One of these was performed by J.J. Thomson in Cambridge using cathode rays. He deflected a beam of cathode rays in an evacuated tube and was able to find from his work the ratio of the charge to the mass for an individual particle in the beam of rays. The ratio of charge to mass is called the **specific charge** for any particle.

A modern television set uses the same principle in producing its picture. J.J. Thomson did not die until 1940 so he was aware before his death of the use to which his important idea had been put, though no doubt he would have been staggered to realise just how many television tubes are produced annually. A photograph of his cathode ray tube is given in Fig 28.4. Although it looks very different from a modern cathode ray tube, it does have electrodes and a deflecting system which are in several respects very similar to the modern tube. However, the practical application of the cathode ray tube was not the only important feature of his experiment.

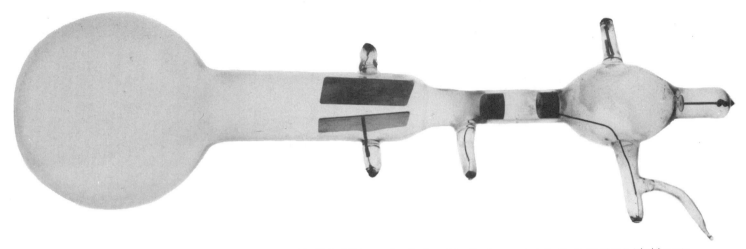

Fig 28.4   J.J Thomson's cathode ray tube. The forerunner of all cathode ray tubes in television sets.

To see the fundamental significance of Thomson's work it is first necessary to examine the theory of his experiment. If an electron is accelerated in a vacuum through a potential difference V then its loss in potential energy is equal to its gain in kinetic energy. Since the potential difference is the potential energy per unit charge we get

$$Ve = \tfrac{1}{2}mv^2$$

where e is the charge on the electron, m is the mass of the electron and v is the electron's final velocity if it is assumed to start from rest. Note that the final velocity depends on the potential difference between the plates but not on how far apart the plates are placed. The equation can be re-arranged to give the speed of the electron as

$$v = \sqrt{\frac{2Ve}{m}}$$

The principle of the experiment is illustrated by Fig 28.5. Electrons are emitted from the cathode and are accelerated in a vacuum towards an anode which is at a potential $+V$ with respect to the cathode. Here most of the electrons collide with the anode but some pass through a small hole in the anode and proceed through another anode at the same potential into the right-hand part of the apparatus which is also all at potential V. In the absence of any field being applied, the electrons coast on in a straight line to the end of the tube where they create a small spot of fluorescence on the glass. If however, as shown, a magnetic field is applied over a region of the tube, then the electrons have a force Bev acting on them in a direction at right angles to their path as shown in Fig 28.5. This causes them to travel in an arc of a circle and to hit the glass tube at P. (Section 18.4 explains why the force is Bev.) If the geometry of the magnetic field, tube and deflection is used to find r, the radius of curvature of the path, then we can apply Newton's second law to the curved part of the electron path.

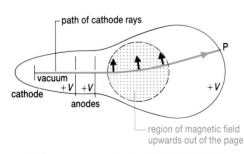

Fig 28.5   Cathode rays are deflected in a circular path when they travel through a uniform magnetic field.

CHARGED PARTICLES

$$\text{force} = \text{mass} \times \text{acceleration}$$

$$Bev = m \times \frac{v^2}{r}$$

This equation can be combined with the equation for v found above. It is impossible to determine v, e and m from these two equations as, although we can measure B, r and V this still gives only two equations for three unknowns. What can be found though is the specific charge e/m by eliminating v.

Since $$Be = \frac{mv}{r} \qquad v = \frac{Ber}{m}$$

But v^2 is given by $$v^2 = \frac{2Ve}{m}$$

so $$\left(\frac{Ber}{m}\right)^2 = \frac{2Ve}{m}$$

or $$\frac{B^2e^2r^2}{m^2} = \frac{2Ve}{m}$$

cancelling $$\frac{B^2e\,r^2}{m} = 2V$$

$$\frac{e}{m} = \frac{2V}{B^2r^2}$$

When J.J. Thomson substituted his readings into the equation, he found that the value he obtained for e/m was over 1800 times greater than for any previously known particle or substance. The value of the specific charge of an electron is now known accurately to be

$$(1.758\ 803 \pm 0.000\ 003) \times 10^{11} \text{ C kg}^{-1}$$

This is a huge figure when compared with the value of 9.65×10^7 C kg^{-1} for hydrogen ions: a value known to Thomson. The implications of this number were not lost on him. He deduced that either there existed a particle of about the mass of a hydrogen ion but with 1800 times the charge or that this was a particle with a negative charge of about the same size as the positive charge on the hydrogen ion but with a mass only one-eighteen hundredth that of the hydrogen ion. Since at that time atoms were **defined** as the smallest possible particles, and the atom of hydrogen was the smallest atom, you should be able to see how revolutionary was the suggestion which Thomson made in 1897 that he had discovered a sub-atomic particle with mass 1/1800 of the mass of the hydrogen atom. This discovery, of what is now called the electron, less than 100 years ago, not only revolutionised academic ideas on matter but has led to the entire electronics industry and much more besides. It is the high value of the specific charge on the electron which makes it so useful because it is so easily manipulated with electric and magnetic fields. It can be given enormous accelerations using only modest voltages. Oscillation frequencies for electrons of more than 1 000 000 000 Hz are commonplace in radio aerials.

While J.J. Thomson was certain that he had discovered a sub-atomic particle, final proof of its existence depended on being able to find the charge and mass separately rather than as a ratio. It took another 15 years before that was achieved.

28.3 THE ELEMENTARY CHARGE

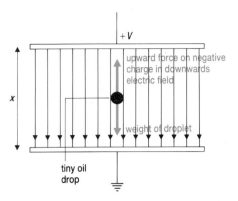

Fig 28.6 The principle of Millikan's experiment to measure the fundamental charge.

One important feature of Thomson's work was that the nature of the gas used did not affect the value for e/m obtained. Until the time of Thomson it had been assumed that there were about 90 different elements and that the atoms of each of these elements were different. Thomson's experiment implied that all these atoms contained within them identical 'cathodic corpuscles' – as Thomson called electrons. Once Thomson had shown that atoms were not the smallest particles of matter, the hunt was on for other sub-atomic particles. That hunt is still going on today.

If cathode rays consist of sub-atomic particles of known fixed specific charge, then finding the charge on one of these particles enables the mass of the particle to be found. The Millikan oil-drop experiment enabled this to be done. The apparatus used was not complicated and many schools and colleges have copies of the original apparatus which you may be able to use. Essentially the apparatus consists of a pair of parallel metal plates across which a potential difference may be applied. In the space between the plates a negatively charged oil droplet is held so that an upward force on it due to the electric field is equal and opposite to its weight. Fig 28.6 shows this in principle.

If the potential difference between the plates is V and their separation is x then the electric field E in the space between the plates is given by

$$E = \frac{V}{x}$$

Since electric field is defined as the force per unit charge, the force F which the field exerts on a charge q is given by

$$F = qE = q\,\frac{V}{x}$$

This is an upward force if q is negative.

The weight of the droplet is $mg = \frac{4}{3}\pi r^3 \rho g$ where r is the radius of the droplet, ρ the density of the oil and g the earth's gravitational field (that is, the gravitational force per unit mass).

This gives

$$\frac{qV}{x} = \tfrac{4}{3}\pi r^3 \rho g$$

$$q = \frac{4\pi r^3 \rho g x}{3V}$$

but for V positive q must be negative. Hence q can be calculated if r, ρ, g, x and V can be measured.

This concludes the principle of establishing the charge on the oil drop but there are some practical problems. A large oil drop cannot be used because its weight will be much greater than any possible upward electrical force on the elementary charge. For this reason an atomiser is used and this produces a cloud of suitable tiny oil droplets. These however are difficult to see unless they are strongly illuminated and are viewed through a microscope. It is then impossible to measure their radius directly because they are so small. A suitable practical arrangement is shown in Fig 28.7(a). The plates are separated by about 5 mm and V is typically about 300 V. Once the microscope is focused on the illuminated space under the hole through which oil droplets can enter, and the p.d. is applied across the plates, a squeeze on the atomiser will cause a shower of droplets to be seen, as shown in Fig 28.7(b). The vast majority of these droplets will be uncharged and will drift downwards through the air with their terminal velocity. A few can be seen to be moving upwards. These are the ones which have somehow acquired a negative charge and will therefore have an upward electrical force on them. V is adjusted to keep a particular drop stationary and its value is measured. This drop is then moved in a controlled way by

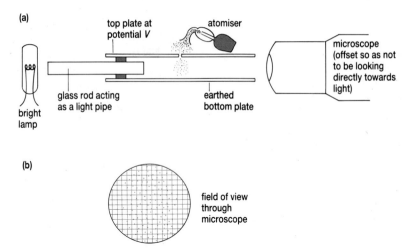

Fig 28.7 Apparatus for Millikan's experiment to measure the fundamental charge.

manipulating V until it is at the top of the graduated scale. From this point the electric field is switched off and the time t taken for the drop to fall across the scale a measured distance y is found. The terminal velocity v, of this drop is given by

$$v = \frac{y}{t}$$

When falling with its terminal velocity, it is not accelerating so the viscous force of the air upwards equals its weight. The viscous force is given by an equation, known as Stokes' equation, with which you are probably unfamiliar. It is

$$\text{viscous force} = 6\pi\eta rv$$

where η is called the coefficient of viscosity of air and has the value 1.8×10^{-5} N s m^{-2} at 20 °C.

This gives

$$\begin{aligned} \text{weight} &= \text{viscous force} \\ \tfrac{4}{3}\pi r^3 \rho g &= 6\pi\eta rv \end{aligned}$$

Cancelling and reorganising this equation gives

$$2\,r^2\rho g \;=\; 9\eta v \;=\; 9\eta\,\frac{y}{t}$$

$$r^2 \;=\; \frac{\eta y}{2\rho g t}$$

$$r \;=\; \sqrt{\frac{9\eta y}{2\rho g t}}$$

This value of r can then be used to find q by substitution into the original equation.

Further practical points to note about this experiment are:
- The microscope inverts the image so everything falling is seen to be rising.
- The microscope can most easily be focused on the correct region of the space between the plates if bright illumination is used and if it is initially focused on a thin wire placed in the hole.

- Particles can go sideways as a result of draughts and they move out of the field of view of the microscope. If a particle is lost this way, that reading has to be abandoned. The light pipe is used to minimise the heating effect of the lamp to try to reduce unwanted movement of the air inside the apparatus.
- This experimental work was started using water droplets but they evaporated too much so Millikan changed to using oil. He also took account of the buoyancy effect of the air but this is only a 0.15% correction and would be unlikely to be one of your larger uncertainties.
- Millikan repeated his readings on a particular drop after irradiating it with a burst of X-rays. This can change the charge of the drop, so different values of q are obtained for the same value of r.

The experiment needs to be repeated many times to show what is meant by the term quantisation of charge. Typical results from the readings with a few droplets are given in the table below, which were taken using oil of density 970 kg m^{-3}, $x = 5$ mm $= 0.005$ m and $y = 0.0025$ m.

V/V	t/s	r /10^{-7}m	q /10^{-19}C
288	60.0	6.29	1.54
240	65.2	6.03	1.64
200	73.6	5.68	1.64
120	109.2	4.66	1.51
265	38.4	7.86	3.28
210	70.0	5.82	1.68
336	51.2	6.81	1.68
370	50.4	6.86	1.56
273	60.0	6.29	1.63
48	92.0	5.08	4.88
285	37.6	7.94	3.14
212	71.3	5.77	1.62

As can be seen from the table of values, the charge on a droplet is not constant but is frequently around 1.6×10^{-19} C. When it does not have this value it has, in this case, values of either twice or three times 1.6×10^{-19} C. If an average is taken therefore, we must assume two charges for the 5th and 11th reading and three charges for the 10th reading. The average is therefore the total of the last column divided by 16 (not 12).

$$\frac{25.80}{16} = 1.61$$

This gives the charge on an electron as -1.61×10^{-19} C though the spread of results here, and the 2 sig. figs value used for the viscosity of air, does not permit 3 significant figures to be quoted. These results therefore give

$$\text{electronic charge} = -1.6 \times 10^{-19} \text{C}$$

The fundamental unit of charge e is taken as the same, but positive, value, namely 1.6×10^{-19}C, and is now known to be the charge on the proton. A more accurate value is 1.602×10^{-19} C.

This experiment has by now been performed accurately many times and using this method no-one has found a charge less than this fundamental unit of charge. This is the meaning of the term quantisation. Charge is not infinitely divisible. It is apparently not possible to have a free charge less than the fundamental charge and all charges are whole number multiples of this charge.

Putting this value for the electron's charge as -1.602×10^{-19} C together with its specific charge, e/m, of -1.759×10^{11} C kg^{-1}, gives the mass of the electron as

$$\frac{1.602 \times 10^{-19} \text{ C}}{1.759 \times 10^{11} \text{ C kg}^{-1}} = 9.11 \times 10^{-31} \text{ kg}$$

This is $\dfrac{1}{1837}$ of the mass of the hydrogen atom.

Example 1

Find the kinetic energy an electron acquires if it falls through a potential difference of 1 V, and its charge is -1.60×10^{-19} C.

This quantity of energy is called an **electron-volt**. It can be found by recalling that

$$1 \text{ volt} = 1 \text{ joule per coulomb} = 1 \text{ J C}^{-1}$$

Therefore for a charge of -1.60×10^{-19} C the kinetic energy acquired as the potential difference rises by one volt

$$= 1 \text{ J C}^{-1} \times 1.60 \times 10^{-19} \text{ C} = 1.60 \times 10^{-19} \text{ J}$$

$$
\begin{aligned}
1 \text{ electron-volt} = \quad 1 \text{ eV} &= 1.60 \times 10^{-19} \text{ J} \\
1 \text{ MeV} &= 1.60 \times 10^{-13} \text{ J}
\end{aligned}
$$

Example 2

A beam of electrons travelling at 1.35×10^7 m s^{-1} enters a uniform electric field between two plates of length l, separated by a distance d. One plate is at a potential of +50 V and the other is at −50 V as shown in Fig 28.8. If $l =$ 0.060 m and $d =$ 0.020 m, find θ, the angular deflection of the beam.

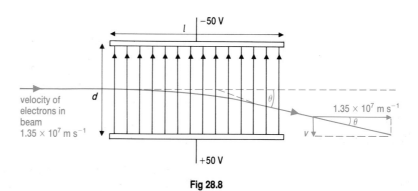

Fig 28.8

The electric field, E, between the plates is the potential gradient

$$E = \frac{100 \text{ V}}{0.02 \text{ m}} = 5000 \text{ V m}^{-1}$$

The field is in a direction towards the top of the page so an electron, which has a negative charge, will have a force on it acting downwards towards the bottom of the page.

$$
\begin{aligned}
\text{Force downwards} &= Ee \\
&= 5000 \text{ V m}^{-1} \times 1.60 \times 10^{-19} \text{ C} \\
&= 8.0 \times 10^{-16} \text{ N}
\end{aligned}
$$

Since this force is always downwards there is no horizontal acceleration and the horizontal velocity is therefore constant. This gives the time taken to travel between the plates as

$$\frac{distance}{velocity} = \frac{0.060 \text{ m}}{1.35 \times 10^7 \text{ m s}^{-1}} = 4.44 \times 10^{-9} \text{ s}$$

While between the plates the downward acceleration is given by

$$\frac{force \ downwards}{mass} = \frac{8.0 \times 10^{-16} \text{ N}}{9.11 \times 10^{-31} \text{ kg}}$$

$$= 8.78 \times 10^{14} \text{ m s}^{-2}$$

(Note how huge this acceleration is. This is possible because of the very small mass of the electron.)

For the downward direction it is now possible to use $v = u + at$ to get

$$v = 0 + (8.78 \times 10^{14} \text{ m s}^{-2} \times 4.44 \times 10^{-9} \text{ s})$$
$$= 3.90 \times 10^6 \text{ m s}^{-1}$$

The forward velocity is unchanged so using the vector diagram shown in Fig 28.8 gives

$$\tan \theta = \frac{v}{1.35 \times 10^7 \text{ m s}^{-1}} = \frac{3.90 \times 10^6 \text{ m s}^{-1}}{1.35 \times 10^7 \text{ m s}^{-1}}$$

$$= 0.289$$
$$\theta = 16° \text{ to 2 sig. figs}$$

In questions similar to this example the normal laws of electricity and mechanics can be applied, provided speeds do not become so large that relativistic effects become important. You will need to pay particular attention to

- keeping all quantities in SI units – distances in metres not centimetres; energies in joules not in electron-volts, etc.
- powers of 10 – remember that 10^7 is put into a calculator as 1 exp 7 and not as 10 exp 7.
- checking answers – think at each stage whether the figure you have found is reasonable; velocities cannot be larger than the speed of light, $3 \times 10^8 \text{ m s}^{-1}$, accelerations are likely to be high, masses to be tiny, charges to be tiny. If you find a number like $m = 3.23 \times 10^{26}$ kg or $q = 1.6 \times 10^{19}$, you have probably omitted the minus sign on a power of 10.
- working from first principles and not from a formula.

QUESTIONS

28.1 What kinetic energy does an electron acquire if it leaves a cathode at zero potential and travels through a vacuum to an anode at +200 V? With what speed does it reach the anode?

28.2 A uniform electric field exists in the space between two uniformly charged parallel plates a distance of 0.014 m apart. An electron travels at right angles to both plates from rest at the negative plate and takes 3.7×10^{-9} s to reach the positive plate. Calculate:
  (a) the acceleration of the electron;
  (b) the force on the electron;
  (c) the electric field;
  (d) the potential difference between the plates;
  (e) the velocity of the electron when it reaches the positive plate.

28.3 An electron beam is generated in a magnetic field of flux density 0.88 mT in a direction at right angles to the field. The beam forms a circular path of radius 6.5 cm. Find the speed of the electrons.

CHARGED PARTICLES

28.4 THE CATHODE RAY TUBE

In section 28.2 Thomson's work on cathode rays was discussed and it was stated that the rays existed in a gas at low pressure when a high potential difference is applied across the tube. This is not the only way of producing electron beams. If a plentiful supply of electrons is required, they can be obtained more readily if the cathode is heated. Emission of electrons from a hot surface is called thermionic emission. It is the process used to produce all the necessary electron beams in a television set. Some materials emit electrons more readily than others and many metals are near their melting points before an appreciable flow of electrons from the surface takes place. Oxides of strontium and barium emit electrons readily at temperatures around 1000 °C and these are used in television sets where they are heated by special heater elements placed in contact with them.

Once a beam of these highly charged, extremely mobile and controllable particles has been produced in a vacuum, it can be directed very readily to any required target. This principle is used not only in the cathode ray tubes of oscilloscopes and televisions but also in electron microscopes and X-ray tubes. Fig 28.9 shows a beam of electrons passing through an evacuated tube until some of them hit a maltese cross. The remaining electrons cause a shadow of the cross to be seen.

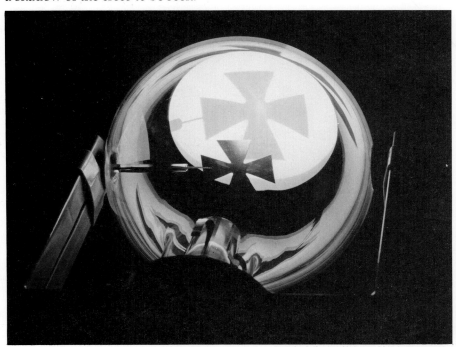

Fig 28.9 In an early classic demonstration of cathode rays, a maltese cross was placed in their path. A clear shadow of the cross was produced, showing that the cathode rays travelled in straight lines in the absence of electric and magnetic fields. When one of these fields is present the shadow is displaced.

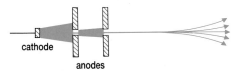

Fig 28.10 A beam of electrons tends to spread out because electrons repel one another.

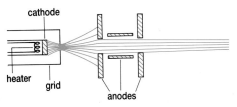

Fig 28.11 An electron gun.

The control of movement of the electrons is sometimes done using magnetic fields and sometimes using electric fields. In most applications it is first necessary to produce a focused beam of electrons. This itself poses problems because electrons repel one another strongly, so a beam which starts off as a fine beam tends to broaden as shown in Fig 28.10. This shows electrons from a heated cathode passing through two small holes lined up on two anodes. The electrons pass through the holes but then spread out because of mutual repulsion.

The focusing is done using an electron gun which consists of a heated cathode, a grid which is a nearly closed cylinder surrounding the cathode and a series of several focusing and accelerating anodes. Fig 28.11 shows three anodes, two of which are circular plates with holes in their centre and one which is cylindrical. The shape of the electric field in the region of the anodes has the effect of deflecting those electrons which are not going straight along the axis of the electron gun. An assembly like this is behaving as an electronic lens and focusing the cathode rays. The amount of focusing is controlled by adjusting the potential of an anode. A high degree of precision is possible nowadays in electron gun technology, so a narrow beam of electrons can be directed at a point less than $\frac{1}{10}$ mm wide on the screen. This is useful for high resolution graphics.

Control of the rate at which electrons pass through the tube is achieved using the grid. The grid is at a potential slightly less than that of the cathode. If it is made more negative it reduces the number of electrons which can pass it to reach the anodes. Making it less negative allows more electrons to pass, so it controls the electron current. This in turn controls the brightness at any point on the screen.

The commercial production of cathode ray tubes is big business and many of them are colour tubes with three electron guns, one for each of the three primary colours, red, green and blue. There are commercial pressures which lead to there being a big choice of style and characteristics. Most television tubes use magnetic deflection. For oscilloscopes, however, electric deflection is more usual because the current requirement for the deflection system is much smaller. A sketch of a typical cathode ray tube in an oscilloscope is shown in Fig 28.12. The electrical connections within the tube and supports to keep anodes and deflection assembly in position are not shown.

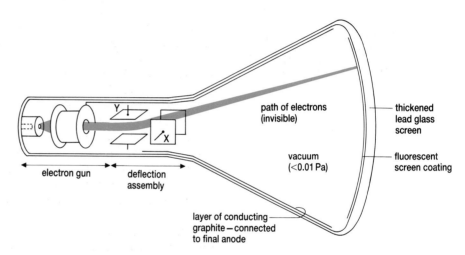

Fig 28.12   A cathode ray tube with electrostatic deflection.

28.5 THE CATHODE RAY OSCILLOSCOPE

The cathode ray oscilloscope has been referred to in Chapter 11 where it was suggested as a versatile instrument for measuring potential difference and time. It was seen to be particularly useful for examining potential differences which vary regularly and rapidly. These are the type of potential differences which occur, for example, as the output from a microphone or within electronic circuits. The cathode ray oscilloscope is able to display on its screen a steady pattern showing how the potential difference is changing. Because of the stability of the pattern on the screen, measurements can be made on it. Often the accuracy of these measurements is not particularly good because the focusing of the trace is poor and the square grid usually placed in front of the screen is rather coarse. However, if more accurate measurements are required, better quality oscilloscopes can be purchased.

Time measurement with an oscilloscope can be done very accurately if the instrument itself is accurately calibrated. No instrument can be better than its calibration.

CHARGED PARTICLES

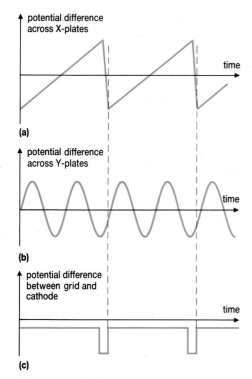

(a) potential difference across X-plates

time

(b) potential difference across Y-plates

time

(c) potential difference between grid and cathode

time

Fig 28.13 The time base on X-plates is synchronised with the signal on the Y-plates. The flyback is suppressed by switching the beam off during flyback.

The basic principle of the cathode ray oscilloscope involves using a cathode ray tube in which the X-plates are used with a sawtooth potential difference called a time base. Its purpose is to move the spot from left to right across the screen at a controllable, constant speed of known value, and then to return the spot to the left-hand side of the screen suddenly, the flyback. Calibrated speeds can vary on different oscilloscopes from around 0.1 cm s^{-1} to 10^6 cm s^{-1}. The potential difference required to cause this spot movement is generated internally within the oscilloscope and is shown in Fig 28.13(a). This voltage causes no vertical movement of the spot. A negative value at the start means that the spot is on the left-hand side of the screen. As the voltage rises to zero, the spot moves uniformly to the centre of the screen and then continues to the right until flyback occurs. An accurate straight line is needed for this increasing voltage if measurements are to be made.

It is usual to suppress the spot while flyback occurs. This is done by applying the voltage shown in Fig 28.13(c). The grid is made more than usually negative so that there is no electron beam through the tube during flyback.

The voltages shown as (a) and (c) on Fig 28.13 produce, by themselves, on the screen of the oscilloscope, what appears to be a succession of spots appearing on the left-hand side of the screen and moving across the screen before disappearing on the right-hand side.

You are recommended to set up an oscilloscope to see this if you are unfamiliar with its operation. No external input is required. Adjust the spot for focus and for brightness, and switch the time base so that the spot sweeps across the screen at a slow rate. If you then increase the time base speed, you will notice that at high speed the moving spot appears as a horizontal line. This is usually due to persistence of vision. Your eye is no longer able to follow an individual spot. Persistence of vision is essential for watching television. It would be most unpleasant if the horizontal motion of the spot could be seen when a television picture is formed. The same time base principle is used for picture formation. The spot moves across the screen 625 times per picture, each sweep being a little lower down the picture than the previous one. Since 25 complete pictures are displayed each second, the time base on a television set needs a frequency of 625×25 s^{-1} = 15 625 Hz.

With an oscilloscope, the voltage variation which it is desired to examine, the signal, is applied to the Y-plates. If the signal is sinusoidal, as shown in Fig 28.13(b), then as the spot moves across the screen it will also be moved up and down to give an appearance on the screen as shown in Fig 28.14. Note that nearly two cycles are displayed, but before the end of the second cycle is reached the flyback has commenced so that the spot is ready to begin its next sweep across the screen. For the pattern on the screen to appear stationary, it is essential that each sweep of the time base begins at the same point on the signal pattern. This is achieved by the signal in Fig 28.13 having a frequency exactly twice the time base frequency.

Keeping a fixed pattern on the screen of the oscilloscope is awkward to do manually because frequencies often drift slightly. To overcome this problem, synchronisation of the time base and signal frequencies is achieved by introducing a pause after flyback. The pause keeps the sweep from starting until a particular Y-plate voltage is reached as it increases. This enables different frequencies to be used on the X- and Y-plates yet for the trace on the screen still to be stationary.

Different oscilloscopes from different manufacturers have different additional facilities. Ranges of time base frequencies are different, so are the outputs and inputs which can be used. Some oscilloscopes are double beam, some have long persistence screens. All oscilloscopes have amplifiers of variable gain connected to the Y-input. This enables signals as small

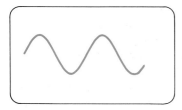

Fig 28.14

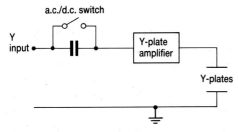

Fig 28.15 The a.c./d.c. switch enables d.c. to be passed to the Y-plates or not, as required.

as a few microvolts or as large as several hundreds of volts to be displayed on the screen.

In addition, most oscilloscopes have the option of putting the signal in through a capacitor as shown in Fig 28.15. If this is done, then any d.c. input is blocked and only a.c. signals can be displayed. It often happens that a signal has an a.c. and a d.c. component, as shown in Fig 28.16. By opening the a.c./d.c. switch to the a.c. position, only the a.c. component of the signal is displayed on the screen and a larger gain can then be used.

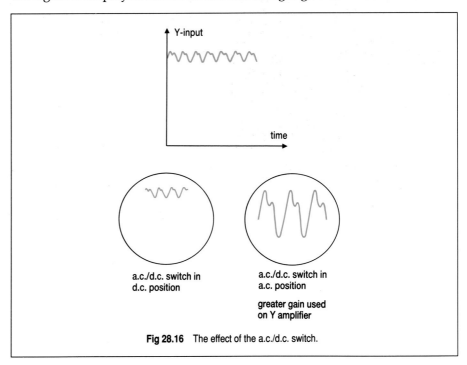

Fig 28.16 The effect of the a.c./d.c. switch.

QUESTIONS

28.4 This can be answered either by reasoning or, preferably, by using an oscilloscope and signal generator and making the required changes.

Draw the appearance of the screen of an oscilloscope when the following changes are made, one at a time, to the potential differences applied to its plates. Assume that you start with potential differences which vary in the way shown in Fig 28.13.

(a) The peak value of the potential difference across the Y-plates is doubled.

(b) The frequency applied to the Y-plates is doubled.

(c) The frequency applied to the Y-plates is halved.

(d) The frequency of the potential difference applied to the Y-plates is reduced to $\frac{1}{8}$ of the value shown.

(e) The frequency of the time base is doubled.

(f) The frequency of the time base is halved.

(g) The amplitude of the time base is halved.

28.5 Sketch the trace seen on an oscilloscope screen when an a.c. input of 2.0 V r.m.s. 50 Hz is applied. The gain on the Y-plates is set at 1 V cm^{-1} and the time base is set to 10 ms cm^{-1}. How will the pattern alter if the time base control is changed to 100 ms cm^{-1}?

28.6 The current through a cathode ray tube is 4.8 μA and the electrons are accelerated by a potential difference of 2000 V. Calculate:

(a) the speed of the electrons through the tube;

(b) the number of electrons passing a point in the tube per second;

(c) the number of electrons per metre of beam.

CHARGED PARTICLES

Ship's radar

A very popular display for ships' radar uses a large screen cathode ray tube as shown in Fig 28.17. The screen is coated with a phosphorescent material so that it glows for a few seconds after it has been bombarded by electrons. In use, a trace on the screen rotates in a way similar to a hand on a watch. One end is fixed at the centre of the screen and the other end rotates in a circle every two seconds. The trace rotates with the same angular velocity as the radar transmitter/receiver aerial on the ship's mast. The voltages applied to the X- and Y-plates to achieve this rotation are shown in Fig 28.18 in which it has been assumed that 16 traces are made

Fig 28.17 A sweep radar screen in use. Reflections of radar waves from targets are shown as dots on the screen.

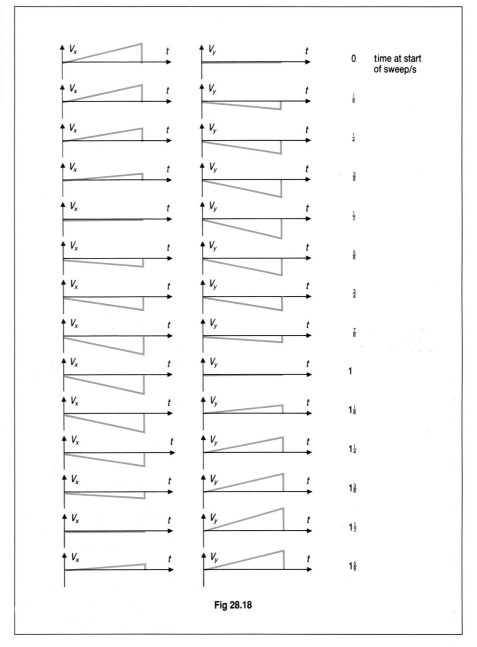

Fig 28.18

per revolution. In practice it is many more, so the trace seems to rotate smoothly, not in 16 jerks per revolution. At the beginning of each radial trace, a short burst of high frequency radio waves is emitted from the aerial. If any ship, or the coast, is close enough, a reflected echo can be received by the aerial before the end of that trace. The echo is amplified and applied to the grid of the cathode ray tube to make the trace

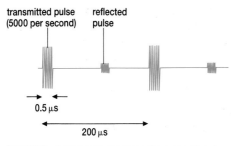

Fig 28.19 A radar system sends out a succession of strong pulses of short wavelength, electromagnetic waves and receives weak reflected echoes.

particularly bright at that point. The brightness lasts until the trace sweeps round on its next revolution. The trace itself has to be visible but is not so bright that its afterglow lasts too long.

The data needed to answer the following questions are given below for a particular radar and is illustrated in Fig 28.19.

Frequency of radio waves	7.23 GHz
Duration of a pulse	0.5 μs
Radius of cathode ray tube screen	10 cm
Speed of radio waves	3.0×10^8 m s^{-1}
Pulse repetition frequency	5000 Hz

(This is the frequency with which pulses are transmitted. It is therefore the number of radial traces in one second, $\frac{1}{2}$ a revolution.)

(a) What is the wavelength of the radio waves used?
(b) How many oscillations take place in each $\frac{1}{2}$ μs pulse?
(c) What is the period of a wave of frequency 5000 Hz?
(d) How far can a wave travel in 200 μs?
(e) What is the maximum range of this radar?
(f) What speed is needed for the spot as it moves outwards along each radius? (Ignore the flyback time.)
(g) A boat is at a distance of 15 km from the radar. What will be the distance from the centre of the screen of its echo?
(h) If an echo is at a distance of 1 cm from the centre of the screen, how far is the ship causing the echo away from the radar?
(i) What is the minimum distance for an object to provide an echo separate from the emitted pulse? What screen distance does this correspond to?
(j) Explain how the graphs of Fig 28.18 give a rotating trace.
(k) Explain the effect on the screen display of the ship carrying the radar changing direction.
(l) How would the display be changed if the sensitivity of the cathode ray tube is altered so that the spot moves at 5000 m s^{-1}? How can this help to overcome a difficulty which part (i) highlights?

28.6 THE MASS SPECTROMETER

Positive rays

In the search for sub-atomic particles which started with J.J. Thomson's discovery of the electron, it was quickly realised that if interesting particles were emitted from the cathode of a discharge tube it was sensible to look for positively charged particles which may be emitted from the anode. J.J. Thomson himself did this but came up against two particular difficulties. In the first place any particles emitted had far smaller charge to mass ratio and so could not be deflected so easily and, secondly, there was a huge variety of different particles. Whereas cathode rays consisted of one type of particle whatever gas or electrodes were used, any particles emitted from the anode did depend very much on the gas within the tube. It was therefore some years before the charge to mass ratios for these positive rays, as they were called could be measured to any great accuracy. Aston, a student of Thomson's, developed a piece of apparatus called a mass spectrometer by 1921 which could be used to determine charge to mass ratios to a high degree of precision.

The mass spectrometer

Modern mass spectrometers are capable of making charge to mass ratios to a high degree of accuracy. Data concerning the mass of atoms are almost always now determined using mass spectrometers.

A diagram of the Bainbridge mass spectrometer is shown in Fig 28.20. It consists of a positive ion source, which may be an ionised gas or a heated crystal of the substance under investigation, a collimator, a velocity

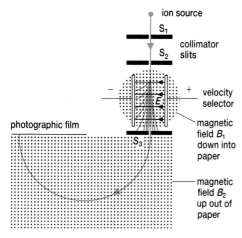

Fig 28.20 The principle of the Bainbridge mass spectrometer.

Fig 28.21 The appearance of a piece of developed film from a mass spectrometer.

selector, a deflecting chamber and a photographic film detector. The collimator is a pair of slits S_1 and S_2 which allow only a narrow beam of positive ions into the space between the plates. In this region exist both strong electric and magnetic fields, arranged so that they exert opposite forces on positive ions.

If an ion, of charge $+q$, is travelling with speed v in this region then

the force exerted on it by the electric field $= qE$,
the force exerted on it by the magnetic field $= B_1qv$

where E and B_1 are the electric and magnetic field strengths respectively.

For a certain velocity, the forces are not only opposite but also equal in magnitude. Charged particles of this velocity are therefore not deflected. This occurs when

$$Eq = B_1qv \quad \text{so } v = \frac{E}{B_1}$$

The charged particles then pass through slit S_3 into a region of uniform magnetic field strength B_2. In this field the force acting on them is B_2qv at right angles to their direction of travel. This causes them to move in a circle of radius r and since

$$\text{force} = \text{mass} \times \text{acceleration}$$

$$B_2qv = m \times \frac{v^2}{r}$$

where m is the mass of the particle. This gives

$$B_2q = \frac{mv}{r} = \frac{m}{r} \times \frac{E}{B_1}$$

so $$\frac{q}{m} = \frac{E}{rB_1B_2}$$

This equation shows that the radius of the path of a particle depends on its charge to mass ratio, and Fig 28.21 shows how a photographic film appears when placed in the path of the particles. The darkest lines indicate large numbers of particles of a particular charge to mass ratio.

Example 3

Work on electrolysis, the conduction of electricity through liquids, shows that a hydrogen ion has a charge to mass ratio of $9.65 \times 10^7 \text{C kg}^{-1}$. A certain mass spectrometer of the type described has $B_1 = 0.93$ T, $B_2 = 0.61$ T and $E = 3.7 \times 10^6 \text{V m}^{-1}$. What will be the radius of the paths of the following ions in the mass spectrometer?

$$H^+, H_2^+, He^+, He^{2+}.$$

For H^+ $$\frac{q}{m} = 9.65 \times 10^7 \text{ C kg}^{-1}$$

Using $$\frac{q}{m} = \frac{E}{rB_1B_2} \text{ gives}$$

$$9.65 \times 10^7 \text{ C kg}^{-1} = \frac{(3.7 \times 10^6 \text{ V m}^{-1})}{r \times (0.93 \text{ T}) \times (0.61 \text{ T})}$$

$$r = \frac{3.7 \times 10^6}{0.93 \times 0.61 \times 9.65 \times 10^7} \text{ m}$$

$$r = 0.0676 \text{ m}$$

An H_2^+ ion is a molecule of hydrogen which has lost one electron. It therefore has the same charge as an H^+ ion but twice the mass. Its charge to mass ratio is 4.82×10^7 C kg^{-1} and this gives $r = 0.135$ m

Continuing in the same way gives

He$^+$:   Charge = +1, Mass = 4, $\dfrac{\text{Charge}}{\text{Mass}} = 2.41 \times 10^7$ C kg^{-1}, radius = 0.270 m

He^{2+}:   Charge = +2, Mass = 4, $\dfrac{\text{Charge}}{\text{Mass}} = 4.82 \times 10^7$ C kg^{-1}, radius = 0.135 m

Note that H^+ and He^{2+} would not be separated.

Example 3 has assumed that the masses of the particles are known so that the radii of their paths in the field B_2 can be found. In practice, the mass spectrometer is used to find the masses by measuring the radii of their paths. The geometry of the apparatus and the position of the lines on the photographic film enable the radius of the orbit to be found.

Before mass spectrometers were used, the relative atomic mass of atoms had been obtained by chemical or electrolytic methods and many atoms were known to have relative atomic masses which were approximately whole numbers. Chlorine, however, seemed an exception with its relative atomic mass of 35.5. If high enough energy is used with a chlorine source in a mass spectrometer, it is possible to create Cl$^+$ ions and to pass them through the instrument. When this is done, no line is found which corresponds to a relative atomic mass of 35.5. Instead a line corresponding to a relative atomic mass of 35 is found together with another line, of one third the intensity, corresponding to a relative atomic mass of 37. It was seen, therefore, that chlorine consists of two types of atom of relative atomic masses 35 and 37. The weighted average of these masses is 35.5. More detail about isotopes, as they are called, will be given in Chapter 29.

QUESTIONS	
	28.7 What is the speed of a positive ion which is undeflected when passing through a region in which the magnetic field strength is 370 mT and the electric field is 8.1×10^4 V m^{-1}. How must the fields be arranged to make zero deflection possible?
	28.8 In a Bainbridge mass spectrometer, the electric field in the velocity selector is 1.19×10^5 V m^{-1} and the magnetic flux density over the whole of the path followed by the ions is 0.53 T. The radii of the paths of several different types of singly charged ions of oxygen were found to be 7.08, 7.53 and 7.97 cm. What is the mass of each of the oxygen ions? The charge on each ion is 1.60×10^{-19} C. What is the relative atomic mass of each of these ions?
	Another line was found corresponding to a path having a radius of 14.16 cm. What ion caused this line?

- Millikan's experiment:

  For oil drop moving at terminal velocity $\quad \frac{4}{3}\pi r^3 \rho g = 6\pi\eta rv \quad$ (to find r)

  For oil drop stationary in electric field $\quad \frac{4}{3}\pi r^3 \rho g = Ee$

  Hence, using r find e. It is found that e has discrete values – that is, charge is quantised.

	Electric field strength E	Magnetic field strength B
stationary charge	Eq	0
charge moving at velocity v	Eq in the direction of the electric field	Bqv at right angles to both velocity and field

- A charge q accelerated through a potential difference V in a vacuum gains kinetic energy equal to the work done on it so, $qV = \frac{1}{2}mv^2$.
- This equation enables the speed of electrons to be found. Deflection experiments enable the speed and the specific charge (q/m) to be found for particles.
- Cathode ray oscilloscope: Y-plate sensitivity is quoted in volts per centimetre, X-plates have time base quoted in milliseconds per centimetre.
- Mass spectrometer: used to find the charge to mass ratio for ions by using deflection techniques.

EXAMINATION QUESTIONS

28.9 A particle of mass m and charge q is accelerated from rest through a potential difference V. Write down an expression for its final speed v in terms of m, q and V.
Find the ratio of the final speeds of a proton and an alpha particle if both are accelerated from rest through the same potential difference. (4)
(UCLES 1983)

28.10 A beam of electrons travelling with speed 1.2×10^7 m s^{-1} in an evacuated tube is made to move in a circular path of radius 0.048 m by a uniform magnetic field of flux density $B = 1.4$ mT.
(a) Calculate, in electron volts, the kinetic energy of an electron in the beam.
(The charge on an electron = -1.6×10^{-19} C and 1 eV = 1.6×10^{-19} J.)
(b) A similar technique is used to accelerate protons to very high speeds. Protons with energies of 500 GeV can be held by magnetic fields in circular orbits of radius 2 km. Suggest why such a large radius orbit is necessary for high energy protons. (7)
(ULSEB 1988)

28.11 The diagram shows a type of cathode ray tube containing a small quantity of gas. Electrons from a hot cathode emerge from a small hole in a conical shaped anode, and the path subsequently followed is made visible by the gas in the tube.

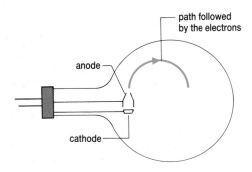

(a) The accelerating voltage is 5.0 kV. Calculate the speed of the electrons as they emerge from the anode.
(b) The apparatus is situated in a uniform magnetic field acting into the plane of the diagram. Explain why the path followed by the beam is circular. Calculate the radius of the circular path for a flux density of 2.0×10^{-3} T.

(c) Suggest a possible process by which the gas in the tube might make the path of the beam visible.
(Specific charge of an electron = 1.8×10^{11} C kg^{-1}.) (9)
(AEB 1987)

28.12 (a) Describe an experiment to determine the charge to mass ratio of electrons, indicating clearly the measurements made. How is the value of e/m calculated from them? What information from the experiment indicates that electrons are *negatively* charged? (7)
(b) A beam of electrons, travelling with a velocity v in the x-direction, enters at point O ($x = 0$, $y = 0$) a region of uniform electric field provided by applying a voltage V between plates A and B, separated by a distance d in the y-direction. The electrons are deflected towards A as shown in the diagram, the point P (x, y) being a point along the trajectory.

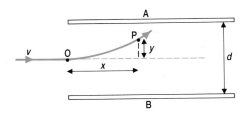

(i) Is the potential of plate A positive or negative with respect to B?
(ii) In terms of the distance x, calculate the angle between the x-direction and the electron beam at P.
(iii) Prove that the path is parabolic, namely $y = ax^2$, and find the value of a. (8)
The position of the electron source is moved so that the direction of the incoming beam at point O is now at an angle θ to the x-direction towards plate B. The initial speed is unchanged.
(iv) Find the distance L along the x-axis at which the beam again has $y = 0$.
(v) Explain how this effect could be used as a basis for a velocity selector for electrons. (5)
(O & C 1984)

28.13 (a) When a small oil drop falls from rest in air, it is acted on by the force of gravity and a resistive force opposing its motion. (Buoyancy may be neglected.)

(i) State how the resistive force depends on the radius and the velocity of the drop.

(ii) Write down the equation for the acceleration of the drop and use it to give a qualitative explanation of how the drop eventually reaches a terminal velocity. (5)

(b) If the oil drop moves between two horizontal conducting plates, an electric field can be applied producing an additional force on the drop if it is electrically charged. Outline an experimental method by which it is possible to determine the charge on the drop using an arrangement of this kind. What further observations would be necessary to determine the charge of the electron? Do *not* give a detailed mathematical analysis of how to calculate the charge. (7)

(c) In an experiment such as the one you have described in (b), explain why

(i) it is important that the conducting plates are horizontal,

(ii) the oil drops should be small,

(iii) the chamber containing the drops is sometimes irradiated with X-rays or gamma rays. (4)

(d) When an electric field of strength E_1 is applied between the two plates, a charged drop moves upwards with a constant velocity, v_1. When the direction of this field is reversed, the terminal velocity of the same drop is v_2 downwards. The drop can be held stationary between the plates by a field of strength E_2. Show that these velocities are related by the equation

$$\frac{v_1}{v_2} = \frac{E_1 - E_2}{E_1 + E_2}$$

Assume that the charge and the mass of the drop remain unchanged. (4)

(JMB 1987)

28.14 A cathode ray oscilloscope has its Y-sensitivity set to 10 V cm^{-1}. A sinusoidal input is suitably applied to give a steady trace with the time base running at such a setting that the trace takes 0.010 s to traverse the screen. If the trace seen has a total peak to peak height of 4.0 cm and contains 2 complete cycles, calculate the r.m.s. voltage and frequency of the input signal? (5)

(ULSEB 1986)

28.15 Explain what is meant by the *time base* of an oscilloscope. Illustrate the wave form associated with a time base and describe what happens to the electron beam in the tube when this time base is switched on. (4)

Describe how an oscilloscope may be used to measure

(i) the peak value of an alternating voltage

(ii) the frequency of an alternating voltage.

Illustrate your explanations with diagrams of what you would see on the screen. (8)

In an oscilloscope the sweep-rate is 100 μs cm^{-1} and the vertical sensitivity is 5 V cm^{-1}. A sinusoidal signal of 20 V peak to peak and frequency 5 kHz is applied to the Y input. Draw a scale diagram of the picture on the screen. (4)

Describe clearly the effect as seen on the screen if the signal remains the same but the sweep rate is increased to 50 μs cm^{-1} and the vertical sensitivity is reduced to 10 V cm^{-1}. (4)

(UCLES 1985)

28.16 (a) Draw a labelled diagram showing the basic structure of the tube of a cathode ray oscilloscope. Explain

(i) how the electron beam is produced

(ii) how the intensity of the beam is varied

(iii) the purpose of the time base. (10)

(b) The diagram below represents an oscilloscope trace for a full-wave-rectified, sinusoidally alternating voltage. The Y-sensitivity is set at 5 V cm^{-1}.

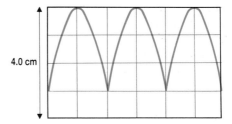

Estimate the mean value of the rectified voltage and calculate its r.m.s. value. (3)

(c) A beam of electrons travelling at 1.0×10^7 m s^{-1} enters the region between the Y-plates of an oscilloscope as shown below. A potential difference of 50 V is applied between the plates.

Calculate each of the following quantities assuming edge effects to be negligible.

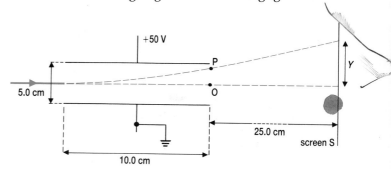

(i) The time a moving electron spends in the region between the plates.

(ii) The vertical acceleration experienced by each electron.

(iii) The vertical displacement OP of the electron beam as it leaves the region between the plates.

(iv) The value of Y, the vertical displacement of the electron beam on the screen S, if the distance between the end of the plates and the screen is 25.0 cm.

(Ratio of charge to mass for an electron = 1.80×10^{11} C kg^{-1}.)   (8)

(AEB 1986)

28.17 (a) The components of a mass spectrometer include an ion source, a velocity selector, and a deflection system. Describe a velocity selector that will select ions of one velocity only from ions with a range of velocities. Derive an expression for the velocity v of those selected.   (7)

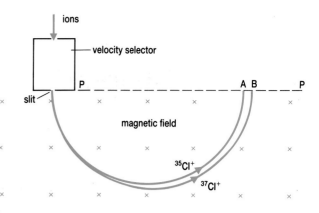

(b) The diagram shows a stream of singly charged positive ions of chlorine isotopes $^{35}Cl^+$ and $^{37}Cl^+$ emerging from a slit into a vacuum at a uniform speed of 500 m s^{-1}. They are then deflected by an extensive magnetic field of uniform flux density 2.0 mT, as shown.

(Take the molar mass of $^{35}Cl^+$ to be 0.035 kg and that of $^{37}Cl^+$ to be 0.037 kg, the electronic charge e to be 1.6×10^{-19} C, and the Avogadro constant N_A to be 6.0×10^{23} mol^{-1}.)

Calculate the linear separation AB of the paths of the $^{35}Cl^+$ and $^{37}Cl^+$ ions as they pass through the plane PP after traversing complete semicircles.   (7)

(OXFORD 1987)

CHARGED PARTICLES

Chapter 29

ATOMIC STRUCTURE

> **LEARNING OBJECTIVES**
>
> At the end of this chapter you should be able to:
>
> 1. use the term nuclear atom correctly;
> 2. use nuclear equations;
> 3. appreciate the random nature of radioactivity;
> 4. understand the term mass defect;
> 5. describe the processes of fission and fusion;
> 6. describe several uses of radioactive nuclides.

29.1 ATOMIC PARTICLES

The discovery of the electron as a sub-atomic particle by J.J. Thomson in 1897 led many people to question what else existed within an atom. It was known that an atom was neutral, so if it was possible for a negatively charged particle to be ejected from the atom, then there must be some positive charge left behind. Thomson himself carried out a long series of experiments on the positive particles which are present in a discharge tube. He came to the conclusion that whereas many of the negative particles present in such a tube are all the same and are independent of the nature of the electrodes and the gas in the tube, there is a huge variety of different positive particles and these do depend on the nature of the gas present. He incorrectly concluded that the positive charge remaining on the atom was distributed evenly within the atom, together with any electrons not ejected.

In 1896, just before the discovery of the electron, Becquerel had discovered the phenomenon which came to be called radioactivity. He found that a photographic plate placed under some crystals containing uranium became exposed even though no light had reached it. Further study of the phenomenon by such people as Marie Curie and Rutherford led to the discovery of more radioactive materials and the isolation of three types of naturally occurring radioactive emissions called alpha, beta and gamma by Rutherford in 1899. More detail on radioactivity is given later in this chapter. It was soon realised by Rutherford that alpha radiation was capable of creating intense ionisation of the air through which it passes owing to its charge and large kinetic energy. It can be detected by allowing it to collide with a fluorescent screen such as one coated with zinc sulphide. On impact with the screen a tiny flash of light, called a scintillation, is produced which can be seen by the eye if a magnifying lens is used. Alpha radiation can therefore be useful in investigating the structure of the atom. By 1908 Rutherford had established that alpha radiation was a stream of positively charged particles identical to helium atoms which have lost two electrons. He did this by collecting a vast number of alpha particles in a discharge tube and then observing the light radiated from the tube. The spectrum he found contained traces of helium, although none had been present before the experiment started. Question 29.1 illustrates how even with strong sources of alpha particles, only a very small mass of helium is produced.

29.1 A radioactive source is emitting 8 000 000 alpha particles per second and the particles are emitted for 50 days. What mass of helium gas is produced and what will be its volume at stp? Assume values for the Avogadro constant and the molar gas constant. The mass of a helium atom is 6.7×10^{-27} kg.

29.2 An alpha particle travels a distance of 7 cm through air and causes the conductivity of the air through which it passes to change. Measurements on the conductivity indicate that the alpha particle causes 50 000 ion pairs per cm along its track. Find the initial speed of the alpha particle if it requires 6 electron volts to produce a single ion pair. The mass of a helium atom is 6.7×10^{-27} kg; 1 eV = 1.6×10^{-19} J.

Beta rays were also found to be high energy particles. They have a negative charge and can be deflected to a much greater extent by a magnetic field than alpha particles. Their charge to mass ratio can be determined from their deflection in magnetic and electric fields, and is found to be identical to that for the electron. These particles are emitted from the *nucleus* of an atom during radioactive decay and are electrons. Beta particles have much less mass than alpha particles and therefore cause much less ionisation. This also means that they can penetrate greater thicknesses of materials than can alpha particles.

Gamma rays are uncharged so cannot be deflected by electric or magnetic fields. They do not interact with matter very strongly and so can penetrate considerable distances through matter before being absorbed. Gamma rays are not particles. They are electromagnetic radiation similar to X-rays of short wavelength.

29.2 THE NUCLEAR ATOM

Experiments done by Marie Curie in 1899 on the amount of ionisation caused by alpha particles gave values for the speed of alpha particles, in the way shown by question 29.2. It was these high speed particles which were used by Rutherford to penetrate the atom. He was using the alpha particle as a bullet to break open an atom and hoped that he could identify any particle which emerged. In no way did he expect what he and his research students, Geiger and Marsden, actually found. The experiment which he suggested that Geiger and Marsden perform was to take a thin layer of massive atoms in a vacuum and to fire alpha particles at them. A thin layer was necessary so that there was less chance of repeated collisions and the hope was that an atom could be burst open. In practice, the thin

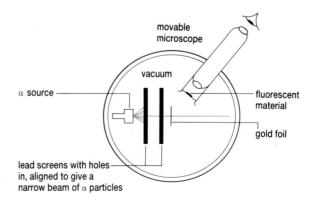

Fig 29.1 An experimental arrangement to detect alpha particle deflections.

layer was a gold foil only a few hundred atoms thick and the alpha particles usually went straight through. Occasionally, however, an alpha particle was deflected through a large angle or even rebounded from a gold atom. The apparatus is shown diagrammatically in Fig 29.1 and the pattern of alpha particle scattering is shown in Fig 29.2 in which thousands of undeflected alpha particles have not been drawn. The rigorous mathematics of why this occasional scattering was so surprising is too complex to be dealt with here but the following example gives some idea of the problem facing Rutherford.

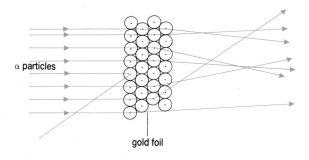

Fig 29.2 Alpha particle deflections by the atoms in a gold foil.

Example 1

An alpha particle of mass 6.7×10^{-27} kg and speed 1.0×10^7 m s^{-1} is to be stopped in a distance equal to the radius of a gold atom, 1.3×10^{-10} m. Find the deceleration of the alpha particle, assumed constant, and the force necessary to cause this deceleration. What electrostatic force exists between the positive charge in a gold atom and an alpha particle when separated by a distance equal to the radius of the gold atom? (The total positive charge within a gold atom is 79 times the charge on a hydrogen ion: the total positive charge within a helium atom is twice the charge on a hydrogen ion.)

Using the equation relating initial velocity u, final velocity v, distance travelled s and constant acceleration a gives

$$v^2 = u^2 + 2as$$
$$0 = (1.0 \times 10^7 \text{ m s}^{-1})^2 + 2a\,(1.3 \times 10^{-10} \text{ m})$$
$$-2a\,(1.3 \times 10^{-10} \text{ m}) = (1.0 \times 10^{14} \text{ m}^2 \text{ s}^{-2})$$

$$a = \frac{-1.0 \times 10^{14}}{2.6 \times 10^{-10}} \text{ m s}^{-2}$$

$$a = -3.8 \times 10^{23} \text{ m s}^{-2}$$

Force necessary to cause this deceleration = ma
$$= 6.7 \times 10^{-27} \text{ kg} \times 3.85 \times 10^{23} \text{ m s}^{-2}$$
$$= 0.0026 \text{ N}$$

The deceleration of the alpha particle is enormous. Also a force of 0.0026 N is extremely large when the mass of the particle it is acting on is considered.

The positive charge on a gold atom is $79 \times 1.6 \times 10^{-19}$ C.

The positive charge on an alpha particle is $2 \times 1.6 \times 10^{-19}$ C so when they are 1.3×10^{-10} m apart the electrostatic force of repulsion

$$= \frac{q_1 \, q_2}{4\pi\varepsilon_0 r^2}$$

$$= \frac{(79 \times 1.6 \times 10^{-19} \text{ C}) \times (2 \times 1.6 \times 10^{-19} \text{ C})}{4\pi \times (8.85 \times 10^{-12} \text{ F m}^{-1}) \times (1.3 \times 10^{-10} \text{ m})^2}$$

$$= 2.2 \times 10^{-6} \text{ N}$$

When the size of this force is compared with the size of the force necessary to cause the alpha particle to rebound, it is seen that the two have completely different orders of magnitude. The value of the electrostatic force is too small by a factor of about 1000. It was the large discrepancy between the two figures which startled Rutherford. There was no way in which his conception of the atom could agree with the results of Geiger and Marsden's experiment. He thought they had been mistaken in taking their readings. Having established that the results were genuine, a complete re-think on the internal structure of the atom was needed. There was no doubt that the force of repulsion had to be at least as large as the figure of 0.0026 N calculated in example 1. This implies that it is the electrostatic force which is too small. The only way in which this force can become larger is for the distance between the two charges to be smaller. This is illustrated in Fig 29.3 where it is shown that if the mass and the positive charge on the gold atom are concentrated into a smaller and smaller particle, then the alpha particle can approach closer to the charge

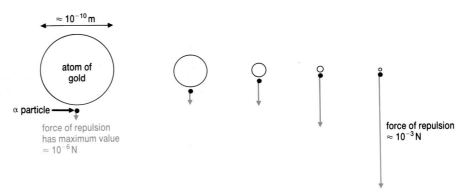

Fig 29.3 A force large enough to deflect alpha particles appreciably as they pass through a gold foil can only be supplied if the alpha particle is very close to all the positive charge in an atom. The diagram shows how the force increases as the volume occupied by the positive charge decreases.

and hence a much larger force of repulsion can be achieved. It was the gold foil experiment, therefore, which resulted in the idea of a nucleus within an atom that contained most of the mass of the atom and all of its positive charge.

More detailed analysis of the results of this experiment shows that the diameter of the atom is of the order of 10 000 times the diameter of the nucleus. The nucleus of an atom occupies a very small fraction of the atom's total volume. This has the effect of making the density of nuclear matter extremely high. Its value was given in Table 22.1 as being of the order of 10^{16} kg m^{-3}. The density of all nuclear matter is approximately constant so a nucleus of twice the diameter of a second nucleus will have 2^3, that is 8, times its mass. The electrons orbit the nucleus but because of their very small mass they do not appreciably deflect alpha particles. Alpha particles do knock electrons out of atoms. This is why alpha particles cause a great deal of ionisation when they travel through air.

ATOMIC STRUCTURE

29.3 The nucleus of uranium has a radius of 9×10^{-15} m. Find the radius of a nucleus (beryllium) which has a mass of $1/27$ of the mass of the uranium nucleus.

The nuclear model of the atom with its orbiting electrons was used by Bohr in 1913 as a basis of his theory of the emission of line spectra. More detail will be given on line spectra in Chapter 30.

Having established that all the positive charge in an atom resides in its nucleus, there still remains the question of what particles exist within the nucleus. Moseley, also in 1913, was working on the frequency of X-rays from many elements and he found the proof that within the atom there is a quantity which increases in regular steps from one atom to the next. He was convinced that this fundamental quantity was the positive charge on the nucleus. His work led directly to the idea that positive charge is quantised and that the number of these positive charges within the nucleus of an atom determine the nature of the atom. Until the work of Moseley, atomic masses had been considered to be the critical quantity to determine the nature of a substance, but from this time onwards the atomic number was seen to be the factor of prime importance. This provided strong support for the periodic table of Mendeleev. The order in which the elements were placed by Mendeleev was in atomic number sequence.

The **atomic number** Z of any element is the number of elementary units of positive charge on the nucleus of the atom. Each of these units of positive charge is associated with a particle called a **proton**, and the atomic number is often called the **proton number**.

Table 29.1 **The proton number is given for the first 10 elements together with the average mass of an atom. These figures have significant differences from those in table 29.2.**

Element	Atomic Number	Atomic Mass/u
hydrogen	1	1.008
helium	2	4.003
lithium	3	6.939
beryllium	4	9.012
boron	5	10.811
carbon	6	12.011
nitrogen	7	14.007
oxygen	8	15.999
fluorine	9	18.998
neon	10	20.183

Table 29.1 lists the first few elements to illustrate the point that the number of protons within an atom determines which element it is. All hydrogen atoms must have 1 proton in their nucleus. If there are two protons in the nucleus of a substance then the substance must be helium; six protons in a nucleus must be a carbon nucleus, etc. Included in the table are the atomic masses. At the beginning of the century these would have been found by chemical analysis. The pattern of numerical values for atomic masses clearly shows that there are a large number of values which are very close to being whole numbers, but the whole numbers are not equal to the atomic number.

The resolution of the problem of what particles exist within a nucleus was not solved satisfactorily until 1932 with Chadwick's discovery of a second nuclear particle, the neutron. The **neutron** is a particle of slightly greater mass than the proton, but it is uncharged. Since the mass of the proton and the mass of the neutron are very nearly equal to one another, it is not surprising that many nuclear masses are integral multiples of this mass. On first consideration it would seem sensible to regard the mass of the proton as the fundamental unit of mass. Indeed this may become the standard of mass at some time in the future, but at present there are practical difficulties in relating this to the mass of large scale bodies. A small unit of mass is needed, however, and it is called the **atomic mass unit, u**. 1 u is one twelfth of the mass of an atom of carbon which contains six protons, six neutrons and six electrons. The conversion factor between the atomic mass unit and the kilogram is found experimentally to be

$$1 \text{ u} = 1.660\ 566 \times 10^{-27} \text{ kg}$$

In these units the masses of the proton, neutron and electron are

$$m_{\text{p}} = 1.007\ 276 \text{ u}$$
$$m_{\text{n}} = 1.008\ 665 \text{ u}$$
$$m_{\text{e}} = 0.000\ 549 \text{ u}$$

As nuclear particles, both protons and neutrons are called **nucleons**. **The number of protons plus neutrons in the nucleus of an atom is called the mass number A of the atom.** The mass number is sometimes called the **nucleon number**. In symbol form the structure of an atom is given by

Mass number (nucleon number) A

                                            Symbol

Atomic number (proton number) Z
for example $^{65}_{29}\text{Cu}$

The example gives the information that the atom of copper (Cu) contains 29 protons. It must therefore contain 29 electrons if it is a neutral atom. Since this particular copper atom contains 65 nucleons altogether and 29 of them are protons, then the other 36 nucleons must be neutrons. An atom of a particular nuclear structure is called a **nuclide**. Fig 29.4 shows diagrammatically the structure of 5 different nuclides. Atoms of three different hydrogen nuclides are shown together with an atom of helium and an atom of bismuth. The bismuth nuclide is $^{209}_{83}\text{Bi}$ and is the largest stable nuclide. Diagrams such as Fig 29.4 are useful for showing certain features of atomic structure, but in the first place they are only two-dimensional

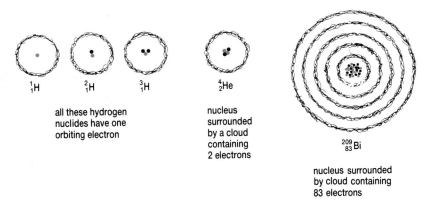

$^{1}_{1}\text{H}$ $^{2}_{1}\text{H}$ $^{3}_{1}\text{H}$ $^{4}_{2}\text{He}$ $^{209}_{83}\text{Bi}$

all these hydrogen nuclides have one orbiting electron

nucleus surrounded by a cloud containing 2 electrons

nucleus surrounded by cloud containing 83 electrons

Fig 29.4 Diagrammatic representation of some atoms from the smallest $^{1}_{1}\text{H}$ to the largest $^{209}_{83}\text{Bi}$ stable atoms.

diagrams of a three-dimensional object, secondly it is impossible to draw them to scale since the nucleus is so small, and thirdly the representation of an electron, even when shown as a cloud of charge surrounding the nucleus, is only a small part of the whole story. Think of the figure purely as a convenient diagram and not as a photograph of an atom.

29.3 ISOTOPES

As described in Chapter 28, work by Thomson on the positive particles present in discharge tubes found that the atoms of each element were not necessarily all the same. At the time he could not explain why this was, but the development of the mass spectrometer showed, for example, that three-quarters of the atoms of chlorine had a mass of about 35 u, whereas the other quarter had a mass of 37 u. This gives the average mass of a chlorine atom as 35.5 u, a figure which had been something of a mystery to chemists for years, since they were unable to explain why, with so many atoms having an atomic mass which is nearly a whole number, chlorine should have an atomic mass definitely not a whole number. The two different types of chlorine atoms are said to be isotopes. **Isotopes** are atoms with the same atomic number but different mass numbers. The discovery of the neutron explained that isotopes were atoms with the same number of protons but with different numbers of neutrons. Modern mass spectrometer determinations enable the mass of about 2500 different nuclides to be measured. Most of these nuclides are unstable but there are about 270 known stable nuclides. Each square on Fig 29.5 represents a stable nuclide.

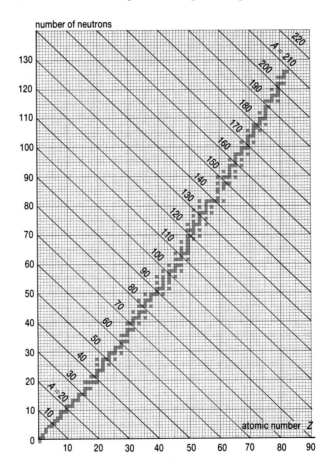

Fig 29.5 A chart showing all the known stable nuclides.

There are some elements which have a large number of isotopes. Tin, for example, has an atomic number of 50. All tin atoms therefore have 50 protons. Together with the 50 protons, stable tin nuclides may have 62, 64, 65, 66, 67, 68, 69, 70, 72 or 74 neutrons. Each of the following nuclides is therefore an isotope of tin and in tin mined on the Earth the percentage abundance of each isotope is remarkably constant whether the tin is mined in Cornwall, Bolivia, Malaysia, Zaire, Zambia, China or Nigeria. This is true for any element. Variation in the abundance of any nuclide is always small. Percentage abundances of each isotope are given.

$^{112}_{50}Sn$	$^{114}_{50}Sn$	$^{115}_{50}Sn$	$^{116}_{50}Sn$	$^{117}_{50}Sn$	$^{118}_{50}Sn$	$^{119}_{50}Sn$	$^{120}_{50}Sn$	$^{122}_{50}Sn$	$^{124}_{50}Sn$
1.0%	0.6%	0.3%	14.2%	7.6%	24.0%	8.6%	33.0%	4.7%	6.0%

Detailed information about all the stable nuclides can be found in reference books but sufficient information about the common nuclides is given in Table 29.2 to be able to see the pattern of atomic structure and for your needs here. The corresponding data for the elementary particles is also given. An asterisk * indicates that the nuclide is unstable. Percentage abundance cannot normally be given for unstable nuclides but in the case of uranium an exception is made because uranium when mined has the proportion stated. This is partly because uranium 238 decays very slowly.

Table 29.2   Giving numerical data about some nuclides

Element	Symbol	Proton number Z	Neutron number	Nucleon number A	Atomic mass / u	% Abundance
proton	p	1	0	1	1.007 28	
neutron	n	0	1	1	1.008 67	
electron	e	−1	0	0	0.000 55	
hydrogen	H	1	0	1	1.007 83	99.99
(deuterium)	(D)	1	1	2	2.014 10	0.01
helium	He	2	1	3	3.016 03	0.000 13
			2	4	4.002 60	~100
lithium	Li	3	3	6	6.015 13	7.4
			4	7	7.016 01	92.6
beryllium	Be	4	5	9	9.012 19	100
boron	B	5	5	10	10.012 94	19.6
			6	11	11.009 31	80.4
carbon	C	6	6	12	exactly 12	98.9
			7	13	13.003 35	1.1
nitrogen	N	7	7	14	14.003 07	99.6
			8	15	15.000 11	0.4
oxygen	O	8	8	16	15.994 92	99.76
			9	17	16.999 13	0.04
			10	18	17.999 16	0.20
fluorine	F	9	10	19	18.998 41	100
neon	Ne	10	10	20	19.992 44	90.9
			11	21	20.993 84	0.3
			12	22	21.991 38	8.8
sodium	Na	11	12	23	22.989 77	100
magnesium	Mg	12	12	24	23.985 04	78.7
			13	25	24.985 84	10.2
			14	26	25.982 59	11.1
aluminium	Al	13	14	27	26.981 54	100
calcium	Ca	20	20	40	39.962 59	97
			22	42	41.958 63	0.6
			23	43	42.958 78	0.1
			24	44	43.955 49	2.1
			26	46	45.953 69	0.003
			28	48	47.952 52	0.2

iron	Fe	26	28	54	53.939 62	5.8
			30	56	55.934 93	91.7
			31	57	56.935 39	2.2
			32	58	57.933 27	0.3
technetium	Tc	43			no naturally occurring nuclides	
lead	Pb	82	122	204	203.973 07	1.4
			124	206	205.974 46	25.2
			125	207	206.975 90	21.7
			126	208	207.976 64	51.7
bismuth	Bi	83	126	209	208.980 42	100
radium	Ra	88	135	223*	223.018 57	
			136	224*	224.020 22	
			138	226*	226.025 40	
			140	228*	229.031 23	
uranium	U	92	142	234*	234.040 90	trace
			143	235*	235.043 93	0.7
			146	238*	238.050 80	99.3

29.4 MASS DEFECT

Table 29.2 can be used to find the mass of any particular nuclide. It can also be used to find the total mass of the particles within that nuclide. An important discrepancy appears as is shown in example 2.

Example 2

Find the mass of the particles contained within the $^{40}_{20}\text{Ca}$ atom and compare it with the atomic mass.

An atom of $^{40}_{20}\text{Ca}$ consists of 20 protons, 20 neutrons and 20 electrons, so the total mass of these particles is

$$
\begin{aligned}
20 \times 1.007\ 28\ \text{u} &= 20.1456\ \text{u} \\
20 \times 1.008\ 67\ \text{u} &= 20.1734\ \text{u} \\
20 \times 0.000\ 55\ \text{u} &= \underline{0.0110\ \text{u}} \\
\text{Total mass} &= 40.3300\ \text{u}
\end{aligned}
$$

From Table 29.2 however the atomic mass of $^{40}_{20}\text{Ca}$ is only 39.962 59 u.

This is a startling result. The law of conservation of mass appears to be broken. The mass of the individual particles is greater than their mass when they are all fitted together into an atom. This would be totally inexplicable were it not for Einstein's theory of relativity in which he equates energy E with an equivalent amount of mass m by the equation

$$E = mc^2$$

where c is the speed of light in free space.

Using the theory of relativity we therefore have in this case

mass of whole atom + mass equivalence of the energy = mass of individual
required to separate the atom particles
into its individual particles

$$
\begin{aligned}
39.962\ 59\ \text{u} + \Delta m &= 40.330\ 00\ \text{u} \\
\Delta m = 40.330\ 00\ \text{u} - 39.962\ 59\ \text{u} &= 0.367\ 41\ \text{u}
\end{aligned}
$$

The difference between the mass of an atom and the mass of its particles taken separately is called the **mass defect**, Δm. The mass defect is small compared with the total mass of the atom. It is only when working to a large number of significant figures that it becomes apparent, but there is a real and crucial difference. The energy ΔE released when an atom is formed from its constituent particles is given by

$$\Delta E = \Delta mc^2$$

In example 2 the mass defect is 0.367 41 u and ΔE, the energy equivalence of this mass, is given by

$$\Delta E = \Delta mc^2 = (0.367\ 41 \times 1.660\ 56 \times 10^{-27}\text{kg}) \times (2.998 \times 10^8 \text{ m s}^{-1})^2$$
$$= 5.484 \times 10^{-11} \text{ J}$$

Note that the mass must be converted from a value in atomic mass units into a mass in kilograms before being substituted into the equation.

The energy of nuclear reactions is often given in electron-volts (eV) or in MeV (see section 28.3).

$$1 \text{ eV} = 1.602 \times 10^{-19} \text{ J}$$
$$1 \text{ MeV} = 1.602 \times 10^{-13} \text{ J}$$

The energy equivalence of the mass defect here is therefore

342.3 MeV

QUESTION	29.4 Show that the energy equivalence of a mass of 1 u is 932 MeV.

If the mass defect is calculated for all nuclides then it is found that, with the exception of the H nuclide, all nuclides have a mass defect. Some of these figures are given in Table 29.3.

Table 29.3 Mass defect of selected nuclides

Nuclide	Mass Defect /u	Mass Defect per nucleon /u
$^{1}_{1}\text{H}$	0	0
$^{2}_{1}\text{H}$	0.002 40	0.001 20
$^{4}_{2}\text{He}$	0.030 40	0.007 60
$^{7}_{3}\text{Li}$	0.042 16	0.006 02
$^{9}_{4}\text{Be}$	0.062 48	0.006 94
$^{11}_{5}\text{B}$	0.081 86	0.007 44
$^{12}_{6}\text{C}$	0.099 00	0.008 25
$^{14}_{7}\text{N}$	0.112 43	0.008 03
$^{16}_{8}\text{O}$	0.137 08	0.008 57
$^{24}_{12}\text{Mg}$	0.212 96	0.008 87
$^{40}_{20}\text{Ca}$	0.367 41	0.009 19
$^{56}_{26}\text{Fe}$	0.528 75	0.009 44
$^{110}_{46}\text{Pa}$	1.010 56	0.009 19
$^{208}_{82}\text{Pb}$	1.757 84	0.008 45
$^{209}_{83}\text{Bi}$	1.761 89	0.008 43
$^{238}_{92}\text{U}$	1.935 38	0.008 13

The mass defect increases with the number of particles within the nucleus, so an additional column has been added to Table 29.3 which gives the mass defect per nucleon. This is therefore giving the amount of mass defect for each of the nuclear particles, and this figure shows an interesting sharp rise at the start to a maximum for the iron nuclide followed by a gradual fall. Fig 29.6 shows the values of the mass defect per nucleon ($\Delta m/A$) plotted against mass number. The fact that the mass defect per nucleon is not the same for all nuclides has important practical applica-

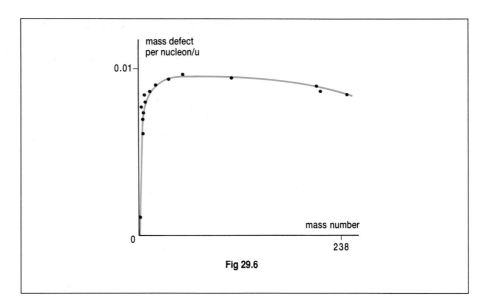

mass defect
per nucleon/u

0.01

0

mass number

238

Fig 29.6

tions. In practice, it is not possible to make atoms by taking protons, electrons and neutrons and forming them into atoms. The fact that there is a very strong repulsion between protons makes union between them and the required number of neutrons impossible at ordinary temperatures. What is possible is a rearrangement of the protons, electrons and neutrons in a particular nuclide to form different nuclides. In questions 29.5, 29.6 and 29.7 you are asked to use the data in Table 29.3 to find how the mass defect changes when certain nuclear changes take place.

QUESTIONS

29.5 What is the change in the mass defect if an $^2_1$H atom joins with a similar atom to make a single atom of the nuclide $^4_2$He? What energy is released in this process?

29.6 What is the change in the mass defect if four $^4_2$He atoms join to make a single $^{16}_8$O atom? What energy is released in this process? What energy would be reuired to separate a single $^{16}_8$O atom into four $^4_2$He atoms ?

29.7 What change in mass defect takes place if a single atom of $^{238}_{92}$U breaks up into two $^{110}_{46}$Pa atoms and several neutrons? What energy is released in this process?

Some of these changes are impossible practically, but they do illustrate the fact that if the light atoms can be joined together to make heavier atoms then the mass defect increases. This can only happen if energy is released. This process is called **fusion** and is the source of the energy of the Sun and the stars.

If the heavy atoms are broken into two roughly equal parts, then again energy is released. This process is called **fission** and is used in all nuclear power stations. On Fig 29.6 any alteration in nuclear structure which results in moving towards the maximum of the graph has the effect of releasing energy. One of the reasons why there is so much iron in the universe is that the iron atom is the atom with the greatest mass defect. Note that an atom with a high mass defect needs a great deal of work to be done on it in order to separate it into its constituent nucleons. It is therefore in a low energy state. The term **binding energy** is used for an atom to indicate how much energy would be released if it were formed from separate protons, electrons and neutrons. The binding energy is therefore the work which would need to be done to separate the atom into individual protons, electrons and neutrons. It is **not** the energy holding the atom together. In practice, the energy required to remove the electrons from an atom is extremely small compared with the energy required to separate the nucleons. For this reason, often only the nucleons are considered.

29.5 NUCLEAR REACTIONS

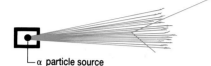

└─ α particle source

Fig 29.7 Occasionally an alpha particle travelling through air hits a nitrogen nucleus and causes a nuclear reaction. The forked track shows that this has happened.

In the previous section, various reorganisations of atomic nuclei were suggested. Some of these transformations can be carried out in practice but others cannot. It was not until 1896, when Becquerel discovered radioactivity, that any possibility existed of changing one element into another. Until that time, creating valuable elements from base elements was merely a dream of the alchemists.

Even after the discovery of radioactivity, it took several more years before the first nuclear transformation was observed in the laboratory. This took place in Rutherford's laboratory in 1919. He used a zinc sulphide screen to detect the particles, but use of a cloud chamber later confirmed his results and enabled pictures to be taken of the particles before and after the reaction. Fig 29.7 shows α (alpha) particles being produced from a source, and most of them simply causing ionisation of the nitrogen gas through which they are passing until they have lost all their energy and stop. One of the α particles, however, enters the nucleus of a nitrogen atom and causes a nuclear reaction to occur. The forked track shows that in the reaction two particles are produced. Measurement of the angles of the tracks and knowledge of the energy of the α particles enabled Rutherford to show that the long track was a proton. This gives a nuclear reaction which can be expressed in equation form:

$$\frac{4}{2}He + \frac{14}{7}N \rightarrow \frac{17}{8}O + \frac{1}{1}H$$

which is sometimes written:

$$\frac{14}{7}N \ (\alpha, p) \ \frac{17}{8}O$$

The presence of the $^{17}_{8}O$ atom cannot be detected chemically because too few atoms are formed in any particular experiment to give enough $^{17}_{8}O$ for analysis. Its presence is deduced from the fact that the equation must balance. There are 9 protons and 9 neutrons on the left-hand side of the equation, so there must be 9 protons and 9 neutrons in the right-hand side. If one proton, $^{1}_{1}H$, is emitted separately, the remaining atom must contain 8 protons and 9 neutrons and so is $^{17}_{8}O$.

Not only must the number of protons and neutrons be the same on both sides of the equation but also there must be equality of mass–energy. If rest masses of the four particles are considered we get, using data from Table 29.2

mass of He	4.002 60	mass of H	1.007 83
mass of N	14.003 07	mass of O	16.999 13
	18.005 67		18.006 96

(The mass of 9 electrons is included on both sides of the equation though in practice, the electrons will usually be lost, at least temporarily, in such a high energy reaction.)

It can be seen that the rest mass of the particles after the reaction is higher than it was before. This implies that the initial α particle must have had a high velocity. The mass associated with its kinetic energy must be at least

$$(18.00696 - 18.00567) \ u = 0.00129 \ u.$$

This is equivalent to

$$932 \; \frac{\text{MeV}}{\text{u}} \times 0.001\ 29 \text{ u}$$

$$= 1.20 \text{ MeV}$$

which is also $1.20 \times 10^6 \times 1.6 \times 10^{-19}$ J $= 1.92 \times 10^{-13}$ J.

If this is equated to the kinetic energy of the α particle, the minimum speed, v, it can have to cause the nuclear reaction is given by

$$\tfrac{1}{2}mv^2 = 1.92 \times 10^{-13} \text{ J}$$

$$v^2 = \frac{2 \times 1.92 \times 10^{-13} \text{ J}}{4 \times 1.66 \times 10^{-27} \text{ kg}} = 5.78 \times 10^{-13} \text{ m}^2 \text{ s}^{-2}$$

$$v = 7.6 \times 10^6 \text{ m s}^{-1}$$

Since the proton emerges with a high kinetic energy, this increases the necessary energy of the α particle.

A nuclear reaction such as this is said to be an **induced** nuclear reaction. It cannot take place spontaneously, as it absorbs energy. It was just such a reaction which led to the discovery of the neutron by Chadwick in 1932. He bombarded a disc of beryllium, placed in a vacuum, with α particles from a polonium source and found an unusual, uncharged radiation coming from the beryllium, Fig 29.8. If a solid material containing many hydrogen atoms was placed in the path of the unusual radiation, paraffin wax was used, then the protons in the wax get knocked on by the radiation. This made it impossible for the unknown radiation to be γ rays as they have no mass. Chadwick's deduction from the quantitative results of his experiment was that the unknown radiation consisted of uncharged particles of mass similar to that of the proton. He called these particles neutrons. The equation of the reaction of the alpha particles and the beryllium becomes

$$^4_2\text{He} + {}^9_4\text{Be} = {}^1_0\text{n} + {}^{12}_6\text{C}$$

Conservation of momentum and energy can be used to find the mass of the neutron. Question 29.8 indicates the way this was done.

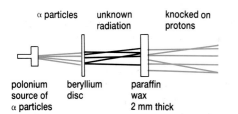

α particles   unknown   knocked on
           radiation   protons

polonium   beryllium   paraffin
source of   disc   wax
α particles   2 mm thick

Fig 29.8 The experiment which resulted in the discovery of the neutron.

QUESTION

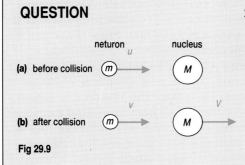

         neutron   nucleus
            u

(a) before collision m M

          v V

(b) after collision m M

Fig 29.9

29.8 A neutron of mass m travelling with velocity u hits a stationary nucleus of mass M head on (Fig 29.9). After the collision, the velocity of the neutron is v and the velocity of the nucleus is V. The collision is an elastic collision.

(a) Write down equations showing conservation of kinetic energy and conservation of momentum for the collision.

(b) Show that eliminating v from these equations to find V in terms of u gives

$$V = \frac{2\,mu}{M + m}$$

(c) Substitute each of the values below into the equation given and use the two resulting equations to eliminate u and find the mass of the neutron. Use the mass of the neutron to find its velocity before the collision.

When the nucleus is a hydrogen nucleus of mass 1 u, V is found to be 3.3×10^7 m s^{-1}. When the nucleus is a nitrogen atom of mass 14 u, V is found to be 4.4×10^6 m s^{-1}.

Until the discovery of the neutron, high energy α particles had been used to cause nuclear reactions. They were not particularly suitable particles for this however. Alpha particles have a positive charge and therefore there is strong repulsive force on them whenever they approach any other nucleus. This makes it difficult for them to penetrate the nucleus of another atom. The uncharged neutron made an ideal missile with which to penetrate the nucleus of atoms to examine their structure.

From Rutherford's first induced nuclear reaction in 1919, the demand has been for higher and higher particle energies. Neutrons are ideal particles for causing nuclear reactions but they cannot be accelerated to high speed in particle accelerators because they have no charge and so are unaffected by electric and magnetic fields. Some of the particle energies and accelerators used in the investigation of nuclear particles are given in Table 29.4.

Table 29.4

1917	8 MeV	α particles from radioactive materials
1932	2 MeV	neutrons after discovery by Chadwick
1932	2 MeV	protons in accelerator of Cockcroft and Walton
1939	32 MeV	positive ions in cyclotron of Lawrence and Livingston
1940	2.3 MeV	electrons in Betatron of Kerst
1950	900 MeV	ions in synchrotron at Brookhaven
1954	5000 MeV	ions in upper atmosphere
1976	300 000 MeV	protons in super synchrotron at CERN, Geneva
1984	800 000 MeV	protons in Fermilab synchrotron, USA
1989		Russian proton accelerator and European electron accelerators nearing completion

Nuclear reactions of considerable complexity are now known. Some important reactions in the development of nuclear physics are listed in the exercise on nuclear reactions. There, brief details on different reactions are given and you are asked to complete the equations and find the necessary energy to balance the equation. Note that any nuclear reaction which releases energy could take place spontaneously – but does not necessarily do so. A reaction which requires energy can only take place if the particle which causes it arrives with sufficient energy.

In all of these reactions, the energy required to balance the equation is measured in MeV. The corresponding figure in a chemical equation is usually of the order of a few electron volts. It is for this reason that whereas a conventional coal-burning power station may burn a million tonnes of coal a year, a nuclear power station requires only a few tonnes of fuel per year. It is worth remembering that each nuclear fission reaction releases of the order of a million times the energy which a single atomic (chemical) reaction releases.

Nuclear reactions

In each of the following reactions you are asked to complete the nuclear equation and to find the quantity of energy, E, necessary to balance the equation. Necessary data can be taken from Table 29.3.

Particle causing reaction	Equation	Notes
alpha	$_2^4\text{He} + 7.7\text{ MeV} + {}_7^{14}\text{N} \rightarrow {}_1^1\text{H} + \quad + E$	The first induced nuclear reaction.
proton	$_1^1\text{H} + {}_3^7\text{Li} \rightarrow {}_2^4\text{He} + \quad + E$	The first experiment to show the possible release of a large quantity of energy.
alpha	$_4^9\text{Be} + {}_2^4\text{He} \rightarrow {}_0^1\text{n} + \gamma + E$ (γ of 4.4 MeV)	Discovery of the neutron.
deuteron	$_1^2\text{H} + {}_1^2\text{H} \rightarrow {}_0^1\text{n} + \quad + E$	Fusion. The source of the Sun's energy.
photon	$_1^2\text{H} + \gamma \rightarrow {}_1^1\text{H} + \quad + E$ (γ of 2.6 MeV)	
neutron	$_5^{10}\text{B} + {}_0^1\text{n} \rightarrow \text{Li} + \text{He} + E$	Control of a nuclear reactor.
none	$_{92}^{238}\text{U} \rightarrow {}_{90}^{234}\text{Th} + \quad + E$	Radioactive decay.
neutron	$_{92}^{235}\text{U} + {}_0^1\text{n} \rightarrow {}_{54}\text{Xe} + {}_{38}^{90}\text{Sr} + 3\,{}_0^1\text{n} + E$	Fission. Energy of around 200 MeV is emitted.

29.6 RADIOACTIVITY

Natural radioactivity has been stated to result in 3 types of radiation, α and β particles and γ rays. It has been seen that α particles are the least penetrating of the radiations but cause the greatest amount of ionisation. Any of these radiations is hazardous and therefore unnecessary exposure to them should be avoided. It is not possible, however, to avoid radiation completely as it is part of our environment. Our own bodies are radioactive and contribute a large fraction of our dose of radiation. Our houses and gardens are radioactive; the Sun is radioactive and we get some very high energy particles from outer space. Besides these sources of radioactivity, there are additional man-made sources. Fall-out from nuclear bombs still exists in the atmosphere, nuclear power stations cause some radiation as do hospitals on account of some of the tests and treatments they use, and from their X-ray machines.

Whatever is written about radiation will find its critics. The problem is knowing how to draw a line between acceptable and unacceptable exposure. A physics textbook is not a suitable place to argue for or against any particular policy. It is nevertheless worth while stating some points about which there is almost complete agreement so that precautions may be taken both within the laboratory and outside it. There is no particular order of importance in these 10 items.

1. Any ingestion of radioactive material should be avoided. Radioactive particles lodged in the lungs, for example, are much more dangerous than if they were outside the body.
2. Hold any radioactive material well away from you and do not point sources at anyone.
3. Keep radioactive materials, when not in use, inside lead containers.
4. Reduce the time of contact to a minimum. Old watches with luminous dials are quite strongly radioactive and people owning them wear them in contact with their wrists for a long time. Usually the backs are metal and so absorb α-radiation.
5. Limit the amount of radiation you receive from X-rays and ultra-violet. There is a unit in a hospital in Brisbane, Australia, with a world-wide

reputation for treating skin cancers. They recommend that people with fair skin should not sunbathe at all, as the risk of getting skin cancer is too high.

6. Keep a gentle draught through your house. Sealing up a room so that it is virtually airtight allows radon gas, which is radioactive, to collect in the room. The radon comes from very slightly radioactive gypsum in the plaster on the walls. By itself this is no great strength, but you are likely to be in your house for a long time and a weak exposure for a long time gives rise to increased risk.

7. A distinction cannot be made between natural and artificial radiation. An 8 MeV α particle from any source is the same.

8. Some radiation treatments in hospitals are extremely successful.

9. Some natural repair takes place within the body after radiation damage has occurred, so that a small rate of exposure for a long time may cause less damage than the same dose given at a high rate of exposure over a short time.

10. There are considerable variations in the background radiation level in different parts of most countries.

29.7 DETECTION OF RADIOACTIVITY

Several detectors have already been mentioned.

A photographic plate was used to discover radioactivity and photographic film is still used for detection of radiation. Figs 29.10 and 29.11 show the effect in photographic emulsion of high energy particles. Fig 29.10 shows an aluminium nucleus colliding with a nucleus in the photographic emulsion and producing 6 alpha particles besides other fragments. In Fig 29.11 a bromine nucleus in the emulsion has been shattered.

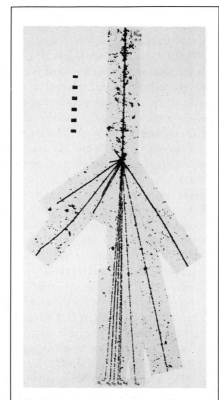

Fig 29.10 An aluminium nucleus collides at high speed with a nucleus in the photographic emulsion and produces, among other particles, a spray of 6 α particles.

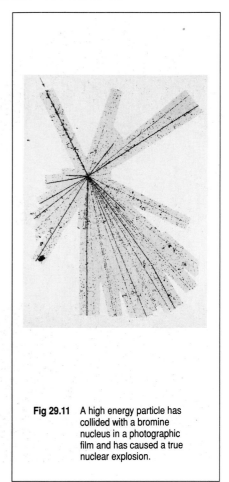

Fig 29.11 A high energy particle has collided with a bromine nucleus in a photographic film and has caused a true nuclear explosion.

ATOMIC STRUCTURE

Scintillation methods, too, have their modern equivalent. Researchers in the early part of the century spent many hours in darkened rooms counting tiny flashes of light from fluorescent screens. Now the same process is used with photomultiplier tubes in which the initial flash of light causes a few photoelectrons (see photoelectric effect, Chapter 30). These constitute a very tiny current which can be amplified to give a pulse of current. The pulses can then be counted electronically.

A cloud chamber has also been mentioned. It depends for its operation on the fact that condensation of a liquid as dew requires something to start it forming. Dew readily forms on dust, for instance, to give fog. Condensation will also take place on ions formed in the air after a radio-active particle has passed through it. A diagram of the diffusion cloud chamber is drawn in Fig 29.12. The solid carbon dioxide cools the black

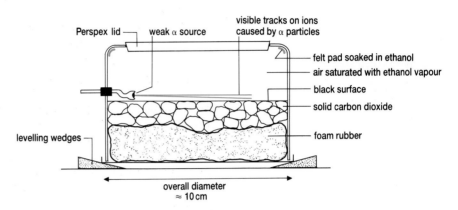

Fig 29.12   A diffusion cloud chamber.

surface to a low temperature and the ethanol vapour near it condenses on the ions caused by the passage of α particles. It is well worth while setting up this piece of apparatus if one is available, and you can get some solid carbon dioxide. You must illuminate the cloud chamber from the side in order to see the tracks, and you must be patient. It sometimes takes a few minutes after putting the lid on before the atmospheric conditions are right for the formation of the tracks. The black surface must be level. Take the Perspex lid off at some stage and you will see clouds of mist forming on the dirt in the air you have let in. Rubbing the Perspex lid creates a static charge on it and this in turn creates an electric field which enables the tracks to be seen more clearly. A photograph of these tracks is shown in Fig 29.13.

Fig 29.13   Alpha particle tracks in a cloud chamber.

Bubble chambers are a modern development of the cloud chamber. These are used with particle accelerators and they can be very large (and very expensive). They use a tankful of liquid hydrogen instead of ethanol vapour. The hydrogen is exceedingly pure and just before a burst of high energy particles enter it the pressure surrounding it is dropped so the hydrogen is above its boiling point. The ions formed by the passage of the nuclear particles act to cause centres on which the hydrogen boils. The tracks caused can be photographed. One such photograph is shown in Fig 29.14. One advantage of the bubble chamber is that, because it is at such a low temperature, very little thermal energy is possessed by any target atom so all the effects caused must be due to the bombarding nuclei.

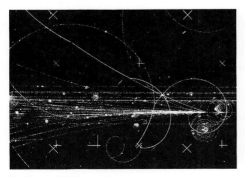

Fig 29.14   Bubble chamber photograph.

The Geiger–Müller tube is a well known radiation detector, though it is being replaced commercially by other ionisation detectors and by solid state detectors. Any detectors of ionisation can be used to detect α particles

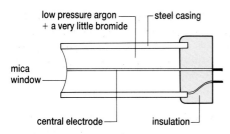

Fig 29.15 A Geiger-Müller tube.

and β particles, and γ rays if it is sensitive enough. A Geiger–Müller tube is illustrated in Fig 29.15. The central electrode is between 400 V and 600 V positive with respect to the case. When an alpha particle enters the argon gas through the mica window, it causes some ionisation and a flow of charge takes place. The bromine present has the effect of preventing this flow becoming a continuous avalanche of charge. The pulse of current caused by the flow of charge can be amplified and the pulses counted electronically. The counter, which also provides the power supply, is called a scaler. If the rate at which pulses are produced is measured by the instrument, then this is called a ratemeter.

Solid state detectors are now available which use semiconductor materials. A p–n junction is connected to a d.c. power supply so that it is not conducting. If some sufficiently energetic radiation is incident on it, charge separation occurs. Holes and electrons are produced. These cause a current to flow which can be amplified and counted.

29.8 RADIOACTIVE DECAY

Any radioactive atom decays spontaneously. What causes it to change is not known. It is known that chemical reactions, temperature and pressure make no difference whatsoever to the likelihood of it undergoing radioactive decay. This is not really surprising when it is realised these changes involve only the electrons within atoms and energies of the order of a few electron volts. Even at 6000 K (surface temperature of the Sun) the energy of thermal vibration is still only about 1 electron-volt whereas binding energies of nuclei are of the order of 10 000 000 eV. What is surprising is that some extremely stable atoms should suddenly decay. Many uranium atoms, for example, were formed with the Earth about 5000 million years ago. They have remained on the Earth ever since and have undergone all sorts of extremes of temperature, pressure and chemical change without altering at all. Yet if you have some uranium metal or uranium oxide in the laboratory, you will find that after all this time some of the atoms suddenly, for no apparent reason, give off an alpha particle and hence decay into thorium. The radioactive decay equation is:

$$^{238}_{92}\text{U} \quad \rightarrow \quad ^{4}_{2}\text{He} \quad + \quad ^{234}_{90}\text{Th} \quad + \quad 4.3 \text{ MeV}$$
$$\qquad\qquad\quad \alpha \text{ particle} \qquad\qquad\qquad\quad \text{energy mostly of emitted}$$
$$\qquad\qquad\qquad\qquad\qquad\qquad\qquad\qquad\qquad \alpha \text{ particle}$$

The only factor known to affect the rate at which α particles are produced by this nuclear reaction is the number of uranium 238 atoms present. We are forced therefore to use the laws of chance to analyse the decay pattern. There are several ways of dealing with the mathematics of this situation, so an actual example will be used.

Example 3
The iron nuclide $^{59}_{28}\text{Fe}$ is radioactive. A solution containing these radioactive iron atoms is used medically in the diagnosis of blood disorders. There is a 50–50 chance that any of the atoms will emit a β particle in a period of 46 days. What fraction of the original total remain after 184 days?

on average $\frac{1}{2}$ of the atoms decay during the first 46 days
on average $\frac{1}{2}$ of the remaining atoms decay during the second 46 days
on average $\frac{1}{2}$ of the remaining atoms decay during the third 46 days
on average $\frac{1}{2}$ of the remaining atoms decay during the fourth 46 days

This is shown in Fig 29.16.

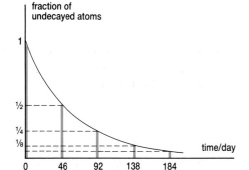

Fig 29.16 Radioactive decay, showing half life.

ATOMIC STRUCTURE

The graph is clearly not a straight line. The time it takes for half of the atoms of any nuclide to decay is called the **half life** of the nuclide. Note that a half life of 46 days does *not* mean that all of the nuclide has decayed after 2×46 days. In this case the fraction remaining after 184 days is 1/16 of the original.

QUESTIONS

29.9 Strontium 90 has a half life of 29 years. If 1 gram of strontium 90 is present in a sample at the start of a series of observations, how long will it be before there is only $\frac{1}{32}$ g of strontium 90 remaining?

29.10 Show that the number of atoms present in 0.0632 g of $^{210}_{82}Pb$ is 1.81×10^{20}. How many $^{210}_{82}Pb$ atoms will be present after 95 years? The half life of $^{210}_{82}Pb$ is 19 years.

29.11 Plot a graph to show how the number of polonium 218 atoms in a sample changes with time using the data in Table 29.5.

Table 29.5

Time /s	Number of $^{21}_{8}Po$ atoms / 10^{15}	Rate of decay / $10^{13}\ s^{-1}$
0	4.38	1.64
50	3.63	1.36
100	3.01	1.13
150	2.50	0.935
200	2.07	0.775
250	1.72	0.643
300	1.42	0.533
350	1.18	0.422
400	0.98	0.366
450	0.91	0.304
500	0.67	0.252

(a) Find the half life of polonium.

(b) Check your answer by starting from some other point on your graph other than when the time is zero.

(c) What relationship is there between the number of polonium 218 atoms present and the rate of decay of the atoms?

(d) What would be rate of decay when the number of polonium atoms present is 1.18×10^{12}?

The methods used for finding the activity of a nuclide have so far either assumed that an exact number of half lives are being used or that a graph can be plotted. Another approach uses a log graph. Consider the data from example 3.

Example 4
Use the data of Table 29.6 for the iron nuclide Fe and plot a sketch graph to show the logarithm of the fraction remaining against time. Use the sketch to find the fraction remaining after 100 days, F_{100} (lg x is the logarithm of x to base 10.)

Table 29.6

Time /day	fraction remaining (F)	lg F
0	1	0
46	$\frac{1}{2}$	−0.301
92	$\frac{1}{4}$	−0.602
138	$\frac{1}{8}$	−0.903
184	$\frac{1}{16}$	−1.204

Because each reading of F is divided by 2, the value of lg F has to have lg 2 (0.301) subtracted from it to get the next reading. The graph must therefore be a straight line. This makes it easy to interpolate to find the value of x, which is the logarithm of the fraction remaining after 100 days.

Using similar triangles

$$\frac{184 \text{ days}}{-1.204} = \frac{100 \text{ days}}{x}$$

$$x = -\frac{1.204 \times 100}{184}$$

$$x = -0.654$$
$$x = \text{lg } F_{100}$$

giving
$$F_{100} = \text{antilog }(-0.654) = 10^{-0.654}$$
$$= 0.22$$

29.12 Use the graph in Fig 29.17 to find the fraction remaining after:
(a) 80 days **(b)** 126 days **(c)** 150 days.

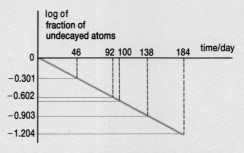

Fig 29.17 A logarithmic graph of radioactive decay.

The data in question 29.11, and the basic principle that the rate of radioactive decay depends only on the number of radioactive atoms present, shows that

the rate of disintegration ∝ number of atoms

or
$$-\frac{dN}{dt} \propto N$$

where N is the number of radioactive atoms present at time t.

This gives the equation

$$\frac{dN}{dt} = -\lambda N$$

where λ is a constant called the **decay constant**. Use of calculus is necessary to solve this equation to get

$$N = N_0 e^{-\lambda t}$$

where N_0 is the number of atoms present at time $t = 0$.

Since $\dfrac{dN}{dt} \propto N$ this also gives

$$\left(\frac{dN}{dt}\right)_t = \left(\frac{dN}{dt}\right)_0 e^{-\lambda t}$$

where dN/dt is the rate of disintegration when $t = 0$. The use of a calculator makes it easy to plot these graphs even if you were unable to see how to get the equation from the initial proportion. Question 29.13 asks you to do this.

QUESTION

29.13 Plot the graph of $N = N_0\,e^{-\lambda t}$ for the range of values $t = 0$ to $t = 10$ s when $N_0 = 6.2 \times 10^{12}$, $\lambda = 0.25$ s^{-1} What is the half life of this decay?

Calculation of the relationship between the half life, τ, and λ can be done by using the fact that if N_0 is the number of radioactive atoms when $t = 0$, $\frac{1}{2}N_0$ will be the number present when $t = \tau$. So

$$\begin{aligned}\tfrac{1}{2}N_0 &= N_0\,e^{-\lambda\tau}\\ \tfrac{1}{2} &= e^{-\lambda\tau}\end{aligned}$$

taking logarithms to base e, (ln), gives

$$\begin{aligned}-0.6931 &= -\lambda\tau\\ \lambda\tau &= 0.6931\end{aligned}$$

or $\quad \tau = \dfrac{0.6931}{\lambda}$

As expected, therefore, when τ is large the decay constant is small; only a small fraction of atoms decay per unit time. When τ is small a large fraction of the radioactive atoms decay per unit time, so λ is large.

INVESTIGATION

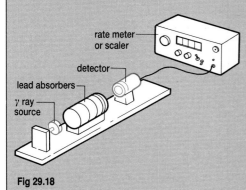

Fig 29.18

Gamma ray absorption

Many schools and colleges have a very weak cobalt 60 source of gamma radiation. Use of such a source by sixth form pupils may be allowed. You should keep the source in its lead container unless you are actually using it and you should handle it with tongs. **NEVER** point it at yourself or anyone else and, since radiation emerges in all directions from this source, position it so that you, and others in the laboratory are all more than a metre away from it. You should also pay particular attention to any special rules which are applied to you, if you are allowed to do this investigation.

The arrangement of apparatus is shown in Fig 29.18. The detector may be a Geiger–Müller tube or a solid state detector, and it must be connected to a suitable power supply and counter. The counter may be used as a ratemeter or a scaler.

Commence the investigation by finding the background count. Monitor the background count for at least 5 minutes before taking any other readings, and make sure that your cobalt 60 source is well out of the way in its lead container while you do this.

Next, put the source into position in its holder. The distance between the source and the counter must be kept fixed, so arrange that there is just sufficient space for all the lead absorbers.

Take a reading of the count rate with no lead absorber in position, and record this value. Use the background count already measured to find the count rate, C, in the detector due to the gamma source alone.

Repeat the experiment with different thicknesses, x, of lead absorbers in position. Plot a graph of the natural logarithm of the count rate against the thickness of the lead. The graph will be a straight line if the absorption is given by the exponential equation

$$C = C_0\, e^{-\mu x}$$

Use your graph to find the value of μ.

An exponential law like this implies that a certain thickness of lead reduces the intensity of the radiation to a half. Twice this thickness reduces it to a quarter, 3 times to an eighth, etc. Find the thickness of lead which will reduce the intensity of gamma radiation to a half.

What thickness of lead is needed to reduce gamma ray intensity from that which gives a count rate of 3.7×10^{10} per second to one of 120 per second?

29.9 USES OF RADIOACTIVE NUCLIDES

Radioactive nuclides are made use of in many completely different ways. In whatever way they are used, it is always necessary to take precautions so that the user is well protected from any harmful radiation.

The techniques used depend on being able to obtain a specific radionuclide. These nuclides are manufactured by specialist firms. They are produced by irradiating different materials in the intense neutron radiation of a nuclear reactor.

To make some radioactive magnesium atoms for use in the tracer experiment which is described later in this section, for example, requires that some magnesium sulphate is placed in a nuclear reactor for a time. The neutron flux penetrates the nucleus of the normal magnesium atoms and some neutrons can be absorbed, so creating radioactive magnesium atoms.

Generally speaking, materials are not made radioactive by being placed near a radioactive source of alpha, beta or gamma radiation. The lead case around a school cobalt 60 source does not become radioactive. This is because the absorption of the gamma radiation by the lead causes no change to the nuclei of the lead atoms.

Five main categories of use are given here. They are as tracers, as penetrating radiation, for medical treatment, for dating archaeological specimens and as power sources. Each use will be treated separately and the examples given are illustrations of some of the many uses now made of radioactivity.

Tracers

Use is made in tracer applications of the ability of detectors to measure extremely small concentrations of a radioactive nuclide. If, for example, a plant research centre wishes to investigate the ability of a tomato plant to take up magnesium through its roots, it is possible to water the plant with a fertiliser containing a known, extremely low concentration of radioactive magnesium 28. If a detector is placed in contact with the leaves of the tomato plant, an increase in countrate soon after watering in the fertiliser indicates that the magnesium has travelled through the plant to the leaves. This can happen within a time as short as 20 minutes.

Tracers are also used for detecting leaks in underground pipes and for measuring wear. Wear measured in a conventional way requires a long experiment until the object wearing away has lost appreciable mass or is appreciably smaller. If the object wearing away is radioactive, the minute quantities of matter worn away in the first few minutes of wear can be detected. This is a much used technique for measuring wear of machinery.

Penetration of radiation

Gamma rays can be used in place of X-rays to photograph solid objects. Fig 29.19 shows a technician placing a gamma ray source inside an aero engine. Film placed around the outside of the engine is exposed by the gamma rays and any imperfections in the metal of the engine, a crack for example, show up as a darker area on the film. This technique has the advantage that high voltage electrical sources are not needed.

Fig 29.19 A gamma ray source being positioned inside an aero engine. The film, which will be exposed by the gamma rays, can be seen on the outside of the engine. Gamma rays will penetrate any crack more readily than sound metal and will give greater film exposure.

The same technique can be used in many ways, choosing a source which is suitable for the thickness of penetration required. The thickness of rolled sheets of paper or plastic can be controlled during manufacture by placing a beta source on top of the sheet and a detector beneath it. If the sheet is too thin, then the count rate will rise. This rise in count rate can be directly coupled to the rollers so that they are separated slightly.

Medical treatment

When radiation is passed through living cells it may do no harm at all. On the other hand, it can kill living cells or cause them to undergo a mutation, that is, to change the biological function of the cell in some way. It is immature cells and cells that are growing or dividing most rapidly which are most sensitive to radiation. The cells in mature insects divide very rarely and for that reason some insects are able to survive extremely large doses of radiation.

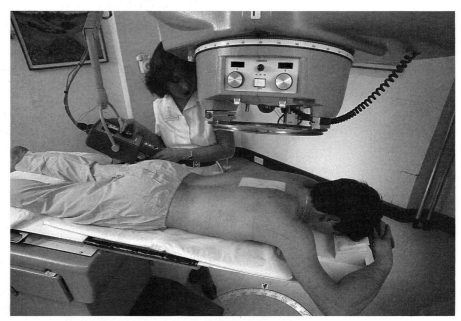

Fig 29.20   A cobalt radiotherapy unit. The cobalt source is housed in the large treatment head and gamma radiation is released through a lead shutter.

This is made use of in the treatment of cancer. Often cancer cells are growing rapidly and are therefore much more likely to be killed by a high dose of gamma radiation from a cobalt 60 source than are normal cells, which divide less frequently. Fig 29.20 is a photograph taken in a hospital cobalt unit and Fig. 29.21 is a pair of photographs showing the dramatic effect that a course of gamma ray treatment can have on a skin cancer. The same is true of other cancers, but photographing them is not possible.

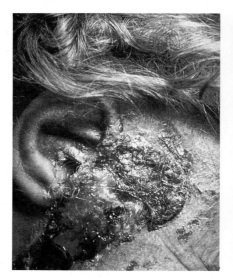 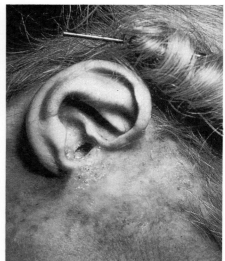

Fig 29.21   A carcinoma of skin by the ear of a patient before and after radiotherapy treatment.

Mutations of living cells are usually harmful, but by irradiating plants it has proved possible to cause mutations which have led to improved varieties of many crops, such as wheat, peas and beans. New strains of plants have been produced which give larger yields and are more resistant to disease.

ATOMIC STRUCTURE

Archaeological dating

The atmosphere contains a small proportion of radioactive carbon 14. During a plant's life carbon dioxide is absorbed, so it too becomes slightly radioactive and the level of radioactivity of plants can be measured. When a tree is cut down and burnt, radioactive decay of any charcoal remaining commences. The longer ago the charcoal was formed, the lower is its radioactivity. This method of dating can now be used with an accuracy of better than to the nearest 100 years. By carrying out experiments on accurately dated events (at Pompeii, for example) it is possible to take into account variations in the level of atmospheric radiocarbon through the centuries. Question 29.24 illustrates this technique further.

Power sources

The obvious example here is in the use of uranium 235 in a nuclear power station. This was briefly referred to earlier. There are, however, other portable power sources which use radioactive materials. Some satellites contain radioactive materials which are decaying naturally and giving enough energy to remain hot throughout a long space flight. The hot material is used in conjunction with thermocouples to supply electricity. Some fire alarms contain a small amount of alpha-emitting substance. This keeps the air in the fire alarm slightly ionised and any alteration in the level of ionisation, caused by smoke, can be detected and used to set off the alarm.

DATA ANALYSIS

Decommissioning a nuclear reactor

A nuclear reactor throughout its life generates power from fission of uranium 235 by neutrons in a reaction of this type:

$$_0^1\text{n} + _{92}^{235}\text{U} \rightarrow _{38}^{90}\text{Sr} + _{54}^{143}\text{Xe} + 3\,_0^1\text{n} + E$$

where E represents about 200 MeV of energy. The fission products, as the strontium (Sr) and xenon (Xe) are called, have very small mass and are mostly contained within the fuel cans. The neutrons, on the other hand, have high speed and are uncharged. They are able to escape from the fuel cans into the graphite (carbon) moderator where they are slowed down. One of the three neutrons is needed to sustain the chain reaction and the others are absorbed by the moderator, the fuel, the control rods and the structural materials of the reactor.

The plans for decommissioning nuclear reactors have to take into account the following information.

- A reactor may have a total mass of 300 000 tonnes and 5% of this may be radioactive.
- The fuel elements can be removed in the normal way. This removes 99.99% of the radioactivity which was on site when the reactor was working.
- The remaining radioactivity is mostly neutron-induced in the steel, concrete and graphite.
- Radioactivity is expressed in becquerels, Bq. 1 Bq is 1 disintegration per second.
- British law defines a radioactive substance as one with an activity greater than 400 Bq kg^{-1}.
- Garden soil contains the radioactive potassium 40 nuclide and frequently has an activity of 800 Bq kg^{-1}; coffee powder is even more radioactive at 1600 Bq kg^{-1}.
- The activity of a source cannot be related directly to a dose of radiation. Some nuclides are more radiologically significant as the dose rate depends not only on the number of radioactive disintegrations but also on the type of particles produced, their range and their energy.

- There are about 2500 known nuclides. Of these, 79 which have a half life longer than 1 year may be present in a reactor. Most of these 79 nuclides do not reach the activity of 400 Bq kg^{-1} to necessitate them being called radioactive.
- Table 29.7 shows the mass and the activity of 13 problem nuclides 10 years after the shutdown of a reactor. A dash indicates an insignificant amount of nuclide.

Table 29.7

Nuclide	Half life / year	Concrete 8000 tonnes /10^{12} Bq	Graphite 2200 tonnes /10^{12} Bq	Steel 3400 tonnes /10^{12} Bq	Total /10^{12} Bq	Particle
$^{3}_{1}$H	12	–	110	–	110	β
$^{14}_{6}$C	5700	–	42	1.7	44	β
$^{36}_{17}$Cl	310 000	–	1.3	–	1.3	β
$^{41}_{20}$Ca	130 000	0.10	0.90	–	1.0	κ
$^{55}_{26}$Fe	2.7	0.82	4.0	2400	2400	κ
$^{60}_{27}$Co	5.2	0.28	11	680	690	γ
$^{59}_{28}$Ni	80 000	–	0.096	2.1	2.2	κ
$^{63}_{28}$Ni	92	–	17	230	250	β
$^{63}_{41}$Nb	20 000	–	–	0.013	0.013	β
$^{108}_{47}$Ag	130	–	0.0017	0.017	0.019	β
$^{151}_{62}$Sm	90	0.15	0.051	–	0.20	β
$^{152}_{63}$Eu	12	0.82	0.11	–	0.93	κ
$^{154}_{63}$Eu	16	0.10	0.61	–	0.71	β
Totals		2.3	190	3300	3500	

All values are given to 2 significant figures. 1 tonne = 1000 kg. κ represents κ capture. This is a mode of decay in which the nucleus captures one of the atom's orbiting electrons.

Answer the following questions using the data supplied.

(a) What is the activity in Bq kg^{-1} of each of the three main sections of the reactor?
(b) Why is it that over a period of 100 years the radioactive iron 55 in the reactor is less of a problem than the radioactive nickel?
(c) Why is it likely that the activity of the long half life nuclides is low and the activity of the short half life nuclides is high?
(d) How long will it be before the activity of the tritium (hydrogen 3) in the reactor is below 0.43×10^{12} Bq?
(e) What will be the activity of the tritium after 420 years?
(f) Using the equation

$$R_t = R_0 \, e^{-0.693t/\tau}$$

where τ is the half life, R_t is the activity at time t, and R_0 is the activity at $t = 0$, find the activity of each isotope 110 years after ceasing to make use of the reactor for power generation.
(Take R_0 to be the values given and t to be 100 years, so

$$R_t = R_0 \, e^{-0.693t/\tau}$$

(g) Which nuclides are the problem ones?
(h) Without making any further calculations, what would you expect the decay graph to look like?

SUMMARY

- Atomic constituents:

Proton	charge +1	mass = 1.007 28 u
Electron	charge −1	mass = 0.000 549 u
Neutron	charge 0	mass = 1.008 67 u

- Nucleon: a proton or a neutron.
- Nuclide: an atom of a particular specification.
- Isotopes: nuclides with the same number of protons but with different numbers of neutrons.
- Atomic number (proton number) Z: the number of protons in the nucleus of an atom.
- Atomic mass number (nucleon number) A: the number of protons + neutrons in the nucleus of an atom.
- Atomic mass unit u: one twelfth the mass of an atom of carbon 12.
- Mass defect, Δm: the difference in mass between the mass of the constituent particles of an atom and the (smaller) mass of the whole atom.
- Binding energy ΔE: the energy equivalence of the mass defect. Found using $\Delta E = \Delta mc^2$. It is the energy required to separate the atom into its constituent particles.
- The table shows the properties of the three naturally occurring types of radioactivity. c is the speed of light in a vacuum.

	Nature	Typical speed	Typical penetration	Charge	Ionisation caused
α alpha	2p + 2n	$0.1c$	6 cm air	$+2e$	a great deal
β beta	1 electron	$0.9c$	few mm Al	$-1e$	some
γ gamma	electromagnetic radiation	c	several cm lead	0	very little

- Nuclear equations: Energy and mass are conserved in any nuclear reaction. The following ways of writing the equations are equivalent

$$^{10}_{5}\text{B} + ^{1}_{0}\text{n} \longrightarrow ^{7}_{3}\text{Li} + ^{4}_{2}\text{He} \quad \text{or} \quad ^{10}_{5}\text{B} \, (\, \text{n}, \, \alpha) \, ^{7}_{3}\text{Li}$$

- Radioactive decay is a random process. The rate of decay is proportional to the number of radioactive atoms present:

$$\frac{dN}{dt} \propto -N \qquad \frac{dN}{dt} = -\lambda N$$

where λ is the decay constant. This gives an exponential decay equation:

$$N = N_0 e^{-\lambda t}$$

For a half life τ, $N = \frac{1}{2} N_0$, which gives $\tau \lambda = 0.693$.

EXAMINATION QUESTIONS

29.14 Explain what is meant by the terms *nucleon, atomic number* and *mass number.* (3)
Briefly describe an experiment which led to an appreciation of the size of the nucleus relative to that of the atom. Explain qualitatively how the results were interpreted. (4)
Experiments indicate that the radius R of a nucleus (which is assumed to be spherical) is given by the approximate relation

$$R/\text{m} \approx 1.2 \times 10^{-15} A^{\frac{1}{3}}$$

where A is the mass number of the nucleus.
(a) Use this relation to estimate the volume occupied by each nucleon in a nucleus. (3)
(b) Hence show that the average distance between nucleons is about 2×10^{-15} m. (2)
(c) Estimate the electrostatic repulsion between neighbouring protons in a nucleus. Comment on your answer. (5)
(UCLES 1987)

29.15 (a) What conclusions about the structure of atoms were made as a result of α particle scattering experiments with thin metal foils? (5)
(b) (i) What is meant by conservation of mass–energy in nuclear reactions? (3)
(ii) Explain how and why the masses of atoms differ from the sum of the masses of their constituent particles. (3)
(c) Radium (Ra) decays to radon (Rn) by the reaction

$$^{226}_{89}\text{Ra} \rightarrow {}^{226}_{86}\text{Rn} + {}^{4}_{2}\text{He} + \gamma$$

The atomic masses are:
radium 3.753×10^{-25} kg; radon 3.686×10^{-25} kg; helium 0.066×10^{-25} kg
(Take the Planck constant h to be 6.6×10^{-34} J s, and the speed of light c in vacuum to be 3.0×10^{8} m s^{-1}.)
(i) Estimate the energy (in joules) released when an atom of ^{226}Ra decays. (4)
(ii) Estimate the wavelength of the gamma (γ) photon emitted during this decay given that 4% of the energy released turns to gamma radiation. (4)
(iii) What happens to the remaining 96% of the energy? (2)
(d) Two units of mass–energy used in nuclear studies are the unified atomic mass unit (u) and the megaelectronvolt (MeV). The unified atomic mass unit is defined as 1/12 of the mass of a carbon 12 atom.
(Take the Avogadro constant N_A to be 6.0×10^{23} mol^{-1} and the electronic charge e to be

1.6×10^{-19} C. The molar mass of carbon12 is 0.012 kg.)
Calculate (i) the value of u in kilograms; (3)
(ii) the equivalent of 1 MeV, in joules; (2)
(iii) the equivalent of 1 u, in MeV. (4)
(OXFORD 1987)

29.16 In an alpha particle scattering experiment, an alpha particle and a gold nucleus (in a piece of gold foil) collide head on and the alpha particle rebounds.
(a) Using the data given, write down a numerical expression for the electrostatic force of repulsion, F, acting on the alpha particle at the instant of collision when the distance between the alpha particle and the gold nucleus is s.
For the alpha particle $A = 4$, $Z = 2$; for the gold atom $A = 197$, $Z = 79$.
Electronic charge, $e = 1.6 \times 10^{-19}$ C.
Permittivity of vacuum, $\varepsilon_0 = 8.85 \times 10^{-12}$ F m^{-1}.
(b) The electric potential energy of the alpha particle and the gold nucleus at the point of impact is Fs.
(i) If the initial kinetic energy of the alpha particle is 1.8 MeV, calculate a value for s. Assume that the gold nucleus has no kinetic energy initially. (1 eV = 1.6×10^{-19} J.)
(ii) What indication does the value of s give about the sizes of the two particles involved? (9)
(ULSEB 1988)

29.17 (a) Sketch a graph of the binding energy E per nucleon against the mass (nucleon) number A for naturally occurring isotopes, indicating an approximate scale for A. (2)
(b) Use your graph to explain how energy may be obtained from fission of heavy nuclei. (3)
(c) The following is a possible nuclear reaction:

$$^{3}_{1}\text{H} + {}^{2}_{1}\text{H} \rightarrow {}_{Z}\text{X} + {}^{1}_{0}\text{n} + 17.6 \text{ MeV}$$

(i) Name the nucleus X.
(ii) Using the data given, calculate the atomic mass of X in u.
(mass of neutron = 1.0087 u; mass of $^{2}_{1}$H atom = 2.0141 u; mass of $^{3}_{1}$H atom = 3.0161 u; 1 u = 931 MeV.)
(O & C 1987)

29.18 Describe an experiment which shows that α particles are helium nuclei. (5)

ATOMIC STRUCTURE

The uranium isotope $^{238}_{92}$U decays to an isotope of thorium (Th) by emitting an α particle of energy 4.2 MeV (6.7×10^{-13} J). Write down an equation for this nuclear transformation. (2)
The α particles from a sample of $^{238}_{92}$U have a maximum range in air. Describe an experiment you would perform to attempt to measure this distance. (4)
The graph shows the number of positive ions and electrons (ion pairs) produced along each millimetre of track of an α particle from $^{238}_{92}$U passing through air at atmospheric pressure and room temperature.

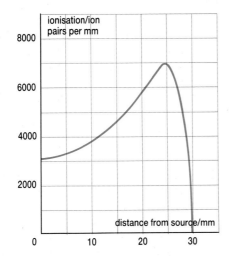

(a) Use the graph to estimate the number of ion pairs produced. Hence, using the energy of the α particle, estimate the energy required to produce an ion pair. (4)
(b) Suggest why the ionisation per millimetre falls to zero at 30 mm. (1)
(c) What is the effect, if any, of a change in the pressure of the air on the range of the α particles? (2)
(d) How would you expect the tracks of β particles of energy 4.2 MeV seen in a cloud chamber to differ from those of the α particles? (2)
(O & C 1985)

29.19 An α particle having a speed of 1.00×10^6 m s^{-1} collides with a stationary proton which gains an initial speed of 1.60×10^6 m s^{-1} in the direction in which the α particle was travelling.
What is the speed of the α particle immediately after the collision?
How much kinetic energy is gained by the proton in the collision?
It is known that this collision is perfectly elastic. Explain what this means.
(Mass of α particle = 6.64×10^{-27} kg; mass of proton = 1.66×10^{-27} kg.) (3)
(ULSEB 1983)

29.20 (a) Explain the nature of alpha, beta and gamma radiations. (6)
(b) A mixture of isotopes is thought to emit alpha, beta and gamma radiations simultaneously all at a low intensity. Describe experimental tests that would (i) confirm all three radiations are present, (ii) measure the relative intensities of the beta and gamma radiations. (9)
(c) A beta particle is moving in a gas along a straight line OA when a uniform magnetic field of flux density 3.0×10^{-4} T is applied. The diagram shows the subsequent path of the particle in a plane perpendicular to the magnetic field.

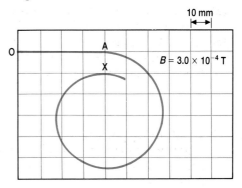

(Take the mass of a beta particle to be 9.1×10^{-31} kg, and its charge to be -1.6×10^{-19} C.)
(i) Explain the form of the path in the magnetic field. (5)
(ii) By estimating the radius of curvature at A, find the speed v of the particle as it passes A. (5)
(iii) Estimate the kinetic energy loss of the particle as it passes from A to X. (5)
(OXFORD 1986)

29.21 (a) Explain what are meant by the *half life* and the *decay constant* of a radioactive isotope.
(b) At the start of an experiment a mixture of radioactive materials contains 20.0 µg of a radioisotope A, which has a half life of 70 s, and 40.0 µg of radioactive isotope B, which has a half life of 35 s.
(i) After what period of time will the mixture contain equal masses of each isotope? What is the mass of each isotope at this time?
(ii) Calculate the rate at which the atoms of isotope A are decaying when the masses are the same.
(Molar mass of isotope A = 234 g; the Avogadro constant = 6.0×10^{23} mol^{-1}.) (9)
(AEB 1987)

29.22 A milk sample is to be tested for evidence of radioactive contamination with the radioactive nuclide strontium 90 (Sr 90), using a Geiger–Müller tube.

(a) What precautions must be taken in order to obtain an accurate value of the activity of the sample?

(b) The radioactive decay process occurring in Sr 90 can be described by the equation $A = A_0 e^{-\lambda t}$ where A is the activity, dN/dt, of the material and λ is its decay constant. The value of λ for the beta decay of Sr 90 is 7.75×10^{-10} s^{-1}. Explain the difficulty involved in attempting to measure its half life experimentally.

(c) The two stages involved in the decay of Sr 90 are described by the equations:

$$^{90}_{38}\text{Sr} \rightarrow {}^{90}_{39}\text{Y} + {}^{0}_{-1}\text{e} : {}^{90}_{39}\text{Y} \rightarrow {}^{90}_{40}\text{Zr} + {}^{0}_{-1}\text{e}$$

At each stage a beta particle is emitted. The half life for the second process is 64 h and the final product, zirconium 90, is stable. How long would it take for the activity of the Sr 90 to fall to 90% of its initial value? Explain why the amount of yttrium 90 (Y 90) present at any time during the period just calculated is likely to be very small indeed compared with the amount of Sr 90 present. (12)

(ULSEB 1988)

29.23 (a) Describe the principle of operation of a solid-state detector of ionising radiation in terms of the generation and detection of charge carriers. (5)

(b) Explain how, when making measurements of radioactivity, practical steps can be taken to overcome problems caused by
  (i) the random nature of radioactivity,
  (ii) background radiation levels. (6)

(c) Radioactive iron, ^{59}Fe, is a radioactive nuclide with a half life of 46 days. It is used medically in the diagnosis of blood disorders. Measurements are complicated by the fact that iron is excreted, i.e. removed, from the body at a rate such that 69 days after administering a dose, half of the iron atoms in the dose have been excreted. That is to say, iron has a biological half life of 69 days.

(i) If the count rate from a blood sample is 960 counts per minute, what will it be from a similar blood sample taken 138 days later? (4)

(ii) How long after the first sample was taken would a further similar sample give a count rate of 480 counts per minute? (4)

(iii) Derive the relation between T the effective half life of the radioactive nuclide in the body, T_b, the biological half life, and T_r, the radioactive half life. (3)

(UCLES 1988)

29.24 The nuclide ^{14}C is used in the radioactive dating of samples of material of biological origin, such as wood. The half life of this isotope is 5570 years. As long as a tree is alive and breathes, the concentration of ^{14}C relative to ^{12}C in the wood will be the same as in the atmosphere, 1.1 part in 10^{12}. But when the tree dies, the ^{14}C is not replenished any more and its decay leads to a gradual decrease in its concentration relative to that of the stable isotope ^{12}C. In a sample of wood taken from an ancient tomb the concentration of ^{14}C relative to ^{12}C was found to be 4.0 part in 10^{13}. How old is the tomb?

29.25 Explain briefly how electrical energy is derived from the heat produced in a nuclear reactor. (2)

In a nuclear reactor the process of nuclear fission is involved. Explain:

(a) how useful energy is released in this process,

(b) why fission generally only occurs for heavy nuclei,

(c) why the release of several neutrons in each reaction is of fundamental importance,

(d) why a nuclear reactor has to be heavily shielded. (8)

A nuclear reactor produces 550 MW of electrical power. Waste heat is carried away by seawater supplied at the rate of 2.7×10^4 kg s^{-1}. If the temperature rise of the seawater is 10.3 K, calculate the mass of nuclear fuel converted into energy in one year.
(1 year $\approx 3.2 \times 10^7$ s; speed of light in a vacuum $= 3.0 \times 10^8$ m s^{-1}; specific heat capacity of seawater $= 4.0$ kJ kg^{-1} K^{-1}.) (8)

(ULSEB 1981)

Chapter 30

THE QUANTUM THEORY

LEARNING OBJECTIVES

At the end of this chapter you should be able to:

1. describe the photoelectric effect and understand why it cannot be explained by classical physics;

2. calculate a photon's energy and relate it to frequency and wavelength;

3. map energy levels and use them to calculate the wavelength of line spectra;

4. describe X-ray spectra;

5. outline the theory of wave particle duality.

30.1 THE EMISSION OF LIGHT

In Theme 3 it was established by several experiments that light can be considered as a wave motion. The theory of the transmission of electromagnetic waves was established by Maxwell in 1865. His work is summarised by four equations relating such quantities as electric field strength, magnetic field strength, electric current, charge, permittivity of free space ε_0 and permeability of free space μ_0. A deduction from Maxwell's theory is that an electromagnetic wave is a combination of two transverse waves which travel through a vacuum with speed c. One of these waves is an electric field wave and the other in the combination is a magnetic field wave at right angles to it. In any electromagnetic wave, the changing magnetic field creates an electric field and the changing electric field creates a magnetic field. The fields are shown in Figs 11.36 and 11.37 (in Chapter 11) for waves travelling in one dimension and two dimensions respectively.

The value of c according to Maxwell is given by

$$ c = \frac{1}{\sqrt{\mu_0\,\varepsilon_0}} $$

The agreement between Maxwell's theory and the experimental value of the speed of light available at the time was very strong evidence for the fact that light is a form of electromagnetic radiation. At the time Maxwell published his theory, the only parts of the electromagnetic spectrum which were known to exist were light, infra-red and ultra-violet. Predictions put forward by Maxwell concerning other electromagnetic radiations were not demonstrated practically until after his death. In 1887 Hertz constructed electrical circuits to produce electromagnetic waves of a different, longer wavelength. The use of these radio waves, as they came to be called, has been of enormous social consequence. Since that time X-rays, γ rays and microwaves have also been shown to be electromagnetic waves.

More recently the definition of the metre has been based on the speed of light so that now:

$$c = 299\ 792\ 458\ \text{ms}^{-1} \text{ exactly by definition of the metre}$$
$$\mu_0 = 4\pi \times 10^{-7}\ \text{N A}^{-2} \text{ exactly by definition of the ampere}$$

$$\text{giving } \varepsilon_0 \quad = \quad \frac{1}{c^2\ \mu_0} \quad = \quad \frac{1}{(299\ 792\ 458\ \text{m s}^{-1})^2 \times (4\pi \times 10^{-7}\ \text{N A}^{-2})}$$

$$= 8.854\ 19 \times 10^{-12}\ \text{C}^2\ \text{N}^{-1}\ \text{m}^{-2}$$

QUESTION

30.1 Show that $\text{C}^2\ \text{N}^{-1}\ \text{m}^{-2}$ is the unit of ε_0, the permittivity of free space and show that $\mu_0\varepsilon_0$ has the unit $\text{s}^2\ \text{m}^{-2}$.

While the transmission of light waves was explained very precisely by Maxwell, there was still a mystery concerning the source of light waves. According to Maxwell, electromagnetic waves are emitted by accelerating charges. All that was known about light production in the middle of the Nineteenth Century was knowledge gained experimentally from the emitted light. Nothing was known about atomic structure so it was not possible to consider what charges were accelerating, but it was known that the emission spectra of hot gases consisted of light of specific wavelengths in a very pronounced pattern. Before any progress could be made on the theory of light emission however, the photoelectric effect was discovered, by Hertz, and this threw doubt on the validity of considering light as a wave motion.

Hertz noticed in 1887 that a spark between two metal spheres could occur more readily if the spheres were illuminated with light from another spark. Light made the space around the spheres conduct electricity more readily. One of the experiments which were done at that time is one which you should repeat. It uses only simple apparatus but it is very instructive.

INVESTIGATION

Place on the cap of a gold leaf electroscope a piece of clean zinc strip as shown in Fig 30.1. The zinc can be cleaned by rubbing it with sandpaper. Charge the gold leaf so that it is negatively charged and ensure that the gold leaf stays at a constant angle to the vertical. If the leaf does fall initially, it is probably because the air inside the electroscope is damp. It can be dried by standing it on a radiator for a

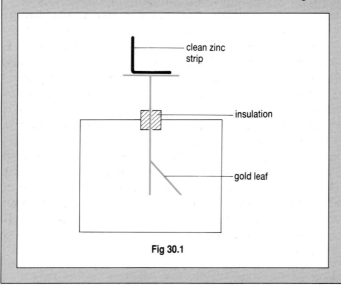

Fig 30.1

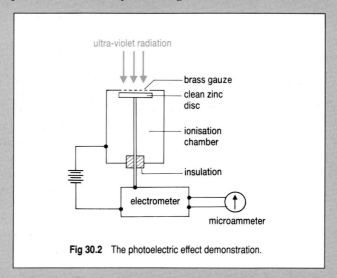

Fig 30.2 The photoelectric effect demonstration.

THE QUANTUM THEORY

while. Then bring an ultra-violet lamp near to the zinc strip. What is the effect on the charge stored on the gold leaf? How does the movement of the gold leaf vary with the intensity of illumination? The intensity of illumination may be changed by moving the lamp closer to, or further from, the zinc strip. Repeat the experiment using a positively charged electroscope. What charge does the ultra-violet light cause to be emitted from the zinc plate? The Sun can be used as the source of ultra-violet light if it happens to be sunny while you are doing this investigation. You must use direct sunlight, however, since ultra-violet radiation is absorbed by window glass.

This investigation can be extended if an electrometer is available. The apparatus is illustrated in Fig. 30.2 and consists of a clean zinc disc held close to a brass gauze. The potential difference between these two items can be varied and the flow of charge measured using the electrometer. The electrometer is a d.c. amplifier with a very high gain so that it can detect very small currents. Investigate how the current varies with
(a) the illumination and
(b) the potential difference between the zinc plate and the gauze.

30.2 THE PHOTOELECTRIC EFFECT

Hertz was not able to explain why the air in the space near his metal spheres became conducting when illuminated with the light from a spark. The radiation from the first spark was clearly assisting the formation of the second. It was also found that visible light from a bright source can initiate sparking provided the metals used for the spheres are sodium or potassium. This is not a recommended experiment because sodium and potassium are so highly reactive. When the early experiments were done with these two metals, they were done in a vacuum and the cleaning of the surfaces was done using an electrically driven scraper in the vacuum.

After the discovery of the electron in 1897, it was immediately suspected that the negatively charged particles which were being emitted by the zinc plate were electrons. By 1899 it had been confirmed that they were electrons by experiments which showed that they had the same specific charge as the electron. This effect, also shown by the investigation, is called the photoelectric effect and the emitted electrons are referred to as photoelectrons.

The photoelectric effect produced some surprising results. It was the energy of the photoelectrons which was surprising. If graphs are plotted of photoelectron current against potential difference between the photo-cathode, the zinc, and the photo-anode, the brass, the result is as shown in Fig. 30.3. As expected, a large photoelectric current is generated when there is greater illumination of the photo-cathode. Also as expected, increasing the potential difference between the electrodes cannot increase the photo-electron current beyond a certain maximum for any level of illumination. This is because the maximum current is limited by the rate at which photoelectrons are produced, and that is controlled by the illumination level, not the potential difference.

The totally unexpected result was that the energy of photoelectrons generated by dim illumination is the same as that generated by bright illumination. This is shown by the **stopping potential**. The stopping potential V_s is the reverse potential which is just sufficient to prevent photoelectrons from reaching the anode. If an electron is emitted with speed v it will stop when it has changed all of its kinetic energy into potential energy. That is when

$$\text{kinetic energy} = \tfrac{1}{2} mv^2 = eV_s$$

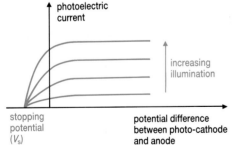

Fig 30.3

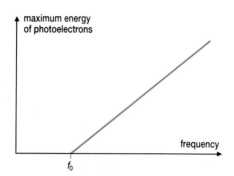

where m is the mass and e the charge of the electron. V_s is therefore directly proportional to the maximum kinetic energy of the photoelectrons and the graph. Fig 30.3, shows that the stopping potential does not depend at all on the brightness of the illumination. Dim illumination can produce photo-electrons of the same energy as bright illumination.

This result is surprising because it contradicts basic wave theory. The energy of any oscillating system is proportional to the square of the ampli-tude (see section 10.3). This implies that if one wave has an amplitude three times that of another wave, the energy of the first wave should be nine times that of the second wave.

Here we have a situation in which the number of photoelectrons pro-duced depends on the intensity of illumination but the maximum energy of these photoelectrons does not. It is for this reason that radio waves, however strong, do not affect photographic film. Likewise, red light does not affect photographic paper but blue light does.

The maximum energy of photoelectrons was found to depend on the frequency of the radiation causing them. This is shown in Fig 30.4. No photoelectrons are produced until the frequency is high enough. This frequency f_0 is called the **threshold frequency**. Further increase in the frequency results in a linear increase in the maximum kinetic energy of photoelectrons. The threshold frequency is such that metals such as zinc will give photoelectrons only if the radiation used to produce them is in the ultra-violet or X-ray region of the electromagnetic spectrum. Sodium, potassium and caesium will give photoelectrons with visible light.

Fig 30.4

The quantum theory

Before the problem concerning the energy of photoelectrons became apparent, Max Planck had given a theory for the emission of radiation from a hot body. He had introduced the idea of discrete energy units. He suggested that in much the same way that matter is not continuous but consists of a large number of tiny particles, so radiation energy from a source of given frequency f is quantised. He further suggested that the basic quantum of energy from a source E is proportional to the frequency of the source. This gives

 $$E = hf$$

where h is a constant of proportionality now called the Planck constant. The value of the Planck constant is found experimentally to be 6.625×10^{-34} J s.

Einstein's photoelectric equation

This concept was adopted by Einstein to give an explanation of the photoelectric effect in 1905. He assumed that radiation was quantised and that to cause a photoelectron a quantum of electromagnetic energy, called a **photon**, is absorbed by an atom in the surface of the cathode. Some of the energy absorbed by this atom releases an electron from it and from the surface of the metal. The rest of the energy becomes kinetic energy, some or all of which the electron may retain when it leaves the photo-cathode. Put as a formula this becomes:

| Energy of incoming photon | = work done to remove electron from metal | + kinetic energy of photoelectron |

If the photoelectron is emitted from the surface with no further interactions with other atoms then it has the maximum kinetic energy corresponding to a speed v_{max}. For such an electron:

$$hf = \phi + \tfrac{1}{2}mv_{max}^2$$

where ϕ is called the **work function** energy of the metal. The threshold frequency is given when v_{max} is zero so

$$hf_0 = \phi + 0$$

and so $\qquad hf = hf_0 + \tfrac{1}{2}mv_{max}^2$

Since, as shown earlier,

$$\tfrac{1}{2}mv_{max}^2 = eV_s$$

the equation can be written

$$hf = hf_0 + eV_s$$

This theory satisfactorily explains the photoelectric effect but it does mean that when light interacts with matter it has to be considered rather differently from before. The wave nature of light is adequate to explain its transmission; the quantum theory, in which light is regarded as a stream of photons, is needed to explain the photoelectric effect. The theories can be shown to be compatible with one another in all circumstances, but at this stage you are expected to use the wave theory when dealing with transmission, diffraction and interference, and the quantum theory when dealing with the photoelectric effect and the emission and absorption of light. One complicating factor which is here being ignored is that there is a so-called contact potential difference between different metals. This has the result that it is difficult to measure stopping potentials reliably, so photoelectric experiments are difficult to do accurately.

Example 1
When ultra-violet radiation of wavelength 319 nm falls on a metal surface, photoelectrons with a stopping potential of 0.22 V are produced. The same surface is then illuminated with ultra-violet radiation of wavelength 238 nm. Find:
(a) the work function of the metal;
(b) the threshold frequency for the metal;
(c) the stopping potential when the wavelength is 238 nm;
(d) the maximum speed of photoelectrons for each wavelength.

$$e = 1.60 \times 10^{-19}\,\text{C} \qquad h = 6.63 \times 10^{-34}\,\text{J s}$$
$$c = 3.00 \times 10^8\,\text{m s}^{-1} \qquad m_e = 9.11 \times 10^{-31}\,\text{kg}$$

First the energy of a photon from each souce of illumination must be found. For $\lambda = 319$ nm:

$$f = \frac{c}{\lambda} = \frac{3.00 \times 10^8\,\text{m s}^{-1}}{319\,\text{nm}} = 9.40 \times 10^{14}\,\text{Hz}$$

Energy of one photon $= hf = 6.63 \times 10^{-34}\,\text{J s} \times 9.40 \times 10^{14}\,\text{Hz}$
$$= 6.23 \times 10^{-19}\,\text{J}$$

For $\lambda = 238$ nm:

$$f = \frac{c}{\lambda} = \frac{3.00 \times 10^8\,\text{m s}^{-1}}{238\,\text{nm}} = 1.26 \times 10^{15}\,\text{Hz}$$

Energy of one photon $= hf = 8.35 \times 10^{-19}\,\text{J}$

(a) For the 319 nm wavelength radiation the stopping potential is 0.22 V. This means 0.22 joules per coulomb, so for a charge of 1.60×10^{-19} C the maximum kinetic energy of the electron is

$$0.22 \; \frac{\text{J}}{\text{C}} \times 1.60 \times 10^{-19} \, \text{C} = 0.35 \times 10^{-19} \, \text{J}$$

The work function is therefore 6.23×10^{-19} J $- 0.35 \times 10^{-19}$ J

$$\phi = 5.88 \times 10^{-19} \, \text{J}$$

(b) The threshold frequency is the frequency for which the quantum of energy of the incoming photon equals the work function so

$$hf_0 = \phi \;\; \text{giving}$$

$$f_0 \;\; = \;\; \frac{5.58 \times 10^{-19} \, \text{J}}{6.63 \times 10^{-34} \, \text{J s}} = 8.87 \times 10^{14} \, \text{Hz}$$

(c) For the radiation of wavelength 238 nm:

$$
\begin{aligned}
hf &= \phi + eV_s \\
8.35 \times 10^{-19} \, \text{J} &= 5.88 \times 10^{-19} \, \text{J} + eV_s \\
V_s &= \frac{(8.35 \times 10^{-19} \, \text{J}) - (5.88 \times 10^{-19} \, \text{J})}{(1.60 \times 10^{-19} \, \text{C})} \\
&= 1.54 \, \text{V}
\end{aligned}
$$

(d) The maximum speed of photoelectrons caused by these photons is given by

$$\tfrac{1}{2} mv^2 \;\; = \;\; eV_s = 0.35 \times 10^{-19} \, \text{J}$$

giving

$$v \;\; = \;\; \sqrt{\frac{2 \times (0.35 \times 10^{-19} \, \text{J})}{(9.11 \times 10^{-31} \, \text{kg})}}$$

$$= \;\; 2.8 \times 10^5 \, \text{m s}^{-1} \; \text{to 2 sig. figs}$$

Repeating for the 238 nm wavelength radiation gives

$$v \;\; = \;\; \sqrt{\frac{2 \times (1.54 \, \text{V}) \times (1.6 \times 10^{-19} \, \text{J})}{(9.11 \times 10^{-31} \, \text{kg})}}$$

$$= \;\; 7.4 \times 10^5 \, \text{m s}^{-1} \; \text{to 2 sig. figs}$$

It is easy to make arithmetical mistakes with problems such as these. Working can be simplified appreciably if energies are expressed in electron-volts. If this is done, it is still necessary to find the energy of a photon but then it can be converted into electron-volts by dividing by the fundamental charge, 1.60×10^{-19} C. Working in the order of the encircled figures in Table 30.1, in fact produces this table.

Table 30.1

Wavelength /nm	Photon energy /eV		Maximum photoelectron kinetic energy /eV		Work function /eV	
319	3.90	②	0.22	①	3.68	④
238	5.22	③	1.54	⑥	3.68	⑤

Note that a stopping potential of 0.22 V means that the photoelectrons have a maximum kinetic energy of 0.22 eV. After ② and ③ are calculated ④ is obtained by direct subtraction and ⑤ must be the same as ④. This enables ⑥ to be obtained.

30.2 What are the frequency and the energy of a photon of red light of wavelength 700 nm and a photon of violet light of wavelength 400 nm?

30.3 What is the threshold frequency of
(a) sodium of work function 2.3 eV
(b) caesium of work function 1.9 eV?
In which region of the electromagnetic spectrum do these frequencies lie?

30.4 What is the maximum kinetic energy of a photoelectron emitted from silver, of work function 4.7 eV, when illuminated with ultra-violet radiation of wavelength 122 nm?

30.5 The maximum energy for photoelectrons emitted from a metal with a work function 2.9 eV is 8.7 eV. What are (a) the maximum energy and (b) the maximum frequency of the incoming photons?

30.6 Assuming that numerical values are available, how can (a) the work function and (b) the Planck constant be obtained from the graph in Fig 30.4? Explain why this graph always has the same gradient whatever metal is used for the photo-cathode.

30.3 SPECTRA AND ENERGY LEVELS WITHIN ATOMS

The quantum theory used to explain the photoelectric effect uses the idea that a photon of electromagnetic radiation has an energy E given by

$$E = hf$$

where h is the Planck constant and f is the frequency. This was quickly realised also as being in some way connected with line spectra.

In section 28.1 it was stated that when a gas is energised sufficiently, by being heated or by having an electric current passed through it, it emits visible light. If the light is examined using a spectrometer, the light is seen to consist of a few colours only. Figs 13.37 and 13.38 show two photographs taken of such spectra. The term 'line spectrum' indicates that only certain specific wavelengths are present, but the lines are created simply because the light entering the spectrometer passes through a narrow slit. Each colour shown in the photograph is really an image of the slit in the particular colour of the light seen. A continuous spectrum is obtained from a heated solid and covers a wide range of frequencies.

The difficulty of explaining line spectra stems again from Maxwell's theory of electromagnetic radiation. If the electrons in an atom are accelerating then, according to Maxwell, they should be emitting radiation, and consequently losing energy. This would imply that the electrons would collapse into the nucleus of the atom and emit radiation of ever-increasing frequency while doing so. This clearly does not happen. Atoms do not collapse and the radiation produced has frequencies with definite values.

It was Niels Bohr who, in 1913, suggested that electrons within atoms could exist in stable energy states without emitting radiation. He suggested that the electrons within atoms could absorb energy in quanta of certain definite amounts when bombarded by other atoms or electrons during such processes as being heated. Once the electron has gained its quantum of energy, the atom is said to be in an **excited state**. Later, often no more than a microsecond later, the electron loses energy as a photon of electro-

---

magnetic radiation as it returns to its former, stable state. This implies that a line spectrum is a map of the changes in energy levels possible within an atom.

The patterns of line spectra for different gases had been measured very accurately during the Nineteenth Century and empirical equations had been found for their wavelengths. Any new theory therefore had to show agreement with the known patterns of wavelengths. Bohr succeeded in doing this for hydrogen to a high degree of accuracy, but later analysis showed that his theory was less successful with multi-electron atoms. The following example shows how the known wavelengths can be converted into energy levels for electrons within a hydrogen atom.

Example 2

Some of the lines present in the hydrogen spectrum have wavelengths of 656, 486, 434, 410, 397, 389 . . . 365 nm. Some of these lines are visible and are shown in the photograph of Fig 13.38. The shorter wavelength lines are in the ultra-violet, and there they get very close to one another as a limit is reached at 365 nm. What energy levels are necessary within a hydrogen atom to allow these wavelengths to be produced?

First it is necessary to convert the wavelengths of the light into their corresponding photon energy. This is done using $E = hf$ and since $c = f\lambda$ we get

$$E = \frac{hc}{\lambda}$$

The calculated values are given in Table 30.2.

Table 30.2

Wavelength / nm	Photon energy / 10^{-19} J	Photon energy / eV
656	3.03	1.90
486	4.09	2.56
434	4.58	2.86
410	4.85	3.03
397	5.01	3.13
389	5.11	3.19
365	5.45	3.41

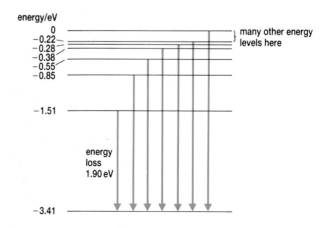

Fig 30.5 Energy levels within the hydrogen atom.

THE QUANTUM THEORY

Whenever energy is measured it is important to state clearly the zero of potential energy being used. Within an atom this is taken to be for an electron just released from the atom. If therefore a photon of 3.41 eV can be emitted from a hydrogen atom there must be an energy level at –3.41 eV. That is 3.41 eV lower than our arbitrary zero. If these levels are mapped for all the photons, we get the horizontal lines shown in Fig 30.5 which is drawn to scale. Note that it has been assumed here that all the electrons fall to the same final energy level. There is a lower energy level within a hydrogen atom. This is at a value of –13.6 eV. When electrons fall to that level, the energy lost as a photon of electromagnetic radiation is much greater and the spectral line is short-wavelength ultra-violet.

QUESTIONS

30.7 Using values taken from example 2, find the wavelengths of a series of spectral lines created when electrons fall (that is, lose energy) from different energy levels to the level at –13.6 eV.

30.8 In a helium–neon laser, electrons are forced from a low energy to a level of – 4.031 eV in the helium. From there, they may fall back to a low energy level in the helium-emitting ultra-violet photons. It is also possible, when a helium atom collides with a neon atom, for electrons in neon atoms to be raised to a level of – 4.026 eV by using the energy of a falling helium atom together with some thermal energy. The energy of helium atoms is continually pumping electrons into the – 4.026 eV state. The electrons in the neon can be induced to fall to a level of – 5.990 eV by photons of an energy equal to this energy fall. This produces the familiar red laser light. Calculate:
(a) the thermal energy required to push an electron into the – 4.026 eV state;
(b) the wavelength of the laser light produced.

30.4 X-RAYS

X-rays were discovered by Röntgen in 1895. (In Germany, X-rays are still called Röntgen rays). He was investigating cathode rays and found that when his cathode rays had high energy, a radiation was produced which was very penetrating and which caused a glow on a fluorescent screen. Because of their ability to penetrate matter they were used for medical investigations almost immediately after their discovery. The dangers associated with X-radiation were realised more slowly. Now there is pressure on doctors and dentists to use X-rays less. There used to be mass screening of the population to check people for tuberculosis of the lung, but this was stopped when it was realised that more harm was probably being done by the X-rays than was being achieved by finding the lung disease. This was especially true when the incidence of tuberculosis of the lungs became so small. When X-rays are used now, the radiographers themselves have to be well screened from the X-rays and the dose for the patient has to be kept to a minimum. This has been made possible by manufacturing photographic X-ray film which is of very high sensitivity.

One reason for having thick glass on the front of a television set is to absorb the X-rays which are produced on the screen by the sudden slowing down of the electrons when they hit the screen. Although the level of radiation is very low from television screens, the dose received from them is built up over many hours. It is not a good idea for young children habitually to sit close to a television screen.

A modern X-ray tube is illustrated in Fig 30.6. If very penetrating, hard X-rays are required, then a high potential difference, typically 120 kV, is used between the cathode and the anode. X-ray tubes used by dentists typically use potential differences of 60 kV. A heated cathode produces

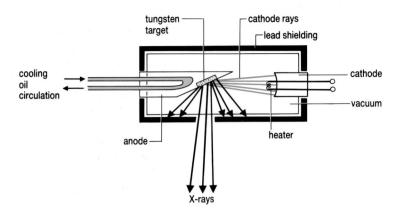

Fig 30.6 A modern X-ray tube.

electrons as in a cathode ray tube, and these can be focused by the shape of the cathode on to an anode, which may need to be cooled by water or oil circulated through it if the power rating is high. As the electrons hit the anode, X-rays are produced. In practice, only about 1% of the power of the electron beam produces X-rays. The rest of the electrical power is wasted as heat.

QUESTIONS

30.9 An X-ray tube operates at 100 kV and the electrons through it constitute a current of 3.0 mA. What is the power of the electron beam? Why does the tube give better X-ray pictures if the electrons are brought to a point focus? Assuming that most of the beam energy is wasted as heat, what problem arises if all the power of the electron beam is brought to a point focus?

30.10 What is the energy of an electron if it is accelerated through a potential difference of 80 kV in an X-ray tube? What is the maximum energy of a photon of X-rays from a tube working with this voltage? What is the maximum frequency of the X-rays the tube produces? What is the wavelength of these X-rays?

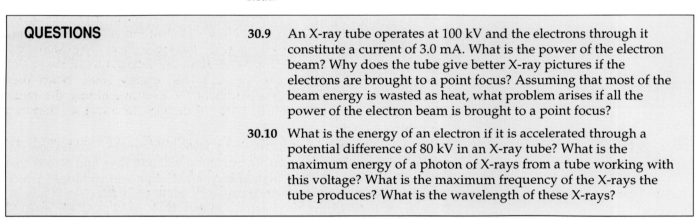

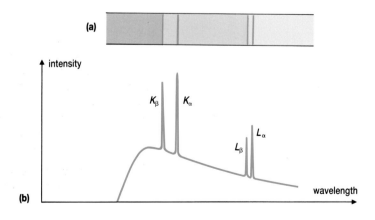

Fig 30.7 (a) An X-ray spectrum showing both a continuous spectrum and a line spectrum.
(b) A graph showing how the intensity of X-rays varies with the wavelength.

THE QUANTUM THEORY

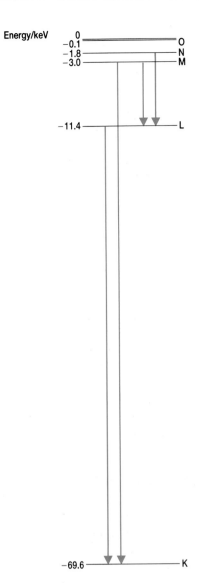

Energy/keV

Fig 30.8 X-ray energy levels.

Question 30.10 shows that the voltage used across an X-ray tube determines a maximum possible frequency, and hence a minimum possible wavelength, for the X rays produced. Usually, however, when an electron strikes the target anode its energy is not lost in creating a single X-ray photon. The electron can lose its energy in a series of encounters with target atoms or it may give several X-ray photons of lower energy. This analysis, by itself, would give rise to a continuous X-ray spectrum with a minimum possible wavelength. This continuous spectrum is found but it is also found that there are some specific wavelengths which are much brighter.

The X-ray spectrum is both a continuous spectrum and a superimposed line spectrum as shown in two ways in Fig 30.7. Fig 30.7(a) is a sketch of an X-ray spectrum and Fig 30.7(b) is a graph showing how its intensity varies with the wavelength. The wavelengths of X-ray line spectra were found to be determined by the element of the target. The photon energies of these lines are very high, so an electron within a target atom must have fallen into one of its lowest possible energy levels. A space will be available in one of these low levels only if the electron originally there has been knocked out by an incoming electron. The following data analysis exercise shows how this is possible.

DATA ANALYSIS

X-Rays

Some of the energy levels within an atom of tungsten are given in Fig 30.8. The energies involved are given in keV so are of the order of a thousand times greater than was the case with hydrogen. You may know that electrons closest to the nucleus in an atom are referred to as the K-shell electrons. These are the ones in the lowest energy state. When these electrons are removed on bombardment by electrons hitting the target, electrons from higher levels fall in to fill the gap. In doing so, they emit high energy X-rays.

(a) Use the data on Fig 30.8 and on the graph of Fig 30.7(b) to identify and find the photon energy of the K_α, K_β, L_α and L_β lines.
(b) What is the wavelength of each of these lines?
(c) At what wavelengths might you expect to find a K_γ and a L_γ line?
(d) The lines found in (c) are possible but in practice their intensity is low. Why do you think this is so?
(e) What is the minimum potential difference which must be across the tube in order to get any K_α lines?

30.5 WAVE PARTICLE DUALITY

It was suggested in section 30.2 that electromagnetic radiation needed to be considered as a wave motion for transmission and as a stream of photons when it interacts with matter. The idea of both wave and particle approaches being applied successfully on an atomic level led de Broglie (pronounced de Broy) in 1924 to suggest that if waves could behave as particles, perhaps particles could behave as waves. He started with the following relationships for an electromagnetic wave, obtained from the quantum theory and Einstein's theory of relativity:

the energy of a photon $\qquad\qquad\qquad\qquad\qquad E = hf$
the mass equivalence (m) of this energy is given by $\quad E = mc^2$
equating these terms gives $\qquad\qquad\qquad\qquad hf = mc^2$
since $f = c / \lambda$ $\qquad\qquad \dfrac{hc}{\lambda} = mc^2$

So the momentum of a photon is given by $mc = h / \lambda$

De Broglie suggested that the wavelength, λ, associated with particles of momentum mv should be given by the same relationship, namely

$$mv = \frac{h}{\lambda}$$

This equation can be tested for electrons. Example 3 shows how the wavelength of an electron can be found.

Example 3

An electron is accelerated by a potential difference of 1500 V in a vacuum. What wavelength does such an electron possess?

$$\text{Energy gained by electron} = Ve$$

$$\text{Kinetic energy of electron} = \tfrac{1}{2}mv^2$$

$$v^2 = \frac{2\,Ve}{m}$$

$$v = \sqrt{\frac{2 \times (1500\ \text{V}) \times (1.6 \times 10^{-19}\ \text{C})}{9.1 \times 10^{-31}\ \text{kg}}}$$

$$v = 2.3 \times 10^7\ \text{m s}^{-1}$$

$$\text{Momentum of electron} = mv$$
$$= 2.3 \times 10^7\ \text{m s}^{-1} \times 9.1 \times 10^{31}\ \text{kg}$$
$$= 2.1 \times 10^{-23}\ \text{N s}$$
$$= \frac{h}{\lambda}$$

$$\text{Wavelength} = \frac{h}{\text{momentum}} = \frac{6.63 \times 10^{-34}\ \text{J s}}{2.1 \times 10^{-23}\ \text{N s}}$$
$$= 3.2 \times 10^{-11}\ \text{m}$$

Example 3 shows that the wavelength of electrons at this speed is of the order of a tenth of the separation of atoms in solids. Davisson and Germer in 1926 used the planes of atoms in a crystal as a diffraction grating to test this theory. They found that electrons could be diffracted in the way de Broglie's theory suggested. An electron diffraction pattern is shown in Fig 30.9. The pattern is circular because, unlike with a diffraction grating, the atoms are not all aligned in the same direction. The corresponding optical pattern to the one shown in Fig 30.9 could be obtained using a diffraction grating with monochromatic light if the grating was spinning round at high speed to give all possible orientations.

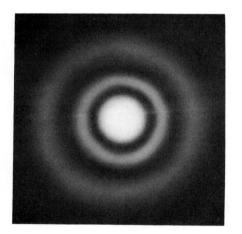

Fig 30.9 When electrons are diffracted by a thin layer of graphite atoms this diffraction pattern is obtained.

QUESTIONS

30.11 What is the wavelength associated with electrons having been accelerated by a potential difference of 56 kV in an X-ray tube?

30.12 Calculate the energy of a photon of light of wavelength 500 nm. What is the momentum of this photon?
The power of a beam of this light is 4.0 W. How many photons pass a point in the beam per second? What is the momentum of all the photons passing a point in the beam per second? What force does the beam exert on an object in its path?

30.13 The first-order electron diffraction pattern occurred at an angle of 22° when electrons, having been accelerated in a vacuum by a potential difference of 250 V, were passed through a thin film of graphite. Assuming you can use, the same equation as for optical diffraction, $n\lambda = d \sin \theta$, find the separation of the atoms in the graphite.

THE QUANTUM THEORY

- Electromagnetic waves can be considered as particles called photons.
- The energy of a photon is proportional to its frequency $E = hf$.
- The photoelectric effect is when a photon of incident light is able to cause an electron to be emitted from the surface of a metal. The energy of emitted electrons is found to depend on the frequency of the radiation and not on its intensity.

- Energy of incoming photon = work done to remove + kinetic energy of
  electron from metal of photoelectron

$$hf = \phi + \tfrac{1}{2}mv_{max}^2$$
$$hf = hf_0 + \tfrac{1}{2}mv_{max}^2$$
$$hf = hf_0 + eV_s$$

- The energy released when an electron falls from energy state E_2 to energy state E_1 equals the energy of the emitted photon of electromagnetic radiation:

$$E_2 - E_1 = hf$$

EXAMINATION QUESTIONS

30.14 Electrons need energy to escape from the surface of a material in the processes of (i) thermionic emission, or (ii) photoelectric emission.
  (a) Explain briefly why energy is needed for the electrons to escape.
  (b) From where *in each case* do the escaping electrons acquire their energy? (5)

(ULSEB 1983)

30.15 In a photoelectric emission experiment a metal surface in an evacuated tube was illuminated with monochromatic light. If the experiment were repeated with light of the *same* wavelength, but of *twice* the intensity, how would this affect
  (a) the photon energy,
  (b) the maximum kinetic energy of the photoelectrons,
  (c) the work function of the metal,
  (d) the photocurrent? (4)

(UCLES 1986)

30.16 (a) (i) State Einstein's equation for the photoelectric effect, and explain the significance of each of its terms (5)
  (ii) Describe a method for the measurement of the Planck constant h, using the photoelectric effect. (9)
  (iii) Describe briefly a device that utilises the photoelectric effect. (4)
  (b) The memory of a computer element can be erased by exposing it to ultra-violet radiation of wavelength 360 nm for a period of 20 minutes. The memory is contained in an insulated silicon film of exposed area 1.5×10^{-9} m^2, and the intensity of the ultra-violet light is 20 W m^{-2}.
  (Take the Planck constant h to be 6.6×10^{-34} J s, the speed of light c in vacuum to be 3.0×10^8 m s^{-1}, and the electronic charge e to be -1.6×10^{-19} C.)
  Calculate:
  (i) the number of photons incident on the film in 1 second; (5)
  (ii) the charge acquired by the film in 20 minutes exposure, if 1% of the incident photons cause photoemission of electrons. (4)
  Explain why erasure would be slower if the memory were exposed to sunlight of the same total intensity. (3)

(OXFORD 1986)

30.17 Describe an experiment to measure the Planck constant using the photoelectric effect. (7)
A strip of clean magnesium ribbon, surrounded by a cylinder of copper gauze maintained at a positive potential of 6.0 V with respect to the magnesium, is connected to the input of an amplifier which measures the p.d. across the resistor R as shown.

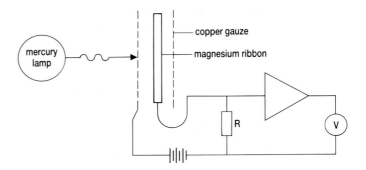

When the magnesium is illuminated with mercury light of wavelength 254 nm the amplifier detects a current flowing in R. State the origin of this current. (1)
Calculate the energies of photons of mercury light of wavelengths 254 nm and 546 nm. (2)
Using these values and the data at the end of the question, describe and explain in detail what you would observe when each of the following experiments is carried out separately.
  (a) The polarity of the 6.0 V battery is reversed.
  (b) The mercury lamp is moved farther away from the magnesium ribbon.
  (c) A filter selects mercury light of wavelength 546 nm in place of that of wavelength 254 nm.
  (d) The e.m.f. of the battery is increased to 10 V.
  (e) The magnesium ribbon is replaced by a strip of copper. (10)
(The photoelectric work function of magnesium $= 4.5 \times 10^{-19}$ J (2.8 eV)
The photoelectric work function of copper $= 8.1 \times 10^{-19}$ J (5.05 eV)
Planck constant $= 6.6 \times 10^{-34}$ J s.)

(O & C 1983)

30.18 (a) What are the dimensions of (i) energy, (ii) momentum, (iii) the Planck constant, h? It can be shown that the total energy E of a particle is related to its momentum p and its mass m_0 when at rest, by the equation

$$E^2 = p^2c^2 + m_0^2c^4$$

where c is the speed of electromagnetic radiation in a vacuum. Show this equation to be dimensionally correct. (4)

THE QUANTUM THEORY

(b) Given that m_0 is zero for a photon, use the above equation to derive an expression for the momentum of a photon in terms of its wavelength. (2)

(c) Estimate the potential difference across which an electron should be accelerated if it is to have the same energy as a photon of ultra-violet radiation. Support your estimate with a calculation, by choosing a suitable value for the wavelength of the electromagnetic radiation. You may assume that the value of the product hc is 2.0×10^{-25} J m. (Charge of an electron = -1.6×10^{-19} C.) (4)

(JMB 1987)

30.19 (a) State **four** ways in which X-rays resemble visible light. (4)

(b) Draw a graph to show the distribution of energy with wavelength in the radiation from an X-ray tube, and explain its features. (8)
Explain the factors that determine:
(i) the short-wavelength limit; (2)
(ii) the intensity of the radiation. (2)

(OXFORD 1987)

30.20 (a) (i) List the main classes of electromagnetic waves in order of ascending frequencies (4)
(ii) Describe **three** properties that all such waves have in common. (3)

(b) Explain the formation of
(i) spectral lines in a gas discharge tube; (4)
(ii) dark absorption lines in the spectrum of sunlight. (4)

(c) (i) Draw a diagram to show the wavelength distribution of the energy of the X-rays emitted by an X-ray tube and explain the principal features of the spectrum. (5)
(ii) Explain how the minimum wavelength λ_{min} is related to the accelerating voltage V applied to the tube. (4)
(iii) Explain how measurements of characteristic X-rays led to an ordering of the elements in terms of their atomic structure. (6)

(OXFORD 1986)

30.21 The diagram shows some of the energy levels of the outermost electron in a mercury atom. Level Q represents the lowest possible energy (the ground state).
(a) Explain why a line spectrum results from an atom with such energy levels.

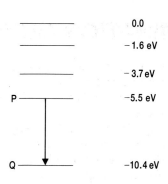

(b) Calculate the energy change in joules when the electron moves from level P to level Q, and determine the wavelength of the spectral line associated with this transition.

(c) Explain what is likely to happen if a moving electron of energy 7.0 eV collides with an isolated mercury atom in its ground state. Explain whether the effect would be different if a photon, also of energy 7.0 eV, were to be incident on the atom.
(The Planck constant = 6.6×10^{-34} J s; speed of light = 3.0×10^8 m s^{-1}; charge on the electron = -1.6×10^{-19} C) (8)

(AEB 1986)

30.22 What is a *photon*? Show that the energy E of a photon and its wavelength λ are related by $E\lambda = 1.99 \times 10^{-16}$ J nm. (4)

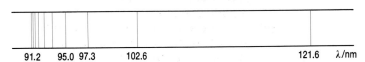

The diagram represents part of the emission spectrum of atomic hydrogen. It contains a series of lines, the wavelengths of some of which are marked. There are no lines in the series with wavelengths less than 91.2 nm

(a) In which region of the electromagnetic spectrum are these lines? (1)

(b) Using the relation between E and λ given above, find the photon energies equivalent to all the wavelengths marked. (1)

(c) Use this information to map a partial energy level diagram for hydrogen. Show, and label clearly, the electron transitions responsible for the emission lines labelled in the diagram above. (4)

(d) Another line in the hydrogen spectrum occurs at a wavelength of 434.1 nm. Identify and label on your diagram the transition responsible for this line. (3)

Emission spectra are often produced in the laboratory using a discharge lamp containing the gas to be investigated. Explain the physical processes occurring within such a lamp which lead to the excitation of the gas and the emission of light.   (4)

(UCLES 1985)

30.23 The ionisation energy for hydrogen is 13.6 eV and the first and second excitation energies are 10.2 eV and 12.1 eV respectively.
Draw an energy level diagram showing the ground state and the next two energy levels for hydrogen. Mark on the diagram the values of the energies corresponding to each level.
Mark on your diagram the transition between two of these levels that would result in the emission of radiation of the longest wavelength. Calculate the value of this wavelength.
(1 eV $= 1.6 \times 10^{-19}$ J;
the Planck constant $= 6.6 \times 10^{-34}$ J s;
speed of light $= 3.0 \times 10^8$ ms^{-1}.)   (7)

(ULSEB 1988)

30.24 Draw a labelled diagram of an apparatus which may be used to demonstrate the phenomenon of electron diffraction.   (3)
Sketch a typical diffraction pattern obtained in this experiment, and explain qualitatively why it has this form.   (4)
Explain very briefly the significance of this experiment.   (2)
Two large plane metal electrodes are arranged parallel to each other in an evacuated tube. One (the collector) is at a positive potential with respect to the other (the emitter). Starting from rest, an electron leaves the middle of the emitter and moves perpendicularly to the plates towards the collector. Sketch clearly labelled graphs showing how the following quantities depend on the distance x from the emitter:
(a) the electric potential energy E_p of the electron,
(b) its kinetic energy E_k,
(c) its speed v,
(d) its associated wavelength λ.   (5)
If the accelerating potential is 150 V, find the wavelength associated with the electron as it reaches the collector.   (3)

(UCLES 1986)

THE QUANTUM THEORY

Appendix A:

ANSWERS TO QUESTIONS

In many of the examination questions physical constants are required in order to answer the question. Different examination boards have different policies regarding the supply of this information. In some cases the required data is given in, or at the end of, the question however some boards supply data booklets containing the information. If therefore you find it necessary to use additional information to that given in the question you will find the required data in the appendix.

Many exam papers are roughly 100 minutes long and carry 100 marks. This implies that the mark given at the end of a question (in brackets) is also a guide to the length of time it should take to answer the question.

In these answers you will often find more significant figures quoted than are strictly justified. Advice is often given not to give answers to more significant figures than are stated in the question, but if this is done with multi-answer questions then inaccuracies can be introduced by rounding up or down too soon. Also the numerical value of the answer affects the number of significant figures which should be given. 1 in 99 is a much smaller uncertainty than 1 in 11 although both 99 and 11 both have two significant figures. Here, therefore, I have attempted to give answers so that any unreliability is limited to the final digit.

Chapter 1

1.1 $[T^{-1}]$; $[MLT^{-2}]$; $[ML^{-1}T^{-2}]$; $[ML^2T^{-2}]$; $[ML^2T^{-3}]$; $[ML^2T^{-3}I^{-1}]$; $[ML^2T^{-3}I^{-2}]$; $[M^{-1}L^{-2}T^4I^2]$

1.2 s^{-1}; $kg\,m\,s^{-2}$; $kg\,m^{-1}\,s^{-2}$; $kg\,m^2\,s^{-2}$; $kg\,m^2\,s^{-3}$; $kg\,m^2\,s^{-3}\,A^{-1}$; $kg\,m^2\,s^{-3}\,A^{-2}$; $kg^{-1}\,m^{-2}\,s^4\,A^2$

1.4 (a) correct, (b) incorrect, (c) correct, (d) incorrect, (e) correct

1.6 $kg^{-1}\,m^{-3}\,s^4\,A^2$

1.7 mass $[E\,v^{-2}]$; length $[E\,F^{-1}]$; time $[E\,F^{-1}\,v^{-1}]$

1.8 kg, m, s; (i) joule (ii) hertz
(i) J s (ii) $[ML^2T^{-1}]$

1.9 (a) potential difference = energy/charge (or power/current), (b) $kg\,m^2\,s^{-3}\,A^{-1}$, (c) Homogenous equation since both sides have the base unit $kg\,m^2\,s^{-3}$

1.10 The definition of potential difference makes use of the quantity *power*. If power is defined in terms of potential difference this is a circular argument.

1.11 (b) ML^2T^{-2}; ML^2T^{-2}

1.12 (c) $[ML^2T^{-3}I^{-1}]$; $[ML^2T^{-2}I^{-2}]$
J; $J\,m^{-3}$; A s; $V\,A^{-1}$ (many others are possible)

1.13 (b) $\varepsilon_0\,\mu_0 = c^{-2}$

1.14 (b) (i) 0.090 Hz; (ii) 0.40
(c) (i) 288 m; (ii) 0.075 Hz
(d) (i) $[ML^2T^{-3}I^{-2}]$; (ii) $[MT^{-2}I^{-1}]$;
(iii) $[M^{-1}L^{-2}T^4I^2]$

1.16 N m

Chapter 2

2.1 (a) 0.0013, 0.13% (b) 0.002, 0.2% (c) 0.0008, 0.08%

2.2 (a) +0.04% (b) +1.3% (c) –0.45% (d) +2% (e) +0.07%

2.3 (68.9 ± 1.4)s, 0.02; [This perhaps should be quoted as (69 ± 2)s]

2.4 Sum: 3.463 mm, 0.003
Difference: 0.303 mm, 0.03

2.5 $(9.73 \pm 0.13)\,m\,s^{-2}$ or $(9.7 \pm 0.2)\,m\,s^{-2}$

2.6 $(1.5 \pm 0.3) \times 10^{-3}\,N\,s\,m^{-2}$. Note that all lengths must be in metres and that $1\,cm^3 = 10^{-6}\,m^3$.

2.8 (a) $(3.75 \pm 0.05)\,m\,s^{-2}$ (b) $6.6\,m\,s^{-1}$ (c) $18.6\,m\,s^{-1}$

2.9 $R = (900 \pm 10)\,\Omega$; $L = (0.380 \pm 0.005)$ H

2.10 $n = -3$ so law is $F \propto 1/d^3$

2.11 greater by 9% of indicated speed

2.12 (i) $(2.0 \pm 0.2) \times 10^{11}$ Pa. (ii) concentrate first on improving the accuracy of measurement of the area of cross section and then on the extension.

2.14 (ii) $k=0.060\,s^{-1}$, $\ln B=0.60$, increasing k decreases time constant.

Chapter 3

3.2 $5.8\,m\,s^{-1}$ in a direction S 31° W

3.3 $70\,m\,s^{-1}$ in new direction

3.4 $75.5\,m\,s^{-1}$ from a direction N 4° W

3.5 (a) 2.6×10^{-6} rad $= 1.5 \times 10^{-4}$ degrees
(b) and (c) $2.6 \times 10^{-3}\,m\,s^{-1}$ towards the centre of the Moon's orbit

3.6 **(a)** 1080 J **(b)** 0 **(c)** 830 J

3.7 **(a)** 0.023 N **(b)** 0 **(c)** 0.0097 N

3.9 10.6 N upwards at 37° to the vertical

3.11 19.2 J

Chapter 4

4.1 Ball $67\,m\,s^{-1}$; Club $36\,m\,s^{-1}$ after hitting ball

4.3 $6.4\,m\,s^{-2}$

4.4 576 m

4.5 $49\,m\,s^{-1}$ in opposite direction; 0.28 m

4.6 (a) 50 s (b) Both travel 2000 m

4.7 1.43 s; $14\,m\,s^{-1}$

4.8 41 m

4.9 $20\,m\,s^{-1}$

4.10 (a) 2.05 s (b) 4.1 s (c) 21 m (d) 91 m

4.11 (a) 68 m (b) $15.8\,m\,s^{-1}$ (c) 68 m
(d) $5.3\,m\,s^{-1}$ upwards (e) 18.5° (f) $16.7\,m\,s^{-1}$

4.12	48 000 N s due East	**6.13**	(b) (i) 2800 N (ii) 1.7×10^5 J
4.13	3.1×10^{-20} N s parallel to field		(c) (i) 80 rad s^{-1} (ii) 1770 N
4.14	**(a)** 3.6×10^{29} N s **(b)** 0		

4.15 765 N s in direction S 79° W

4.16 2.9 m beyond stop line

4.18 (a) 0; 9.8 m s^{-2} (b) 2.5 m s^{-1} (c) 0.76 m^{-1}

4.19 1.4 mm

4.20 1.03 s after being served (b) 30.8 m

4.21 (a) 5.57×10^6 m s^{-1} at 90° to PQ
(b) 2.67×10^9 s; 2.09×10^{15} m s^{-2} at 90° to PQ

4.22 63°

4.23 30 N s; 25 000 m s^{-2}

Chapter 5

5.1 All three diagrams are identical. The force the road exerts on the car is equal and opposite to the weight of the car with the resultant force zero in each case.

5.2 61 000 N

5.3 (a) 1.5 m s^{-1} (b) 45 000 N (c) 55 000 N
(d) 20 500 N

5.4 1.47 m s^{-2}; 2.6 s

5.5 0.61 m s^{-1}; A exerts force of 4900 N forwards on B; C exerts force of 5900 N forwards on B

5.6 55 N

5.7 Acceleration; 1.1 m s^{-2} upwards. Force; 270 N downwards

5.8 (a) 1800 N (b) 2.7

5.9 17 m s^{-2}; 9 m s^{-2}; 3 m s^{-2}
(a) 1.4 N s; 1.4 N
(b) 0.96 Ns; 0.75 N

5.10 (a) 3500 N (b) 0.35 m s^{-2}

5.11 (b) 5 N; 2.5 m s^{-2} to the right

5.12 (a) 4.0 m s^{-2} at start (b) 7.1 m s^{-1} at end

5.13 (a) 4.2 m (b) 18 m s^{-1} upwards (c) 3.2 m
(d) (i) – 10 m s^{-2} (ii) – 0.64 N s (iii) – 0.80 N
(e) 2.2 s (f) 6.4 m s^{-1}

5.14 (c) (i) 5 N (ii) 5 N (iii) 3.5 m s^{-1}

5.15 5.0 s

5.16 (a) 320 m (b) 5000 N

5.17 30 m s^{-1}

5.18 (a) 2.5 m s^{-1} (b) 1200 N (c) Extra 80 N to accelerate slab

5.19 (ii) 11.6 ms
(iii) 51 s^{-1}

Chapter 6

6.1 (a) 75 N (b) 350 N (c) 81 N (d) ≈ 100 N
(e) 800 N

6.2 0.030 N m anticlockwise

6.3 P = 1.4×10^5 N
Support at B = 0, at C = 5.6×10^5 N

6.4 3.6 N m anticlockwise

6.5 Force at A = 4.3×10^5 N, at B = 4.2×10^5 N

6.6 6.0×10^7 N m anticlockwise

6.8 17.5 m.

6.9 300 N

6.10 (a) 59 N (b) (i) 54 N

Chapter 7

7.1 2400 N

7.2 7200 N

7.3 66 m s^{-1}; 3.1 N s

7.4 8200 N; 29 m s^{-1}

7.5 2.1×10^{-20} N s; 2.1×10^{-12} N

7.6 5.7 m s^{-1} in direction van was travelling

7.7 0.63 m s^{-1}

7.8 4 mm s^{-1}

7.9 6500 J

7.10 2.9×10^6 J; a counterpoise weight is used

7.11 46 000 J

7.12 1.6×10^{-15} J

7.13 7 J

7.14 0.027 J

7.15 12.5 J per spring

7.16 2200 J

7.17 9.7 J

7.18 93 kW

7.19 41 kW; P = Fv

7.20 (a) 6250 m (b) 3.75×10^8 J (c) 3.75×10^8 J

7.21 15.0 m s^{-1}; 43.4 m s^{-1}

7.22 16 000 N

7.23 Breaking force greater than 8.1 N

7.24 602 J; 5.0 m s^{-1}

7.25 31 m

7.26 0.96 J; 3.6 m s^{-1}

7.27 1.5×10^{-13} J on ebb and 1.5×10^{-13} J on flow: 660 MW

7.28 3.6×10^3 m s^{-1}

7.29 1.5v

7.30 (a) 0.156 m s^{-1} (b) 0.056 J

7.31 θ=60°; $u = 8.7 \times 10^5$ m s^{-1}; $v = 4.3 \times 10^5$ m s^{-1}

7.32 1.42 m s^{-1} sθ; 60

7.33 alpha particle 0.983E; nucleus 0.017E

7.34 multiplication by a factor of 2.8

7.35 (b) 330 N

7.36 $\frac{1}{2}u$; KE at $t_0 = \frac{1}{4}mu^2$; KE at $t_s = \frac{1}{2}mu^2$

7.37 After 1.5 s velocity of container (v) is 0.20 m s^{-1}; after 2.0 s $v = 0$

7.38 (i) 1.5×10^6 J (ii) 1.0×10^5 N (iii) 60 s

7.39 (a) (i) $\frac{\pi}{2}\sqrt{\frac{I}{8}}$ (i) $\frac{1}{2}mgl\theta^2$ (iii) $m\theta\sqrt{\frac{gl}{2}}$

7.40 E/4

7.41 (b) (iii) 4 (iv) alpha particle

7.42 1.2×10^6 kg s^{-1} = 1200 tonnes per second

7.43 3000 N

Chapter 8

8.1 (a) 4π (b) $\frac{3}{4}\pi$ (c) $\frac{5}{2}\pi$ (d) 100π (e) $\frac{1}{18}\pi$
(f) 48π

8.2 0.105 rad s^{-1}; 5.2×10^{-3} m s^{-1}
1.745×10^{-3} rad s^{-1}; 8.7×10^{-5} m s^{-1}
1.454×10^{-4} rad s^{-1}; 5.8×10^{-6} m s^{-1}

8.3 (a) 1.99×10^{-7} rad s^{-1}

(b) 7.27×10^{-5} rad s^{-1}
(c) 2.6×10^{-6} rad s^{-1}
(d) 2.6×10^{-6} rad s^{-1}

8.4	5.95×10^{-3} m s^{-2}
8.5	(a) 367 rad s^{-1} (b) 84 m s^{-1} (c) 3.1×10^4 m s^{-2}
8.6	1.18×10^4 m s^{-2}
8.7	340 rad s^{-1}
8.8	4.46 N
8.9	For period of 6 s: tension = 275 N and support force = 480 N
	For period of 3 s: tension = 1100 N and support has to push down with a force of 160 N.
8.10	5.4 m s^{-1}
8.11	920 N downwards
8.12	27 m s^{-1}; 2760 N
8.13	Weight 343 N; 424 N along line of tilt **or** vertical component 343 N up and horizontal component 250 N towards centre of circle. 36°
8.14	18.3 cm
8.15	(a) (i) 3.1 m s^{-1} (ii) 2.2 m s^{-1} (b) 0.20 N
8.16	(b) 0.14 revolutions per second
8.17	(c) 11.5°
8.18	(iii) 0.31
8.19	(i) 1.62 m s^{-2} (iii) 9.4°
8.20	(c) (i) 1.68 m s^{-1} (ii) 0.566 N (iii) 0.084 N s

Chapter 9

9.1	(a) 1.0×10^{-42} N (b) 9.8 N (c) 3.6×10^{22} N
9.2	(a) 6×10^{-5} N (b) 20 N (c) 3×10^{-6} rad (= 1.7×10^{-4} degrees)
	(d) $M_e = mR^2/x^2 \tan\theta$
9.3	3.4×10^8 m from the centre of the Earth
9.4	9.8 J kg^{-1}
9.5	-6.26×10^7 J kg^{-1}; 1.28×10^7 m; 6.26×10^7 J kg^{-1}
9.6	6.2×10^5 m s^{-1}
9.7	1.08×10^{11} m; 2.28×10^{11} m; 4.51×10^{12} m;
9.8	4×10^{23} kg
9.9	-3.0×10^{10} J; 1.5×10^{10} J
9.10	2.3×10^{10} J
9.11	Loss of P.E. = 4.0×10^8 J; gain in K.E. = 2.0×10^8 J; total energy loss = 2.0×10^8 J
	Initial speed = 7300 m s^{-1}; final speed = 7600 m s^{-1}
9.12	13.9 N; 0.154 m s^{-2}
9.13	24.9 N kg^{-1}; 373 N; (a) 2.25 m s^{-2} (b) 33.7 N (d) at pole 373 N, at equator 339 N
9.14	(b) 2.71×10^{-3} m s^{-2}
9.16	(i) 5.5×10^{26} kg (ii) 6.9×10^{-11} N m^2 kg^{-2}
9.17	(a) 10 N kg^{-1} (b) 30 J kg^{-1} (c) 20 N (d) 60 J
9.18	(b) (i) kinetic energy of spacecraft (ii) thermal energy (c) (i) 1.3×10^7 J kg^{-1} (ii) 1.26×10^{14} J
9.19	1690 m s^{-1}; 1.41×10^6 J kg^{-1}
9.20	(c) 0.0123
9.21	(b) 8000 m s^{-1}
9.22	7000 m s^{-1} (a) 19.6 N kg^{-1} (b) 16000 m s^{-1}

Chapter 10

10.3	9.4 m s^{-1} 1.5×10^5 m s^{-2}
10.4	(a) 5.3 hours (b) 0.35 (c) 0.67 m h^{-1} (d) – 6 cm, – 16 cm, – 16 cm, – 10 cm, – 8 cm, 0 (e) 1.6 m
	(f) 12 h (g) height = 1.6(1 + sin $\pi t/6$), height in metres, t in hours.
10.5	0.246 s
10.7	(a) 6.0 cm (b) 5.1 cm
10.8	(b) (ii) $T = 2\pi\sqrt{\dfrac{m}{k}}$ (iv) 0.059 m (v) 1.75 J
10.9	(c) (i) 3.8 mm (ii) 7.5 mJ (iii) 0.27 m s^{-1}
10.10	0.62 mm
10.11	0.44 s: 0.037 m; 0.45 m s^{-1}; –7.4 m s^{-2}
10.12	(a) 2.27×10^4 m s^{-2} (b) 30.2 m s^{-1}
10.13	(c) (i) 31 m s^{-1} (ii) 4.9×10^5 m s^{-2} (iii) 6.0 W
10.14	(a) 0.045 m (b) 1.01 s
10.16	(b) 0.050 m; 2.0 s (c) 0.16 m s^{-1}
10.17	(i) 0 (ii) 0.99 m s^{-2} (iii) 0.67 s
10.18	(a) 0.71 s
	4th bounce is 1.66 m, 5th bounce is just less than 1.5 m

Chapter 11

11.1	Red; 4.3×10^{14} Hz: violet; 7.5×10^{14} Hz
11.2	Not usually, but there can be a need when pulses of waves are being considered.
11.4	At 40 m; intensity = 1.1×10^{-3} W m^{-2}, amplitude = 0.80 mm, at 100 m; intensity = 1.8×10^{-4} W m^{-2}, amplitude = 0.32 mm.
11.5	See diagram below

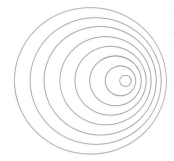

Wavelength in front = 0.30 cm; frequency in front = 130 Hz
wavelength behind = 1.3 cm; frequency behind = 31 Hz.
This effect is known as the Doppler effect. The apparent frequency, f', is related to the true frequency, f, by the equation

$$f' = f\left(\frac{c-u}{c}\right)$$

where c is the speed of the wave and u the speed of the source.

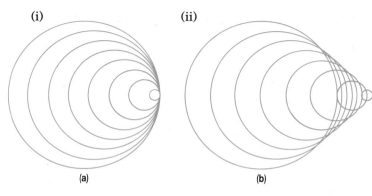

(i)　　　　(ii)

(a)　　　　(b)

11.7 Parallel straight lines before hitting the mirror, parts of circles after reflection.

11.8 6.7 μs

11.9 2.7;　0.37

11.10 (c) (i) 80 m (ii) 28°, critical angle is 26.4°

11.11 13.1°

11.12 (a) (i) 1.33 m (ii) 5 μm (iii) A leads B by 2.4 rad

11.13 (a) C (b) 1.88×10^{-3} s

11.15 60°; $\frac{1}{4}$

11.16 13.4%

11.17 2.2×10^{8} m s^{-1}

11.18 (b) 1.4 mW m^{-2};　4.2 m

11.19 (i) $\theta = 0.40$ rad (ii) $\phi = 0.20$ rad (iii) 2.97×10^{8} m s^{-1} (iv) 0.05 cm

Chapter 12

12.1 0.02 rad

12.3 33.8°

12.4 82.3°;　97.7°

12.5 53°, this is called the Brewster angle and is the angle at which the reflected ray is polarised most.

12.6 1.515

12.7 1.514

12.8 144 m

12.9 (a) 20 dioptre (b) 56.3 mm (c) 10 mm (d) 0.125

12.10 118 mm

12.11 Light is parallel after passing through the lens. This can be seen as an image at infinity.

12.12 Virtual image, $v = -5.5$ cm;　$m = 0.45$

12.13 Lens power = −1.25 dioptre;　range 18.5 cm − ∞

12.14 Angular magnification;　50, 100, 200
Brightness;　2500 times, 10 000 times, 40 000 times

12.16 (b) 35°　34%

12.17 Angle of emergence 77°, taking $n = 1.5$

12.18 1.41

12.19 41.4°

12.20 (a) (ii) 800 mm, 200 mm (iii) 1.0 mm, 16 mm (b) (iii) − 0.5 dioptre

12.21 From $u = 41.7$ mm ($m = 6$) to $u = 50.0$ mm ($m = 5$)

12.22 −15 cm

12.23 110 mm

12.24 (b) (i) 0.0092 rad (ii) 2.8 mm (iii) 0.011 rad (iv) 1.2　(c) 6.0

12.25 (c) (i) 200 mm (ii) 150 mm, 7.6° (iii) 1.15°, 6.6

(iv) 6.7　(v) 9.4

12.26 (a) (ii) 1.0 ms

Chapter 13

13.1 275 N;　52.6 Hz;　2.4 Hz

13.2 Taking speed of sound = 344 m s^{-1} at 20 °C gives 0.59 m

13.3 444 Hz or 436 Hz

13.4 (a) increase (b) decrease (c) increase

13.5 4.52 mm

13.6 Ship must lie at some position on four curved lines so that it is either half a wavelength (1500 m) or one and a half wavelengths (4500 m) nearer one of the beacons

13.7 3.3 s

13.8 5.9×10^{-7} m

13.9 3. At 25.0°, 39.3° and 57.7°

13.10 First 10.0°;　second 23.5°;　incomplete third after 54.8°

13.12 (b) Beat frequency of 100 Hz; two beat periods visible on screen

13.13 (c) (i) 50 N (ii) 42 mm

13.14 (b) 3.5 cm (c) (i) 0.3 m

13.15 1.0 m

13.16 (a) (i) 340 m s^{-1} (ii) 696 mm

13.18 (c) $\theta_1 = 3.25 \times 10^{-3}$ rad;　$\theta_2 = 6.5 \times 10^{-3}$ rad

13.19 (a) 6.16×10^{5} m^{-1} (b) yes for $\phi = 105°$

13.20 (b) First order for $\sin \theta_1 = 0.3846$;　$\theta_1 = 22.6°$; second order for $\sin \theta_2 = 0.7692$;　$\theta_2 = 50.3°$; $\sin \theta_3$ would need to be 1.1538

13.21 (a) (iii) third order blue, second order red; $\lambda_R = 681$ nm (b) 589 nm

Chapter 14

14.1 (a) 5200 C (b) 25 mA (c) 700 s

14.2 (a) 140 C (b) (i) 1.05 mC (ii) approximately the same as (i) (c) zero. The charge in the wire oscillates but no charge moves permanently past any point in the wire.

14.3 4.0×10^{-4} m s^{-1}

14.4 1.40×10^{5} mol; 8.44×10^{28} mol^{-1}; approximately 1

14.5 (a) 0.94 mA

14.6 (a) 600 000 J (b) 2500 C (c) 0.25 A

14.7 B, 8 V: E, 10 V: F, 10 V: W, 12 V: X, 20 V: Y, − 2 V: Z, − 10 V

14.8 There is one deliberate mistake in this table. Can you find it?

	Battery	A	B	C
charge/C	800	100	700	800
power/W	40	2	14	24
energy/J	16000	800	5600	9600
current/A	2.0	0.25	1.75	2.0
potential difference/V	20	8.0	8.0	4.0

14.9 6900 Ω in dark:　113 Ω in light

14.10 (a) (i) ≈1000 Ω (ii)100 Ω (iii) 20 Ω (iv) 250 kΩ
(b) 0.73 V

14.11 (a) (i) 301 Ω (ii) 297 Ω (iii) 293 Ω (iv) 179 Ω
(b) when V is small V^2 is very small and
$R \approx 1/b$
(c) 130 Ω

14.12 2.8×10^{-8} Ω m; aluminium

14.13 Regard the coil as 190 circular turns; 15 Ω

14.14 1010 °C

14.15 8.4 V; 0.75 A

14.16 Answer in terms of the potential difference across the internal resistance not in terms of the motor taking so much current that there is not enough for the lights

14.17 21.1 V; 0.56 Ω

14.18 5.2 V; 1.76 Ω

14.19 12 A; 2880 W; 4900 s

14.20 12.1 MJ; 2.4 Ω; 240 W

14.21 2.8 kW

14.22 £181.86

14.23 (a) 0.985 (b) (i) 0.050 mm s^{-1} (ii) 10 000 s

14.25 (a) 77 Ω (b) 3.5 m

14.28 (i) 4.9 Ω (ii) 1.7 Ω;
2.0 A; greater than half

Chapter 15

15.1 $a = 3.1$ A; $b = 3.2$ A; $c = 1.1$ A; $d = 4.3$ A

15.2 170 μA; 64.08 mA

15.3 9 Ω; 1.0 W, 0.75 W, 2.25 W and 0.50 W to internal resistance

15.4 (a) 5.4 V (b) 0.6 V (c) 5.0 V (d) 0.33 mW;
100 mW (e) 120 mW (f) 20 mW

15.5 (a) 1.5 A (b) C, –1.2 V; D, –7 V; E, 8 V;
(c) 11 V (d) 1.6 Ω (e) 52.5 W; 16.5 W
(f) 24.5 W, 2.7 W, 1.8 W, 40 W

15.6 (a) 80 kΩ (b) 20 kΩ (c) and (d) 36 Ω

15.7 42 k Ω in parallel (e.g. 39 kΩ + 1.8 kΩ + 1.2 kΩ)

15.8 If the fine control resistance is too small it has too little effect. It is most suitably about $\frac{1}{10}$ of the coarse control resistance.

15.9 (a) 8.98 V (b) 0.69 V 860 Ω

15.10 0.61 V; 95 kΩ

15.11 (a) 0.0375 Ω (b) 0.225 V

15.12 300 V, 119 920 Ω; 30 V, 11 920 Ω; 3 V, 1120 Ω;
0.3 V, 40 Ω

15.13 900 kΩ, 400 kΩ, 233 kΩ, 150 kΩ, 100 kΩ,
67 kΩ, 43 kΩ, 25 kΩ, 11 kΩ, 0

15.14 (a) 320 Ω, (b) 1.33 kΩ

15.15 $R = 1.58$ Ω; $l_1 = 13.6$ cm

15.16 (a) Potential at X = 5.989 V: potential at Y = 6.000 V (b) $V_0 = 10$ V

15.17 (1) 0.0080 V (2) 0.0080 V (3) $RE/(r + R)$ (4) E
(6) gradient is internal resistance

15.18 (b) 6.0 Ω (c) 500 % (d) bulb filament breaks

15.19 (i) in internal resistance 0.29 A; in 5 Ω and 10 Ω 0.13 A; in 4 Ω and 8 Ω 0.16 A
(ii) 1.9 V (iii) zero (iv) (a) no change
(b) increase of 0.089 A

15.20 (ii) 0.30 A (iii) 2.0 Ω. Component x= 20 Ω

15.21 (a) 360 Ω (b) 0.96 V

15.22 (i) 10 V read as 10 V: (ii) 10 V read as 6.0 V
40 V f.s.d: yes; 40 V, 30 V: 25 kΩ/V

15.23 (b) (i) 0.90 mA (ii) 0.09 Ω; 0.81 Ω; 8.1 Ω;
81 Ω

15.24 (a) 6.0 V (b) 4.0 V

15.25 239 Ω becomes 239.5 Ω when under stress;
0.02 V

15.27 (b) 2 Ω

Chapter 16

16.1 9.0 MN

16.2 230 N; 1.38×10^{29} m s^{-2}

16.3 5.7×10^{-9} N

16.4 3.4×10^5 N C^{-1}

16.5 (a) 1.37×10^6 C (b) 2.7×10^{-9} C m^{-2}

16.6 (a) zero (b) 1.6×10^{-23} N m

16.7 (a) 1400 V (b) 2.24×10^{-16} J (c) 2.24×10^{-16} J
(d) 2.2×10^7 m s^{-1}

16.8 $V_A = -1300$ V; $V_B = 2200$ V; $V_C = -1300$ V;
$V_D = -4800$ V; $V_E = -1300$ V

16.9 (b) (i)8.2×10^{-8} N (ii) 2.2×10^6 m s^{-1},
1.52×10^{-16} s

16.10 $\approx 5 \times 10^{21}$ N C^{-1}

16.14 (b) (i) 60 000 N C^{-1} (ii) 6000 V

16.16 (b) (i) zero (ii) 3 mJ

16.17 (a) both forces of attraction = 5.4×10^{-9} N
(b) 6.4×10^{-9} N downwards

Chapter 17

17.1 0.024 μF; Capacitance is doubled

17.2 Charge on 22 μF = 198 μC; charge on 47 μF = 423 μC; Total capacitance = 69 μF

17.3 Charge on 0.050 μF = 1.0 μC; charge on 0.010 μF = 0.20 μC; charge on 0.030 μF = 1.20 μC; e.m.f. = 60 V; total capacitance = 0.020 μF

17.4 (a) 6 mC, 0.018 J (b) 12 mC, 0.072 J

17.5 92 V

17.6 (a) 2000 μC (b) 165 μC on both (c) 20 mJ by 100 μF; 1.4 mJ by 10 μF; 0.29 mJ by 47 μF
(d) 20 V across 100 μF; 16.5 V across 10 μF;
3.5 V across 47 μF

17.7 Maximum charge is 90 μC

17.8 Maximum current is 4.5 mA at $t = 0$

17.9 (a) 4.5 mA (b) 90 μC (c) 0.020 s (d) 0.020 s
(e) second (f) 63.2%, 99.3%

17.10 36.8, 13.5, 5.0, 1.8, 0.67

17.13 5.96 s; 24 s; time constant = 8.6 s; half life = 0.693 × time constant

17.14 (b) (i)25 V, 1000 μC (ii) 8.3 V

17.15 (a) 2 in series (b) 2 in series in parallel with 2 in series

17.16 (a) 4 V (b) 4 μC (c) 8 μJ

17.17 0.37

17.20 400 s

17.21 20 μA, (32 ± 2) μA

17.22 (b) (i) 0.028 mm

Chapter 18

18.1 1.77×10^{-4} T
18.2 0.0034 N
18.3 0.39 m
18.4 4.8×10^7 C kg^{-1}
18.5 (a) 0.40 mm s^{-1} (b) 3.1×10^{-24} N
(c) 7.8×10^{22} m^{-1} (d) 0.24 N
18.6 (a) 43.6 m s^{-1} (b) 0.26 T (c) positive
18.7 1.04×10^{25} m^{-3}; 0.15 m s^{-1}
18.8 (a) reduced (b) increased (c) reduced
18.9 1.8 rad; 0.36 V; 0.20 V rad^{-1}
18.10 1.85 mT
18.11 13 mT
18.12 1.6 mT; 0.20 mT
18.13 8.5×10^{-5} N m^{-1}; 100 Hz
18.14 6.3×10^{-8} V
18.16 (c) (ii) 0.20 m s^{-1}; 1.22×10^{-5} V
18.17 (a) 2.5 mV (b) 0.0048 N m
18.19 (d) (i) Series resistor, 9000 Ω
18.20 (b) $4\pi \times 10^{-7}$ NA^{-2}
18.21 (b) 4 mΩ in parallel

Chapter 19

19.1 0.45 m s^{-1}
19.2 0.37 V
19.4 520 A s^{-1}
19.5 (a) 313 W (b) zero (c) 213 J (d) zero
9.2 A; 12.5 s
19.6 (a) 0.50 s (b) 0.25 s (c) 0.048 s (d) 0.22 s
(e) ± 3.2 V (f) $+ 6.6$ A s^{-1}
19.7 (a) 0.50 A; 3.0 W (b) 0.43 W; 0.091 W
(c) 1860 revolutions per minute
19.8 Idealised graph is a straight line with zero
speed when current is 6 A
19.9 2.9×10^{-4} N
19.11 (b) (ii) 15 H; 6.0 Ω
19.12 11.3 J; 150 V
19.13 (c) (i) 375 000 C; (ii) 88 N m
19.14 (b) (i) 1.96×10^{-3} Wb (ii) 0.029 V
19.15 4.0 mH
19.16 (i) 8 W; 12.5 W (ii) 92 W (iii) 1700 r.p.m.
(iv) 216 W (v) 50.5 A (vi) max 220 W

Chapter 20

20.2 4.0 mA; one leads by $\pi/4$; one lags by $\pi/4$
20.4 (a) 6.00 A (b) 8.49 A (c) 339 V (d) 2880 W
(e) 1440 W
20.5 (a) 4.0 A (b) 4.0 A (c) 2.55 A
20.6 0.28 A at $t = 0$; $- 0.23$ A at $t = 2$ ms;
$- 0.28$ A at $t = 2.5$ ms; 0.11 A at $t = 4$ ms
20.7 Values for checking; at 100 Hz, $X_C = 1590$ Ω;
at 500 Hz, $X_C = 318$ Ω; at 1000 Hz, $X_C = 159$ Ω;
at 3000 Hz, $X_C = 53$ Ω
20.8 480 Ω; 0.24 A
20.9 85 Ω; 0.24 A
20.10 0.49 mA; $V_C = 3.5$ V; $V_R = 4.9$ V; 0.6 mA

20.11 7.7 µF
20.12 $I_R = 0.91$ mA; $I_C = 1.26$ mA; 1.82 mW
20.14 (a) 198 Ω (e) $V^2 = V_L^2 + V_R^2$ (g) 54 mA;
222 Ω (h) 0.29 W (i) 0.45
20.15 (a) 1.1 A (b) 55 V (c) 234 V (d) 213 Ω (e) 0.68 H
(f) a capacitor is placed in parallel
20.16 (a) $X_L = 72.3$ Ω; $X_C = 67.7$ Ω;
(b) $I_L = 0.249$ mA; $I_C = 0.266$ mA (d) 0.017 mA
(e) X_C rises, I_C falls; X_L falls, I_L rises
20.17 25.2 MHz
20.18 0.37 A; 1.74 V; 1.18 V
20.19 250 turns
20.20 1960 A; 4350 A, $N_s/N_p = 0.451$; 3MW
20.21 (a) 1.0 A (b) 2.0 A (c) 2.0 A
20.22 (b) (i) 1000 turns (ii) 0.050 A
20.23 (d) (i) 0.40 A (ii) 375 Ω (iii) 225 Ω (iv) 36 W (v)
469 Ω (vi) 6.8 µF
20.24 (c) (i) 9.9×10^{-11} F (ii) 4.2
20.25 (b) 4.0 V; 3.0 V; $V_L = 2.5$ V; $V_C = 2.0$ V
20.26 900 µJ
20.27 (b) (i) 3600 Hz (ii) 0 (iii) 10 V (iv) 20 mA (v) 0
20.28 (b) 0.99 µF; 4.9 mH
20.29 50%

Chapter 21

21.1 1.0 V to 5.6 V
21.2 720 Ω
21.3 600 Ω
21.4 0.99
21.5 (b) 0.4 V (c) 5.6 V
21.6 12 µA
21.8 Warning light comes on when temperature is
too low
21.9 See table below

Inputs			Outputs			
A	B	C	AND	NAND	OR	NOR
0	0	0	0	1	0	1
0	1	0	0	1	1	0
1	0	0	0	1	1	0
1	1	0	0	1	1	0
0	0	1	0	1	1	0
0	1	1	0	1	1	0
1	0	1	0	1	1	0
1	1	1	1	0	1	0

21.10 (a) NOR (b) EX-NOR
21.11 Least significant bit, $S_0, = 1$, $S_1 = 1$, $S_2, = 0$,
$S_3 = 1$, $C_4 = 1$
21.12 See table below

R	S	CP	R'	S'	Q	$\overline{Q}$
0	0	0	0	0	1 or 0	0 or 1
0	1	0	0	0	1 or 0	0 or 1
1	0	0	0	0	1 or 0	0 or 1
1	1	0	0	0	1 or 0	0 or 1
0	0	1	0	0	1 or 0	0 or 1
0	1	1	0	1	1	0
1	0	1	1	0	0	1
1	1	1	1	1	0	0

When CP = 0 no change of Q or $\overline{Q}$ is possible.
When CP = 1, Q and $\overline{Q}$ can change over. Q = $\overline{Q}$ = 0 is not used

21.14 0.186 V; $I_{in} = I_f = 21.7$ μA

21.15 gain is 20 but output cannot be greater than + 5 V or less than – 5 V

21.17 (i) (a) voltage follower (b) voltage adder (c) precision rectifier (d) current to voltage converter (e) integrator (f) differentiator (ii) to change to different impedances (iii) so the gain is 1 (v) 360 s

21.19 Top box EX–OR; bottom box AND; Adder; P is sum Q is carry

21.21 (a) (ii) S_1; 00011: S_2; 11100

21.23 (b) 0.15 mV

21.24 (a) (i) 5 (ii) 250

21.25 – 0.9 V

21.27 (b) There are several possibile modifications. One, using 4 diodes, gives (i) –10 V (ii) 0

Chapter 22

22.1 13.6; 13.6 g

22.2 0.375; 0.402

22.3 7.1×10^{-3} m s^{-1}

22.4 (a) 2.0×10^{-11} N (b) 8.7×10^{-22} J (c) 1.67×10^{-21} J (d) 0.38 nm (e) 5.2×10^{-12} N

22.5 3.6×10^{-9} m

22.6 2.28×10^{-10} m if in cubical pattern 2.56×10^{-10} m if atoms are close packed. (The number of atoms per unit volume is fixed) Separation of layers = 2.09×10^{-10} m

22.7 0.30 g s^{-1}

22.8 1.0×10^9 kg s^{-1} (1 000 000 tonnes per second!)

22.9 4.5×10^5 kg

22.10 (a) 0.94 N (b) 747 N

22.11 24.7 m

22.12 zero; Antarctic ice is on land; Arctic ice is largely floating.

22.13 2.3×10^{17} kg m^{-3}

22.14 $r_{Pb} = 7.1 \times 10^{-15}$ m; $r_O = 3.0 \times 10^{-15}$ m

22.15 (a) 0.024 m^3 (b) 2.0×10^{-4}

22.16 3×10^{-9} m

22.17 (c) (iii) 2.5×10^6 J kg^{-1}; (iv) 2.23×10^6 J kg^{-1}

22.19 720 000 N

22.20 147 000 N

22.21 255 kPa; 250 kPa; 5 kN m^{-2} upwards

22.22 6.0×10^7 m^3; 5.5×10^{10} kg

22.23 0.21%; added

Chapter 23

23.1 2.1×10^5 Pa; 1.4×10^7 Pa

23.2 (a) 7.8×10^{-7} m^2 (b) 3.7×10^{-4} (c) 4.1×10^7 Pa (d) 32 N

23.3 32 000 N; 22 m s^{-2}

23.4 (a) (i) 1.14×10^{11} Pa (ii) 1.11×10^8 Pa

23.5 (a) 4.21 mm (b) 0.0364 J

23.6 (a) (ii) Stress is only 5.0×10^7 Pa so no permanent extension occurs; extension =

1.3 mm

23.8 (a) 5×10^6 Pa; 5.6×10^{-4} (b) 8.9×10^9 Pa (c) 0.056 J

23.9 (b) (i) 3.3×10^5 Pa (ii) 0.10 (iii) 3.3×10^6 Pa (c) 0.16 J

23.10 2.8×10^{-10} m (a) 2.0×10^8 Pa (b) 3 J (c) 1.4×10^{10} Pa

23.12 1.2 J

23.13 Strain = 0.187; 0.187 > 0.167; survives. Suitable since 0.187 < 0.200; no break.

23.14 (b) 1 000 000 Pa; mass of giant/mass of man = 8; 0.085 m; Y is larger

Chapter 24

24.1 69 °C

24.2 – 56.3 °C

24.3 (a) 304.2 K; 31.0 °C. (b) the gas is not an ideal gas

24.4 342.8 K

24.5 1.05 kg m^{-3}; 1200 m^3

24.7 223 K

24.8 290.0 K

24.9 b =2810 K; 1410 Ω is the value of P

24.10 (a) 118 700 Pa at 51 °C; 118 300 Pa at 50 °C: $\pm\frac{1}{4}$°C (b) 650 Ω at 51 °C; 677 Ω at 50$\frac{1}{4}$°C: ±0.2 °C

24.11 881°C

24.12 – 183°C

Chapter 25

25.1 391 °C

25.2 8.24×10^{-6} m^3

25.3 314 kPa; 214 kPa above atmospheric pressure

25.4 5.6×10^{-21} J

25.5 (a) 500 m s^{-1} (b) 290 000 m^{2}s^{-2} (c) 539 m s^{-1}

25.6 6.2×10^{-21} J

25.8 2.6×10^{-7} m

25.10 (a) (ii) $2mc \sin\theta$ in a direction at right angles to the wall and away from the wall (c) 3750 J

25.11 (a) (i) 40 mol (ii) 14 kg (iii) 14 kg m^{-3} (iv) 150 m s^{-1}

25.12 (a) 0.120 mol He; 0.060 mol Ne (b) (ii) 2.24 (c) (i) 0.060 mol (ii) 0.030 mol

25.13 (a) 0.60 m (b) 25 cm (c) 18 kg (d) 20 kg (e) wire snaps before piston is pulled out   (f) piston pulled out at load of 20 kg. 72 kg would have to be suspended to break wire.

25.14 1.04×10^{20} J for both; hydrogen 2500 m s^{-1}; oxygen 620 m s^{-1}

25.15 (a) 6.0×10^{-20} N

25.16 (c) (i) 2.93×10^{-10} kg (ii) 6.07×10^{12} (iii) 550 m s^{-1} (iv) 5.0×10^{-9} m

25.17 (a) 1.24×10^{-20} J (b) 7460 J : 483 m s^{-1}

Chapter 26

26.1 (a) 100 J (b) 160 J: Increase in internal energy is the same. Since less work is done on the gas

(≈ 60 J), more heat has had to be supplied to it (≈ 100 J)

26.2

−50	−70	20
25	25	0
100	140	−40
−75	−75	0

26.3 (b) (i) $P \propto 1/V$ at constant temperature; (ii) $V \propto T$ at constant pressure; (iii) $P \propto T$ at constant volume

(c) (i) 26 200 J (ii) 18 700 J (iii) 5200 J (**not** zero; see answer g)

(d) (i) 0 (ii) −7500 J (e) −5200 J (f) (i) 18 700 J (ii) 18 700 J

(g) 5200 J of work are done by the gas

(h) ∞

(i) Assuming a straight line, if the final temperature is ≈ 650 K work done by the gas = 5100 J and the decrease in internal energy is also 5100 J so no heat has been supplied or taken away.

26.4 0.375

26.5 (a) 0.090 mol

(b)

1480	0	1480
1880	2630	−750
−2520	0	−2520
−845	−845	0

(d) 179 kW (f) 5.8 g s^{-1}

26.6 20.4%

26.7 at 370 °C; 55.2%: at 645 °C; 68.6%

26.8 (b) 3270 J kg^{-1} K^{-1}

26.9 85 W; 0.42 K s^{-1}

26.10 (i) 1290 J kg^{-1} (ii) ≈ 75 K

26.11 (b) ≈ 1 MJ

26.12 (b) (i) 1.2×10^{-3} m^3 (ii) 187 J (c) (i) 31 J (ii) 21 J (iii) 52 J

26.13 (a) 800 J; 800 J of heat transferred from gas (b) 500 J (c) 1300 J; 1300 J of heat supplied to gas

26.14 (b) (i) T_C = 2800 K, T_D = 2300 K (ii) Work done is zero; +300 J +1840 J −440 J −1700 J (iv) 880 J (v) 0.34

26.15 (c) (i) 1.7×10^5 J (ii) 2.1×10^6 J (iii) 6.3×10^{-20} J

26.16 (c) 37 400 J (d) 3740 J; 1370 m s^{-1}

26.17 (a) 8.8×10^8 W (b) 39%

Chapter 27

27.1 0.98 °C

27.2 (a) 1100 W (b) 209.95 °C (this is 0.05 °C below the temperature of the steam to 1 sig fig)

27.4 (a) 4000 W

27.5 (a) greater (b) same (c) less

27.6 (c) (i) 88.5 °C, 11.5 °C (ii) 1.93 W

27.7 7.8 W

27.8 (ii) 150 °C m^{-1}; 29 W (iii) at x = 0.25 m

27.9 (a) $(2.40 \pm 0.05) \times 10^{-8}$ W Ω K^{-2} (b) 2.44×10^{-8} J^2 C^2 K^{-2}. Both figures and units agree

Chapter 28

28.1 3.2×10^{-17} J; 8.4×10^6 m s^{-1}

28.2 (a) 2.05×10^{15} m s^{-2} (b) 1.86×10^{-15} N (c) 11 600 V m^{-1} (d) 163 V (e) 7.6×10^6 m s^{-1}

28.3 1.01×10^7 m s^{-1}

28.4 (a) amplitude doubled (b) twice number of cycles (c) half number of cycles (d) one eighth number of cycles (e) half number of cycles (f) twice number of cycles (g) same number of cycles in smaller distance across screen

28.5 Amplitude of 2.8 cm; 1 cycle occupies 2 cm: 5 cycles occupy 1 cm

28.6 (a) 2.7×10^6 m s^{-1} (b) 3.0×10^{13} s^{-1} (c) 1.13×10^6 m^{-1}

28.7 2.2×10^5 m s^{-1}

28.8 2.67×10^{-26} kg; 2.84×10^{-26} kg; 3.00×10^{-26} kg; 16.1; 17.1; 18.1. A singly charged oxygen molecule.

28.9 1.41

28.10 (a) 400 eV

28.11 (a) 4.2×10^7 m s^{-1} (b) 0.12 m

28.12 (b) (i) positive (ii) $\tan \phi = eVx/mdv^2$ (iii) $a = eV/2mdv^2$ (iv) $2mdv^2 \cos \theta \sin \theta/eV$

28.14 14.1 V; 200 Hz

28.15 initially 1 cycle occupies 2 cm peak to peak height 4 cm

28.16 (b) 9 V; 10.6 V (c) (i) 1.0×10^{-8} s (ii) 1.8×10^{14} m s^{-2} (iii) 9.0 mm (iv) 54 mm

28.17 (b) 1.1 cm

Chapter 29

29.1 2.32×10^{-13} kg; 1.3×10^{-12} m^3

29.2 1.0×10^7 m s^{-1}

29.3 3×10^{-15} m

29.5 0.0256 u; 23.8 MeV

29.6 0.0155 u; 14.4 MeV 14.4 MeV

29.7 0.0858 u; 80 MeV

29.8 (c) 1.0 u; 3.3×10^7 m s^{-1}

29.9 145 years

29.10 5.7×10^{18}

29.11 (a) (184 ± 2) s (c) Rate of decay/ number of atoms = 3.74×10^{-3} s^{-1} (d) 0.44 $\times 10^{10}$ s^{-1}

29.12 (a) 0.30 (b) 0.15 (c) 0.104

29.13 2.8 s

29.14 (a) 7.2×10^{-45} m^3 (c) 57 N

29.15 (c) (i) 9.0×10^{-12} J (ii) 5.5×10^{-13} m (d) (i) 1.67×10^{-27} kg (ii) 1.6×10^{-13} J (iii) 930 MeV

29.16 (b) (i) 1.26×10^{-13} m

29.17 (c) (i) Helium 4 (ii) 4.0026 u

29.18 (a) 135 000 ion pairs; 31 eV per ion pair

29.19 6.0×10^5 m s^{-1}; 2.12×10^{-15} J

29.20 (c) (ii) 1.6×10^6 m s^{-1} (iii) 6.3×10^{-19} J

29.21 (b) (i) 70 s; 10 μg; (ii) 2.5×10^{14} s^{-1}

29.22 (c) 1.36×10^8 s (4.3 years)

29.23 (c) (i) 30 counts per minute (ii) 27.6 days (iii) $1/T = 1/T_b + 1/T_r$

29.24 8100 years

29.25 fuel converted = 0.59 kg

Chapter 30

30.2 Red: 4.29×10^{14} Hz; 2.84×10^{-19} J;
Violet: 7.50×10^{14} Hz; 4.97×10^{-19} J

30.3 (a) 5.55×10^{14} Hz; visible (b) 4.59×10^{14} Hz; visible

30.4 8.8×10^{-19} J

30.5 (a) 1.86×10^{-18} J (b) 2.8×10^{-15} Hz

30.7 122 nm, 103 nm, 97.5 nm, 95.3 nm, 94.0 nm, 93.3 nm, 92.9 nm .....

30.8 (a) 8.0×10^{-22} J (b) 633 nm

30.9 300 W

30.10 1.28×10^{-14} J; 1.28×10^{-14} J; 1.93×10^{19} Hz; 0.0155 nm

30.11 5.2×10^{-12} m

30.12 3.96×10^{-19} J; 1.32×10^{-27} N s; 1.01×10^{19} s^{-1}; 1.33×10^{-8} N; 1.33×10^{-8} N

30.13 2.1×10^{-10} m

30.15 (a) same (b) same (c) same (d) twice

30.16 (b) (i) 5.5×10^{10} (ii) 0.11 μC

30.17 The voltmeter shows (a) no reading (b) lower reading (c) no reading (d) no change (e) no reading

30.18 (a) (i) $[ML^2T^{-2}]$ (ii) $[MLT^{-1}]$ (iii) $[ML^2T^{-1}]$ (c) for $\lambda = 300$ nm potential difference = 4.2 V

30.21 (b) 7.8×10^{-19} J; 254 nm

30.22 (a) ultra-violet (b) 2.18, 2.09, 2.04, 1.94, 1.64 aJ (1 aJ = 10^{-18} J)

30.23 651 nm

30.24 (d) 1.0×10^{-10} m

Appendix B:
USEFUL DATA

Physical Quantities

Quantity	Defining Equation	Unit	Abbreviation
mass	base unit	kilogram	kg
time	base unit	second	s
frequency	number of cycles/time	hertz	Hz
length	base unit	metre	m
angle	arc length/radius	radian	rad
angular velocity	angle/time	radian per second	rad s^{-1}
area	length $\times$ breadth	metre^2	m^2
volume	length $\times$ breadth $\times$ height	metre^3	m^3
density	mass/volume	kilogram per metre^3	kg m^{-3}
velocity	distance in given direction/time	metre per second	m s^{-1}
acceleration	increase in velocity/time	metre per second^2	m s^{-2}
momentum	mass $\times$ velocity	kilogram metre per second	kg m s^{-1}
force	mass $\times$ acceleration	newton	N
moment of force, torque	force $\times$ distance from axis	newton metre	N m
pressure, stress	force/area	pascal	Pa
strain	extension/original length	no unit	
work, energy	force $\times$ distance in direction of force	joule	J
power	work/time	watt	W
gravitational field	gravitational force/mass	newton per kilogram	N kg^{-1}
gravitational potential	gravitational energy/mass	joule per kilogram	J kg^{-1}
thermodynamic temperature	base unit	kelvin	K
heat capacity	energy/temperature change	joule per kelvin	J K^{-1}
specific heat capacity	heat capacity/mass	joule per kelvin kilogram	J K^{-1} kg^{-1}
specific latent heat	energy/mass	joule per kilogram	J kg^{-1}
electric current	base unit	ampere	A
electric charge	current $\times$ time	coulomb	C
electric field strength	force/charge	newton per coulomb	N C^{-1}
specific charge	charge/mass	coulomb per kilogram	C kg^{-1}
potential difference	energy/charge	volt	V
resistance	potential difference/current	ohm	Ω
resistivity	resistance $\times$ area/length	ohm metre	Ω m
conductance	current/potential difference	siemens	S
capacity	charge/potential difference	farad	F
magnetic flux density	force/current $\times$ length	tesla	T
magnetic flux	magnetic flux density x area	weber	Wb
inductance	e.m.f/rate of change of current	henry	H
radioactive activity	counts/time	becquerel	Bq

Physical Data

Physical Quantity	Symbol	Numerical value and unit
Speed of light in a vacuum	c	2.998×10^8 m s^{-1}
Planck constant	h	6.626×10^{-34} J s
Gravitational constant	G	6.672×10^{-11} N m^2 kg^{-2}
Permeability of free space	μ_0	$4\pi \times 10^{-7}$ H m^{-1}
Permittivity of free space	ε_0	8.854×10^{-12} F m^{-1}
Boltzmann constant	k	1.381×10^{-23} J K^{-1}
Molar gas constant	R	8.314 J K^{-1} mol^{-1}
Electron charge	$-e$	-1.602×10^{-19} C
Specific charge of electron	e/m	1.759×10^{11} C kg^{-1}
Electron rest mass	m_e	9.110×10^{-31} kg
Proton rest mass	m_p	1.673×10^{-27} kg
Neutron rest mass	m_n	1.675×10^{-27} kg
Avogadro constant	N_A	6.022×10^{23} mol^{-1}
Atomic mass unit	u	1.661×10^{-27} kg
Electron volt	eV	1.602×10^{-19} J
Acceleration of free fall	g	9.81 m s^{-2}
Standard temperature and pressure (s.t.p.)		
standard temperature, ice point, 0 °C		273.15 K
standard atmospheric pressure		$1.013\,25 \times 10^5$ Pa
Density of air at s.t.p.	ρ	1.29 kg m^{-3}
Molecular mass of air		$0.028\,98$ kg mol^{-1}
Molar heat capacity of air at constant pressure	$C_{p,m}$	29.1 J K^{-1} mol^{-1}
Molar heat capacity of air at constant volume	$C_{v,m}$	20.8 J K^{-1} mol^{-1}
Density of water	ρ	1000 kg m^{-3}
0 °Celsius		273.15 K
Triple point of water		0.01 °C = 273.16 K
Specific heat capacity of water		4190 J kg^{-1} K^{-1}
Mass of the Earth		5.98×10^{24} kg
Equatorial radius of the Earth		6.378×10^6 m
Mean density of the Earth		5520 kg m^{-3}
Period of rotation of Earth on its axis ($\approx$ 1 day)		8.616×10^4 s
Mean distance of Earth from Sun		1.50×10^{11} m
Period of revolution of Earth (1 year)		3.16×10^7 s
Mass of the Sun		1.99×10^{30} kg
Radius of Sun		6.96×10^8 m
Power output of Sun		3.9×10^{26} W
Power intensity received at Earth (solar constant)		1.4 kW m^{-2}
Mass of the Moon		7.35×10^{22} kg
Radius of the Moon		1.74×10^6 m
Mean distance of Moon from Earth		3.84×10^8 m
Period of rotation about axis = period of revolution in orbit (1 lunar month, 27.3 days)		2.36×10^6 s
Acceleration of free fall on the Moon		1.62 m s^{-2}

Prefixes for Units

Prefix	Multiplying Factor	Symbol
atto	10^{-18}	a
femto	10^{-15}	f
pico	10^{-12}	p
nano	10^{-9}	n
micro	10^{-6}	μ
milli	10^{-3}	m
centi	10^{-2}	c
deci	10^{-1}	d
deca	10	da
hecto	10^{2}	h
kilo	10^{3}	k
mega	10^{6}	M
giga	10^{9}	G
tera	10^{12}	T
peta	10^{15}	P
exa	10^{18}	E

Mathematical Relationships

Algebra

$$a^2 = a \times a \qquad a^{\frac{1}{2}} = \sqrt{a} \qquad a^{-2} = \frac{1}{a^2} \qquad a^{-\frac{1}{2}} = \frac{1}{\sqrt{a}}$$

If $\ a = 10^x \qquad \lg a = x \qquad$ If $a = e^x \qquad \ln a = x$

$\lg(ab) = \lg a + \lg b \qquad\qquad \ln(ab) = \ln a + \ln b$

$\lg(a^n) = n \lg a \qquad\qquad\qquad \ln(a^n) = n \ln a$

$\lg\left(\frac{a}{b}\right) = \lg a - \lg b \qquad\qquad \ln\left(\frac{a}{b}\right) = \ln a - \ln b$

$(1 + x)^n \approx 1 + nx$ if x is small

The two solutions to the equation $ax^2 + bx + c = 0$ are given by $\ x = \dfrac{-b \pm \sqrt{b^2 - 4ac}}{2a}$

Trigonometry

$$\sin \theta = \frac{o}{h} \qquad \cos \theta = \frac{a}{h} \qquad \tan \theta = \frac{o}{a}$$

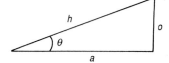

In any right angled triangle, the Pythagoras theorem gives $\ o^2 + a^2 = h^2$

In the circle drawn the radius is 1 unit of length
the opposite side, PR, is $\sin \theta$ units of length
the adjacent side, OR, is $\cos \theta$ units of length
and the arc length PQ is θ units of length.

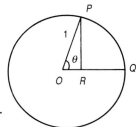

From the right angled triangle we get $\ \sin^2 \theta + \cos^2 \theta = 1$
For small angles $\sin \theta \approx \theta \approx \tan \theta$, with angles measured in radians.

In any triangle

$$\frac{a}{\sin A} = \frac{b}{\sin B} = \frac{c}{\sin C}$$

and $a^2 = b^2 + c^2 - 2bc \cos A$

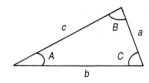

Geometry

Circumference of a circle of radius r $= 2\pi r$
Area of a circle of radius r $= \pi r^2$
Volume of a sphere of radius r $= \frac{4}{3}\pi r^3$
Surface area of a sphere of radius r $= 4\pi r^2$
Volume of a cylinder of radius r and height h $= \pi r^2 h$

For a straight line graph $y = mx + c$ where m is the gradient of the graph and c is the intercept on the y - axis

Calculus

The gradient of a graph is found by differentiating

if $y = x^n$ the gradient, $\frac{dy}{dx}$ is given by $\frac{dy}{dx} = nx^{n-1}$

if $y = \sin x$ $\qquad \frac{dy}{dx} = \cos x$

if $y = \cos x$ $\qquad \frac{dy}{dx} = -\sin x$

If $y = e^x$ $\qquad \frac{dy}{dx} = e^x$

The area beneath a graph can be found by integrating. This is the reverse process to differentiating

If $y = x^n$ the integral, $\int y dx$ is given by $\int y dx = \frac{x^{n+1}}{n+1} + c$

if $y = \cos x$ $\qquad \int y dx = \sin x + c$

if $y = \sin x$ $\qquad \int y dx = -\cos x + c$

if $y = \frac{1}{x}$ $\qquad \int y dx = \ln x + c$

if $y = e^x$ $\qquad \int y dx = e^x + c$

where c is a constant

Appendix C:

GUIDELINES FOR DATA ANALYSIS EXERCISES

Many answers are given in this section, but not all of them. You should treat the answers as checks which enable you to see if you are progressing in the right direction. In places hints are more appropriate than answers. You will need to use the numerical values of physical constants from time to time. These are given in the appendix. As with answers to questions, the final digit of an answer should be treated with caution. It is given to help with checking and to avoid introducing too many rounding up errors. It may not be justified by the accuracy of the data supplied.

Chapter 3
Forces in frameworks

Note that throughout this exercise the forces shown are the forces exerted by the struts on the joints. The diagram does not show the forces on the struts.

(d) This will be a rectangle
(e) 150 kN
(f) 600 kN
(g) CD is in *tension*. It pulls on C to the right and it pulls on D to the left, so it is being stretched. Similarly JB, IC and DF are also in tension. IB and IH are two of nine in compression.
(h) JI, DE, BC and HC

Chapter 4
Accelerometer graphs

Great accuracy is not attainable here so it is not worthwhile taking too long by using too many points. The graph given, and your own will be different, gives (b) 40 m s^{-1} (c) 60 m s^{-1} after about 210 s and (e) 11 000 m; 6000 m while accelerating and a further 5000 m while slowing down

Chapter 5
Passenger loading of European Airbus

(a) 3000 km   (b) 600 km   (c) 75 000 kg   (d) 1.88 m s^{-2} (e) 141 000 N   (f) 176 000 N   (g) 60 000 kg
(h) Range: 3000 km;   400 km;   800 km;   1200 km;   3000 km   (i) 100

Chapter 6
Building design

(a) A mass of 1 kg has a weight of 9.8 N
(b) Because of the angle of the roof
(c) 0.73 m
(e) The total loading is the weight supported per unit area; the purlin loading is the weight supported per linear metre. 4.5 m is the horizontal distance from the centre of the building to the outside, of this distance the left hand wall supports 2.0 m so the purlin supports the remaining 2.5 m.
(f) The hanger supports half the weight of the ceiling between the two walls.
(g) Because the hanger is suspended from a point immediately next to the purlin.
(h) Roof loading 7500 N m^{-1};   ceiling loading 2500 N m^{-1};   so $M = 10\,000 \times 7.6^2 /8 = 72\,000$ N m^{-1}. This requires the heavier, and hence stronger, purlin.

Chapter 7
Energy and momentum of balls used in various sports

(a) 120 J;   165 J;   110 J;   16 J;   38 J;   75 J
(b) 6.2 N s;   11.8 N s;   3.2 N s;   1.4 N s;   1.6 N s;   3.0 N s
(c) Answers *must* be identical to (b)
(d) Answers also identical to (b)
(e) (i) 1.55 kg (ii) 0.25 kg (iii) 0.60 kg
(f) 120 J
(g) Each mass is held by hand and so collisions are not between a single object and a ball
(h) Taking "most nearly elastic" to mean the collision in which the velocity of separation is the highest proportion

of the velocity of approach gives the answer - golf ball.
(i) 4500 N; 1470 N; 2400 N; 1000 N; 520 N; 740 N

Chapter 8
Behaviour of aircraft tyres when rotating at high speeds

Speed v / m s^{-1}	Clearance /mm	Angular velocity ω /rad s^{-1}	$r\omega^2$ /m s^{-2}	Force per unit mass /N kg^{-1}
70	21	156	11 000	11 000
80	25	178	14 200	14 200
90	29	200	18 000	18 000
100	34	222	22 200	22 200
110	42	244	26 900	26 900

(h) This data is suitable for tyres between 850 and 1000 mm. For a constant speed the acceleration of larger radii tyres is less than for small radii ($a = v^2/r$) but the mass is greater. These two factors partially cancel one another out.

Chapter 9
Satellite Motion

(a) 8.0   (b) 6.2 MJ kg^{-1} (found as minus half the potential energy or, more accurately from $GM/2r$)

(c) approximately 60 MJ kg^{-1} (d) 1.8×10^{11} J   (e) 3×10^{10} J; 4500 m s^{-1} (f) 480 MJ greater potential energy at greater distance   (g) 6100 m s^{-1}(loss of 480 MJ of kinetic energy)

Chapter 10
Energy in SHM

(a) 2.45 cm   (b) 20 s^{-1} (c) 0.52 m s^{-1} (d) 0.24 J   (e) 7 = 8 = 9 = 0.24 J;   1 + 4 = 2 + 5 = 3 + 6 = 0.24 J
(f) 0.012 J;   0.36 J;   1.19 J   (g) as (f)   (h) –0.36 J   (i) 0.59 J   (j) 0.23 J;   –0.95 J

Chapter 11
Seismic surveying

(b) SD$_8$ = 1.15 km;   SXD$_8$ = 1.77 km;   depth = 0.67 km
(d) $A = 2.6 \times 10^{10}$ kg s^{-2} m^{-1} = 2.6 Pa
(f) The further seismometers give the same amplitude but will receive weaker vibrations; direct waves should show larger amplitude than reflected waves
(g) If a wave strikes the boundary at the critical angle it can be refracted with an angle of refraction of 90°. Re-entry of this wave back to D$_8$ at the critical angle, is possible, presumably as a result of irregularities in the boundary layer.

Chapter 12
Camera

(a) 6.25 mm   (c) $\sqrt{32}$ = 5.657   (e) (1/500) s; f/16   (g) 600 mm – 1200 mm

Chapter 13
Synthesisers

For Fig 13.18 (a) $3f_0$ (b) $5f_0$

Chapter 14
Light bulb construction

(b) $\frac{1}{4}$A; 960 Ω   (c) 0.0121 mm   (d) A is long life, C is high temperature   (e) and (f) see table

	A	B	C
length/ m	0.158	0.14	0.121
surface area/ m^2	6.4×10^{-6}	5.3×10^{-6}	4.3×10^{-6}
power lost per unit area/ W m^{-2}	0.94×10^7	1.13×10^7	1.4×10^7
temperature/ °C	2420	2600	2730

(g) Cost for A £10.10; cost for C £11.60; Light energy from A 3.9×10^7 J; light energy from C 5.2×10^7 J; Cost of a unit of light energy from A/cost of unit from C = 1.2

(h) If you want less light use a lower wattage bulb not a long life one.

Chapter 17
Charging a capacitor through a resistor

$V_C = 10$ V; $V_R = 0$ V; $I = 0$; $q = 0$; $Q = 4000$ μF

Chapter 18
Meter sensitivity

Type	Wire diameter / mm	Approx. s.w.g size	Number of turns	Resistance /Ω	Current sensitivity /mm mA^{-1}	Voltage sensitivity /mm mV^{-1}
A	0.200	36	24	0.78	50	64
B	0.100	42	96	12.5	200	16
C	0.050	47	384	200	800	4.0
D	0.025	50	1536	3200	3200	1.0

Table 18.1 should contain these figures when completed. Note that as the current sensitivity increases the voltage sensitivity decreases

Chapter 19
Electromagnetic flow metering

(a) To avoid shorting X and Y   (d) Hydrocarbons   (e) Because of its resistence to corrosion
(f) Because there is no changing flux through it   (g) E_1 is not constant   (i) 0.35 m s^{-1} (k) 18 m
(l) If the conductivity of the liquid is low the internal resistance between X and Y is high. The resistance of the cable adds to this internal resistance and may reduce the output to too small a value.

Chapter 20
Filter circuits

(a) - (d)

frequency/ Hz	50	5000
reactance/ Ω	9950	99.5
impedance/ Ω	9997	1005
V_C/ V	1.99	0.198
V_R/ V	0.200	1.99

(e) To have high reactance for high frequency and therefore to have the high frequency a.c. across it.
(g) 16 Ω   (h) 0.85 mH   (i) S_2

Chapter 22
Blood pressure

(a) 16 000 Pa;   10 700 Pa
(b) 116 000 Pa;   110 700 Pa
(c) Extra pressure as a result of 1.2 m height between heart and feet = 12 500 Pa
(d) 7300 Pa
(e) Normal pressure in brain is $(0.4 \times 1060 \times 9.8)$ Pa = 4200 Pa below systolic pressure, i.e. 11 800 Pa above atmospheric pressure. When accelerating the drop in pressure across 0.4 m of body = $(0.4 \times 1060 \times (28 + 9.8))$ Pa = 16 000 Pa below systolic pressure = 0

Chapter 24
Temperature scales

Absolute zero becomes –468 °F and the triple point 32 °F.
Any other temperature on scale X becomes $T \times 500/273.16$ giving the boiling point of water as 683.02 °X or 215 °F. Scale F has an uncanny similarity to the Fahrenheit scale.

Chapter 25
Kinetic theory

Component of velocity in a direction towards the wall $= c \cos \theta$
Change in momentum at a collision $= 2mc \cos \theta$
Time between collisions with wall $= 2r \cos \theta / c$
Rate of change of momentum $= mc^2/r$
Pressure on wall due to one molecule $= mc^2/4\pi r^3$

Molecular speeds

(a) 400 m s^{-1} (b) 332 m s^{-1}
(e) (i) 0.20 (ii) 0.040 (iii) 0.11 (iv) 0.053 (v) 0.43
(f) (i) 320 m s^{-1} (ii) 370 m s^{-1} (iii) 410 m s^{-1} The fastest 10% of molecules are travelling at over 1000 m s^{-1}.
(g) Numerical values are given to the nearest whole number. The height of the graph for molecules travelling at 4000 m s^{-1} may only be 0.005 but this still represents a large number of molecules.

Chapter 26
Behaviour of gases in a turbofan jet engine

(a) 6290 mol   (c) 664 K   (d) 2 610 000 Pa   (e) 96.3 MJ   (f) 1.82 kg s^{-1} (g) 157 tonnes   (h) 386 K

(j) 8 400 000 J   (l) Increase in internal energy is always $nC_{vm}\Delta T$ so the four values are 57.4 MJ;   96.3 MJ;   −132.7 MJ and −21.0 MJ. Efficiency = (96.3 − 29.4) / 96.3 = 0.69   (m) 55 MW   (n) 0.57.

Chapter 27
Heat Losses from a House

(a) e.g. from the living room

	Temperature difference/K	Area /m^2	U– value /W m^{-2} K^{-1}	Power loss /W
outside wall	25	25 – 2.4	0.51	288
inside wall	5	25 – 2	3.5	402
window	25	2.4	5.7	342
door	5	2	5.2	52
floor	25	24	0.6	360
ceiling	25	24	0.35	210
			Total	1654

Working in a similar way for the other rooms gives:-
bathroom; 360 W: hall;   850 W: kitchen; 960 W: bedroom 1, 700 W: bedroom 2, 790 W. A total of 5310 W. None of these figures are very accurate. They give a guide to the requirements and the heating system is made large enough to cope with increased demand. It is not worthwhile dealing with small details.
Extra load for air changes about 1300 W so whole system may have any output larger than 7 kW.

Insulation provided by clothing

(a) Radiation is absorbed by a metal and no fluid exists in a metal in which convection can occur
(b) The insulation is done mostly by the still air trapped between the fibres
(d) (i) 0.0208 m^2 K W^{-1} (ii) 0.208 tog

Chapter 28
Ship's radar

(a) 4.15 cm   (b) 3615   (c) 0.200 ms   (d) 60.0 km   (e) just under 30 km   (f) 10 cm in 0.2 ms so speed is 500 m s^{-1} (g) 5.0 cm   (h) 3.0 km   (i) 37.5 m; 0.25 mm

Chapter 29
Nuclear reactions

The energies necessary to balance each of the equations are as follows:-
6.5 MeV;   17.3 MeV;   1.3 MeV;   3.26 MeV;   0.39 MeV;   2.8 MeV;   4.3 MeV

Decommissioning a nuclear reactor

(a) 2.9×10^5 Bq kg^{-1}; 8.6×10^7 Bq kg^{-1}; 9.7×10^5 Bq kg^{-1}

(c) Long half life materials have their long half life because there is a low probability of any particular nucleus decaying per unit time.

(d) 96 years (e) 3200 Bq

(f) All in 10^{12} Bq. H, 0.34;   C 43.5;   Cl 1.3;   Ca 0.9995;   Fe 1.7×10^{-8};   Co 1.1×10^{-3};   Ni 2.198;   Ni 117.7;   Nb 0.012 96;   Ag 0.011;   Sm 0.93;   Eu 2.9×10^{-3};   Eu 9.3×10^{-3}

(g) Nickel 59, carbon 14, nickel 59, chlorine 36, calcium 41

(h) The graph will not be exponential; it falls fairly quickly but because of the long half life isotopes then flattens out.

Chapter 30

X-rays

(a) K_{α}, 58.2 keV; K_{β}, 66.6 keV, L_{α}, 8.4 keV, L_{β}, 9.6 keV

(b) K_{α}, 2.13×10^{-11} m; K_{β}, 1.86×10^{-11} m; L_{α}, 1.48×10^{-10} m; L_{β}; 1.29×10^{-10} m

(c) K_{γ}, 1.83×10^{-11} m; L_{γ}, 1.10×10^{-11} m,

(e) While it is possible for an electron in the K-shell to be given exactly 58.2 keV to lift it to the L-shell this is in practice unlikely. Inner electrons are normally dislodged from the atom and this requires a potential difference of at least 69.6 kV.

Index

Bold page references denote key entries.